THE AMERICAN CASSIINAE

A Synoptical Revision of Leguminosae tribe Cassieae subtribe Cassiinae in the New World

Howard S. Irwin and Rupert C. Barneby

T DEN

This material is based upon research supported by the National Science Foundation under Grants No. GB 2293, GB 6458, BMS 74-19289, BMS 76-24079 and DEB 78-18365 and is published with support of National Science Foundation Grant No. DEB 80-05535. The Foundation provides awards for research and education in the sciences. The awardee is wholly responsible for the conduct of such research and preparation of the results for the publication. The Foundation, therefore, does not assume responsibility for such findings or their interpretation. Any opinions, findings, conclusions, or recommendations expressed in this publication are those of the authors and do not necessarily reflect the views of the National Science Foundation.

Library of Congress Cataloging in Publication Data

Irwin, Howard S.
The American Cassiinae.
(Memoirs of the New York Botanical Garden; v. 35)
Bibliography: p.
Includes index.
1. Cassia—Classification. 2. Senna—Classification.
3. Chamaecrista—Classification. 4. Botany—America—
Classification. I. Barneby, Rupert C. II. Title.
III. Title: Cassiinae. IV. Series.
QK1.N525 vol. 35 581s [583'.323] 82-8282

International Standard Book Number 0-89327-241-8

DECEMBER 1982 MEMOIRS OF THE NEW YORK BOTANICAL GARDEN **35**: 455–918

D. Sectio SENNA

Senna P. Miller sect. **Senna**. *Senna* P. Miller, 1754, l.c. supra, sensu parum ampliato.—Typum jam sub genere indicavimus.—*Cassia* sect. *Senna* (P. Miller) DeCandolle ex Colladon, Hist. Casses 92. 1816. *C.* subgen. *Senna* (P. Miller) Bentham in Martius, Fl. Bras. **15**(2): 96. 1870 & Trans. Linn. Soc. London **27**: 518. 1871.

Cassia sect. *Senna* sensu Vogel, 1837, p. 35, ex parte, quoad spp. eglandulosas (*C. browniana* exclus.); Symon, 1966, p. 76 (typo erroneo exclus.).

Corolla zygomorphic along its vertical axis; stamens 4-morphic; 3 adaxial staminodes; 4 median (sometimes sterile), the anthers short and straight; 2 latero-abaxial much larger, elevated sideways into the plane horizontal to the floral axis of symmetry, the incurved anthers facing each other like the arms of ice-tongs; 1 centric abaxial, diminished or sterile.—Petiolar glands 0; pod mostly plano-compressed, with seeds compressed parallel to the valves, in 1 sp. 4-angular, winged lengthwise along valves, with seeds compressed parallel to interseminal septa, never terete, the valves either plane, or low-corrugate, or raised over each seed as a pyramid, an umbo, or a vertically aligned crest; seeds 1-seriate transverse, areolate.—The section consists of 4 series, 1 circumtropical, 2 monotypic Neotropical, and 1 Paleotropical represented by 2 adventive spp. in the New World.

As redefined herein, the section *Senna* is essentially consubstantial with Bentham's series *Pictae* and *Brachycarpae* of *Cassia* subgen. *Senna* sect. *Chamaesenna,* though deprived of the anomalous *S. paradictyon* and enriched by a few species discovered since 1871. Its nucleus is *Cassia senna* L., which included the two medicinal sennas of commerce, *S. alexandrina* and *S. italica,* and its one constant differential character is androecial. In sect. *Senna* the two long stamens do not project directly or obliquely forward, as in sect. *Chamaefistula,* in a plane opposed to the vexillar petal, but are lifted sideways into a plane horizontal to the flower's vertical axis, bringing their long, incurved anthers, that face each other like twin arms of tongs, into position to deposit pollen on the sides rather than on the venter of visiting bees. Historically *Senna* was thought to differ from other cassias in the long-funiculate seeds suspended at middle of the cavity in locules narrower than the pod itself. A few individual species of the section have, it is true, pods in some respect peculiar either to themselves or to their series, but the pod of others can be closely matched either in the whole series or in individual species of our enlarged and modified sect. *Chamaefistula*.

The American species of sect. *Senna* are divided here into four series. The two larger correspond with ser. *Pictae* and ser. *Brachycarpae* of Bentham's revision except that we have segregated from the *Pictae* a monotypic ser. *Aculeatae* and established for the wholly anomalous *S. paradictyon* a monotypic section. The fourth series consists of the recently discovered *S. spinigera,* unique among Cassieae for its thornlike stipules but retaining most other features of sect. *Senna.*

Key to the Series of sect. *Senna*

1. Stipules herbaceous, dorsoventrally bifacial, not modified into thorns, sometimes deciduous.
   2. Stems and lf-stalks smooth, not armed with epidermal prickles; axis of racemes and floral bracts not capitate-glandular.
      3. Hypanthium turbinate, much shorter than the pedicel; pod (5.5–)7–19 cm long, 10–58-ovulate; seeds smooth or obscurely rugulose and the areole not elevated on a ridge; spp. 8 native + 1 cultivated in the Americas, few in Africa and Australia. Di. ser. *Pictae* (p. 456).
      3. Hypanthium slenderly infundibuliform, as long or longer than the pedicel; pod 2.5–5(–7) cm long, 6–10-ovulate; testa of seeds elaborately rugulose and the areole engraved into a raised ridge; Paleotropic, 2 spp. naturalized in the Americas. Div. ser. *Senna* (p. 479).
   2. Stems and lf-stalks armed with recurved epidermal prickles; axis of racemes and dorsal face of floral bracts capitate-glandular; sp. 1, widely interruptedly dispersed form Cuba to Paraguay. Diii. ser. *Aculeatae* (p. 478).
1. Stipules modified into stout, laterally subcompressed straight thorns set at right angles to the stem, persistent. Rachis of lvs rigid, terminating in a vulnerant spinule, the midrib of lfts also spinulose-excurrent; lfts 4–6 pairs, to 4.5–6 cm; sp. 1, local in e.-centr. Espirito Santo, Brazil. Dii. ser. *Spinigerae* (p. 476).

Di. ser. PICTAE (Bentham) Irwin & Barneby

Senna sect. **Senna** ser. **Pictae** (Bentham) Irwin & Barneby, comb. nov. *Cassia* sect. *Chamaesenna* ser. *Pictae* Bentham in Martius, Fl. Bras. **15**(2): 126. 1870 & Trans. Linn. Soc. London **27**: 550. 1871, max. ex parte.—Sp. lectotypica: *C. picta* G. Don ≡ **Senna pistaciifolia** var. **picta** (G. Don) Irwin & Barneby.

Cassia sect. *Chamaesenna* DeCandolle ex Colladon, Hist. Casses 95. 1816, ex parte, quoad typ.—Sp. lectotypica: *C. reticulata* Willdenow ≡ **Senna reticulata** (Willdenow) Irwin & Barneby.—*Cassia* sect. *Senna* subsect. *Chamaesenna* (Colladon) Bentham in Bentham & Hooker, Gen. Pl. **1**(2): 572. 1865.—Equated with *Cassia* subgen. *Senna* sect. *Senna* by Symon, 1966, p. 76, a concept that included our ser. *Senna* and ser. *Pictae.*

Cassia sect. *Herpetica* [Rumphius] DeCandolle ex Colladon, Hist. Casses 91. 1816.—Sp. typica: *C. alata* Linnaeus ≡ **Senna alata** (Linnaeus) Roxburgh.—*Herpetica* (Colladon) Rafinesque, Sylva Tellur. 123. 1838, pro gen.

Chamaesenna Pittier, Arb. y Arbust. Legum. 130. 1928.—Sp. lectotypica (Britton & Rose, 1930, p. 250; Britton & Killip, 1936, p. 178): *Ch. reticulata* (Willdenow) Pittier ≡ **Senna reticulata** Willdenow.—Although by intention possibly based on sect. *Chamaesenna* DC. (v. supra), and so interpreted by Cowan in Index Nom. Generic. **1**: 335. 1979, this genus was attributed to Rafinesque, Sylva Tellur., sine pag., in which work the name does not occur; here interpreted as a *genus novum* without antecedent history or typus.

Corolla and androecium of sect. *Senna*; hypanthium turbinate, much shorter than the pedicel; fl-buds asymmetrically obovoid; sepals little graduated in length, membranous or subherbaceous, green or colored, faintly nerved or, if prominently so, the nerves few and distant; style linear 4–12 mm, gently or scarcely incurved distally, the stigmatic cavity obliquely terminal; ovules (10–)12–58; body of pod (7–18 cm) usually plano-compressed, the valves then either faintly corrugate or umbonate over the seeds and these broadside to the valves, but in *S. alata* 4-angular, the valves undulately winged lengthwise and the seeds broadside to the interseminal septa; testa of seeds smooth or faintly rugulose.—Shrubs and trees, some precociously flowering; stipules commonly herbaceous and at base amplexicaul on side further from petiole, sometimes cordate on both sides; lfts

4–30 pairs, ample when few, diminishing with increased number; racemes 10–70-fld, the buds often subtended by enlarged, upwardly imbricated colored bracts.—Spp. 9 with us, one of them adventive, perhaps 3–4 more in Africa and Australia but these requiring systematic review; ours widespread from s. Mexico to s.-e. Brazil and n. Argentina, 1 (*S. alata*) cultivated in warm temperate regions of both hemispheres and adventively circumtropical.

While we have somewhat altered its definition, we adopt in practice very nearly the limits set to ser. *Pictae* by Bentham, excluding only the peculiar *S. aculeata* and *S. paradictyon,* each transferred to a monotypic series, and provisionally omitting some Australian species of which we know too little. The series is very closely related to Paleotropical ser. *Senna,* which Bentham (as ser. *Brachycarpae*) distinguished by a shorter, broader and few-ovulate pod, to which difference may now be added the slenderly funnelform hypanthium longer than the pedicel and the elaborately corrugated seed-testa.

The androecium of ser. *Pictae* follows the general plan of sect. *Senna,* but there is some variety in the dehiscence of the two long anthers. In the better- and longer-known species, of which the young flowering spike takes the form of a cone of colored bracts, dehiscence occurs through two slits looking upward and inward from the porrect pollen-cup which stands opposed, at a wide angle, to the bosses that terminate the adaxial side of the anther. In the closely related North Andean *S. pistaciifolia, S. cushina,* and *S. viarum,* as in the Mexican *S. centranthera,* of which the bracts are small, greenish and conceal only the very young, still sessile flower-buds, the angle between cup and bosses is closed or nearly so, the anther-tip becoming in consequence minutely two-lipped, and the pollen discharged by lateral pores, one, it seems, sometimes permanently closed. We have hesitated over the advantages of recognizing this group as coordinate with the nucleus of bracteose *Pictae,* but the close relationship between *S. nicaraguensis* and *S. centranthera,* which have the same umbonate pod but would otherwise be referred to distinct series, argues conclusively against it.

Key to American Species of ser. *Pictae*

1. Floral bracts petaloid or subpetaloid, usually yellow or orange, in any case dilated and 12–30 mm long.
    2. Seta of lf-rachis dilated into an herbaceous or subcarneous, deltate or lance-triangular blade 1.5–5 × 1–4 mm; lfts mostly 7–14 pairs and the distal pair 7–20 × 3–10 cm; mature pedicels 4–11 mm.
        3. Petiole including pulvinus (2.5–)3.5–13(–16) cm; lft-blades always thinly (sometimes inconspicuously) pilosulous on upper face, more densely so on lower; pod plano-compressed, not winged, the cavity linear-elliptic in cross section; seeds narrowly oblong-claviform, compressed parallel to the valves and bearing an areole ±3–5 mm long on each broad face. 147. *S. reticulata* (p. 458).
        3. Petiole including pulvinus 1–3.5(–5) cm, the true petiole often subobsolete, the first pair of lfts inserted next to or close to the pulvinus; lft-blades usually glabrous on upper face and only thinly or remotely pilosulous dorsally on veins and veinlets, sometimes glabrous throughout; pod sharply 4-angular, carinate by the sutures and winged lengthwise along the valves, the cavity rhombic in cross section; seeds rhomboid, compressed parallel to the interseminal septa and bearing an areole ±2–3 mm long on each narrow face. 148. *S. alata* (p. 460).
    2. Seta of lf-rachis subobsolete, subulate or linear-claviform, much less than 1 mm wide; either lfts more numerous or smaller (or both), or the pedicels longer.
        4. Stipules dimidiately cordate-acuminate 6–18 mm wide; filaments of 2 large stamens 5.5–7 mm; s. Mexico to Panama. Valves of pod umbonate over each seed. 149. *S. nicaraguensis* (p. 463).
        4. Stipules either (of native Brazilian *S. martiana*) obliquely ovate-lanceolate 2–5 mm wide or (of cultivated African *S. didymobotrya*) cordate and yet wider but then symmetrically so; filaments of 2 large stamens 2–4 mm; valves of pod either corrugate or

sharply crested; native in e. Brazil, seldom cultivated in Centr. America, Cuba, s.-e. United States.

5. Stipules obliquely lanceolate or ovate, dilated on side further from petiole and auriculately semiamplexicaul; tertiary venulation of lfts sharply prominulous, the blade distinctly reticulate; valves of pod pyramidally distended over each seed, the pyramids ± confluent into a crest; e. Brazil. 150. *S. martiana* (p. 465).
5. Stipules subsymmetrically cordate, auriculate at base on both sides; tertiary venulation of lfts immersed, the blade essentially penninerved; valves of pod corrugately elevated over each seed as a simple ridge horizontal to the pod's long axis; native to Africa, cultivated and sparingly naturalized in the Americas. 151. *S. didymobotrya* (p. 467).

1. Floral bracts not petaloid, ovate, lanceolate or subulate 2–10 mm.
6. Lfts 3–12 pairs, the distal pair 4–15 × 1.5–5 cm; n.-w. S. America, Galápagos Is., s. along the Andes to Bolivia.
7. Lfts less than 8 pairs, or obovate-oblong-elliptic and obtuse, or glabrous dorsally and marginally flat.
8. Stipules early reflexed, persistent; lfts (4–6 pairs) ovate- or broadly lance-acuminate; pedicels mostly 20–42 mm; filament of 2 long abaxial stamens 5–8 mm; Andean Ecuador and Colombia. 154. *S. viarum* (p. 474).
8. Stipules erect-spreading, caducous; lfts mostly broadest above the middle and commonly obtuse mucronate; pedicels mostly 12–20 mm; filament of 2 long abaxial stamens 3–5 mm; Andes from Colombia to Bolivia and lowlands of n.-w. Peru and Ecuador; Galapagos Is. 153. *S. pistaciifolia* (p. 471).
7. Lfts at once 8–12 pairs, lance-acuminate in outline, loosely pilosulous dorsally and marginally revolute; n.-centr. Peru (Amazonas and Huánuco). 155. *S. cushina* (p. 475).
6. Lfts (15–)20–33 pairs, the distal pair ±1.5–4 × 0.4–1.2 cm; s. Mexico. 152. *S. centranthera* (p. 468).

147. Senna reticulata (Willdenow) Irwin & Barneby, comb. nov. *Cassia reticulata* Willdenow, Enum. pl. hort. reg. Berol. **1**: 443. 1809.—"*Habitat in* Para Brasiliae."—Holotypus, *Hb. Willd. 7994,* collected by Sieber and acquired from Hoffmannsegg, B (2 sheets, one = F Neg. *1740*)!—*Chamaesenna reticulata* (Willdenow) Pittier, Arb. y Arbust. Legum. II Cesalp. 131. 1928, without bibliographic reference.

Cassia strobilacea Humboldt, Bonpland & Kunth, Nov. Gen. & Sp. **6**(qu): 347. 1824.—"Crescit in sylvis ad ripam fluminis Magdalenae prope La Bocca de San Bartholome [Antioquia, Colombia]."—Holotypus, labelled "n. 1630. Río Magdalena-San Bartolomé," P-HBK! isotypus, B (*hb. Willd. 7982*)!—Equated with *C. reticulata* by Vogel, 1837, p. 35.

Cassia tarantan Humboldt, Bonpland & Kunth, Nov. Gen. & Sp. **6**(folio): 348. 1824.—"Crescit prope Cumanam [Sucre, Venezuela]."—Holotypus, labelled "n. 1178. Cumaná," P-HBK! isotypus, B (*hb. Willd. 7981*)!—Equated with *C. reticulata* by Bentham, 1871, p. 550.

Cassia dumetorum Bertero ex DeCandolle, Prod. **2**: 499. 1825.—". . . [Colombia:] ad Sanctam-Martham. Bertero."—Holotypus, G-DC! = F Neg. *33458*.—Equated with *C. reticulata* by Bentham, 1871, p. 551.

Cassia annunciata E. H. L. Krause, Beih. Bot. Centralbl. **32**(2, heft 3): 345. 1914.—"Banquito in Nicaragua . . . [*Jean Borntraeger* in hb. Krause] No. 11828."—No typus found at STR (Gagnieu in litt., 1979).

Cassia reticulata sensu Colladon, 1816, p. 115; Vogel, 1837, p. 35; Bentham, 1870, p. 127; 1871, p. 550; Standley, 1922, p. 411; Standley & Steyermark, 1946, p. 124; Schery, 1951, p. 52; Romero-Castañeda, Fl. Centro Bolívar 114, fig. 44, 45. 1967; García-Barriga & Forero-González, 1968, p. 59, fig. 18; Croat, Fl. Barro Colorado I. fig. 275. 1978.—*Chamaesenna reticulata* sensu Britton & Rose, 1930, p. 250.

Coarse, amply leafy, potentially arborescent shrubs of rapid growth, at anthesis (1–)2–8(–9) m, precociously flowering as a softly woody bush but sometimes becoming a tree with trunk up to 8 cm diam few-branched into an irregular crown, pilosulous-puberulent throughout or almost so with fine erect, incumbent or sinuous hairs up to 0.1–0.5 mm, the vesture of hornotinous branchlets usually composed of erect shorter and longer hairs (the latter sometimes 0), that of the foliage either spreading or subappressed, the lvs ± bicolored, the lfts deep green drying olivaceous, dull on both faces, paler beneath, the racemes at first axillary to developed lvs, later forming an exserted candelabriform panicle.

Stipules herbaceous obliquely triangular-lanceolate (5–)6–10 mm, at base dilated on side further from petiole and auriculately amplexicaul, the auricle or the whole blade reflexed in age, the blade strongly parallel-nerved distally, becoming dry and deciduous before the lf.

Lvs (2–)2.5–7(–8.5) dm; petiole including the discolored, when dry shrunken pulvinus (2.5–)3.5–13(–16) cm, at middle 2–6(–7) mm diam, terete or faintly 2-ridged ventrally; rachis (1.1–)1.3–4.5(–5) dm, tapering and progressively more trigonous distally, the flat, narrowly margined ventral face of each interfoliolar segment gradually dilated upward, the terminal seta modified into a subcarneous (often eaten) lance-triangular or deltate conduplicate blade 1.5–5 × 1–4 mm; petiolar glands 0; pulvinules (when fresh orange-yellow) 2.5–6 mm; lfts (6–)7–13(–15) pairs, either conspicuously or scarcely accrescent distally (but the proximal 1–3 pairs always smallest), the distal or largest pair oblong, oblong- or exceptionally subrhombic-obovate (6.5–)7–18(–19) × (2–)3–7(–8.5) cm, 2–3.3 times as long as wide, at apex most commonly broadly rounded mucronulate, less often shallowly emarginate or deltately subacute, at base inequilaterally cuneate or rounded, the margin plane, the straight centric midrib immersed or almost so above, cariniform beneath, the 10–18 pairs of camptodrome secondary veins usually prominulous on both faces, sometimes faintly so above, the tertiary and reticular venulation sometimes finely raised on both, sometimes only on dorsal face, sometimes faint on both.

Racemes incurved to vertical many-flowered, in early anthesis capped by a blunt cone of bright yellow or orange-yellow (when dry fuscous) bracts, early and often much elongating, the stout axis including the short peduncle becoming (1–)1.5–5.5(–6) dm; bracts subpetaloid, amply obovate or rhombic-obovate, abruptly acuminulate (12–)15–22(–24) mm, minutely but densely puberulent dorsally, caducous as the pedicel begins to lengthen; pedicels at and after full anthesis (3–)4–9(–10) mm; fl-buds obliquely obovoid obtuse, minutely puberulent; sepals subpetaloid, little graduated, oblong-obovate-concave 10–14 mm, often disjointing before the petals; petals glabrous, yellow drying whitish or ochroleucous brown-veined, all conspicuously clawed, the blades little heteromorphic, oblong-obovate or obovate-flabellate, the longest petal (12–)14–22 mm; androecium glabrous, functionally 2-merous, the filaments of 4 median stamens 1.5–2.5 mm, of 2 latero-abaxial ones thickened and 2.5–5 mm, of the central abaxial one 4–5.5 mm, the anthers of 4 median stamens sterile or almost so, lance-oblong from shortly sagittate base, including the porrect 2-porose beak 3–5 mm, the anther of the centric abaxial one sterile 2.5–4.5 mm, its basal lobes hastately recurved, the 2 large fertile anthers lunately lanceolate in outline, including the spoutlike, porrect, 2-porose beak 9.5–12 mm; ovary densely fuscous-velutino-pilosulous, sometimes in addition pilose with longer pale hairs; style filiform or almost so, gently incurved 4.4–6 mm, at apex 0.2–0.3 mm diam; ovules 28–40.

Pod stiffly widely ascending or declined, broadly linear straight (8–)10–16 × (1.2–)1.3–1.75 cm, plano-compressed, bicarinate by the scarcely thickened sutures, cuneately contracted at base into a stipe ±3–6 mm, the stiffly papery lustrous, finely cross-venulose, nigrescent glabrate valves corrugately elevated over the seeds in low transverse ridges (not crested lengthwise), the locules 3–4 mm long but as wide as the pod itself; seeds transverse, narrowly oblong-claviform compressed parallel to the valves 5.1–7.6 × 1.9–2.3(–2.5) mm, the testa light brown sometimes faintly darker-mottled, often obscurely pitted, dull or moderately lustrous, crackled in age, the linear-elliptic, highly lustrous areole 3.1–5.1 × 0.4–0.7 mm; $n = 14$.—Collections: 145.

Forest margins and clearings, riverbanks, strands and shores, often in standing water, essentially a transient denizen of open places in wet macrothermic woodland, sometimes weedy but, unlike related *S. alata,* seldom forming extensive thickets, mostly below 400 m but ascending to 1000 m in Chiapas, 600 m in Honduras, and 1100 m along río Cauca in Colombia, discontinuously widespread from 18°30′N in s.-e. Mexico (Veracruz, Tabasco, Yucatán, s. Oaxaca, Chiapas) through Central America in the inter-Andean valleys of Colombia as far as Valle de Cauca and along the Pacific lowlands to Gulf of Guayaquil, Ecuador; again widespread through the Orinoco and Amazon basins from Portuguesa and Barinas, Venezuela e. to Trinidad, s.-e. through the Guianas and the Amazon delta to n. Piauí, Brazil and s.-w. on the s. affluents of the Amazon to Pasco in Peru, Rondônia in Brazil, and on Río Beni s. to 15°30′S in the Yungas of La Paz, Bolivia. At higher latitudes cultivated only; commonly infested with small stinging ants.—Fl. almost throughout the year, but most freely at the end of rainy seasons.—*Yaaxhabin* (Yucatan Maya); *barajo* (Mexico and C.A., from the resemblance of imbricated sleeping leaflets to a spread deck of cards); *palo de jiote, varajita* (Oaxaca); *mariloa* (in Chanantec); *huevo de tortuga* (Chiapas); *sambrán* (El Salvador); *laureño* (Panama); *bajagua* (C.A., Colombia); *galbecillo, dorancé* (Colombia); *tarantán* (Venezuela); *majagua* (Venezuela); *majagüillo* (Colombia, Venezuela); *mucuteno* (Venezuela); *carrion crow bush* (Trinidad); *srabitie* (Surinam); *dartrier* (French Guiana); *María mole* (Maranhão); *retama* (upper Amazonia); *tampush, shmashut* (Peru); *matapasto grande* (Brazilian Amazonia), some of these names applied indiscriminately to *C. alata.*

Senna reticulata and *S. alata* are strikingly similar at anthesis and as a pair are readily distinguished from other *Pictae* by their ample, distally accrescent, oblong-obovate leaflets and the massive cone or cylinder of petaloid, orange-yellow bracts that surmounts the several simultaneously expanded flowers. Of the two, *S. reticulata* is potentially the taller, more commonly arborescent and more pubescent plant, but the one differential character that is truly reliable at anthesis is the length of the petiole; in *S. reticulata* the first pair of leaflets is inserted upward of 3.5 cm distant from the pulvinus whereas in *S. alata* it is less than 3.5 cm distant, often subcontiguous to the pulvinus. In applying this criterion, care must be taken to note the true insertion of the first pair, which are readily deciduous in some specimens but leave a tell-tale scar on the leaf-stalk. The flat nigrescent pod containing seeds that are conformably flattened parallel to the valves are extremely different from the winged pod and obcompressed seeds of *S. alata,* so different in fact that Britton & Brown referred the species to different genera. The ingenuous opinion of the common people, who often have only one vernacular name for the two species and use them for the same medicinal purposes, appears nevertheless to acknowledge the close kinship between them expressed in multiple similarities of form, habit and properties.

148. Senna alata (Linnaeus) Roxburgh, Fl. Indica **2**: 349. 1824, based on *Cassia alata* Linnaeus, Sp. Pl. 378. 1753.—"*Habitat* in America *calidiore.*"—Holotypus (Brenan, 1967, p. 64), BM (hb. Cliffort., *Cassia* No. 3, a single leaf)!—Lectotypification by LINN *528.26,* proposed by De Wit (1955, p. 232), cannot be sustained, the specimen only acquired by Linnaeus from P. Browne in 1758; cf. Isely, 1975, p. 197.—*Cassia herpetica* Jacquin, Observ. Bot. **2**: 24, t. 45, fig. 2. 1767, nom superfl. based on *Cassia no. 9* P. Browne, Nat. Hist. Jam. 224. 1756, which explicitly = *C. alata* L.—*Herpetica alata* (Linnaeus) Rafinesque, Sylva Tellur. 123. 1838.

(?) *Cassia bracteata* Linnaeus fil., Suppl. Pl. Syst. Veg. 232. 1781.—"*Habitat in* Surinamo. C. G. Dalberg."—Holotypus, S (hb. Linn.), seen in microfiche (IDC 5075. 166: II.4)!—Provisionally equated with *C. alata* by Lamarck, Encycl. Méth. **1**: 648. 1785, and decisively so by Wight & Arnott, Prod. fl. pen. Ind. or. 287. 1834; this disposition, accepted by Vogel (1837, p. 21) and confirmed by Bentham (1871, p. 550), subsequently traditional. The typus, consisting of a half a leaf and a flowering raceme with ovaries just developing, is ambiguous between *S. alata* and *S. reticulata,* and the protologue, which calls for "foliola . . . subtus mollia, cana" would suggest the latter. By following tradition, we evade the consequences of replacing the well-known epithet *reticulata* (Willdenow, 1809) with the long buried *bracteata* (Linnaeus fil., 1781).

Cassia alata Linnaeus var. *rumphiana* DeCandolle, Prod. **2**: 492. 1825.—"In Java. *Herpetica* Rumph., [Herb. Amboin.] 7, [=Auctuarium, ed. Burmann: 35,] t. 18. [1755]."—Holotypus, the cited plate!

Cassia alata var. *perennis* Pampanini, Nuovo Giorn. Bot. Ital., n. ser. **14**: 595. 1907.—". . . in India orientali, probabiliter in insula Penang."—Described from a plant "nato spontaneamente nell Orto botanico di Firenze nel estate 1905," this surmised to have arisen from seed sown in 1901.—Holotypus, coll. *23.XI.07,* FI!

Cassia alata sensu Colladon, 1816, p. 91; Vogel, 1837, p. 21; Wight, Ic. Pl. Ind. Or. **1**(13): t. 253/890. 1840; Bentham, 1870, p. 126 & 1871, p. 550; Hooker, Curtis's Bot. Mag. **105**: t. 6425. 1879; Sessé & Mociño, 1888, p. 60 & 1893, p. 101; Prain, J. Asiatic Soc. Bengal **66**(2, pt. 1): 161. 1897; Perkins, Contrib. U.S. Nat. Herb. **10**: 160. 1907; Merrill, Philip. J. Sci. Bot. **5**: 50. 1910; Rock, Legum. Pl. Hawaii 83, t. 32. 1920; Standley, 1922, p. 410; Blake Contrib. U.S. Nat. Herb. **24**: 90, pl. 30. 1922; Merrill, Enum. Philip. Pl. **2**: 262. 1923; Pio Corrêa, Dicc. Pl. Ut. Bras. **2**: 518 + fig. 1931; Zingg, Philip. J. Sci. **54**: 256. 1934; Britton & Killip, 1936, p. 178; Standley & Steyermark, 1946, p. 109; Boelcke, Darwiniana **7**(2): 301, t. VIII, fig. e (seed); Pellegrin, Les Lég. du Gabon 96. 1948; León & Alain, 1951, p. 267, fig. 111; Schery, 1951, p. 53; Steyaert, 1952, p. 507; Dimitri & Rial Alberti, 1954, fig. 12; De Wit, 1955, p. 231; Symon, 1966, p. 95; Brenan, 1967, p. 64; Aubréville, 1968, p. 53, pl. 8, fig. 6; Henderson, Malay Wild Fls. **1**: 98, fig. 96. 1974; Isely, 1975, p. 63, map 16; Peixoto et al., 1978, 247, estampa 2.—Merian, Hist. Gén. Insectes t. 58. 1647, often referred here, =*Inga* sp!

Cassia bracteata sensu Willdenow, Sp. pl. **2**: 525. 1799; Poiret, Encycl. Méth. Suppl. **2**: 129. 1811; Colladon, 1816, p. 92.

Herpetica alata sensu Cook & Collins, Contrib. U.S. Nat. Herb. **8**(2): 159, t. 39. 1903; Britton & Rose, 1930, p. 243; Degener, New Ill. Fl. Hawaiian Is. pl. s.n. 1932.

Rankly leafy shrubs of rapid growth, either monopodial branched distally or several-stemmed from the ground and ± as broad as tall, at anthesis commonly 1–4 m, less often (in wet tropical Venezuela, Ecuador, the Guianas) arborescent and 4–6(–10) m with trunk to 7 cm diam, the older branches stout, pith-filled, the terete striate annotinous ones with lf-stalks, at least the major veins of the dorsal face of lfts and all axes of inflorescence densely minutely puberulent and often also pilosulous with erect pale lustrous hairs up to 0.1–0.25(–0.3) mm, the ample membranous lfts subconcolorous, dull or moderately lustrous, usually glabrous on upper face except for a line of thickened trichomes along midrib, less often thinly minutely pilosulous near the pulvinus or exceptionally overall, the vertical racemes at first axillary, sometimes later forming a candelabriform panicle, each capped in early and mid-anthesis with a blunt cone of imbricated yellow bracts.

Stipules firmly herbaceous deltate-trigangular or lance-ovate, dilated and auriculate-amplexicaul on the side further from the petiole, excluding the auricle 6–16(–20) mm, the coarsely nerved blade commonly reflexed or replicate in age, becoming dry, deciduous or subpersistent.

Lvs (2–)3–7(–7.5) dm; petiole including the discolored, when dry shrunken pulvinus 1–3.5(–5) cm, the true petiole often subobsolete, the first, relatively small pair of lfts inserted next to or shortly distant from the pulvinus and turned backward toward the stem; rachis ±1–6 dm, the segments rounded dorsally, on ventral side widely openly sulcate and slightly dilated upward, the sulcus interrupted between each pair of lfts, the terminal seta modified into a recurved deltate-subcordate acute conduplicate herbaceous blade 2.5–4.5 mm; petiolar glands

0; pulvinules (orange when fresh, then blackish) 2.5–6 mm; lfts (6–)7–14 pairs, ± strongly accrescent distally but either the distal or the penultimate pair, rarely some pairs near middle of blade largest, the terminal pair usually broadest, these broadly oblong or obovate from inequilaterally rounded or proximally semicordate base, 7–19(–21) × (3–)3.5–9.5(–13.5) cm, 1.2–2.5 times as long as wide, at apex usually broadly obtuse or subemarginate mucronulate, rarely depressed-deltate subacute, the margin plane or subrevolute, the centric straight midrib immersed or shallowly depressed above, cariniform beneath, the 10–16 pairs of camptodrome with (few, random) intercalary secondary veins and all tertiary and reticular venules finely prominulous on both faces, often more sharply so beneath.

Peduncles stout, incurved to vertical, 6.5–17 cm, near base 2.5–7 mm diam; racemes many-fld, dense before anthesis but the tapering axis much elongating, in fruit becoming 1.5–6 dm; bracts firm, yellow or orange with pallid submembranous margin, fuscous when dry, broadly rhombic-oblanceolate or obovate-flabellate, deltately subacute (1.3–)1.7–3 cm, densely puberulent dorsally, nidulating the young fl-buds, caducous as the pedicel elongates; pedicels at and after anthesis (4–)5–11 mm; fl-buds obliquely obovoid obtuse, puberulent (but inner sepals glabrate); sepals petaloid, deep chrome- or orange-yellow, little graduated, oblong-obovate, cucullate at tip, 11.5–16 mm, finely reticulate-venulose, the venation immersed; petals glabrous, bright yellow drying stramineous brown-veined, conspicuously clawed, subheteromorphic, the vexillar one longest (15–)16–23(–24) mm, its blade subquadrately obovate obtuse or emarginate, sometimes lyrately constricted near middle, the rest more broadly obovate, the 2 abaxial sometimes of slightly different lengths, all concave at full anthesis forming a bowl-shaped perianth; androecium glabrous, functionally 2-merous, the filaments of 4 median stamens 2.6–4 mm, of 2 large fertile ones dilated and 3.5–5 mm, of 1 abaxial one dilated and 6–10 mm, the anthers of 4 median and 1 centric abaxial stamens sterile or almost so, narrowly oblong straight, including the porrect, biporose beak 2.6–4.6 mm, those of the fertile abaxial ones stoutly lunately lanceolate 9.5–13 mm, unequally sharply sagittate at base, at apex abruptly truncate 2-umbonate and contracted into an abruptly divaricate biporose beak ±0.5 mm; ovary densely fuscous-puberulent and sometimes in addition minutely white-pilosulous; style filiform or gently tapering distally, strongly evenly incurved, 4–5.5(–6) × 0.2–0.4 mm; ovules ±44–58.

Pod stiffly widely ascending, subsessile, broadly linear in outline, straight or a trifle decurved, sharply tetragonal, carinate by the sutures, winged lengthwise down the middle of each valve, 11–18(–19) cm long, including the wings 2–2.8 cm wide, 9–12 mm thick from suture to suture, the papery, regularly crenulate wings 4–9(10) mm wide, the green, thinly fleshy valves becoming blackish and stiffly papery, puberulent or glabrate, transversely venulose, the cavity rhombic in cross section, the well-developed membranous interseminal partitions 2.5–3 mm apart, the locules occupying the whole cavity; dehiscence tardy, primarily through the inertly gaping ventral suture; seeds rhomboid widest above the middle, compressed parallel to the septa, their wide faces 5.5–6.8 × 4.5–5.6 mm, the testa smooth or obscurely rugulose, tan or dark brown, dull or moderately lustrous, crackled in age, the areole on the narrow faces elliptic or linear-elliptic 2.1–3 × 0.5–0.9 mm. n = 12.—Collections (Neotropical only): 126.

Riverbanks, lake shores, margins of ponds and ditches, seasonally wet savannas, becoming colonially weedy in disturbed brush-woodland, pastures, orchards and plantations, along roads and in waste places, mostly below 500 m but ascending to 1050 m on the Brazilian Planalto and up to 2000 m in New Guinea, native probably along rivers of the Guianas and perhaps around the periphery of the Orinoco and Amazon basins in Venezuela, Brazil and Colombia, now fully but erratically naturalized in the wet and seasonally humid Neo- and Paleotropics, in the New World from s. Mexico (Nayarit e. through Balsas Depression to s. Oaxaca and the Gulf lowlands) to n.-centr. Colombia, and through the Greater Antilles (Cuba, Jamaica, Hispaniola, Puerto Rico) and the Leeward and Windward Is. to Trinidad; from the Maracaibo and Orinoco basins in Venezuela s. and s.-e. through the Guianas and the Amazon delta region to Ceará, coastal Bahia, the highlands of s.-centr. Goiás, and less commonly on the sources of the Amazon from s.-e. Ecuador to n.-e. Bolivia, thence s., probably or certainly adventive, to Paraguay, n. Argentina (Corrientes) and coastal s.-e. Brazil (Paraná, Santa Catarina); cultivated in warm temperate United States, especially on the Gulf Coastal Plain and in s. California, locally spontaneous in peninsular Florida; fully established by mid-XVII century in Java and now widespread, both as a weed and in cultivation for medicine and ornament, through equatorial Africa and from s. India and Sri Lanka to s. China and the Philippines, thence extending through Malesia to n. Australia and e. through Micronesia to the Society and Hawaiian Is.—Fl. in N. and n. S. America mostly IX–III, in equatorial latitudes nearly throughout the year, s.-ward mostly I–VI.—*Flor del secreto* (s. Mexico); *tulipán* (s. Mexico); *guacamaya francesa* (Cuba); *taratana, talentro* (Antilles, variants of *tarantan*); *mocote* (Venezuela), *dartrier* (French colonial America); *soroncontil* (Nicaragua); *guajavo* (Dominican Republic); *Carrion Crow bush* (Guyana); *taperibá guazú* (Argentina); *ringworm shrub, candlesticks* (U.S.); *seven golden candlesticks* (Australia); *acapulco* (Philippines, with several variants, denoting origin in Mexico), in s. America known by the same names as *S. reticulata,* q.v.; Aubréville, Roxburgh, Zingg, Steyaert, De Wit, Merrill, ll. cc. quote many vernacular terms in use in Africa and Indo-Malesia.

149. Senna nicaraguensis (Bentham) Irwin & Barneby, comb. nov. *Cassia nicaraguensis* Bentham, Trans. Linn. Soc. London **27**: 552. 1871.—"Tropical America, Mexico, and Central America, *Seemann, Oersted.*"—Lectoholotypus, ticketed "In provincia Segovia, Nicaragua, 1851, *Œrsted, Cassia no. 37.*" K (hb. Benth.)! = NY Neg. *1421*; paratypus, *Œrsted, Cassia no. 38,* K (hb. Benth.)!—*Chamaesenna nicaraguensis* (Bentham) Britton & Rose, N. Amer. Fl. **23**(4): 250. 1930.

Cassia seleriana Harms, Bull. Herb. Boiss. **7**: 551. 1899.—"Habitat in Mexico, in prov. Chiapas in fruticeto apud Tonala: Sel[er] n. 2065; et in Guatemala, in dept. Santa Rosa ad Frajanes in 1000 m altitud.: J. Donn[ell] Smith n. 6116 (leg. Heyde et Lux)."—Lectoholotypus, *Seler 2065,* presumably †B; isotypi, GH, K!—Correctly equated with *C. nicaraguensis* by Standley, Contrib. U.S. Nat. Herb. **23**: 411. 1922.

Cassia nicaraguensis sensu Schery, 1951, p. 51, fig. 151.

Amply leafy arborescent shrubs at anthesis 1.5–7(–9) m, with stout subterete smooth or in age lenticellate branches clothed when young with large herbaceous semicordate stipules, except for sometimes minutely residually puberulent lf-stalks and (exceptionally) dorsal face of lfts glabrous throughout, the long plurifoliolate lvs sublustrously olivaceous either concolorous or slightly paler beneath, the inflorescence terminal paniculate, the panicle either leafy-bracteate or naked distally but not much exserted from foliage, the young racemes terminating in a blunt cone of imbricated petaloid bracts.

Stipules early caducous (present only to early anthesis, absent from fruiting spms), dimidiately cordate-acuminate 10–35 × (6–)7–18 mm, the plane blades amplexicaul by their obtuse, greatly dilated contrapetiolar basal angle, flabellately many-nerved from base, reticulate distally, the midrib falcately curved outward into an acute or caudate apex.

Lvs (disregarding depauperate ones in inflorescence) 16–40(–52) cm; petiole including discolored but little dilated pulvinus 3–9 cm, at middle (1.2–)1.5–2.8 mm diam, bluntly trigonous, obscurely sulcate, the lf-stalk distally becoming more sharply trigonous and openly shallow-sulcate; rachis 8–36 cm, its longer interfoliolar segments 10–21(–24) mm; petiolar glands 0; lfts (7–)9–17(–19) pairs, subequal or a little accrescent upward, ascending from rachis face upward on turgid, dorsally keeled, when dry wrinkled pulvinule (1–)1.4–2.8 mm, in outline varying (sometimes on one lf) from elliptic-, lance- or ovate-oblong to elliptic-oblanceolate, the larger (3.5–)4–8 × 1–2.8 cm, all either obtuse or deltately subacute and abruptly mucronate by the costa excurrent by (0.5–)1–3 mm, at asymmetric base broadly cuneate to shallowly cordate on proximal side, cuneate to rounded on distal one, the margin plane, the slender straight centric midrib cariniform beneath, less prominent or immersed above, the 10–16(–19) pairs of major camptodrome (with random intercalary) secondary veins and fine tertiary and reticulate venulation prominulous on both faces, the ultimate defined areoles of dorsal one much < 1 mm diam.

Racemes incurved to vertical, at first densely soon loosely (8–)15–70-fld, the tapering axis becoming (6–)9–40 cm; bracts petaloid yellow or submembranous brownish-yellow, obovate-cuneate or broadly obovate obtuse 18–25 mm concave, enveloping the young fl-buds, thrown off as the pedicel begins to elongate; pedicels at full anthesis and afterward ascending 12–21 mm; buds obovoid-ellipsoid obtuse; sepals submembranous yellowish, orange or brownish, little graduated, all broadly obovate or oblong-obovate obtuse, the innermost (13–)14–21 mm; petals (glabrous) yellow drying pale or stramineous, strongly arborescently brown-venulose, all obovate or oblong-obovate contracted at base into a slender claw, little heteromorphic except the vexillary one sometimes broadest, the longest 16–28 mm, commonly ⅓ longer than sepals, sometimes scarcely exserted; androecium glabrous 4-morphic: 3 adaxial staminodes with cordate blades; 4 median stamens with filaments 2–3.5 mm, anthers oblong, slightly incurved, the body 4–5 mm, contracted into a porrect truncate beak ±1 mm; 1 sterile abaxial stamen opposed to the vexillum, cruciform, the basal tails of the anther-sacs divaricate at nearly 90°; on either side of the last 2 long stamens with filament 5.5–7 mm, the anthers linear-oblong, strongly dorsoventrally compressed, incurved through ⅓ circle, (10.5–)11–18 mm, truncate at apex and produced as an obscure divergent beak ±0.5 mm; ovary glabrous; style filiform, hamately incurved at tip, 6.5–8.5 mm; ovules (15–)18–32(–35).

Pod randomly spreading-declined from stiff arched pedicels, short-stipitate, the stipe 3–6 mm, the linear-oblong, strongly flattened body 7–12.5 × 1–1.9 cm, bicarinate by the sutures, at base cuneately contracted into stipe, at apex more abruptly rounded-truncate into the small persistent style-base, the thin reddish-green or atropurpureous valves becoming papery lustrous, finely transverse-venulose, expressed over each seed in the form of an obtuse mound, these elevations forming a pronounced crest down the length of the valve; dehiscence through both sutures, inert; seeds uniseriate compressed-pyriform or subrhombic-pyriform 4–4.7 × 2.1–2.6 mm, on each face above the areole elevated into a mound conformable with the external protuberances of the valves, the ochraceous sublustrous testa rugulose-reticulate, the elliptic smooth areole 1.1–1.3 × 0.5–0.9 mm.—Collections: 62.—Fig. 11 (androecium), 13 (pod, seed).

Dry and seasonally moist mixed forest, tropical deciduous forest, disturbed woodlands and thicket remnants of forest, sometimes abundant locally in disturbed habitats along streams and highways, 30–1200 m, interruptedly dispersed between Panama Canal and s. Mexico, chiefly along the Pacific foothills and coastal plain, but reaching the Caribbean slope in w. Honduras and adjoining Guatemala and the Gulf slope in centr. Chiapas, in Mexico scattered along the seaward slope of Sa. Madre del Sur from Istmo de Tehuantepec to w. Jalisco.—Fl. IX–XII(–IV), fr. XI onward.—*Huevo de iguana* (Chiapas).

A handsome tall shrub or bushy treelet that deserves the attentions of gardeners in monsoon climates, *S. nicaraguensis* is recognized, in the context of sect. *Senna,* at anthesis by broad foliaceous, dimidiately cordate-amplexicaul stipules and petaloid bracts shielding the young flower buds, and in fruit by the lustrous, narrowly oblong, umbonately crested pod. These features, seldom coetaneous, coincide with long narrow multifoliolate leaves that are characteristic in the range of the species. In *S. nicaraguensis* the crest running lengthwise along the pod's valves is composed of simple mammillate cones raised over the seeds; in allied Brazilian *S. martiana,* similar in foliage and petaloid bracts but different in the narrow stipules, these projections are keeled both vertically and transversely and take the form of T-shaped studs. The Mexican *S. centranthera* resembles *S. nicaraguensis* in form and ornamentation of the pod but differs in the narrow stipules and yet more numerous, smaller leaflets. Where the ranges of this pair overlap in southern Mexico they are separated by life-zone and altitude.

150. Senna martiana (Bentham) Irwin & Barneby, comb. nov. *Cassia martiana* Bentham in Martius, Fl. Bras. **15**(2): 127. 1870.—". . . in provincia Bahiensi prope Joazeira [=Juàzeiro] in vicinia fluminis S. Francisco: Martius."—Holotypus, mounted with Martius's manuscript *Observatio 2349* which employs the preoccupied epithet *vogeliana,* M! = NY Neg. *10473*; isotypi, one labelled Vila Nova da Rainha [=Jacobina], M!

Cassia martiana sensu Bentham, 1871, p. 551; Rizzini, Rodriguesia **28**(41): 180. 1976–1977; Peixoto et al. 1978, passim & estampa 4.

Amply leafy, softly woody shrubs of rapid growth with spreading-incurved hornotinous branches, at anthesis 1.5–4 m, usually finely softly pilosulous throughout with fine erect shorter and longer hairs, the latter up to 0.2–0.35 mm but sometimes lacking, one or both faces of lfts sometimes glabrous, the dull olivaceous foliage sub-bicolored, the firmly membranous reticulate lfts a trifle paler beneath, the axillary long-pedunculate many-fld racemes forming in bud a cone of yellow bracts, early elongating, either immersed in or exserted from foliage.

Stipules erect or spreading, obliquely lance- or ovate-acuminate 8–20 × 2–5 mm, at base auriculate-amplexicaul on side further from petiole, the stiffly herbaceous blades becoming dry but not prickly, distally strongly striate-nerved, the margins loosely revolute, the whole deciduous before the lf.

Lvs 1.5–4 dm, the expanded blade linear-oblong-elliptic in outline, the lfts a little decrescent toward either end; petiole including little dilated pulvinus 6–30 mm, often not or scarcely longer than first interfoliolar segment of rachis; rachis 12–35 cm, the longer segments up to 7–18(–20) mm, rounded dorsally, sulcate laterally, thick-margined and openly shallow-sulcate ventrally; petiolar glands 0; pulvinules 0.5–1.3 mm, the lft-blades appearing sessile against lf-stalk; lfts

(10–)12–24(–26) pairs obliquely oblong, oblance- or narrowly ovate-oblong, deltate- or triangular-acute mucronate, the longest of a lf 3–5 × 0.6–1.7 cm, 2.7–4(–5) times as long as wide, at base semicordate on proximal and cuneate on distal side, the margin plane, the marginal nerve scarcely thickened, the centric straight midrib, (4–)5–8(–9) pairs of camptodrome and random intercalary secondary veins, tertiary connecting and reticular venulation all prominulous on both faces, sometimes more sharply so beneath.

Racemes stiffly erect ±20–70-fld, the early elongating axis together with stout peduncle becoming 2–5(–8) dm; bracts either firm or submembranous, yellow or fuscous-yellow, obovate acute or deltately acuminulate 12–20 × 8–14 mm, nidulating the young fl-buds, deciduous as pedicel elongates; pedicels at anthesis 3–12 mm, in fruit 6–21 mm; fl-buds very obliquely obovoid or almost semi-obovoid obtuse, densely puberulent, or distally glabrate; sepals yellow or the outer ones fuscous, little graduated, obovate concave, the longest 9.5–15 mm; petals vivid yellow, when dry whitish brown-veined, all obovate beyond the short narrow claw, the vexillar one slightly smaller and the 2 abaxial often slightly larger and more oblique than the rest, the longest 15–25 mm; androecium glabrous, functionally 2-merous, the filaments of 4 median stamens 1.5–2.5 mm, of 2 latero-abaxial ones 2–3 mm, of the centric abaxial one 4–5 mm, the anthers of 4 median stamens ovate-oblong nearly straight 2.8–5.5 mm, contracted into a porrect 2-porose beak ±0.7 mm but sterile or almost so, those of 2 large fertile stamens lunately lanceolate in outline 8–14 × 2–2.3 mm, inequilaterally sagittate at base, gradually contracted distally into a scarcely defined shortly 2-lipped 2-porose beak, that of the centric abaxial one staminodal or much reduced and sterile, 1.5–5 mm; ovary densely but sometimes very minutely pilosulous-puberulent; style filiform 6–11 mm, slightly incurved and 0.15–0.3 mm diam near apex; ovules 10–16.

Pod obliquely ascending-spreading, narrowly oblong plano-compressed 7–10.5 × 1.6–2.4 cm, contracted cuneately at base into a stipe 3–6 mm, bicarinate by the scarcely thickened sutures, the stiffly papery or subcoriaceous valves fuscous or nigrescent, umbonately expressed over the seeds in the form of 1 vertical and one transverse ridge forming a T-shaped stud, the seed-locules 4.5–8 mm long, nearly as wide as the pod's cavity; seeds (little known) broadly obovoid compressed parallel to the valves ±5–6 mm long and nearly as wide, colliculate or transversely ridged on each face above the areole, the atropurpureous testa intricately rugulose sublustrous, the broadly elliptic areole ±1 × 0.8 mm.—Collections: 13.

Caatinga thickets, riverbanks and around ephemeral ponds, becoming colonial and weedy along roadsides, 20–450 m, apparently local in n.-e. Brazil between 5° and 11°S, from Rio Grande do Norte (Angicos) s. through Paraíba and Pernambuco to extreme n. Bahia (Bonfim and Joàzeiro upstream along R. São Francisco to the lower R. Pardo).—Fl. irregularly through the year.—*Canafístula caixão*; *c. brava*.

In habit and foliage this opulently decorative senna resembles the African *S. diplobotrya,* but differs in the obliquely semiamplexicaul (not symmetrically cordate) stipules, the subsessile, intricately reticulate leaflets, the crested pod, and the relatively broad rugulose seeds with small areole. This species is variable in length of pedicels and petals, in size of the long anthers, and length of the individual seed-locules in the pod. These variable features show, in our too small sample, some degree of correlation, both between themselves and collectively with dispersal. The following contrasts will need testing as material accumulates; if found viable, they will define taxonomic varieties:

1. Pedicels at anthesis ±12 mm, in fruit 18–21 mm; longer petals 20–25 mm; long abaxial anthers ±13–14 mm; locules of pod 6.5–8 mm long; n. Bahia (foliage pubescent).—Includes the typus of *S. martiana*.
1. Pedicels at anthesis 3–5 mm, in fruit ±6–8 mm; petals to ±15 mm; long anthers 8–9.5 mm; locules of pod ±4.5–5.5 mm; Pernambuco n.-ward (foliage either pubescent or glabrous).

151. Senna didymobotrya (Fresenius) Irwin & Barneby, comb. nov. *Cassia didymobotrya* Fresenius, Flora (Regensburg) **22**: 53. 1839.—". . . in Abyssinia a cl. Rüppell . . ."—No holotypus survives at FR (H. J. Conert in litt. 21.I.1979); our interpretation follows, with some reluctance, that of Bentham and all subsequent literature.

Cassia verdickii de Wildeman, Ann. Mus. Congo, Bot. Sér. 4, fasc. **2**: 49, t. 16, fig. 6–11.—"Lukafu [Katanga], juillet 1900 (Ct. Verdick, n. 590)."—No typus seen.—Equated by Steyaert, 1952, l.c. infra, with *C. didymobotrya,* but the plate does not show the enlarged floral bracts typical of the species and the (9–11-jugate) leaflets are fewer than normal.

Cassia nairobensis L. H. Bailey, Hortus Second 146. 1941, nom. sine descr. lat.—Cf. Brenan, 1967, l.c.

Cassia didymobotrya sensu Bentham, 1871, p. 551; Robyns, Fl. Spermat. Parc Nat. Albert **1**: 373, pl. 37. 1948; Schery, 1951, p. 50; Steyaert, 1952, p. 504, pl. XXXVI; De Wit, 1955, p. 241; Brenan, 1967, p. 66, fig. 12; Isely, 1975, p. 80.

Shrubs, usually several-stemmed and of broadly rounded outline, at anthesis (0.6–)1–5 m, reportedly sometimes arborescent to 9 m, the terete striate, densely leafy annotinous branchlets with foliage and inflorescence pilosulous throughout or almost so with erect or spreading-ascending white hairs up to 0.2–0.6 mm often (especially on raceme axis) mixed with random granular or longer ± flexuous thickened trichomes, the mustily foetid foliage subconcolorous, the dull-olivaceous lfts equally pubescent on both faces or rarely glabrous above, the racemes incurved to vertical from distal axils, forming a panicle at length ± exserted.

Stipules thinly herbaceous, early reflexed, symmetrically ovate-cordate-acuminate 6–17 mm, palmately several-nerved from base, tardily deciduous.

Lvs mostly 1–3.5(–5) dm, in expanded outline narrowly oblong-elliptic; petiole including discolored but not much swollen pulvinus (1–)1.5–4.5(–5.5) cm, at middle 1.3–3 mm diam, rounded dorsally, very narrowly margined and shallow-sulcate ventrally; rachis ±0.7–2.8(–4) dm; petiolar glands 0; pulvinules 1.2–2.7 mm; lfts (8–)9–16(–18) pairs, slightly decrescent at either end of rachis, from asymmetric base elliptic-oblong ±2–5.5(–6.5) × 0.8–2(–2.5) cm, 2.3–3 times as long as wide, at apex obtuse or broadly deltate-subacute, aristate-mucronate, the pallid margin plane, the centric straight midrib immersed above, cariniform beneath, the (6–)7–11 pairs of slender camptodrome secondary veins finely prominulous beneath, not or scarcely so above, the tertiary venulation faint, not forming a raised reticulum.

Peduncles with tapering axis of many-fld raceme together 1–4 dm, the several fls simultaneously expanded overtopped by a cone of firm brownish or blackish-green bracts, these broadly ovate, abruptly subacute or acuminulate concave (8–)10–17 mm, caducous; pedicels at anthesis and afterward 4–10 mm; fl-buds obliquely obovoid, densely puberulent; sepals firm, oblong-obovate, little graduated, the longest (9–)10–14 mm; petals glabrous, yellow, fading stramineous brown-veined, little heteromorphic, beyond the short slender claw oblong or obovate, the longest 17–25(–27) mm, at anthesis all concave, forming a bowl-shaped perianth; androecium glabrous, functionally 2-merous, the anthers of 4 median stamens 2.5–3.3 mm, of 2 large abaxial ones 3–4 mm, of 1 centric abaxial one 4–5 mm, the anthers of 4 median and 1 (slightly longer) centric abaxial stamens including the pronounced porrect beak 4–5.5 mm, those of 2 large fertile ones lunately lanceolate in outline 9–11 mm, the short porrect biporose beak 0.6–1 mm; ovary densely yellowish- or pallid-pilosulous; style linear, gently incurved distally 7.5–12 mm, just below stigma 0.2–0.4 mm diam; ovules ±12–17.

Pod stiffly ascending or randomly spreading-declined, the stipe 7–11 mm, the linear-oblong, strongly compressed body 7–12 × (1.5–)1.7–2.3(–2.5) cm, bicarinate by the sutures, the fuscous-green, finally papery nigrescent valves finely pilosulous with short sometimes mixed with longer white hairs, corrugate by low obtuse ridges elevated over each seed, finely transverse-venulose, the locules between the narrow membranous interseminal septa 4–5.5 mm long, occupying the whole width of cavity; seeds oblong narrowed proximally, compressed parallel to the valves, 6–8.2(–"9") × 3–4.6(–"5") mm, the testa pale tan sometimes overlain with fuscous-green, smooth or obscurely pitted, moderately lustrous, crackled in age, the oblong-elliptic areole 3–4 × 0.7–1.1(–"1.5") mm.—Collections: 30.

Savanna thickets, margin of forest galleries, lake shores and riverbanks, becoming colonial and weedy in old fields, coppiced woodland and on roadsides, widespread in tropical Africa, whence introduced into tropical and warm temperate horticulture, sporadically naturalized in s. India, Sri Lanka, Malesia, s. peninsular Florida, apparently only cultivated in s. California, Cuba, Mexico, Panama.—Fl. in N. America mostly XII–II.

This coarsely handsome but ill-scented African senna, occasionally cultivated in North America and liable to become naturalized in frostless regions, resembles our native *S. nicaraguensis* and *S. martiana* in the narrow multifoliolate leaves and the massive erect racemes capped by a cone of dilated bracts. At anthesis it differs from both in the symmetrically cordate stipules and the less highly colored, usually fuscous rather than orange-yellow floral bracts, further from *S. nicaraguensis* in the short filaments of the two large abaxial stamens, and further from *S. martiana* in the essentially simple pinnate rather than reticulate venulation of the leaflets. The simply corrugate, not crested or umbonate pod of *S. didymobotrya* is decisively characteristic.

152. Senna centranthera Irwin & Barneby, sp. nov., *S. nicaraguensi* affinis eique quoad perianthium ac androecium persimilis, sed stipulis parvis (6–10 nec 10–35 mm longis) haud dimidiatis nec semicordato-amplexicaulibus, bracteis parvis (2–7.5 nec 18–25 mm longis), foliolis simul duplo numerosioribus (15–33 nec 10–16-jugis) ac dimidio brevioribus (15–40 nec 35–80 mm longis), sepalisque minoribus (9–11.5 nec 13–21 mm longis) manifeste diversa.—MEXICO. Jalisco: w. slope of mountains 11 mi. s. of Autlán, 1350 m, 25.IX.1966 (fl), *W. R. Anderson & C. W. Laskowski 3745*.—Holotypus, NY; isotypi, GH, US.

Arborescent shrubs with trunk to 3(–?) cm diam, becoming 4–5 m tall, the hornotinous branches densely clad in long multifoliolate lvs and terminating in an ample panicle of many-fld racemes, except for the glabrous upper or both faces of the dull olivaceous, either subcon- or discolorous lfts and for the dorsally glabrous sepals commonly pilosulous throughout with spreading-ascending pale hairs to 0.2–0.45 mm, sometimes glabrate except for residually puberulent lf-stalks, the panicle of racemes at least distally leafless and exserted from foliage.

Stipules early reflexed, herbaceous, deltate-ovate-acuminate or broadly lanceolate 6–12 × 2.5–5(–6) mm, deciduous.

Lvs 15–35 cm; petiole including little swollen but discolored pulvinus 2–3.5 cm, at middle 1–1.6 mm diam, rounded dorsally, narrowly winged and open-sulcate ventrally, the wings of lf-stalk segments slightly widened upward and the sulcus there deeper; petiolar glands 0; lfts 20–33 pairs, the proximal ones smaller but the rest not or little graduated, all ascending from rachis, face upward, on scarcely dilated pulvinule 0.7–1.3 mm, in outline lance- to narrowly ovate-oblong or lance-elliptic, obtuse mucronulate (8–)15–40 × (3–)4–12 mm, 2.7–4 times longer than wide, at strongly assymmetric base rounded on both sides or broadly cuneate distally, the slender midrib with ±8–10 pairs of major camptodrome and often some intercalary secondary veins finely prominulous on both faces, the tertiary venulation yet finer, scarcely raised.

Racemes crowded along a straight primary axis 1–2.5 dm, the lower ones leafy-bracteate, at first densely then loosely 10–45-fld, the 1–2 simultaneously expanded fls standing well below the unopened buds, the axis including peduncle becoming 7–20 cm; bracts submembranous lance- or ovate-acuminate 2–4.5 mm, caducous from below young buds at first elongation of pedicel; pedicels at and after full anthesis 9–16 mm; fl-buds obovoid-ellipsoid obtuse, except for sometimes minutely ciliolate sepals glabrous; sepals submembranous, brown except where exposed in vernation, the inner ones subpetaloid, all broadly oblong-elliptic concave, finely many-nerved from base, subequal in length but the inner a little broader, the longest 9–11.5 mm; petals yellow drying stramineous brown-veined, all glabrous, little heteromorphic but varying from broadly oblanceolate to obovate-cuneate beyond the slender claw, the 3 adaxial a trifle longer than the abaxial pair, the longest 13–20 mm; androecium glabrous, functionally 2-merous, the filaments of 4 median stamens 1.8–2.5 mm, of the 3 abaxial ones 3.3–5 mm, the anthers of 4 median and the centric abaxial one sterile or greatly reduced in size, shallowly sigmoid 3–4.5 mm, with porrect beak 0.7–1.3 mm, those of the 2 large abaxial ones lunately incurved 10–14 mm, almost imperceptibly narrowed distally into an erect beak dehiscent by minute terminal, ultimately confluent slits, all anthers strongly dorsoventrally compressed, sagittate at base, the acute descending spurs of the larger anthers 1.5–2.5 mm; ovary glabrous, narrowed into a filiform incurved or hooked style 4.5–6.5 mm; ovules ±24–30.

Pod obliquely pendulous, the stipe ±3 mm, the broadly linear body 7–11 × ±1.4 cm, strongly plano-compressed and carinate by the sutures but the lustrous green, at length stiffly papery valves raised over each seed as a prominent fuscous umbo, the seed-locules ±4 mm long and twice as wide, occupying a little more than half the pod's width; seed unknown.—Collections: 12.—Fig. 28.

Seasonally wet open woodland and semi-deciduous scrub thickets in hill-savanna, sometimes on pedregal and probably confined to volcanic bedrock, 1350–2000 m, local along the Pacific slope and crest of Sa. Madre del Sur in s. Jalisco (mun. Autlán), w. Michoacán (mun. Tsitsio, Quitupán, Cotija de la Paz and Coalcomán), Guerrero (mun. Tixtla and Chilpancingo), and w. Oaxaca (distr. Juxtlahuaca), Mexico.—Fl. VI–XI.—*Palo María.*

Senna centranthera appears to be a submontane derivative of lowland *S. nicaraguensis,* morphologically modified by reduction of stipules and bracts and by simultaneous doubling in number and halving in size of the leaflets. It might appear a priori that this evolutionary sequence is reversed, for foliaceous stipules and ample colored bracts must surely be advanced features in *Senna* as a whole. In the narrower context of ser. *Pictae,* however, the petaloid bracts, unless they have made independent parallel appearances in North America, South America, Africa and Australia, which seems unlikely, are antecedent to the smaller, more conventional ones found in *S. centranthera* and its Andean kindred described below.

The range, habitat and superficial gross aspect of *S. centranthera* roughly coincide with those of *S.* (sect. *Chamaesenna*) *multifoliolata* and possibly may give

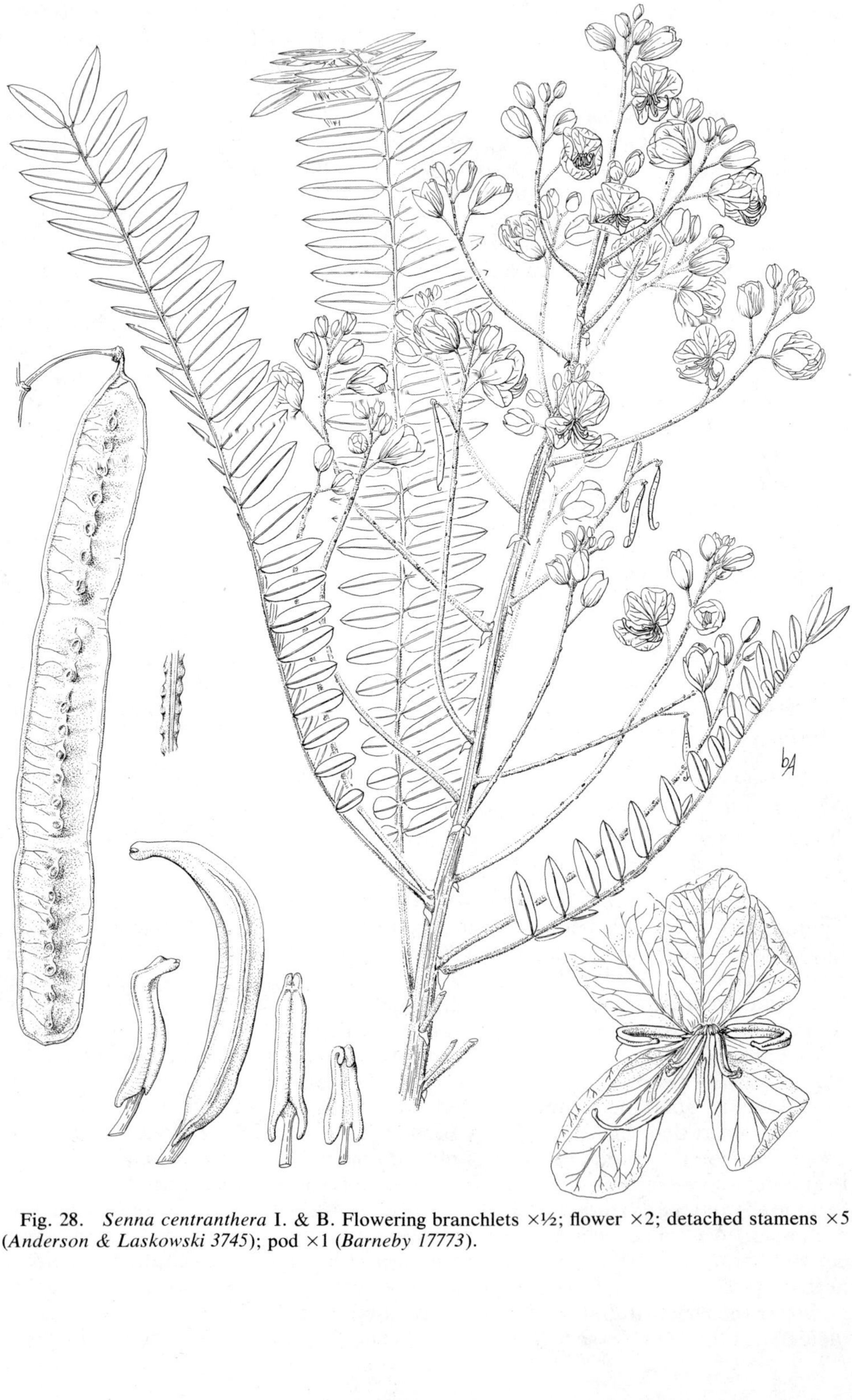

Fig. 28. *Senna centranthera* I. & B. Flowering branchlets ×½; flower ×2; detached stamens ×5 (*Anderson & Laskowski 3745*); pod ×1 (*Barneby 17773*).

rise to misidentification. The glandless leaf-stalk, the exserted panicle, and most importantly the androecium proper to sect. *Senna,* not to mention the cristate pod, are the fundamentally distinctive characters of *S. centranthera.*

153. Senna pistaciifolia (Humboldt, Bonpland & Kunth) Irwin & Barneby, comb. nov. *Cassia pistaciifolia* Humboldt, Bonpland & Kunth, Nov. Gen. & Sp. 6(folio): 349 ("*pistaciaefolia*"). 1824.—Typus infra sub var. *pistaciifolia* indicatur.

Amply leafy shrubs with smooth terete hornotinous branches and branchlets, at anthesis mostly 0.8–2.5 m, reportedly sometimes arborescent to 10 m, often appearing and sometimes truly glabrous but the young stems, lf-stalks, lower face of lfts and axes of inflorescence sometimes puberulent or pilosulous with weak hairs up to 0.4–0.7 mm, the stiffly membranous lfts dull olivaceous on both faces, paler beneath, the loosely many-fld racemes at first axillary to distal lvs, then either immersed in foliage or exserted, some smaller later ones often forming a terminal depauperately leafy or leafless panicle.

Stipules (caducous, poorly known) firm erect obliquely lance-acuminate 3–12 × 1.5–3 mm, ± dilated at base on the side further from petiole, strongly striate-nerved.

Lvs (0.7–)1–4 dm; petiole including moderately swollen pulvinus (1–)2.5–7 cm, at middle 0.7–2.3 mm diam, rounded dorsally, very narrowly thick-margined and plano-sulcate ventrally; rachis (2–)3–24 cm; petiolar glands 0; lfts 3–10(–11) pairs accrescent distally but the penultimate one sometimes largest, this and the distal pair broadly oblanceolate, oblong-obovate or elliptic (3.5–)4.5–10.5 × (1.2–)1.5–4.7 cm, ±2–3 times as long as wide, at apex broadly rounded, deltate or (if damaged) emarginate, commonly conspicuously mucronate, at base (when relatively broad) inequilaterally rounded or subcordate or (when narrower) cuneate on the distal side, the midrib centric straight, the 10–16 pairs of camptodrome (and often random intercalary) secondary veins with tertiary and reticular venulation all prominulous on both faces, a trifle more sharply so dorsally, there going out into a raised mesh of areoles variable in size but the smallest 0.5 mm or less diam.

Racemes loosely 10–45 fld, the axis together with peduncle becoming (0.5–)0.7–3 dm; bracts firm, not or scarcely colored, ovate apiculate or ovate- or lance-acuminate, together forming around the very young fl-buds a small cone, caducous as the pedicels begin to elongate, absent from many flowering and all fruiting spms; pedicels at and after full anthesis 12–20 mm; fl-buds obliquely obovoid or oblong-obovoid obtuse, glabrous or thinly puberulent; sepals little graduated, oblong or obovate obtuse 8–14.5 mm, either yellowish or fuscous, membranous-margined, the tips cucullate; petals yellow with darker veins, when dried whitish brown-veined, glabrous, heteromorphic, all 3 adaxial or only the vexillar petals oblanceolate to narrowly obovate, the 2 lateral sometimes and 2 abaxial always obovate, the latter obliquely so and usually largest, the longest petal 15–22 mm; androecium glabrous, the filaments of 4 median stamens 1.3–2.5 mm, of 3 abaxial ones 3–5.5 mm, those of 2 latero-abaxial ones much thickened, the anthers of 4 median stamens oblong, dorsoventrally compressed, little incurved 3–5 mm, abruptly contracted into a porrect biporose beak 0.7–1.4 mm, those of the 2 latero-abaxial ones lunately incurved 13–16 mm, at base inequilaterally sagittate (the longer descending spur 2.5–3 mm), at apex very slightly contracted below a short 2-lipped beak, the upper lip 2-porose, the central abaxial anther sterile, 2.5–3 mm, its basal lobes spreading-decurved ±1 mm; ovary either glabrous or strigulose; style 4.5–6.5 mm, subfiliform slightly dilated toward the gently incurved apex, 0.2–0.4 mm diam just below the oblique stigma; ovules (17–)18–30.

Pod obliquely declined, the stipe 3.5–5 mm, the oblong plano-compressed body (7–)8–12 × 1.6–2.4 cm, the papery glabrous or early glabrate valves greenish with fuscous stripe down middle, finely venulose, tardily separating along the scarcely thickened sutures, the interseminal septa 3.5–5 mm apart, forming locules 7–16 mm wide extending transversely across ⅔ to nearly the whole width of the cavity; seeds (not seen ripe) horizontal to the pod's long axis, apparently narrowly obovate compressed parallel to the valves, the areole oblong-elliptic occupying more than half the seed-face.

The material that has accumulated in herbaria under the names *Cassia pistaciifolia, C. fraxinifolia* (considered by Bentham only varietally distinct from the preceding), *C. pearcei* and *C. picta* appears to represent a single major specific entity. All have essentially the same perianth and characteristic androecium, and the same pod except for some variation in width of the seed-locules and in degree to which the valves are elevated over the seeds. These minor features of the pod, however, are sufficiently correlated with variation in pubescence, number of leaflets, venation of the bracts and dispersal to justify the recognition of three varieties.

Key to the Varieties of *S. pistaciifolia*

1. Bracts puberulent dorsally, faintly keeled but not prominently several-nerved; lfts (3–)4–6(–7) pairs, their midrib ± depressed-sulcate ventrally, their margin subrevolute; centr. Ecuador n. through Cauca valley to n.-w. Colombia (Antioquia).
   153a. var. *pistaciifolia* (p. 472).
1. Bracts glabrous, prominently 5–several-nerved dorsally; lfts variable in number, but the midrib prominulous ventrally and the margin plane; s. Ecuador to Bolivia and Galápagos Is.
   2. Lfts 6–11 pairs, at least in most larger lvs; pod plane, the valves indistinctly elevated over the seeds as faint corrugations; n. Peru (Lambayeque, Cajamarca and Amazonas) to Bolivia, on the Amazon slope except marginally at the n.-w. limit.
      153b. var. *glabra* (p. 473).
   2. Lfts 3–8, in most upper lvs not over 6 pairs; valves of ripe pod colliculately elevated over each seed, the colliculi confluent into a low crest along middle of pod; Pacific n.-w. Peru (Piura), Pacific s.-w. Ecuador and Galápagos Is. 153c. var. *picta* (p. 473).

153a. Senna pistaciifolia (Humboldt, Bonpland & Kunth) Irwin & Barneby var. **pistaciifolia**. *Cassia pistaciifolia* Humboldt, Bonpland & Kunth, 1824, l.c., sens. str.—"Crescit locis montosis, inter urbem Popayan et Meneses, alt. 700 hex."—Holotypus, consisting of 1 lf + pods, ticketed "n. 2133. Popayan," P-HBK! isotypus, B (hb. Willd. *7946*)!—*Chamaesenna pistaciifolia* (Humboldt, Bonpland & Kunth) Britton ex Britton & Killip, Ann. N.Y. Acad. Sci. **35**(3): 179. 1936.

Chamaesenna colombiana Britton & Killip, Ann. N.Y. Acad. Sci. **35**(3): 179. 1936.—"Vicinity of Medellín, Colombia. Type from Boquerón, January 5, 1928, [*Rafael A.*] *Toro 866*, type . . . Cucaracho, Antioquia, December 25, 1927, *Toro 839*. Antioquia, *Jervise*; *Triana*."—Holotypus, *Toro 866*, NY! = NY Neg. *4958*; paratypi, *Toro 839*, NY! *Jervise s.n.*, K! *Triana s.n.*, K!

Lfts either glabrous or softly pilosulous beneath; otherwise as given in varietal key.—Collections: 7.

Habitat little known, to be expected in disturbed upland forest or residual thickets in formerly forested valleys, ±1200–2000 m, local in the w. Andes of Colombia, especially in the Cauca valley between Medellín and Popayán (Antioquia and Cauca), and apparently disjunct in central Andean Ecuador (Chimborazo).—Fl. X–II(–?).

The populations in the Cauca valley that differ from the holotypic phase of *Cassia pistaciifolia* in having leaflets glabrous on both faces do not differ otherwise, and we cannot maintain even varietally the two species of *Chamaesenna* postulated by Britton & Killip. The disjunct record from Ecuador (Huigra, *Rose 23505*) is based on plants bearing year-old pods and newly expanding leaves, and may represent a distinct form. Rose recorded it as a shrub, with hard wood used for house-building.

153b. Senna pistaciifolia (Humboldt, Bonpland & Kunth) var. **glabra** (Bentham) Irwin & Barneby, comb. nov. *Cassia pistaciifolia* var. *glabra* Bentham, Trans. Linn. Soc. London **27**: 551. 1871, based on *C. fraxinifolia* Humboldt, Bonpland & Kunth, Nov. Gen. & Sp. **6**(folio): 349. 1824.— ". . . inter flumen Amazonum [=Marañón] et oppidum Jaen de Bracamoros [Cajamarca, Peru], alt. 250 hex."—Holotypus, so labelled, P-HBK! former isotypus, †B, survives as F Neg. *1731*!

Cassia pearcei Bentham, Trans. Linn. Soc. London **27**: 552. 1871.—"Bolivia, La Bama, at an elevation of 9000–10000 feet, *Pearce*."—Holotypus, *Pearce s.n.* from "La Banca," K! = NY Neg. *1563*; isotypus, BM!
Cassia cuspidata sensu Macbride, 1943, p. 162.
Cassia pearcei sensu Macbride, 1943, p. 177.

Foliage usually quite glabrous, exceptionally the lfts strigulose beneath; otherwise as in key.—Collections: 12.

Evergreen and deciduous brush-woodland (?450–)900–2800 m, scattered along the e. slope of the Peruvian Andes, on headwaters of Marañón and Ucayali rivers, from Cajamarca and immediately adjoining Lambayeque to Cuzco and (presumably adjoining n.-w.) Bolivia, at its extreme n.-w. limit in Lambayeque crossing the crest to the edge of the Pacific slope.—Fl. IX–IV(–VI).

The var. *glabra* differs from var. *pistaciifolia* most visibly in the more numerous, usually quite glabrous leaflets. These are also the supposed differential characters of *C. pearcei* which we cannot distinguish from south Peruvian var. *glabra*. We have not found on any map the type-locality of *C. pearcei,* written by Bentham (perhaps by typographic error) "La Bama," but appearing on the label as "La Banca." The place was said by Bentham to be in Bolivia, but this does not appear on the original label; it may have been in Peru.

The name *C. cuspidata,* published by Willdenow in 1809, which has obviously several years' priority over *C. fraxinifolia* or *C. pistaciifolia,* was equated with the former by Vogel (1837, p. 37) and taken up by Macbride in *Flora of Peru* for the plant here called *S. pistaciifolia* var. *glabra*. The holotypus consists of a single leaf from a plant grown in the Botanic Garden at Berlin, preserved as *Herb. Willdenov. 7979* (B). We are unable to identify this with certainty and are obliged to follow Bentham in treating *C. cuspidata* as a doubtful synonym.

153c. Senna pistaciifolia (Humboldt, Bonpland & Kunth) Irwin & Barneby var. **picta** (G. Don) Irwin & Barneby, stat. nov. *Cassia picta* G. Don, Gen. Hist. Dichl. Pl. **2**: 444. 1832.—"Native of Guayaquil [Ecuador]."—No typus found at BM or OXF; presumed isotypi, "Ex Guayaquil," *hb. Pavon. 14/27,* MA! and *s.n.,* FI (*hb. Webb.*)!

Cassia applanata N. J. Andersson, Kongl. Svensk. Vet.-Akad. Handl. **1853**: 254. 1854.—"[Galápagos Is.:] . . . locis graminosis regionis mediae insulae Chatham."—Holotypus, S = photo at US! presumed isotypi, †B = F Neg. *1728*, BR, K, W. (*Andersson 237*)!—Equated by Bentham (1871, p. 552) with *Cassia picta*, a disposition accepted by subsequent students of Galápagos flora.

Cassia picta sensu Robinson, Proc. Amer. Acad. Sci. **38**: 149. 1902; Rudd in Wiggins & Porter, Fl. Galápagos 605, map. 1971.

Lfts always glabrous, otherwise as given in key.—Collections: 19.

Scrub thickets, forest clearings, roadsides, on Galápagos a pioneer in malpais, locally abundant on the Pacific lowlands and Andean foothills below 600 m in s.-w. Ecuador (Guayas), n.-w. Peru (Piura) and the Galápagos archipelago (Is. Fernandina, Isabela, San Salvador, Sta. Cruz, Sta. María, San Cristóbal).—Fl. III–VII(–XII), probably throughout the year unless drought-inhibited.

154. Senna viarum (Little) Irwin & Barneby, comb. nov. *Cassia viarum* Little, Phytologia **18**: 470, fig. 16. 1969, based on *Chamaesenna velutina* Britton & Killip, Ann. N.Y. Acad. Sci. **35**(3): 179, pl. II. 1936.—"[In the] Plaza [of] Mutiscua, Norte de Santander, Colombia, 2600 m . . . February 20, 1927, *Killip & Smith 19655*, type . . . Bogotá, *Popenoe*."—Holotypus, NY! isotypi, K, US!—*Cassia velutina* (Britton & Killip) García-Barriga & Forero-González, Cat. Ilustr. Pl. Cundinamarca **3**: 61, fig. 21. 1968; non *C. velutina* Vogel ≡ **Senna velutina** (Vogel) Irwin & Barneby.

Cassia velutina sensu García-Barriga, Rev. Acad. Colomb. Ci. Exact. Fis. Nat. **13**(50): 273. 1968.

Trees with amply leafy rounded crown, at anthesis (3–)4–15 m, the smooth gray trunk up to ±1–1.5 dm diam, the annotinous stems and lf-stalks densely pilosulous with straight erect or commonly flexuous and ± entangled, pale shining hairs up to (0.3–)0.4–0.8 mm, the pubescence extending outward ± equally densely along the midrib and (especially dorsally) the secondary veins, sometimes also to the intervenium of the lfts, the thin-textured foliage bicolored, the dull olivaceous lfts paler beneath, the glabrescent racemes at first axillary to and shorter than cauline lvs but often distally forming a small exserted panicle.

Stipules early deflexed, lanceolate or obliquely ovate-caudate 6–20 × 4–8 mm, the herbaceous blades revolute, finely striate-nerved, pubescent adaxially as well as dorsally, persistent often until fall of lf.

Lvs 15–35 cm; petioles including scarcely swollen pulvinus (2–)4.5–8(–9) cm, at middle 1.5–4 mm diam, subterete but very narrowly thick-margined and openly sulcate ventrally; rachis 6–16 cm; petiolar gland 0; pulvinules 3.5–7(–8) mm; lfts 4–6 pairs, accrescent distally but the penultimate longest, the distal pair a little smaller, all broadly lance- or narrowly ovate-acuminate, the largest 7–15 × 2.3–5 cm, 3–4 times as long as wide, at base inequilaterally rounded or on proximal side shallowly cordate, the margin revolute, the straight centric midrib deeply impressed above, cariniform beneath, the 9–15 pairs of camptodrome (with random intercalary) secondary, connecting tertiary and subsequent open reticular venulation all sharply finely prominulous on both faces.

Racemes loosely ±10–30-fld, the elongating axis together with peduncle (0.7–)1–3 dm; bracts firm, broadly ovate acute 4–7 mm, nidulating the very young fl-buds but cast off as soon as the pedicel elongates; pedicels ascending, at full anthesis and afterward (1.5–)2–4.2 cm; buds obliquely obovoid obtuse, fuscous-yellowish or atropurpureous; sepals firm, striate-nerved, oblong-obovate obtuse, little graduated, the longest (10–)11–17 mm; petals yellow, when dry stramineous brown-veined, of nearly equal length but heteromorphic, the vexillar one cuneate-oblanceolate, the rest obovate, the 2 abaxial ones more obliquely and broadly so, the longest 15–23 mm; androecium glabrous, functionally 6-merous, the filaments of 4 median stamens 1.5–2 mm, of 3 abaxial ones 5–8 mm, those of the 2 latero-abaxial much dilated, the anthers of 4 median stamens oblong nearly straight 4.5–5.5 mm abruptly narrowed into a porrect 2-porose beak 0.7–1 mm, that of the central abaxial one sterile, 5.5–8 mm, prolonged at base into falcately descending lobes 1–1.5 mm, those of 2 latero-abaxial ones much the largest, lunately incurved, 14–18 mm, the theca on the abaxial side prolonged backward into a sharp tail 2–3.5 mm, that on the other much shorter, the whole anther gradually tapering at apex, minutely 2-lipped at apex; ovary subglabrous, less commonly pilosulous; style (5–)6–8 mm, slightly incurved and tapering into the oblique stigmatic cavity, near apex 0.35–0.5 mm diam.

Pod stiffly obliquely descending or pendulous, the stipe 2–3 mm, the narrowly oblong plano-compressed body mostly 10–13 × 1.8–2.1 cm (shorter only due to abortion of some ovules), ±2 mm thick, the obtuse or subacute sutures little thickened, the stiff pithy valves becoming atrofuscous along the middle, paler at the edges, finely transverse-venulose, scarcely elevated over the seeds, the cavity divided into transverse chambers several times wider than long; seeds transverse, in outline narrowly oblanceolate 5.5–7 × 2–2.6 mm, the testa smooth brown sublustrous, ultimately crackled, the linear-elliptic areole 3.5–4 × 0.6–1 mm.—Collections: 15.

Scrub thickets and disturbed brush-woodland, 2600–2900 m, described as native on and near the Andean crest in centr. Ecuador and s. Colombia (Nariño) but also cultivated there in parks, gardens and as a street tree, also planted and occasionally escaping into hedges and waste places in n.-e. Colombia (Cundinamarca, especially in and around Bogotá, and Norte de Santander, to be expected elsewhere).—Fl. VIII–IV, perhaps intermittently through the year.—*Chalán* (Ecuador); *alcaparra de Bogotá.*

A coarse but handsome, truly arborescent senna, related to *S. pistaciifolia* sens. lat., but distinct in its reflexed and revolute, long-persistent stipules, in leaflets broadest below rather than at or above the middle, and in the relatively long filaments of the abaxial stamens. Field observations, so far as they go, suggest that *S. viarum* is indigenous only in Ecuador and extreme southern Colombia, but even there is locally cultivated. Northward in the eastern cordillera it is both cultivated and naturalized; the typus was collected in a village square.

155. Senna cushina (Macbride) Irwin & Barneby, comb. nov. *Cassia cushina* Macbride, Field Mus., Bot. **13**(Fl. Peru), part 3(1): 161. 1943, a legitimate substitute for *C. andina* Rose ex Macbride, Field Mus., Bot. **8**: 93. 1930; non *C. andina* Philippi, 1858.—"Peru . . . Cushi, Dept. of Huánuco, June 19–23. 1923, *Macbride 4825* . . ."—Holotypus, F! = F Neg. *53945*.

Trees 6–10 m, with habit and pubescence almost of *S. viarum,* but the lfts much more numerous and pedicels shorter.

Stipules obliquely lanceolate 10–16 × 3–4 mm, reflexed and coiled horizontally around stem, deciduous.

Lvs ±3–4 dm; petiole 5–8 × 0.2–0.3 cm; rachis 16–24 cm; lfts 7–16 pairs, all broadly lance-acuminate, the longest (usually penultimate) pair mostly 8–11 × 1.5–3 cm, all dull olivaceous, glabrous above except sometimes for a line of hairs along the depressed midrib, beneath pallid and pilosulous with loose shining hairs up to 0.7–1 mm, the margin revolute, the ±14–16 pairs of camptodrome secondary veins immersed or faintly raised above, sharply prominulous beneath.

Inflorescence of *S. viarum,* but pedicels at full anthesis only 11–20 mm, like those of *S. pistaciifolia*; bracts deltate-flabellate ±2.5–3 × 2.5–3 mm, beyond middle several-nerved, thinly pilosulous dorsally; fl-buds almost or quite glabrous, the brownish-yellow sepals up to ±11 mm; long petals ±16 mm; androecium of *S. viarum,* the filament of 2 long anthers ±4 mm; ovary either densely velutinous or thinly puberulent; style ±4.5 mm; ovules 16–24.

Pod unknown.—Collections: 2.

Humid woodland of the *monte* formation, ±2000–2300 m, known only from 2 localities on the Amazonian slope of the Andes in n.-e. and centr. Peru: Marañón valley in Amazonas (above Pomacochas, prov. Bongará); and on a fork of Río Pachitea in e. Huánuco (Cushi).—Fl. VI–IX.

Our inchoate concept and description of *S. cushina* are derived from two flowering specimens, the typus which has mature flowers but has lost bracts and stipules, and *Ferreyra & Ancleto 15183* from Amazonas, which retains both bracts and stipules at the earliest stage of anthesis. Although to some degree complementary they fall far short of providing conclusive evidence of a species distinct from the already highly variable *S. pistaciifolia,* widespread at similar elevations in the Peruvian Andes. The relatively numerous lance-acuminate leaflets, marginally revolute and loosely pilosulous beneath, provisionally separate the species both from *S. pistaciifolia* and the closely allied *S. viarum*; but the taxonomy of the whole group will need reevaluation as more is learned about the details of dispersal, about variation in shape and number of leaflets, and particularly about the configuration of the mature pod.

Dii. ser. SPINIGERAE Irwin & Barneby

Senna sect. **Senna** ser. **Spinigerae** Irwin & Barneby, ser. nov., racemo ante anthesin bracteis amplis imbricatis strobilaceo cum nonnullis ser. *Pictarum* speciebus congruae, sed ab eis imprimis stipulis in spinam validam horizontalem vulnerantem transformatis, ulterius foliorum rachi rigida subspinescenti, foliolisque pauci (4–6)-jugis spinuloso-mucronatis diversae.—Sp. unica: **S. spinigera** (Rizzini) Irwin & Barneby ≡ *Cassia spinigera* Rizzini.

Resembling ser. *Pictae* in form of inflorescence, floral morphology, pod and seeds but the stipules transformed into narrowly lanceiform persistent, horizontally divaricate spines and the 4–6 pairs of apically spinulose lfts borne on a stiff lf-stalk sharply spinulose at tip.—Sp. 1, of local dispersal in Atlantic s.-e. Brazil.

We are obliged to describe an independent series for this truly remarkable senna, unlike any other in the long straight stipular spines that stand out at right angles from the stem, and are echoed upward by a spinulose seta terminating the leafstalk and a pinlike mucro formed by the excurrent midrib of the leaflets, both more or less vulnerant. Despite these unique features, the flowers and pods of *S. spinigera,* as pointed out in the protologue, are essentially like those of ser. *Pictae,* and the enlarged colored bracts gathered into a cone at apex of the young flowering raceme are not different from those of *S. alata, S. reticulata, S. martiana* or *S.* (ser. *Aculeatae*) *aculeata.*

156. Senna spinigera (Rizzini) Irwin & Barneby, comb. nov. *Cassia spinigera* Rizzini, Rev. Brasil, Biol. **32**(3): 319, fig. 1–4. 1972.—". . . Baixo Guandú, Espírito Santo, ab Apparicio P. Duarte n. 13.973(20-V-1971) lecta florifera et fructifera."—Holotypus, RB! isotypus, NY!

Cassia spinigera sensu Peixota et al., 1978, passim & estampas 1, 3.

Amply leafy subshrubs with stout diffuse and incurved-ascending subterete purplish-castaneous stems up to 1 m armed at the nodes with stout stipular spines and by spinescent lf-stalks, glabrous except for minutely puberulent axes of inflorescence and ovary, the chartaceous foliage subconcolorous, scarcely lustrous, the few axillary and pseudoterminal racemes loosely many-flowered, far exserted. Stipules horizontally divaricate, in outline narrowly lanceolate, 1–2 cm, slightly compressed laterally, vulnerantly spinulose, persistent.

Lvs 8–13 cm; petiole including moderately swollen pulvinus ±1.5–2 cm, at middle ±1 mm diam, rounded dorsally, narrowly thick-margined and shallowly sulcate ventrally; rachis 3.5–5.5 cm, the terminal seta modified into a vulnerant spinule 5–6 mm; petiolar glands 0; pulvinules 2–2.5 mm, wrinkled when dry; lfts (4–)5–6 pairs, accrescent distally, the distal pair broadly obovate-cuneate, shallowly widely emarginate ±4.5–6 × 2–3 cm, spinulose-mucronate at apex, at base decurrently rounded on proximal and cuneate on distal side, the straight centric midrib with ±13–15 pairs of camptodrome secondary and reticular venules all immersed or almost so above, sharply prominulous beneath; smaller proximal lfts proportionately wider, obovate, less asymmetric at base.

Peduncle together with tapering raceme-axis ±2–3 cm, at early and mid-anthesis bearing at tip young fl-buds enclosed in a cone of yellow (when dry fuscous) bracts; pedicels at and after full anthesis 3–5 mm; bracteoles 0; bracts broadly ovate concave ±10–12 mm, caducous as the pedicel elongates; fl-buds obliquely obovoid obtuse; sepals firm but probably yellow when fresh, oblong-obovate, little graduated, 8.5–12 mm, faintly nerved from base; petals yellow, brown-veined when dried, elliptic-obovate or obovate-flabellate, little heteromorphic, the longest ±13–14 mm; androecium glabrous, functionally 2-merous, the filaments of 4 median and 2 large fertile stamens ±1.5 mm, of the centric abaxial staminodal one 4–5 mm, the anthers of 4 median stamens and of the centric abaxial one sterile or almost so, narrowly oblong straight, including the short porrect 2-fid beak 3.5–4 mm, those of the 2 fertile ones lanceolate, lunately incurved, ±7 mm, sagittate at base, gradually narrowed at apex and produced into a very small divaricate 2-porose beak; ovary densely but very minutely velvety-puberulent; style linear ±4 mm; ovules ±16.

Pod declined, the stipe ±1–1.5 mm, the plano-compressed, broadly linear-oblong body 5.5–7.5 × 1.1–1.4 cm, straight or very slightly incurved, bicarinate by the scarcely thickened sutures, the stiffly chartaceous, rather coarsely transverse-venulose, lustrously atropurpureous or nigrescent valves low-corrugate over the seeds but not crested lengthwise, tardily separating along both sutures, the seed-locules 3.5–4 mm long, the partitions between them closed but not elevated into septa; seeds oblong-obovoid compressed parallel to the valves ±4–6 mm, the testa lustrous, the areole oblong-elliptic ±4 × 2 mm.—Collection: 1.

On rocks in dry woodland near 100 m, known only from Rio Doce valley on the border of Minas Gerais and Espírito Santo (Baixo Guandú, near lat. 19°30′S), s.-e. Brazil.—Fl. V–VI.

Rizzini has described and figured the seedling plant of *S. spinigera,* the stem of which, immediately above the cotyledons, swells into a fusiform orange axis up to 5–7 mm diameter bearing the first foliage leaves and accompanying stipules, these already spinuliform but soft. Fertilization, mostly by *Xylocopa* sp., dehiscence of pod and germination of the seed have been studied by Peixoto et al. on plants cultivated at Jardim Botánico, Rio de Janeiro.

Diii. ser. ACULEATAE Irwin & Barneby

Senna sect. **Senna** ser. **Aculeatae** Irwin & Barneby, stat. et nom. nov. *Echinocassia* Britton & Rose, N. Amer. Fl. **23**(4): 251. 1930, pro gen.—Sp. unica: *E. aculeata* (Bentham) Britton & Rose ≡ **Senna aculeata** (Bentham) Irwin & Barneby.

Resembling ser. *Pictae* in form of inflorescence, floral morphology, pod and seeds, but the stems and lf-stalks armed with acicular or compressed, vulnerant epidermal prickles, and the raceme-axis together with dorsal face of the ample colored floral bracts and margin of some lfts charged with capitate-glandular hairs.—Sp. 1, of seasonally flooded savanna habitats in n. and e.-centr. South America and Cuba.

The only member of its genus at once prickly and glandular, *S. aculeata* is further notable for the coriaceous texture of its broad, cordately amplexicaul stipules and of its lance-oblong, marginally thickened leaflets, all of which are tipped with a sharp spinule. The straight flat unappendaged pod is essentially like that of *S.* (ser. *Pictae*) *reticulata*. The species represents a specialized offshoot of ser. *Pictae* adapted to a habitat shunned by most sennas.

157. Senna aculeata (Bentham) Irwin & Barneby, comb. nov. *Cassia aculeata* Pohl ex Bentham in Martius, Fl. Bras. **15**(2): 128, t. 39. 1870.—". . . in provincia Goyaz prope Curralinho: Pohl; in uliginosis districtus Paranagoa provinciae Piauhy: Gardner n. 2551."—Lectoholotypus (Britton & Rose, 1930, p. 251, by implication): *Pohl 120,* K! isotypi, NY, W! paratypi, *Gardner 2551,* BM, FI, NY, W!—*Chamaesenna aculeata* (Bentham) Pittier, Arb. y Arbust. Legum. II. Cesalpin. 131. 1928. *Echinocassia aculeata* (Bentham) Britton & Rose, N. Amer. Fl. **23**(4): 251. 1930.

Cassia aculeata sensu Bentham, 1871, p. 551; Pio Corrêa, Dicc. Pl. Ut. Bras. **2**: 587, fig. 1931.

Coarse prickly diffuse or assurgent shrubs at anthesis 0.6–3 m, with terete purplish-brown stems armed, like the lf-stalks, with scattered straight acicular or deltate-acuminate and laterally compressed, either straight or recurved, vulnerant epidermal prickles up to 2–7 mm, the axis of inflorescence and often the young branchlets, some lf-stalks, and margin of some lfts charged with erect capitately thickened glandular trichomes up to 0.2–0.5 mm, the foliage olivaceous subglaucescent, the stiff subcoriaceous spine-tipped lfts subconcolorous glabrous on both faces, the erect racemes axillary to distal lvs or, by inhibition of the leading shoot, appearing terminal.

Stipules erect foliaceous broadly ovate acute or acuminate (1.5–)2–6 × (0.8–)1.2–3.3 cm, at base cordate-amplexicaul, the plane blades many-nerved from base, tardily deciduous.

Lvs (9–)12–31 cm, the blades narrowly oblong or oblong-elliptic in outline, the proximal lfts smallest, the rest subequal or slightly decrescent distally; petiole including the discolored but little swollen pulvinus (1–)1.5)–3(–3.5) cm, at middle (1.2–)1.5–2.5 mm diam, terete except for 2 narrow approximate ventral ridges enclosing the narrow sulcus; rachis 7–22 cm; petiolar gland 0; pulvinules 1–2 mm; lfts 6–14, of lvs subtending peduncles seldom over 9 pairs, oblong- or lance-oblong acuminate, the longest of each lf 2.5–7 × 1–1.8(–2) cm, 2.7–4.2 times as long as wide, at apex contracted into a vulnerant spinule 0.8–2 mm, at base inequilaterally rounded or semicordate, the margin bordered with a thickened pallid, smooth or glandular-verruculose nerve, the subcentric midrib with ±9–14 pairs of camptodrome (and random intercalary) secondary nerves faintly raised above, sharply prominulous beneath, an irregular tertiary-reticular venulation finely prominulous dorsally, invisible above.

Racemes shortly pedunculate many-fld, before anthesis of first fl taking the form of a clavate cone of petaloid bracts, the axis quickly elongating and elevating the bracteal cone beyond the open fls, at length 1–4 dm; bracts membranous yellow, dorsally capitate-glandular, obovate or oblong- or rhombic-obovate from a broadly cuneate base, 11–22 × 9–14 mm, at apex obtuse or deltately subacute, folded over the young fl-bud, deciduous as the pedicel elongates; pedicels at full anthesis and afterward 8–20 mm; fl-buds obliquely obovoid, at least at base capitate-glandular; sepals membranous petaloid, not much graduated, oblong-obovate or obovate obtuse, the longest 8.5–12.5 mm; petals glabrous, yellow with darker yellow veins, drying whitish brown-veined, the vexillar one oblong or oblong-obovate, the rest ± obliquely obovate, the 2 abaxial slightly longer, the longest petal 13–16 mm; androecium glabrous, functionally 2-merous, the 3 adaxial members staminodal, the anthers of 4 median and one central abaxial stamens greatly reduced and sterile or almost so, the filaments of 4 median stamens 1.5–2.5 mm, of 2 fertile ones thickened and 2–3 mm, of the centric abaxial one filiform 4–4.5 mm, the anthers of 4 median and of the centric abaxial stamens narrowly ovate, shortly porrectly beaked 2.5–4 mm, of the 2 fertile ones lunately lanceolate strongly obcompressed 8.5–10 mm, sagittate at base, contracted at apex into a porrect beak ±0.8–1 mm, this forming an introrsely oriented pollen-cup sometimes divided by a filiform septum; ovary glabrous; style filiform 4–5 mm, gently incurved and 0.2–0.25 mm diam just below the subacute apex; ovules 20–28.

Pod obliquely ascending, broadly linear-oblong, plano-compressed, 8.5–11(–12.5) × 1.9–2.7(–3) cm, cuneately contracted at base into a stipe ±3–5 mm, bicarinate by the scarcely thickened sutures, the papery valves fuscous except toward the pallid margins, low-corrugate by simple transverse ridges expressed over the seeds, when ripe separating along both sutures and ± recurved from the ventral side; seed-locules 3–4 mm long, extending transversely almost from suture to suture; seeds transverse, oblong-obovate or obovate, compressed parallel to the valves, 6.3–8.5 × 3–4 mm, the testa brown lustrous, the linear areole 3.8–5.3 × 0.6–0.9 mm.—Collections: 36.

Annually flooded savannas, pantanal thickets, lagoon margins, becoming weedy in disturbed grassland, wet pastures and on roadsides, 10–500 m, of tricentric dispersal: locally abundant in scattered stations around the s. edge of Amazon Basin from the upper forks of R. Guaporé in e. Bolivia and adjoining Brazil e. to s. Piauí and s.-centr. Goiás, thence s. in the Paraguai Basin through Mato Grosso to centr. Paraguay; río Apure basin in Portuguesa, Barinas, w. Guárico and Apure, Venezuela, very likely naturalized; both cultivated and escaped in w. Cuba (La Habana).—Fl. in Brazil and adjacent countries mostly IV–VIII, in Venezuela IX–I, in Cuba I–III, but governed probably by water supply rather than by season.—*Taperiba guarú* (Paraguay); *Flor de Barinas* (Venezuela).

Div. ser. SENNA

Senna sect. **Senna** ser. **Senna**. *Senna* P. Miller, 1754, l.c., sensu str.—Typus supra sub genere indicatur.

Cassia subgen. *Senna* sect. *Chamaesenna* ser. *Brachycarpae* Bentham, Trans. Linn. Soc. London **27**: 552. 1871 (non sect. *Chamaefistula* ser. *Brachycarpae* Bentham, op. cit. p. 529).—Sp. lectotypica: *C. angustifolia* Vahl = **Senna alexandrina** P. Miller.

Cassia sect. *Senna* sensu Vogel, 1837, p. 35, ex parte, quoad spp. 101–103, caeteris exceptis.
Senna sensu Batka, Bot. Zeit. **7**(11): 185–194. 1849; C. Martius, Versuch Monogr. Sennesbl. 1–159. 1857; Batka, Monogr. Cassien gruppe Senna 1–152. 1866; Britton & Rose, N. Amer. Fl. **23**(4): 245. 1930.

Corolla and androecium of sect. *Senna*; hypanthium slenderly funnelform, much elongated (1.5–4 mm), usually longer than the true pedicel, solid for most of its length, concave distally; ovules 6–10; pod plano-compressed, broadly oblong in profile, straight or incurved, the papery valves low-convex or crested over the seeds, these compressed parallel to the valves, emarginate at apex, the testa rugulose, the areole engraved into a raised ridge.—Spp. 7, of tropical savanna and warm temperate desert climates, potentially frutescent but flowering precociously and often monocarpic, primarily African, 2 extending e. into s.-w. Asia and India, these naturalized in West Indies and Caribbean S. America.

Our narrow definition of ser. *Senna,* differentiated from allied and otherwise similar series by the elongate hypanthium, is coextensive with the original genus *Senna* and with Bentham's *Cassia* ser. *Brachycarpae* (of sect. *Chamaesenna,* not of sect. *Chamaefistula*), except that we exclude the Australian *pleurocarpa* F. Mueller, which seems best transferred to ser. *Pictae.* The peculiar, narrowly funnelform hypanthium of ser. *Senna* is solid through most of its length, that is to say fused with the vascular column of the pistil, and concave only near the orifice. It simulates a pedicel, but is differentiated from the very short genuine pedicel by a joint and a change in coloration or texture. The corolla and androecium, and in particular the pincerlike pair of long anthers displayed horizontal to the flower's vertical axis, are essentially those of ser. *Pictae,* but the seeds are different in the pyriform-obcordate broad profile, in the elaborately rugulose testa and in the elevation of the areole on a salient ridge.

The two species of ser. *Senna* adventive in the Neotropics are those known historically from remote antiquity for the purgative properties of their leaves and pods. The strictly botanical and voluminous medicinal literature concerning them goes back far beyond Linnaeus and we have made no effort to summarize it or to compile a full synonymy. The interested reader should turn to the monographs of Carl Martius (1849) and Batka (1866), the latter especially for its superbly detailed pictures of the species known up to that time. For the systematics of the African members of the series Brenan's paper (1958) precursory to *Flora of Tropical East Africa* is indispensable, but the group as a whole still needs a comprehensive monographic review. This, however, can hardly upset the nomenclature, at least within genus *Senna,* of the species naturalized in the Americas, which are represented in our territory by typical and average races of their sort. In genus *Cassia* (sens. lat.) the nomenclature is somewhat less securely established, due to an intricately confused historical development. The Alexandrian or narrow-leaved senna with plane pods and the Italian or broad-leaved senna with crested pods were both included by Linnaeus in his *Cassia senna,* but shortly thereafter opinion became divided about which element was typical and deserved the epithet *senna.* Lamarck in 1785, perhaps following N. L. Burman (Fl. ind. 96, t. 33, fig. 2. 1768), interpreted the Italian as typical, a viewpoint that had some historical justification, although it had formed for Linnaeus a variety of β distinct from the typical. Colladon, working under the aegis of De-Candolle, abandoned *C. senna* Linn. as a mixture, and called the Italian and Alexandrian sennas *C. obovata* and *C. acutifolia* respectively. While these names are now, by retroactive law, illegitimate or superfluous, they were adopted by Vogel and by Bentham (1871) and became standard usage. It is only in modern times that *C. senna* has reemerged as the correct name for the narrow-leaved species with plane pods. Yet the Lamarckian viewpoint survives in Symon (1966), who considered the Italian senna to be the type-species of genus *Senna* Miller, mistakenly dated to the eighth edition of Gardeners' Dictionary rather than to the abridged fourth of 1754.

Key to the Neotropical Species of ser. *Senna*

1. Terminal lfts lanceolate 4–7 times as long as wide; 2 longest anthers (with us) 9.5–10 mm; valves of pod shallowly elevated over the seeds, plane when half-formed; established on Hispaniola and in s. Mexico, perhaps elsewhere. 158. *S. alexandrina* (p. 481).
1. Terminal lfts obovate 1.4–2.5 times as long as wide; 2 longest anthers (with us) 7.5–8.5 mm: valves of pod elevated over each seed, parallel to the pod's long axis, as an obtuse crest 3–4 mm long and ±1.5 mm tall, this developing long before the pod reaches full size or maturity; circum-Caribbean weed. 159. *S. italica* (p. 482).

158. Senna alexandrina P. Miller, Gard. Dict. ed. 8, *Senna* no. 1. 1768.—". . . *Senna alexandrina sive foliis acutis* C[aspar] B[auhin] P[inax 397.] 1623 . . . grows naturally in Egypt."—No typus seen. The name *S. alexandrina,* unaccountably overlooked by Bentham (1871), has since Colladon (1816, p. 94) been equated with *Cassia senna α* Linnaeus, *C. lanceolata* Forsk. and *C. acutifolia* Delile, while the specific epithet is the earliest available in genus *Senna* for *Cassia senna* Linnaeus, sens. str. and the historic pre-Linnaean one for the Egyptian senna of the pharmacopeias.

Cassia senna Linnaeus, Sp. Pl. 377. 1753, ex parte, exclus. var. *β*.—"*Habitat in Ægypto.*"—No typus found at LINN or BM (hb. Cliffort.); lectotypus, Morison, Pl. Hist. Univ. Oxon. **2**(2): t. 24, fig. 1. 1715, cited in the protologue.

Cassia angustifolia Vahl, Symb. bot. **1**: 29. 1790.—"Habitat in Arabia. Inter Forskalaei plantas . . . inveni."—No typus examined. Referred to *C. senna* by Brenan, 1958, p. 243.—*Senna angustifolia* (Vahl) Batka, Bot. Zeit. **7**(11): 193, t. 2, fig. 2. 1849.

Cassia acutifolia Delile, Fl. d'Egypte, Expl. Pl. 75. 1813 + t. 27, fig. 1. 1826.—"Cette plante croît dans les vallées du désert, au midi et à l'est de Syène."—No typus examined. Referred to *C. senna* by Brenan, 1958, l.c.—*Senna acutifolia* (Delile) Batka, Bot. Zeit. **7**(11): 193. 1849.

Senna alexandrina Garsault, Fig. pl. méd. **1**: pl. 42, fig. B & Expl. abr. pl. 33. 1765.—". . . vient dans la Perse, l'Arabie, la Syrie."—Name unacceptable, because binary nomenclature not uniformly adopted.—*Cassia alexandrina* (Garsault) Thellung, Bull. Herb. Boiss. II, **8**: 788. 1908.

Cassia lanceolata sensu Colladon, 1816, t. 15, fig. C.

Senna angustifolia sensu C. Martius, Versuch Monogr. Sennesbl. 64, 73–75 (bibliography). 1857; Batka, Monogr. Cassien gruppe Senna 30 + tab. (optima!). 1866. *Cassia angustifolia* sensu Bentham, 1871, p. 553; Baillon, Nat. Hist. Pl. ed. angl. **2**: 120, fig. 101–102. 1872.

Cassia senna sensu Andrews, Fl. pl. Anglo-Egypt. Sudan **2**: 118, fig. 49. 1952; Keay in Hutchinson & Dalziel, Fl. W. Trop. Afr. ed. 2, **1**(2): 453, fig. 148c. 1958; Isely, 1975, p. 124.

Ours erect-ascending, precociously flowering, ultimately suffrutescent herbs at anthesis 3–10 (in Paleotropics potentially –30) dm, sparsely strigulose with fine, truly appressed straight hairs up to 0.15–0.3 mm, the foliage pallidly olivaceous, concolorous or yellowish above, the axillary and finally subterminal loose racemes surpassing their lf.

Stipules ascending or spreading lance-subulate 1.5–3 mm, at base dilated on side further from petiole, there ±1 mm wide, early dry deciduous.

Lvs 5–16 cm; petiole including wrinkled but not much dilated pulvinus 1–2.5 cm, at middle 0.5–1.2 mm diam, shallowly sulcate ventrally; rachis (2.5–)3–13 cm; petiolar glands 0 (but spicules present between lfts); pulvinules ±1 mm; lfts (3–)4–8 pairs, ± accrescent distally, in outline lanceolate acuminulate or acute mucronulate, the longest (distal or penultimate) ±2.5–5 × 0.5–1 cm, 4–7 times as long as wide, at base subsymmetrically cuneate, the margin plane, the midrib and 3–5(–6) pairs of narrowly ascending secondary veins prominulous on both faces, an open irregular tertiary venulation raised or not.

Racemes (7–)10–30-fld, the fl-buds elevated well beyond the 2–4 simultaneously open flowers, the elongating axis together with peduncle becoming 1–2 dm; bracts membranous or membranous-margined, yellowish or fuscous, obovate 5–7 × 4.5–6 mm, at base shortly auriculate on both sides, enveloping the young fl-buds, cast off as the pedicel elongates; pedicels at and after anthesis 1–2(–2.5) mm, surmounted by a narrowly funnelform hypanthium 2.5–4 mm (easily and often in the past mistaken for part of pedicel); fl-buds ascending, obliquely obovoid or semi-obovoid obtuse; sepals submembranous fuscous or yellowish pallid-margined, glabrous or microscopically papillose-puberulent, in outline obovate or oblong-oblanceolate and of subequal length ±9–11 mm; petals glabrous yellow or orange-yellow drying stramincous or whitish brown-veined, in outline obovate or oblong-obovate beyond the short claw, ±13.5–15 mm; androecium glabrous, the 3 adaxial members minute or sterile (anthers 1.1–2 mm), the filaments of 4 median ones 2–2.6 mm, of 2 latero-abaxial ones 2.2–2.5 mm, of the centric abaxial one 3.5–4 mm, the anthers of 4 median and the centric abaxial one 3.3–4.3 mm, straight or almost so, of the 2 latero-abaxial lunately incurved, narrowly lance-attenuate 9.5–10 × 1.3–1.5 mm, at base inequilaterally sagittate, all 7 fertile anthers at truncate apex bluntly 2-umbonate on ventral side and 2-porose on dorsal one; ovary densely strigulose; style glabrous linear-filiform, evenly incurved from base, sometimes a trifle thickened distally, at or just below the obliquely terminal stigmatic cavity 0.2–0.25 mm diam; ovules 6–10.

Pod spreading or declined stipitate, the true stipe (beyond orifice of hypanthium) 2.5–3 mm, the broadly oblong or oblong-elliptic, straight or gently incurved, plano-compressed body with us 4–5.5 × 1.6–1.7 (in paleotropic races potentially up to 7 × 2.6) cm, rounded at both ends and at base abruptly cuneately contracted into stipe, the style-base becoming infraterminal, the thinly papery dull or moderately lustrous valves livid-brown becoming paler along the slender sutures, slightly elevated but not longitudinally crested over ripe seeds, the cavity dilated only along middle and there narrowly but completely membranous-septate; dehiscence inert but complete, along both sutures: seeds compressed parallel to the valves, in outline oblong-paddle-shaped, at apex truncate-emarginate, 5.2–6.2 × 3–3.6 mm, the ivory-stramineous testa sharply elaborately sinuous-rugulose overall, the oblanceolate areole 1.5–2.4 × 0.4–0.5 mm engraved into a vertical ridge at proximal end of seed; n = 14 (?13).—Collections: 30 (4 neotropical).—Fig. 10 (androecium), 13 (pod, seed), both under synonym *angustifolia*.

Waysides and waste places in the lowlands, adventive and naturalized in Hispaniola (s.-centr. Dominican Republic and Haiti) and in the Balsas valley in s. Mexico, cultivated and possibly casual elsewhere in the New World; native of grasslands and deserts of Tropical Africa mostly n. of the Equator, extending e. through Near East to the plains of peninsular India.—Fl. in the West Indies year around.—*True senna*; *Alexandrian s.*; *Tinnevelly s.*; *sen de España* (Guerrero).

The exareolate seed illustrated as that of *Cassia angustifolia* by Malick & Krishna, 1978, fig. 1, is certainly misidentified.

159. Senna italica P. Miller, Gard. Dict. ed. 8, *Senna* no. 2. 1768.—". . . *Senna italica sive foliis obtusis* C[aspar] B[auhin] P[inax 397. 1623] . . . grows naturally in India."—No typus found at BM by Brenan (Kew Bull. 1958, p. 240) or by us (1978); our interpretation follows Brenan, l.c.

and 1967, p. 65 (subsp. *italica*). *Cassia italica* (P. Miller) Sprengel, "Bot. Gart. Univ. Halle 21. 1800" fide Mabberley, Taxon **30**(1): 9. 1981, q.v. for history of name and early synonymy.

Cassia senna β Linnaeus, Sp. Pl. 377. 1753.—"*Senna italica s. foliis obtusis* Bauh. . . . Habitat [not distinguished from that of typical *C. senna*] in Ægypto."—LINN *528/19*! is authentic for the variety.

Cassia ligustrina P. Miller, Gard. Dict. ed. 8, *Cassia* no. 12. 1768.—". . . sent me from the Havannah [Cuba] by the late Dr. Houston."—A spm of *S. italica* labelled 'Havanna' in hb. Miller. (BM!) may be authentic, but lacks the characteristic ticket in Houston's hand.—Referred to *C. obovata* by Bentham, 1871, p. 533.—Non *C. ligustrina* Linnaeus, 1753.

Cassia obtusa Roxburgh, Hort. Bengal. 31. 1814.—No typus seen, but the identity established by Wight, Ic. pl. Ind. Or. **3**: 757. 1843.—*Senna obtusa* (Roxburgh), Roxburgh, Fl. Ind. ed. Carey **2**: 344. 1832.

Cassia obovata Colladon, Hist. Casses 92, t. 15, fig. A. 1816, nom. illegit. *Senna obovata* (Colladon) Link, Handbuch **2**: 140. 1831; Batka, Bot. Zeit. **7**: 192, t. 2, fig. 3. 1849.

Cassia porturegalis Bancroft, "Jamaica Courant, Nov. 1827 & Roy. Gaz. 49, 46: 22. 1827" fide Mabberley, Taxon **30**(1): 9. 1981.—Protologue not seen, but the type-locality evidently Port Royal, Jamaica.—Spm. authent., K (hb. Hook.)!—Equated with *Cassia obovata* by MacFadyen, Fl. Jam. 340. 1837.

Cassia obovata sensu Vogel, 1837, p. 101; Bentham, 1871, p. 553; Baillon, Nat. Hist. Pl., ed. angl. **2**: 118, figs. 96, 97. 1872; Dict. Pl. **1**: 647 + fig. 1876.

Senna obovata sensu C. Martius, Versuch einer Monogr. der Sennesblätter 58, 96–97 (bibliography). 1857; Batka, Monogr. Cassien gruppe Senna 32 + tab. (optima!). 1866; Britton & Rose, 1930, p. 245; Font Quer, Pl. medicin. 351 + fig. 1962; Malick & Krishna, 1978, fig. 3 (seed).

Cassia italica sensu Keay in Hutchinson & Dalziel, Fl. W. Trop. Afr. ed. 2, **1**(2): 453, fig. 148D. 1958; Adams, 1972, p. 325; Täckholm, Students' Fl. Egypt pl. 90, fig. B. 1974; Jafri, Fl. Libya **61**: 3, fig. 1. 1978; Mabberley, Taxon **30**(1): 9–10 (review of nomenclature). 1981.

Ours erect or diffuse, precociously flowering, sometimes monocarpic, in age weakly suffruticose herbs at anthesis 2–7.5(–10) dm, the young stems, lf-stalks, dorsal or rarely both faces of lfts and axes of inflorescence strigulose with appressed or narrowly ascending hairs up to 0.15–0.2 mm, the thick-textured foliage pale green or glaucescent, the erect axillary racemes either longer or shorter than their subtending lf.

Stipules early spreading-deflexed, lanceolate or triangular-lanceolate ±3.5–5 × 1–2 mm, the firm blades auriculately dilated at base on side further from petiole, coarsely 3–5-nerved, tardily deciduous.

Lvs 6–11 cm; petiole including livid pulvinus 16–27 mm, at middle 0.6–1.3 mm diam, coarsely 3-costate dorso-laterally, shallowly sulcate ventrally; rachis 4–6.5 cm; petiolar gland 0; pulvinules livid 1–1.7 mm; lfts 4–6(–7) pairs, ± accrescent distally but either the terminal or the penultimate pair longest, the terminal pair almost always broadest, all lfts oblong-obovate to broadly obovate obtuse mucronulate or subemarginate apiculate, the longest pair 1.5–2.6 × (0.8–)0.9–1.7 cm, 1.4–2.2(–2.5) times as long as wide, the margin plane, the midrib and 4–7 pairs of camptodrome with random intercalary secondary veins and erratic tertiary venulation subequally prominulous on both faces.

Racemes loosely openly ±7–20-fld, the ascending fl-buds elevated well beyond expanded fls, the axis together with short peduncle 7–20 cm; bracts firm fuscous rhombic-ovate-apiculate 2–4 × 1–3 mm, early caducous; true pedicels 0.8–1.8 mm, the narrowly funnel-form hypanthium 2.5–3.5 mm; fl-buds obovoid-ellipsoid glabrous or minutely puberulent; sepals fuscous pallid-margined, of subequal length, all oblong-elliptic obtuse ±7.5–10 mm: petals glabrous yellow drying stramineous brown-veined, of equal length or the 2 abaxial ones a trifle longer than the rest, all obovate or oblanceolate obtuse, the longest 10–12.5 mm; androecium glabrous, the anthers of 3 minute staminodes 1.2–1.8 mm, the filaments of 4 median stamens 2–2.5 mm, of 2 long latero-abaxial ones 2.8–4 mm, of the centric abaxial one 3.5–4 mm, the anthers of 4 median and of the centric abaxial one linear-oblong straight or almost so 3.5–5.5 × 0.7–0.8 mm, those of the 2 long stamens 7.5–8.5 × 0.8–1.1 mm, lunately incurved, all anthers abruptly truncate at apex, obtusely 2-umbonate on ventral and 2-porose on the dorsal side; ovary densely minutely strigulose; style glabrous linear-filiform 4.5–6 mm, evenly incurved from ovary, very slightly dilated at the obliquely truncate apex, there 0.25–0.3 mm diam; ovules 6–10.

Pod horizontal or declined, the stipe (above hypanthium) 1–1.5 mm abruptly expanded into the body, this in profile lunately oblong, broadly rounded at both ends 2.5–4.5 × 1.3–1.6 cm, plano-compressed, bicarinate by the slender, shallowly undulate sutures, the gray- or brownish-glaucescent, medially fuscous valves papery, transversely venulose, finely puberulent, elevated over each seed as an obtuse crest 3–4 mm long, ±1.5 mm tall, oriented parallel to the pod's long axis, the cavity membranous-septate between seeds, the valves internally contiguous and eseptate toward the sutures; seeds compressed parallel to the valves, in broad profile subquadrately pyriform, truncate-emarginate at apex, the moderately lustrous testa intricately pallid-sinuous-rugulose overall against a yellowish-brown, gray or fuscous field, the lustrous livid areole engraved into the distal end of a ridge ±3 × 0.35–0.5 mm occupying the proximal end of seed-face.—Collections (from neotropics): 16.

Roadsides, waste places about towns and country dwellings, abandoned gardens, sometimes on seashores, adventive and fully established in the lowlands of the Greater and Lesser Antilles (Jamaica, Hispaniola, St. Martin, Anguilla, Union I., Curaçao, Aruba and Bonaire) and arid coastal Venezuela (Golfo de Venezuela and Penísola de Paraguaná, Isla de Margarita); native of tropical and warm temperate Africa and Near East.—Fl. with us throughout the year when not drought-inhibited.—*Italian senna*; *Port Royal s.*; *Aleppo s.*

E. Sectio PARADICTYON Irwin & Barneby

Senna sect. **Paradictyon** Irwin & Barneby, sect. nov., petalo abaxiali uno heteromorpho obliquo margine crenulato sect. *Peiranisiae* notulisque plurimis sect. *Sennae* similis, sed inter congeneres omnes pedicellis abbreviatis basi 2-bracteolatis, alabastris anguste lanceolatis, bracteisque cum sepalis glumaceis prominule striato-nervosis anomala.—Herbae macrophyllae e xylopodio; caulium internodia profunde excavato-sulcata; foliola ampla chartacea 2–5-juga; legumen nutans oblongum plano-compressum 14–18-ovulatum; semina areolata.—Sp. 1, **S. paradictyon** (Vogel) Irwin & Barneby ≡ *Cassia paradictyon* Vogel.

Corolla irregular, 1 abaxial petal heteromorphic asymmetric, marginally crenulate; sepals subequilong, like the enlarged deciduous bracts glumaceous, finely densely parallel-nerved; pedicels very short, 2-bracteolate at base; androecium almost of sect. *Senna,* but the 2 long anthers less curved and more erect (their orientation vis-à-vis one another uncertain).—Herbs from xylopodium, the internodes of stems deeply sulcate; petiolar gland 0; lfts ample 2–5 pairs; pod nodding, oblong-elliptic in outline, plano-compressed, vestigially septate; seeds transverse, compressed parallel to valves, areolate.

The preeminently singular *S. paradictyon* suggests an improbable intergeneric liaison between *Senna* ser. *Pictae* and *Chamaecrista* sect. *Xerocalyx*. Its bibracteolate pedicels, lanceolate straight flower-buds and striately veined bracts and sepals, nowhere repeated in *Senna,* are reminiscent of *Xerocalyx,* but the highly modified zygomorphic androecium, the inertly dehiscent plano-compressed pod and the areolate seeds are all sterling characters of *Senna*. It was described by Vogel, who lacked the pod, as a cassia of unknown affinity somewhat resembling *S. reticulata* in foliage. Bentham, who still had only flowering material for study, associated it with *S. aculeata* in his ser. *Pictae* of sect. *Chamaesenna,* probably influenced by a general similarity in the enlarged herbaceous stipules, ample obovate leaflets and sharply sagittate long anthers to various members of that series. The pod, however, has turned out to be short and broad, consistent rather with Bentham's concept of ser. *Brachycarpae* (our ser. *Senna*), which incidentally has similarly short (but of course ebracteolate) pedicels. The superficially chamaecristoid character of *S. paradictyon* is such that Bentham was deceived into misidentifying as this species a sterile specimen (*Martius s.n.* from Serro Frio, Minas Gerais, M) of *Chamaecrista choriophylla* var. *latifolia*. In its perianth, androecium, pod and seed it is pure *Senna,* but we see no alternative to describing for this perplexing hippogriff a monotypic section. Observation of the fresh flower is needed to determine accurately the orientation of the long anthers around which the narrow flower-buds take their characteristic form.

160. Senna paradictyon (Vogel) Irwin & Barneby, comb. nov. *Cassia paradictyon* Vogel, Syn. Gen. Cass. 45 & Linnaea **11**: 686, descr. ampliat. 1837.—"In Brasilia: Sellow leg."—Holotypus, labelled "São Paulo," received by Kunth from Humboldt, †B = F Neg. *1724*! neoholotypus, *Sellow s.n.,* K! = NY Neg. *1470.*

Cassia paradictyon sensu Bentham, 1870, p. 128, exclus. pl. Martian. e Serro Frio quae = *Chamaecrista choriophylla* var. *latifolia* (M!); Bentham, 1871, p. 551.

Coarse, amply leafy herbs with stout, few-lvd stems procumbent and incurved-ascending from vertical lignescent rootstock up to 3 cm diam, glabrous except for minutely puberulent raceme-axes, the internodes prominently angulate and sulcate by 3 ribs descending from each lf, at anthesis 3–6 dm, the few large plane chartaceous, prominently finely reticulate lfts dull olivaceous subconcolorous, the long-pedunculate racemes arising vertically from lf-axils, in fruit either shorter or a little longer than the lf.

Stipules erect herbaceous, subsymmetrically ovate obtuse or deltately acuminulate 1–3 × 0.7–1.8 cm, many-nerved from base, the sharply prominulous venules branched and anastomosing distally, the blades persistent.

Lvs 1.5–3 dm; petioles stout (5–)6.5–10 cm, obtusely 5-ribbed, at middle (1.3–)2–3 mm diam, deeply sulcate ventrally; rachis (3–)4.5–14 cm, terminating in a stiff linear or sometimes claviform seta; petiolar glands 0; pulvinules 4–6.5 mm, discolored when dry; lfts (2–)3–5 pairs, accrescent distally, their blades decurrent on the pulvinule at least proximally and appearing sessile or almost so on rachis, the distal pair amply obovate or obovate-flabellate, 6.5–12 × 5–9 cm, 1.2–1.6 times as long as wide, very obtusely rounded or obscurely depressed-deltate-acuminate at apex, the subcentric, slightly incurved midrib immersed above, cariniform beneath, the ±10–16 pairs of camptodrome (with random intercalary) secondary, connecting tertiary and fine reticular venules all sharply prominulous on both faces, the camptodrome arches confluent well within the plane, sharp-edged margin and giving rise toward it to a very fine and dense reticulum.

Peduncles 7–21 cm; racemes loosely (10–)15–35-fld, the yellowish glumaceous, prominently multistriate bracts and emergent fl-buds crowded at apex into a narrow cone, the axis early elongating, in fruit 7–21 cm; bracts ovate-elliptic obtuse or deltately subacuminulate 9–14 mm, covering only the very young fl-buds, persistent until just before anthesis; pedicels at and after anthesis 2–3.5 mm, subtended at very base by a pair of narrowly lance-subulate membranous-margined bracteoles 3–5 mm; fl-buds symmetrically lanceolate in profile, obtuse at narrow apex, glabrous; sepals resembling bracts in texture and venulation, scarcely or little graduated, narrowly lanceolate to linear- or lance-elliptic or (the outer) ovate-elliptic, all obtuse, the longest 16–23 × 1.8–4 mm; petals yellow or orange-yellow fading stramineous (not brown-veined), 4 obovate-cuneate, 1 more oblique and crenulate-margined, all of about equal length, the longest 18.5–27 mm; androecium glabrous, functionally 2-merous, the filaments of 4 median and 2 large abaxial stamens 1.8–2.5 mm, of the centric abaxial staminodal stamen 5–6.5 mm, the anthers of 4 median stamens lance-oblong straight, including the short porrect 2-porose beak 5–5.5 mm, those of 2 fertile abaxial ones lanceolate-attenuate, gently incurved from sharply hastate base, 11–14 mm, not beaked but tapering into the terminally minutely 2-porose apex; ovary glabrous or minutely puberulent laterally; style gently incurved, sublinear, 4–6 mm, 0.2–0.3 mm diam; ovules 14–18.

Pod declined, the stipe 1.5–2 mm, the body oblong-elliptic triangular-acuminate at both ends, plano-compressed 35–47 × 14–19 mm, bicarinate by the scarcely thickened sutures, the stiffly papery, green then pale brown and ± lustrous valves finely transverse-venulose, not at all raised over the seeds, the cavity continuous (vestigial membranes between seeds but no defined locules); dehiscence primarily through the ventral suture, the valves there gaping in age; seeds transverse, pyriform, subemarginate at apex, compressed parallel to the valves, 7.5–8 × 4.8–5 mm, the testa smooth atropurpureous lustrous, the linear-elliptic areole ±5 × 1.2 mm.—Collections: 14.

Campo, below 400 m, uncommon but locally plentiful in the foothills of the Paraguai and Paraná valleys in lat. 20°–26°S, in s. Mato Grosso and (presumably adjoining) São Paulo, Brazil and s.-e. Paraguay.—Fl. XI–III.—*Suliman* (guaraní); *fedegoso do campo*.

It is by inadvertence that Vogel described in *Synopsis Cassiae* the leaflets of *C. paradictyon* as 6–8 *pairs,* for this is corrected to 6–8 *leaflets* in the extended description in *Linnaea*. The epithet *paradictyon* alludes to the close reticulation around the outer periphery of the leaflets, lying between the margin itself and the camptodrome arches of the major secondary veins.

According to Balansa, who collected the species near Caaguazú in 1876, the Guaranis prepared a poisonous extract from the foliage.

F. Sectio PEIRANISIA (Rafinesque) Irwin & Barneby

Senna sect. **Peiranisia** (Rafinesque) Irwin & Barneby, stat. nov. *Peiranisia* Rafinesque, Sylva Tellur. 127. 1838.—Sp. typica (unica): *P. aversiflora* (Herbert) Rafinesque (the basionym erroneously attributed to Hooker) ≡ *Cassia aversiflora* Herbert ≡ **Senna aversiflora** (Herbert) Irwin & Barneby.

Corolla irregular, 1 abaxial petal, that opposed to the displaced pistil, obliquely dilated, its claw commonly shorter and thicker than that of the rest; androecium consisting of 3 adaxial staminodes and 7 perfect stamens, these either homomorphic or modified into 2 sets of 4 shorter median and 3 abaxial, the centric abaxial one variable in relation to its 2 neighbors.—Petiolar glands present or 0;

pod mostly plano-compressed, in ser. *Excelsae* terete-subquadrangular; seeds transverse areolate, usually compressed parallel to the valves, in ser. *Excelsae* turned broadside to interseminal septa.—The section consists of 7 series endemic to the wet and dry tropics, extratropical savannas, pampas and warm temperate deserts of N. and S. America.

Under the title of sect. *Peiranisia* we bring together all American sennas, except the peculiar *S. paradictyon,* in which one adaxial petal is modified to perform a specialized function, either as a shield over the androecium, as a platform for visiting pollinators, or as a flag. This odd petal, always opposed to the asymmetrically displaced pistil, is analogous to the so-called cucullus of genus *Chamaecrista,* but must have evolved by an independent path. The concept of such a group cuts across the grain of Bentham's basically carpological classification of *Cassia,* emphasis being shifted from the pod to organization of the corolla. Assembled from distant pages of Bentham's monograph are: from sect. *Chamaefistula* the ser. *Excelsae*; and from sect. *Chamaesenna* a) *S. wislizeni* (our ser. *Deserticolae*), b) the infraserial group 'Molles' (our ser. *Isandrae*), these both excised from eglandulose ser. *Floridae,* c) the large interglandulose ser. *Rostratae* (our ser. *Interglandulosae*), and d) the junceous or prickly *Aphyllae.* The Paleotropical *Cassia* ser. *Auriculatae,* which should be redefined (in *Senna,* and shorn of Brazilian *S. reniformis*) so as to include *C. singueana* Del. and *C. petersiana* Bolle, is technically referable to *Peiranisia* because of its irregular corolla. Its status as a series of sect. *Peiranisia* or as a coordinate but certainly kindred section requires evaluation in the context of African *Senna.*

On evidence from the androecium sect. *Peiranisia* divides into two circles of affinity. In one, consisting of ser. *Isandrae* and *Excelsae,* there are seven subisomorphic, thin-walled, short-beaked or submuticous fertile anthers. In the other, comprising ser. *Interglandulosae, Deserticolae, Aphyllae* and the monotypic *Chlorocladae* and *Egregiae,* the seven firm-walled fertile anthers are differentiated into two sets of four and three respectively, the latter abaxial, longer and often strongly beaked. Other characters correlate imperfectly with this androecial dichotomy. The second group as delimited by the androecium contains three series with and two without petiolar glands. The pods of ser. *Interglandulosae, Deserticolae* and their allies are plano-compressed, papery or coriaceous, dry, with valves flat or variously elevated over the seeds. In ser. *Isandrae* the pod varies from plane and papery (*S. haughtii*), when technically indistinguishable from the preceding, to subsucculent lignescent (*S. atomaria*), while that of ser. *Excelsae* is ligneous and subterete. The variety of form assumed by the pod of sect. *Peiranisia* is, however, not greater than described above for sect. *Chamaefistula* sens. lat.

Key to the Series of sect. *Peiranisia*

1. Plants leafless at anthesis, the junceous or spinulose (sometimes elaborately ramified) stems green and photosynthetic; stem-nodes and peduncles subtended by a single fuscous entire bractlet (stipules apparently 0); Argentina and adjacent Bolivia and Paraguay.
Fiv. ser. *Aphyllae* (p. 567).

1. Plants normally leafy at anthesis (exceptionally the lvs tardily deciduous); stipules present (if caducous their attachment-point marked by a small scar).
    2. Petiolar gland between proximal and often between subsequent pairs of lfts.
        3. Thecae of 2–3 abaxial stamens at once confluent distally and contracted into a tubular beak; lfts 1–many pairs, but 1–2 pairs only n. of Amazonia.
        Fi. ser. *Interglandulosae* (p. 488).
        3. Thecae of 3 long abaxial anthers separate to apex, the beakless or shortly beaked anther 2-porose; lfts 1–2 pairs. Microphyllous shrubs of s. Bolivia and w. Argentina.

4. Stipules setiform; stiffly erect subjunceous shrubs ±5–10 dm of tropical n.-w. Argentina (Salta, extreme n.-w. Formosa) and s. Bolivia.
Fii. ser. *Chlorocladae* (p. 563).
4. Stipules glanduliform, tumidly convex dorsally; shrublets 2.5–4 dm of temperate w.-centr. Argentina (Mendoza and Neuquén). Fiii. ser. *Egregiae* (p. 566).
2. Petiolar gland 0. Thecae of abaxial anthers either separated distally into 2 independent beaks or, if confluent, the composite beak ±2-lipped.
5. Microphyllous, the longest lvs 0.8–4.5(–5) cm, the longest lfts at most 15 mm; centric abaxial anther resembling its neighbors in form but raised well beyond them on a longer filament; deserts and deciduous tropical woodlands of Mexico and s.-w. United States. Fv. ser. *Deserticolae* (p. 570).
5. Macrophyllous, either the longer lvs well over 4 cm, or the longer lfts over 2 cm, or (usually) both; centric abaxial anther as long as or shorter than its immediate neighbors; widespread.
6. All 7 fertile anthers at once of nearly the same size and isomorphic in respect to their beaks, these uniformly latero-extrorse; lfts (1–)2–12 pairs; inflorescence various, composed of axillary racemes only, or of racemes clustered on leafless brachyblasts, or terminally paniculate, but if the last the axis of panicle contracted and not or scarcely exserted from foliage; pod strongly compressed, the seed locules occupying the thickness of the cavity. Fvi. ser. *Isandrae* (p. 578).
6. All 7 fertile anthers of nearly same length but heteromorphic in respect to their beak, that of 4 median stamens and of the centric abaxial one latero-extrorse, that of the 2 latero-abaxial (antepetalous) stamens porrect; lfts 11–19 pairs; inflorescence an ample thyrsoid panicle elevated well above foliage; pod terete-subquadrangular and each seed-cavity suspended between empty locules.
Fvii. ser. *Excelsae* (p. 599).

Fi. ser. INTERGLANDULOSAE (Bentham) Irwin & Barneby

Senna sect. **Peiranisia** ser. **Interglandulosae** (Bentham) Irwin & Barneby, comb. nov. *Cassia* sect. *Chamaesenna* ser. *Interglandulosae* Bentham in Martius, Fl. Bras. **15**(2): 120. 1870.—Sp. lectotypica: *C. biflora* sensu Bentham = **Senna pallida** (Vahl) Irwin & Barneby.—*Cassia* sect. *Chamaesenna* ser. *Rostratae* Bentham, Trans. Linn. Soc. London **27**: 252. 1871, nom. substitut. illegit.

Peiranisia Rafinesque, 1838, l.c., sens. str.—Typum supra sub sect. *Peiranisia* indicavimus.
Panisia Rafinesque, Sylva Tellur. 128. 1838, pro gen.—Sp. typica: *P. biflora* (Linnaeus) Rafinesque ≡ *Cassia biflora* Linnaeus.—Identity of typus not established, but probably either *Senna pallida* (Vahl) Irwin & Barneby sens. lat. or *S. angustisiliqua* (Lamarck) Irwin & Barneby, certainly a member of ser. *Interglandulosae*.
Peiranisia sensu Pittier, 1928, pp. 126–130; Britton & Rose, 1930, pp. 259–267, max. ex parte, spp. nonnullis exclus.

Corolla of sect. *Peiranisia*; sepals strongly graduated; androecium functionally 7-merous, the fertile members differentiated into sets of 4 shorter median and 3 usually longer, always longer-beaked abaxial ones, the anthers all firm 4-sulcate lengthwise, the beak of the 4 median ones sublateral or infraterminal, dehiscent by 2 short subparallel slits or by one U-shaped one, the beak of the abaxial anthers tubular (sometimes a trifle swollen distally), at apex horizontally or obliquely truncate or subbilabiate, the orifice looking upward or inward and either simple or divided by a slender septum; pod narrowly oblong or linear in outline, plano-compressed, the ripe valves stiffly or pliantly papery, remaining flat or either corrugately or pyramidally elevated over each seed, shallowly depressed between them and there sometimes developing a line of fracture; seeds 1-seriate transverse areolate, in outline narrowly oblong, obovate-pyriform or rhombic.—Shrubs and trees; stipules subulate-setiform or sometimes asymmetrically dilated, early deciduous; petiolar glands always between proximal pair of lfts, often between some succeeding pairs, the terminal seta of rachis sometimes glandiform;

lfts 2–63 pairs, their size varying inversely with number; racemes 1–16(–25)-fld, often subumbellately 2–3(–4)-fld, the pedicels in some spp. subtended by a fusiform gland; style nearly always short (1–3.5 mm), but in *S. aristeguietae* elongate (5–7 mm), the minute stigmatic cavity obliquely terminal; ovules 16–76.—Spp. 19, widespread almost throughout the Neotropics, absent only from the w. slope of the Andes s.-ward from Ecuador, most numerous and diverse in climates with a pronounced dry season, 1 extending n. in Mexico into warm temperate desert.

The ser. *Interglandulosae* is sharply and definitively characterized by an asymmetric, peiranisioid corolla coincident with petiolar glands and three abaxial anthers contracted into a tubular beak. Similarly long-beaked anthers occur also in sect. *Senna,* notably in ser. *Trigonelloideae* and *Basiglandulosae,* and are not of themselves diagnostic of any one group.

The series first took precise form in Bentham's monographs (1870, 1871), where it was called successively *Interglandulosae* (derived from a descriptive term of Vogel's) and *Rostratae,* the latter nomenclaturally superfluous. As defined by Bentham it was marred only by introduction of an admittedly peripheral element in *S.* (*Trigonelloideae*) *leiophylla* (including *S. cobanensis* as a variety), similar in the androecium but quite other in the zygomorphic corolla. The genus *Peiranisia* as elaborated by Britton & Rose (1930) was largely equivalent, wholly so as to the first 26 species described, but was injudiciously structured to admit elements (spp. 27–34 incl.) of our ser. *Basiglandulosae, Brachycarpae* and *Pachycarpae,* collectively different in their zygomorphic perianth.

The *Interglandulosae* fall into two circles of affinity characterized primarily by seed-form. In *S. multijuga* and the three species next in order the seeds are narrowly oblong, subtetragonally compressed into crowded narrow locules much wider than long; in the remainder the locules are broadly oblong or nearly square and accommodate compressed or rhomboid, emphatically bifacial seeds. In all members of the first group the racemes are several-flowered and lack fusiform glands at base of the pedicels. In many but not all others the raceme is subumbellately few-flowered, in several quasi-dichotomously two-flowered, the flowers opening simultaneously and enantiostylous, and in all except *S. aristeguietae* the pedicels, or some of them, are subtended by a gland. Setules that secrete a viscid sap occur in some but not all members of the narrow-seeded group, but also in the broad-seeded *S. aristeguietae,* which is unique in ser. *Interglandulosae* because of its elongate style. Throughout the series the three abaxial anthers of any flower are isomorphic as between themselves, but are often elevated on filaments of different lengths. When this occurs it is the abaxial stamen nearest the asymmetrical petal, or this accompanied by its (centric) neighbor, that lengthens, and these one or two long stamens are more or less loosely invested and shielded from the style by the incurved petal juxtaposed. Meanwhile the abaxial stamen further from the asymmetrical petal, oriented more or less in line with the displaced pistil, is drawn back toward the flower's center on a filament not or scarcely longer than those of the short median stamens. This androecial asymmetry, constant and highly developed in *S. multijuga,* is less fixed in other species, in which particular flowers may display it in different degrees. Moreover in some species where the three abaxial stamens are normally equal we have found either one longer than the rest or less often the centric one longer than its two neighbors, repeating a pattern that has become stabilized in the closely related ser. *Deserticolae.* While the group with narrow seeds forms a somewhat detached cluster within the orbit of the series it lacks supporting characters adequate for independent serial status.

Key to Species of ser. *Interglandulosae*

1. Lfts, at least of longer lvs, more than 14 pairs.
  2. Pedicels not subtended laterally by a fusiform or stipitate gland.
    3. Axes of inflorescence viscid-setulose; Bahia (upland caatinga); multifoliolate forms of: 164a. *S. acuruensis* var. *Acuruensis* (p. 508).
    3. Axes of inflorescence glabrous, simply pilosulous (always so in Bahia), or if setulose the setules non-secretory.
      4. Stems nearly always hispid with lustrous horizontal setae (3–)4–9 mm (these exceptionally wanting); pod 3.5–10.5 mm wide; long inner sepals 8.5–14 mm; species local either in the e. Cordillera of Colombia-Venezuela or of upland Costa Rica-Panama; seeds ovate, with ovate or ovate-elliptic areole well over 1 mm wide.
        5. Petioles 4–11 mm; axis of individual racemes 8–23 mm; petals relatively small, the long innermost 16–20 mm; beak of 3 abaxial anthers poorly developed, 0.35–0.5 mm; pod 7–10.5 mm wide, each seed-cavity wider than long; seeds 4.5–5 mm, their areole 2.2–3 mm; e. Cordillera of n.-e. Colombia and adjoining Venezuela. 166. *S. mutisiana* (p. 512).
        5. Petioles 11–22 mm; axis of individual racemes 1–10 mm; petals larger, the longest 23–29 mm; beak of abaxial anthers 0.9–1.5 mm; pod 3.3–5 mm wide, each seed-cavity longer than wide; seeds 2.6–3.1 mm, their areole 1.3–1.5 mm; Costa Rica and Panama. 167. *S. williamsii* (p. 513).
      4. Stems lacking hispid setae; long sepals 3.5–8 mm; pod 15–21 mm wide; species widespread from s. Mexico to Bolivia and s.-e. Brazil, extensively cultivated and naturalized (if petiole 4+ cm cf. *S. formosa*). 161. *S. multijuga* (p. 492).
  2. Pedicels subtended on one side by a fusiform, sessile or stalked gland resembling that between proximal pair of lfts.
    6. Racemes all or mostly exactly 2-fld: stipules linear-attenuate erect or simply incurved to erect; Mexico and Central America; multifoliolate vars. of: 177. *S. pallida* (p. 531).
    6. Racemes subumbellately 4–8-fld; stipules at base divergent from lf-pulvinus across the stem, the blade thence falcately or subsigmoidally incurved to erect; Andes of Ecuador and Peru; multifoliolate forms of: 174. *S. incarnata* (p. 526).
1. Lfts of longer lvs up to but not over 14 pairs, often much fewer.
  7. Stems, lf-stalks and commonly all axes of inflorescence charged with erect viscid setules mixed with short villi and sometimes also some long lustrous non-viscid setae; e. Brazil and Maracaibo Basin in Venezuela.
    8. Style 1.5–3 mm; racemes (2–)3–12(–18)-fld, the axis 2–70 mm; sepals glabrous dorsally or the outermost sometimes thinly puberulent and ciliolate, the inner always glabrous dorsally; e. Brazil (Maranhão to Pernambuco and Bahia, perhaps n. Minas Gerais).
      9. Ventral suture of pod bluntly 3-nerved but not winged; lfts variable in number (4–20 pairs), but if in the range of 5–9 pairs then the axes of inflorescence glabrous and of Bahia. 164. *S. acuruensis* (p. 506).
      9. Ventral suture of pod narrowly winged (wing 0.8–1 mm wide); lfts 5–9 pairs; all axes of inflorescence setulose; Maranhão, Piauí, Ceará, and Paraiba. 165. *S. trachypus* (p. 509).
    8. Style 5–7 mm; racemes 2–5-fld, the axis not over 1 mm; sepals all gray-puberulent dorsally; full range uncertain, known from Zulia, Venezuela, and from Minas Gerais, Brazil. 163. *S. aristeguietae* (p. 502).
  7. Stems and lf-stalks varying from glabrous to densely pilosulous but nowhere viscid-setulose (some minute thickened but non-secretory trichomes often mixed with the filiform hairs and rarely, in e. Brazil, some horizontal lustrous setae).
    10. Beak of 4 median anthers dehiscent by one U-shaped aperture. Circum-Caribbean to s. Mexico and Costa Rica, Andean n. Colombia, n. and interior Venezuela, disjunct in Pacific lowland Ecuador.
      11. Filaments of median anthers and usually of 3 adaxial staminodes confluent at base; terminal lfts of larger lvs mostly 3–10 cm; absent from Antilles except e. Cuba.
        12. Lfts at once relatively large (4–10 × 1–3 cm) and acuminate distally; s. Mexico (Chiapas) and Costa Rica. 173. *S. tonduzii* (p. 525).
        12. Lfts of larger lvs sometimes up to 6 cm long but often smaller and in any case not acuminate, in outline oblong, oblanceolate, elliptic-obovate, or obovate, at apex obtuse, deltately acute, or emarginate; allopatric.

13. Pod coarsely venose, the pallid sinuous venules superficial and prominulous, in age detached from the pericarp; racemes 2–7 (few exactly 2)-fld; local in n.-w. Venezuela (Lara, Aragua). 172. *S. saeri* (p. 524).

13. Pod faintly and delicately transverse-venulose, the venules neither discolored nor separating in age; racemes mostly 1–4 (often exactly 2)-fld; widespread in Venezuela to e. Cuba, n. Colombia, Pacific Ecuador.

14. Long sepals 5–6.5 mm; petals externally pubescent; anthesis coeval with vegetative growth, the racemes associated with recent lvs or gathered into a ± leafless panicle terminal to annotinous branchlets; range as given. 170. *S. robiniifolia* (p. 520).

14. Long sepals mostly 7–10.5 mm; petals glabrous; anthesis during arrest of vegetative growth, the racemes borne on brachyblasts axillary to old or fallen lvs, often on wholly leafless branchlets; Venezuela only. 171. *S. viciifolia* (p. 522).

11. Filaments all free from base; lfts up to 3(–3.3) cm, mostly less than 3 cm; Greater Antilles (Hispaniola, Puerto Rico & Virgin Is., s. Bahamas & Caicos Is., Jamaica, absent from Cuba), some cultivated and naturalized elsewhere.

15. Terminal lfts of some larger lvs at least 11 mm long, racemes mostly arising directly from axils of coeval lvs or from developed axillary branchlets; range as given except for Puerto Rico and Virgin Is. 168. *S. angustisiliqua* (p. 514).

15. Terminal lfts of largest lvs 3.5–9(–10) mm long; racemes borne on spurlike brachyblasts axillary to early deciduous primary cauline lvs; n.-e. Dominican Republic, Puerto Rico & Virgin Is. 169. *S. polyphylla* (p. 517).

10. Beak of 4 median anthers dehiscent by 2 parallel slits. Continental N. and S. America to s. Windward Is., Swan Is. in w. Caribbean, and Bahamas with extreme e. Cuba.

16. Plants of e. Brazil (e. Pernambuco s. e. through Bahia to Minas Gerais and s. Goiás).

17. Lvs 1.5–3.5 dm, the lfts large, up to 4–6 cm; inflorescence and ample branched leafless panicle of 3–12-fld racemes; no glands at base of pedicels; trees. 162. *S. formosa* (p. 501).

17. Lvs 4.5–11.5 cm, the lfts less than 3.5 cm; inflorescence of mostly 2-fld racemes axillary to normal foliage lvs; a fusiform gland at base of each pedicel; shrubs.

18. Stems (commonly) and stipules (always) hispid with lustrous horizontal setae 2–7 mm; filaments of at least one pair of median stamens connate at base; beak of median anthers opening by one U-shaped aperture; pod 4.5–5.5 mm wide, the valves raised over each seed in the form of a 4-armed pyramid, the distance between interseminal septa 3.3–4 mm; seeds compressed-rhomboid, ± as long as wide; e.-centr. Bahia to Pernambuco. 176. *S. aversiflora* (p. 529).

18. Stems and stipules strigulose-pilosulous with incurved or appressed hairs much less than 1 mm (sometimes mixed with minute claviform hairs); filaments all free; beak of median anthers opening by 2 parallel slits; pod 6.5–8 mm wide, the valves elevated over each seed as a simple ridge transverse to the long axis of pod, the interseminal septa 1.5–2 mm apart; seeds narrowly obovoid, much longer than wide; n.-centr. and extreme s. Bahia to s.-w. Goiás. 175. *S. rostrata* (p. 528).

16. Plants of extreme n. Brazil (Roraima), Peru, n. to Mexico and Bahamas.

19. Range limited to inter-Andean valleys and Pacific slope in Peru and Ecuador.

20. Racemes subumbellately (3–)4–9-fld; stipules 0.5–2.5 mm wide, falcately incurved to erect; Pacific slope in lat. 2–12°S. 174. *S. incarnata* (p. 526).

20. Racemes 1–3, mostly exactly 2-fld; stipules linear-setiform erect; local on both slopes of Andes in lat. 3–6°S.

21. Lvs 3.5–10 cm; lfts (5–)6–9 pairs. 177f. *S. pallida* var. *nemorosa* (p. 543).

21. Lvs (1–)1.5–3.5 cm; lfts 3–4 pairs. 179. *S. huancabambae* (p. 561).

19. Range extending from Colombia to Mexico, n.-e. Brazil and the Bahamas.

22. Lfts of larger lvs 4–several pairs; range as given. 177. *S. pallida* (p. 531).

22. Lfts of larger lvs 2 or 3 pairs (if brought hither by plants of Yucatán

Peninsula, cf. 177i. *S. pallida* var. *goldmaniana*, p. 547); s. Mexico (México, Morelos, Guerrero, Oaxaca, Chiapas) to El Salvador and Honduras. 178. *S. holwayana* (p. 559).

161. Senna multijuga (L. C. Richard) Irwin & Barneby, comb. nov. *Cassia multijuga* L. C. Richard, Actes Soc. Hist. Nat. Paris **1**: 108. 1782. —Typus infra sub. subsp. *multijuga* indicatur.

Trees and treelets at maturity attaining 6–40 m with trunk 1–6 dm diam, often precociously flowering as shrubs only (3–)4–6 m, most commonly encountered as slender trees ±6–15 m, varying from fully glabrous to densely pilosulous throughout (as described for vars. following), the annotinous branchlets subterete verruculose canaliculate, the hornotinous ones ribbed, densely leafy, the long narrow multifoliolate lvs strongly bicolored, dark green (when dried often brunnescent) and dull or sublustrous above, pallid beneath, the inflorescence a fully exserted or at base partly leafy-bracteate, either simple thyrsiform or complexly branched pyramidal or pseudo-corymbose panicle of racemes, its axes commonly finely pilosulous, sometimes glabrous, sometimes setulose.

Stipules 3–14 mm, either linear-setiform or asymmetrically dilated at base and the blade there undulate or replicate on itself, in any case caducous, absent from many mature flowering and all fruiting specimens.

Lvs often ± heteromorphic, those low on annotinous branchlets and those in and near the panicle often shorter and simpler than the rest, the larger ones (6–)7–31(–35) cm (these only described hereafter); petiole including discolored, variably dilated pulvinus 4–32 mm, at middle 0.5–1.8 mm diam, shallowly margined and shallowly open-sulcate ventrally; gland between proximal pair stipitate or sessile, the stipe when present either glabrous or pilosulous, the whole in profile (0.8–)1–4.5(–5) mm, the ovoid or fusiform, sometimes lanceiform or hornlike body 0.3–0.65(–0.8) mm diam, a similar gland sometimes at second pair and (in var. *multijuga*) sometimes also at a few distal pairs, minute subuli- or verruculiform glands commonly present between all other pairs, these sometimes represented by a tuft of glandiform spicules or sometimes 0; pulvinules (as seen in dorsal view) 0.3–2(–2.4) mm; lfts (13–)15–56 pairs, inserted along rachis at points 1.5–13(–16) mm apart, very gradually decrescent upward and downward from a point below middle of rachis, the whole blade in expanded outline therefore very narrowly lanceolate, the larger lfts varying from linear or linear-oblong to lance-oblong, oblong or oblong-elliptic, deltately subacute, obtuse or truncate-emarginate, the largest 0.75–5.3 cm, usually conspicuously mucronulate by the excurrent costa, at base strongly inequilateral, the margins usually revolute at least near the pulvinule, sometimes plane, the midrib cariniform beneath, immersed or impressed above, the secondary venation of 5–11 major camptodrome (sometimes with intercalary) pairs either fully immersed or prominulous, but only delicately so, on one or both faces, tertiary venulation 0 or faint and irregular, sometimes immersed but discolored beneath.

Peduncles 1–4.5(–6) cm; racemes 3–16(–25), the more vigorous of any panicle mostly at least 5-fld, the axis becoming (0.2–)0.4–4(–8) cm, but some distal racemes of a complex panicle often depauperate and 2–5-fld with axis 0.1–0.4 cm; bracts lance-elliptic or ovate cymbiform (1–)1.5–6 mm, shielding the very young bud but early caducous; pedicels becoming (13–)14–32 mm, not subtended by a gland; buds globose when very young, opening long before true anthesis, glabrous or basally puberulent; sepals when mature subpetaloid, yellow, greenish or brownish, strongly graduated, obovate or orbicular very obtuse, the smaller firmer outer ones ±½–⅓ as long as the inner, the longest (3.5–)4–7.5(–8) mm; petals yellow, commonly glabrous dorsally, sometimes puberulent externally along veins, heteromorphic, the 3 adaxial and 1 abaxial similar (except the latter often larger), obovate-oblanceolate 7–21 mm, contracted into a stalklike claw, the fifth (abaxial, alternately right and left up the raceme) sessile, very obliquely semi-ovate or boomerang-shaped, coarsely veined, commonly largest of all and (measured along the mid-vein) 14–23(–25) mm; androecium usually glabrous, the free filaments rarely and anthers exceptionally pilosulous, the filaments of 4 median and 1 latero-abaxial stamens 0.6–2.3 mm, those of 2 abaxial ones (1.8–)2.5–7(–9) mm, the anthers of 4 median stamens slightly incurved, at apex truncate or obscurely strangulated into a short neck, the lateral beak 0.3–0.5 mm, dehiscent by 2 parallel slits, the body of 3 abaxial stamens lunately incurved (4.5–)5–9 mm, contracted into a porrect, ventrally 2-porose beak (0.8–)1–1.8(–2.2) mm; ovary glabrous, glabrous laterally but ciliolate, pilosulous overall, or both pilosulous and remotely pilose; style 1–2 × 0.3–0.6 mm, glabrous; ovules 48–76.

Pod erratically geotropic in attitude, stipitate, the stipe 2–9 mm cuneately dilated upward, the broadly linear, abruptly obtuse body (6.5–)8–20 × 1.3–2.1(–2.5) cm, bordered by prominulous sutures undulately constricted only at aborted ovules, the very flat reddish or purple-castaneous, seldom greenish, ultimately papery, nigrescent, transversely venulose valves obscurely expressed over seeds as narrow transverse mounds, the interseminal septa 2–4 mm apart, the individual seed-chamber much wider than long; seeds narrowly oblong or oblong-ellipsoid, bifacially or subquadrately compressed (4.3–)4.8–8(–9) × 1.5–2.1 mm, the pale brown or fawn-gray testa dull or sublustrous crackled, the linear or linear-elliptic areole 2.3–4.5 × 0.25–0.5 mm, 5–11 times as long as wide.

Our concept of *S. multijuga* is the traditional one established in Bentham's monograph, modified only so as to accommodate *Cassia verrucosa* (as predicted necessary by Bentham himself) and the more recently described *C. doylei*. So defined it differs from other *Interglandulosae*, except *S. mutisiana*, *S. williamsii* and some plurifoliolate forms of *S. pallida* and *S. acuruensis*, in the numerous leaflets (at least 16 pairs in larger leaves), from the first two just mentioned in the smaller sepals (the large inner ones ±4–7.5, not 8–14 mm) and lack of lustrous hispid setae on the young stems, from all of *S. pallida* in the several-flowered racemes and absence of fusiform glands at base of the pedicels, and further from all these three together in the very broad pod and narrowly oblong seeds. Its closest relative, *S. acuruensis*, which is vicariant in the seasonally arid caatinga country of northeastern Brazil, resembles *S. multijuga* in the pod and seeds, but in those cases where the leaflets (in var. *acuruensis* only) attain equality in number the leaf-stalks and axes of inflorescence are densely viscid-setulose.

The species varies much in stature, sometimes flowering as a softly woody arborescent shrub but capable of attaining the port and girth of a genuine forest tree. The vesture varies erratically in density, in quality, and independently in color from whitish to sordid or golden-yellow, the filiform hairs being mixed in some populations both north and south of the Amazonian Hylaea with erect orange setules. Bentham already commented on the range of flower-size, which now appears to be both seasonal and racial, although the determining factor is often open to conjecture. Most arresting visually is the variation in size and number of leaflets, the two modes of variation being mutually adjusted to stabilize the total photosynthetic area exposed to sunlight and simultaneously governing the distance between pairs of leaflets along the rachis, smaller leaflets being always more crowded. Finally the stipules, though inconveniently caducous and lacking from most mature specimens, vary with significant geographic correlation from linear-setiform to basally dilated and falcately hemilanceolate in outline, the dilated blade being nearly always undulate or folded on itself.

Analysis of the phenetic variation has shown that *S. multijuga* can be divided into three major geographic blocs by the character of the stipules, blocs that coincide with recognized patterns of floristic dispersal. The dilated stipule is circum-Hylaean and Guayanan, with a detached outpost in the wet coastal forest of Bahia. The setiform stipule replaces this in southeastern Brazil southward from central Bahia and is the only type encountered in Mexico. Within each of the two South American blocs, which we term subspecies, are encountered parallel modifications of size and number of leaflets, taxonomically recognized at the varietal level. A random variation in vesture is found almost throughout South America but is feebly, if at all, correlated with dispersal or with other phenetic characters.

Senna multijuga is one of the handsomest of the Neotropical sennas, an ornamental tree of rapid growth, free of disease and prolific in flower. Its var. *multijuga* was early taken into cultivation and by mid-XIX century was already naturalized in Malesia and southern India, probably having been brought originally by the Dutch from Surinam to Java. Other introductions, from Trinidad to Puerto Rico, of a setulose form of var. *multijuga* into Martinique, of var. *lindleyana* into Fiji, very likely from Rio de Janeiro, and into California from São Paulo, are a matter of record, but the origin of much of the material planted in the American tropics as a garden or street tree is lost to view. The literature contains records of *Cassia multijuga,* sens. lat., from Argentina (Tucumán), Central America, and West Indies, but we have found no evidence that it is native in any of these countries.

Key to Subspecies and Varieties of *S. multijuga*

1. Stipules at base inequilaterally dilated, the herbaceous blade there 0.8–2.5 mm wide and undulately crimped or folded on itself; Amazon Basin from s. Goiás, Brazil and Yungas, Bolivia n. to s.-e. Colombia, Guayana Highland, the Guianas, Trinidad, and the s. affluents of the Orinoco in Venezuela; disjunct in coastal s.-e. Brazil (Bahia); cultivated or naturalized or both in Colombia (Meta), Panama, Costa Rica, West Indies, East Indies, and widely planted in tropical botanical gardens: subsp. *multijuga*:
    2. Lfts of larger lvs 16–31 pairs, up to ±2–5 cm long, inserted at points up to (5–)6–11(–15) mm apart along rachis; range as given except coastal s.-e. Brazil. 161/Ia. var. *multijuga* (p. 495).
    2. Lfts of larger lvs 26–56 pairs, up to 0.8–1.4 cm long and inserted at points 1.5–4 mm apart along rachis; s.-e. Brazil (Bahia, Espírito Santo). 161/Ib. var. *verrucosa* (p. 496).
1. Stipules linear-attenuate or setiform, equilateral at base and there not over 0.6 mm wide, both margins plane; native either in s.-e. Brazil (coastal centr. Bahia to centr. Minas Gerais, s. to Santa Catarina) or s.-e. Mexico (Veracruz, Oaxaca, Chiapas); naturalized in s.-e. Venezuela (Bolívar) and n.-w. Colombia (Bolívar); subspp. *lindleyana* and *doylei*:
    3. Lfts at once relatively long or relatively broad (or both), the larger ones at least 2 cm long and 5.5 mm wide, and spaced at points up to (5–)6–11(–15) mm apart along rachis; native in s.-e. Brazil, cultivated in United States, Fiji and probably elsewhere. 161/IIa. var. *lindleyana* (p. 498).
    3. Lfts smaller and narrower, the largest not over 15 mm long or over 4.5 mm wide and spaced up to 2.5–5 mm apart along rachis; one sympatric, one remotely allopatric.
        4. Pairs of lfts in larger lvs 15–32; coast ranges of s.-e. Brazil (Rio de Janeiro to Santa Catarina), sympatric with var. *lindleyana,* naturalized in Venezuela and Colombia. 161/IIb. var. *peregrinatrix* (p. 499).
        4. Pairs of lfts in larger lvs 33–53; s.-e. Mexico (Veracruz, Oaxaca, Chiapas). 161/III. subsp. *doylei* (p. 500).

161/I. Senna multijuga (L. C. Richard) Irwin & Barneby subsp. **multijuga**. *C. multijuga* L. C. Richard, 1782, l.c., sens. str.—". . . e Cayenna . . .

missarum a domino LeBlond."—Holotypus, *LeBlond s.n.*, P (hb. Richard., 2 sheets)! isotypus, P-LAM!

161/Ia. Senna multijuga (L. C. Richard) Irwin & Barneby subsp. **multijuga** var. **multijuga**. *Cassia multijuga* L. C. Richard, l.c., sens. str.—Typus supra sub subsp. *multijuga* indicavimus.—*Peiranisia multijuga* (L. C. Richard) Britton & Rose ex Britton & Wilson, Sci. Surv. Porto Rico and the Virgin Is. **6**(3): 350. 1926.

Cassia calliantha G. F. W. Meyer, Prim. Fl. Essequib. 169. 1818.—"In sylvis plantationis Königsberg [near Georgetown, Guyana]."—Collected by E. K. Rodschied; no typus found at GOETT (1977), but the detailed description, as recognized by Bentham (1870, p. 123), decisive.

Cassia richardiana Kunth, Mimoses 139, t. 42. 1823.—"Crescit in ripa fluminis Atabapo (Alto Orinoco) . . . bords du Rio Atabapo près de son confluent avec l'Orenoque [near lat. 4°N in Amazonas, Venezuela]."—Holotypus, labelled "*n. 1180,* Rio Atabapo," collected by Humboldt & Bonpland in April, 1800, P-HBK! isotypus, B (*hb. Willd. 7990*)!—Reduced by Bentham l.c. to *C. multijuga*.

Cassia fulgens Wallich ex Vogel, Syn. Gen. Cass. 44 & Linnaea **11**(6): 685 (descr. ampliat.). 1837.—". . . 5310 . . . Hbt. in India orient."—Holotypus, B (hb. Kunth.), not seen; isotypi, *Wallich Cat. 5310* ex hort. Calcutt., K, NY (misit Bentham)!—Equated by Bentham, l.c., with *C. multijuga*.

Cassia ampliflora Steudel, Flora **26**: 760. 1843.—"[Hostmann & Kappler . . . im Innern von Surinam gesammelt haben . . .] nr. 829."—Holotypus, P (hb. Steud.), 3 sheets! isotypi, K, NY!—Equated by Bentham, l.c., with *C. multijuga*.

(?) *Peiranisia aristulata* Britton & Killip, Ann. N.Y. Acad. Sci. **35**(3): 182. 1936.—"Villavicencio, Meta, Colombia, December, 1928, *Apollinaire Marie,* type (Herb. Instituto de la Salle, Bogotá)."—Holotypus †1948, survives as photograph, K! clastotypus (fragm), NY!—Certainly referable to *C. multijuga* sens. lat., but, lacking stipules, ambiguous between var. *multijuga* and var. *lindleyana*.

Cassia multijuga sensu Bentham, 1807, p. 123, ex parte, exclus. syn. *C. selloi, C. magnifica* et *C. semifalcata* cum locis pluribus austrobrasiliensibus, t. 37 (legumen male depictum); Schery, 1951, p. 72, fig. 127; Dimitri & Rial Alberti, 1954, fig. 2; De Wit, 1955, p. 253; Worthington, Ceylon Trees 190. 1959.

Trees attaining 25 m with trunk 3(–4) dm diam, flowering precociously as shrubs as low as 3(–2) m, most commonly seen as slender trees with rounded crown ±6–15 m, the annotinous growth sometimes quite glabrous but the branchlets with lf-stalks and lfts (on dorsal or both faces) and axes of inflorescence commonly strigulose, pilosulous or subtomentulose with straight appressed, spreading, or curly and entangled, pallid or lutescent hairs up to 0.1–0.7 mm, the inflorescence sometimes in addition charged with erect orange or discolored setules mixed with the filiform trichomes.

Stipules 4–12 mm, unilaterally dilated at base and ± falcately incurved, the base of the firm, herbaceous blade 0.8–2.5 mm wide, undulately crimped or replicate on itself.

Larger mature lvs 13–31(–35) cm; gland at proximal pair (1.2–)1.5–4.5 × 0.3–0.8 mm, sometimes similar glands at a few distal pairs, between all other pairs commonly a tuft of spicules, sometimes a minute subulate gland; pulvinules (0.8–)1–2(–2.4) mm; lfts 16–37 pairs, inserted up to (4–)5–13(–16) mm apart along rachis, in outline oblong, oblong-elliptic, lance- or exceptionally linear-oblong, mostly (20–)23–43(–53) × 5–12(–13) mm and (2.7–)3–5.2(–5.8) times as long as wide, exceptionally (in upper Amazonian Brazil and adjacent Peru), up to 15–20 × (2.6–)3–4 mm, the secondary camptodrome venation sometimes fully immersed on both faces but the (5–)6–11 pairs of major secondary veins usually prominulous beneath, sometimes on both faces, tertiary venules then erratic.

Long inner sepals 4–6.5(–7) mm; longest petal 16–26 mm; androecium and pod of the species.—Collections: 296.

Riverine forest, both on várzea and terra firme, river- and creek-banks in rain forest, entering savanna and cerrado by way of forest-islands and gallery-forest, becoming locally abundant in disturbed forest or capoeira, from near sea level in the Guianas up to 850 m on the Guayana Highland, to 1100 m in Ecuador, 750 m in Bolivia and 1000 m on the Planalto of s.-centr. Goiás, widespread around the periphery and along rivers within the Amazonian Hylaea, the Guayana Highland and the Guianas, from s.-e. Colombia and Venezuela s. of the Orinoco through the foothills of the Ecuadorian and Peruvian Andes to the Yungas in Bolivia, e. in Brazil to coastal Amapá, the Amazon Delta in Pará, and coastal Maranhão, s. in Brazil to the Amazon-Paraguay divide near 17°S in s.-centr. Goiás; widely dispersed in tropical gardens both within and outside its native range (Trinidad, Martinique, Puerto Rico and Virgin Is.; s. Florida; tropical Africa; s. India, Java, Sumatra), reported from Central America and Mexico but the only specimens seen (from Honduras and the Canal Zone in Panama) are from cities and probably either street- or weed-trees.—Fl. irregularly throughout the year.—*Marimari* or *San Francisco Marimari* (Guayana); *riariadan, kashadan* (Guyana); *palaka* (Surinam), *kabana-fau* (Taki-Taki in French Guiana); *arrepillo, uña de gato, marupa* (Amazonas, Venezuela); *jarupari* (Terr. do Roraima); *sabueira, camunze* (Pará); *mocairey* (Huitoto of Loreto, Peru); *quillocaspi, quillosisa* (Loreto and San Martín, Peru).

The commonest and most widespread form of *S. multijuga* is characterized by the syndrome of basally dilated stipules and large membranous leaflets well spaced along the rachis. The populations differ much in pubescence and in flower-size, but erratically so. Trees with setulose panicle and sometimes a few setules along the leafstalks are known from the segment of the whole range that extends from Sierra Imataca and Mount Roraima in southeastern Venezuela southwestward to the upper Río Negro and (interruptedly) to Peruvian Amazonia, but does not replace the ordinary, simply pilosulous or glabrate form. Locally around Iquitos in Loreto, Peru and on the sources of Rio Juruá in Acre, Brazil, the species is represented by a notable variant with 20–36 pairs of exceptionally narrow leaflets up to (13–)15–22 × 2.6–4(–4.3) mm relatively crowded along the rachis at points (2–)3–4.5 mm apart. This might suggest passage into distantly allopatric var. *verrucosa,* but has substantially fewer and (even though narrow) shorter leaflets; we interpret it as a minor, but visually striking variant.

All but the last of the synonyms listed above were recognized as such by Bentham and require no further comment. We refer here *Peiranisia aristulata* with diffidence, for it is known only from fragments of the holotypus lost with the collection of Instituto de la Salle in Bogotá. These fragments seem certainly to represent *S. multijuga* sens. lat., but lacking stipules we cannot be sure that they are not var. *lindleyana*. The species, except for var. *peregrinatrix* introduced into Bolívar, has been collected in Colombia only on the Putumayo River and can hardly be expected as a native element of the llano flora in Meta. Perhaps a topotypus from Villavicencio will one day settle the question.

It seems likely that the earliest name applicable to *S. multijuga* is *Cassia marimari* Aublet (1775), for which there is a voucher in the Aublet herbarium at BM (=MO Neg. *2271*). As published, however, *C. marimari* is merely a technical synonym of *C. galegifolia* Linn.

161/Ib. Senna multijuga (L. C. Richard) subsp. **multijuga** var. **verrucosa** (Vogel) Irwin & Barneby, stat. nov. *Cassia verrucosa* Vogel, Syn. Gen. Cass. 38 & Linnaea **11**: 682 (descr. ampliat.). 1837.—"In Brasilia; Sellow leg. inter Victoria et Bahia."—Holotypus, B (ex hb. Kunth.), not seen = F Neg. *1767*! isotypi, FI, K (hb. Hook.) = NY, Neg. *1467* = IPA Neg. *990,* P, W!

Cassia centijuga Wawra, Flora **47**(n.r. **22**): 248. 1864.—"Peckolt Coll. n. 228; Urwaldbaum; Februar; Schoten im Marz."—Holotypus, divided into *228* (fl) and *228b* (fr), collected at Cantagallo, prov. Rio de Janeiro, W!—Lfts ±100, not 100 pairs!

Cassia verrucosa sensu Bentham, 1870, p.p., quoad pl. Blanchet., caeteris exceptis; 1871, eadem ex parte.

Trees 10–40 m, with trunk to 12–60 cm diam, the annotinous branchlets, ventral angles of lf-stalks and axes of inflorescence densely pilosulous-tomentulose with fine sordid spreading or flexuously incumbent hairs to 0.15–0.4 mm.

Stipules herbaceous, obliquely dilated at base, semi-ovate or falcately lanceolate 3–8 × 0.7–1.6 mm.

Lvs 7–16 cm; gland at proximal (and sometimes at second) pair 1–2.5 × 0.25–0.45 mm, smaller subulate glands (often obscured by vesture) at all following pairs; pulvinules 0.3–0.7 mm; lfts of major lvs 26–56 pairs, inserted along rachis at points 1.5–4 mm apart, the larger in outline linear, linear-lanceolate or -oblong, 8–14 × 1.6–3.2 mm, (3–)3.5–6 times as long as wide, the secondary camptodrome venation immersed on both faces or faintly raised on upper and visibly discolored on lower face.

Long inner sepals 4–5.5 mm; petals glabrous, the adaxial ones 11–16 mm, the falcate abaxial one 14–21 mm; androecium glabrous; ovary puberulent along sutures or overall.—Collections: 17.

Humid forest of the coastal plain, locally inland along rivers, below 300 m, often preserved for shade in cocoa plantations, local in s.-e. Bahia and n. Espírito Santo from the lower Contas valley to the vicinity of Linhares (±14°–19°30′S); Organ Mts. Rio de Janeiro.—Fl. VI–VII(–VIII).—*Angico branco.*

In its relatively narrow area of dispersal var. *verrucosa* is easily recognized by the high number and small size of the leaflets combined with basally dilated herbaceous stipules, a character noted by Vogel in the protologue but afterwards overlooked by Bentham. Around Ilheus and Itabuna and in the middle Pardo valley it has been collected close to var. *lindleyana,* here at its northern limit and possibly introduced from further south. The var. *lindleyana* has setiform stipules and leaflets at once fewer and larger than those of var. *verrucosa,* although not as large in this segment of its range as ordinarily in Rio de Janeiro or Minas Gerais. The more distantly related *S. acuruensis,* with similar (but narrower) pod and similar narrow seeds, has at the very most 20 pairs of leaflets and a viscid-setulose inflorescence; it is a weakly arborescent shrub adapted to different habitats in the caatinga formation of the foothills and uplands.

Bentham, treating *Cassia verrucosa* as a distinct species, suggested that it might prove to be only a variant of *C. multijuga* with many small leaflets, a view readily adopted herein. Bentham's concept of *C. verrucosa* was, however, somewhat more inclusive than our own, embracing elements of our subsp. *lindleyana* var. *peregrinatrix,* superficially similar in small size (though different in lower number) of leaflets and technically distinct in having the setiform stipules of subsp. *lindleyana.*

161/II. Senna multijuga (L. C. Richard) subsp. **lindleyana** (Gardner) Irwin & Barneby, stat. nov. *Cassia lindleyana* Gardner, London J. Bot. **2**: 341. 1843.—". . . from the Organ Mountains [Rio de Janeiro, Brazil . . . *Gardner*] 367 . . . in woods, common."—Holotypus, collected IV.1837, K (hb. Hook.)! isotypi, FI (hb. Webb.), K (hb. Benth.), NY!

In all respects like subsp. *multijuga*, but stipules setiform, not asymmetrically dilated at base, and range (except as modified by man) more southern and eastern, from Bahia and Minas Gerais e. of Rio São Francisco to Santa Catarina, locally sympatric with subsp. *multijuga* var. *verrucosa* in e.-centr. Bahia.

161/IIa. Senna multijuga (L. C. Richard) subsp. **lindleyana** (Gardner) Irwin var. **lindleyana**. *Cassia lindleyana* Gardner, 1843, l.c., sens. str. sed char. mut.—Typus supra sub subsp. *lindleyana* indicavimus.—*C. multijuga* var. *lindleyana* (Gardner) Bentham in Martius, Fl. Bras. **15**(2): 123. 1870, quoad nom., exclus. pl. Pohl.; Bentham, 1871, p. 546.

Cassia scabrella Hoffmannsegg, Zweiter Nachtrag Verzeichn. Pflanzenkult. 81. 1826.—"Hab. in Bras[ilia]."—No typus known to survive, but the protologue detailed and suggestive of no other Brazilian senna.

Cassia semifalcata Vellozo, Fl. Flumin. 167 ("*semi-falcata*"). 1825 & Ic. **4**: t. 68. 1835.—"Habitat silvis maritimis [presumably in vicinity of Rio de Janeiro]."—No typus survives; lectoholotypus, the cited plate.—Equated with slight reservation by Bentham, 1870, p. 123, with *C. multijuga* sens. lat., but the figure and habitat suggest no other known cassia.

Cassia selloi G. Don, Gen. Hist. Dichl. Pl. **2**: 442. 1832.—"Native of Brazil, Sello."—Holotypus, presumably seen by Don in herb. Lambert., not found at BM or OXF (1980); presumed isotypus, K! = NY Neg. *1465*.

Cassia magnifica Martius, Flora **20**, Beibl. 2 (Herb. Fl. Brazil.): 106. 1837.—"[Martius Herb.] 141 . . . In silvis Capocabanae [presumably Copacabana, Guanabara, for Martius himself was in s. Minas Gerais during March and April when the plant was said to flower]."—Holotypus, collected by B. Luschnath, BR! isotypi, K = NY Neg. *1466*, NY (2 sheets)!

Cassia trachypus sensu Bentham, 1870, p. 122, pro parte, quoad *Mart. Herb. 1072* ex S. Paulo.

Cassia chrysoclada Taubert ex Glaziou, Bull. Soc. Bot. Fr. **53**, Mém. 1(3): 167, nom. nud. 1906.—"Serra dos Orgãos, à Theresopolis, Rio-Jan., no. 8416."—Equated with *C. multijuga* by Harms, Fedde's Repert. **20**: 124. 1924.

Essentially like var. *multijuga* in aspect, and in detail except (as noted under subsp. *lindleyana* supra) for the setiform stipules, the pubescence equally variable except the axes of inflorescence never setulose.

Lfts of larger lvs 16–36(–46) pairs, the larger ones (1.8–)2–4.6 × (3.5–)4.5–13(–14) mm, (2.8–)3–5 times as long as wide, the longest interfoliolar segments of rachis (4–)5–13(–15) mm.

Fls and pods of var. *multijuga*.—Collections: 93.

Forest clearings, margins of humid woodland, often along streams, inland in gallery forest, thriving in disturbed and regenerating brush-woodland, sometimes preserved as a shade-tree in pastures and plantations, mostly 50–950 but on Sa. do Caparaó ascending to 1800 and on Sa. da Caraça to 1400 m, locally plentiful along the Atlantic slope and crest of the coast ranges of s.-e. Brazil between s. Bahia and n.-e. Santa Catarina, sparingly n. (?native, in the range of *S. multijuga* var. *verrucosa*) on the coastal plain of Bahia to Ilheus and Itajuipé, inland in São Paulo to the headwaters of Rios Tieté and Paranapanema and in Minas Gerais to the s. Sa. do Espinhaço (Catas Altas, Sabará etc.) and n. along the Velhas fork of Rio São Francisco to Várzea da Palma; cultivated in s. Brazilian cities and gardens (Brasília, Rio de Janeiro, São Paulo), in s. California, in Fiji, and probably elsewhere in the Neotropics.—Fl. I–IV(–V), in California gardens X–XII.—*Farinha senca* (Minas Gerais); *alleluia, amarellinho* (S. Paulo).

Cassia lindleyana was separated in the first instance from *C. selloi* because of the copious yellow vesture of the young branches and inflorescence and maintained by Bentham for the same reason as a variety of *C. multijuga*. The unnumbered variants of *S. multijuga* sens. lat. that differ in color and density of pubescence are here considered taxonomically (though not necessarily horticulturally) insignificant and var. *lindleyana* is now characterized by the coincidence of distant ample leaflets with setiform stipules. It is the south Brazilian analogue of var. *multijuga* and, except for stipules, essentially like it. In the part of its range that extends along the coastal ranges southwestward from the Organ Mountains var. *lindleyana* is sympatric with var. *peregrinatrix,* differing from this in the less crowded, always longer and nearly always much broader leaflets. In southern Minas Gerais, especially around Viçosa (where repeatedly collected), var. *lindleyana* is represented locally by a form with leaflets approaching var. *peregrinatrix* for narrowness but at the same time longer and more numerous. It appears strictly parallel to that form of var. *multijuga* with narrow leaflets that is locally dominant around Iquitos and the sources of Rio Juruá in upper Amazonia. We take this to be equivalent to *C. selloi* γ Vogel (1837, p. 38) which seems to be represented at K by a Sello collection labelled "*C. selloi* var. *polyphylla* Vog." (nom. nud.).

The three synonyms listed above, all based on plants from the neighborhood of Rio de Janeiro, were already referred by Bentham to *C. multijuga.* In our account of *C. trachypus,* q.v., a partial confusion with var. *lindleyana* on Bentham's part is reviewed.

161/IIb. Senna multijuga (L. C. Richard) subsp. **lindleyana** (Gardner) var. **peregrinatrix** Irwin & Barneby, var. nov., a var. *lindleyana* foliolis simul brevioribus et angustioribus (majoribus 9–15 × 2–4 mm) a subsp. *multijuga* var. *verrucosa,* cum qua olim confudit Bentham., imprimis stipulis setiformi-linearibus basi haud dilatatis foliolisque minus numerosis (foliorum majorum saepissime 15–32, nec 33–53-jugis) diversa.—BRAZIL. Paraná: Banhado, mun. Piraquara, 31.1.1971 (fl, fr jun), *G. Hatschbach 26189.*—Holotypus, NY: isotypi, NY, MBM.

Cassia verrucosa sensu Bentham, 1870, p. 124, p.p. quoad pl. hilariana; non Vogel.

Trees (3–)4–10 m, the hornotinous branchlets, lf-stalks and axes of inflorescence ± densely puberulent-tomentellous with incurved hairs up to 0.1–0.25 mm or with these mixed with spreading-ascending longer hairs to 0.3–0.5 mm, the pubescence sordid or lutescent, the axes of inflorescence in addition (together with some lf-stalks) nearly always ± densely charged with erect setules to 0.2–0.5 mm, the lfts either glabrous on both faces and ciliolate or commonly pilosulous beneath or overall.

Stipules setaceous or very narrowly linear-attenuate (3.5–)5–13 × 0.2–0.4 mm. Lvs below the panicle 6–13 cm; gland between proximal pair of lfts 2–5 × 0.25–0.6 mm; pulvinules 0.5–1 mm; lfts 15–32 (of some lvs low on annotinous branchlets as few as 14–9) pairs, inserted at points 2–5 mm apart along rachis, in outline lance-oblong or narrowly oblong up to 9–15 × 2–4 mm, 3–5 times as long as wide, a faint venation of ±5–8 pairs of camptodrome secondary veins commonly prominulous above, sometimes delicately so beneath, the tertiary venulation fully immersed on both faces, sometimes discolored beneath.

Long inner sepals 5–7.5 mm; petals glabrous or exceptionally puberulent dorsally along veins, the long abaxial ones up to (14–)16–21 mm; androecium and ovary glabrous, or the latter remotely pilosulous.—Collections: 20.

Moist forest and slopes of morros along the Atlantic and high interior slopes of the coast ranges of s.-e. Brazil, from near sea-level up to ±900 m, certainly native below the Tropic line between s.-e. São Paulo and n.-e. Santa Catarina; one record from the Organ Mountains in Rio de Janeiro (Novo Friburgo); distantly allopatric and presumably naturalized in savanna woodland at ±250–300 m on the lower Paragua and adjoining Caroní rivers in Bolívar, e.-centr. Venezuela, and on the Caribbean lowlands of n.-w. Bolívar (Turbaco; Palmito) in Colombia.—Fl. in Brazil X–II, in Venezuela and Colombia VIII–X.

In southeastern Brazil var. *peregrinatrix* is somewhat precariously distinguished from var. *lindleyana* only by the shorter, narrower and more crowded leaflets, in this respect standing in relation to var. *lindleyana* as does var. *verrucosa* to var. *multijuga*. The cases are different, however, in that the native range of var. *peregrinatrix* appears to lie entirely within (though is far from coextensive with) that of its close relative. The majority of specimens differ further from var. *lindleyana* in the setulose pubescence of the panicle, never encountered in *lindleyana* and lacking from var. *peregrinatrix* only at the extremities of its range in Rio de Janeiro (Novo Friburgo) and Santa Catarina (Moro de Cambirella). In its Venezuelan and Colombian stations, where var. *peregrinatrix* has passed as a form of *S. mutisiana* lacking lustrous hispid setae, its inflorescence is setulose as always in Paraná and São Paulo. Curiously the Venezuelan stations lie within the relatively restricted range of setulose var. *multijuga,* different not only in the broad distant leaflets but also in the dilated stipules. In absence of the fruit and in ignorance of the habitat we cannot be absolutely certain that the Venezuelan and Colombian plants, which appear certainly conspecific in a narrow sense, are genuine var. *peregrinatrix* and do not represent an independently derived, autochthonous form of *S. multijuga* linking southeast Brazilian subsp. *lindleyana* with Mexican subsp. *doylei,* all being similar in their small leaflets and setiform stipules. The bicentric occurrence on the lower Caroni in Venezuela and on the densely populated and long inhabited coastal plain of northwestern Colombia is suggestive of random dispersal from a southern focus, and it is this hypothesis which suggested the epithet.

161/III Senna multijuga (L. C. Richard) subsp. **doylei** (Britton & Rose) Irwin & Barneby, stat. nov. *Peiranisia doylei* Britton & Rose, N. Amer. Fl. **23**(4): 265. 1930.—"Type from Pichucaleo [=Pichucalco], Chiapas, June 19, 1907, *C. B. Doyle 259.*"—Holotypus, US! clastotypus (fragm + phototypus), NY!—*Cassia doylei* (Britton & Rose) Lundell, Phytologia **1**: 214. 1937.

Cassia doylei sensu Pennington & Sarukhan, Arb. Trop. Méx. 192 + fig. 1968.

Trees 6–17 m with trunk up to 2.5 dm, the hornotinous branchlets, ventral angles of lf-stalks and axes of inflorescence minutely pilosulous with pale or commonly lutescent hairs to 0.1–0.25 mm.

Stipules 4–6 × 0.3–0.6 mm.

Lvs 9–19 cm; gland at first pair of lfts 0.8–2 × 0.3–0.6 mm, minute subulate or verruculiform glands at all other pairs; pulvinules 0.3–0.6 mm; lfts 33–53 pairs, inserted 2–4 mm apart along rachis, in outline linear-oblong or narrowly lanceolate, up to 7.5–14 × 1.8–3 mm, (2.8–)3.6–6.2 times as long as wide, the venation of upper face immersed, the secondary camptodrome veins of lower face discolored and immersed or faintly promunilous.

Long inner sepals 3.5–4.8 mm; petals glabrous dorsally, the longest ±17–23 mm; androecium glabrous.—Collections: 24.

Humid evergreen forest, below 600(–800) m, flourishing in disturbed and cutover woodlands, sometimes persisting in hedges and pasture thickets, locally abundant on the Gulf coastal plain and bordering hill country in Veracruz and adjoining s. Hidalgo, n. Oaxaca, n.-e. Chiapas and Tabasco.—Fl. VI–XI.—*Cachimbo, guaje cachimbo, guajillo*; *dormilón, d. blanco*; *chipilin*; *hormiguerilla*; *picho guacamayo*.

While *S. doylei* has appeared very distinct from all *Interglandulosae* in the Mexican flora because of its numerous and small leaflets combined with a broad pod and long narrow seeds, these distinctive characters are all shared by one or another form of South American *S. multijuga*. In the high number and small size of the leaflets it resembles southeast Brazilian var. *verrucosa*, but differs in the setiform, not dilated stipules; in stipules it resembles subsp. *lindleyana*, likewise southeast Brazilian, but differs from var. *lindleyana* in the much smaller and from var. *peregrinatrix* in the more numerous leaflets. Ultimately it is the geographic isolation from other populations of *S. multijuga* rather than any morphological peculiarity that prompted us to treat *Cassia doylei* as an independent subspecies of *S. multijuga* rather than as a third variety of subsp. *lindleyana*. The only Central American member of ser. *Interglandulosae* that could be confounded with subsp. *doylei* is *S. pallida* var. *foliolosa*, comparable in its many small leaflets but readily distinguished at anthesis by the exactly 2-flowered racemes and later on by the narrow pod and compressed-obovoid seeds.

The subspecies was first discovered by the Danish botanist F. M. Liebmann in 1842. According to Pennington & Sarukhan (1968, l.c.) the wood has no industrial use, but the plant is sometimes grown for ornament.

162. Senna formosa Irwin & Barneby, sp. nov., inflorescentia florisque forma cum *S. multijuga* arcte congrua et ei certe affinis sed foliis longipetiolatis (petiolo 4–6 nec 0.3–3.2 cm), foliolisque maximis, foliorum majorum 5–11(–15)- nec 15–56-jugis 4–6 cm usque longis et 1.2–2.2 cm usque latis secus rachin distantius (15–28 nec 2.5–11 mm) semotis, adspectu ideo absimillimo diversa.—BRAZIL. Bahia: Santa Cruz Cabrália, Reserva Biológica Pau-Brasil, 2.XII.1971 (fl), *A. Eupunino 45*.—Holotypus, CEPEC; isotypi, NY, R.

Amply leafy trees 6–15 m, trunk 1–3 dm diam, with pale brown lenticellate older and smooth atropurpureous young branchlets, these with ventral angles of lf-stalks, pulvinules and midrib of lfts, and all axes of inflorescence residually pilosulous with incurved-ascending hairs 0.15–0.4 mm, the foliage bicolored, the thinly chartaceous lfts brownish-olivaceous sublustrous above, paler dull beneath, the inflorescence an amply branched leafless panicle of racemes terminal to branchlets, not or scarcely exserted.

Stipules caducous, none seen.

Lvs (1.5–)2–3.5 dm; petiole including livid wrinkled pulvinus 4–6 cm, at middle 1.7–2.7 mm diam, proximally subterete, distally like rachis openly sulcate ventrally; rachis (5–)12–26 cm; glands between all pairs of lfts sessile or substipitate, in profile 1–2.5 mm, the livid lance-ellipsoid or ovoid head 0.4–0.7 mm diam; pulvinules 3–5 mm; lfts (3–)5–12(–15) pairs, inserted along rachis at points (1–)1.5–2.8 cm apart, slightly decrescent both up- and downward from above middle of blade, in outline broadly oblong or lance-oblong from asymmetrically cuneate or semicordate base, 4–6.2 × 1.2–2.2 cm, 2.5–3.5 times as long as wide, mucronulate at obtuse-emarginate apex, the margins revolute, the centric midrib depressed and puberulent above, cariniform beneath, the 9–13 pairs of major camptodrome (with some intercalary) secondary veins immersed or almost so above, bluntly prominulous beneath, tertiary venulation faint and irregular.

Peduncles 3.5–6.5 cm; racemes shortly 3–12-fld, the axis 0.2–4.5 cm, the fls at anthesis raised beyond level of unopened buds; bracts membranous ovate-cymbiform 2–2.5 mm, caducous as pedicels begin to elongate; pedicels at full anthesis ±2–3.5 cm; fl-buds globose when very young, the sepals separating long before anthesis; sepals ovate obtuse, moderately graduated, the outer 3–5 mm, the innermost 6–8.5 mm, all glabrous or the outer ones thinly puberulent ciliolate, all yellowish or yellowish-margined, delicately reticulate-venulose; petals glabrous, bright yellow drying pale yellow brown-veined, 4 subhomomorphic, in outline obovate or oblong-obovate contracted into a slender claw, the longest 17–24 mm, one abaxial petal (alternately left and right up the raceme-axis) short-clawed, coarsely veined, in outline obliquely oblong incurved away from vertical axis of fl through ±90°, measured along curvature ±3–3.5 cm; androecium glabrous, the filaments of 4 median stamens thickened 2–2.5 mm, of 1 latero-abaxial one 3–3.5 mm, of 2 abaxial ones 8–12 mm, the anthers of 4 median stamens slightly incurved 5.5–7 mm, constricted just below truncate apex, the very short divaricate beak dehiscent by parallel slits, the anthers of 3 abaxial ones lunately incurved, the body 6–10.5 mm, narrowed into a slender porrect introrsely 2-porose beak 1.5–2 mm; ovary glabrous; style ±2 × 0.5 mm, gently incurved and attenuate into the minute stigmatic cavity; ovules ±70.

Pod seen only in its incipient stages, then not distinguishable from that of *S. multijuga*.—Collections: 6.—Fig. 29.

Humid coastal forest below 100 m, surviving in capoeira, very local near the coast of s. Bahia in lat. ±16–17°S.—Fl. XII–IV.

An exceptionally handsome arborescent senna and a promising candidate for tropical horticulture on the grand scale. We suppose that *S. formosa* represents an offshoot of widespread and multiracial *S. multijuga* in which the leaves have undergone that reciprocal loss of numbers and gain in size which has obviously occurred in several independent lines of inheritance in *Senna*. The flower of *S. formosa* is essentially that of *S. multijuga* and its incipient pod foreshadows a similar ripe one. Its small area of dispersal around Santa Cruz Cabrália and Pôrto Seguro lies within the shared range of *S. multijuga* vars. *verrucosa* and *lindleyana,* which exhibit different stages of multiplication and diminution of the leaflets. The extent to which these taxa may now be or, before disturbance of the coastal forest, may have been ecologically segregated requires study in the field.

The earliest collection of *S. formosa* known to us was distributed under Glaziou's number *8413,* (BR, G, K), without indication of exact locality. It is not mentioned by Glaziou in his Liste, and was perhaps not collected by him personally.

163. Senna aristeguietae Irwin & Barneby, sp. nov., ab affini *S. acuruensi* quoad habitum pubemque viscido-setulosam congrua imprimis stylo elongato 5–7 (nec 1.8–3) mm longo sepalisque et intimis dorso dense puberulis, ulterius a var. *acuruensi* foliolis minus numerosis 8–11 (nec 14–20)-jugis, a var. *interjecta* petiolis brevibus 8–13 (nec 15–30) mm longis diversa.—VENEZUELA. Zulia: cerca de Cabimas, carretera de la 'Sideroca,' XI.1968 (fl, fr submaturo), *L. Aristeguieta* (with *C. Blanco & L. E. Carrillo*) *6856*.—Holotypus, NY (2 sheets); isotypi, MICH, US.

Diffusely bushy shrubs ±1.2–2 m with terete castaneous or livid hornotinous branchlets, pubescent with 4 sorts of trichomes, viscid-hispidulous nearly throughout with orange excretory setules ±0.2–0.3 mm, the stems and lf-stalks in addition minutely retrorse-puberulent with fine recumbent hairs ±0.2–0.4 mm, the lfts on both faces antrorsely pilosulous, and the stems, lf-stalks and valves

Fig. 29. *Senna formosa* I. & B. Flowering branchlet ×½; petals ×1; detached stamens + staminode ×3 (*Eupunino 45*).

of pod furthermore hispid with lustrous horizontal setae 0.8–1.8 mm, the foliage strongly bicolored, the membranous lfts when dry dull olivaceous-brownish or dark brown above, pallid beneath, the inflorescence of 2–5-fld umbelliform racemes axillary to coeval lvs, either immersed in foliage or shortly exserted.

Stipules erect, linear-attenuate or flexuously setiform 5–8 mm, deciduous. Lvs 6–12.5 cm; petiole including livid pulvinus 8–16 mm, at middle 0.4–0.7 mm diam, rounded dorsally, narrowly or sometimes obscurely sulcate ventrally, ± twice as long as the first interfoliolar segment of rachis; rachis 4.5–9.5 cm; petiolar glands between proximal and often between the next 2–3 pairs of lfts, shortly stipitate, in profile 1–2.2 mm, the slenderly fusiform acute glabrous head 0.15–0.3 mm diam; lfts 8–11 pairs, ascending from rachis on livid pulvinule 0.6–1 mm, slightly accrescent upward, in outline oblong-obovate or oblong, obtuse mucronulate, the distal pair 12–24 × 5–10 mm, ±2.2–3 times as long as wide, all at base rounded-subcordate on proximal and broadly cuneate on distal side, the margin plane, the venation of upper face fully immersed, the slender centric midrib prominulous beneath, the 5–7 pairs of camptodrome secondary veins there faintly raised or almost imperceptible, the intervenium delicately reticulate by immersed but discolored veinlets.

Peduncles 2–7 cm; bracts caducous, not seen; gland at base of pedicels 0; racemes 2–5-fld, the axis not over 1 mm; pedicels when 2 quasi-dichotomous, when 3–5 umbellate, at and after anthesis 21–30 mm, in fruit commonly bent beyond middle; fl-buds subglobose, densely puberulent and sometimes also setulose; sepals submembranous greenish or livid-tinged, obovate or oblong-obovate obtuse concave, the outermost 6–11 mm, the innermost 11–16 mm; petals yellow, puberulent dorsally along and between veins or glabrate, 4 broadly obovate narrowed to a slender claw, the cucullus broadly obliquely boomerang-shaped concave, nidulating the 2 long abaxial stamens, the longest petal 2.7–3.5 cm; filaments glabrous or minutely pilosulous, of 4 median stamens 1.4–2.3 mm, of the centric abaxial one 1.5–2.5 mm, of 2 latero-abaxial ones (5–)6–11 mm, the anthers of 4 median stamens including the short sublateral beak 5.5–7 mm, the beak dehiscent by 2 fully separated parallel slits, those of 3 abaxial ones more strongly incurved, their body 6–8 mm, abruptly contracted into an erect tubular 1-porose beak 2.6–4 mm; ovary densely white-setose and setulose; style filiform 5–7 mm, at incurved, sometimes slightly dilated apex 0.3–0.4 mm diam; ovules 20–26.

Pod randomly spreading-declined, the stipe ±9 mm, the linear, plano-compressed, straight or slightly decurved body 8.5–11 × 0.75–0.8 cm, the papery pale brown valves scarcely elevated over seeds, becoming rather coarsely pallidly transverse-venulose and reticulate, the narrow interseminal septa ±4 mm apart; seeds not seen ripe, but a narrow areole evident on testa of immature ones.—Collections: 4.—Fig. 30.

Thickets and disturbed brush-woodland, near sea level in Venezuela and up to near 750 m in Brazil, the natural range enigmatic, collected once on the e. shore of Lago Maracaibo in Zulia, Venezuela, where possibly introduced, and twice in interior n.-w. and s.-centr. Minas Gerais (Virgem da Lapa; Rio Manso), Brazil.—Fl. in Venezuela X–XI, in Brazil III–VI.

This morphologically clearcut but otherwise contentious senna was first collected in Brazil by Auguste de Saint-Hilaire before 1822, about the turn of the turn of the century by Glaziou in south-central Minas Gerais (Rio Manso, ±20°20′S, 44°20′W), again in the northwest of the same state (between Virgem da Lapa and Barra de Salmas) in 1959 by Mendes Magalhães, and in 1968 it inexplicably turned up in the Maracaibo Basin in northwestern Venezuela. It is a strongly characterized species, perhaps most closely related to the Brazilian *S. acuruensis,* of which it has the complex vesture of retrorse villi (on stems), antrorse villi (on leaflets), viscid setules, and long non-secretory setae, but differs from all forms of it, as from all habitally similar *Interglandulosae,* in the elongate style. The shaggy-setose pistil and dorsally pubescent sepals distinguish it further

Fig. 30. *Senna aristeguietae* I. & B. Flowering branchlet ×1; detail of stem pubescence ×6; two stamens + staminode ×3; two segments of pod ×3. (*Aristeguieta 6856*).

from *S. acuruensis* sens. lat., of which the typical variety has more numerous (14–20, not 8–11) pairs of leaflets and the var. *interjecta,* similar in leaflet number, has substantially longer petioles. The seeds of *S. acuruensis* are narrowly oblanceolate in outline, whereas those of *S. aristeguietae,* not yet known fully ripe, appear to have the broad outline prevalent elsewhere in the series. Assuming that the Venezuelan plant is truly conspecific with the Brazilian ones, from which it differs only inconsequentially in length of sepals and longer filaments, we suppose that it was introduced in Zulia. A bicentric dispersal of this pattern is unprecedented and it is easier to look to eastern Brazil, the focus of speciation for viscid-setulose *Interglandulosae,* for a source of origin, than to Venezuela. The kindred Venezuelan species most like *S. aristeguietae* is *S. saeri,* found in the hill country of Lara at least 200 km east of Lago Maracaibo, which is instantly distinguished by fewer (4–6, not 8–11) pairs of larger, dorsally venulose leaflets, a narrower, more coarsely venose pod (4.5–6, not 7.5–8 mm wide), and a brachystylous ovary. Our choice of the Venezuelan plant as typus of what we believe will turn out to be a species endemic to Minas Gerais was dictated simply by the superior quality and relative completeness of the specimens at hand.

The species is named in honor of our eminent Venezuelan colleague Leandro Aristeguieta.

164. Senna acuruensis (Bentham) Irwin & Barneby, comb. nov. *Cassia acuruensis* Bentham in Martius, Fl. Bras. 15(2): 122 (''*açuruensis*''). 1870, sens. ampliat.—Typus infra sub var. *acuruensi* indicatur.

Profusely handsomely floriferous arborescent shrubs, at anthesis (1–)2–5 m, with densely leafy pliant annotinous branchlets, these with all lf-stalks and (commonly but not always) axes of inflorescence viscidulous with erect yellowish or orange tapering or capitate setules up to 0.2–0.4 mm mixed with minute incurved whitish villi and (rarely) in addition pilose with filiform spreading-ascending hairs to 1 mm, the foliage bicolored, the thinly chartaceous lfts dull dark brownish-olivaceous above, pallid beneath, varying (as described for the vars.) from glabrous to densely pilose on one or both faces, the inflorescence of simple racemes arising directly from axils of distal lvs and about as long or longer than them, sometimes by reduction of foliage distally becoming paniculate.

Stipules erect setiform 3–8(–9) mm caducous.

Lvs of primary stems (5–)6–15 cm (of some branchlets shorter, not further described), highly variable in number and size of lfts, these varying (acc. to var.) from 4 to 19(–20) pairs, when fewer larger; petiole including firm, moderately dilated pulvinus 9–30 mm, at middle 0.35–0.6 mm diam, shallowy sometimes obscurely sulcate ventrally; rachis 2–12.5 cm; petiolar glands between several proximal and often, but diminished upward, between all pairs of lfts, stipitate, the lower in profile 1–3 mm tall, the stipe nearly always puberulent, the glabrous discolored head slenderly ovoid-acuminate or claviform 0.2–0.45 mm diam; lfts ascending from rachis, face upward, on slender livid pulvinule 0.6–1.1 mm, in outline ovate-, obovate- or oblong-elliptic to broadly obovate or oblong-obovate, obtuse mucronulate or subemarginate (exceptionally incipiently deltate-acute), the largest of the larger lvs 1–3.7 × 0.4–1.8 cm, at base asymmetrically rounded, the straight centric midrib with (4–)5–7 pairs of camptodrome secondary veins usually all finely raised above, always so beneath, the midrib there cariniform, the tertiary and reticular venulation variable, above immersed, faint, or evident, beneath usually finely prominulous.

Peduncles 2.5–5.5 cm; racemes (2–)3–9(–17)-fld, the fls either subequidistant or (when few) subumbellate or subverticillate, the axis 0.2–7 cm; bracts oblanceolate or obovate-elliptic concave 1.5–3 mm, nidulating the very young fl-buds, promptly caducous; pedicels 21–32 mm; buds subglobose obtuse, glabrous beyond immediate base or the outer sepals puberulent and ciliolate; sepals greenish or yellowish, often petaloid-margined, finely several-nerved from base, strongly unequal, the inner ± twice as long as outermost, broadly obovate-suborbicular

(6–)6.5–11(–12) mm; petals golden-yellow, dorsally glabrous or at base puberulent along major veins, highly dissimilar in form, the 3 adaxial relatively small, long-clawed, oblance-obovate in outline, the 2 abaxial larger but dissimilar, one asymmetrically ovate-obovate-subcordate beyond the short claw, the opposed one obliquely reniform scoop-shaped subsessile, this nidulating 2 of the 3 beaked anthers; androecium glabrous, the filaments of 4 median stamens 1.5–3 mm, of one abaxial (1.5–)2–3 mm, of 2 abaxial 5.5–10.5 mm, the anthers of 4 median ones oblong little incurved 4–7 mm, with abruptly incurved beak ±0.5 mm dehiscent by parallel or almost confluent slits, those of 3 abaxial ones lunately incurved 6–8 mm, abruptly contracted into an erect tubular beak 3–4 mm, this slightly enlarged at apex and its orifice divided by a filiform septum; ovary glabrous, thinly ciliolate especially below middle or thinly pilosulous; style 1.8–3 mm, at apex 0.4–0.55 mm diam; ovules 38–50.

Pod spreading or pendulous, the stipe 4.5–8 mm, deltately dilated into the valves, the narrowly oblong flat body 8–11 × 1.1–1.5 cm, the 3 sutural nerves all cordlike or the middle one narrowly winged, the plane papery nigrescent valves finely transverse-venulose, scarcely elevated over the seeds; seeds linear-oblong in outline, obtusely 4-angular, 5.6–7 × 1.1–1.6 mm, the testa pale brown, sometimes faintly rugulose, marked on each face by a linear-elliptic areole 1.5–2.3 × 0.25–0.4 mm.

The typus of *Cassia acuruensis,* the only example known to Bentham, was at once geographically and morphologically marginal to its species. In the light of modern collections it takes its due place in a series of Bahian sennas very constant in characters of flower and pod but extremely variable in size and number of leaflets and, largely independently of these, in quality and distribution of the mixed pilose and viscid-setular vesture. Until recently the typi of *Cassia acuruensis,* with 14–20 pairs of small leaflets pilose on both faces, and of *C. catingae,* with only 4–7 pairs of substantially larger glabrous ones, the latter further characterized by glabrate inflorescence, appeared safely and distantly distinct; but they are now united by numerous intermediate forms which we group here into three varietal categories, characterized primarily by leaflet number. These accord fairly well with dispersal patterns, the multifoliolate types being largely confined to Chapada Diamantina and those with fewer leaflets to the more eastern ranges and the coastal lowlands, but all coinciding in the lower and middle Contas valley where the essential unity of the species as redefined herein becomes especially evident.

The species is very closely related to the northwardly vicariant *S. trachypus* and now no longer separable from it by number of leaflets although reliably distinct in the thickened, not winged sutures of the pod. In flowering condition only *S. acuruensis* var. *catingae* is likely to be confused with *S. trachypus*; its fully disjunct range and glabrescent axes of inflorescence then provide for rapid discrimination. Viscid-setular pubescence of stem and leaf-stalk, absence of stipitate glands at base of the pedicels and linear seeds enclosed in a broad pod together sharply distinguish both *S. acuruensis* and *S. trachypus* from all otherwise similar Brazilian *Interglandulosae* except the rare *S. aristeguietae,* different in the elongate style and pubescent sepals.

Key to the Varieties of *S. acuruensis*

1. Lvs at once shortly petiolate and plurifoliolate, the petiole 9–16(–19) mm, the lfts relatively many and small, in larger lvs 14–19(–20) pairs, the amplest of them 10–18 × 4–8 mm; upland centr. Bahia to centr. Pernambuco. 164a. var. *arcuruensis* (p. 508).
1. Lvs at once relatively long-petiolate and paucifoliolate, the petiole 15–30 mm, the lfts fewer

and larger, in larger lvs 4–13 pairs, the amplest of them 14–37 × 8–18 mm; e.-centr. and s.-e. Bahia.

2. Lfts 8–13 pairs; all axes of inflorescence densely viscid-setulose. 164b. var. *interjecta* (p. 508).

2. Lfts 4–7 pairs; axes of inflorescence glabrous or almost so, the pedicels sometimes residually setulose. 164c. var. *catingae* (p. 509).

164a. Senna acuruensis (Bentham) Irwin & Barneby var. **acuruensis**. *Cassia acuruensis* Bentham in Martius, Fl. Bras. **15**(2): 122. 1870, sens. str.—"Habitat in silvis Capoeiras deserti s. d. Sertão do Rio S. Francisco in Serra Acuruá; Blanchet n. 2851."—Holotypus, K (hb. Hook.)! = NY Neg. *1464*; isotypi, †B = F Neg. *1643,* BR, F, FI, K (hb. Benth.), NY, W!

Cassia acuruensis sensu Bentham, 1871, p. 546.

Stems, lf-stalks and all axes of inflorescence setulose and villosulous, the stems rarely in addition pilose; lvs as described in the varietal key, the lfts strongly bicolored, usually pilosulous on both faces with pale or lutescent hairs, rarely glabrous above, the tertiary venulation of upper face commonly immersed.—Collections: 20.

Caatinga woodland, surviving after disturbance in hedges and thickets, mostly between 300 and 1000 m, locally plentiful in scattered stations on both slopes of Chapada Diamantina in centr. Bahia, lat. 10–14°S, extending rarely e. toward the coastal lowlands along Rio de Contas; apparently somewhat isolated in e. Pernambuco (Sa. do Genipapo) and centr. Alagôas.—Fl. IX–III.

164b. Senna acuruensis (Bentham) var. **interjecta** Irwin & Barneby, var. nov., quoad foliolorum numerum inter var. *acuruensem* et var. *catingae* interludens, foliolis foliorum majorum 8–13-jugis ±15–30 × 9–14 mm notabilis, inflorescentia dense viscido-setulosa var. *acuruensem* sed petiolis elongatis var. *catingae* simulans.—BRAZIL. Bahia [or possibly adjoining Pernambuco]; 25 km de Salgueiro, 18.VIII.1964 (fl), *L. Duarte 446 e A. Castellanos in hb. Brade. 33392*.—Holotypus, BRADE; isotypus, NY.

Cassia trachypus Mart. Herb. (BR!); sensu Bentham, 1870, p. 122, minore ex parte, quoad specimina Martiana e Minas Gerais.

Stems, lf-stalks and axes of inflorescence setulose as var. *acuruensis* (but not known to be pilose); lvs as described in the varietal key, the lfts not strongly bicolored, varying from glabrous to thinly pilosulous dorsally along midrib or also rarely on upper face, the tertiary venulation of upper face either prominulous or immersed.—Collections: 5.

Habitat not recorded, to be expected in lowland hill country, in thickets or caatinga woodland, 350 m or lower, known with certainty only from scattered stations on the lower S. Francisco, Paraguaçú and Contas rivers in e. Bahia and probably adjacent Pernambuco, possibly in n. Minas Gerais.—Fl. VIII–X, XII–III, perhaps intermittently through the year.

As inferred in the synonymy, it was to specimens of var. *interjecta* that Martius gave the new name *Cassia trachypus,* taken up by Bentham for a mixed concept involving elements of three species. We have lectotypified *C. trachypus* Benth. by fruiting specimens collected by Gardner, thereby excluding the original, unpublished *C. trachypus* Mart. ms. The vague locality given on Martius's label

(BR), '*in prov. Minarum*' is associated with the date July 1818. At this time Martius was collecting in the general vicinity of Grão Mogol and the Jequetinonha-São Francisco divide in northern Minas Gerais. The var. *interjecta* is known by exact data no further south than the Contas valley in southern Bahia and Martius's record for Minas needs confirmation.

164c. Senna acuruensis (Bentham) var. **catingae** (Harms) Irwin & Barneby, stat. nov. *C. catingae* Harms in Engler, Bot. Jahrb. **42**: 210. 1908.—"Bahia: Catinga bei Calderão [located, op. cit. p. 196, in "Gebiet des Rio das Contas," i.e. the village of Caldeirão ±25 km s.-w. of Maracás near 13°35′S, 40°35′W] (Ule n. 7250.—Okt. 1906)."—Holotypus, †B = F Neg. *1669*! neoholotypus, former isotypus, HAMB! = NY Neg. *8912*; isotypus, K! = IPA Neg. *1145* = NY Neg. *1546*.

Stems and lf-stalks setulose and villosulous but the main axis or this and the pedicels smooth and glabrous; lvs as described in the varietal key, the lfts bicolored but not strongly so, glabrous on both faces or thinly pilosulous along midrib beneath, the tertiary venulation finely prominulous on both faces.—Collections: 10.

Habitat scarcely known, sometimes (perhaps always) in caatinga, ±150–700 m, both slopes of Sa. Geral in the drainage of Rios Paraguaçú, Contas and Pardo, centr. & s.-e. Bahia.—Fl. IX–II.

165. Senna trachypus (Bentham) Irwin & Barneby, comb. nov. *Cassia trachypus* Bentham in Martius, Fl. Bras. **15**(2): 122. 1870, sens. restr., exclus. *C. trachypode* Mart. in herb.—"Habitat in silvis humidis prope Crato prov. Ceará: Gardner n. 1575; in collibus circa Oeiras prov. Piauhy frequens: Gardner n. 2119; in sepibus silvarumque marginibus prov. Minas Geraës: Martius; ad S. Carlos prov. S. Paulo: Patricio da Silva Manso (Mart. Herb. Fl. Bras. n. 1072.)."—Lectotypus, *Gardner 2119,* collected V.1829 (fl, fr), K (hb. Hook.)! = IPA Neg. *983*; isotypi, BM, F, FI, K (hb. Benth.), NY (hb. Meissner.), W! paratypi conspecifici, *Gardner 1575,* BM, K, W! The paratypi *Martius sn* ex Minas Gerais, BR! and *Silva Manso 269* distrib. *Martius. Herb. 1072,* †B = F Neg. *1758,* BR, K = NY Neg. *1463,* NY! represent respectively *S. acuruensis* var. *interjecta* and *S. multijuga* var. *lindleyana*.

Cassia trachypus sensu Bentham, 1871, p. 546, exclus. loc. S. Paulo et Minas Gerais.

Arborescent shrubs at anthesis 1.5–4 m, with slender, densely leafy reddish-brown annotinous branchlets, these with all lf-stalks and axes of inflorescence at once finely minutely villosulous with filiform incurved whitish hairs and viscidly setulose with orange or yellow peg- or tack-shaped setules up to 0.1–0.2 mm, the thinly chartaceous, notably bicolored lfts glabrous on both faces, above dark dull green brunnescent when dry and pallidly venulose, beneath pallidly olivaceous-glaucescent, the inflorescence composed of axillary simple racemes immersed in or distally exserted from foliage.

Stipules erect setiform 2–5.5 cm caducous, lacking from fruiting spms.

Lvs (4.5–)6–16(–18) cm; petiole including moderately dilated firm pulvinus (1.5–)2–4.5(–5) cm, at middle 0.4–0.8 mm, rounded dorsally, very shallowly sulcate ventrally; rachis (1.5–)2–8.5(–10) cm; petiolar glands between 2 proximal or commonly (diminishing distally) between all pairs of lfts stipitate, the lowest in profile 1.3–2.5 mm, the body slenderly ovoid-fusiform acute 0.2–0.6 mm diam glabrous; lfts 5–9 pairs, moderately accrescent upward, ascending from rachis face upward on pulvinule 0.8–1.6 mm, in outline obovate or oblance-obovate obtuse or subtruncate, mucronulate by the excurrent midrib, the distal pair (16–)20–42 × 8–21 mm, 1.6–2.5 times as long as wide, at asymmetric base rounded on proximal and cuneate on distal side, the margin plane, the centric straight midrib with 5–8 pairs of major camptodrome and intercalary secondary veins finely prominulous above, sharply so beneath, a tertiary reticulation commonly visible but immersed above, finely but evidently raised beneath.

Peduncles 1.5–4 cm; racemes loosely (2–)4–12(–18)-fld, the axis becoming (0.2–)1–6(–8) cm; bracts obovate cymbiform ±1.5–2 mm, early dry and caducous from below young fl-buds as pedicels elongate; pedicels 2.3–3.7 cm, glandless at base; buds obovoid obtuse glabrous or almost so; sepals brownish pale-margined, faintly several-nerved from base, strongly graduated, the outermost scarcely ½ as long as inner ones, these ovate- or obovate-suborbicular very obtuse concave 10–13 mm; petals golden-yellow glabrous, resembling those of *S. pallida* in outline and asymmetry, the obliquely reniform abaxial one longest, 2.5–3.4 mm; androecium glabrous, the filaments of 4 median stamens 1.5–2.8 mm, of one abaxial one 2–3.5 mm, of the one exactly abaxial and of one contiguous to it 4–9.5 mm, the anthers of 4 median stamens little curved, the body 5.2–6.5 mm, the divaricate biporose beak scarcely 0.5 mm, the anthers of 3 abaxial ones more strongly incurved, the body 5.5–9 mm, contracted into an erect tubular beak 2–3 mm, the orifice 1-porose or divided by filiform septum; ovary glabrous, the stout, gently incurved style 1.5–2 mm, not dilated distally, ±0.5 mm diam; ovules 44–50.

Pod pendulous, the stipe 5–8 mm dilated upward, the body linear-oblong straight or almost so, strongly compressed, 7.5–11 × 1.1–1.6 cm, the papery brunnescent or nigrescent valves flat or obscurely colliculate down the middle, finely transverse-venulose, dull or sublustrous in age, the sutures narrowly sharp-winged, the ventral wing ±0.8–1 mm wide; seeds linear-oblong, obtusely 4-angular 5.4–6.5 × 1.1–1.6 mm, grooved on each side, the testa pale brown dull crackled, the faces marked with a linear-elliptic areole 1.1–1.6 × 0.25–0.4 mm.—Collections: 18.—Fig. 31.

Seasonally dry woodlands (caatinga, cerradão), 100–370 m, locally plentiful over Atlantic n.-e. Brazil n. of Rio São Francisco, from the Paraíba valley in s.-centr. Maranhão to n.-centr. and s. Ceará and w. Paraíba and Pernambuco.—Fl. III–VI(–XI), the empty pods sometimes long-persistent.

A handsomely floriferous senna, notable for the combination of strongly bicolored, pale-venulose foliage, viscid-setulose leaf-stalks and inflorescence, beaked abaxial anthers and linear seeds, a syndrome repeated in northeastern Brazil only by the southwardly vicariant *S. acuruensis*. The latter is very closely related, differing ultimately only in the bluntly three-nerved, not acutely narrow-winged sutures of the pod but further from var. *acuruensis,* found immediately south of Rio S. Francisco, in the fewer and larger leaflets and from geographically more distant *S. acuruensis* var. *catingae,* which has almost the same foliage, in the presence of viscid setules throughout the inflorescence as well as on the petioles. Superficially similar Brazilian *S. rostrata* and *S. aversiflora* lack viscid setules entirely and differ in their umbellately 2–3-flowered racemes and in the stalked glands subtending the pedicels.

Our concept of *S. trachypus* is materially narrower than that of Bentham, who included in the protologue of *Cassia trachypus* an element of *S. multijuga* var. *lindleyana* in the shape of *Martius Herbarium No. 1072* collected in São Paulo by Silva Manso. This collection, of which two likenesses have been distributed (see above) as phototypes of *C. trachypus,* disagrees with the protologue in its

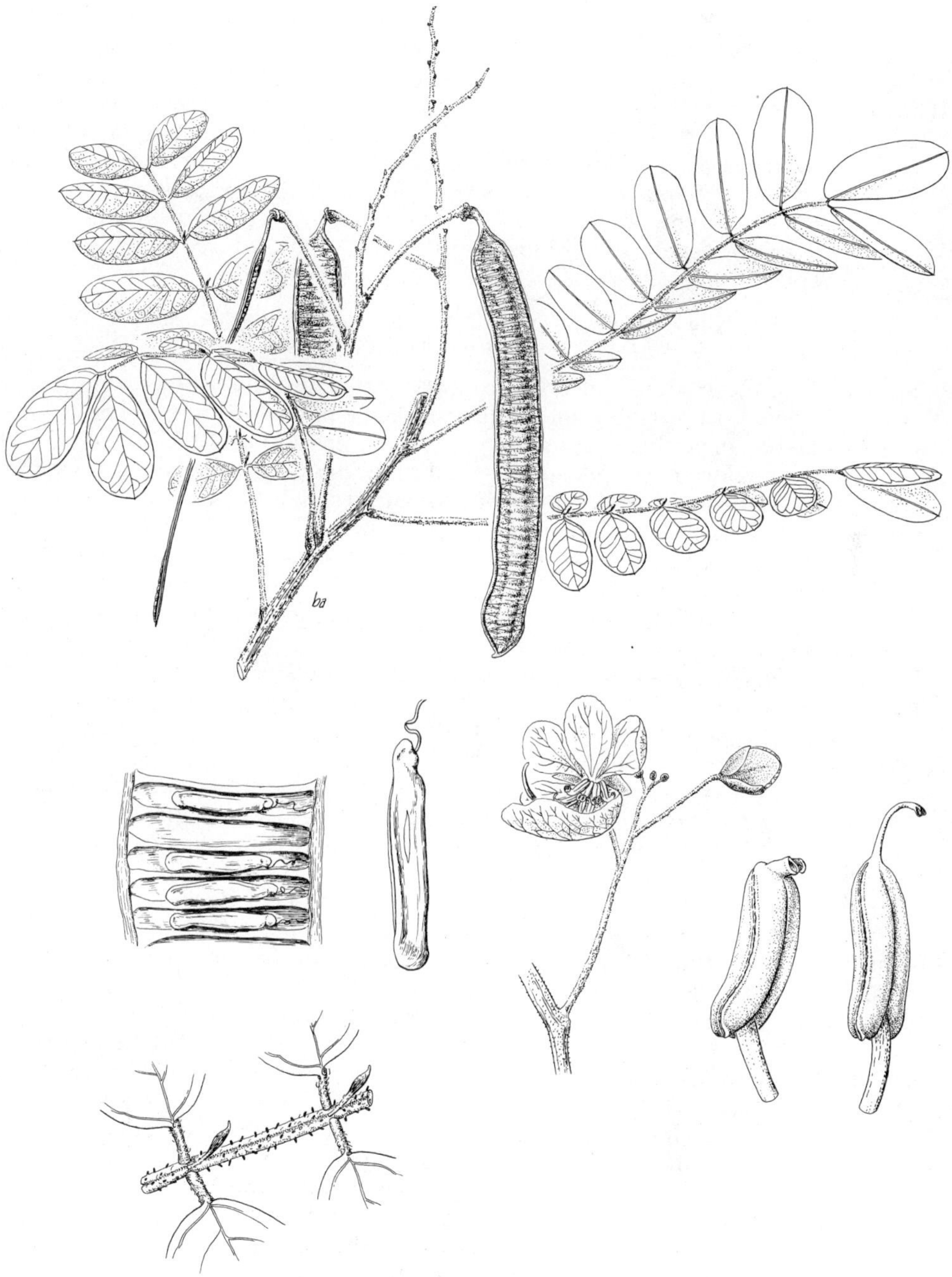

Fig. 31. *Senna trachypus* (Benth.) I. & B. Fruiting branchlet ×½; segment of leafstalk, showing petiolar glands ×4; detached stamens ×4; seed-locules ×2; seed ×4 (*Eiten 4287,* fl; *Drouet 2433,* fr).

glandless foliage and inflorescence and could have suggested neither Bentham's diagnosis (''racemis . . . viscido-puberulis'') nor the epithet attributed to Martius, apposite only to the plants with scabrously setulose pedicels. The specimens from which Bentham derived the epithet *trachypus* are also not conspecific with the lectotypus of *C. trachypus* but represent the related and, at anthesis, deceptively similar *S. acuruensis* var. *interjecta*.

166. Senna mutisiana (Kunth) Irwin & Barneby, comb. nov. *Cassia mutisiana* Kunth, Mimoses 142, pl. 43. 1824 (jan).—"Crescit in Regno Novo-Granatensi . . . communiquée par le célèbre Mutis, professeur de botanique à Santa Fe de Bogota."—Holotypus, B (not seen) = F Neg. *1713*! isotypi, P-HBK! P (hb. Bonpland.)!—*Peiranisia mutisiana* (Kunth) Pittier, Bol. Ministerio RR. EE. **1–3** (Arb. Arbust. Legum.): 128. 1928.

Cassia mutisiana sensu Humboldt, Bonpland & Kunth, Nov. Gen. & Sp. **6**(folio): 359. 1824 (jul); Bentham, 1871, p. 546. *Peiranisia mutisiana* sensu Britton & Killip, 1936, p. 181.

Shrubs at anthesis 1.5–3 m with densely leafy, distally ribbed annotinous branchlets, these with lf-stalks and often the foliage pilosulous in varying density with spreading or incumbent, commonly lutescent hairs up to (0.2–)0.3–0.6(–0.7) mm, the subconcolorous lfts pilosulous on both faces or only beneath, then sometimes merely at one basal angle of midrib, sometimes glabrous except for ciliolate margins, the stems in addition nearly always hispid with horizontally spreading pale or yellowish shining setae up to (3–)4–9 mm (these exceptionally quite lacking), the inflorescence of short but loosely several-fld racemes at first axillary to long coeval lvs and immersed, sometimes by reduction or atrophy of distal lvs becoming corymbose and exserted.

Stipules herbaceous, falcately lance-attenuate or unilaterally dilated and then semi-ovate-acuminate, 4–14 × 0.5–4.5 mm, deciduous before associated lf.

Lvs (below the panicle, where often much reduced) 6–17 cm; petiole including moderately thickened pulvinus 4–11 mm, at middle 0.6–1.5 mm, narrowly thick-margined and grooved ventrally; rachis 5–15.5 cm, the ventral wings constricted at each pair of pulvinules; glands between proximal 1–3 pairs of lfts subsessile or stipitate 1–3 mm, the fusiform acute body 0.3–0.7 mm diam, shorter spiculiform glands between all other pairs; pulvinules 0.3–0.5(–0.8) mm; lfts (25–)30–52 pairs inserted on rachis 1.5–3.5(–4.5) mm apart, very gradually decrescent both up- and downward from a point below middle of rachis, the larger ones in outline oblong or lance-oblong from asymmetrical base, obtuse mucronulate or subemarginate, 6–13 × 1.6–4 mm, dorsally carinate by the midrib, the margins (at least proximally) revolute, the venation other than costa immersed.

Peduncles 1.5–5 cm; racemes 5–9-fld, the axis 8–23 mm; bracts submembranous lance-elliptic cymbiform 2.5–4 mm, caducous; pedicels 13–28 mm, a fusiform stipitate gland beside each; buds globose glabrous or puberulent at base; sepals very unequal, the small outer ones firm, the inner subpetaloid, broadly obovate-suborbicular 8.5–11 mm; petals glabrous, 4 subequiform obovate beyond the narrow claw, the fifth obliquely or falcately oblanceolate, the longest 16–20 mm; androecium glabrous, the filaments of 4 median and 1 or 2 of the 3 abaxial ones 0.5–1.5 mm, that of 1 or 2 abaxial ones 2–4 mm, the anthers of 4 median stamens 3.3–5.2 mm subtruncately rounded at apex, the very short beak lateral 2-pored, those of 3 abaxial 3.2–5 mm, gently contracted to a very short erect 2-porose beak 0.35–0.5 mm; ovary thinly pilosulous; style 2–3.5 mm, at incurved apex 0.3–0.35 mm diam; ovules 20–30.

Pods randomly ascending or pendulous, the stipe 3.5–5 mm, the flat body (5–)7–10.5 × 0.7–1.05 cm, the thinly papery valves castaneous nigrescent glabrate, the interseminal septa 4–5 mm apart; seeds broadly or narrowly compressed-obovoid 4.5–5 × 2.4–3.7 mm, the testa dull brown, the obovate or elliptic-obovate areole 2.2–3 × 1.2–1.7 mm.—Collections: 17.

Open hillsides and forest margins, 900–2260 m, scattered along the Eastern Cordillera of the Colombian Andes from Huila n. to Norte de Santander and thence n.-e. to Cordillera de Mérida in Venezuela, lat. 3°–8°30′N.—Fl. I–VIII, perhaps intermittently through the year.

Senna mutisiana is usually strongly characterized by the coincidence of lustrously setose stems with narrow multifoliolate leaves, glands at base of pedicels slender pod, features all apparent in Turpin's exquisite colored engraving in Kunth's *Mimoses*. The species is closely related to northwardly vicariant *S. williamsii*, but sufficiently different in its short proper petioles, looser racemes of somewhat smaller flowers, short-beaked anthers and broader pods that accommodate larger seeds in wider locules. The setose indumentum of the stems, so striking to the eye in both these species, is here as elsewhere in Cassieae an unreliable character. One of the several collections of *S. mutisiana* known from Cordillera de Mérida in Venezuela (*López-Figueiras & Rodríguez 9078*) lacks it entirely and superficially resembles some microphyllous race of *S. multijuga*; and Pittier encountered in Chiriquí, Panama forms of *S. williamsii* growing close together, one with (no. *5407*) and one without (no. *5406*) cauline setae.

167. Senna williamsii (Britton & Rose) Irwin & Barneby, comb. nov. *Peiranisia williamsii* Britton & Rose, N. Amer. Fl. **23**(4): 265. 1930.—"Type from Bismark, above Penomene, 2000–3000 ft elevation, Panama, March 18, 1908, R. S. Williams 567."—Holotypus, NY! isotypi, NY = NY Neg. *9398*, US!—*Cassia williamsii* (Britton & Rose) Standley, Field Mus. Bot. **18**(Fl. Costa Rica): 519. 1937.

Cassia williamsii sensu Schery, 1951, p. 70, fig. 126.

Weakly arborescent shrubs 2–3.5 mm with trunk up to 3.5 cm diam and subterete annotinous branchlets densely leafy distally, these with ventral angle of lf-stalks pilosulous with fine spreading-incumbent pallid hairs up to 0.1–0.2 mm mixed with minute thickened orange or livid trichomes, the stems in addition nearly always hispid with stiff lustrous white setae 4.5–9 mm, the very numerous small lfts subconcolorous glabrous, the corymbose-paniculate inflorescence composed of shortly or subumbellately few-fld racemes arising from axils of coeval lvs and shorter than them.

Stipules herbaceous, falcately lance-attenuate 5–12 × 1–2 mm, deciduous.

Lvs 9–22 cm; petiole including little-swollen pulvinus (8–)11–22 mm, at middle 0.6–1.2 mm diam, bluntly narrow-margined and shallowly sulcate ventrally; rachis 7.5–19 cm; glands between proximal (1–)2–3 pairs stipitate, in profile (1.6–)2–4.2 mm, the fusiform acute body 0.3–0.6 mm diam, between all other pairs a much shorter and narrower acicular gland; lfts 36–63 pairs, inserted 1.5–3.5(–4) mm apart along rachis, very gently decrescent both up- and downward from near or below middle, pitched forward from rachis on pulvinule 0.4–0.8 mm, the blades of larger ones linear-oblong from asymmetrical base, obtuse mucronulate, 7–16 × 1.3–2.5 mm, ±4.5–6 times as long as wide, the smaller distal ones often linear-oblanceolate and proportionately wider, all appearing veinless except for slender centric midrib prominulous beneath, a secondary venulation of 5–7 pairs of camptodrome nerves sometimes faintly discolored beneath but fully immersed.

Peduncles 1–5 cm; racemes (2–)3–7-fld, the axis 1–10 mm; bracts submembranous lanceolate 2.5–5.5 mm, caducous; pedicels 18–32 mm; buds globose glabrous; sepals strongly graduated, the 2 inner ones obovate-suborbicular 10–14 mm; petals glabrous, except for one all obovate to broadly oblanceolate beyond the slender claw, the odd abaxial one obliquely or falcately lanceolate, the longest 23–29 mm; androecium glabrous, the filaments of 4 median stamens 1–1.7 mm, those of 3 abaxial ones 1.5–3 mm, the anthers of 4 median stamens 5–5.5 mm, gently incurved, the very short biporose beak lateral, those of 3 abaxial stamens slightly more incurved 5–5.7 mm, the porrectly erect biporose beak 0.9–1.5 mm; ovary puberulent; style 2–2.5 mm, near apex 0.35–0.4 mm diam; ovules 36–44.

Stipe of pod 5–6 mm, the narrowly linear straight compressed body 10.5–15 × 0.33–0.48 mm, the interseminal septa 2.5–4(–4.5) mm apart, the seed cavity as long or longer than width of valves; seeds rhombic-obovate 2.6–3.4 mm, the testa pale brown dull, the oblong-elliptic areole 1.3–1.5 × 0.5–0.6 mm.—Collections: 10.

Habitat little known, sometimes in disturbed rainforest ±550–950 m, scattered along the Pacific slope of the Continental Divide from s.-e. Costa Rica (Cordillera de Talamanca) through Chiriquí to Coclé (Valle de San Antón; Penonomé) and Panamá (Cerro Campana) in Panama.—Fl. XI—III.

A highly ornamental senna, closely akin to *S. mutisiana,* which see for commentary. The proportionately narrow, pluriovulate pod distinguishes the species from otherwise comparable *Interglandulosae*. Schery's figure (l.c.) incorrectly shows some flowers solitary in leaf-axils.

168. Senna angustisiliqua (Lamarck) Irwin & Barneby, comb. nov. *Cassia angustisiliqua* Lamarck, Encycl. Méth. **1**: 649. 1785.—Typus infra sub var. *angustisiliqua* indicatur.

Shrubs and treelets at anthesis 1–4.5 m, with twiggy, often tortuous branches and densely leafy, virgately erect, ascending-incurved or plagiotropic hornotinous branchlets, varying from quite glabrous below the ovary to variably pilosulous with fine spreading-incurved, gray or yellowish hairs up to 0.2–0.65 mm commonly mixed with short clavate-glanduliform ones, the foliage bicolored, the lfts dark dull green and glabrous above, paler but either glabrous or pubescent beneath, ciliolate or not, the subumbellately (1–)2–4-fld racemes mostly axillary to coeval lvs and shorter than them, sometimes crowded together on short leafy axillary sideshoots.

Stipules erect, narrowly linear-attenuate or setiform (1.5–)2.5–7(–10) × 0.15–0.5 mm, early deciduous.

Major lvs (deciduous, sometimes lacking from specimens in late fl or from arid sites) 5–12(–13) cm; petiole including discolored pulvinus 6–28 mm, 1–3 times as long as first segment of rachis, rounded dorsally, narrowly grooved ventrally; rachis 2–9.5 cm; petiolar glands between the proximal or 3 proximal pairs of lfts, erect, stipitate or subsessile, including stipe 1.3–3.5 mm tall, the orange or yellowish body compressed or (when narrow) subterete, acute 0.4–0.8(–1) mm diam; pulvinules 0.5–1 mm; lfts 4–14 pairs, scarcely or markedly accrescent upward, in outline oblong-elliptic to -obovate, abruptly acute, obtuse mucronulate or rarely emarginate, the distal pair (10–)11–33 mm (further described under vars.), at base asymmetrically cuneate or rounded, the margins revolute, the straight centric midrib immersed above, cariniform beneath, giving rise to (4–)5–8(–9) pairs of fine camptodrome (and often a few intercalary) secondary veins prominulous beneath, not or scarcely so above, the tertiary venulation immersed or faintly visible beneath only.

Peduncles (0.7–)1–3 cm; racemes subumbellately (1–)2–4, many exactly 2-fld, the axis 0–4.5 mm; pedicels at anthesis subfiliform, in fruit little thickened (8–)10–28 mm, subtended on one side by a sessile or short-stipitate subuliform-ellipsoid gland; bracts oblanceolate cucullate 1.3–3.5 mm, shielding the very young bud but deciduous with first elongation of pedicel; buds subglobose, glabrous or thinly pilosulous; sepals green or yellowish with pallid membranous margins, obovate-orbicular, strongly graduated, the large inner ones 5–8 mm; petals yellow, dorsally pilosulous or at least near claw puberulent, erratically heteromorphic, 3 adaxial smaller, symmetrically or asymmetrically oblanceolate or obovate, among themselves either similar or dissimilar in size and shape, the 2 abaxial larger, 16–28 mm, one obliquely obovate, opposed to the divergent pistil, one (often largest of all) obliquely reniform, loosely nidulating the long stamens; androecium glabrous except for sometimes puberulent filaments; filaments of 4 median stamens ± dilated 1.1–1.8 mm, of 3 abaxial ones filiform (1–)2.2–4.7 mm, all free to base; anthers of 4 median stamens little incurved 2.5–5 mm, the depressed beak 0.2–0.5 mm dehiscent by one U-shaped pore, of 3 abaxial ones lunately incurved, the body 4–5.5 mm, the tubular beak erect 1–1.8 mm, slightly dilated at 1-pored apex; ovules 16–26.

Pod pendulous or obliquely spreading, the stipe 3–5.5 mm, the linear body straight or slightly decurved (3–)4.5–8 × 0.5–9.5 cm, the chartaceous or castaneous, finally nigrescent, thinly pilosulous or glabrate valves slightly depressed between each seed and simply mounded over them; seeds (of var. *angustisiliqua* and *inaguensis,* those of var. *fulgens* not seen) compressed-obovoid or subquadrately ovoid 3–4.6 × 1.8–3.6 mm, the testa brown, dull or sublustrous, the ovate-elliptic areole 1–1.4 × 0.4–0.7 mm.

As mentioned in our discussion of *S. pallida* sens. lat., *Cassia angustisiliqua* with *C. crista* and *C. fulgens,* all of which were reduced by Bentham to *C. biflora,* are more closely related to *S. viciifolia* and *S. robiniifolia,* forming with them and Central American *S. tonduzii* a vicariant series collectively characterized by the U-shaped dehiscence of the four median anthers. As here defined so as to accommodate the plants of Hispaniola and Jamaica that have passed latterly as *C. biflora,* to which we here add the local south Bahaman *Peiranisia inaguensis, S. angustisiliqua* differs from *S. robiniifolia* of eastern Cuba and Caribbean South America in the smaller leaves usually more shortly petioled and lacking glands beyond the second or third pair of leaflets, in the shorter, commonly obtuse leaflets, in the free median and adaxial filaments, and in the shorter pod of which the ripe valves remain continuous until after dehiscence, not fragile along the lines of the interseminal septa. The Jamaican *S. angustisiliqua* var. *fulgens,* with more than one petiolar gland and leaflets rather larger than those of genuine var. *angustisiliqua* of Hispaniola, is to that extent intermediate to Cuban *S. robiniifolia,* expressing a relationship that might be anticipated by study of the dispersal of these probably recently divergent forms.

Key to the Varieties of *S. angustisiliqua*

1. Beak of 4 median anthers lateral, divergent from the theca at ±90°; petioles 6–15(–18) mm, mostly 1–2 times as long as the first interfoliolar segment of rachis; Hispaniola, Jamaica.
   2. Petiolar gland between proximal pair of lfts only; pod (3–)4.5–7.5(–8) × 0.5–0.7(–0.8) cm; Hispaniola. 168a. var. *angustisiliqua* (p. 515).
   2. Petiolar glands between 2–3 lowest pairs of lfts; pod 10–12 × 0.8–0.95 cm; Jamaica. 168b. var. *fulgens* (p. 516).
1. Beak of 4 median anthers terminal suberect, projecting beyond the theca; petioles 13–28 mm, mostly 2–3 times all long as the first interfoliolar segment of rachis; s.-e. Bahamas (Inagua, Caicos). Glands between 1, 2 or 3 pairs; pod 6–9 × 0.7–1.1 cm. 168c. var. *inaguensis* (p. 516).

168a. Senna angustisiliqua (Lamarck) Irwin & Barneby var. **angustisiliqua**. *Cassia angustisiliqua* Lamarck, 1785, l.c.—". . . dans l'Isle de St. Domingue [Hispaniola]."—Holotypus, P-LA!

Cassia crista Jacquin, Collectanea **1**: 82 & Ic. Pl. Rar. **1**: 8, pl. 74. 1787.—"Apud nos in caldario . . . Sponte crescit in Florida [doubtless an error for Hispaniola]."—Holotypus, cult. in hort. Schoenbr., W (hb. Jacqu.)!—*C. frondosa* var. *crista* (Jacquin) G. Don, Gen. Hist. Dichl. Pl. **2**: 442. 1832, comb. illegit., the varietal epithet anterior. *Peiranisia crista* (Jacquin) Britton & Rose, N. Amer. Fl. **23**(4): 261. 1930.—Illegally reduced by Willdenow, Sp. Pl. 524, to *C. frondosa* (*Hb. Willd. 7995,* B!).

Cassia frondosa Aiton, Hort. Kew. **2**: 53. 1789.—"*Nat.* of the West Indies. *Introd.* [to Great Britain] about 1769."—Holotypus, "Hort. Kew. 1782," BM!

Cassia semperflorens DeCandolle, Cat. Hort. Monspel. 90. 1813.—". . . in plurimis hortis occurrit . . ."—Holotypus, from a plant cultivated at Montpellier, G-DC! spms retained at MPU, cult. there 15.V.1808, 7.X.1812 and in 1813, are authentic; clastotypus (+ drawing of holotypus, misit Hochreutiner anno 1929), NY!—*C. biflora* var. (β) *semperflorens* (DeCandolle) Colladon, Hist. Casses 103. 1816; DeCandolle, Prod. **2**: 496. 1825.

Cassia frondosa var. (γ) *pubescens* Colladon, Hist. Casses 106. 1816.—"in Sto. Domingo."—No typus found at G or P (or passed over unrecognized).

Cassia crista var. *oligophylla* Urban, Arkiv Bot. **22A**(8): 33. 1929.—"Haiti, Plaine Cul-de-Sac, Croix-des-Bouquets, prope Manneville . . . [Ekman] n. H 3069 . . ."—Holotypus, collected 20.I.1925, S! = NY Neg. *10485*.—Referred to *P. crista* by Britton & Rose, 1930, p. 261.

Peiranisia haitiensis Britton ex Britton & Rose, N. Amer. Fl. **23**(4): 261. 1930.—"Haiti. Type from vicinity of Ennery, l'Artibonite, . . . January 13, 1906 [properly 1926], *E. C. Leonard 8799.*"—Holotypus, NY! isotypus, US!—*Cassia haitiensis* (Britton & Rose) Jiménez, Rhodora **62**: 236. 1960.

Peiranisia fitchiana Britton & Rose, N. Amer. Fl. **23**(4): 264. 1930.—". . . Santo Domingo City, Santo Domingo, March 9–10, 1913, *Rose, Fitch & Russell* 3795a."—Holotypus, NY! isotypus, US!—*Cassia fitchiana* (Britton & Rose) Jiménez, Rhodora **62**: 236. 1960.

Lfts of major lvs 8–14 pairs, the terminal pair (10–)11–26 × (3.5–)4–9.5(–10) mm, 1.8–3.6 times as long as wide; otherwise as given in description and varietal key.—Collections: 91.

Thickets, roadsides, sometimes on limestone or serpentine barrens, from sea level up to 700(–1050) m, widespread and locally abundant over Hispaniola w.-ward from a line drawn through Dominican Republic between San Juan and Puerto Plata.—Fl. abundantly IX–III, sporadically through the year.—Early cultivated on Guadeloupe (Basseterre, P, hb. Richard.) and from mid-XVII cent. in w. Europe.

168b. Senna angustisiliqua (Lamarck) var. **fulgens** (Macfadyen) Irwin & Barneby, stat. nov. *Cassia fulgens* Macfadyen, Fl. Jam. **1**: 342. 1837.—"Common, Port-Royal mountains."—Lectoholotypus, 'Jamaica, *Macfadyen s.n.,*' K! = NY Neg. *10479*.—Non *C. fulgens* Wallich ex Vogel, 1837.

Peiranisia jamaicensis Britton ex Britton & Rose, N. Amer. Fl. **23**(4): 262. 1930.—"Jamaica. Type from below Flamstead, about 900 m. altitude, October 20, 1905, [*William*] *Harris 9031.*"—Holotypus, NY! isotypi, F, US!—Non *Cassia jamaicensis* (Britton) Adams, 1970.

Peiranisia oxyadena Britton & Rose, N. Amer. Fl. **23**(4): 263. 1930, ex parte, exclus. basionym. qui *C. pallidae* pertinet. *Cassia biflora* sensu Linnaeus, Pug. Jam. Pl. in Amoen. Acad. **5**: 397. 1760, equated with *Cassia* no. 6 of Browne, Hist. Jam. 223. 1756; non *C. biflora* Linnaeus, 1753, nom. dub.—Spm. authent., LINN *528.21*! —sensu Lindley in Bot. Reg. **16**: t. 1310. 1830 & auct. plur. jamaic.; Adams, Fl. Pl. Jam. 325. 1972.

Lfts of major lvs mostly 7–9 pairs, the terminal pair obovate 18–33 × 7–15 mm, 2–2.5(–2.8) times as long as wide; otherwise as in description and varietal key.—Collections: 9.

Thickets on open hillsides and in second-growth woodland, 600–1250 m, local in e.-centr. Jamaica (St. Andrews and St. Thomas parishes).—Fl. VII–XI.

168c. Senna angustisiliqua (Lamarck) var. **inaguensis** (Britton & Rose) Irwin & Barneby, stat. nov. *Peiranisia inaguensis* Britton ex Britton & Rose,

N. Amer. Fl. **23**(4): 261. 1930.—"Inagua Island, Bahamas. Type from Salt Point Hill, October 12, 1904, *Nash & Taylor 971.*"—Holotypus, NY!

Lfts of major lvs 4–7(–8) pairs, the terminal pair obovate obtuse mucronulate 12–21 × 6–12 mm, 1.6–2.4 times as long as wide; otherwise as in description and varietal key.—Collections: 7.

Coppice scrub and coral limestone pavement below 30 m, known only from Providenciales and Great Inagua islands, s.-e. Bahamas.—Fl. VI–XII.

Britton separated *Peiranisia inaguensis* from *P. crista* by the broad pod (said to be 8–10 not 4–7 mm wide) and from *P. jamaicensis* by glabrate foliage. Modern collections show that the pod varies on Great Inagua from 6.5 to 9 (none available to Britton were actually 10) mm wide, and in the context of ser. *Rostratae* no reliance should be placed on pubescence characters, everywhere fallible. The orientation of the median anther-beaks, erect and obviously terminal in *P. inaguensis* but turned at rightangles to the body of the theca in *S. angustisiliqua* and its var. *fulgens,* is on our view the best, even though somewhat precarious differential character. In general var. *inaguensis* differs further from var. *angustisiliqua* in the longer petioles and peduncles and fewer (mostly 5–8, not 8–14) pairs of leaflets, but our three collections from Nord-Ouest province, Haiti (*Leonard 11070, 11726, 13704,* all NY, US), referred to var. *angustisiliqua* because of the androecium, have essentially the foliage of var. *inaguensis*. In foliage (except for vesture) and a relatively broad pod the vars. *inaguensis* and *fulgens* are closely similar, differing, as already noted, in the anther-beak. On South Caicos the range of var. *inaguensis* coincides with the southernmost outpost of closely related *S. pallida* var. *bahamensis* (*P. Wilson 7639,* NY, US), instantly separated by the 2-pored median anthers, lack of dense resinous punctuation of the leaflet blades, and a really narrow pod only 4–5.5 mm wide.

169. Senna polyphylla (Jacquin) Irwin & Barneby, comb. nov. *Cassia polyphylla* Jacquin, Collectanea **4**: 104. 1791.—Typus infra sub var. *polyphylla* indicatur.

Stiffly much-branched microphyllous shrubs and treelets at anthesis 1.5–8 m, with long, densely leafy, widely ascending, horizontal or geotropic branches, the young branchlets and lf-stalks commonly strigulose-pilosulous with appressed or widely ascending pallid hairs to 0.2–0.4(–0.5) mm, sometimes subglabrous, the small firm leaflets commonly glabrous but sometimes thinly pubescent beneath and ciliolate, on both sides olivaceous brunnescent but a little paler beneath, the lvs dimorphic, the longer primary ones subtending condensed brachyblasts or spurs clad in fasciculate shorter lvs mixed with 1–3-fld peduncles, the longer lvs early deciduous, the inflorescence thyrsiform, the relatively large and showy fls exserted from the small foliage.

Stipules (of primary lvs) erect, subulate-setaceous 2–5.5 mm, tardily deciduous, those of spur-lvs shorter, more persistent.

Lvs of primary nodes of long-shoots solitary 1.2–5.5(–6) cm; petiole including moderately dilated pulvinus 2–5.5(–6.5) mm, at middle 0.45–0.7 mm diam, rounded dorsally, openly sulcate ventrally; rachis 0.6–4.5(–5) cm; gland 1 sessile between proximal pair (0.4–)0.6–1 mm, in profile obliquely ovate, semiovate or incurved-lanceolate, acute or acuminulate, glabrous (much eaten and then apparently 0); lfts 3–13 pairs not or scarcely accrescent upward, ascending from rachis face upward on wrinkled pulvinule 0.2–0.4(–0.5) mm, in outline obovate, oblong-obovate or broadly elliptic, obtuse-mucronulate or rarely subretuse,

3.5–9(–10) × 2–4(–4.5) mm, ±1.7–2.8(–3.0) times as long as wide, asymmetrically rounded at base, the margin plane, the centric, dorsally cariniform midrib with 3–4(–5) pairs of camptodrome secondary and rare or random connecting tertiary venules all immersed above, the secondary venation faintly or decisively prominulous beneath; spur-lvs similar but mostly shorter, composed of fewer and slightly smaller lfts.

Peduncles (4–)6–21 mm; racemes 1-3-fld, the axis 0–1(–2) mm; bracts oblanceolate concave, cucullate at tip, 1.7–2.4 mm, caducous; pedicels (9–)10–19 mm, commonly but not always subtended by stipitate glands; buds globose, glabrate above the base; sepals greenish with petaloid margins, strongly graduated, all obovate-suborbicular obtuse, the largest inner one 5–7 mm; petals yellow, either glabrous or thinly puberulent dorsally, not strongly heteromorphic, the 3 adaxial slightly smaller, obovate-oblanceolate, the 2 abaxial similar but one more oblique and more shortly clawed, this (12–)13–22(–26) mm; androecium glabrous, the filaments of 4 median anthers free, 0.9–2 mm, those of 3 abaxial ones 1.8–3.5 mm, the anthers of 4 median stamens including very short, slightly oblique beak 3–5 mm, this opening by U-shaped slit, the body of 3 abaxial ones lunately incurved 3–4.6 mm, contracted into a slightly porrect tubular 1-porulose beak 0.7–1.2 mm; ovary either glabrous or densely strigulose-pilosulous; style 1–1.8 mm, 0.2–0.35 mm diam at apex; ovules (18–)20–46.

Pod spreading or pendulous, the stipe 5–7.5 mm, the linear compressed, commonly decurved or irregularly or undulately contorted body (4.5–)6–14 × 0.6–0.9(–1) cm, the thin, finally papery nigrescent sublustrous valves low-mounded over the seeds, faintly and delicately transverse-venulose, when dry easily broken across through the interseminal septa; seeds oblong- or obovate-paddle-shaped, 2.7–3.7 × 2–2.5 mm, the testa chocolate- or chestnut-brown, scarcely lustrous, crackled, the elliptic or oblong-elliptic areole 1–1.5 × 0.4–0.6 mm.

An ornamental senna, notable in ser. *Rostratae* for the arrangement of the 1–3-flowered racemes among fascicles of diminished leaves borne close together along stiff wandlike branchlets, these being long shoots produced during rainy periods whereas the fascicles are drought-adapted brachyblasts arising from axils of the primary leaves. While *S. polyphylla* usually has a marked individual facies it is only precariously distinct from *S. pallida* sens. lat. A quite similar differentiation of the branchlets occurs in some populations of *S. pallida* on the seasonally arid limestones of Yucatan as also in some winter-flowering, drought-deciduous *S. pallida* in northwestern Mexico. Here, however, the primary leaves when grown out become much larger and coarser, while the asymmetry of one abaxial petal (the cucullus) becomes more emphatic and the beak of the four median stamens is infraterminal rather than obliquely terminal. Although the origin of *S. polyphylla* by reduction and xeromorphic adaptation from the stock of *C. pallida* seems certain, the species is now abruptly and easily distinguished from Antillian forms of the latter by its small leaves, small plane leaflets and the peculiar architecture of the inflorescence.

Revision of the material that has passed in recent times as *Cassia polyphylla* reveals that xeromorphic diminution of the foliage has not stopped with the typical phase of the species, now thought to be endemic to southern Puerto Rico and the United States Virgin Islands, but has proceeded to further stages in distantly disjunct populations localized on northern Hispaniola and the island of Anegada, described below as new varieties. In using the varietal key, care must be taken to distinguish the genuine primary leaves that subtend each brachyblast but are caducous (and absent from many fruiting specimens) from the smaller, more persisent leaves associated with the flowers.

Key to the Varieties of *S. polyphylla*

1. Lfts of primary lvs 5–13 pairs, of some spur-lvs at least 4 pairs; Puerto Rico, Virgin Is., Hispaniola.
   2. Primary lvs 2.5–5.5(–6) cm, the lfts (6–)7–13 pairs; ovary canescently strigulose-pilosulous; Puerto Rico and Virgin Is. 169a. var. *polyphylla* (p. 519).
   2. Primary lvs 1.5–2.5 cm, the lfts 5–7(–8) pairs; ovary glabrous or transiently puberulent along the sutures; n.-w. Dominican Republic. 169b. var. *montis-christi* (p. 519).
1. Lfts of primary lvs 3–4 pairs, of spur-lvs mostly 2 pairs. Ovary glabrous. Anegada I. 169c. var. *neglecta* (p. 520).

169a. Senna polyphylla (Jacquin) Irwin & Barneby var. **polyphylla.** *C. polyphylla* Jacquin, Collectanea **4**: 104. 1791 & Ic. Pl. Rar. **3**(11): t. 460. 1792.—"In insula Porto rico sponte crescit; in caldario horti Caesarei Schönbrunnensis floret Aprili & Majo . . ."—Holotypus, Hb. Jacqu. ex hort. Schoenbrun., W! = NY Neg. *10487*.—*Peiranisia polyphylla* (Jacquin) Britton & Rose ex Britton & Wilson, Sci. Survey Porto Rico & Virgin Is. **5**(3): 373. 1924.

Cassia marginata Willdenow, Enum. pl. hort. bot. Berol. 444. 1809.—"*Habitat in* Surinamo."—Holotypus, a spm. cult. at Berlin, *Hb. Willd. 8001*, B!

Cassia tenuissima Sessé & Mociño, Fl. Mex. ed. 2, 102. 1894.—No locality given.—Presumed holotypus, labelled "*Cassia tenuissima*" and agreeing with the short diagnosis, Hb. S. & M. *1133*, MA! = F Neg. *44228*; clastotypus, F!—"*C. microphylla* S. & M.," s. loc., BM, MA (hb. S. & M. *1192*) = F Neg. *44246* may be isotypic.

Cassia polyphylla sensu Colladon, 1816, p. 104; Vogel, 1837, p. 31; Bentham, 1871, p. 544; Britton, Addisonia **1**: pl. 2. 1916; Little, Woodbury & Wadsworth, Trees Puerto Rico, Sec. Vol. 280, pl. 373. 1974.

Lfts of primary lvs 7–13 pairs, 3.5–8(–10) × 2–4(–4.5) mm, commonly glabrous on both faces, sometimes ciliolate, sometimes thinly pilosulous dorsally; n = 14 (Irwin & Turner, 1960).—Collections: 44.

Dry hillsides, thin stony woodland, thickets, sometimes on outcrops or rock-ledges, on either eruptive or sedimentary (calcareous) bedrock, below 600 m, locally plentiful along the coast and foothills of s. Puerto Rico from Cabo Rojo e. to Guayama (lat. 66°5′–67°13′W), and on the Virgin Is. of St. Thomas and St. Croix; cultivated in U.S. (Florida, California), Windward Is. (Grenada), Guyana (Georgetown), Surinam (Paramaribo), s.-e. Brazil (Campinas, S. Paulo), and formerly under glass in Europe, probably elsehwere.—Fl. in Puerto Rico most abundantly X–III, but intermittently through the year, the fruits long persisting.—*Retama prieta*.

169b. Senna polyphylla Jacquin var. **montis-christi** Irwin & Barneby, var. nov., a var. *polyphylla* foliis minoribus paucifoliolatis, primariis (macroblastorum) 1.5–2.5 cm usque, foliolis plerumque 5–7-jugis, ovarioque glabro vel minutissime ciliolato (nec canescenti-pilosulo) absimilis.—DOMINICAN REPUBLIC. Monte Cristi: hillside, in thickets, 100 m, 57 km e. of Montecristi on Santiago road, 14.VI.1969 (fr), *Alain H. Liogier 15707*.—Holotypus, NY; isotypi, GH, K, US.

Lfts of primary lvs 5–7(–8) pairs, mostly 4–7 × 1.6–3.5 mm, glabrous, of spur-lvs 3–5 pairs.—Collections: 5.

Dry hillside thickets, locally plentiful, 100–300 m, known only from valley of Río Yaque in prov. Valverde and adjoining Montecristi, n.-w. Dominican Republic.—Fl. XII–VI.

169c. Senna polyphylla Jacquin var. **neglecta** Irwin & Barneby, var. nov., ab aliis speciei varietatibus foliorum omnium abbreviatorum (12–20 mm) primariorum foliolis 3–4 jugis caeterum 2–3-jugis, a var. *polyphylla* ulterius ovario glaberrimo distans.—BRITISH VIRGIN ISLANDS: rocky plain near settlement, Anegada I., 19–20.II.1913 (fr), *N. L. Britton & W. C. Fishlock 1034.*—Holotypus, NY; isotypi, F, US.

Lfts of primary lvs 3–4 pairs, the distal pair 6–9 × 3–4.5 mm, glabrous on both faces. Collection: 1.

Stony ground below 50 m, known only from Anegada I., British Virgin Islands.—Fl. (?) XII–I.

Described by Britton as a tree 4 meters tall.

170. Senna robiniifolia (Bentham) Irwin & Barneby, comb. nov. *Cassia robiniifolia* Bentham, Trans. Linn. Soc. London **27**: 545 (''*robiniaefolia*''). 1871.—''Columbia, Ocaña, *Schlim,* n. 2; Cuba, *Wright,* n. 2370.''—Lectoholotypus (Britton & Killip, 1936, p. 181, by implication sub *Peiranisia santanderensi*): *Wright 2370,* collected (acc. original pencilled ticket at NY) from a tree cultivated at El Retiro, prov. Oriente, Cuba), K! = NY Neg. *1429*: isotypi, G, GH, NY!—*Peiranisia robiniifolia* (Bentham) Britton & Rose, N. Amer. Fl. **23**(3): 263. 1930.

Peiranisia macrochlamys Pittier, Arb. y Arbust. nuev. Venezuela, Decad. **6–8**: 92. 1927.—''Portuguesa: Vegas del río Portuguesa en la carretera a Guanare . . . Deciembre 28, 1925 (Pittier 12026, tipo).—Lara: Alrededores de Barquisemeto . . . (J. Saer d'Héguert 254).''—Holotypus to be sought at VEN; isotypi, GH, MAD, NY, US! paratypi (*Saer 254*), NY, US!

Peiranisia santanderensis Britton & Killip, Ann. N.Y. Acad. Sci. **35**(3): 181. 1936.—''. . . between Pamplonita and Chinácota, Rio Pamplonita valley, Norte de Santander, Colombia, 1300–1800 m altitude, March 17, 1927, *Killip & Smith 20750* . . . Ocaña, *Schlim 2* . . .''—Holotypus, *Killip & Smith 20750,* NY! isotypi K, US!—Non *Cassia santanderensis* (Britton & Killip) Lasseigne, based on *Adipera santanderensis* Britton & Killip.

Slender trees and arborescent shrubs 4–8 m, with terete striate or ribbed-canaliculate younger and strongly lenticellate annotinous branchlets, strigulose-pilosulous with forwardly subappressed and often some, less often mostly widely incurved-ascending, sometimes lutescent hairs up to 0.2–0.6 mm, the foliage bicolored, the lfts above dark green (brunnescent in drying) and either glabrous or puberulent-pilosulous, beneath paler and either glabrous, strigulose, or densely pilosulous, always ciliolate, the terminal or subterminal, basally leafy-bracteate but often distally leafless and exserted panicle composed of mostly 2- or 2–4-fld racemes.

Stipules falcately linear-attenuate 4.5–11 × 0.2–0.45 mm caducous (lacking from fruiting spms).

Lvs (disregarding reduced distal ones) 8–18(–24) cm; petiole including firm, moderately dilated pulvinus 1.5–3(–3.5) cm, at middle 0.6–1 mm diam, bluntly carinate-ribbed dorsally, openly grooved ventrally; rachis (3.5–)4.5–12(–14) cm; petiolar glands ascending from between all or several, exceptionally only the proximal pairs of lfts, stipitate or subsessile, (1.3–)1.8–3.2 mm, the slender stipe either glabrous or puberulent, the glabrous, laterally compressed body 0.2–0.4(–0.7) mm diam, commonly incurved or even hamate distally; lfts (of major lvs) 5–9(–10) pairs, strongly accrescent upward, the ultimate pair elliptic or elliptic-oblanceolate, triangular- to deltate-acute or obtuse mucronate (2.5–)3–6 × 0.9–1.7(–1.9) cm, 2.5–4 times as long as wide, at base rounded on proximal and cuneate on distal side, the margin revolute, the straight centric midrib cariniform beneath, with (8–)9–14 pairs of major camptodrome secondaries finely prominulous usually on both faces, rarely only beneath, the reticular venulation prominulous either on both faces or only beneath.

Peduncles 1.5–3 cm; racemes subumbellately or very shortly (1–)2–4-fld, the axis 0–5(–8) mm; bracts elliptic-oblanceolate concave 1.7–2.7 mm, very early caducous; glands at base of pedicels either 0 or random and caducous; pedicels 20–33 mm; buds subglobose glabrous or at base puberulent, the sepals sometimes ciliolate; sepals brownish with pale margins, strongly graduated, all broadly obovate-suborbicular, the inner 5–6.5 mm; petals yellow, puberulent dorsally along and often between the veins, 3 adaxial smaller, oblanceolate-obovate narrowed to slender claw, 1.2–2 cm, the 2 abaxial larger, one short-clawed and obovate-cordate, one subsessile reniform-scoop-shaped, the longer 1.5–2.4 cm; androecium glabrous; filaments of staminodia and 4 median stamens 1–2 mm, connate at base or up to middle, those of 3 abaxial ones free, 2–3 mm; anthers of 4 median stamens 3–4 mm usually bottle-shaped by constriction just below the very short divaricate beak which dehisces by one U-shaped slit, those of 3 abaxial ones 3–5 mm, contracted into a tubular porrect beak 1.8–2.5 mm slightly enlarged and 1-porose at apex; ovary strigulose-pilosulous; style 1–1.4 mm, ±0.2 mm diam; ovules 24–34.

Pod obliquely ascending or pendulous, the stipe 4–8 mm, the linear compressed, slightly curved or straight body (8–)10–16 × 0.55–0.75 cm, 2-carinate by the obtuse sutures, the papery atrocastaneous valves becoming low-convex over seeds, faintly transversely venulose, finally fragile and easily broken along lines of interseminal septa; seeds compressed-obovoid 2.8–4 × 1.6–3 mm, the pale brown testa sublustrous crackled, the elliptic areole 1.2–1.8 × 0.5–0.75(–1) mm.—Collections: 57.

Open woods, thickets, hedges, sometimes along streams, of highly discontinuous, montane and lowland dispersal in e. Cuba, Venezuela, n.-e. Colombia and Pacific lowland Ecuador, long cultivated and perhaps in some places naturalized: lowlands and foothills of e. Cuba (Oriente) from Holguín and Manzanillo s.-e. to s. foothills of Sa. Maestra, formerly cultivated at La Habana: Cauca valley and Cordillera Oriental of n. Andes at middle elevations (±850–2100 m) in Colombia (Antioquia, Norte de Santander and Santander) and adjoining Venezuela (Táchira); mountainous n. Venezuela at 1000–1700 m in Falcón (Sa. San Luis) and Distrito Federal (Parque del Avila and cultivated in Caracas); lowland n. Venezuela, perhaps only cultivated but status not recorded, in Lara (Barquisemeto) Yaracuy (Yaracuy) and Portuguesa (Upper río Portuguesa); disjunct near 500 m on lower Paragua river near 7°N in Bolívar; again widely disjunct in lowland disturbed forest at 20–350 m in s.-w. Ecuador in Manabí (Porto Viejo, Manta) and Guayas (Guayaquil, Isla Piñá, Pedro Carbo).—Fl. in Cuba (VII–)X–III, in Venezuela VII–XII, in Colombia III–V (but data doubtless incomplete), in Ecuador V–VII(–IX).—*Urquía* (Cuba); *orimaco* (Venezuela).

Our concept of *S. robiniifolia,* embodied in the foregoing description, is essentially that of Bentham except as modified or amplified by a century of collecting; and dissents from that of Britton who believed the species to be endemic to Cuba. The Cuban material acquired since 1871 shows that Bentham's key character of "4–10"-flowered racemes was mistakenly derived from a succession of 2–3-flowered ones, not of single flowers; in all modern collections the basic racemose elements of the inflorescence are mostly 2-flowered, only a few early ones becoming 3–4-flowered by elongation of the axis beyond the first pair of pedicels. In this respect, as in the union of the anterior filaments of the androecium, *S.*

robiniifolia agrees with *S. viciifolia* (which see for comment and differential characters) and the distantly allopatric Central American *S. tonduzii,* different chiefly in the elliptic-acuminate, prominently venulose leaflets. The inflorescence, except for the 1-pored dehiscence of the four median anthers, is indistinguishable from that of polymorphic *S. pallida* which, however, within or near the range of *S. robiniifolia,* has shorter, proportionately wider leaflets, glands more persistent at insertion of the pedicels and fewer on the leaf-stalks, and a pod at once narrower (3–4.5, not 5.5–7.5 mm wide) and when ripe not fragmenting along lines of interseminal septa.

The known range of *S. robiniifolia* is difficult to account for except under the supposition that it has been modified by man. In eastern Cuba, where even in the first instance it was seen by Charles Wright only as a cultivated tree, it is reported from the intensively settled and disturbed lowlands; conceivably it may be merely introduced in Cuba as an ornamental. It appears to be truly native in the northern Andes, but the few examples from this region are more densely pubescent than those from Cuba or Venezuela and can hardly represent the source of the Cuban plant. In Venezuela *S. robiniifolia* has been found in relatively cool mountain forest, in Falcón and near Caracas; in the hot, seasonally dry llanos of the upper Portuguesa River; near (or in) the cities of Yaracuy and Barquisimeto, where it may be only planted; and distantly disjunct on the Paragua fork of the Caroní south of the Orinoco. The Ecuadorian populations differ slightly in the more obtuse leaflets and perhaps larger seeds (known from too few samples) and will perhaps deserve some form of taxonomic recognition. Careful observation in the field will doubtless contribute to a solution of this problem in plant geography.

Senna robiniifolia was collected first in Cuba by Ramón de la Sagra (P, hb. A. Richard., nos. *133/1* and *231/7*).

171. Senna viciifolia (Bentham) Irwin & Barneby, comb. nov. *Cassia viciifolia* Bentham, Trans. Linn. Soc. London **27**: 544 ("*viciaefolia*"). 1871.—"Columbia, Fendler, n. 2230, 2231."—Lectoholotypus, collected "between Turmero and La Victoria," Aragua, Venezuela, 14.III.57 (fl, fr), *Fendler 2230,* K (hb. Benth.)! = NY Neg. *1561*; paratypi, collected "between La Victoria and Biscaina," 1.II.57 (fl), *Fendler 2231,* K (hb. Benth., hb. Hook.)!—*Peiranisia viciaefolia* (Bentham) Pittier, Bol. Ministerio de RR. EE. **1–3**: 128. 1928, sine cit.; Cat. Fl. Venezol. **1**: 376. 1945.

Peiranisia magniflora Britton & Rose ex Pittier, Bol. Ministerio de RR. EE. **1–3**: 130. 1928.—". . . Venezuela; el tipo es Pittier no. 8678a de Valencia (Carabobo)."—Lectoholotypus, labelled "Carabobo: upper Guaremales, road from Puerto Cabello to San Felipe . . . 100 to 500 meters, XII.1919," NY (annotated by Britton)! = NY Neg. *4951*; isotypus, labelled "Vicinity of Valencia, 8.XII.1919," US!—Overlooked by Pittier, Cat. Fl. Venezol., 1945.

Arborescent shrubs and slender trees at anthesis 1.5–8(–10) m, closely resembling and related to *C. robiniifolia* but the drought-deciduous foliage proteranthous and the inflorescence of 2–3-fld racemes borne on brachyblasts developing after or only shortly before fall of the suffulcrant lf, the hornotinous branchlets and foliage pilosulous with loosely ascending-incurved hairs up to 0.3–0.75 mm, the bicolored lfts above dark green (brunnescent in drying) and puberulent or exceptionally glabrate, below paler and densely pilosulous, when young often canescent.

Stipules erect linear-attenuate 3–14 × 0.2–0.45 mm caducous.

Larger lvs (8–)10–18(–21) cm; petiole including dilated, when dry wrinkled pulvinus (1.7–)2.2–3.7 cm, at middle 0.6–1 mm diam, openly grooved ventrally; rachis (3.5)4–10 cm; glands between 2–3(–5) proximal pairs of lfts, short-stipitate or sessile (1–)2–3.2 mm, the compressed or subclaviform body 0.3–0.7 mm diam; lfts (5–)6–9(–10) pairs, strongly accrescent upward, the distal pair usually oblanceolate or narrowly obovate obtuse, rarely elliptic-oblanceolate and deltately subacute 3.6–6 × (0.9–)1–2.3 cm, (2.2–)2.5–4.2 times longer than wide, the base, margin and venation as in *C. robiniifolia*.

Axis of flowering brachyblasts up to 2 cm, often condensed into a short spur or knot; racemes all or almost all (1–)2-fld, the axis 0–1 mm or rarely elongating to bear a third fl 4–5 mm above the pair; peduncles 9–23 mm; pedicels 16–35 mm, straight in fruit, at base subtended laterally by a compressed-fusiform, early divaricate and deciduous gland; buds subglobose, glabrous or almost so; sepals all obovate-suborbicular, strongly graduated, the largest innermost one (6.5–)7–10.5 mm; petals (of *Interglandulosae*) glabrous, the longest (20–)22–33 mm; androecium glabrous, the filaments of 4 median stamens and 3 staminodia fused at base, 0.8–2 mm, those of 3 abaxial ones free 2–3.8 mm; anthers of 4 median stamens 4.5–5.6 mm, commonly constricted just below the very short divaricate beak, this dehiscent by one U-shaped slit, those of 3 abaxial ones 4.5–7 mm, contracted into a tubular, distally a little dilated and 1-porose beak 2.5–4.2 mm; ovary strigulose or glabrous; style and ovules of *C. robiniifolia*.

Pod of *C. robiniifolia,* 6–7 mm wide; seeds (little known) paddle-shaped 3.2 × 2.4 mm, the testa castaneous sublustrous, the areole oval 1.2 × 0.8 mm.—Collections: 37.

Deciduous woodland, xeromorphic brush-woodland, thickets in savanna and along highways, 300–700 m, locally plentiful in intermontane valleys and around n. edge of the Orinoco valley in n.-w. Venezuela, from e. Lara to n. Aragua and n.-w. Anzoátegui, n. Cojedes and n. Guárico.—Fl. XI–II, sporadically later.

A species closely related to the marginally sympatric *C. robiniifolia,* which it resembles in form of foliage (except leaflets more commonly obtuse), in the flower, and in the connate filaments of the staminodia and median fertile stamens, but technically different in the somewhat larger sepals, the glabrous petals, and more importantly in the rhythm of anthesis and form of the inflorescence. In *C. robiniifolia* the racemose elements of the paniculate inflorescence are borne on annotinous branchlets axillary to recently developed leaves or, as the season advances, paniculately at the leafless apex of the current year's growth, anthesis coinciding with the period of greatest vegetative activity. By contrast in *C. viciifolia* the abbreviated axis of the panicle takes the form of a brachyblast arising from the axil of an old leaf soon due to fall or of one already fallen; and anthesis coincides with a season of drought and vegetative quiescence. Where they are sympatric or closely vicariant, *C. robiniifolia* flowers from July to December when in full leaf, *C. viciifolia* from December to January, sporadically later, abundantly and characteristically only from leafless branchlets. It might be thought that these two sennas, so similar in most respects, are merely seasonal aspects of one climatically adaptable species; but the perfect correlation in *C. viciifolia* of few petiolar glands, larger sepals, glabrous petals, protantherous foliage and condensed panicle suggests that they are biologically distinct entities adapted to different climatic regimes. If our still rather incomplete data on flowering seasons are accurate, the two species, by virtue of dyschronous anthesis, are unlikely or unable to exchange pollen.

The typi of *C. viciifolia* and *P. magniflora,* which originated from the east and west ends, respectively, of Lago de Valencia, are unquestionably conspecific in the most exact sense. At the time Pittier described *P. magniflora* he appears to have interpreted as *C. viciiflora* the Venezuelan form of *C. biflora* with pubescent foliage.

171. Senna saeri (Pittier) Irwin & Barneby, comb. nov., *Peiranisia saeri* Britton & Rose ex Pittier, Bol. Ministerio RR. EE. **1–3**: 129. 1928.—". . . Venezuela . . .," but no collection cited.—Lectoholotypus, *J. Saer 8,* collected "on shores of La Ruesca, Barquisimeto, Lara, Venezuela, VIII.1923" (fl), NY (annotated by Britton as "cotype")! isotypi, US (2 sheets)!

Shrubs of bushy outline, at anthesis 1.5–4 m, with purplish-brown, in age pallid-lenticellate branchlets, densely pilosulous throughout with fine spreading-incurved sordid gray or brunnescent hairs up to 0.2–0.45 mm mixed with minute reddish or orange thickened but eglandular ones, the foliage bicolored, the lfts dull brownish-olivaceous and papillate above, paler beneath, the inflorescence of few-, rarely mostly 2-fld racemes at first axillary to and ± immersed in foliage, distally becoming corymbose-paniculate.

Stipules incurved-erect, linear-lanceolate or subulate 2–6 mm, firm but early caducous.

Lvs (disregarding depauperate distal ones) 5–11.5 cm; petiole including the livid, moderately dilated pulvinus 1.1–2.3 cm, at middle 0.45–1 mm diam, rounded dorsally, narrowly grooved ventrally; rachis 2–5 cm; petiolar glands (caveat: much eaten) between all pairs of lfts, stipitiate or subsessile, in profile (1–)1.2–2.2 mm, the stipe puberulent, the livid glabrous laterally compressed acute body 0.3–0.6 mm diam; lfts (3–)4–6 pairs, strongly accrescent upward, ascending face up on terete, moderately swollen pulvinule 0.9–2.4 mm, the distal pair obovate, elliptic-obovate, -oblanceolate, or elliptic, obtuse mucronulate, shallowly emarginate or subacute, (20–)23–46 × (7–)10–19 mm, ±1.8–3.1 times longer than wide, at asymmetric base rounded on proximal and either rounded or cuneate on distal side, the margin revolute, the straight centric midrib narrowly canaliculate on ventral and cariniform on dorsal face, the 6–10 pairs of major camptodrome secondary with subsequent reticulate venulation always finely sharply prominulous beneath, less so or almost immersed above; proximal lfts progressively smaller downward along rachis and usually proportionately broader and somewhat fewer-veined.

Peduncles 1–3.5 cm; racemes 2–7 (but seldom all exactly 2)-fld, the fls often subumbellately clustered or pseudoverticellate, the axis (0–)1.5–12 mm; bracts broadly oblanceolate concave 2–8.5 mm, covering the young fl-buds, promptly caducous; a stipitate gland (often eaten) beside each pedicel; pedicels 17–30 mm; fl-buds subglobose, either glabrous or proximally puberulent; sepals greenish or brownish with pallid margins, ovate- or obovate-suborbicular obtuse, strongly graduated, the outermost not over half as long as the innermost, these delicately several-nerved from base, 5.2–6.7 mm; petals yellow, puberulent dorsally, 3 adaxial relatively small and long-clawed, oblance-obovate, the 2 abaxial larger, one short-clawed and ovate-cordate, the other obliquely reniform-scoop-shaped, the longer one 1.6–2.4 cm; androecium functionally 7-merous, the filaments thinly puberulent, those of 4 median stamens 0.8–1.6 mm, of 3 abaxial ones 1.8–3 mm, the anthers glabrous or exceptionally puberulent, of 4 median stamens gently incurved 3.3–4.6 mm, the infraterminal beak scarcely 0.3 mm dehiscent by a single U-shaped slit, those of 3 abaxial ones more strongly incurved, the body 3.7–5.5 mm, contracted into an erect tubular beak (2–)2.3–3.3 mm, this slightly enlarged and 1-porose at apex; ovary densely yellowish-strigulose; style 0.9–1.5 mm, not dilated distally 0.25–0.4 mm diam; ovules ±25.

Pod spreading and pendulous (from obliquely geotropic branchlets appearing erect), the stipe 4–7 mm, the linear, strongly compressed, straight or irregularly decurved body 6.5–15.5 × 0.45–0.6 cm, bluntly 2-carinate by the sutures, the papery atro-castaneous coarsely transverse-venulose valves bullately elevated between septa, the sinuous pallid venulation superficial and elevated, finally detached from the epicarp, the truly ripe valves apparently fracturing along the lines of septation, hence not truly dehiscent; seeds (few seen) paddle-shaped, 2.5–3.7 × 2.2–3 mm, the pale brown, scarcely lustrous testa cross-crackled, bearing on each face an elliptic areole 1.2–1.5 × 0.5–0.6 mm.—Collections: 21.

Dwarf, drought-decidous scrub-woodland of dry hillsides and valley floors, 450–900(–1450) m, sometimes forming extensive thickets, locally plentiful about the sources of Ríos Tocuyo and Yaracuy in Lara and (perhaps disjunct) in Parque Nac. Rancho Grande in n. Aragua, Venezuela.—Fl. II–V, VII–XI, the full range unknown.—*Plantanico.*

Senna saeri is closely akin to *S. robiniifolia,* within the range of which it occupies a restricted enclave. It is a shrub or treelet smaller at maturity than *S. robiniifolia*; its leaves are mostly shorter, its pairs of leaflets one or two fewer, and its racemes have on the average slightly more numerous flowers. The latter differ only to the extent that the beak of the long anthers tends to be more pronounced. The two species are difficult to distinguish at anthesis, but the sharply raised venulation of the pod of *A. saeri,* which stands out pallid against a dark ground and becomes detached from it in age, is characteristic.

172. Senna tonduzii (Standley) Irwin & Barneby, comb. nov. *Cassia tonduzii* Standley, Contrib. U.S. Nat. Herb. **20**: 215. 1919.—Type in the U.S. National Herbarium, no. 471718, collected along the Río Tiliri, Costa Rica, November 23, 1892, by A. Tonduz (no. 7213) . . ."—Holotypus, US! = NY Neg. *9685*; isotypi, BR, G, GH, US!; paratypi, *Tonduz 1484, 3197, E. W. Nelson 3127,* US!—*Peiranisia tonduzii* (Standley) Britton & Rose, N. Amer. Fl. **23**(4): 265. 1930.

Peiranisia (?) *verbenensis* Britton ex Britton & Rose, N. Amer. Fl. **23**(4): 266. 1930.—"La Verbena, San José, Costa Rica . . . January 29, 1924, *Standley 32215.*"—Holotypus, NY! isotypus, US!—*Cassia verbenensis* (Britton & Rose) Standley, Field Mus., Bot. **18**(Fl. Costa Rica pt. 1): 519. 1937.

Potentially arborescent shrubs, sometimes flowering at 1–3 m but capable of attaining 10 m, with ultimately gray-barked trunk, densely pallid-lenticellate annotinous and prominently ribbed hornotinous branchlets, strigulose-pilosulous with fine appressed or ascending hairs up to 0.2–0.5 mm, the dense ample foliage bicolored, the lfts glossy dark green and glabrous above, paler duller and strigulose beneath, the inflorescence a leafy panicle of mostly 2-fld racemes arising directly from axils of developed foliage lvs or from axillary leafless branchlets, the fls immersed in foliage, some rarely forming a small, shortly exserted terminal panicle.

Stipules erect, falcately linear-attenuate 3–7 × 0.3–0.45 mm, at first green, early dry caducous (lacking from most mature specimens).

Lvs (disregarding small ones associated with axillary flowering branchlets) 9–23 cm; petiole including dilated, discolored pulvinus (10–)12–27 mm, at middle 0.5–1.2 mm diam, rather coarsely but narrowly wing-margined; rachis (4–)5–10.5

cm; glands between all pairs of lfts (caveat: much eaten) stipitate, in profile oblanceolate acute 1.7–4 mm tall, the stipe glabrous, the dilated red-brown body laterally compressed 0.4–0.7 mm wide; pulvinules 1–2.6 mm; lfts 4–7 pairs, strongly accrescent upward along the rachis, in outline elliptic or rarely ovate-elliptic 4–10.2 × 1.1–3.2 cm, (2.3–)2.7–4 times as long as wide, acuminate at apex and usually inversely acuminate at base, at insertion of pulvinule inequilaterally cuneate or (when broad) subcordately rounded on proximal side, the margins revolute, the centric straight midrib impressed-canaliculate above, cariniform beneath, giving rise on each side to ±7–11 pairs of camptodrome (and often some weaker intercalary) secondary veins, these faintly prominulous above, sharply so beneath, the tertiary connecting venules irregular, faint on both faces or finely prominulous beneath.

Peduncles 7–24 mm; racemes 1–3-, mostly 2-fld, the axis 0–1 mm, the fls when paired opening successively; bracts obovate cymbiform ±1.5 mm, nidulating the very young buds but caducous as the pedicel starts to elongate; pedicels becoming 16–22 mm, in fruit bent at a point ±3–4 mm below hypanthium, at base subtended laterally by a stipitate or subsessile ellipsoid gland; buds globose, glabrous overall or minutely puberulent at base; sepals strongly graduated, the outer broadly obovate, the inner suborbicular, all greenish with pallid margins, the longest 7.5–10 mm; inner perianth highly irregular, the yellow petals all pubescent dorsally, 3 adaxial long-clawed oblanceolate 8–16 mm (sometimes hardly longer than androecium), early deciduous, the 2 abaxial subsessile much larger but unlike each other, one broadly flabellate-obovate or suborbicular ±2–2.5 cm, the other (alternately right and left) obliquely reniform (scoop-shaped) ±2.5–3 cm, folded over 2 longer stamens; androecium glabrous, or the filaments sometimes thinly puberulent, the filaments of 4 median stamens 1–1.8 mm, of 3 abaxial ones 1.5–4 mm, the anthers of 4 median stamens oblong, slightly incurved 3.5–4.2 mm, contracted into an erect but obliquely 2-porose beak 0.5–0.8 mm, those of 3 abaxial ones lunately incurved 5.5–8 mm, contracted into a lanceolate beak (1-porose or septate) 3–5.5(–6.5) mm; ovary densely strigulose; style gently incurved, not dilated, 1–1.4 mm; ovules 30–48.

Pod pendulous, the stipe 3.5–8 mm, the body linear compressed 12–18 × 0.65–0.8 cm, the papery valves lustrous brown nigrescent, remotely strigulose; seeds unknown.—Collections: 30.

Forest margins, wooded ravines and hillsides, thickets in the oak-belt, (650–)900–1950 m, locally plentiful but highly localized and bicentric in centr. America: central upland and highland Chiapas, s. Mexico (near 16°30′–17°30′N, 92°30′–93°30′W); highland centr. Costa Rica (s. Alajuela and adjoining n. San José; perhaps also Guanacaste).—Fl. (VI–)VIII–I.—*Candelillo* (Brenes).

Senna tonduzii belongs to the circum-Caribbean group of *Interglandulosae* that includes the better known *S. robiniifolia* and *S. angustisiliqua,* from which it differs principally in the ample elliptic-acuminate leaflets. The plants from the centers of its bicentric range in Mexico and Costa Rica appear indistinguishable and the discovery of its presence in Guatemala or Honduras would not be unexpected. In the light of Costa Rican material now available for comparison the typus of *Peiranisia verbenensis* emerges as nothing remarkable.

173. Senna incarnata (Bentham) Irwin & Barneby, comb. nov. *Cassia incarnata* Pavón ex Bentham, Trans. Linn. Soc. London **27**: 545. 1871.—"Tropical America: Peru . . . Pav! in herb. Lamb. nunc Mus. Brit. . . . and Bolivia."—Holotypus, ticketed by Pavón "*Cassia incarnata* sp. nov.

de Mexico'' BM! = NY Neg. *110*; isotypi, labelled 'in collibus Pachacana [=Pachacamac], MA (*hb. Pavon. 14/8*) and s. loc. FI! and ''Pérou, misit Pavón 1827,'' G (hb. Moricand.)! = F Neg. (+ fragm) *28006*.

Cassia versicolor sensu Bentham, Pl. Hartweg. 128 [*Hartweg 729* from Malacatos, Ecuador, K! = NY Neg. *1562*]. 1843; non Meyen ex Vogel, 1837.
Cassia incarnata sensu Macbride, 1943, p. 169.

Shrubs, either erect and finally arborescent or diffuse and bushy, at anthesis 0.5–5 m, the angulately ribbed, finally brown and lenticellate branchlets and the lf-stalks thinly or densely either pilosulous with spreading or strigulose with subappressed weak shining hairs to 0.2–0.8 mm, these mixed with random short thickened livid ones, the firm dull olivaceous subconcolorous lfts above glabrous, beneath varying from densely pilosulous to glabrous except for ciliolate midrib, the inflorescence of subumbellate racemes axillary to normal or distally reduced foliage, not or shortly exserted.

Stipules linear-elliptic attenuate at both ends (4–)5–15 × (0.5–)0.8–2.5 mm, at base divergent from pulvinus across the stem, thence falcately or subsigmoidally incurved to erect, the firm herbaceous blades early dry brunnescent, deciduous long before lf.

Lvs below inflorescence (4–)5–14 cm; petiole including little dilated pulvinus (10–)12–25 mm, at middle 0.5–1.2 mm diam, bluntly 3-ribbed latero-dorsally, openly shallowly sulcate ventrally; rachis (3–)3.5–11.5 cm; gland 1 sessile or shortly stipitate between proximal pair of lfts (1–)1.8–3.6 mm, the stipe when present either glabrous or puberulent, the livid glabrous body narrowly fusiform to obliquely rhombic acute, when broad laterally compressed; lfts (6–)7–14(–17) pairs, of almost equal size or commonly a little decrescent both up- and downward from near middle, ascending from rachis face upward on scarcely dilated pulvinule (0.5–)0.7–1.1(–1.4) mm, in outline oblong-elliptic to elliptic-obovate obtuse mucronulate or deltately subacute, those (largest) near middle of rachis (9–)11–24 × 3.5–8.5 mm, 2.2–4 times as long as wide, at base asymmetrically rounded or subcordate, the margin plane, the centric midrib cariniform beneath, the venation otherwise immersed or 5–8 pairs of secondary camptodrome venules faintly raised beneath.

Peduncles 1.5–5 cm; racemes subumbellately or very shortly (3–)4–8(–9)-fld, the axis becoming 1.5–5(–7) mm; bracts obovate or oblanceolate cucullate 1.5–4.5 mm, caducous; pedicels 16–27 mm, the lower ones subtended laterally by an early divergent or deflexed stipitate gland 2–5 mm, this sometimes rudimentary or 0; fl-buds globose, glabrous above immediate base; sepals submembranous greenish with petaloid margins, faintly nerved, strongly graduated, all broadly obovate-suborbicular but the outer scarcely half as long as the innermost, these 7–10 mm; petals yellow, glabrous or dorsally puberulent near claw, 3 adaxial shorter and narrower, oblong- or obovate-oblanceolate beyond the short claw, the 2 abaxial longer, one subsymmetrically obovate, the other boomerang-shaped 15.5–21 mm, its blade nidulating the longest stamen; androecium glabrous but the anthers granular-papillose, the filaments of 4 median stamens 1–1.4 mm, of 2 abaxial ones 2–3.5 mm, of one abaxial one 3–5.5 mm, the anthers of 4 median stamens oblong, slightly incurved 2.7–3.8 mm, with lateral biporose beak scarcely 0.4 mm, the anthers of 3 abaxial lunately incurved, the body 3.2–4.5 mm, the tubular beak 1.2–1.6 mm, dehiscent by a pore usually interrupted by a slender septum; ovary canescently strigose-pilosulous, the glabrate style linear sometimes slightly enlarged upward 1.8–3.5 × 0.2–0.35 mm; ovules 20–28.

Pod erratically pendulous, the stipe 5–7 mm, the linear straight ribbon-like body

6.5–11 × 0.45–0.65 cm, the papery fuscous continuous valves pallid along the sutures, elevated over each seed as a low-convex or low-pyramidal mound; seeds fuscous or brownish, oblong- or rhombic-obovoid (2.3–)3.1–3.9 × (1.4–)2.2–2.7 mm, the testa moderately lustrous crackled, the elliptic areole (0.8–)0.9–1.3 × 0.5–0.9 mm.—Collections: 31.

Brushy and stony hillsides and rocky bluffs, 1500–2800 m, sometimes forming thickets in overgrazed pasture or in hedges, locally abundant in widely separated stations along the crest and Pacific slope of the Andes in Ecuador (Chimborazo; Loxa) and Peru (Cajamarca; Ancash; Lima), between 2°15′ and 12°S, in Lima descending locally (at Atocongo) to the foothills (loma formation) at 300–400 m.—Fl. almost throughout the year.

Senna incarnata is readily distinguished from other Andean *Interglandulosae* by its falcate stipules and umbelliform racemes of about 4–8 flowers, these contrasting with the mostly 2-flowered racemes of *S. pallida* sens. lat. All superficially similar sympatric sennas lack stipitate glands at base of the pedicels and have truncate, not tubular-beaked long anthers. This is a clearly differentiated, little variable and seldom misidentified species.

175. Senna rostrata (Martius) Irwin & Barneby, comb. nov. *Cassia rostrata* Martius, Verhandl. Vereins Beförd. Gartenbaues Königl. Preuss. Staaten **3**: 99, t. 7. 1827.—Cultivated at Munich from seed collected by Martius "in dem Fermo von Minas Novas und in der Wüste der Provinz Minas Geraës."—Lectoholotypus, cult. VII.1827 in hort. monac. from seed collected in Brazil (*Martius 1544, Cassia saltuum* nom. nud.), M! paratypus, *Martius 1381,* "in deserto Minarum Generalium ad Contendas . . . in caldario floruit mense Jan. 1822," M!—*C. biflora* var. *rostrata* (Martius) Bentham in Martius, Fl. Bras. **15**(2): 121. 1870, max. ex parte, exclus. pl. Blanchet. hirta. *C. bicapsularis* var. *rostrata* Bentham ex Isely, Mem. N.Y. Bot. Gard. **25**(2): 198 (lapsu pro *C. biflora* var.). 1975.

Cassia rostrata sensu Voigt in Sylloge Pl. Nov. Min. Cognit. Ratisbon. **2**: 55. 1828; Vogel, 1837, p. 28. *C. biflora* var. *rostrata* sensu Bentham, 1871, exclus. syn. *C. aversiflora*.

Shrubs 1–5 m, in aspect and most details resembling coarse, large-fld *S. pallida,* the smooth green, angulately ribbed, freely forking hornotinous branchlets and lf-stalks at least thinly, sometimes densely pilosulous with fine whitish spreading-ascending hairs up to 0.2–0.5 mm mixed with random thickened livid ones, the pallid glaucescent foliage subconcolorous and appearing glabrous but the lfts often minutely ciliolate and rarely pilosulous beneath, the inflorescence of 2-fld racemes mostly axillary to recent lvs, late in season becoming subcorymbose and exserted.

Stipules erect herbaceous narrowly linear-attenuate 4–11 × 0.25–0.6 mm, glabrous, caducous.

Lvs 4.5–11.5 cm; petiole including moderately dilated pulvinus 12–22(–26) mm, at middle 0.5–11.5 mm diam, prominently 3-ribbed dorso-laterally and deeply 2-sulcate between the ribs, openly sulcate ventrally; rachis 3–5.5 cm; petiolar gland between proximal pair sessile or shortly stipitate 2–3.5 mm, the stipe when present puberulent, the glabrous compressed livid or orange body in profile obliquely obovate acute (1–)1.1–2.5 mm wide; sometimes similar smaller glands at 1–2 pairs next following; lfts 4–8 pairs, ascending from rachis face upward on linear pulvinule 1.2–2 mm, slightly accrescent upward, in outline obovate, oblong-obovate or broadly oblanceolate, obtuse or shallowly emarginate, muticous or minutely mucronulate, the distal pair (14–)17–32 × (6–)8–15 mm, 1.7–2.4 times as long as wide, at base asymmetrically rounded, the membranous pallid margin plane or incipiently revolute, the midrib immersed above, cariniform beneath, the 6–8 pairs of camptodrome secondaries either immersed or faintly prominulous on either face.

Peduncles 1.8–3.8 cm; racemes (1–)2-fld, the axis 0, the pedicels quasi-dichotomous; bracts lanceolate 2–4 mm caducous; pedicels 1.8–3 cm, subtended laterally at base by a sessile gland, in fruit abruptly bent shortly below hypanthium; fl-buds globose glabrous; sepals fuscous-green with submembranous or petaloid margins, strongly graduated, all obovate-orbicular, the innermost at least twice as long as outermost, 10.5–14 mm; petals yellow, puberulent externally at claw or to middle, strongly heteromorphic, 3 adaxial relatively short, ovate or oblong-obovate beyond the slender claw, the 2 abaxial heteromorphic, one like the adaxial but ampler, the other (alternately right and left) sessile very obliquely reniform 27–33 mm, folded to enclose the longest stamens; androecium glabrous, the filaments free, of 4 median stamens 1–1.8 mm, of 2 abaxial 3.3–7 mm, of one (lateral) abaxial 6–9 mm, the anthers of 4 median stamens oblong, slightly incurved 5–8 mm, with divaricate or infraterminal, very short beak opening by 2 parallel slits, those of 3 abaxial ones lunately incurved, the body 6.5–9 mm abruptly contracted into an erect tubular beak 3–4 mm, this opening by a terminal pore often divided by a slender septum; ovary white-strigulose; style 1–1.5 mm, at incurved apex ±0.35 mm diam; ovules 44–56.

Pod erratically spreading-ascending or pendulous, the stipe (6–)7–11 mm, the linear compressed straight or slightly curved body (4.5–)6.5–11 × 0.65–0.8 cm, the papery valves becoming fuscous or nigrescently castaneous, delicately cross-venulose, elevated over the seeds in the form of low transverse ridges 1.5–2 mm wide; seeds narrowly obovoid, irregularly compressed or compressed-subquadrangular 3.5–4.4 × 1.4–2.3(–2.6) mm, the testa pale brown dull crackled, the elliptic areole (sometimes on the narrower faces of the seed) 1.2–1.5(–1.8) × 0.4–0.6 mm.—Collections: 27.

Stream-banks, pebbly riverbeds, woodland margins, becoming locally plentiful in disturbed cerrado or cerradão and persistent in pasture thickets and hedges, ±250–700 m, local, best known from the middle São Francisco valley in lat. 15°–19°S, in centr. and n.-w. Minas Gerais and extreme s.-centr. Bahia (Urandi), extending e. of the crest of Sa. do Espinhaço to upper Rio Araçuaí (Minas Novas) and interruptedly w. into the valley of Rio Paranaíba in the Triângulo Mineiro (Tupaciguara) and s.-w. Goiás (Jataí); an old record from Paraguay, without exact data of provenance (*Fleischer s.n.*, P), requires confirmation.—Fl. IV–VIII(–IX).

Senna rostrata resembles some broad-fruited North American form of *S. pallida* and was interpreted by Bentham as an outlying Brazilian variety of the equivalent *Cassia biflora*. Its pod, however, differs from that of *S. pallida* sens. lat. in having about fifty seeds crowded into a pod only about 7–11 cm long. In consequence the seed-locules become at least twice as wide as long and each seed, confined within its narrow stall, assumes a narrowly oblong-obovoid, compressed-subquadrangular form never encountered in *S. pallida*. This peculiar fruit, which suggests that of sympatric *S. acuruensis* rather than that of distantly allopatric (trans-Amazonian) *S. pallida*, entitles *S. rostrata* to the specific rank in which Martius originally described it. Within its own range *S. rostrata* can be mistaken only for the species next following, *S. aversiflora*, which see for comment.

176. Senna aversiflora (Herbert) Irwin & Barneby, comb. nov. *Cassia aversiflora* Herbert in Sims, Curtis's Bot. Mag. **53**: t. 2638. 1826.—". . . raised at Spofforth [Dean Herbert's home in Yorkshire, England] . . . from

Brazilian seed.''—No typus known to survive, but the protologue decisive.

Cassia aversiflora sensu Vogel, 1837, p. 43 (the name wrongly attributed to W. J. Hooker). *C. biflora* var. *rostrata* sensu Bentham, 1870, p. 121, ex parte (quoad pl. Blanchet. e Jacobina et Martianam hirtam in Obs. 1 notatam), non *C. rostrata* Mart. sens. str.
Cassia aversifolia Britton & Rose, N. Amer. Fl. **23**(4): 259 (lapsu). 1930.

Erect shrubs of bushy outline, reportedly sometimes arborescent, closely resembling *S. rostrata* in habit, with virgate, obtusely angulate-ribbed hornotinous branches either sparsely or densely hispid with fine lustrous, horizontally spreading setae up to (2–)3.5–7 mm and in addition charged with random minute thickened livid trichomes, the lf-stalks, some lft-margins and ovary furthermore minutely white-strigulose, the setae sometimes very sparse or confined to the base of some branchlets or to the stipule-margins, the foliage glaucescent bicolored, paler beneath.

Stipules flexuously incurved-ascending, linear-attenuate 4–11 × 0.2–0.5 mm, 1-nerved, caducous.

Lvs 5–9.5 cm, slender-petioled; petiole including moderately dilated livid pulvinus (13–)15–28 mm, at middle 0.45–0.8 mm diam, 3-ribbed latero-dorsally and sulcate between the ribs, openly sulcate ventrally; rachis 2.5–4.5 cm; petiolar gland between proximal pair of lfts sessile or shortly stipitate 2–4 mm, the reddish body elliptic- or semi-obovate obtuse or acute glabrous, strongly compressed, in profile (0.8–)1–1.4 mm wide; rarely a smaller gland at second pair; lfts (4–)5–8 pairs, by day ascending from rachis, turned half face to face on linear livid pulvinule 1–1.4 mm, in sleep sharply tilted backward, strongly accrescent distally, in outline obovate to broadly oblanceolate obtuse or shallowly emarginate, the distal pair 15–28 × 4.5–12 mm, (2–)2.2–3.4 times as long as wide, all at base asymmetrically rounded to broadly cuneate or (when broad) subcordate, the margin plane, the slender centric midrib immersed on upper and prominulous on lower face, the 4–7 pairs of major camptodrome secondary veins very slender, sometimes finely raised beneath, the tertiary venulation imperceptible.

Racemes numerous axillary to young lvs, capitately 2-fld, the peduncles 1.5–3.7 cm, the quasi-dichotomous pedicels 1.8–3 cm, in fruit bent shortly below hypanthium; bracts varying from lanceolate to linear-oblanceolate 1.5–2.5 mm, caducous; a gland sessile beside base of each pedicel; buds either globose or angulate by distinctly keeled outer sepals, glabrous or almost so; sepals submembraneous, fuscous with pale or petaloid margins, strongly graduated but all broadly obovate-suborbicular obtuse, the large inner ones 7–10 mm; petals yellow, externally puberulent along (and often between) the major veins, heteromorphic, 3 adaxial and 1 (larger) abaxial obovate or oblong-obovate from a narrowly cuneate claw, the fifth (abaxial) largest, very obliquely semi-obovate or reniform sessile or almost so, concave and nidulating the long stamens 21–34 mm; androecium usually glabrous, the thecae rarely setose in the lateral grooves, the filaments of 4 median anthers 1.4–2.6 mm, on both or at least on one side of the fl united at base, those of 2 abaxial ones 3–4 mm, of 1 (lateral) abaxial 4.5–6 mm, the thecae of 4 median stamens oblong slightly incurved 4–5 mm, the divaricate subterminal beak ±0.5 mm opening by one U-shaped slit, those of 3 abaxial ones more strongly incurved, the body 4.4–6.5 mm abruptly contracted into an erect tubular beak 2.4–3.8 mm, opening by a single, ±2-lipped pore; ovary strigulose; style 1–1.8 mm, at incurved tip 0.3–0.4 mm diam; ovules 26–52.

Pod erratically oriented, the stipe 6–10 mm, the narrowly linear straight or slightly curved body 9.5–16.5 × 0.45–0.55 cm, the brownish-nigrescent, thinly strigulose valves expressed over each seed in the form of a 4-ribbed pyramid almost as long as wide, the distance between interseminal septa 3.3–4 mm; seeds (few seen truly ripe) compressed-rhomboid, ±3.2 mm diam, the testa dull blackish, the areole elliptic ±1 × 0.4 mm.—Collections: 21.—Fig. 32.

Caatinga thickets, ±80–700, on Chapada Diamantina up to 1000 m, apparently local in the taboleiro hill-country of n. and e.-centr. Bahia from Rio Paraguaçú w. to Sa. Açuruá, n.-e. interruptedly to e. Pernambuco (Tapera; Pau D'Alho) and Paraíba (Areia).—Fl. irregularly through the year.

Senna aversiflora was not distinguished from *S. rostrata* by Bentham, who treated the Brazilian *Interglandulosae* with twin-flowered racemes as a single variety of *S. pallida* (see next following). At a time when only a few incomplete and mostly cultivated specimens were available for comparison this was hardly surprising, but the now substantial material accumulated from the eastern Brazilian states is found to fall into two perfectly distinct series, treated herein as species independent both of each other and of *S. pallida*. Differences between them in pubescence, androecium, pod and seed have been brought out in our key to ser. *Interglandulosae* (couplet 18/18, q.v.). In the context of this series, as indeed in *Senna* as a whole, the lustrous setose cauline pubescence of *S. aversiflora,* which recalls the multifoliolate *S. mutisiana* and *S. williamsii,* is a precarious differential character; it is in fact sometimes reduced to a rudiment concentrated on the stipules. Rare calvescent plants of *S. aversiflora,* among them the specific typus, are deceptively similar at anthesis to *S. rostrata*. But the longer and narrower pod, with valves elevated over each square or broadly quadrilateral seed-locule into a decussately folded pyramid, is decisively different.

The epithet *aversiflora,* as explained by Dean Herbert, referred to the "singular manner in which the pair of flowers appear to shrink from each other, by the incurvation of the petals that come in contact." It incorporates the earliest known mention of enantiostyly in *Senna*.

177. Senna pallida (Vahl) Irwin & Barneby, comb. nov. *Cassia pallida* Vahl, Ecl. Amer. **3**: 12. 1807.—Typus infra sub var. *pallida* indicatur.

(?)*Cassia biflora* Linnaeus, Sp. Pl. 378. 1753, nom. ambig.—"*Habitat in* Indiis."—Described from a plant grown in Clifford's garden at Hartekamp = *Cassia* (No. 12) *foliolis octo parium ovalibus emarginatis, glandula acuta inter infima* Linnaeus, Hort. Cliff. 159. 1737.—No typus extant either in hb. Cliffort., BM, or at LINN. The protologue (1737 + 1753), strongly suggestive of some member of *Interglandulosae,* has traditionally been interpreted in the sense of our *S. pallida* sens. lat., but could apply with equal probability to *S. angustisiliqua*. If indeed based on a form of *S. pallida,* exact application of the name would still be impossible, for var. *bahamensis* and var. *pallida* were equally as likely to have reached Europe in early XVIII century.

(?)*Cassia tenuissima* Linnaeus, Sp. Pl. 378. 1753.—"*Habitat in* Havana."—Based on a manuscript description of Dr. Houston, for which no corresponding spm survives at BM or LINN.—This obscure name, listed here because equated with a form of *Cassia biflora* by Vogel (1837, p. 29) and Bentham (1871, p. 543), is indecipherable from the seven words of original description.

Cassia dispar Willdenow, Enum. pl. hort. bot. Berol. 441. 1809, nom. ambig.—"*Habitat in* America meridionali."—Holotypus, *Hb. Willd. 7966,* B! consists of sterile branchlets from a plant grown at Berlin. While almost certainly representing *S. pallida* sens. lat. and probably its var. *pallida,* precise identification is not possible without flowers.

Precociously flowering shrubs and treelets variable in stature and habit of growth, at anthesis commonly 0.5–5 m, in exceptional forest environment up to 7(–10) m, varying from fully glabrous (the ovary seldom so) to densely pilosulous with pallid or sometimes lutescent, variably spreading or accumbent hairs up to 0.2–1(–1.3) mm commonly mixed with random minute thickened (but not secretory) trichomes and resinous dots, the lfts greatly variable in number (2–31 pairs),

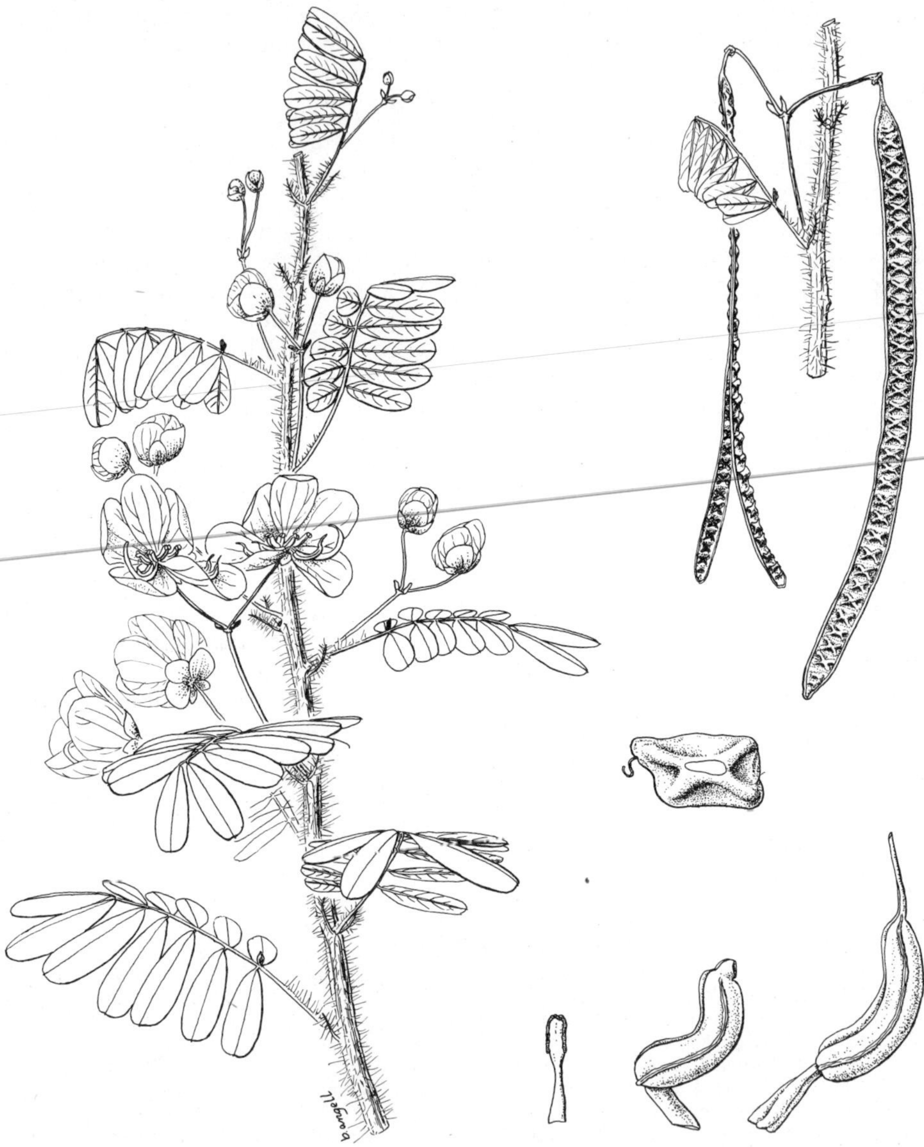

Fig. 32. *Senna aversiflora* (Herbert) I. & B. Flowering branchlet ×½; detached stamens + staminode ×3 (*Santos 1968*). Fruiting branchlet ×½; seed ×4 (*Duarte 10552*).

size (0.6–6 cm) and venation, either accrescent upward along rachis or only the proximal somewhat smaller than the remainder, the usually strongly bicolored foliage seasonally dimorphic, the primary lvs often drought-deciduous and replaced by diminished, simpler lvs of axillary brachyblasts, the inflorescence composed of mostly 2 (some rarely 1- or 3–4)-fld racemes arising either solitary from axils of coeval cauline lvs, or from axils of smaller lvs on axillary branchlets, or from highly condensed, leafless or microphyllous brachyblasts, the fls thus all or mostly immersed in foliage or all exserted, then sometimes forming a showy pseudocorymbose or thyrsoid panicle.

Stipules subulate, setiform or linear, less often somewhat expanded when commonly falcate, (1–)1.5–12(–15) × (0.1–)0.2–2 mm, caducous, absent from most fruiting specimens.

Primary cauline lvs (3.5–)4.5–16(–18) cm; petiole (3.5–)5–40 mm; lf-stalk prolonged beyond distal lfts into a setiform deciduous or less often a glandiform, more persistent appendage; petiolar gland always present between or (adnate to rachis and apparently) shortly above first pair of lfts, sessile or more often stipitate, (1–)1.4–4(–4.3) mm long, the reddish, orange or fuscous head ± laterally compressed, in profile 0.3–1.3(–1.5) mm wide; less commonly smaller glands at 1–2(–3) succeeding pairs; lfts varying in outline from narrowly oblanceolate, oblong-oblanceolate or oblong-elliptic to broadly obovate, emarginate, obtuse or deltately acute, muticous or conspicuously mucronulate, the margins usually revolute at least near pulvinule, sometimes all around, rarely quite plane; midrib cariniform beneath, giving rise to 4–10(–11) pairs of major camptodrome secondary veins visible or prominulous on lower or both faces (rarely altogether immersed) and sometimes a tertiary or quaternary reticulate venulation, prominulous or defined by color, this developed on lower or both faces.

Peduncles (2–)4–30(–36) mm; axis of racemes 0–1.5(–3) mm; pedicels 8–27(–32) mm, subtended on one side by a sessile or stipitate fusiform gland, in fruit ± abruptly bent above middle; sepals broadly obovate or orbicular very obtuse, greenish, brownish or subpetaloid with usually membranous margins, strongly graduated, the large inner ones (4.5–)5–11(–13) mm, the smaller outermost ± half as long; petals golden yellow, usually puberulent externally especially on claws and major veins, heteromorphic in the manner of ser. *Interglandulosae,* the 2 larger abaxial ones (13–)14–32(–36) mm; filaments glabrous or puberulent, those of 4 median stamens dilated, sometimes confluent at base, 0.7–2.2 mm, those of 3 abaxial ones filiform, free 2–5.5 mm, the anthers of 4 median stamens 2.8–6 × 1–1.8 mm, at apex either abruptly truncate or smoothly contracted into a lateral or obliquely terminal, very short beak dehiscent by 2 parallel (very exceptionally confluent) slits, those of 3 abaxial ones 3–8(–9) mm, abruptly contracted into an erect or porrect tubular beak 1–3.5(–3.8) mm opening by a single pore; ovary strigulose, pilosolous, rarely glabrous, the style 0.7–2.2(–2.5) mm, incurved but not dilated distally; ovules mostly 28–50, in var. *bahamensis* 20–30.

Pod pendulous or spreading-declined at random angles, the stipe (2.5)3–9 mm, the linear, straight, decurved or somewhat contorted body mostly 5–15(–17, in one var. 17–26) × 2.5–8.5(–10) mm diam, the ultimately brown or nigrescent papery valves raised over each seed as a simple transverse mound or a more or less sharply defined 4-armed pyramid; seeds in broader pods obovate (rarely oblong-obovate), in narrower ones rhombic-paddle-shaped in lateral view, 2.3–4.8(–6) × (1.5–)1.7–3.4(–3.7) mm, the testa pale amber or brownish to yellowish-olivaceous, exceptionally castaneous, dull or exceptionally sublustrous, smooth becoming crackled or exceptionally pitted overall, the areole round, elliptic or oblong-elliptic 0.4–2.4(–3.5) × 0.3–1.2 mm.—Fig. 1 (petiolar nectary).

Apart from questions of nomenclature, discussed below, our concept of *S. pallida* is equivalent to *Cassia biflora* sensu Bentham (1871, p. 543) with these differences: the specific limits are narrowed by loss of var. *rostrata* (=*S. rostrata* + *S. aversiflora*) and of those Antillean and South American *Interglandulosae* characterized by U-shaped dehiscence of the median anthers (=*S. angustisiliqua* and varieties); but augmented in another direction by annexation of *Cassia foliolosa* and by numerous forms which Bentham a century ago had no opportunity of knowing. The division of the biflorous *Interglandulosae* into major groups differentiated by anther dehiscence is somewhat artificial and technical, the more so because in some Guatemalan populations of *S. pallida* var. *quiedondilla* and again in some individual flowers of var. *gaumeri* the twin beaks of one or more median anthers are so shortened as to become confluent. Moreover the overall phenetic similarity between *S. angustisiliqua* and some forms of *S. pallida* demonstrates an intimate relationship. However, once *S. angustisiliqua* is admitted to the already forbiddingly polymorphic *S. pallida,* it becomes logically indefensible to exclude the also close kindred *S. polyphylla, S. robiniifolia* and *S. viciifolia* which have the androecium of *S. angustisiliqua* and hardly more decisive morphological differences than the varieties of *S. pallida* recognized herein. The Brazilian *S. rostrata* and *S. aversiflora,* of which the diagnostic fruits and vesture were unknown to Bentham, are in reality much further removed from the core of *S. pallida,* both in form and in geographical dispersal, despite their similar androecium, than are *S. angustisiliqua* and its immediate allies.

Bentham called attention to the wide variation in flower-size in his *C. biflora* and modern observers (Schery, 1951, p. 74; Isely, 1975, p. 71) have commented on instability of pubescence and noted the need of revision and the probable segregation of infraspecific taxa. The latter prophecy has been fully borne out in our study which is based on herbarium material many times greater than that available to Bentham a century ago, but flower-size and pubescence have proved obstructive rather than helpful in the definition of what appear to be, for less immediately obvious reasons, genuine self-perpetuating entities. The stems and foliage of *S. pallida* vary from perfectly glabrous to densely pilosulous throughout, but the imperceptibly gradual intergrades in type of vesture, which itself varies independently in density, orientation, length and color, are only very locally correlated with other morphological features of significance and against a wider background emerge as erratic. Thus, following var. *pallida* across the northern states of Venezuela we find the plants in Carabobo glabrous, in Aragua variably pilosulous, in Distrito Federal, Miranda and Yaracuy randomly pilosulous and subglabrous, and in Nueva Esparta again prevailingly glabrous, as in distant Carabobo or on the nearby island of Grenada. The small range of the exceptionally pilose form (*Peiranisia velutina* sens. str.) that represents var. *pallida* on the Cauca-Patía divide in southern Colombia is bracketed by populations of glabrous plants, immediately to the north (Huila) and south (Nariño). In our var. *bahamensis,* endemic to the Bahamas and far eastern Cuba, the prevailing condition is glabrous below the ovary, but our one collection from Atwood Cay, otherwise identical (even to the diagnostic androecium) to those from islands to the north and south, is pilosulous. Pilosulous *Cassia trichocraspedon,* originally thought to be exceptional in the long-ciliate sutures of the pod, and glabrate *C. tortuosa* share a broad pod that is foreign to vicariant forms of var. *pallida* and except for pubescence are essentially identical; we refer these without hesitation to a collective var. *trichocraspedon*. However visually arresting, pubescence in this group contributes nothing to the definition of useful taxonomic categories.

A marked feature of *S. pallida* is seasonal dimorphism of the foliage, perhaps potentially present in all forms of the species but becoming especially evident in regions subject to annual dry seasons such as the Sonoran Desert and Yucatan Peninsula. Here a given plant may produce in wet months solitary two-flowered racemes from axils of ample primary stem leaves and, during drought, when those primary leaves have been shed, continue to flower from brachyblasts, either leafless or clad in much diminished and often much simplified leaves. The drought phases of varieties which, in days of prosperity, are obviously different in number or venation of the leaflets, or in their pods and seeds, are disconcertingly similar.

As for flower-size, it seems impossible to derive any useful criteria from this source. The seemingly random irregularity of the corolla is difficult to measure exactly, and we have found that flowers from two stems in one collection, from two collections from one site, and from sympatric collections taken at different seasons of the year, may span most of the range of variation in the whole series *Interglandulosae*. It is true that the flower of var. *bahamensis* is consistently on the small side of the norm and that of var. *macdougalliana* near or at its upper limit, but no discontinuities of dimension were found between any two closely allied forms. The same can be said of the anthers and their filaments.

The characters found most useful as taxonomic markers within *S. pallida* are emphasized in the varietal key following. Most of them are derived from aspects of the foliage: the condition of the seta, the length of the petiole in relation to the interfoliolar segments of the leafstalk, the number, size, accrescence, and in particular the venulation of the leaflets. These are reinforced in some cases by features of the pod or seed, rarely of the always inconveniently caducous stipules, and in almost all cases by considerations of range, indispensable for identification of incomplete specimens and those representing the inhibited drought-phase of their sort. The ranges of most varieties are mutually exclusive, but the differences, especially in southern Mexico and Central America, are as often altitudinal as not. In all cases where our knowledge is reasonably detailed we find that the varieties adapted to cool uplands are different from those in tierra caliente at the same latitude. In Pacific Mexico var. *pallida* and var. *geminiflora* are sympatric in the lowlands; we suspect that they are ecologically separated, but not totally for intermediate forms are frequent. Exceptionally two or three varieties of var. *pallida* occur in one place. At the same season in different years, near Lachao (municipio of Juquila) in the Sierra Madre of southern Oaxaca MacDougall found populations of var. *geminiflora* (with about 5 pairs of large, upwardly accrescent, intricately venulose leaflets), var. *macdougalliana* (similar except for smooth leaflets and glandiform seta), and var. *triquetripes* (with 11–15 pairs of small, non-accrescent leaflets, a normal petiolar seta and dilated rachis). These specimens, taken out of context of *S. pallida* sens. lat., might pardonably be attributed to distinct species. Unfortunately we have no information on micro-habitats or possibly critical differentiation by soil, exposure or elevation. All have the same flower adapted to pollination by bumblebees and they flower simultaneously at midwinter. A combined genetic-ecological study of this and similar situations in highland Guatemala would surely be rewarding.

For more than two centuries the epithet constantly applied to the present species is the descriptive *biflora,* which dates from *Species Plantarum* (1753) but is based on a plant seen by Linnaeus in the conservatory of George Clifford at Hartekamp in 1736. With the passage of years *Cassia biflora* became stabilized in the literature, but at the same time it acquired by accretion an increasingly vague and comprehensive meaning. The broad concept of a polymorphic *C. biflora* (excluding var. *rostrata*) developed by Bentham is coextensive with our *Senna pallida* and *S. angustisiliqua* combined, including such of the now recognizable infraspecific taxa as had reached Europe at that time. Britton & Rose (1930) disassembled a number of the component elements of *C. biflora* sensu Bentham into independent species of *Peiranisia,* but their analysis was incomplete and the residue left in the synonymy of *P. biflora* was still a disorderly mixture of *S. pallida* (several vars.) and *S. angustisiliqua.* Precise identification of *Cassia biflora* L. is not possible, because no typus was preserved. The protologue (Linnaeus, 1753) and the antecedent description in *Hortus Cliffortianus* (1738) are the only evidence to be considered. The description is a full one and clearly indicates a *Senna* of ser. *Interglandulosae* with two-flowered racemes but the critical features that distinguish *S. angustisiliqua* from *S. pallida* or, assuming for sake of argument that it was indeed our *S. pallida,* that would determine the particular variety of it, are not mentioned. Linnaeus attributed *C. biflora* to the Indies, meaning the West Indies (and not India, as Britton & Rose, 1930, p. 260, erroneously supposed), but this might indicate either Jamaica, or the Bahamas, or Hispaniola, or Cuba and consequently a different taxon. We mention under *S. pallida* var. *bahamensis* the supposition of N. E. Brown that Linnaean *C. biflora* came from the Bahamas, but there is no trace of hard evidence that such was the case.

The identity of *C. tenuissima* L., which was at first doubtfully (Colladon, 1816) and later by many authors without qualification referred to the synonymy of *C. biflora,* is equally mystifying. It was based by Linnaeus on a manuscript phrase-name of William Houston, but no corresponding specimen can be found in either the Linnaean collections or among Philip Miller's at British Museum, and the perfunctory protologue defies interpretation.

Precise application of these two Linnaean names, based on convincing typifications, has eluded us. We regard them as insolubly ambiguous, and regretfully fall back for an epithet on the next earliest valid binomial, *C. pallida* Vahl.

Key to the Varieties of *S. pallida*

1. Appendage terminal to lf-stalk subulate or setiform, not modified and dilated into a gland; range of whole species.
    2. Bahama and Caicos Is. and coastal n.-e. and far e. Cuba; anthers slender, the 4 median ones 1–1.2 mm diam, their tubules smoothly contracted at apex into an obliquely terminal beak. 177j. var. *bahamensis* (p. 548).
    2. Continental N. and S. America with Lesser Antilles, s. Windward Is. and Swan Is. in s. Caribbean; anthers nearly always stouter (1.2–)1.3–2 mm diam, abruptly truncate at apex and the tubules there commonly umbonate at the angle opposed to the laterally divaricate, often slightly infra-terminal beak, if whole anther less than 1.25 mm diam the umbo pronounced.
        3. Reticulate tertiary venulation emphatically developed on one or both faces of lfts; range of the whole species short of Peru, Yucatan Peninsula and the Bahamas with e. Cuba.
            4. Tertiary venulation forming a reticulum either distinctly prominulous or strongly differentiated by color (or both) on dorsal face only, the ventral face simply penninerved or apparently veinless; n.-e. Brazil, Venezuela and Colombia, Lesser Antilles, lowlands of Central America and Mexico, extending n. on Pacific slope to Baja California Sur and Sinaloa, on the Gulf slope to Veracruz, the common xermorphic form of the species. 177a. var. *pallida* (p. 537).
            4. Tertiary venulation finely prominulous on both faces of lfts; Pacific slope of Mexico from Oaxaca to Sonora, sympatric with the last but more mesophytic and less widespread. 177c. var. *geminiflora* (p. 540).
        3. Reticulate tertiary venulation imperceptible or almost so on both faces of lfts, these superficially either penninerved or veinless.
            5. Lfts of larger primary cauline lvs 2–8(–9) pairs, perceptibly and usually conspicuously accrescent distally, the terminal pair longest; areole of seeds 0.4–0.9 mm in greatest diam.
                6. Isolated in middle Marañón valley in trans-Andine Peru. 177f. var. *nemorosa* (p. 543).
                6. Scattered over s. Mexico and Guatemala.
                    7. Plants of Yucatan Peninsula, Mexico.
                        8. Lfts of primary cauline lvs 4–6 pairs, of brachyblast lvs (2–)3–4 pairs; seeds smooth; Campeche, Yucatán and Quintana Roo, relatively widespread and common. 177h. var. *gaumeri* (p. 546).
                        8. Lfts of primary lvs 2–4, of brachyblast lvs exactly 2 pairs; local in Quintana Roo; seeds pitted. 177i. var. *goldmaniana* (p. 547).
                    7. Plants of Pacific Mexico to Chiapas and highland Guatemala.
                        9. Pod 6.5–9 mm wide; lowland Michoacán and Guerrero, in matorral at 0–800 m. 177e. var. *palmeri* (p. 543).

9. Pod 4–5.5 mm wide; either Sonora–Sinaloa, or Oaxaca–Guatemala.
 10. Habitat in desert and thorn-forest below 600 m in Sonora and Sinaloa; valves of pod simply mounded over the seeds. 177d. var. *shreveana* (p. 542).
 10. Habitat and dispersal different, in Oaxaca, Chiapas and Guatemala, if below 1000 m the valves of pod raised over each seed into a sharply defined decussate pyramid.
  11. Matorral and thorn-forest at 5–700 m in e. Oaxaca and adj. Chiapas; valves of ripe pod pyramidally elevated over each seed. 177b. var. *isthmica* (p. 539).
  11. Pine-oak forest at 1200–2150 m; valves of ripe pod plane or weakly bullate over each seed.
   12. Primary lvs 6–13 cm, their rachis 6–13 cm and lfts 6–9 pairs, the longest 18–35 mm; Guatemala. 177g. var. *longirostrata* (p. 544).
   12. Primary lvs 3–6.5 cm, their rachis 1–3 cm and lfts 3–7 pairs, the longest 9–17(–20) mm; Oaxaca and Chiapas. 177g(bis). var. *brachyrrhachis* (p. 545).

5. Lfts of primary lvs 9+ pairs, at least in larger lvs (caveat: deciduous, and absent from some specimens with ample panicle which may be accompanied by only diminished distal lvs), the pairs not or scarcely accrescent upward, the terminal pair not obviously the largest; areole of seeds mostly 1.1–2.4(–3.5) mm in greatest diam.
 13. Lfts of largest lvs 9–15(–17) pairs (if led hither by plants of lowland s.-e. Mexico, cf. 177b. var. *isthmica*).
  14. Lf-stalk sharply carinate dorsally and conspicuously green-winged latero-ventrally, the wings constricted at insertion of lfts; where sympatric with the next, the petioles 6–15 mm; local in s. Guerrero and s. Oaxaca. 177m. var. *triquetripes* (p. 552).
  14. Lf-stalk either sharply or bluntly carinate dorsally but only narrowly winged latero-ventrally, the wings not constricted at insertion of lfts.
   15. Petiole of major lvs 15–30 mm, 3–5 times as long as first interfoliolar segment of lf-stalk; Mexico to Nicaragua. 177k. var. *quiedondilla* (p. 549).
   15. Petiole of major lvs 6–13(–15) mm, 1–2 times as long as first interfoliolar segment; local in Guatemala. 177l. var. *tuerckheimiana* (p. 551).
 13. Lfts of larger lvs (12–)15–31 pairs; local in Guatemala. 177n. var. *foliolosa* (p. 553).

1. Appendage terminal to lf-stalk modified into a gland resembling that between proximal pair of lfts; local either in highland centr. Costa Rica or in s. Mexico (México, Oaxaca, Veracruz).
 16. Lfts of larger lvs 9–12 pairs; ovules 26–52; either the larger lfts over 2 cm long or the pod over 12 cm long.
  17. Plants of highland Costa Rica. 177o. var. *cordillerae* (p. 554).
  17. Plants of s. Mexico.
   18. Larger lfts 22–32(–35) mm long; ovules 48–52; pod extremely long and narrow, 17–26 × 0.35–0.4 cm, twisted on its vertical axis and shallowly undulate along the sutures; lowlands of n. Oaxaca and adj. Veracruz. 177p. var. *lemniscata* (p. 555).
   18. Larger lfts 25–60 cm long; ovules ±25–32; pod 7–13 × 0.6–0.85 cm; either Pacific lowland Oaxaca or s. México.
    19. Margin of lfts plane; venation of lfts immersed on upper face; Sa. Madre del Sur in s. Oaxaca. 177q. var. *macdougalliana* (p. 556).
    19. Margin of mature lfts strongly revolute; venation of lfts sharply prominulous on upper face; s.-w. México. 177r. var. *trichocraspedon* (p. 557).
 16. Lfts of larger lvs 14–24 pairs and not over 2 cm long; ovules 16–22; pod 6.5–11 cm long; highland n.-w. Oaxaca at 1530–1700 m. 177s. var. *delgadoana* (p. 558).

177a. Senna pallida (Vahl) Irwin & Barneby var. **pallida**. *Cassia pallida* Vahl, 1807, l.c., sens. restr.—"Habitat ad St. Martham [Colombia]. *von Rohr.*"—Holotypus, C (hb. Vahl.)! = IDC Microfiche 8/1.7.8/11/1 = NY Neg. *10483*.

Cassia discolor Desvaux, J. Bot. **3**(2): 72. 1814.—"Habitat in Jamaica."—Holotypus, s. loc., P (hb. Desvaux.), as to lvs only, the attached pod alien.—Referred by Bentham, 1871, p. 555 to *Cassia glauca*, but the pod alone belonging there.

Cassia venustula Humboldt, Bonpland & Kunth, Nov. Gen. & Sp. **6**(qu): 352. 1824.—"Crescit locis aridis prope Cumanam [Sucre, Venezuela]."—Holotypus, labelled "n. 529. Cumaná," P-HBK! isotypus, B (*hb. Willd. 7959*)!

Cassia acapulcensis Humboldt, Bonpland & Kunth, Nov. Gen. & Sp. **6**(qu): 353. 1824.—"Crescit prope Acapulco Mexicanorum [Guerrero]."—Holotypus, P-HBK! isotypus, B (*hb. Willd. 7970/5*)!

Cassia xiphoidea Bertoloni, Nov. Comment. Acad. Sci. Instituto Bonon. **4**: 415. 1840.—"Habitat in Volcan de Pacaya [dept. Amatitlán, Guatemala] . . . Velasquezius dedit nomine *Cassia species nova decussata*."—No typus known to survive, but the description and locality persuasively suggest some pubescent phase of var. *pallida*.

Cassia pallidior Rose, J. Wash. Acad. Sci. **17**: 167, quoad typum, paratypis costaricensibus exclusis. 1927.—"Type . . . [US] 676583, collected by H. Pittier near Alajuela, Panama, January, 1914 (no. 2343) . . ."—Holotypus, US! isotypus, NY!

Peiranisia deamii Britton ex Britton & Rose, N. Amer. Fl. **23**(4): 261. 1930.—"Gualán, dept. Zacapa, Guatemala, January 14, 1905, *Charles C. Deam 291*."—Holotypus, NY! isotypus, MO!—*Cassia gualanensis* Lundell, Carnegie Inst. Wash. Publ. **478**: 212. 1937 (non *C. deamii* (Britton & Rose) Lundell, 1937).

Peiranisia velutina Britton & Killip, Ann. N.Y. Acad. Sci. **35**(3): 181. 1936.—"Patía Valley, Nariño, Colombia, 200–500 m altitude, [F.C.] Lehmann B.T. 822 . . ."—Holotypus, NY! clastotypus (fragm), US! paratypi, *Lehmann 7792*, F, K; *Lehmann B.T. 827*, NY!

Potentially arborescent shrubs at anthesis (0.6–)1–3(–4) m, with ribbed but usually not deeply sulcate annotinous branchlets, highly variable in pubescence, either glabrous to (sometimes including) the ovary or thinly to densely pilosulous with forwardly incumbent or spreading, pallid or rarely lutescent hairs to 0.3–0.8(–1.3) mm, the bicolored lfts glabrous on both faces, puberulent beneath and ciliolate, or pilosulous overall, the inflorescence when young consisting of simple 2-fld racemes axillary to coeval lvs, in age often becoming ± paniculate and exserted.

Stipules linear-lanceolate or -oblanceolate 3–9(–11) × 0.3–1(–1.2) mm, carinate dorsally.

Major lvs 5–15(–18) cm; petiole (7–)9–22 mm, 1–2.8 times as long as first interfoliolar segment of rachis; rachis (2–)2.5–8(–10.5) cm; glands between proximal 1–3 pairs of lfts, stipitate or sessile, the lowest 1.5–3.5 mm, the ovate acute or elliptic compressed body 0.4–1.2 mm wide; lfts commonly 5–8, rarely in none more than 4 or in some up to 9–10(–11) pairs, accrescent upward along rachis, the distal pair mostly obovate obtuse mucronulate, sometimes oblanceolate obtuse or abruptly deltate-acute, 18–42(–53) × (5.5–)7–21(–22) mm, 1.5–3.3 times as long as wide, the venation of 5–9(–11) pairs of camptodrome secondary veins nearly always finely prominulous (exceptionally shallowly engraved) on upper face, always so beneath, there giving rise to a well defined tertiary reticular venulation either prominulous or if almost immersed strongly defined by color.

Peduncles (7–)9–22(–27) mm; racemes mostly 2-, rarely some 3–4-fld, the pedicels (9–)11–23 mm; long sepals (5.5–)6–10(–11) mm; long petals (pubescent or rarely glabrous dorsally) (15–)17–27(–29) mm; filaments glabrous or puberulent, those of 4 median stamens ± dilated free 0.8–1.8 mm, those of 3 abaxial ones 2.2–4.5(–5) mm, the anthers of 4 median stamens 2.8–5.3 mm, abruptly truncate, the lateral beak ±0.2–0.4 mm, those of 3 abaxial ones 4–6.4 mm, the beak (1.4–) 1.6–3 mm; ovary commonly strigulose, rarely glabrous or pilose; style 1.2–2.4 mm.

Stipe of pod 2.5–7 mm, the body (5–)6–12(–15) × (0.25–)0.33–0.45(–0.5) cm, the ripe valves raised over each seed as a distinct 4-armed pyramid; seeds 2.3–3.5 × 1.5–2.8(–3) mm, the areole 0.4–0.8 × 0.4–0.8 mm.—Collections: 434.—Fig. 11 (androecium).

Forest borders and clearings, glades in deciduous woodland, brushy hillsides, benches of barrancas, stream-banks and savanna thickets, thriving in disturbed habitats and becoming locally weedy along highways and in hedges, widely dispersed, mostly in xeromorphic plant-communities, from sea level up to 1100 m, exceptionally higher (in Michoacán to 1440, in Guatemala to 1480, in Honduras and Nicaragua to 1200 m), over much of extreme n. South and Central America and the tierra caliente of s. and s.-w. Mexico, there extending on the Gulf slope to Veracruz (but absent from Yucatan Peninsula) and on the Pacific slope to Sinaloa and the Cape region of Baja California Sur, in Colombia s. in inter-Andean valleys to Río Patía in Cauca and Nariño, e. through Caribbean Venezuela to Nueva Esparta and the Windward Islands (Grenada; Martinique, perhaps introduced); Swan Islands in the w. Caribbean; one station in the middle Orinoco drainage in Guárico, Venezuela; apparently much isolated in savannas of n. Território do Roraima, Brazil.—Fl. abundantly X–IV, sporadically throughout the year, in dry seasons often from brachyblasts with greatly diminished foliage.—*Abejón*; *vara prieta*; *flor de San José* (Mexico, Central America); *montenegro* (Costa Rica); *brusca*; *güeregüere* (Venezuela).

This is the common lowland continental phase of *S. pallida* characterized by relatively few (commonly 5–9, rarely up to 11) pairs of distally accrescent leaflets intricately reticulate beneath and by a narrow pod with valves raised when ripe over each seed in the form of a four-armed pyramid. It is dispersed over the whole range of the species short of Peru, extratropical northwestern Mexico, Yucatan and the Bahamas, where it is replaced by closely analogous forms differing in simpler venulation of the leaflets commonly associated with a pod simply mounded over each seed. The var. *pallida* varies much in amplitude of the leaflets and flowers, the variation due partly to seasonal or climatic influences and partly, no doubt, to genetic control (not experimentally demonstrated); and conspicuously in pubescence, as described in our discussion of *S. pallida* sens. lat. The holotypi of *Cassia pallida, C. pallidior* and *C. acapulcensis* represent glabrate or glabrescent, those of *C. xiphoidea* (probably) and *Peiranisia deamii* moderately pilosulous, and that of *P. velutina* densely pilosulous states. In a general but imprecise fashion our var. *pallida* is equivalent to the *C. biflora* of floristic works covering the northern periphery of South America, Central America, and Mexico.

The identity of *C. xiphoidea* listed in the synonymy of var. *pallida* was obscure to Britton & Rose who mentioned it (1930, p. 268) nevertheless as a doubtful member of genus *Peiranisia*. No typus has survived in Bertoloni's herbarium at Bologna (1977) and unless a duplicate can be found the name can be interpreted only through the description. This agrees very well with our concept of pilosulous Guatemalan var. *pallida,* especially in characterization of the pod as sword-shaped and torulose (i.e. expressed over the seeds) and in the telling adjective *decussata* quoted from the collector Velásquez, which we interpret as a reference to the sharp X-shaped pyramidal form of each torulose segment of the pod valves.

177b. Senna pallida (Vahl) var. **isthmica** Irwin & Barneby, var. nov., a var. *pallida,* caeterius congrua, foliolorum venatione dorsali penninervia (nec reticulata), a var. *shreveana* et *longirostrata,* quoad venationem similibus, leguminis valvulis maturis supra semina eximie decussatim pyramidatis, ulterius a priori stipulis expansis (0.5–2 mm latis) diversa.—MEXICO. Oaxaca: Paso Alicia–Guiengola–Las Tejas, near Tehuantepec, 2.IX.1969, (fl), *T. MacDougall s.n.*—Holotypus, NY.

In aspect quite like var. *pallida,* except for venation of lfts, at anthesis 2–3 m, the lfts usually pubescent beneath, sometimes glabrous above, rarely so on both faces except for costa ciliate beneath, the hairs up to 0.25–0.55 mm, the racemes mostly borne on leafless branchlets axillary to coeval lvs, in dry springs borne on brachyblasts furnished with greatly diminished foliage.

Stipules herbaceous narrowly oblanceolate or linear-lance-acuminate 3.5–12 × (0.5–)0.6–2 mm.

Major lvs 5.5–12 cm; petiole 9–20(–22) mm, at middle 0.4–0.8 mm diam, 2–3 times as long as first segment of lf-stalk; rachis (3–)3.5–6 cm; gland between proximal (exceptionally also second) pair of lfts stipitate, the stipe puberulent, the whole in profile 2–3.3 mm, the head 0.5–1 mm wide; lfts (6–)7–10 pairs accrescent upward, the distal pair obovate, oblong-obovate or oblanceolate 15–30 × 6–13 mm, 1.9–3.1 times as long as wide, the secondary venation of ±6–8 pairs of camptodrome veins fully immersed above, discolored but not raised beneath, tertiary venulation essentially 0.

Peduncles 7–17 mm; pedicels 10–18 mm; long sepals 5–7 mm; long petals 14–22 mm; filaments pilosulous or glabrous, those of 4 median stamens free 0.7–1.2 mm, of 3 abaxial ones 2–3.8 mm, the anthers of 4 median stamens 4–5 mm, the beak 1.3–2 mm; style 1–1.7 mm.

Stipe of pod 3–4 mm, the body 5–10.5 × 0.27–0.33(–0.35) cm, the interseminal septa 2.5–3 mm distant, the valves elevated over each seed as a sharply defined 4-armed pyramid; seeds 2.3–3.1 × 2.1–2.7 mm, the areole 0.4–0.6 × 0.3–0.4 mm.—Collections: 11.

Disturbed thorn-forest, brushy hillsides and pasture thickets, 5–600 m, locally plentiful on the Coastal Plain and foothills of the Pacific slope of Isthmus of Tehuantepec in s.-e. Oaxaca, e. sparingly to the source valleys of Río Grijalva on the Gulf slope in w. and n.-w. Chiapas.—Fl. X–V, sporadically later.

The var. *isthmica* is precariously distinguished from var. *pallida* by the simple pinnate venation of the leaflets, but is peculiar for the syndrome of dilated stipules, relatively small flowers and very narrow pods, characters that occur, it is true, elsewhere in var. *pallida* but nowhere else, it seems, so constantly associated. The variety resembles distantly allopatric var. *longirostrata* and var. *shreveana* in leaflet venation, but differs in the broad stipules and in the sharply decussate pyramidal boss that develops early over each seed, a feature characteristic of var. *pallida*. It differs from var. *longirostrata* in ecology and therefore in dispersal, var. *isthmica* being an element of the xeromorphic lowland scrub and thorn-forest of the Isthmus coastal plain and neighboring foothills there and in lowland valleys of Chiapas, whereas var. *longirostrata* is adapted to the cooler, primitively wooded uplands of central and southern Guatemala. The var. *quiedondilla,* another form of *S. pallida* with similar foliage, has a wider flat pod and large seeds with correspondingly large areole. A few flowering specimens from middle elevations in Chiapas appear, nevertheless, for the present ambiguous between the two.

177c. Senna pallida (Vahl) var. **geminiflora** Irwin & Barneby, var. nov., a var. *pallida* foliolis amplis tenuiter membranaceis utraque facie subaequaliter subtillimo venularum rete ornatis dissimilis. MEXICO. Colima: 13–15 km w. of Santiago, 100 m, 5.XII.1970 (fl, fr), *R. McVaugh 25011*.—Holotypus, MICH = NY Neg. *10491*.

Cassia geminiflora Colladon, Hist. Casses 103, t. 3. 1816.—"*Cassia* sp. nova. *Moc. et Sessé pl. mex. ined. ic.* Hab. in Mexico."—Lectotypus, *Colladon, pl. 3,* l.c.—Typotypi probabiles: *Pl. Novae Hispaniae a Sessé aliisque lectae 1194,* F! = NY Neg. *10492*; "*C. biflora*" *Sessé & Mocino s.n.,* BM! "*C. biflora,*" hb. S. & M. *1194,* MA! = F Neg. *44429*.

Weak shrubs with terete albescent hornotinous and very slender pliant annotinous branchlets, at anthesis 0.5–2 m, pilosulous throughout or the ample, thin-textured lfts pilosulous beneath only and sometimes there only along midrib, the 2-fld racemes all axillary to coeval lvs and shorter than them, rarely becoming paniculate distally late in season.

Stipules setiform or narrowly linear-attenuate 2–6.5 × 0.1–0.3 mm.

Major lvs 7–13 cm; petiole (12–)15–30 mm, ±1.5–2.5 times as long as first segment of rachis; rachis 2.5–7(–8) cm; gland 1 at or (adnate and apparently) above first pair of lfts 2–3.5 mm, the stipe often pubescent, the body 0.3–0.7 mm wide; lfts (3–)4–7 pairs, the distal ones elliptic, elliptic-oblanceolate or obovate, deltate- to triangular-acute or obtusely rounded at apex, (27–)30–44(–50) × (9–)10–19(–21) mm, 1.9–3.1 times as long as wide, the midrib with 7–11 pairs of major camptodrome and some intercalary secondary veins prominulous on both faces and giving rise on both to a finely prominulous 3- and 4-nary mesh of veinlets.

Peduncles 8–18(–27) mm; pedicels (12–)16–25 mm, subtended by usual glands; long sepals 7–10 mm; long petals 17–28 mm; androecium of var. *pallida,* the 4 median anthers (their filaments free or rarely united at base) 3.3–5 mm, the 3 abaxial ones (4.5–)5–7.5, the beak 1.8–2.5 mm; ovary white-strigulose; style 0.7–1.4 mm.

Stipe of pod 5–9 mm, the body 7–13.5 × 0.42–0.55(–0.6) cm, the valves scarcely expressed over the seeds, the interseminal septa 2–2.5 mm apart; seeds (of var. *pallida*) 2.7–3 × 1.7–2.6 mm, the areole ±0.5 mm diam.—Collections: 42.

Tall subdeciduous forest, wooded bottomland, canyon slopes, becoming abundant in regenerating brush-woodland, 30–600 m, coastal lowlands and foothills of Sierras Madre Occidental and Sur, apparently common in Jalisco and Colima, n. interruptedly to Río Mayo in s. Sonora and immediately adjoining Chihuahua, not recorded from coastal Michoacán but reappearing in s. Guerrero (mun. La Unión) and s.-centr. Oaxaca; in Jalisco ascending e. and up to 1200 m along Río Grande de Santiago to the barrancas of Guadalajara and the extreme s.-w. prong of Zacatecas; at its n. limit on Río Mayo and n. to the middle Bavispe valley in Sonora represented by populations variably intergradient into var. *shreveana.*—Fl. (IX–)X–IV.

This elegant variety differs from sympatric but probably ecologically differentiated var. *pallida* in the ample, thin-textured leaflets delicately and intricately reticulate on both faces and is the only form of *C. pallida* so characterized. The lax habit of the very slender pliant branchlets and the ample thin foliage might suggest that var. *geminiflora* consists merely of a selection of shade-forms of var. *pallida*; but we have many examples of mesomorphic var. *pallida* from Central America with leaflets as large and thin-textured as those of var. *geminiflora* which nevertheless retain the reticulation normal for their kind. The variety is strongly characterized at the middle of its range and southeastward, but at its northern limit on Ríos Mayo and Bavispe is represented by forms with leaf-reticulation increasingly faint and the leaves themselves subject to the seasonal dimorphism by means of which *S. pallida* sens. lat. is adapted to survive temporary droughts. Most of the collections from Alamos, Sonora, and vicinity are of this ambiguous type, seemingly passing into the yet more xeromorphic var. *shreveana* closely adjacent.

The variety is newly described here, the epithet being deliberately chosen to coincide with that of *Cassia geminiflora* Collad., in the belief that they are one and the same in a taxonomic sense. Colladon described his species from a drawing made in Mexico by an artist of the Nueva España expedition; we lectotypify it herein by the engraved reproduction in *Histoire des Casses*. A specimen that agrees very well with the picture in having 6–7 pairs of elliptic leaflets survives in the Sessé & Mociño herbarium and we suspect but cannot be absolutely certain that it is equivalent to a typotype. The specimen itself is labelled "*C. biflora*," but cannot be the *C. biflora* described by Sessé (Fl. Mex. 101; Pl. Nov. Hisp. 59) from Apatzingán which had 8–9 pairs of leaflets and which must have been, for geographical reasons, our var. *pallida,* although the leaflet number is not fully in harmony with what is known of var. *pallida* in the Balsas valley. We hazard a guess that Sessé's *C. biflora* (but not his specimen) may belong to var. *quiedondilla,* described from the same general region in Michoacán (near Ario de Rosales), but this is no more than speculation.

177d. Senna pallida (Vahl) var. **shreveana** Irwin & Barneby, var. nov., quoad foliolorum numerum ac legumen latum cum var. *pallida* necnon var. *geminiflora* congrua, ab ambabus foliolis simpliciter penninerviis nec reticulatis, a var. *palmeri,* quoad foliolorum venationem numerumque praesimili legumine angusto 4–5.5 (nec 6.5–9) mm lato diversa.—MEXICO. Sonora: 7 mi. n.-e. of Cajamé, 2.III.1933 (fl), *Forrest Shreve 6139.*—Holotpyus, NY; isotypi, F, MICH.

Shrubby or potentially arborescent, xeromorphic, glabrous or residually strigulose-pilosulous, with habit and stipules of var. *palmeri,* the firm pallid glaucescent foliage ample after rain, the 2-fld racemes then axillary and shorter than the subtending lf, in dry spring months flowering again from depauperately microphyllous brachyblasts arising from lvs already fallen, the fls then well exserted.

Primary lvs (absent from many specimens) 5.5–9 cm; petiole 12–24 mm, ±2.5–3 times as long as first interfoliolar segment of lf-stalk; rachis 2.5–5.5 cm; glands between proximal and rarely also the second pair of lfts 1–2.2 mm, the head 0.3–0.5 mm diam; lfts 4–7(–8) pairs accrescent upward, obovate obtuse to oblanceolate deltate-acute, the distal pair 13–27 × (4–)6–10 mm, mostly 1.8–3, rarely 4–4.5 times as long as wide, the secondary venation of 5–8 pairs of camptodrome (with rare and random faint connecting venules) visible on one or both faces; lvs of brachyblasts much smaller, 1–5.5 cm, the longest lfts up to 6–16 mm.

Peduncles (4–)6–26 mm; pedicels 8–25 mm; long sepals 5.5–8.5 mm; long petals 16–22 mm; anthers of 4 median stamens 3.5–4.8 × 1.4–1.8 mm, the beak of 3 abaxial ones 1.3–2.3 mm.

Stipe of pod 4–6 mm, the body (5–)6–11.5 × 0.4–0.55 cm; seeds 3–3.4 × 2.3–2.7 mm, the areole 0.4–0.7 × 0.4–0.55 mm.—Collections: 27.

Thickets and stony hillsides along draws and dry washes in the first foothills of Sierra Madre Occidental and adjoining coastal plain, commonly in drought-deciduous thorn-forest, below 600 m, locally abundant between 26° and 28°30′N in s.-w. Sonora and n.-w. Sinaloa; in the middle foothills to the east, on Ríos Bavispe and Mayo, passing freely into var. *geminiflora* and in lowland centr. Sinaloa into var. *pallida*.

The range of var. *shreveana* lies in the zone of transition between the Sonoran Desert proper and the thorn forest of the Sinaloan coastal plain. It seems likely to represent a recent offshoot of the otherwise intertropical var. *pallida* which has invaded a genuine desert climate and become somewhat modified in the process, the leaflets being here of thicker texture so as to reveal superficially only the costa and its pinnate secondary venation. It has adapted further, like var. *gaumeri* on the seasonally desiccated karst limestones of Yucatan, to an abundant flowering in the dry spring season, January–March, from microphyllous brachyblasts, as well as a conventional fall one, September to January, from axils of primary leaves. In the latter stage var. *shreveana* closely resembles south-Mexican var. *palmeri,* but this becomes readily distinguished in due course by the broad pod. In the middle foothills of Sonora var. *shreveana* passes gradually also into the more mesophytic var. *geminiflora,* which see for further comment.

177e. Senna pallida (Vahl) var. **palmeri** (Britton & Rose) Irwin & Barneby, stat. nov. *Peiranisia palmeri* Britton ex Britton & Rose, N. Amer. Fl. **23**(4): 261. 1930.—"Vicinity of Acapulco, Guerrero, October 1894 to March 1895, *E. Palmer 260.*"—Holotypus, US! isotypi, F (2 sheets), NY = NY Neg. *9396,* US (2 sheets)!—Non *Cassia palmeri* Wats., 1887.

Arborescent shrubs 1–5(–8) m, with terete whitish annotinous and virgate, deeply sulcate hornotinous branchlets, thinly strigulose distally but the lfts glabrous, of firm texture and subglaucescent, the 2-fld racemes at first axillary to coeval lvs, ultimately corymbose-paniculate and ± exserted.

Stipules linear-setiform 2–6 × 0.2–0.3 mm.

Major lvs 5.5–12(–61) cm; petiole 16–40 mm, ±2–3 times as long as first segment of rachis; rachis (2–)2.5–8(–10) cm bluntly carinate dorsally; gland between proximal pair stipitate 1.4–2.8 mm, the body 0.5–1 mm wide; lfts 4–8(–9) pairs, the distal pair narrowly or broadly obovate obtuse or deltately subacute 15–25(–30) × 7–12(–15) mm, 1.7–2.3 times as long as wide, simply penninerved, the 5–6 pairs of camptodrome secondary veins above fully immersed or almost so, beneath finely prominulous or merely discolored, subsequent venulation invisible.

Peduncles 1.2–2.8 cm; pedicels 1.6–2.7 cm; long sepals 6.5–9.5 mm; long petals 2.4–3.1 cm; androecium glabrous, the filaments free, those of 4 median stamens 1.3–1.8 mm, of 3 abaxial ones 3.5–5 mm; anthers of 4 median stamens (as var. *pallida*) 4–6 mm, of 3 abaxial ones 5–7 mm, the beak (1.8–)2.2–3.5 mm.

Stipe of pod 6–9 mm, the body 7–13(–15) × 0.65–0.8(–0.9) cm, the valves shallowly convex over seeds; seeds 3.1–3.7 × 1.8–3.1 mm, the areole 0.8–0.9 × 0.55–0.7 mm.—Collections: 9.

Brushy flats and hillsides, from near sea-level behind the beach up to 800 m in the foothills, apparently local along the Pacific slope of Sa. Madre del Sur in Michoacán (n. of Playa Azul) and Guerrero (between ríos Tecpan and Papagayo), and inland, in the hill-country w. of Iguala, Guerrero.—Fl. I–IV.

The var. *palmeri* is in some ways intermediate morphologically between sympatric lowland var. *pallida* and allopatric upland var. *quiedondilla,* combining the relatively few pairs of leaflets of the former with the deeply sulcate young branchlets, the simply venulose leaflets, the corymbose-paniculate inflorescence and the broad pod of the latter. Although the pod is much wider (6.5–9 not 2.5–5 mm) than that of var. *pallida,* the seeds are not or scarcely larger and bear on the faces the same small areole. The strong resemblance in most technical features between var. *palmeri* and the Peruvian var. *nemorosa,* which has seeds of a size to match their pod, is no doubt due to convergent evolution.

177f. Senna pallida (Vahl) var. **nemorosa** (Humboldt, Bonpland & Kunth) Irwin & Barneby, stat. nov. *C. nemorosa* Humboldt, Bonpland & Kunth, Nov. Gen. & Sp. **6**(folio): 353. 1824.—"Crescit in nemoribus Provinciae Jaen de Bracamoros [Cajamarca, Peru], alt. 238 hex."—Holotypus, labelled "n. 3662. *Querocotillo,*" P-HBK! isotypus, †B (hb. Kunth.) = F Neg. *1765.*

Weakly arborescent, at anthesis 1–2.5 m, sparsely pilosulous with subappressed hairs to 0.3–0.5 mm, the lfts glabrous on both faces, the inflorescence and general facies of var. *pallida*.

Stipules subulate-setiform 1.5–4 × ±0.2 mm.

Primary cauline lvs 3.5–9(–10) cm; petiole 5–15 mm; rachis 2–7 cm; gland sessile or short-stipitate, 2–2.6 × 0.5–0.9 mm; lfts (5–)6–9 pairs, oblong-elliptic or obovate, not or only slightly accrescent distally, the largest 8–18 × 5–9 mm, the pinnate venation finely prominulous above, reticular venulation 0.

Peduncles 1.2–3 cm; pedicels 15–20 mm; large sepals 6–8.5 mm; long petals 16–22 mm; 4 median anthers ±4 mm, with pronounced conic points at the truncate apex, the 3 abaxial ones 4.5–5 mm with beak 1.8–2.5 mm.

Stipe of pod 5–6 mm, the body 6–11 × 0.6–0.75 cm; seeds oblong or obovate in profile 3.9–4.5 × 1.8–2.5 mm, the areole 0.45–0.6 × 0.4–0.5 mm.—Collections: 10.

Xeromorphic scrub-thickets and algarrobal, 400–800 m, highly localized in the middle Marañón valley, lat. 5°35′–5°50′S, in Cajamarca (e. prov. Jaén) and adjoining Amazonas (prov. Bagua), Peru.—Fl. III, VI, X, perhaps intermittently through the year.

The var. *nemorosa* is greatly isolated geographically from the rest of its species, its compact range lying some 750 km distant from the southernmost outpost of var. *pallida* in southern Colombia. In general appearance and in details of the simply penniveined leaflets and the broad pod it very closely mimics the Mexican var. *palmeri,* from which it differs feebly in the shorter petioles, smaller corollas and larger seeds with a proportionately much smaller aerole on either face. Macbride (1943, p. 158), following Bentham (1871), included this endemic Peruvian senna in a comprehensive *Cassia biflora*. The specimens from Lima and Ancash, however, that Macbride refers to *C. biflora* actually represent the related *Senna incarnata*.

177g. Senna pallida (Vahl) var. **longirostrata** (Britton & Rose) Irwin & Barneby, stat. nov. *Peiranisia longirostrata* Britton & Rose, N. Amer. Fl. **23**(4): 264. 1930.—"Type from Solola, Guatemala, January 27, 1915, *E. W. D. Holway 134*."—Holotypus, US! clastotypus (fragm) + phototypus, NY!—*Cassia longirostrata* (Britton & Rose) Lundell, Phytologia **1**: 214. 1937.

Shrubs 1–3 m with deeply sulcate hornotinous branchlets, varying from glabrous below the always strigulose-pilosulous ovary to densely pilosulous up to the glabrous fl-buds, the vesture of stems and lf-stalks composed of stiff, spreading-ascending hairs up to 0.4–0.8 mm but the lfts either pubescent on both faces or only beneath, the mature inflorescence corymbosely or virgately paniculate.

Stipules linear-attenuate or narrowly linear-oblanceolate 5–15 × 0.4–1 mm.

Major lvs 6–13 cm; petiole 2–3.4(–4) cm, at middle (0.5–)0.6–1(–1.1) mm diam, (2–)2.3–5 times as long as first segment of rachis; gland between proximal pair of lfts sessile or stipitate (1–)1.4–3(–3.5) mm, the stipe when present either glabrous or pilosulous, the body 0.4–1 mm wide; lfts up to 6–9 pairs, accrescent upward along rachis, the distal pair largest, obovate, oblong-obovate or oblanceolate obtuse or deltately subacute 18–35 × (7–)8–14 mm, 1.8–3 times as long as wide, the simply pinnate venation of (4–)5–7(–8) pairs of camptodrome and some few faint intercalary secondary veins immersed and scarcely or not perceptible above, beneath either finely prominulous or immersed and discolored, tertiary venulation essentially 0.

Peduncles (4–)8–22 mm; racemes 2(–4)-fld, the axis 0–1.5(–4) mm; pedicels 10–

21 mm; long sepals (6–)6.5–9 mm; long petals (16–)18–21 mm; filaments glabrous or pilosulous, free to base, those of 4 median stamens 0.9–1.8 mm, those of 3 abaxial ones 3–5.5 mm; anthers of 4 median stamens 3.2–5.3 mm, bluntly truncate, the lateral beak not over 0.4 mm, dehiscent by parallel pores, those of 3 abaxial ones 5–7 mm, contracted into beak 1.6–2.8 mm; style 1.7–2.5 mm.

Stipe of pod 3–5 mm, the body 6.5–9.5 × 0.4–0.5 cm, the valves simply mounded over the seeds, the interseminal septa 1.5–2 mm apart; seeds not seen ripe, but the areole small, <1 mm diam.—Collections: 24.

Thickets and brushy hillsides in oak-pine, cypress- and mixed forest, (1200–) 1400–2000 m, scattered through the mountains of Guatemala from s. Huehuetenango s.-e. to Jalapa (±89°30′–91°30′).—Fl. XI–II.

Due to great superficial variation in pubescence and in outline of leaflets var. *longirostrata* is difficult to characterize neatly. It is closely related to vicariant and partly sympatric var. *quiedondilla* and resembles it in the relatively long petioles, but differs in the fewer leaflets accrescent upward along the rachis, the distal pair largest, and in the narrow pod and (presumably) smaller seed with smaller areole. Both leaflets and pods of var. *longirostrata* resemble those of lowland var. *pallida* except that the leaflets are simply penniveined, as in other highland varieties, and not reticulate; and the petiole is actually and relatively longer, 2–5, not ±1–2 times as long as the first segment of the rachis. The rarer var. *tuerckheimiana,* the third form of *S. pallida* found in the Guatemalan highlands at the same elevations and in the same latitudes, resembles pubescent var. *longirostrata* at anthesis, but has the broad pod, large seeds and numerous (10–13-jugate) leaflets of var. *quiedondilla* combined with the short petiole of var. *pallida.*

The species *Peiranisia longirostrata* was described by Britton & Rose from a fragmentary specimen and contrasted in *North American Flora* with *P. oxyadena, P. shaferi* and *P. robiniifolia,* all distantly allied. The report in the protologue from Veracruz was based on a specimen of pubescent var. *quiedondilla* (*Purpus 2342*), mentioned under the latter heading. Despite the epithet, the beak of the long anthers is not exceptionally long in the context of either *S. pallida,* sens. lat., or of the whole series *Interglandulosae.*

177g(bis). Senna pallida (Vahl) var. **brachyrrhachis** Irwin & Barneby, var. nov., omnibus cum var. *longirostrata* conveniens nisi foliis primariis abbreviatis 3–6.5 (nec 6–13) cm longis, rachi 1–3 tantum (nec 3–9) cm longo, foliolisque plerumque 4–7 (nec 6–9)-jugis, longioribus 9–17 (nec 18–35) mm longis.—MEXICO. Oaxaca, distr. Sola de Vega: en encinar-pinar, 1800 m, 19.X.1976 (fl), *M. Sousa et al. 6297.*—Holotypus (2 sheets) MEXU.

Shrubs or treelets ±2 m tall, sparsely minutely strigulose distally, the lfts ciliolate and their midrib puberulent beneath, otherwise glabrous, the annotinous branchlets pallid, striped lengthwise, the racemes mostly terminal to brachyblasts, forming a narrow leafy thyrse.

Stipules subulate setiform 2.5–5 × 0.15–0.3 mm.

Primary lvs 3–6 cm; petioles (1.2–)1.5–3 cm; rachis 1–2.3 cm, mostly shorter than petiole, the longer interfoliolar segments 2–5.5 mm; leaflets (3–)4–6 pairs, strongly accrescent distally, obovate or oblong-elliptic obtuse or emarginate, the distal pair 9–15 × 4–9 mm, ±2–2.5 times as long as wide, the margins plane, the midrib prominulous beneath, the 4–6 pairs of secondary venules discolored but immersed beneath; gland between proximal pair 2–3 × 0.2–0.5 mm.

Peduncles 9–19 mm, 2–3-fld; pedicels 10–13 mm; sepals glabrous dorsally, the broad inner ones 7–8 mm; longer petals 15–17 mm; filaments of long stamens pilosulous 4–5 mm, their anther, including tubular beak, 5.5–7 mm; ovules ±38.

Stipe of pod (not seen fully ripe) ±3 mm, the linear body 6.5–7 × 4 mm, the valves plane.—Collections: 7.

Pine-oak woodland, 1300–2150 m, locally plentiful in Sa. Madre del Sur in Oaxaca (distr. Sola de Vega) and in the highlands of Chiapas (mun. Tzimol, Totolapa, La Trinitaria, Las Margaritas and Amantenango del Valle).—Fl. VII–I.

This montane variety is closely akin to var. *longirostrata,* but differs in the abbreviated leaf-rachis and smaller, more crowded leaflets, the dimensions given in our diagnosis. Representative material of it was seen by us only when our manuscript was, as we thought, finished, but in time to intercalate the description next following the closely related Guatemalan var. *longirostrata*.

177h. Senna pallida (Vahl) var. **gaumeri** (Britton & Rose) Irwin & Barneby, stat. nov. *Peiranisia gaumeri* Britton & Rose, N. Amer. Fl. **23**(4): 292. 1930.—"Yucatán, *G. F. Gaumer 391.*"—Holotypus, collected at Izamal, date unknown, NY! = NY Neg. *9395*; isotypi, F, US, W!—*Cassia gaumeri* (Britton & Rose) Lundell, Carnegie Inst. Publ. **478**: 212. 1937.

Peiranisia yucatensis Britton & Rose, N. Amer. Fl. **23**(4): 263. 1930.—"Yucatan, 1917–1921, *C. F. Gaumer & Sons 24066.*"—Holoptypus, NY! = NY Neg. *9399*; isotypi, F, MA, US!—*Cassia yucatensis* (Britton & Rose) Lundell, Bull. Torr. Club **69**: 391. ("*yucatanensis*"). 1942.

Shrubs 1–3 m with stiffly ascending or virgate pallid terete annotinous and prominently ribbed hornotinous branchlets, seasonally dimorphic in aspect due to replacement during drought of longer lvs by smaller-lfd axillary brachyblasts, varying from glabrous below the strigulose ovary to densely pilosulous everywhere except for glabrous sepals with spreading or incurved-ascending hairs up to 0.2–0.6 mm, the foliage moderately bicolored, the lfts often distinctly resinous-dotted above, paler beneath, the 2-fld racemes borne on either inhibited or shortly developed but then leafy brachyblasts, surpassing the lvs of these but surpassed by the major cauline ones.

Stipules of primary lvs subulate-setiform 1.5–6 × 0.2–0.5 mm, those of brachyblasts shorter and more persistent.

Primary cauline lvs (absent from most fruiting and some flowering spms) 4–7(–10) cm; petiole (6–)9–16 mm, 0.4–0.7 mm diam, 1.2–2 times as long as first interfoliolar segment of lf-stalk; gland at proximal (exceptionally also at second) pair, stipitate or subsessile (1–)1.5–2.5 × 0.3–0.7 mm; lfts 4–6 pairs strongly accrescent upward, obovate or cuneate-obovate obtuse or deltately acute mucronulate, the distal (largest) pair 1–2(–2.8) × 0.6–1.1 cm, 1.7–2.5 times as long as wide, the midrib and 4–6 pairs of camptodrome (with random intercalary) secondary veins immersed and imperceptible above, faintly prominulous beneath, tertiary venulation not developed; lfts of brachyblasts mostly (2–)3–4 pairs, smaller.

Peduncles 4–16 mm; pedicels 8–18 mm; long sepals (4.5–)5–7 mm; long petals 14–18 mm; filaments glabrous free to base, those of 4 median stamens 1–1.5 mm, of 3 abaxial ones 2–4.7 mm, the anthers of 4 median stamens 2.8–4.5 mm, abruptly truncate, the very short lateral beak of all, or of at least the 2 longer ones opening by parallel slits, these sometimes confluent into a U-shaped slit in one or two of the shorter ones, the anthers of 3 abaxial stamens 3–6.5 mm, their beak 1.3–2 mm; style 0.9–1.3 mm.

Stipe of pod (3–)4–7 mm, the body 5.5–8.5 cm × (4.5–)5–6.5 mm, the septa 2.5–3 mm apart; seeds broadly obovate-suborbicular or paddle-shaped in lateral view, 2.9–3.5 × 2.4–3.5 mm, the areole 0.5–0.7 × 0.3–0.5 mm.—Collections: 16.

Thickets and glades in the drought-deciduous woodland, becoming abundant in disturbed scrub and brush, below 100 m, widespread over the n.-w. half of Yucatán Peninsula, from coastal Campeche (Chapotón) n.-e. to e. Yucatán (Tizimin) and n. Quintana Roo.—Fl. I–VI, then sparingly through the year.

Peiranisia yucatensis and *P. gaumeri,* here evaluated as more and less pubescent aspects of *S. pallida* var. *gaumeri,* were characterized by Britton & Rose (1930, p. 259, in key) as having only 3–4 pairs of leaflets not over 1 cm long. The typi of both species were collected in the drought phase and have all complete leaves and flowers issuing from axillary brachyblasts, the leaves therefore much smaller than the primary cauline ones, all shed except in part of *P. gaumeri,* where long but naked leaf-stalks with up to 6 pairs of scars remain to tell the tale. Even though they were originally segregated from *P. biflora* sensu Britton & Rose (which covered our *S. pallida* var. *pallida* and var. *bahamensis*) for the wrong reason, these xeromorphic populations from Yucatan Peninsula, of which we now have many examples of the more amply leafy, wet-season phase, seem genuinely to differ from var. *pallida* in the simpler venation of the leaflets, markedly so from the races of var. *pallida* native to Veracruz shortly to the west. In other respects, especially in the relatively short and wide pod elevated over the seeds as a simple mound or very obscure pyramid and in the resinous-dotted leaflets, var. *gaumeri* simulates var. *bahamensis,* although retaining the stout anthers of var. *pallida*. But the erratic dehiscence of some of these anthers by one U-shaped rather than two discrete parallel slits suggests a line of affinity leading toward *S. angustisiliqua* distantly allopatric on the Greater Antilles and in northern South America.

177i. Senna pallida (Vahl) var. **goldmaniana** Irwin & Barneby, var. nov., omnes inter speciei formas foliolis 2–4-jugis seminumque testa castanea dense minutim foveolata praestans.—MEXICO. Quintana Roo: La Vega [on coast opposite Isla Cancun near 21°06′N; cf. Goldman, Smithson. Miscell. Coll. **115**: 24, 242. 1951], 14–26.III.1901 (fl, fr), *E. A. Goldman 632*.—Holotypus, US = NY Neg. *10493*.

Slender shrubs, in habit and resinously dotted foliage resembling var. *bahamensis,* with terete branchlets, glabrous except for minutely puberulent ovary, the mature foliage pallid-glaucescent beneath, the 2-fld racemes borne on depauperately leafy branchlets axillary to coeval lvs.

Stipules subulate or setaceous 1–5 × 0.2–0.3 mm.

Primary cauline lvs 4.5–7.5 cm; petiole 12–20 × 0.5–0.6 mm, 1–1.5 times as long as first (or only) interfoliolar segment of lf-stalk; rachis 1–4.2 cm; gland between proximal pair, shortly stipitate, 1.1–2 mm, the head 0.4–0.8 mm diam; lfts 2–4 pairs, accrescent distally, the terminal pair broadly obovate obtuse 2–2.8 × 1–2 cm, 1.3–2 times as long as wide, the venation of 5–6 pairs of camptodrome secondary nerves weakly prominulous on both faces, reticular venulation 0; lvs of branchlets shorter, the lfts always 2-jugate.

Peduncles 12–17 mm; pedicels 1–2.2 cm; long sepals ±6 mm; long petals ±13–17 mm; filaments glabrous, of 4 median stamens ±1 mm, of 3 abaxial ones ±2.5 mm, the anthers of 4 median stamens abruptly truncate 4.5–4.8 × 1.5 mm, of 3 abaxial ones 5–5.5 mm, with beak ±1.5 mm; style 1.5 mm.

Stipe of pod 5–7 mm, the body 5.5–9 × 0.55–0.7 cm, the valves simply mounded over seeds; seed 3 × 1.7–2.8 mm, the castaneous testa densely finely pitted overall, the areole 0.7–0.9 × 0.6–0.7 mm.—Collections: 5.

Habitat not recorded, known only from the n.-e. shoulder of Yucatan Peninsula, Terr. Quintana Roo, Mexico.—Fl. II–IV, probably later.

A remarkable plant, clearly distinct from var. *gaumeri,* vicariant immediately to the east, in its few large leaflets, those of primary leaves two to four (not 4–6), those of leaves associated with flowering brachyblasts exactly two (not 3–4) pairs; and unique among known forms of *S. pallida* in its pitted seeds. The same characters distinguish it from var. *bahamensis,* which has in addition characteristically slender anthers; and from var. *pallida,* of which the nearest known stations lie some 800 km southwestward in Veracruz and 500 km southeastward on Swan Islands in the western Carribean. In general aspect var. *goldmaniana* is reminiscent of *S. holwayana* var. *absoidea* which has leaves similarly simplified down to two pairs of leaflets, but differs, inter alia, in the proportionately very long and slender petiole, 5–9 (not 1–1.5) times as long as the rachis proper.

177j. Senna pallida (Vahl) var. **bahamensis** Irwin & Barneby, var. nov., ab aliis speciei formis antheris angustis ±1–1.2 mm diam, mediis 4 rostro oblique terminali (nec laterali) dehiscentibus diversa.—BAHAMA ISLANDS. New Providence: near Nassau, 26.I.1903 (fl, fr), *A. H. Curtiss 50.*—Holotypus, NY; isotypus, US.

Slender shrubs 1.5–4 m with flexuous terete lustrous lenticellate annotinous branchlets, the hornotinous branchlets with lf-stalks and pallid lower face of strongly 2-colored lfts scattered with lustrous resinous dots, usually glabrous up to the thinly pilosulous (exceptionally also glabrous) ovary, rarely (Atwood Cay) pilosulous with fine incumbent hairs ±0.4 mm, the inflorescence axillary and incipiently paniculate, the (1–)2(–3)-fld racemes mostly borne on very slender, usually leafless branchlets axillary to coeval lvs, these (especially early in season) sometimes reduced to 1 or 2 pseudofasciculate peduncles.

Stipules linear-attenuate or falcately linear-oblanceolate, 3–6 × 0.2–1 mm.

Primary lvs 4.5–9(–11) cm; petiole 10–22 mm, 2–3 times as long as first interfoliolar segment of rachis; rachis (1.5–)2–5(–7) cm; gland between proximal pair 1.8–3 mm, the head 0.5–1.1 mm diam; lfts (3–)4–7(–8) pairs, accrescent upward, the distal pair obovate or broadly oblanceolate obtuse or emarginate, (1.6–)1.9–3.5 × (0.6–)0.8–1.5(–1.6) cm, 1.5–2.8 times as long as wide, the venation of (5–)6–8(–9) pairs of camptodrome (and often fainter intercalary) secondary veins either faintly raised beneath or on both faces or fully immersed, a fine tertiary venulation sometimes faintly discolored beneath but not prominent.

Peduncles 2–12(–15) mm; pedicels 11–18 mm; long sepals 5–8 mm; long petals 14–19(–21) mm; androecium glabrous, the filaments of 4 median stamens 1–2.2 mm, of 3 abaxial ones 2–3.5 mm, the anthers of median ones 3–4 mm, of 3 abaxial ones 3–4.5, the beak 1–1.4 mm.

Stipe of pod 4–7 mm, the body (5–)6–9.5(–10) × 0.4–0.55 cm, the valves simply raised over seeds, not pyramidally so, the interseminal septa 3–3.7 mm apart; seeds 2.5–3.2 × 2–2.7 mm, the areole 0.5–0.9 × 0.3–0.6 mm.—Collections: 37.

Coppice and beach scrub, locally abundant especially in disturbed and ruderal habitats, below 70 m, widespread in the Bahamas from New Providence and Eleuthera s.-e. to Samana Cay and Acklin I.; South Caicos I.; n.-e. and extreme e. Cuba in Camaguey (Cayos Romano and Paloma) and Oriente (Baracoa to Cabo Maisí).—Fl. (IX–)XI–II(–III).—Cultivated in U.S.A., Hawaii, and perhaps elsewhere.—*West Indian Gorse.*

The var. *bahamensis* is separated physically from other forms of its species by discontinuities, west to Yucatan, southwest to Honduras and south to Colombia, nowhere less than 900 km wide; and morphologically by the slender anthers gently contracted at apex rather than premorsely truncate. On Caicos Islands its range narrowly overlaps that of *S. angustisiliqua* var. *inaguensis* and on Cuba impinges on that of *S. robiniifolia,* but it is distinguished from both of these by the biporose dehiscence of the median anthers and by the relatively short and narrow pod. In general appearance and detail var. *bahamensis* is essentially monomorphic, relatively small-flowered by comparison with some continental *S. pallida,* and glabrous except for the exceptional population on Atwood Cay (*Wilson 7370,* F, NY) already mentioned in our discussion of the species.

Writing from Kew to Dr. Britton in 1924, N. E. Brown (NY, ms) mentioned a specimen of *Cassia biflora* preserved in the Dale Herbarium at BM, collected in the Bahamas before 1739, possibly the oldest specimen of *S. pallida* extant. Brown suggested that this might be conspecific with the plant seen by Linnaeus in Clifford's garden at Hartekamp in 1736 which was to become the typus of *Cassia biflora* L. Britton in his reply to Brown wrote that "in working out the Bahama Flora . . . I came to the conclusion that the Bahama plant would prove to be [*C. biflora*], more by inspiration perhaps than for any other reason." There seems to be no substantial connection, however, between the two.

Old specimens collected by Macfadyen (K), presumably on Jamaica, apparently represent this variety. We have no modern evidence of its occurrence on the island.

177k. Senna pallida (Vahl) var. **quiedondilla** (Micheli) Irwin & Barneby, stat. nov. *Cassia quiedondilla* Micheli, Mém. Soc. Phys. Sci. Nat. Genève **34**(3): 272, pl. 19. 1903.—"[MEXICO. Michoacán:] . . . Los Frenos [properly Los Fresnos], 1850 m, mars 1898, [*Langlassé*] no. 52."—Holotypus, G! isotypi, G, GH, MEXU, NY (fragm), US (=NY Neg. *9088*)!—*Peiranisia quiedondilla* (Micheli) Britton & Rose, N. Amer. Fl. **23**(4): 262. 1930.

Cassia tapahuascana A. Molina, Ceiba **18**(1–2): 98. 1874.—"Honduras: Dept. El Paraíso: . . . Quebrada Honda, Montaña Tapahuasca, entre San Lucas y Morolica . . . *Molina* 14234; . . . pine forest of Tapahuasca, between Manzaragua and San Lucas, alt. 1400 m., Nov. 24, 1966, *Molina* 18734 . . ."—Holotypus, designated by Molina, l.c., *Molina 18734,* F! = NY Neg. *9684*; isotypus (not seen), EAP; paratypus, *Molina 14234,* F (2 sheets)!

Weakly arborescent shrubs at anthesis commonly 1–4.5 m, in undisturbed forest sometimes becoming slender trees up to 7(–10) m, with terete annotinous and older branches and usually deeply sulcate hornotinous branchlets, varying from quite glabrous below the ovary to thinly pilosulous except for always glabrous sepals, the lfts commonly quite glabrous on both faces or glabrous except for a tuft of hairs dorsally in the anterior basal angle of midrib, sometimes pilosulous over whole lower (and exceptionally also upper) face, the inflorescence of 2(–4)-fld racemes either leafy-bracteate and formed of either single racemes or of 2 (several) together on condensed brachyblasts axillary to coeval lvs, or more commonly, at least at maturity, corymbose- or thyrsoid-paniculate and ± exserted.

Stipules linear-attenuate (2.5–)4–7.5(–10) × 0.2–0.5(–0.7) mm, rarely expanded and up to 1 mm wide.

Major lvs 5.5–11(–12) cm, those of branchlets (especially late in season, not further described) shorter and simpler; petioles (12–)15–30 mm, 3–5 times as long as first segment of rachis; rachis 4–8(–9.5) cm; gland between proximal pair 1.5–3(–4.2) × 0.4–1 mm; petiolules 0.6–1.8 mm; lfts (8–)9–15(–17) pairs [caveat:

of summer-flowering branchlets often fewer!], the distal pair no longer or a trifle shorter than their neighbors, in outline oblong, oblong- or obovate-oblanceolate obtuse or deltately subacute, 11–26(–36) × 3–10(–11) mm, (1.8–)2–4 times as long as wide, the simply pinnate venation of midrib and 5–9(–11) pairs of major camptodrome secondary veins immersed and usually imperceptible above, visible but scarcely or not prominulous beneath.

Peduncles (8–)10–26(–35) mm; racemes 2(–4)-fld; pedicels 10–16(–20) mm; long sepals (6–)7–11(–12) mm; long petals (16–)18–29(–32) mm; androecium usually glabrous, the filaments of 4 median stamens either free or partly united 1–2.2 mm, those of 3 abaxial ones free 2–3.8 mm, the anthers of 4 median stamens 3.8–5.5 mm, conic-pointed distally, the body of 3 abaxial ones (3–)3.6–7(–8) mm, the beak (1.2–)1.8–3.6 mm; style 1.2–2 mm.

Stipe of pod 4–7 mm, the body 5–10 × (0.5–)0.6–0.75(–1) cm, the valves simply mounded over the seeds; seeds either ovate or oblong in outline, 3–4.5(–6) × 2–2.5 mm, the testa brown-olivaceous, the ovate-elliptic or (in longer seeds) linear-oblong-elliptic areole (0.9–)1.2–2.4(–3.5) × 0.4–0.7 mm.—Collections: 63.

Glades, clearings and margins of oak-, pine- and mixed deciduous upland forest, persisting and sometimes locally abundant in regenerating brush, 900–2100(–2300) m and occasionally lower (in Honduras down to 600, in Chiapas to 700 m), common along and near the Continental Divide from s. Mexico (highlands of Chiapas) through Guatemala into w. El Salvador (Ahuachapán, Sta. Ana, Sonsonate), thence e., becoming rarer and scattered, through w.-centr. Honduras (Cortés, Distrito Central, El Paraíso) into Nicaragua (Jinotega); unknown from Oaxaca, but reappearing in scattered stations around the periphery of the Balsas Depression in s. Mexico, on and near the summit of Sa. Madre del Sur in Guerrero and the s. slope of Cordillera Neovolcánica in Michoacán.—Fl. X–III, occasionally in summer.

As defined by the forcgoing description, var. *quiedondilla* is the commonest and most widely dispersed representative of the *S. pallida* complex in the relatively moist and cool, primitively wooded uplands of southern Pacific Mexico and northwestern Central America, a zone occupied in Costa Rica by the closely related var. *cordillerae,* different in the glandiform seta terminating the leafstalk, and shared locally, in Guatemala only, with var. *longirostrata* and var. *foliolosa*. The two last-named resemble var. *quiedondilla* in their penniveined, not reticulate leaflets and are collectively distinguished thereby from var. *pallida* which, largely vicariant in the tierra caliente at the same latitudes but ascending occasionally, perhaps only as a weed, into the uplands, differs further in its short petiole and from all but var. *longirostrata* in the strongly accrescent distal leaflets. Common characters of the highland varieties are a broad pod and a relatively large seed, large areole, or both. The var. *quiedondilla* differs from var. *longirostrata* in the more numerous, less strongly graduated leaflets and from var. *foliolosa* in its less numerous (9–17, not 17–31 pairs) of larger ones.

The differences readily apparent between specimens of var. *quiedondilla,* though very likely in part genetic, are certainly often merely edaphic or seasonal. Long primary cauline leaves that at early anthesis may subtend single 2-flowered racemes, or leafless branchlets sometimes contracted into a brachyblast bearing a fascicle of 2–3 racemes, give way distally, early or late in the season, to much shorter and simpler ones, these finally reduced to bracts of an exserted, loosely or compactly corymbose or thyrsoid panicle. Much fruiting material, indeed, lacks the long leaves entirely, as do leafy branchlets of the second order of ramification leading to the panicle; such specimens in fruit may be recognized by their relatively broad flat pod and by the individual seed. Transition between a leafy thyrse and a naked panicle of racemes is very gradual; both types occur in the same intermediate area but are seldom seen in any one collection from a given place. The typus of *Cassia quiedondilla* does, however, well illustrate the coincidence of two inflorescence types, the terminal panicle as illustrated in the protologue and the series of brachyblasts subtended by long leaves of mature branchlets illustrated by the isotypus at US. Phenetically identical branches have been collected in Chiapas and Huehuetenango; of the virgate leafy type cf. *Ton 1865* (NY) and *Holdridge 2284* (US), of the exserted paniculate one numerous examples at NY, F, and US.

In Sierra de Cuchumatanes in Quiché and Huehuetenango, Guatemala and in highland central Chiapas the beak of one, two, three or exceptionally of all four median anthers of var. *quiedondilla* dehisce by a U-shaped slit, exactly in the style of *S. angustisiliqua*. The plants so characterized are otherwise identical with sympatric ones with biporose median anthers and we regard this anomalous dehiscence as taxonomically vexatious, because it destroys the technical distinction we have drawn between *S. pallida* and *S. angustisiliqua,* but not taxonomically significant. A notable plant, ostensibly from Veracruz (Zacuapan, *Purpus 2342,* F, NY, but not the sheet so numbered at US) appears to represent a densely pubescent state of var. *quiedondilla* not otherwise exactly matched; but there being some question of its provenance and the correctness of the label we have omitted Veracruz from our statment of dispersal pending modern verification of var. *quiedondilla* on the Gulf slope in Mexico.

The pollination of var. *quiedondilla* by bees, as seen in the mountains of Guerrero between Tixtla and Chilapa, is described in detail by Buchmann (Bull. S. Calif. Acad. Sci. **73**(3): 171, fig. 1. 1974).

177l. Senna pallida (Vahl) var. **tuerckheimiana** Irwin & Barneby, var. nov., cum var. *quiedondilla* necnon var. *longirostrata,* ambabus in montanis guatemalensibus incolis foliolisque penninerviis (nec reticulatis) congruis, comparanda, ab ea, quoad foliola numerosa (10–13-juga usque) et legumen latum (7–8 mm) simili, petiolo abbreviato 6–13 (nec 13–30) mm longo, hac eadem notula et insuper legumine lato a var. *longirostrata* diversa.—GUATEMALA. Baja Verapaz: Santa Rosa, IX.1888 (fl), *H. von Tuerckheim in Donnell Smith 1422.*—Holotypus, US; isotypi, MO, NY, US.

Weak, sometimes subsarmentose shrubs 1.5–3 m, densely pilosulous with sordid or lutescent hairs to 0.4–0.6 mm, the leaflets densely pilosulous beneath, at least puberulent above, the 2-fld racemes both axillary and forming a small terminal panicle.

Stipules erect setiform 2–6 × 0.2–0.3 mm.

Major cauline lvs 10–15 cm (those near and in the inflorescence, not further described, smaller and simpler); petiole 6–13(–15) mm, at middle 0.6–1.3 mm, ±1.2–1.8(–2.1) times longer than first segment of rachis; rachis 7–11 cm; gland between proximal pair 2–2.6 mm, sessile or stipitate, the stipe pilosulous, the body strongly compressed, in profile 0.7–1.2 mm wide; lfts up to 10–13 pairs, not or scarcely accrescent upward, in outline elliptic, elliptic-oblanceolate or oblanceolate, obtuse-mucronulate or deltately acute (1.8–)2.2–3.7 × 0.7–1 mm, (2.2–)3–4.2 times as long as wide, the margins strongly revolute, the venation of 5–8(–9) pairs of camptodrome and some faint intercalary secondary veins either immersed or prominulous above, finely prominulous beneath, tertiary venulation imperceptible.

Peduncles 18–36 mm; pedicels 15–19 mm; long sepals 11–13 mm; long petals

22–32 mm; filaments glabrous or pilosulous, of 4 median stamens commonly united at base, 1.5–2.1 mm, of 3 abaxial ones free, 3.5–5 mm, the anthers of 4 median stamens 4.5–5.5 mm, sharply truncate, the lateral beak 0.3–0.5 mm, dehiscent by parallel slits, those of 3 abaxial ones 7–8.5 mm, the tubular beak 2.5–3.8 mm; style 1.5–1.8 mm.

Stipe of pod 5–7 mm, the body 8–13 × 0.7–0.8 cm, the interseminal septa 2.5–3.5 mm apart; seeds (few seen) obovate ±4 × 2.8 mm, the areole 1.8–2 × 1.1 mm.—Collections: 7.

Cool humid forest, 1300–1700 m, local, known only from the highlands of Quezaltenango (Volcanes Zunil and Santa María) and Santa Rosa (Sta. Rosa; Salamá), Guatemala.—Fl. IX–I.

This variety is closely related to var. *quiedondilla,* with which it has in common relatively many, simply penniveined leaflets (in larger leaves up to 10–13 pairs) not accrescent upward along the rachis, and a broad, large-seeded pod, but is readily distinguished by the short petioles, only 1–2, not 3–5 times longer than the first interfoliolar segment of the leaf-stalk. The dense yellowish or sordid pilose pubescence seems further characteristic, but in the context of ser. *Interglandulosae* it would be imprudent to put diagnostic emphasis on this feature. The also related var. *longirostrata* is similar in ecology and in the simply veined leaflets, but these are fewer (up to 6–9 pairs) and strongly accrescent distally, and the narrow, small-seeded pod is only 4–5.2 (not 7–8) mm wide.

Donnell Smith distributed the type-collection of var. *tuerckheimiana* as *Cassia biflora,* but later determined the holotypus (US) as *C. viciifolia.* The latter (distantly allopatric in Venezuela) is superficially similar in ample foliage, large flowers, and width of pod, but the median anthers dehisce by one U-shaped slit and the pod becomes fragile when ripe, breaking along transverse lines of interseminal septa.

177m. Senna pallida (Vahl) var. **triquetripes** Irwin & Barneby, var. nov., foliolis numerosis (plerumque 10–15-jugis) subtillime penniveniis (nec dorso reticulatis) cum var. *quiedondilla* proxime congrua, ab ea foliorum petiolo breviori rachique dorso acute carinato de latere alis ad pulvinulos constrictis ornato ideo triquetro, ulterius verosimiliter seminis areola majori diversa.—MEXICO. Guerrero: distr. Montes de Oca [=mun. La Unión], San Antonio Buenos Aires, 20.X.1937 (fl), *G. B. Hinton 11512.*—Holotypus, NY; isotypi, K, MICH, US.

Arborescent shrubs 3–5 m, with very slender, angulate-sulcate hornotinous branchlets, appearing glabrous but the young stems and lf-stalks minutely puberulent, the strongly bicolored lfts either glabrous or bearing a tuft of hairs dorsally at the anterior basal angle of midrib, the 2-fld racemes arising singly from axils of primary coeval lvs and shorter than them or some ultimately forming a small leafless terminal exserted panicle.

Stipules linear-setiform or -subulate 1.5–3.5 × 0.15–0.35 mm.

Major lvs 7–13.4 cm; petiole (3.5–)6–15(–18) mm, at middle 0.4–1 mm diam, 2–3(–5) times as long as first interfoliolar segment of rachis; rachis 5–9.5 cm, sharply keeled dorsally and relatively broadly winged laterally, the wings constricted at insertion of lfts; gland between proximal pair 1.5–3 × 0.5–1 mm; lfts 10–15 pairs, not (except in some smaller distal lvs) accrescent upward, in outline oblong-elliptic obtuse or deltately acute, conspicuously mucronulate, the longer ones (16–)18–24(–27) × 4.5–8(–9) mm, 3–4 times as long as wide, the delicate pinnate secondary venation of ±5–7 pairs of camptodrome veins fully immersed or rarely obscurely prominulous above, faintly discolored but not prominulous beneath, the tertiary venulation, when visible, faint and random.

Peduncles 11–27 mm; pedicels 16–25 mm; long sepals 6.5–9.5 mm; long petals 23–36 mm; filaments of 4 median stamens (glabrous or puberulent, free or basally confluent) 1.2–1.7 mm, of 3 abaxial ones 2.5–5.5 mm, the anthers of 4 median stamens 4–5.2 mm, of 3 abaxial ones 6.5–8, the beak 2.6–3 mm; style 1–1.4 mm.

Stipe of pod ±6 mm, the body 13–19 × 0.5–0.6 cm, resembling that of var. *lemniscata* but not twisted; seeds (few seen) ±4–5 × 3 mm, the areole 2.7–3.5 × 1.5 mm.—Collections: 6.

Cool oak-pine and alder woodland, 1000–1550 m, apparently local along crest of Sa. Madre del Sur in Guerrero (mun. La Unión, Ajuchitlán) and s. Oaxaca (distr. Pochutla, Juquila), Mexico.—Fl. X–II.

This still imperfectly known variety is sympatric in Guerrero with var. *quiedondilla* and resembles it in number and venulation of the leaflets; but clearly differs in the short petiole and the transversely dilated rachis of the leaf-stalk. If we are not mistaken in associating with Hinton's two flowering collections (nos. *11512, 14915*) from Guerrero two in fruit from Oaxaca (*M. Sousa 5586,* MEXU, from Pochutla; *Reko 3699,* US, from an unidentified locality), the pod tends to be narrower than that of var. *quiedondilla* and the areole of the seeds larger, but these characters require confirmation from more complete material.

177n. Senna pallida (Vahl) var. **foliolosa** (Bentham) Irwin & Barneby, stat. nov. *Cassia foliolosa* Bentham, Trans. Linn. Soc. London **27**: 544. 1871.—"Guatemala, *Skinner in herb. Lindley.*"—Holotypus, *Skinner 18,* CGE! = NY Neg. *9691.—Peiranisia foliolosa* (Bentham) Britton & Rose, N. Amer. Fl. **23**(4): 265. 1930.

Weak shrubs at anthesis 1–3 m with subterete annotinous and strongly sulcate, virgate or weakly plagiotropic annotinous branchlets, appearing glabrous but the stems and lf-stalks commonly minutely strigulose-pilosulous, the many small crowded lfts glabrous at least on both faces, the inflorescence axillary or corymbose and exserted.

Stipules linear-attenuate 2–5.5 × 0.2–0.25 mm.

Major lvs 5–14 cm; petiole (7–)10–18 mm, 3–6 times as long as first segment of rachis; rachis 3.5–11.5 cm; gland between proximal pair stipitate 1.6–3 mm, the body 0.3–0.6 mm diam, rarely a smaller gland at second pair; petiolules 0.2–0.5 mm; lfts (12–)15–31 pairs, not accrescent toward summit of rachis, narrowly lance-oblong, oblong, oblong-obovate or -elliptic obtuse or deltately acute 6–15 × 1.7–3.7 mm, 3–5 times as long as wide, the venation invisible above, the midrib prominulous beneath but the simple pinnate secondary venation of 4–7 pairs of nerves either immersed and faintly colored or altogether imperceptible.

Inflorescence and flower of var. *quiedondilla*; style 1.5–2.5 mm.

Pod (little known) like that of var. *quiedondilla,* the stipe 4–7 mm, the body 6.5–8.5 × 0.5–0.6 cm, the valves plane or nearly so; seeds 2.7–3.5 × 1.7–2 mm, the areole 1.1–1.2 × 0.5–0.7 mm.—Collections: 10.

Quebradas and rocky hillsides in seasonally dry, often semideciduous brush-woodland, 900–1800 m, local in the highlands of s.-e. Guatemala (deptos. Zacapa, Chiquimula, Guatemala, Jalapa, Jutiapa and Sta. Rosa), to be expected in adjoining El Salvador.—Fl. X–II, perhaps also in summer.

In the protologue of *C. foliolosa,* which was described from a single branchlet in young flower from an unrecorded locality in Guatemala, Bentham remarked that it had precisely the inflorescence of his *C. biflora* (=our *S. pallida*, p. max. p.) but leaflets much more numerous, smaller, and more simply veined, thereby suggesting some microphyllous form of *S. multijuga*. We now have matching material with fruit and seed which further emphasize the similarity to *S. pallida* var. *quiedondilla*, of a type unknown to Bentham, with essentially similar even though less numerous leaflets. Leaflets of holotypic *Cassia foliolosa* vary from 12 to 21 pairs, those of other, more modern collections 17 to 31 pairs; and the number rises in some var. *quiedondilla* to 17 pairs. The syndrome of diminished surface and augmented number of leaflets is a common modification of the senna leaf, the loss of photosynthetic area being compensated by reduplication, and we have no hestitation, in the absence of supporting differential characters, in reducing *C. foliolosa* to varietal status.

177o. Senna pallida (Vahl) var. **cordillerae** Irwin & Barneby, var. nov., a var. *pallida* late per Americae Centralis regionem inferiorem dispersa seta petiolari in glandulam transformata, foliolorum venatione simpliciter subtillime penninervia (nec dorso reticulata), legumine latiori 5–6.5 (nec 3–4.5) mm lato, seminumque areola 1.8–2.2 (nec 0.5–0.8) mm diam valde diversa; differunt varietates mexicanae seta glanduliformi praeditae *lemniscata* legumine longissimo (17–26 cm), *macdougalliana* foliolis majoribus 3–6 (nec 2–4) cm usque longis utrinque nec dorso tantum glaucescenti-pallidis, et *trichocraspedon* foliolis margine eximie revolutis necnon utraque facie prominule venulosis.—COSTA RICA. Alajuela: buissons au bord du río Virilla, XII.1895 (fl, fr), *A. Tonduz 9824* distrib. *Donnell Smith 7007*.—Holotypus, US; isotypi, F, K, NY, US.

Cassia pallidior Rose, J. Wash. Acad. Sci. **17**: 167. 1927, pro parte, quoad pl. costaricens, cit., exclus. typ.

Amply leafy arborescent shrubs 1.5–5 m with pliant, bluntly angulate, commonly plagiotropic hornotinous branchlets, the thin-textured foliage strongly bicolored, the lfts dull dark green and glabrous above, pallid-glaucescent and at least along costa thinly strigulose beneath, ciliolate, the inflorescence of 2-fld racemes axillary to coeval lvs and shorter than them, rarely 2 together on a short leafless axillary branchlet.

Stipules (caducous, little known) falcately linear-acuminate 4.5–8 × 0.5–1.5 mm.

Major lvs 8–16(–18) cm; petiole 10–20(–24) × 0.6–1.1 mm, a little more or less than twice as long as first interfoliolar segment of rachis; rachis 6–11 cm, the terminal seta (caveat: sometimes deciduous, often eaten) transformed into an ellipsoid gland; petiolar gland 1 between proximal pair, sessile or shortly stipitate, 1.7–2.5 × 0.6–0.9 mm; lfts (8–)9–12(–13) pairs, oblong-elliptic or -oblanceolate obtuse mucronate, the larger pairs (18–)20–42 × 6–14 mm, 2.5–3.3 times as long as wide, the simply pinnate venation of (5–)6–8 pairs of major camptodrome secondaries immersed or almost so above, delicately raised or merely colored beneath.

Peduncles (8–)10–21 mm; pedicels 17–23 mm, in fruit bent well above middle; long sepals 6.5–8.5 mm; long petals 21–30 mm; filaments glabrous or puberulent, all free, those of 4 median stamens 1.2–2, of 3 abaxial ones 2.7–4.5 mm; anthers of 4 median stamens 4–5.2 mm, obliquely truncate, each tubule conic-pointed distally, the very short declined beak lateral, those of 3 abaxial ones strongly incurved, the body 6.5–7.5 mm, the erect beak 2.6–3.1 mm.

Stipe of pod 5.5–8.5 mm, the body (8–)11–17 × 0.5–0.65 cm, the papery nigrescent valves continuous in age, simply expressed over seeds, the interseminal septa 2.5–4 mm distant; seeds compressed-obovoid 3.2–3.5 × 1.7–2 mm, the testa brown sublustrous, the areole 1.8–2.2 × 0.8–1.2 mm.—Collections: 20.

Margins and openings in seasonally moist forest, persisting in pasture thickets, along hedges and on stream banks, 850–1525 m, locally abundant on both slopes of Cordillera Central in San José and Alajuela, Costa Rica.—Fl. VIII–II, sporadically through the year.

In the flora of Costa Rica there are two abruptly distinct biflorous *Interglandulosae,* one in the seasonally dry lowlands of Guanacaste and Puntarenas, the other in the central highlands of Alajuela and San José. The former has mostly 4–7, rarely 9 pairs of leaflets reticulate beneath and a pod 3–4.5 mm wide sharply raised over the seeds in decussate pyramids; it is the widespread var. *pallida,* of which the seeds have an areole less than 1 mm diam. The highland senna described above as var. *cordillerae* has mostly 9–12 pairs of simply penniveined leaflets, pods 5–6.5 mm wide simply mounded over the seeds, and an areole about 2 mm long. It differs further from other highland Central American varieties of *S. pallida* (all allopatric northward) in the glandiform seta between the distal pair of leaflets. In this last feature it resembles the Mexican vars. *lemniscata, macdougalliana* and *trichocraspedon,* of which the diagnostic characters are brought out under the next three varieties following.

The protologue of *Cassia pallidior* Rose, even though the peculiar seta went unnoticed, is based largely on material of var. *cordillerae,* but the specified typus represents glabrate var. *pallida.*

177p. Senna pallida (Vahl) var. **lemniscata** Irwin & Barneby, var. nov., inter speciei varietates omnes hucusque notas legumine longissimo simul ac angustissimo 17–26 × 0.35–0.42 cm insignis, his notulis a sympatrica var. *pallida,* ulterius foliolis dorso reticulatim nervulosis diversa, statim distinguenda.—MEXICO, Veracruz: 10 km e. de Cedillo, camino a la laguna, Hidalgotitlan, 16.III.1974 (fr), *Brigada Dorantes 2581.*—Holotypus, F = NY Neg. *10490*; isotypus, MEXU.

Weak shrubs with slender flexuous subterete branchlets, at anthesis 1–4 m, glabrous below the puberulent ovary, the ample foliage dark green above, pallid-glaucescent beneath, the 2-fld racemes mostly axillary to and much shorter than the coeval lvs.

Stipules (little known) linear-lanceolate 3–4 × 0.2–0.4 mm.

Major lvs 9–14.5 cm; petiole 13–19 mm, ± twice as long as first segment of rachis; rachis 6–9.5 cm, bluntly keeled dorsally; gland between proximal pair of lfts, shortly stipitate, 1.6–2.5 mm, the body 0.5–0.9 mm wide; lfts 9–12 pairs, the longest (often not the distal pair) elliptic-oblong or -oblanceolate obtuse or emarginate 22–32 × 6–10 mm, ±3–3.5 times as long as wide, delicately penninerved, the 5–8 pairs of fine camptodrome secondary nerves immersed above, faintly prominulous beneath, tertiary venulation imperceptible.

Peduncles 13–23 mm; pedicels 15–18 mm; sepals dorsally resin-dotted where exposed in bud, the long innermost ones 7.5–10 mm; long petals 27–30 mm; androecium glabrous, the filaments free, of median stamens ±1.2 mm, of 3 abaxial ones 3–3.5 mm; anthers of 4 median stamens 4.5–5 mm, truncate at apex but the angles rounded, body of 3 abaxial ones 4.5–5 mm, the beak 1.6–1.9 mm; style ±2 mm.

Stipe of pod 4–5 m, the body narrowly linear ribbon-like, straight, decurved or contorted, 17–26 × 0.35–0.42 cm, the lustrous reddish, delicately venulose valves closely fitting over seeds, the sutures obscurely strangulated at the intraseminal septa, these 4–5 mm apart; seeds dull olivaceous, like those of var. *pallida* 3.3–3.4 × 2.2–2.4 mm, the areole 1.1 × 0.6–0.7 mm.—Collections: 9.

Savanna thickets and regenerating or disturbed, tall evergreen humid forest of the Gulf Coastal Plain and first foothills to the w., apparently local between ríos Blanco and Coatzacoalcos in s. Veracruz (mun. Zongolica and Hidalgotitlan), adjoining Oaxaca (mun. Tuxtepec, n.-w. Juchitán) and the s.-e. corner of Puebla (Bosque de Ajenjibre).—Fl. X–II.—*Zanca de tordo*.

This variety is clearly and obviously distinct from the var. *pallida* common on the Gulf Coastal Plain in Veracruz, different in the glandiform seta terminating the leafstalk, in the simple dorsal venation of the leaflets, in the resinous outer sepals, and especially in the very long and narrow pod (±17–26, not 6–15 cm long), unknown elsewhere in the species. The number of leaflets (9–12 pairs in longer leaves) is higher than average for var. *pallida* in this region, but not reliably diagnostic. The modified seta, the ample thin-textured foliage and the general habit of var. *lemniscata* forcibly recall the Costa Rican var. *cordillerae*, which differs, of course, in the more ordinary dimensions of the pod and less obviously in the large areole (±2, not ±1 mm diam) on the faces of the seed. The epithet refers to the often contorted pods, fancifully likened to the ribbons of a Roman wreath.

177q. Senna pallida (Vahl) var. **macdougalliana** Irwin & Barneby, var. nov., adspectu toto racheosque seta in glandulam fusiformem transformata var. *trichocraspedon* simulans, ab ea imprimis foliolis planis superne impresse penniveniis diversa.—MEXICO. Oaxaca: Rancho "El Belén" a 21 km al n. de Pochutla, 310 m, 15.II.1976 (fl, fr jun), *Mario Sousa et al. 5363*.—Holotypus, US = NY Neg. *10496*; isotypus, MEXU.—Non *Cassia macdougaliana* Rose (1909).

Treelets to 3.5 m with flexuous subterete hornotinous branchlets, glabrous or almost so up to the ovary, the ample blue-green lfts completely so, the 2-fld racemes borne solitary or commonly 2–3 together on leafless nodiform brachyblasts axillary to coeval lvs, the large fls immersed in foliage.

Stipules (little known), lance-subulate 2–2.5 × 0.5 mm.

Major lvs 12–17 cm; petiole (15–)17–32 mm, at middle 0.7–1.2 mm diam, rounded dorsally, 1.8–2.3 times as long as first interfoliolar segment of rachis; rachis 7–11.5 cm; petiolar gland between proximal pair subsessile, broadly oblanceolate in profile, 2–2.8 × 0.8–1.5 mm; lfts up to 7–10 pairs, accrescent upward along rachis, in outline elliptic obtuse to deltately subacute, the distal pair 32–60 × 12–21 mm, 2.3–3 times as long as wide, the venation of 7–10 very slender pairs of camptodrome secondary veins immersed above, faintly raised beneath, a weak irregular tertiary venulation discolored but immersed.

Peduncles 10–29 mm; pedicels 18–32 mm; sepals apparently early reflexed, the long inner ones 8.5–10 mm; long petals 26–31 mm; androecium glabrous, the filaments of 4 median stamens either free or confluent at base 1.4–1.8 mm, those of 3 abaxial ones 4–5.5 mm, the anthers of 4 median stamens 4.5–5.5 mm, those of 3 abaxial ones 6–9 mm, the beak 2–2.5 mm; style 1–1.5 mm.

Stipe of pod (poorly known) 7–9 mm, the body ±11–12 × 0.7 cm; seeds unknown except as withered ovules but the areole early becoming ±2 mm long.—Collections: 5.

Habitat not recorded, but to be expected in the oak belt, 300–1200 m, apparently very local, known only from the Pacific slope of Sa. Madre del Sur in s. Oaxaca (Juquila; n. of Pochutla; Candelaria Loxicha).—Fl. II–III. *Flor de Nixtamal* (Sousa).

In the diagnosis we have compared this handsome floriferous variety with var. *trichocraspedon,* similar in the glandiform seta of the leafstalk and in the relatively ample, prevailing elliptic leaflets; but these are plane, not conspicuously revolute, and smooth on the upper face, the secondary pinnate venation and midrib being fully immersed into the epidermis. The general aspect of var. *macdougalliana* instantly and compellingly suggests var. *geminiflora,* which has been collected both at Juquila and in the neighborhood of Pochutla, and must be actually or very nearly sympatric; but this differs in its normal setiform appendage to the leaf-stalk, in the sharply reticulate upper face of the leaflets, and in the small areole of the seed-faces, ±0.5 not 1–2 mm diameter. The also very closely related but distantly allopatric Costa Rican var. *cordillerae,* which has the same glandiform seta and the same large areole, differs precariously in its strongly bicolored leaflets, dull dark green above and pallid beneath, not pallidly glaucescent on both faces, and perhaps in the (still too little known) narrower pod.

The variety was collected first in March, 1849, somewhere in Oaxaca, by Galeotti (no. *3469,* NY), very likely in the Sierra Madre del Sur, but the station not recorded. We dedicate the variety to Thomas MacDougall, whose collections of this group of *Interglandulosae* from remote places in Oaxaca have been especially helpful and challenging.

177r. Senna pallida (Vahl) var. **trichocraspedon** (Sandwith) Irwin & Barneby, stat. nov. *Cassia trichocraspedon* Sandwith, Kew Bull. **1936**: 6. 1936.—"MEXICO. State of Mexico: District of Temascaltepec; Ypericones . . . fl. and fr. Feb. 1934, *Hinton 5582* . . ."—Holotypus, K! = NY Neg. *1426*; isotypi, F, NY! paratypi, *Hinton 355, 2972, 5742,* all F, K, and all but *355,* NY!

Cassia tortuosa Smith & Schubert, Contrib. Gray Herb., n. ser. **127**: 21, t. 1, fig. 16–18. 1939.—"MEXICO: Mexico: distr. Temascaltepec, Nanchititla . . . June 11, 1934, *Hinton 6144* . . ."—Holotypus, GH! isotypi, F, MEXU, NY, US!

Slender, ultimately arborescent shrubs at anthesis 1–5 m with terete or faintly ribbed, often abruptly flexuous but stiff branches, either pilosulous throughout with spreading or spreading-incurved hairs up to 0.3–1 mm, the lfts then pubescent on both faces but more densely so beneath, or only the young stems and lf-stalks pilosulous and the lfts then glabrous ciliolate, the inflorescence formed of 1–several, mostly 2-fld racemes pseudofasciculate or shortly racemose on leafless brachyblasts axillary to coeval or lately fallen lvs, some rarely terminal to branchlets.

Stipules setiform 1.5–6 mm.

Major lvs 7–12.5(–14) cm; petiole (7–)10–20 mm; rachis (3–)4–8.5 cm; gland between or immediately above first pair of lfts stipitate, including glabrous or hispidulous stipe 3–4 mm, the lance-elliptic body 0.55–0.7 mm diam; sometimes a similar gland between second pair, and the seta similarly modified but more slender; lfts (6–)7–10(–11) pairs, commonly but not always heteromorphic, those of lvs low on the branchlets often proportionately broader and more obtuse than the rest, the majority lance- or oblong-elliptic acute, some oblong-obovate subobtuse, always accrescent upward along rachis, the distal pair (2–)2.5–4.5 ×

0.6–1.2 cm, ±2–3 times as long as wide, the margins strongly revolute at maturity, the centric midrib with 6–10 pairs of major camptodrome secondary nerves finely prominulous above, sharply and more strongly so beneath, tertiary reticulation 0 or weak and irregular.

Peduncles (3–)8–20 mm; pedicels 12–24 mm; long sepals ±6–8 mm; long petals up to 20–28 mm; androecium glabrous, the filaments of 4 median stamens 1.1–1.6 mm, of 3 abaxial ones 2.5–5 mm, the anthers of 4 median stamens 4–5 mm, of 3 abaxial ones 5–7, their beak 2–2.8 mm; ovary ciliate or ciliolate along sutures, glabrous or almost so laterally; style 1.2–1.5 mm; ovules 26–34.

Stipe of pod 4.5–9 mm, the body 7.5–13 × 0.6–0.85 cm, the papery valves nigrescent glabrous, the sutures either permanently pilosulous-ciliolate or glabrescent; seeds 3.4–4.8 × 3–3.7 mm, the areole oblong-elliptic 1–1.7 × 0.5–0.9 mm.—Collections: 9.

Oak woods and brushy hillsides in the oak-pine belt, ±1400–1700 m, local in the s. and s.-w. foothills of Volcán Toluca in mun. Temascaltepec, Amatepec, and Almoloya de Alquisiras, México.—Fl. XII–III.

In the protologue of *C. trichocraspedon* Sandwith emphasized as diagnostic features the copious pubescence of the leaves and especially the fringe of hairs persistent along the pod sutures, but he recognized the relationship to *Peiranisia palmeri* (our *S. pallida* var. *palmeri*), very similar in the pod although different in the glabrate, broader and blunter, less venulose leaflets. The pubescence of the typus and several paratypi of *C. trichocraspedon,* all of which are derived from two localities (Ypericones and Telpintla) about 20 km apart, is indeed peculiar in the group, but the typus of *C. tortuosa,* which has the same glandiform seta and broad pod as *C. trichocraspedon,* and which was described from a station immediately contiguous to Cerro de Ypericones, has glabrate ciliolate foliage and only residual cilia along the sutures of the pod. We feel certain that these two are merely aspects of one taxon, the diagnostic value of pubescence being, as elsewhere in ser. *Rostratae,* illusory. Collectively *C. trichocraspedon* and *C. tortuosa* differ significantly from other Mexican forms of *S. pallida* only in the terminal gland of the leaf-stalk, analogous to that of Costa Rican var. *cordillerae,* which differs chiefly in its thin-textured, very markedly bicolored, less venulose leaflets and those characters brought out in our varietal key. The highly irregular corolla, thought by Smith & Schubert to distinguish *C. tortuosa* from other *Interglandulosae,* is actually, to the contrary, characteristic of the series and in no detail different from the corolla of sympatric *C. trichocraspedon.*

177s. Senna pallida (Vahl) var. **delgadoana** Irwin & Barneby, var. nov., foliorum seta terminali glandula obsita cum vars. *macdougalliana* et *lemniscata* comparanda, ab ambabus foliolis numerosis (14–24, nec 6–12-jugis), ovulisque paucis (16–22, nec 30–52), patriaque montana diversa.—MEXICO. Oaxaca, distr. Juxtlahuaca: a 2 km al SE. de Copala, 1700 m, 3.II.1977 (fl, fr jun), *M. Sousa et al. 6978.*—Holotypus, MEXU (2 sheets).

Subarborescent, the annotinous branchlets gray lenticellate, the young ones and the foliage loosely strigulose, the lfts bicolored, dull olivaceous above, pale beneath, the inflorescence composed of solitary or paired axillary 1–2-flowered racemes.

Stipules erect linear-setiform deciduous.

Lvs 6–10 cm; petiole 11–15 mm; rachis 3–8 cm, the longer interpetiolular segments 3–6 mm; gland between proximal pair 2–2.3 mm, oblanceolate compressed acute, and a gland at tip of rachis collateral with the seta; lfts (7–)9–24 pairs, narrowly elliptic or elliptic-oblanceolate, the longer ones (near middle of rachis) 14–20 × 3–6 mm, 3–4 times as long as wide, the margin revolute, the upper face veinless, the midrib prominent beneath and 5–8 pairs of secondary veins discolored but immersed.

Peduncles 5–9 mm; fruiting pedicels ±1 cm; stipe of pod 5–6 mm; body linear (4.5–6–11 × 0.6–0.9 cm, the valves a little elevated over the seeds but not decussately ridged; seeds 16–22, paddle-shaped 3.5–4.5 × 3–4 mm, the areole 1.3–2 × 1 mm.—Collections: 3.

Oakwoods, 1530–1700 m, locally plentiful but known only from the vicinity of Copala, distr. Juxtlahuaca, Oaxaca.—Fl. I–II, fr. VIII.

The var. *delgadoana,* marked by the combination of many pairs of relatively small, revolute, simply penniveined leaflets, a gland between the distal pair collateral with the seta, and a short few-seeded pod, has a notably distinct facies and perhaps deserves specific status. We dedicate it to Dr. A. Delgado S., a perceptive student of Cassieae and Mexican legumes in general.

178. Senna holwayana (Rose) Irwin & Barneby, comb. nov. *Cassia holwayana* Rose, Contrib. U.S. Nat. Herb. **8**: 301. 1905.—Typus infra sub var. *holwayana* indicatur.

Shrubs and treelets, potentially drought-deciduous, at anthesis 1–5 m, with terete smooth or striate hornotinous and pale lenticellate, often stiffly contorted older branchlets, glabrous except for often (not always) dorsally puberulent petals and always strigulose ovary, the foliage dull olivaceous often pallidly glaucescent, the relatively ample lfts either concolorous or paler beneath, often brown-venulose against a green intervenium, the inflorescence composed of mostly 2-fld racemes arising singly or 2–3 together from leafless brachyblasts axillary to coeval or lately fallen primary lvs, these brachyblasts rarely elongating and depauperately leafy, the individual racemes then axillary.

Stipules of primary cauline lvs erect linear-subulate or rarely a little dilated and subfalcate, 1.5–5(–6) × 0.25–0.6(–0.8) mm, early dry setiform caducous, those of brachyblasts shorter and more persistent.

Lvs (3–)4–13(–14.5) cm; petiole including ovoid pulvinus (1–)1.3–4 cm, at middle 0.25–0.9(–1) mm diam; rachis 3–35(–38) mm; gland between proximal pair of lfts stipitate or subsessile, (1.3–)1.8–4 mm, the body ± compressed, fusiform or oblanceolate acute or less often obtuse (much eaten and often mutilated or deformed), in profile 0.4–1.2 mm wide; lfts 2–3 pairs, the distal larger; pulvinules 0.8–2.2(–2.8) mm; distal pair of lfts of larger mature lvs symmetrically obovate, elliptic, or elliptic-oblanceolate obtuse or shallowly emarginate, exceptionally rhombic-obovate and deltately acute 1.5–7 × 1.1–3(–3.4) cm, the margin plane or incipiently revolute near pulvinule, either simply or reticulately venulose.

Peduncles (7–)9–25 mm, almost all exactly 2 (exceptionally a few 1 or 3)-fld, the axis 0–1 mm, the pair of fls opening simultaneously and enantiomorphic; bracts subtending nascent fl-buds ovate-deltate acute 0.8–2 mm, caducous as pedicels elongate; pedicels becoming 1–2.4 cm, subtended at base on one side by a fusiform or fusiform-compressed gland similar to the petiolar one; buds globose, glabrous (sepals sometimes minutely ciliolate); sepals green or commonly brownish with pallid margins, strongly graduated, all obovate obtuse or the inner suborbicular, the longest (6–)7–10.5(–13) mm, the outermost ±½ shorter; petals yellow, highly heteromorphic, 3 adaxial smaller, subsymmetrically obovate-oblanceolate cuneately contracted into a slender claw, 2 abaxial longer, up to (1.7–)2–2.8 cm, one opposed to the pistil obliquely obovate, the other obliquely reniform and loosely involute over the androecium; androecium glabrous, or the filaments exceptionally puberulent; filaments of 4 median stamens thickened (1–)1.2–2 mm, of abaxial ones filiform 3–6 mm, the anthers (either smooth or minutely papillate) of 4 median stamens oblong obtusely truncate slightly incurved 3.3–6 mm, with very short lateral beak opening by parallel slits, of 3 abaxial ones lunately incurved, the body 5–8 mm, the narrowly tubular 1-porose erect or porrect beak (1.6–)2–2.9(–3.6) mm; ovary strigulose, the glabrate style 1.1–3.5 × 0.25–0.35(–0.5) mm; ovules (26–)30–48.

Pod variably spreading or pendulous, the stipe 4–8 mm, the linear compressed body 6.5–14 × 0.4–0.6 cm, the glabrescent brownish papery valves nigrescent in age, expressed over each seed in the form of an oblong transverse mound or more sharply defined 4-lobed pyramid; seeds (poorly known) irregularly paddle-shaped or rhomboid-compressed, 2.6–4.2 × 2.1–3.1 mm, the testa brown-olivaceous or atrocastaneous dull, smooth or minutely pitted, the broadly elliptic or subcircular areole 0.4–0.9 mm diam.

As remarked by Bentham (1871, p. 543, sub. *C. multiflora*), this species is extremely close to *S. pallida* sens. lat. (=*C. biflora* sensu Benth.), of which it has precisely the flower, pod and seed. Its range lies entirely within that of *S. pallida* var. *pallida,* with which it is sometimes closely associated; but the fewer and at the same time larger leaflets lend it an individual aspect so readily recognized that specimens of it are almost never misidentified except in the drought-deciduous phase when flowers and pods appear on leafless branchlets. In most populations of *S. holwayana* the leaflets vary in number from two pairs in smaller leaves to three pairs in larger ones, and their venation is decisively reticulate on both faces, especially on the lower. In a small part of the species range in southern Guatemala and El Salvador a striking variant with all leaves constantly and exactly 4-foliolate and the leaflets themselves simply penniveined can be readily distinguished as var. *absoidea*.

Key to the Varieties of *S. holwayana*

1. Lfts 2–3, and in some larger lvs almost always 3 pairs, the rachis of all lvs mostly 8–35 mm; lfts reticulately venulose, with 8–14 pairs of secondary veins giving rise to tertiary venulation distinctly prominulous on the dorsal or both faces; range of the species.
178a. var. *holwayana* (p. 560).
1. Lfts exactly 2 pairs in all lvs, separated by a rachis only 3–6 mm; lfts simply penniveined, with 5–6 pairs of major secondary veins; local in s. Guatemala and El Salvador.
178b. var. *absoidea* (p. 561).

178a. Senna holwayana (Rose) Irwin & Barneby var. **holwayana.** *Cassia holwayana* Rose, l.c., a legitimate substitute for *C. multiflora* Martens & Galeotti, Bull. Acad. Roy. Bruxelles **10**(2): 307. 1843.—". . . au bord des ruisseaux des régions alpines de la cordillère orientale d'Oaxaca, de 6 à 7,000 pieds, surtout près de Capulalpan et de San Pedro Nolasco."—Holotypus, collected XII.1840 (fl), *Galeotti 3145*, BR! isotypus, K (hb. Hook.)! = NY Neg. *1417*.—Non *C. multiflora* Vogel, 1837.—*Peiranisia holwayana* (Rose) Britton & Rose, N. Amer. Fl. **23**(4): 264. 1930.

Peiranisia salvadorensis Britton ex Britton & Rose, N. Amer. Fl. **23**(4): 264. 1930.—"El Salvador. Type from San Francisco [Gotera, dpto. Moranzàn], 1929, *Salvador Calderón.*"—Holotypus, *Calderón 2485,* NY! isotypus, F!

Cassia multiflora sensu Bentham, 1871, p. 543.

Lvs (3–)4.5–13(–14.5) cm; petiole (1–)1.3–4 cm, at middle (0.45–)0.5–0.9(–1) mm diam; rachis (5–)8–35(–38) mm, the longer interfoliolar segments (6–)8–17(–21) mm; distal pair of lfts symmetrically obovate, elliptic or elliptic-oblanceolate obtuse or shallowly emarginate, exceptionally rhombic-obovate and deltately acute (2.7–)3–7 × (1.2–)1.4–3(–3.4) cm, ±(1.5–)2–3 times as long as wide; venation as given in key.—Collections: 81.

Cactus-grassland, scrub thickets, regenerating pine-oak forest, semideciduous woodland and thorn-forest, sometimes plentiful on road-banks, adapted to xeromorphic plant-communities but occurring in exposed rocky places and along barrancas within climax forest, near sea level on Istmo de Tehuántepec up to 1900(–2200) m in interior valleys of s. Mexico, common and locally abundant in centr. and s. Oaxaca, n. and n.-w. in scattered stations to the Tehuacán Desert in Puebla and the e. arm of Balsas Depression in México, Morelos and e. Guerrero, e. through the Pacific slope of highland Chiapas just into Guatemala (Huehuetenango); apparently disjunct in e. El Salvador (San Vicente, Morazán) and s.-centr. Honduras (Morazán, Choluteca), to be expected in adjoining Nicaragua.—Fl. (VII–)IX–III.

The criteria adduced by Britton & Rose in support of *Peiranisia salvadorensis* no longer seem either decisive or significant, for leaflets and pods as long and wide as those of *P. salvadorensis* occur at random in Oaxacan populations of the species. Specimens collected on Gulf of Fonseca, Honduras by George Barclay and Dr. Sinclair (K, US) have many 6-foliolate leaves of var. *holwayana* but the slender habit and venulation of var. *absoidea*.

178b. Senna holwayana (Rose) var. **absoidea** Irwin & Barneby, var. nov., a var. *holwayana* habitu gracillimo, petiolis subfiliformibus 0.25–0.4 (nec 0.45–0.9) mm diam, foliolisque semper 2-jugis, paribus inter se rachi brevissimo 3–6 mm tantum longo semotis, laminis insuper penninerviis nec reticulatis diversa.—GUATEMALA. Zacapa: brushy rocky slope, 200–500 m, hills above village of La Fragua, 14.X.1940 (fl), *P. C. Standley 74800*.—Holotypus, F = NY Neg. *9689*; isotypus (fragm), NY.

Lvs 4–7 cm; petiole 2–3.6 cm, at middle 0.25–0.4 mm diam, terete or almost so, 5–9 times as long as rachis; rachis 3–6 mm; distal pair of lfts broadly obovate 15–36 × 11–24 mm, ±1.3–1.6 times as long as wide; venation as given in key.—Collections: 6.—Fig. 33.

Dry brushy hillsides, stream-beds, quebradas, 75–600 m, very local in two widely separated localities; valley of río Motagua in Zacapa and immediately adjoining Chiquimula, Guatemala; and about Lago Olomega in San Miguel, s.-e. El Salvador.—Fl. X–II.

A visually striking plant, curiously simulating in habit and foliage *Chamaecrista* (sect. *Absus*) *zygophylloides* var. *deamii*, but with the axillary two-flowered racemes and beaked anthers of *Senna* ser. *Rostratae*. The bicentric dispersal of var. *absoidea*, especially in relation to the disjunct Salvadorian populations of var. *holwayana*, is difficult to understand; further collections from El Salvador may, of course, establish a more continuous pattern.

179. Senna huancabambae (Harms) Irwin & Barneby, comb. nov. *Cassia huancabambae* Harms in Fedde, Repert. Sp. Nov. **18**: 94. 1922.—"Peru: Tal des Flusses Huancabamba, südl. Sondòn [=Sondor], Dep. Piura,

Fig. 33. *Senna holwayana* (Rose) var. *absoidea* I. & B. Flowering branchlet ×1; distal segment of leafstalk, showing petiolar gland ×3; detached stamens + staminode ×3 (*Standley 74800*). Pod ×1; seed ×5 (*Tucker 929*).

Prov. Huancabamba, 5°20′S . . . 1700–1800 m (Weberbauer no. 6285—V.1912 . . .).''—Holotypus, †B = F Neg. *1693*! isotypi, F, GH, NY!

Cassia huancabambae sensu Macbride, 1943, p. 169.

Sparsely leafy subshrubs with blackish knotty caudex and stiff, diffuse or virgately ascending, simple or few-branched, angulately ribbed stems 0.5–1 m, thinly strigulose with subappressed hairs 0.2–0.3 mm, the foliage subconcolorous, the lfts glabrous above, the inflorescence of exserted, umbellately 2–3-fld racemes narrowly racemose-paniculate.

Stipules subulate-setiform 0.8–2 mm deciduous.

Lvs (1–)1.5–3.5 cm; petiole 5–12 mm, at middle 0.3–0.5 mm diam, openly shallow-grooved; rachis 4–13 mm; petiolar gland 1 between proximal pair of lfts, stipitate, (1.5–)2–3 mm, the slenderly fusiform acute body 0.3–0.5 mm diam; pulvinules 0.4–0.6 mm; lfts 3–4 pairs, slightly accrescent distally, obovate or oblong-obovate obtuse or retuse, at base subsymmetrically rounded, the distal pair 6–12 × (2–)3–6 mm, the midrib slightly elevated beneath, the venation otherwise immersed.

Peduncles directly axillary (no brachyblasts) to primary cauline lvs, 10–21 mm; racemes subumbellately 2–3-fld; bracts elliptic-oblanceolate ±2 mm caducous; raceme-axis 0–1 mm; glands at base of each pedicel stipitate, resembling those of the petiole; pedicels 10–13 mm; buds subglobose puberulent; sepals strongly graduated, 2 outer firm ovate ±½ as long as inner ones, these broadly obovate-suborbicular membranous-margined 5–6.5 mm; petals yellow, puberulent dorsally, not strongly heteromorphic except the 2 abaxial a little longer than the 3 adaxial and one of them more oblique, the blades beyond the short claw broadly obovate or oblong-obovate, the longest (including claw) 16–23 mm; androecium glabrous, the filaments of 4 median stamens 1–2.5 mm, of 3 abaxial ones 2–3.5 mm, the body of the 4 median anthers 3–3.5 mm, contracted into a divaricate 2-porose beak 0.4–0.6 mm, that of 3 abaxial ones 3.3–3.8 mm, scarcely more incurved than the median but contracted into an erect beak 1.2–2 mm, this 1-porose or the orifice divided by a filiform septum; ovary densely silky-puberulent, the glabrescent style ±1 × 0.2–0.3 mm; ovules 34–38.

Pod widely spreading-recurved, the stipe 5–7 mm, the linear compressed body 7–11.5 × 0.45–0.5 cm, the papery valves turning brown or castaneous, elevated over the seeds in the form of simple mounds; ovules ±20–40; seeds unknown.—Collections: 5.—Fig. 34.

Dry rocky hillsides and riverbanks, in drought-deciduous scrub ("rainy-green formation") with arborescent cactaceae, 1300–2000 m, local, known only from near the crest of the Andean divide on the headwaters of río Jubones in Azuay, Ecuador and río Huancabamba in e. Piura (prov. Huancabamba) and adjoining Cajamarca (prov. Jaén), Peru.—Fl. VIII–XII.

The circumstances of its Andean dispersal and the racemes of 2–3 flowers subtended by a stipitate gland neatly distinguish *S. huancabambae* from all *Interglandulosae* except *S. pallida* var. *nemorosa*. From this it differs in stiff subjunceous stems clad in short 6–8-foliolate leaves that scarcely equal the internodes in length. Its habit of growth is idiosyncratic, but approaches that of distantly allopatric *S. chloroclada*. The similarity is certainly coincidental, for *S. chloroclada* has the beakless anthers and glandless raceme of the closely related and sympatric ser. *Aphyllae*.

Fii. ser. CHLOROCLADAE Irwin & Barneby

Senna sect. **Peiranisia** ser. **Chlorocladae** Irwin & Barneby, ser. nov., habitu toto *Aphyllis* rigidicaulibus subspinosis praesimiles, sed ramulis hornotinis saltem in anthesin usque foliis perfectis abbreviatis 2–4-foliolatis instructis, antherisque fertilibus longitrorsus 4-sulcatis apice distincte etsi breviter rostratis diversae.—Sp. typica (unica): *S. chloroclada* (Harms) Irwin & Barneby.

Fig. 34. *Senna huancabambae* (Harms) I. & B. Flowering stem ×½ (*Hutchinson 6729*). Detached stamens + staminode ×3 (*Holm-Nielsen 4886*). Fruiting branchlet ×½ (*Steyermark 53739*).

In habit and in perianth resembling the stiffly junceous members of ser. *Aphyllae,* but small 2–4-foliolate lvs with petiolar gland present at all nodes into and through anthesis, then deciduous; fertile anthers sulcate both laterally and dorsoventrally, distinctly though shortly beaked, the beak of the 4 median ones lateral, that of the 3 abaxial ones porrect, all 2-porose; pod essentially of ser.

Aphyllae, but the valves (little known in really mature state) apparently tumescent.—Sp. 1, of w. Gran Chaco and bordering Andean foothills in s. Bolivia and n. Argentina.

Burkart (1952, l.c.) has alluded to *Cassia chloroclada* as transitional between the leafy and leafless sennas of Argentina and, following this lead, we interpolate the species between our ser. *Interglandulosae* and *Aphyllae*. The thrysiform inflorescence of depauperate racemes and the marked asymmetry of the corolla suggest an affinity to ser. *Aphyllae* which is borne out by the general appearance of the plant, especially after the leaves have fallen. But the laterally sulcate fertile anthers contracted at apex into a short biporose beak forcibly recall the androecium of *S*. (ser. *Trigonelloideae*) *mucronifera*. The pod, scarcely known in fully mature state, appears similar to that of ser. *Aphyllae* except that the valves are more tumid. The genuine kinship of *S. chloroclada* is in our view ambiguous and will require reappraisal as more material accumulates; its morphological peculiarities, however, will in any case entitle it to independent serial status. Bravo (1981, p. 273) refers the species to a modified ser. *Pachycarpae* uncritically characterized.

180. Senna chloroclada (Harms) Irwin & Barneby, comb. nov. *Cassia chloroclada* Harms in Herzog, Meded. Rijks-Herb. Leiden **27**: 37. 1915.—"[Bolivia. Tarija, Azero:] Monte bei Cumbarute . . . 800 m ([*Herzog*]) no. 1149—Dez. 1910)."—Holotypus, †B = F Neg. *1671*; neoholotypus, former isotypus, G! isotypus, W!

Cassia chloroclada sensu Burkart, Darwiniana **4**: 324. 1942; 1952, p. 168; Bravo, 1981, p. 273, fig. 2, 5 (E–H).

Microphyllous subshrubs at anthesis 0.5–1 m, with stout terete striate stems ±1.5–3.5 mm diam freely branched distally, some branchlets tapering into a stout thornlike leafless apex, the stems and foliage minutely pilosulous throughout with incurved hairs ±0.1–0.2 mm, the distant 2–4-foliolate lvs shorter than the internodes, the lfts subconcolorous gray-green dull, the 1–4-fld racemes axillary from distal lvs forming a sparse narrow thyrse.

Stipules erect appressed setiform 0.3–2 mm, tardily deciduous.

Lvs 1–2 cm; petiole including obscurely differentiated pulvinus 3–6 mm, at middle 0.3–0.45 mm diam, shallowly grooved ventrally; rachis 0–3.5 mm; gland between proximal (or only) pair, ovoid or conic-subulate 0.2–0.8 mm, puberulent; pulvinules 0.4–0.7 mm; lfts of most lvs 1 pair, of a few lvs 2 pairs, the distal (or only) pair obliquely elliptic or narrowly obovate, abruptly acute or subobtuse 6–16 × 2.5–9 mm, at base cordate on proximal and rounded to cuneate on distal side, the midrib with 5–7 pairs of fine camptodrome (and few weak intercalary) secondary nerves immersed or faintly raised above, bluntly prominulous beneath, a few connecting tertiary veinlets visible beneath only.

Peduncles 2–13 mm, the raceme-axis 0–7 mm; bracts ovate subscarious 1.2–2 mm; pedicels 5–12 mm; fl-buds obovoid obtuse puberulent; sepals submembranous, often fuscous at middle but otherwise greenish-yellow pallid-margined unequal, the long inner ones oblong-ovate ±6.5 mm, the outermost ± half as long; longest petal ±15 mm; androecium glabrous, the filaments of 4 median stamens ±0.5 mm, of 3 abaxial ones ±3 mm, the anthers of median stamens ±4 mm, abruptly truncate, the divaricate beak 0.5 mm, the anthers of abaxial ones lunately incurved 7–9 mm, rounded at base, the very short beak obliquely truncate 2-porose; style 1.5 × 0.15–0.2 mm erect.

Pod ascending, the stipe ±5 mm, the turgidly linear body ±10 × 0.55 cm, bluntly carinate by the ventral suture (this 0.5–0.8 mm wide), the finely rugulose evenulose valves becoming papery brownish; seeds obliquely basipetal, compressed-pyriform ±5.5 × 3 mm, the smooth testa lustrous black, the oblong-elliptic areole ±2.5–3.5 × 1 mm.—Collections: 8.

Open brush-woodlands and gullied barrancas, ±280–850 m, along the e. foothills of the Andes and w. margin of Gran Chaco drained by ríos Parapetí, Pilcomayo and Bermejo between 20° and 23°S latitude in Tarija, Bolivia and in n.-e. Salta and extreme n.-w. Formosa, Argentina.—Fl. XI–IV.

Fiii. ser. EGREGIAE Irwin & Barneby

Senna sect. **Peiranisia** ser. **Egregiae** Irwin & Barneby, ser. nov., ser. *Aphyllis* et *Chlorocladis* certe affines, ab iis imprimis foliis glandulaque petiolari sua bene evolutis, ab his imprimis stipulis minusculis glanduliformibus atrorubentibus dorso tumidis diversae.—Sp. typica (unica): **S. kurtzii** (Harms) Irwin & Barneby.

Inflorescence, perianth, androecium and pod almost of ser. *Aphyllae* but small leaves charged with a petiolar gland present at all nodes; stipules glandiform ovate-apiculate in outline, dorsally tumid, livid-reddish.—Dwarf microphyllous subshrubs, the (2–)4-foliolate lvs mostly shorter than the internodes; anthers firm, those of 3 long abaxial stamens beakless 2-porose, sulcate dorsoventrally but not along the lateral sutures.—Sp. 1, of w.-centr. sub-Andean Argentina.

Like ser. *Chlorocladae* immediately preceding, the monotypic *Egregiae* seem clearly related to sympatric *Aphyllae,* which they resemble in almost everything except the retention of small microphyll foliage. The peculiar minute but swollen and discolored stipules are unique in *Senna.* The species is referred by Bravo (1981, p. 259) to a modified ser. *Pachycarpae.*

181. Senna kurtzii (Harms) Irwin & Barneby, comb. nov. *Cassia kurtzii* Harms, Feddes Repert. Sp. Nov. **18**: 94. 1922.—"*Argentina*: Mendoza, oberer Rio Salado, La Cienegita . . . (F. Kurtz no. 7093 u. 7666–I. u. II. 1892–1893)."—Lectoholotypus, *Kurtz 7093,* †B = F Neg. *1696*; no isotypus seen; paratypus, *Kurtz 7666,* NY!

Cassia egregia Sandwith, Kew Bull. **1927**: 181. 1927.—"[Argentina. Gob. Neuquen:] Near Chos Malal . . . at 1200 m., November 20th, 1925, [H. F.] Comber 163 . . ."—Holotypus, K! = NY Neg. *1572.*—Equated by Sandwith himself (K, annotation, 1929) with *C. kurtzii,* the synonymy accepted by Burkart, 1952, p. 166, in clave.

Cassia conjugata sensu Kurtz, Verhandl. Bot. Ver. Brandenburg **35**: 102. 1894; non Ruiz & Pavón.

Cassia kurtzii sensu Bravo, 1981, p. 259, fig. 1, 5 (A–D).

Stiffly many-stemmed xeromorphic microphyllous suffrutex ±2.5–4 dm, with woody caudex and ascending or diffusely incurved-ascending terete striate pale green hornotinous and basally persistent brown leafless annotinous branches, finely inconspicuously strigulose with appressed straight hairs up to 0.2–0.3 mm, but one or both faces of lfts glabrous and the calyx sometimes thinly hirsute, the lvs mostly 4-foliolate shorter than the internodes, the concolorous thick-textured lfts commonly livid-speckled, the pseudo-racemose inflorescence consisting of 1(–2)-fld, shortly pedunculate racemes arising from axils of small distal lvs.

Stipules livid-reddish glanduliform, tumidly convex dorsally, in outline ovate-apiculate 1–1.6 mm, persistent.

Lvs 7–23 mm; petiole including slender yellowish pulvinus 1–7 mm, at middle 0.25–0.5(–0.6) mm diam, obscurely sulcate; gland between proximal pair, slenderly fusi- or subuliform 0.4–0.7 × 0.1–0.4 mm, yellowish or orange; pulvinules 0.5–0.8 mm; lfts of almost all lvs 2 pairs, occasionally only 1 pair or subpalmately 3-foil, the distal pair longer, these narrowly or broadly obovate obtuse apiculate or shallowly emarginate 4–13 × 2.2–6 mm, ±1.5–2.2 times as long as wide, at base inequilaterally rounded, the margin plane pallid or discolored, the midrib bluntly prominulous beneath, the venation otherwise fully immersed.

Peduncles ascending (1.5–)5–26 mm; racemes all 1- or partly 2-fld, the axis then 1–7 mm; bract subtending the first (or only) fl resembling a stipule in shape and texture 1.5–2.2 mm, of the second fl obsolete; pedicels 12–20 mm; fl-buds globose, either glabrous above the puberulent base or erratically setose-hirsute with weak straight hairs up to ±1 mm; sepals submembranous, obovate or suborbicular, the small outer ones often dark at middle, the inner membranous-margined, well graduated, the smallest outer one 3–5 mm, the largest inner one 6–9.5(–10) mm; petals glabrous golden- then orange-yellow, 4 oblong-obovate obtuse, one abaxial one with thickened claw and oblong-elliptic blade turned almost at right-angles to it, this petal a little longer than the rest ±15.5–17 mm; androecium glabrous, the staminodes ±2 mm wide, the filaments of 4 median anthers 1.8–2 mm, of 3 abaxial ones very unequal 3–9(–10) mm, that of the centric stamen shortest, the anthers of 4 median stamens 3.6–4 mm up to the very short (±0.3 mm) ascending beak, those of the 3 abaxial stamens subequal lunately lanceolate in outline 4–5.5 × 1.3–1.7 mm, tapering at summit but not beaked, dehiscent by oblique short slits; ovary strigulose and sometimes also loosely ascending-pilose; style glabrous 1–1.4 × 0.25–0.3 mm almost straight; ovules 20–22.

Pod (not well known) ascending and curved outward, the stipe 6–9 mm, the body 6.5–9 × 0.8–1 cm compressed linear bicarinate by the sutures, the firmly papery valves elevated over seeds in the form of simple transverse ridges, the interseminal septa very narrow, the locules ±3 mm long; seeds transverse, narrowly ovate in outline 4.7–5 × 2.7–2.8 mm, compressed parallel with the valves, the moderately lustrous brown dark-spotted testa rugulose, the elliptic or almost round areole 0.4–0.8 × 0.2–0.35 mm; $2n = 24$.—Collections: 9.

Dry rocky hillsides and along intermittent stream beds, ±650–1600(–1700) m, locally plentiful in the e. foothills of the Andes drained by ríos Atuel, Colorado and Neuquén, lat. 35°15′–39°S in Mendoza and Neuquén, Argentina.—Fl. XI–III.—*Cacho carnero*; *yerba loca*.

Fiv. ser. APHYLLAE (Bentham) Irwin & Barneby

Senna sect. **Peiranisia** (Rafinesque) ser. **Aphyllae** (Bentham) Irwin & Barneby, comb. nov. *Cassia* sect. *Chamaesenna* ser. *Aphyllae* Bentham, Trans. Linn. Soc. London **27**: 542. 1871.—Sp. lectotypica: *C. aphylla* Cavanilles ≡ **Senna aphylla** (Cavanilles) Irwin & Barneby.

Cassia ser. *Aphyllae* sensu Burkart, Darwiniana **7**(2): 235–239. 1946; Bravo, Darwiniana **21**(2–4): 343–391, fig. 1–15. 1978.

Corolla of *Peiranisia*; sepals strongly graduated; androecium functionally 7-merous, the fertile members differentiated into sets of 4 shorter median and 3 longer abaxial ones, the anthers firm, 4-sulcate 2-porose, essentially beakless, the pores of the 4 median ones lateral-infraterminal, those of the 3 long ones obliquely terminal; pod linear plano-compressed, 2-carinate by the ± thickened sutures, the valves stiffly chartaceous, faintly corrugated but not otherwise differentiated over the seeds, these obliquely transverse, broadside to the valves, areolate.—Xeromorphic shrubs and subshrubs from deeply penetrating woody roots sometimes thickened into a xylopodium, leafless except for fleeting 2–4-foliolate eglandular eophylls of seedlings (and exceptionally some transient similar lvs on new branchlets during periods of rapid growth), the lvs of adult branchlets represented by minute fuscous triangular or subulate scales (perhaps homologous with stipules), the stiffly junciform or pliantly flexuous, sometimes elaborately ramulose branchlets often spine-tipped, green and photosynthetic, the 1–5-fld racemes either obviously lateral or terminally thyrsiform; style short, slightly incurved distally, the ciliolate stigmatic orifice looking up- and inward; ovules 14–36.—Spp. 11, primarily of the temperate and subtropical Monte phytogeographic province of Argentina, few of (or extending into) the Chaco and sub-Andean (Prepuna) provinces in n.-w. Argentina, s.-e. Bolivia and adjoining Paraguay, in lat. ±20–44°S.

This primarily Argentine group of leafless sennas has recently been described in minute monographic detail by Dr. Lilia D. Bravo (1978, l.c.); it would be superfluous and indeed impertinent to attempt in this place any revision of her work. For the convenience of the reader who may not have immediate access to the source monograph we present here only a) a key, translated and slightly modified from that of Bravo, in which a brief statement of range is incorporated for each species; and b) a conspectus of the species, which we feel obliged to transfer from *Cassia* to the genus *Senna*. The taxonomic formulations are taken intact from Bravo's monograph.

The immediate relationship of ser. *Aphyllae* appears to lie with ser. *Interglandulosae,* with which it agrees in form of the corolla, pod and seed, and perhaps with the xeromorphic North American *Deserticolae,* but less directly. The suppression of functional leaves after the seedling stage, and the consequent (or concurrent) loss of petiolar gland is the most obvious differential feature of ser. *Aphyllae,* but monotypic ser. *Chlorocladae,* which has almost the same junciform habit but retains normal, even if small and precociously deciduous leaves provided with a small petiolar gland, preserves what appears to be an intermediate stage of xeromorphic reduction.

The mature stem-nodes of the *Aphyllae* bear a minute deltate or subulate entire scalelike bract which Bravo interprets as a rudimentary leaf. We question whether these are not in fact rudimentary stipular structures; if they are not, the *Aphyllae* are exstipulate and in this respect unique in *Senna*. The composite inflorescence of ser. *Aphyllae* is composed of depauperate, sometimes corymbiform, one- or few-flowered racemes resembling those of many *Interglandulosae* but lacking glands at base of the pedicels. Bravo made a key distinction between a so-called terminal (compound) inflorescence found in *S. acanthoclada* and *S. nudicaulis* and a lateral inflorescence common to the rest. It appears to us that the individual racemose elements of the inflorescence are always axillary in origin. When arising serially toward the tip of flowering branchlets they assume the appearance of an exserted thyrse and it is this that is, we think inaccurately, called a terminal inflorescence.

Key to the Species of ser. *Aphyllae*

1. Mostly erect shrubs and subshrubs 4–25 dm tall, less often smaller but if so the stems glabrous; ultimate branchlets more than 2 cm long.
    2. Branchlets mostly spreading at ±90° from parent axis, stiff and spine-tipped. Chaco phytogeographic province.
        3. Stems and sepals glabrous; racemes borne along the length of the flowering branchlets; n.-w. Argentina (s. Salta and w. Chaco to n. Córdoba.
        182. *S. chacoënsis* (p. 569).
        3. Stems finely velvety-puberulent and sepals finely pilosulous; racemes crowded toward tip of branchlets; middle Pilcomayo valley in Paraguay.
        183. *S. spiniflora* (p. 569).

2. Branchlets either less abruptly divaricate or, if not so, the whole axis conspicuously zigzag; Monte and Prepuna phytogeographic provinces.
   4. Branchlets stout, at once fastigiately ascending and narrower at both ends than at the middle, either 3–9 mm diam or, if less, bearing toward base fine setae ±5 mm long.
      5. Adult branches mostly more than 4 mm diam at middle, glabrous at base; peduncles 5–10 mm.
         6. Pistil strigose; calyx puberulent and sometimes finely setose, the larger sepals 5–7 × 4–5 mm; pedicels 8–13 mm; local in s.-w. Salta, Argentina. 184. *S. rigidicaulis* (p. 569).
         6. Pistil glabrous; calyx glabrous, the larger sepals 8–10 × 5–7 mm; pedicels 14–16 mm; n.-w. Argentina (n.-ward from Catamarca) and adjacent Bolivia. 185. *S. crassiramea* (p. 569).
      5. Adult branches ±4 mm diam at middle, finely setose at base; peduncles 10–25 mm; n.-w. Argentina (Salta to La Rioja). 186. *S. rigida* (p. 570).
   4. Branchlets neither fastigiately erect nor narrowed toward point of origin, 1–5 mm diam, always glabrous.
      7. Young branches, bracts, peduncles and pedicels pubescent or, if glabrous, then the calyx silky-setose.
         8. Ovary and pod glabrous; sepals charged with setae 3–3.5 mm; anthers of long abaxial stamens 3–4 mm; w. Argentina (Catamarca and s.-w. Santiago del Estero to San Luis). 187. *S. trichosepala* (p. 570).
         8. Ovary densely and pod thinly pubescent; sepals merely puberulent, especially proximally; anthers of long abaxial stamens 5–7 mm; n.-w. Argentina (Catamarca to San Juan). 188. *S. fabrisii* (p. 570).
      7. Plant wholly glabrous or the pedicels and calyx rarely thinly puberulent, the latter never setose.
         9. Dwarf shrublet 1.5–2(–3) dm, the stems decumbent from a massive xylopodium, the ultimate branchlets spinose; seeds 4–4.4 mm; local in Catamarca. 189. *S. pachyrrhiza* (p. 570).
         9. Shrubs and shrublets variable in stature, 4–25 dm, the primary branches erect-ascending, the ultimate branchlets pliant, commonly zigzag, often pendulous; seeds 3.5–4 mm; widespread over centr. and n.-w. Argentina. 190. *S. aphylla* (p. 570).

1. Diffuse or prostrate, intricately branched shrublets, the stems silky-puberulent and the ultimate branchlets at once sharply divaricate from their parent axis, less than 2 cm, rigid, and thornlike.
   10. Primary stems 2–5(–7) dm, prostrate, repeatedly and intricately branched (suggestive of *Ulex*); racemes corymbiform, mostly 3–5-fld, the peduncle 4–10 mm, the pedicels 15–18 mm; w. Argentina (s. Catamarca to w. San Juan and n. San Luis). 191. *S. acanthoclada* (p. 570).
   10. Primary stems suberect 1–2 dm, little branched; racemes 1-fld, their peduncle 0.5–4 mm, the pedicel 10–15 mm; very local in n.-e. Neuquén, Argentina. 192. *S. nudicaulis* (p. 570).

182. Senna chacoënsis (Bravo) Irwin & Barneby, comb. nov. *Cassia chacoënsis* Bravo, Darwiniana **21**(2–4): 359, fig. 4. 1978, based on *C. aphylla* var. *robusta* Burkart, Darwiniana **7**(2): 237. 1946. *C. rigida* (Hieronymus) var. *robusta* Burkart, Leg. Argent. ed. 2, 168, 544. 1952.

183. Senna spiniflora (Burkart) Irwin & Barneby, comb. nov. *Cassia spiniflora* Burkart, Darwiniana **7**(2): 235. 1946.

Cassia spiniflora sensu Bravo, 1978, p. 361, fig. 5.

184. Senna rigidicaulis (Bravo) Irwin & Barneby, comb. nov. *Cassia rigidicaulis* Burkart ex Bravo, Darwiniana **21**(2–4): 364, fig. 6. 1978.

185. Senna crassiramea (Bentham) Irwin & Barneby, comb. nov. *Cassia crassiramea* Bentham in Hooker, Icones Pl. t. 1063. 1870.

Cassia crassiramea sensu Bentham, 1871, p. 542; Bravo, 1978, p. 366, fig. 7.

186. Senna rigida (Hieronymus) Irwin & Barneby, comb. nov. *Cassia aphylla* var. *rigida* Hieronymus, Bol. Acad. Nac. Ci. Córd. **4**(1): 24. 1881. *C. rigida* (Hieronymus) Burkart, Leg. Argent. ed. 2, 544. 1952.

Cassia aphylla var. *rigida* sensu Burkart, 1946, p. 239. *C. rigida* sensu Bravo, 1978, p. 369, fig. 8.

Bravo (1978, p. 373, fig. 9) recognized in addition to the typical one a var. *inermis,* different in its more pliant branchlets and slightly shorter pedicels.

187. Senna trichosepala (Chodat & Wilczek) Irwin & Barneby, comb. nov. *Cassia aphylla* var. *trichosepala* Chodat & Wilczek, Bull. Herb. Boiss. II, **2**(5): 475. 1902. *C. trichosepala* (Chodat & Wilczek) Bravo, Darwiniana **21**(2–4): 373, fig. 10. 1978.

188. Senna fabrisii (Bravo) Irwin & Barneby, comb. nov. *Cassia fabrisii* Bravo, Darwiniana **21**(2–4): 377, fig. 11. 1978.

189. Senna pachyrrhiza (Bravo) Irwin & Barneby, comb. nov. *Cassia pachyrrhiza* Bravo, Darwiniana **21**(2–4): 380, fig. 12. 1978.

190. Senna aphylla (Cavanilles) Irwin & Barneby, comb. nov. *Cassia aphylla* Cavanilles, Ic. Pl. **6**: 41, t. 561. 1801.

Cassia aphylla sensu Bentham, 1871, p. 542; Burkart, 1952, p. 168, fig. 29 (b, c); in Cabrera, Fl. prov. Buenos Aires **4**(3): 451, fig. 141. 1967; Bravo, 1978, p. 382–389, fig. 13–15; Darwiniana **21**(2): 393–399, fig. 1–3 (ontogeny of seedlings and infraspecific differentiation). 1978.

Bravo (1978, ll. cc.) recognized two subspecies, one of them divided into two varieties, differing in branching patterns, number of flowers to the raceme, and dispersal.

191. Senna acanthoclada (Grisebach) Irwin & Barneby, comb. nov. *Cassia acanthoclada* Grisebach, Symb. Fl. Argent. 116. 1879.

Cassia acanthoclada sensu Bravo, 1978, p. 355, fig. 3.

192. Senna nudicaulis (Burkart) Irwin & Barneby, comb. nov. *Cassia nudicaulis* Burkart, Bol. Soc. Argent. Bot. **6**(3–4): 226. 1957.

Cassia nudicaulis sensu Bravo, 1978, p. 354, fig. 2.

Fv. ser. DESERTICOLAE Irwin & Barneby

Senna sect. **Peiranisia** ser. **Deserticolae** Irwin & Barneby, stat. et nom. nov. *Palmerocassia* Britton ex Britton & Rose, N. Amer. Fl. **23**(4): 253. 1930, pro gen.—Sp. typica: *P. wislizeni* (A. Gray) Britton & Rose ≡ **Senna wislizeni** (A. Gray) Irwin & Barneby.

Cassia sect. *Chamaesenna* ser. *Rigidae* *Parvifoliae (partim) Bentham, Trans. Linn. Soc. London **27**: 547. 1870 (quoad *C. wislizeni,* caet. exclus.).

Corolla of sect. *Peiranisia*; sepals strongly graduated; androecium functionally 7-merous, the fertile anthers subisomorphic but their filaments graduated in length, the 4 median shortest, the centric abaxial one much the longest, its neighbor on either side intermediate, the beaks of all short, those of 3 abaxial anthers a little more porrect than the rest but scarcely longer, all 7 dehiscent alike by 2 subparallel slits; pod linear plano-compressed, bicarinate by and dehiscent through the somewhat dilated sutures, the stiffly papery or coriaceous valves plane or corrugately elevated over the seeds; seeds 1-seriate subobliquely transverse, rhombic-trapeziform, compressed parallel to the valves, areolate.—Microphyllus xeromorphic drought-deciduous shrubs and treelets, the branchlets commonly differentiated into long-shoots and knotlike brachyblasts axillary to fallen primary lvs; stipules setiform; petiolar gland 0; lfts 1–5(–6) pairs; inflorescence paniculate, of few- but proportionately large-fld racemes; style poorly differentiated from ovary 0.7–2 mm, the stigmatic cavity minute terminal; ovules (10–)12–40.—Spp. 3, of tropical thornforest and arid interior plateau and desert environments in Mexico and s.-w. United States.

It is the rule in *Senna* that, wherever the three abaxial stamens are differentiated in length it is the centric one, that opposed to the vexillar petal, that is shorter and slimmer than its neighbors on either side. In ser. *Deserticolae,* nevertheless, it is the centric stamen that is longest of all, an arrangement duplicated nowhere else in the genus. The flower of ser. *Deserticolae* is otherwise that of ser. *Isandrae,* except that the beak of all anthers is a little better developed and more porrect. The narrowly linear pod is not different from that of ser. *Interglandulosae,* a group that differs consistently in presence of petiolar glands and generally in the distal confluence of the thecae of each long anther into a common tubular beak.

The series was first systematically assembled by Britton & Rose who segregated the species then known into the genus *Palmerocassia,* placing this next following *Gaumerocassia* (our ser. *Isandrae*) in a very plausible sequence. Only *S. wislizeni* was known to Bentham (1871), who inserted it in his heterogeneous eglandulose ser. *Floridae* associated with members of our ser. *Galeottianae.* The morphology and dispersal of all these largely Mexican sennas are now much better known and ser. *Galeottianae,* while habitally similar to some peiranisias, and like them lacking petiolar glands, have a fundamentally different zygomorphic corolla and specialized style and anther-tips unlike anything encountered in sect. *Peiranisia.*

Key to the Species of ser. *Deserticolae*

1. Lfts (2–)3–5(–6) pairs; dry interior valleys and plateaus of Mexico n.-ward from Oaxaca to Sonora and Coahuila and of adjoining s.-w. United States. Ovules 22–40. 193. *S. wislizeni* (p. 571).
1. Lfts of all lvs exactly 1 pair; local in Chihuahuan and Tehuacán deserts in Mexico.
    2. Lfts obovate or obovate-cuneate ±3–9 × 2–6 mm, the margins plane; ovules 10–16; w.- and w.-centr. Chihuahua. 194. *S. monozyx* (p. 575).
    2. Lfts broadly flabellate-obovate or -obcordate 5–12 × 5–11 mm, the margins strongly revolute; ovules 20–24; s.-e. Puebla and adjacent Oaxaca. 195. *S. unijuga* (p. 576).

193. Senna wislizeni (A. Gray) Irwin & Barneby, Phytologia **44**(7): 500. 1979. *Cassia wislizeni* A. Gray, Pl. Wright. **1**: 60. 1852, sens. ampliat.—Typus infra sub var. *wislizeni* indicatur.

Stiffly branched xeromorphic microphyllous and drought-deciduous, potentially arborescent shrubs at anthesis 1–5(–6) m, the branches of 3 sorts: a) pliant hornotinous long-shoots with developed internodes, floriferous at least distally; b, present in only some plants) shorter, stiffly divaricate branchlets lateral to long-shoots, usually floriferous, early leafless and functioning as thorns; and c) knotlike brachyblasts axillary to fallen primary lvs and consisting of a tuft of stipules and a fascicle of usually smaller lvs; old stems gray or castaneous, commonly lenticellate, the young stems, lf-stalks and inflorescence commonly strigulose with appressed or subappressed gray hairs to 0.15–0.45 mm or (locally) more densely pilosulous throughout with erect lustrous hairs to 0.3–0.5 mm (exceptionally few random setae to 1 mm), the firm subconcolorous lfts either olivaceous or pallidly subglaucescent, varying from glabrous on both faces to dorsally strigulose or pilosulous on both faces, the racemes of proportionately large showy fls arising mostly from axils of lvs on long-shoots, forming an open or contracted panicle, a few rarely in the dry season from leafless branchlets.

Stipules erect, narrowly lanceolate or setiform (1–)2–7 mm, those of primary lvs often early caducous, those of fasciculate lvs subulate or acicular 0.7–3.5 mm, early dry but persistent.

Primary lvs (0.9–)1.2–4.5(–5) cm; petiole including slender wrinkled pulvinus 2.5–11 mm, at middle 0.2–0.5 mm, narrowly sulcate; rachis (2.5–)4–25(–37) mm, the longer interfoliolar segments (1–)2–7 mm; gland 0; pulvinules 0.4–1 mm; lfts (2–)3–5(–6) pairs, little or not accrescent distally, broadly or narrowly obovate, obtuse mucronulate (aristulate) or shallowly emarginate, the distal pair 4–15 × 2.5–7.5 mm, 1.6–2.3 times as long as wide, semicordate or inequilaterally rounded at base, the margin plane, the venation above fully immersed or faintly raised, the midrib prominulous beneath either its whole length or only proximally, the (3–)4–6 pairs of camptodrome with rare intercalary secondary veins and sometimes tertiary connecting venules immersed or weakly raised above, either immersed and discolored or prominulous beneath; fasciculate lvs similar but mostly smaller, the lfts 2–4 pairs.

Racemes loosely (2–)3–13(–17)-fld, 1–3 fls expanding on a given day, the axis together with peduncle becoming (0.5–)1–5(–7) cm, sometimes persistent and ultimately thorny; bracts subulate, narrowly lanceolate or ovate 1.5–3 mm caducous; pedicels 12–24(–26) mm; young fl-buds globose, glabrous, puberulent or rarely pilosulous; sepals yellowish, greenish or purplish-brown, strongly graduated, the smallest outermost one ovate or oblong-elliptic 3–5(–5.5) mm, the innermost amply obovate-suborbicular membranous-margined 6.5–10 × 5.5–8 mm; petals glabrous yellow, fading yellow brown-veined, conspicuously clawed, 4 oblong-obovate or -elliptic unequal, one abaxial one thick-clawed and highly asymmetric, its blade turned at ±90° to claw, either longer or shorter than the rest, the longest petal (caution: fls with expanded petals but immature androecium may be smaller) 16–28 mm; androecium glabrous, the 3 staminodes 1–1.8 mm wide, the filaments of 4 median stamens 0.9–2 mm, of 2 latero-abaxial ones 2–3.5 mm, of the centric abaxial one (4.5–)5.5–9 mm, the 7 fertile anthers of similar girth and except for direction of beak of similar outline, all gently incurved, the 4 median ones 4–6.2 × 1.4–2 mm, their divaricate beak 0.35–0.6 mm, the body of 3 abaxial anthers 4.8–6.2 × 1.7–2 mm, its porrectly incurved beak 0.7–1.5 mm, the fertile anthers all dehiscent by 2 short slits; ovary and style glabrous, the latter 0.7–2 × 0.2–0.4 mm, very slightly incurved distally, the minute stigmatic cavity terminal; ovules 22–40.

Pods obliquely ascending and randomly declined from ascending pedicels, the stipe 5–8 mm, the linear plano-compressed straight or gently decurved body (7–)8–24 × 0.55–0.75(–0.8) cm, bicarinate by the cordlike sutures, the firmly papery or subcoriaceous dull or highly lustrous purplish-green or livid, finally brown or nigrescent valves becoming delicately transverse-reticulate, at maturity either plane or raised over each seed in the form of 1 or 2 transverse ridges, the interseminal septa thickened but very narrow, the locules 4–7.5 mm long, commonly as long or a little longer, rarely a little shorter than wide; seeds strongly compressed parallel to the valves, in outline rhombic, rhombic-paddle-shaped or almost square 4–5.5 × 3–4 mm, the brown or pinkish-brown testa smooth dull, the lustrous elliptic areole 0.6–1 × 0.3–0.7 mm.

In a study of *Cassia* for the Chihuahuan Desert Flora we developed (Irwin & Barneby, 1975, p. 16, fig. 2) a comprehensive concept of *C. wislizeni* that encompassed, as three vicariant varieties, three species of Britton's genus *Palmerocassia*. We are now obliged to stretch the definition of the species a little further so as to accommodate *Cassia pringlei,* thought to differ decisively from *C. wislizeni* and all close allies in the long pod and in development of divaricate thorny branchlets lateral to annotinous long-shoots. These branchlets are, however, an uncertain feature of *C. pringlei,* striking when present but absent from some specimens, and the length of the pod, 16–24 cm when all 31–40 ovules develop successfully, is not decisively diagnostic, since the pod of other varieties, with 22–32 ovules, can reach a length of 16 cm. It should be emphasized that the calyx, the corolla, and the androecium, with its centric abaxial anther elevated beyond its neighbors on a long filament, remain essentially uniform and are exactly shared only by the closely related *S. monozyx* next following.

In the context of its genus, which is of tropical mesophytic derivation, we suppose that the southern var. *pringlei* represents the primitive core of *S. wislizeni* sens. lat. from which the other varieties, as they moved north into progressively colder and drier climates, were derived by selection. The order of presentation follows this hypothesis.

Key to the Varieties of *S. wislizeni*

1. Stipules of primary lvs promptly caducous, usually before maturity of the associated lf; valves of pod dull when ripe, not raised over seeds or, if so, then only as a single low transverse ridge; secondary venulation of lower face of lfts not or obscurely prominulous; s. Mexico n. to the s. lobe of Chihuahuan Desert (to n. Zacatecas and s. Nuevo León), wholly s. of lat. 25°N.
   2. Body of pod 16–24 cm; ovules 31–40; thorny secondary branchlets often but not always present; s. Mexico in lat. 16–20°N (Colima to Puebla s. to Oaxaca). 193a. var. *pringlei* (p. 573).
   2. Body of pod 9–15(–16) cm; ovules 25–32; no thorny secondary branchlets; s. Chihuahuan Desert in lat. 21–25°N (Querétaro to n. Zacatecas and s.-w. Nuevo León). 193b. var. *painteri* (p. 574).
1. Stipules of primary lvs more persistent, usually into or through maturity of the associated lf; valves of pod highly lustrous when ripe, elevated over each seed in the form of 2 transverse ridges, these sometimes joined at middle by a shorter vertical ridge; secondary venulation of lower face of lfts bluntly or sharply prominulous; centr. and n. Chihuahuan Desert n.-ward from lat. 25°N, extending feebly n.-w. into Sonora.
   3. Vesture of the whole plant appressed; desert n. from 27°N. 193c. var. *wislizeni* (p. 574).
   3. Vesture of the whole plant pilosulous; Mapimí Depression in centr. Chihuahuan Desert in lat. 25–26°N (n.-w. Durango and adjoining Coahuila). 193d. var. *villosa* (p. 575).

193a. Senna wislizeni (A. Gray) var. **pringlei** (Rose) Irwin & Barneby, stat. nov. *Cassia pringlei* Rose, Contrib. U.S. Nat. Herb. **5**: 194. 1899.—"Collected by Mr. C. G. Pringle, at Tomellin Station, State of Oaxaca, altitude 615 meters, July 14, 1897 (No. 7478); and by Mr. E. W. Nelson between Petlatcingo [=Petlalcingo] and Acatlan, State of Puebla, November 20, 1894 (No. 1996)."—Lectoholotypus, *Pringle 7478,* US! isotypus, GH!—*Palmerocassia pringlei* (Rose) Britton ex Britton & Rose, N. Amer. Fl. **23**(4): 254. 1930.

Cassia morelensis Greenman, Proc. Amer. Acad. Sci. **39**: 79. 1903.—"MEXICO. State of Morelos: near Jojutla . . . 12 May, 1901, C. G. Pringle, n. 9642 . . ."—Holotypus, GH! isotypus,

M!—Correctly equated with *C. pringlei* by Standley, 1922, p. 409 and with *Palmerocassia pringlei* by Britton & Rose, 1930, p. 254.

Strigulose with appressed hairs 0.15–0.3 mm or (locally) more densely pilosulous with erect hairs 0.3–0.5 mm; primary lvs mostly 1.2–4.5 cm, with (3–)4–5(–6) pairs of lfts up to 4–12.5 mm long; otherwise as given in key.—Collections: 49.

Thorn-forest and drought-deciduous brush-woodlands, sometimes on cliffs, on either volcanic or calcareous bedrock, (150–)350–1850 m, discontinuously widespread over interior arid valleys of s. Mexico, from Colima e. through Balsas Depression to Puebla, there extending e. to the forks of río Papaloapan on the Gulf slope, s. to centr. and s. Oaxaca.—Fl. (IV–)VI–X(–XI).—*Mulato*; *rompebota*(*s*); *tecahuil*; *huesillo*.

Normally var. *pringlei* is inconspicuously strigulose with appressed or subappressed hairs, in this respect agreeing with var. *painteri* and var. *wislizeni*. An otherwise identical plant pilosulous overall with erect and slightly longer hairs has been collected twice on the north fork of río Papaloapan in Puebla (near Coxcatlán and San Luis Tultitlanapa) and once in southeastern Oaxaca (near Yautepec). Morphologically this variant has the same claim to taxonomic recognition as has var. *villosa,* but it has not established an independent area of dispersal. The type-locality of *Cassia pringlei* is Tomellín on the south fork of the Papaloapan and several collections of typical var. *pringlei* have been made in the valley of Oaxaca, these together lying on a direct line between the known stations for the pilosulous form.

193b. Senna wislizeni (A. Gray) var. **painteri** (Britton & Rose) Irwin & Barneby, Phytologia **44**(7): 500. 1979. *Palmerocassia painteri* Britton ex Britton & Rose, N. Amer. Fl. **23**(4): 254. 1930.—"Type from Higuerillas, Querétaro, August 23, 1905, *Rose, Painter & Russell 9807*."—Holotypus, NY! isotypus, US!—*Cassia wislizeni* var. *painteri* (Britton & Rose) Irwin & Barneby, Sida **6**(1): 16, fig. 2. 1975.

Like var. *pringlei,* but thorny lateral branchlets 0, the vesture always strigulose, the inflorescence tending to be more compactly paniculate; primary lvs mostly 1.5–4 cm, with 3–5 pairs of lfts up to 5–15 mm; otherwise as given in key.—Collections: 22.

Larrea scrub and chaparral, 1450–2100 m, commonly on limestone, locally plentiful over interior upland San Luis Potosí, thence extending into n.-e. Zacatecas, extreme s. Coahuila, s. Nuevo León, extreme s.-w. Tamaulipas and extreme n. Querétaro, the range nearly coextensive with the s. lobe of the Chihuahuan Desert.—Fl. V–VIII.—*Consa* (Querétaro).

The var. *painteri* is both morphologically and geographically intermediate between var. *pringlei* and var. *wislizeni,* combining the former's indistinct venulation of the leaflets and dull texture of the almost plane pod valves with the latter's relatively short pod, fewer ovules and lack of divaricate thorny branchlets.

193c. Senna wislizeni (A. Gray) Irwin & Barneby var. **wislizeni**. *Cassia wislizeni* A. Gray, Pl. Wright. **1**: 60. 1852, sens. str.—"[Chihuahua:] Carizal and Ojo Caliente, south of El Paso, *Dr. Wislizenus*."—Holotypus (Isely, 1975, p. 210), GH!—*Palmerocassia wislizeni* (A. Gray) Britton ex Britton & Rose, N. Amer. Fl. **23**(4): 254 ("*Wislizenii,*" an impossible genitive). 1930.

Cassia wislizeni sensu Bentham, 1871, p. 574; Wooton & Standley, 1915, p. 334; Turner, 1959, p. 76, map 37 (Texas only); Kearney & Peebles, 1960, p. 405; Isely, 1975, p. 133, map 61 (U.S. only). *Cassia wislizeni* var. *wislizeni* sensu Irwin & Barneby, 1975, p. 16 (in clave), fig. 2 (map).

Secondary thorny branchlets 0, but the raceme axes sometimes stiffly or almost pungently persistent; vesture strigulose, the hairs up to 0.2–0.45 mm; primary lvs 12–34 mm, with 3 or 3–4 pairs of lfts, the largest 4–10 × 2.5–7.5 mm; ovules 22–25; otherwise as given in key.—Collections: 25.

Larrea scrub in foothills of desert mountains, often but not exclusively on limestone, 850–1570 m, widespread over the n. lobe of the Chihuahuan Desert from s.-e. Chihuahua and n.-w. Coahuila to the Rio Grande valley in trans-Pecos Texas (upstream from Presidio Co.), far s.-w. New Mexico and extreme s.-e. Arizona, thence feebly s.-w. and down to ±500 m into the Sonoran Desert in n.-centr. Sonora.—Fl. (IV–)VI–XI.—*Pinacate*; *yerba del pinacate*.

This variety and the next differ collectively from the two foregoing in the glossy texture of the pod's valves which become elevated over each seed in the form of two parallel transverse ridges across the proximal and distal edges of the locule. These ridges are sometimes connected by a vertical one, forming over the seed a depressed tent or blunt-angled pyramid such as is characteristic of some varieties of *S. pallida*.

193d. Senna wislizeni (A. Gray) var. **villosa** (Britton & Rose) Irwin & Barneby, Phytologia **44**(7): 500. 1979. *Palmerocassia villosa* Britton ex Britton & Rose, N. Amer. Fl. **23**(4): 254. 1930.—"Mapimi, Durango, October 21–23. 1898, *E. Palmer 518*."—Holotypus, NY! = NY Neg. *9393*; isotypi, GH, US!—*Cassia wislizeni* var. *villosa* (Britton & Rose) Irwin & Barneby, Sida **6**(1): 16. 1975.

Like var. *wislizeni* except for the pilosulous vesture; primary lvs 10–25 mm, with 2 or 3 pairs of lfts, the largest 3–8 mm; body of pod 7.5–16 × 0.6–0.8 cm, 24–30-ovulate, otherwise as in var. *wislizeni*.—Collections: 5.

Larrea scrub and matorral, in canyons and foothills of desert mountains, strongly calci (or gypso)-phile, 1250–1860 m, local around the s.-w. periphery of Mapimí Depression and the lower Nazas valley in far s.-w. Coahuila and adjacent Durango.—Fl. VI–IX.

This variety is weakly distinguished from var. *wislizeni* by its loose vesture but so definitely segregated geographically that we think it useful to recognize it formally. The leaflets are on the average fewer by one pair than in var. *wislizeni* and lead on to the related *S. monozyx*, its vicariant neighbor in south-central Coahuila, in which all leaves are bifoliolate.

194. Senna monozyx (Irwin & Barneby) Irwin & Barneby, Phytologia **44**(7): 500. 1979. *Cassia monozyx* Irwin & Barneby, Sida **6**(1): 16. 1975.—"MEXICO. Coahuila: limestone ridge, 1800 m, Valle Seco in Sierra de la Paila, mun. General Cepeda, 7.VI [properly 6.VI]. 1944, *G. Hinton 16565*."—Holotypus, NY!

Stiffly branched subspinescent microphyllous shrubs up to 1(–?) m with crooked fuscous trunks, resembling *S. wislizeni* vars. *wislizeni* and *villosa* in habit, fl and pod but the primary and fasciculate lvs homomorphic and all 2-foliolate, the young branchlets and sometimes the dorsal face of lfts strigulose with appressed gray hairs to 0.25 mm, the foliage subconcolorous, the firm lfts yellow-green or olivaceous-subglaucescent, the few-fld racemes forming a small exserted panicle.

Stipules of both primary and fasciculate lvs setiform 1–4 mm, persistent into or past the maturity of associated lf.

Lvs 5–15 mm; petiole including scarcely differentiated pulvinus (1–)1.5–5 mm, at middle 0.2–0.4 mm diam, rounded dorsally, indistinctly sulcate ventrally, the terminal appendage setiform, early dry; pulvinules 0.3–0.7 mm; lfts 1 pair obovate or cuneate-obovate 3–9(–10) × 2–6(–6.5) mm, (1–)1.1–1.6 times as long as wide, at apex apiculate or less often retuse-emarginate mucronulate, the margin plane, the midrib and 3–5(–6) pairs of secondary veins bluntly prominulous on both faces.

Racemes mostly 1–5(–7), rarely to 15-fld, the axis including peduncle (0–)1–3(–5) cm, sometimes thornily persistent into the second year; bracts ovate acute or subulate 1–1.5 mm caducous; pedicels 12–20 mm; fl-buds globose glabrous; sepals well graduated, the firm brownish outermost one 4–4.5 mm, the broadly obovate-suborbicular pallid-margined innermost one 6.5–8 mm; petals of *S. wislizeni,* the longest 18–21 mm; androecium of *S. wislizeni*; ovary glabrous, the stout style 1–2 mm, gently incurved and tapering toward apex, ±0.25 mm diam just below the minute terminal stigmatic cavity; ovules (10–)12–16.

Pod irregularly ascending-spreading, the stipe ±5 mm, the body 8–10 × 0.65–0.8 cm, in compression, texture and color like that of *S. wislizeni* var. *wislizeni,* the seed-locules 4.5–6 mm long; seeds strongly compressed parallel to the valves, in outline rhombic-paddle-shaped 4.5–5 × 3 mm, the dull or sublustrous brown valves smooth, the glossy elliptic areole 0.9–1 × 0.4–0.5 mm.—Collections: 12.—Fig. 35.

Rocky ridges, bajadas and canyon-slopes in and around the foothills of calcareous desert mountains, in the matorral-chaparral transition zone, 750–1800 m, locally frequent in w.- and s.-centr. Coahuila (Sa. Mojada e. to Sa. de la Madera, s. to Sa. de la Paila).—Fl. V–IX.

Under *S. wislizeni* we have traced northward from southern Mexico to southern Arizona a series of xeromorphic modifications involving progressive diminution of foliage, loss of leaflets and decline in ovule-number. *Senna monozyx* represents the climax of this series. While it retains essentially unaltered the flower and the glossy corrugated pod of vicariant *S. wislizeni* var. *wislizeni* and var. *villosa,* its leaves have shrunk to a single pair of diminutive leaflets, its ovules have declined from 22 or more to about 12–16, and it has established its independent range of dispersal in the extremely arid desert mountains of south-central Coahuila. The similarly but more grossly bifoliolate *S. unijuga* is a parallel but not directly related product of similar selective forces operative in the Tehuacán Desert far to the south.

195. Senna unijuga (Rose) Irwin & Barneby, comb. nov. *Cassia unijuga* Rose, Contrib. U.S. Nat. Herb. **5**: 195. 1899.—"Collected by Mr. C. G. Pringle in the eastern part of the State of Puebla [Mexico], near Tehuacan, August 5, 1897 (No. 6773)."—Holotypus, US! isotypi, GH, MEXU, MO, NY, US!—*Palmerocassia unijuga* (Rose) Britton ex Britton & Rose, N. Amer. Fl. **23**(4): 254. 1930.

Cassia unijuga sensu Standley, 1922, p. 409.

Stiffly branched but not or scarcely thorny, often gnarled and crooked microphyllous shrubs and treelets at anthesis (0.3–)0.5–3 m, with gray or fuscous terete, ultimately lenticellate trunks and branches, the hornotinous stems, lvs, and in-

Fig. 35. *Senna monozyx* I. & B. Flowering branch ×1; detached stamens ×4; pod ×1; seed ×3 (*Pinkava 4212*).

florescence up to the glabrate interior sepals densely pilosulous with erect or ascending, mostly straight hairs up to 0.4–1 mm, the short bifoliolate lvs homomorphic, all solitary on long-shoots or a few fasciculate on axillary brachyblasts but these scarcely smaller, the racemes of few, proportionately large fls lateral to long-shoots, irregularly paniculate.

Stipules erect linear-attenuate or setiform, often livid, those of primary lvs mostly 2–4.5(–6) mm, tardily deciduous, those of brachyblasts shorter persistent.

Lvs 8–15 mm; petiole including little differentiated pulvinus 1–4.5(–5.5) mm, at middle 0.3–0.5 mm diam, obscurely depressed-sulcate ventrally; gland 0; pulvinules 0.4–0.6 mm; lfts 1 pair, broadly flabellate-obovate or obcordate, emarginate mucronulate or subtruncate acuminulate, 5–12 × 5–11 mm, 0.8–1.3 times as long as wide, at base broadly rounded or cordate on proximal and more narrowly so on distal side, the margin strongly revolute, the upper face veinless, the midrib always and often 3–5 pairs of weak, widely ascending secondary veins prominulous beneath.

Racemes subumbellately 1–3-fld, the very short axis together with peduncle (4–)8–27 mm; bracts submembranous purplish or livid, lance- or elliptic-acuminate 4–5.5 × 1.2–2 mm deciduous; pedicels (7–)10–15 mm; young fl-buds globose pilosulous, the glabrate inner sepals early emergent; sepals firm livid except for pallid margins, well graduated, the smallest outer one elliptic obtuse 4.5–6 mm, the broadly obovate-suborbicular innermost one 7.5–9 × 5.5–7 mm; petals glabrous yellow fading brownish-yellow dark-veined, 4 subhomomorphic except for size, oblong-obovate beyond the claw, one abaxial thick-clawed and heteromorphic, its blade turned at ±90° to claw, the longest petal (fully expanded) 18–26 mm; androecium glabrous, the staminodes 1.1–1.4 mm wide, the 7 fertile stamens only a little differentiated, the anthers all oblong from bluntly sagittate base a little incurved and ±1.4–2 mm diam, somewhat dorsoventrally compressed, the filaments of 4 median stamens 1.2–1.5 mm, of 2 latero-abaxial ones 1.5–2.5 mm, of the centric abaxial one 3–4.5 mm, the beaks of 4 median anthers latero-terminal very short, those of 3 abaxial more porrect but only 0.5–0.6 mm, all dehiscent by short slits; ovary glabrous; style (sometimes indistinctly demarcated from ovary) ±1.2–1.5 mm, gently incurved, 0.25–0.5 mm diam just below the terminal minute stigmatic cavity; ovules ±20–24.

Pod (little known) ascending, the stipe 6–9 mm, the plano-compressed straight or slightly decurved body ±9–13 × 0.6 cm, bicarinate by the broad sutures, the stiffly papery, brown or atropurpureous glossy, faintly venulose valves scarcely raised over the seeds, their thickened margins in age cracking transversely the whole length of the pod, the seed-locules ±4.5 mm long; seeds (poorly known) compressed parallel to the valves, obtusely rhombic in outline, ±4.3 × 3.1 mm, the dull purplish-castaneous testa minutely pebbled, the lustrous elliptic areole ±1 × 0.4 mm.—Collections: 13.

Thorn-forest and thin oak woodland, on calcareous, gypseous and metamorphic bedrock, 1300–2000 m, narrowly localized in the Tehuacán Desert in s.-e. Puebla and closely adjoining n.-w. Oaxaca, Mexico.—Fl. VI–X(–XI).

A very distinct species, readily recognized by its villous-pilosulous stems and foliage, small bifoliolate leaves, broadly flabellate-obcordate, marginally revolute leaflets, and large asymmetric corolla. The transverse cracks that appear along the thickened margins of the ripe pod-valves recall *S. atomaria*.

Fvi. ser. ISANDRAE Irwin & Barneby

Senna sect. **Peiranisia** ser. **Isandrae** Irwin & Barneby, nom. et stat. nov. *Isandrina* Rafinesque, Sylva Tellur. 126. 1838, pro gen.—Sp. typica (unica): *I. arborescens* Rafinesque (nom. illegit.) ≡ *Cassia emarginata* auct., non Linnaeus ≡ *Senna emarginata* Rafinesque = **S. atomaria** (Linnaeus) Irwin & Barneby.

Cassia sect. *Chamaesenna* ser. *Floridae* Bentham, Trans. Linn. Soc. London **27**: 547. 1871, minore ex parte, quoad "**Molles" (omnes), sed typ. exclus.

Gaumerocassia Britton ex Britton & Rose, N. Amer. Fl. **23**(4): 252. 1930, pro gen.—Sp. typica: *G. peralteana* (Humboldt, Bonpland & Kunth) Britton & Rose (quoad nom., non descr.)

≡ *Cassia peralteana* Humboldt, Bonpland & Kunth ≡ **Senna peralteana** (Humboldt, Bonpland & Kunth) Irwin & Barneby.

Pseudocassia Britton & Rose, N. Amer. Fl. **23**(4): 230. 1930, ex parte, quoad spp. 2 et 3, exclus. typ.

Isandrina sensu Britton & Rose, 1930, p. 268.

Corolla of *Peiranisia*; sepals strongly graduated; 7 fertile stamens subisomorphic, the thin-walled anthers bluntly truncate, their thecae not confluent distally, the short beaks latero-terminal or infraterminal, dehiscent by separate slits.—Shrubs, some vinelike, and trees; stipules erect narrow caducous; petiolar glands 0; lfts (1–)2–12 pairs; pedicels in 1 sp. (*S. peralteana*) bracteolate; ovules 8–120; pod linear or linear-oblong plano-compressed, the ripe valves either papery or coriaceous and dehiscent through the sutures, or woody and indehiscent, then tardily fragmenting through interseminal septa; seeds transverse, broadside to the valves, areolate.—Spp. 7, of tropical Mexico (1 n. to s. Sonora), Central America, the Antilles and n. South America, 2 s. to n. Peru.

The series *Isandrae,* of which all members except the local *S. sousana* and *S. haughtii* had been collected prior to 1871, was defined by Bentham as a subsidiary group, descriptively termed Molles, of a heterogeneous sect. *Chamaesenna* ser. *Floridae,* herein typified by a chamaesenna, *S. siamea.* It is characterized by an unmistakable androecium, for the fruit varies greatly in length correlated with ovule-number, in texture of valves and in dehiscence. The pod of *S. racemosa, S. andrieuxii* and *S. haughtii* is relatively short, few-ovulate, stiffly papery and dehiscent through the sutures, whereas that of *S. atomaria, S. mollissima, S. sousana* and *S. peralteana* is much elongate, lignescent or stiffly coriaceous and indehiscent except by tardy transverse fracture. This dichotomy was evaluated by Britton & Rose at generic level, and doubtless does correspond with divergent evolutionary paths. It is here reflected in the order of presentation, the first three species, equivalent to *Isandrina,* having the more generalized sennoid pod, the last four, equivalent to *Gaumerocassia,* having the more specialized pod that leads on to the climactic *Excelsae,* next below.

The *Isandrae* are semi-xerophilous, withstanding long periods of desiccation, the more advanced species shedding their leaves in the dry months and sometimes flowering from leafless branchlets. The bracteolate pedicels of *S. peralteana* are a rare anomaly in the genus and are known in no other species of sect. *Peiranisia.*

Critical reevaluation of the types of the two commoner species known to Bentham and subsequent generations as *Cassia emarginata* and *C. atomaria* show that the names of both have been misapplied. The epithet *emarginata* disappears into the synonymy of *S. bicapsularis* and is confusingly replaced by *atomaria,* which in turn gives place to an unfamiliar *mollissima.* The following equations encapsulate the changes:

Cassia emarginata Linn. = **Senna bicapsularis** (L.) Roxb.
Cassia emarginata sensu Benth. = **Senna atomaria** (L.) Irwin & Barneby
Cassia atomaria Linn. ≡ **Senna atomaria** (L.) Irwin & Barneby
Cassia atomaria sensu Benth. = **Senna mollissima** (Willd.) Irwin & Barneby

Key to the Species of ser. *Isandrae*

1. Pod 7–19.5 cm long, plano-compressed, papery, inertly dehiscent along both sutures; ovules 8–28; either lfts of larger lvs 6 or more pairs, or the largest lfts only 2–6 cm long, or both.
   2. Widespread in Mexico and w. Cuba; ovules 14–28.

3. Lfts of larger lvs 4–12 pairs, but if less than 6 pairs then only thinly subappressed-pilosulous and the ovary and pod glabrous; range as just indicated. 196. *S. racemosa* (p. 580).

3. Lfts of larger lvs 3–5 pairs, densely white- or gray-lanate beneath; ovary lanate and pod rusty-pilosulous; local in s. Puebla and adj. Oaxaca. 197. *S. andrieuxii* (p. 585).

2. Local in n.-w. Peru; ovules 8–10. 198. *S. haughtii* (p. 587).

1. Pod 2–4 dm long, compressed but thick and fleshy when first formed, becoming woody, indehiscent along the sutures but tardily breaking transversely along lines of interseminal septa; ovules 46–120; lfts (1–)2–6 pairs, the largest (maturely expanded) lfts (3–)5–16 cm long.

4. Free-standing trees and shrubs, the old stems never armed with stipular thorns; pedicels ebracteolate; venulation of upper face of lfts commonly immersed or almost so; widespread in Mexico, West Indies, n. S. America; Pacific Ecuador and n. Peru.

5. Lfts nearly always obovate or elliptic-obovate, broadest near or beyond middle of blade, and obtuse; racemes commonly less than 20-fld and the fls themselves relatively small, the longest petal 13–23 mm, but if more than 20-fld the longest petal less than 20 mm; range as indicated except Ecuador-Peru. Lvs often but not consistently hysteranthous, the fls mostly arising directly from annotinous branchlets. 199. *S. atomaria* (p. 588).

5. Lfts ovate-acuminate, broadest below middle; either the racemes well over 20-fld, or the fls themselves large (the longest abaxial petal at least 25 mm), or if the inflorescence that of *S. atomaria* then arising from contemporary branchlets and axillary to newly expanding lvs; of bicentric range in Pacific Mexico (s.-ward from Sinaloa) and Guatemala, and in dry coastal Ecuador and Peru. Lvs always synanthous.

6. Racemes pliantly pendulous from lf-axils, and their axis (at least in Mexico) elongating prior to anthesis of first fls, the fl-buds consequently racemosely projected well above expanded fls; fls large, the longest abaxial petal (measured along its arcuate midvein) 19–36 mm. Pacific N. and S. America, almost exclusively in lowland tierra caliente, in s. Mexico at 0–250, exceptionally up to 800 m. 200. *S. mollissima* (p. 592).

6. Racemes erect from lf-axils, their axis not or scarcely elongating prior to anthesis of first fls, these and the buds subcorymbose; fls small, the longest abaxial petal ±15–17 mm. Local in montane woodlands of Sa. Madre del Sur in Oaxaca, at 1300–1850 m. 200bis. *S. sousana* (p. 596).

4. Scandent or sarmentose shrubs and vines of Yucatán Peninsula and adjacent Belize, Chiapas and Guatemala, the annotinous and older stems armed with stipular thorns (not present in all herbarium spms); pedicels minutely 2-bracteolate; secondary and tertiary veins of lfts commonly prominulous on upper face. Lvs synanthous. 201. *S. peralteana* (p. 597).

196. Senna racemosa (P. Miller) Irwin & Barneby, comb. nov. *Cassia racemosa* P. Miller, Gard. Dict. ed. 8, *Cassia* no. 19. 1768.—Typus infra sub var. *racemosa* indicatur.

Round-headed subarborescent shrubs and trees of rapid growth with smooth gray trunks, at anthesis 3–10 m, the trunk sometimes reaching 3 dm diam but commonly much more slender, variably pilosulous or strigulose throughout or almost so with straight, forwardly subappressed or (more often) with some or all spreading or loosely ascending incurved or sinuous hairs up to 0.2–0.8 mm, the vesture of the foliage gray or when young golden, of the inflorescence golden or rusty, the lvs bicolored, dull or lustrous green above, paler beneath, the several or many pluriflorous racemes axillary to lvs crowded toward the tips of hornotinous branchlets and together forming a broadly round- or flat-topped, not or scarcely exserted corymbiform panicle.

Stipules erect, subulate, linear- or lance-attenuate or linear-oblanceolate 2.5–9 mm, less than 1 mm wide, caducous.

Lvs (7–)8–26 cm; petiole including firm, moderately dilated pulvinus 2–4.5 cm, at middle (0.7–)0.8–1.8 mm diam, subterete, very shallowly or obscurely sulcate ventrally; rachis (3–)3.5–17 cm; pulvinules 1–3.2 mm; lfts of larger lvs 4–12 pairs, moderately accrescent upward, the distal pair elliptic, elliptic-oblong, -obovate or broadly obovate, at apex deltately acute or obtuse to subemarginate and mucronulate, 2–6 × 0.9–2.4(–2.7) cm, 1.8–3.2 times as long as wide, at base when broad asymmetrically rounded-subcordate, when narrow broadly inequilaterally cuneate, the midrib and 5–9(–10) pairs of camptodrome (with occasional intercalary) secondary veins either immersed or finely prominulous above, commonly prominulous beneath, the tertiary venulation either expressed or invisible.

Racemes densely 15–60(–120)-fld, the fl-buds in praefloration nodding and crowded into a blunt cone surrounded by several simultaneously expanded fls, the axis together with peduncle becoming 4–13(–20) cm; bracts firm, early spreading or reflexed, lance- or triangular-acuminate or -caudate 3–7 × 0.8–2.8 mm, deciduous at or soon after anthesis; pedicels 16–30 mm; fl-buds obovoid-subglobose, rusty-pilosulous when young but the inner sepals often glabrate marginally; sepals moderately or strongly graduated, obovate-suborbicular cymbiform, the outer ones 2.5–7 mm, the innermost 3.5–10 mm; petals rich dark yellow, usually glabrous dorsally beyond the puberulent claw, 4 subhomomorphic broadly obovate beyond the short slender claw, obtuse or emarginate, (9–)10–16.5 mm, 1 abaxial strongly heteromorphic subsessile obliquely oblong as long or slightly longer than the rest (9–)10–18 mm; staminodes glabrous, the small quadrate or cordate blade 0.8–1.2 mm wide; fertile stamens 7, homomorphic except slightly diminished abaxially, the filaments 1.2–2.5 mm, the oblong, slightly incurved or gently sigmoid glabrous or thinly weakly pilosulous anthers (2.6–)3–5.5 × 1.1–1.7, each theca 2-sulcate lengthwise, obtuse-subtruncate distally, the very small beaks lateral, opening by subparallel slits; ovary either glabrous or densely pilosulous; style 1.3–4 mm, gently incurved and slightly narrowed distally, 0.25–0.4 mm diam near the minute terminal stigma; ovules (16–)18–28.

Pod obliquely pendulous, the stipe 2.5–7 mm, the linear plano-compressed body straight or slightly arched downward, 8–19.5 × 1–1.6 cm, broadly flat-keeled by the sutures, the stiffly papery, reddish-brown or finally nigrescent valves prominently transverse-venulose, separating along both sutures, the seed-locules 4–10 mm long, as wide as the cavity; seeds transverse, compressed parallel to the valves, in outline oblong-obovate 4.4–7 × 2.6–4 mm, the testa beneath the cracking or flaking waxy outer integument pale fawn- or pinkish-brown smooth sublustrous, the areole linear or narrowly oblong-elliptic 1.4–3.6 × 0.5–0.9 mm.

The ample material of Mexican *Senna* that has been referred in herbaria, either by Britton directly or by reference to Britton's conspectus of *Gaumerocassia* (1930, p. 252), to *Cassia liebmannii, C. ekmaniana* and *C. peralteana* is essentially uniform in habit, inflorescence, flower-structure, pod and seed, and can readily and conveniently be accommodated within one species. Vicariant groups of populations differ between themselves in pubescence of foliage and ovary, in leaflet-number, in size of sepals, and in length of style and pod, but the differences are neither great nor absolute and fail to counterbalance the overall similarities. For this composite species we adopt the old epithet *racemosa,* misapplied by Bentham to what is here called *Senna silvestris* and never hitherto satisfactorily identified. Within the species we can make out five varieties, neatly separated geographically but only weakly so morphologically.

As redefined, *S. racemosa* has obvious close relations in *S. andrieuxii,* which see for comment, and *S. peralteana,* which is easily recognized by the reflexed stipular thorns of the mature stem and the highly asymmetric, narrowly racemose flowers. More numerous leaflets and a dehiscent pod not greatly thickened along the sutures separate *S. racemosa* from the partly sympatric *S. atomaria.* The dense corymbiform panicle of racemes, each of which bears simultaneously several expanded, deeply campanulate, rich yellow flowers elevated to or beyond the cone of nodding buds, combined with plurifoliolate glandless leaves and longitudinally sulcate anthers distinguish *S. racemosa* from all other Mexican and Central American sennas.

Key to the Varieties of *S. racemosa*

1. Ovary and pod glabrous; mature lfts reticulately venulose on upper face; e. Mexico (s.-e. San Luis Potosí to Yucatan Peninsula), n.-e. Guatemala (Petén) and far w. Cuba; local in s.-w. Mexico (w. Michoacán and adjoining Colima).
   2. Lfts 4–8 pairs, their blades subappressed-pubescent, the upper face only minutely puberulent or glabrate; pod 8–12 cm, the seed-locules relatively narrow, 4–6 mm long; Yucatan Peninsula (Campeche to Quintana Roo) and adjacent Guatemala, w. to n. Oaxaca; far e. Cuba. 196a. var. *racemosa* (p. 582).
   2. Lfts of larger lvs 8–10 pairs; pod (unknown in s.-w. Mexican var. *coalcomanica*) 11–19.5 cm, the seed-locules 6–10 mm long; distantly allopatric.
      3. Lfts densely softly pilosulous on both faces; local at ±1000–1400 m in valley of río Moctezuma in s.-e. San Luis Potosí and closely adjoining Querétaro and Hidalgo. 196b. var. *moctezumae* (p. 583).
      3. Lfts subappressed-strigulose; local near 140–500 m in w. Michoacán and adjoining Colima. 196c. var. *coalcomanica* (p. 584).
1. Ovary and pod densely rusty-pilosulous; pinnate venulation of lfts prominulous beneath only and tertiary venulation of both faces fully immersed; s. Mexico (s.-w. México to Chiapas).
   4. Lfts of larger lvs 9–12 pairs; s.-e. México, Morelos and adjoining Guerrero. 196d. var. *sororia* (p. 584).
   4. Lfts of larger lvs 6–9 pairs; e. Oaxaca and Chiapas. 196e. var. *liebmannii* (p. 585).

196a. Senna racemosa (P. Miller) Irwin & Barneby var. **racemosa**. *Cassia racemosa* P. Miller, l.c., sensu str.—". . . sent me from Carthagena, by the late Mr. Robert Millar."—Holotypus, labelled in Miller's hand, BM! = NY Neg. *161* = Bailey Hort. Neg. *05169*.

Cassia ekmaniana Urban, Symb. Antill. **9**: 442. 1928.—"Cuba: Prov. Pinar del Río in peninsula Guanahacabibes . . . inter Remates et Yayales, m. Mart. [1924] flor. et fruct.: [*Ekman*] n. 18738 . . ." –Holotypus, presumably †B, not seen; isotypi, G, GH, NY (2 sheets)! paratypi, *Millspaugh 1481* ex insula mexic. Cozumel, F, NY!—*Gaumerocassia ekmaniana* (Urban) Britton ex Britton & Rose, N. Amer. Fl. **23**(4): 253. 1930.

Gaumerocassia peralteana sensu Britton & Rose, 1930, l.c., quoad descr., basionym. exclus.

Foliage strigulose-pilosulous, the leaf-hairs commonly subappressed, the lfts minutely puberulent or glabrous above; lvs (7–)9–17(–20) cm; lfts 4–8 pairs, the largest (2.1–)2.6–6 × (1–)1.2–2.4(–2.7) cm, varying (sometimes on one plant) from elliptic or elliptic-oblanceolate subacute to obovate obtuse, the secondary veins 7–10 pairs from midrib, the mature blades reticulate on upper face; calyx relatively small, the outer sepals 2.3–3.5 mm, the glabrate inner ones 3.5–5.5 mm; 4 petals (9–)10–14 mm, the odd petal (9–)10.5–15.5 mm; anthers glabrous; ovary glabrous; style 1.3–2.2(–2.4) mm; body of pod 8–12 × 1–1.2(–1.5) cm.—Collections: 39.

Low hills and plains, in drought-deciduous brush savanna below 200 m, widespread over Yucatán Peninsula in e. Campeche, Yucatán and Quintana Roo (including Isla Cozumel), Mexico and n.-e. Guatemala (Petén), extending weakly w. into n. Oaxaca (distr. Tuxtepec) and localized on Guanahacabibes Peninsula across the Yucatan Channel in w. Pinar del Río, Cuba.—Fl. I–III, VIII, the fr. long persisting after dehiscence.—*Yaax-haben* (Maya).

The relatively widespread var. *racemosa* is characterized by thinly pubescent foliage that becomes reticulately venulose when mature, a glabrous brachystylous pistil, and a relatively short pod. The latter, however, is no fewer-seeded than the long pod of other varieties, but accommodates each seed in a narrower locule.

The history of the name *Cassia racemosa,* which has only twice been associated correctly with a biological reality, has been clouded by vain efforts to apply Miller's description to a plant native to northern Colombia. The name was as early as 1797 reduced by Martyn, in his revised edition of Miller's Dictionary, to the synonymy of *C. emarginata,* but this was an error. The type-collection was examined by Vogel and briefly described in 1841 (Linnaea **15**: 70) but not identified with any species recognized in Vogel's revision of the genus (1837), where the name had been overlooked. Bentham (1870, p. 126; 1871, p. 549) revived *C. racemosa* for the taxon known to us as *Senna silvestris* which, while widely dispersed over much of forested tropical South America, has not been traced further northwestward than the Amazon-Orinoco divide in southeastern Colombia and the Maracaibo Basin in Venezuela. Britton & Killip (1936, p. 169) were unable to recognize Miller's species in any Colombian cassia, and the supposed origin at Cartagena seems certainly to be mistaken, either due to a lapse of memory on Miller's part or to mixture of materials at Chelsea. Miller's plant (BM) was critically examined once again by Amshoff (1939, p. 23), who rejected Bentham's interpretation but was unable to match it with any known species. In 1898 Pollard (in Millspaugh, Field Mus. Bot. **1**: 366) had applied the name *Cassia racemosa* correctly to the arborescent senna of Yucatan Peninsula (Izamal, *Gaumer 354*; Mérida, *Schott 974*), but without explanatory comment; his identification was accepted by Standley (1922, p. 411) but either overlooked or repudiated by Britton & Rose (1930). The attempt by Britton (sub *Gaumerocassia*) to distinguish between a larger-leaved *G. ekmaniana* from a smaller-leaved, sympatric *G. peralteana* is clearly ineffective even when applied to the specimens actually annotated by Britton himself, and involves a misinterpretation of genuine *Cassia peralteana* H.B.K., the earliest name for the species described in North American Flora as *Pseudocassia anisopetala* and *Ps. petensis*.

196b. Senna racemosa (P. Miller) var. **moctezumae** Irwin & Barneby, var. nov., pistillo glaberrimo brachystylo cum var. *racemosa* congruens sed ab ea pube omnium partium densa pilosula aureola, foliorum majorum foliolis 8–10 (nec 4–8)-jugis, legumine longiore 10–19.5 (nec 8–12) cm longo patriaque remota submontana abstans.—MEXICO. Hidalgo: mountains along Rte. 85 ±1 mile n.-e. of Jacala, 23.III.1961 (fl), *Robert Merrill King 4212*.—Holotpyus, NY; isotypi, MICH, US.—Ibid., 1.VII.1948 (fr), *H. E. Moore & C. E. Wood 3673*.—Paratypi, NY, MICH = U.S.D.A. Plant Introduction *260707,* cult. at LASCA (*A. Griffiths 4990,* NY).

Cassia ekmaniana sensu Isely, 1975, p. 82, quoad descr., basionym. exclus.

Foliage densely loosely pilosulous with hairs up to 0.3–0.5 mm, the vesture of young lvs and inflorescence golden; lvs 11–25 cm; lfts (6–)8–10 pairs, the largest 2.5–5.3 × 0.9–2.1 cm, in outline elliptic, either acute or obtuse mucronulate, the secondary veins 6–8 pairs, the mature blades reticulate on both faces; calyx a trifle larger than that of var. *racemosa,* the outer sepals 3–4.5 mm, the inner 4.5–6 mm; 4 petals 12–16 mm, the odd petal 14–16 mm; anthers glabrous or minutely puberulent with a few incurved hairs up to 0.2 mm; ovary glabrous; style 1.5–2 mm; body of pod (10–)11–19.5 × 1.2–1.6 cm.—Collections: 15.

Rocky hillsides, cliff ledges and road-cuts, on limestone bedrock, 1000–1400 m, local on the Gulf slope of Sa. Madre Oriental drained by río Moctezuma, lat. 20°45′–21°25′N in extreme s.-e. San Luis Potosí and closely adjacent Querétaro and Hidalgo.—Fl. III–V, occasionally later. Cultivated (as *Cassia ekmaniana*) in s. California and s. peninsular Florida, fl. in summer.

The handsomely floriferous var. *moctezumae,* strongly isolated from the rest of its species on the Gulf slope of Sierra Madre Oriental, is morphologically intermediate between var. *racemosa,* of which it has the reticulate mature leaflets and glabrous brachystylous pistil, and var. *sororia,* like it in the relatively numerous pairs of leaflets and the copious loose vesture but unlike in the immersed tertiary venulation of the leaflets and in the rusty-pilosulous ovary. The remotely allopatric var. *coalcomanica* is closely similar in facies, but has finely strigulose leaflets and a style about 3 mm long.

196c. Senna racemosa (P. Miller) var. **coalcomanica** Irwin & Barneby, var. nov., foliolorum numero cum var. *moctezumae* et var. *sororia* congrua, ab ambabus pube foliolorum appressa, ab hac ulterius ovario glabro diversa.—MEXICO. Michoacán: Aquila, distr. Coalcomán, 140 m, 21.IX.1938 (fl), *G. B. Hinton 12640.*—Holotypus, NY; isotypi, K, MICH, US.—Ibid., 6.XI.1941 (fl), *Hinton 16105.*—Paratypi, K, MICH, NY, US.

Foliage strigulose with straight subappressed hairs up to 0.4–0.6 mm, the vesture of inflorescence a little looser and rusty; lvs 16–20 cm; lfts 8–9 pairs, up to 2.8–4.8 × 1.2–2 cm, obovate or elliptic-obovate obtuse mucronulate, the 6–8 pairs of secondary veins and sometimes the tertiary venulation of mature blades finely prominulous above; outer sepals 3.5–4.5 mm, the inner ones only thinly pubescent dorsally 6–7 mm; 4 petals 13.5–15 mm, the odd one 15–17 mm; anthers thinly pilosulous toward apex with weak hairs up to ±0.4 mm; ovary glabrous; style ±3 mm; pod unknown.—Collections: 3.

Rocky hillsides and stream banks, in drought-deciduous brush-woodland, 140–500 m, local, known only from the Pacific foothills in Colima (mun. Coquimuatlán) and s.-w. Michoacán (mun. Aquila) in s.-w. Mexico.—Fl. X–I.

For differential commentary see under the preceding and following.

196d. Senna racemosa (P. Miller) var. **sororia** Irwin & Barneby, var. nov., ob ovarium dense pilosulum cum var. *liebmannii* praesertim comparabilis, ab ea foliolis magis numerosis 9–12 (nec 6–9)-jugis patriaque aliena, ab aliis speciei varietatibus ovarii pube foliolisque maturis supra eveniis diversa.—MEXICO. Guerrero: 6 km s.-e. of Taxco, 27.X.1977 (fl), *R. C. & G. K. Barneby 17772.*—Holotypus, ENCB; isotypus, NY.

Foliage densely loosely pilosulous with hairs up to 0.4–0.8 mm, the vesture of lvs gray or yellowish, of inflorescence rusty-fuscous; lvs 11–26 cm; lfts 9–12 pairs, the largest 2.3–5.4 × 1–2.6 cm, either elliptic or obovate, obtuse or deltately acute, the secondary veins 5–7 pairs, fully immersed above, prominulous beneath, the tertiary reticulation faintly prominulous beneath only or merely discolored; outer sepals 4–7 mm, inner ones 6.5–10 mm; long petals ±15 mm; androecium of var. *liebmannii,* the anthers sparsely pilosulous distally with weak hairs ±0.5–0.7 mm; ovary densely rusty-pilosulous; style glabrous 2.5–4 mm; body of pod (little known) 13–18 × 1.1–1.3 cm, the seed locules 6–7 mm long.—Collections: 9.

Ipomoea-thorn-forest and brush-grassland transition in the hill country around the n.-e. margin of Balsas Depression at ±1100–1600(–2100) m, local, s.-w. México (mun. Tejupilco and Ixtapan de la Sal), Morelos (mun. Yautepec, mun.

Jiutepec) and n.-centr. Guerrero (mun. Chichihualco, mun. Zumpango del Río, mun. Taxco), ±99°–100°20′W, 17°30′–19°S, in s. Mexico.—Fl. X–III.

Hinton's collections of var. *sororia* from Temascaltepec District were determined at Kew as *Cassia liebmannii* and *C. andrieuxii,* at the time the most nearly appropriate alternatives. The variety is related to both, but different in the numerous pairs of leaflets, in this respect approaching the more northern and western, fully and distantly disjunct populations that we have described above as var. *moctezumae* and var. *coalcomanica* respectively, both distinguished by their glabrous pistils.

196e. Senna racemosa (P. Miller) var. **liebmannii** (Bentham) Irwin & Barneby, stat. nov. *Cassia liebmannii* Bentham, Trans. Linn. Soc. London **27**: 549. 1871.—". . . Mexico, prov. Oaxaca. *Liebmann.*"—Lectoholotypus, *Liebmann Pl. Mex. 5017 = Caesalp. Mex. 41,* collected XI.1842 at Santiago Astata ['S. Iago Estata'], Oaxaca, C! = NY Neg. *10484*; isotypi, C, K! = NY Neg. *1420*.—*Gaumerocassia liebmannii* (Bentham) Britton ex Britton & Rose, N. Amer. Fl. **23**(4): 253. 1930.

Cassia liebmannii sensu Standley, 1922, p. 411.

Densely pilosulous throughout with fine weak spreading-ascending hairs to 0.4–0.8 mm, the vesture of lvs gray, of inflorescence rusty; lvs (7–)8–16 cm; lfts 6–9 pairs, the largest 2–3.7 × 1–1.7 cm, obovate, elliptic-obovate or -oblong, obtuse or subemarginate mucronulate, the 5–7 pairs of secondary veins immersed above, prominulous beneath, tertiary venulation faintly visible but scarcely raised beneath only; outer sepals 3.5–5.5 mm, inner ones 5.5–8(–9) mm; 4 petals 14–16.5 mm, the odd one 15–18 mm; anthers 3–5.5 mm, the thecae thinly pilose distally with weak hairs to 0.5–0.7 mm; ovary densely pilosulous; style glabrous 2.5–4 mm; body of pod (not well known) 11–19 × ±1 cm, the seed-locules 6–7 mm long.—Collections: 10.

Brushy hillsides ±600–1200 m, locally plentiful on the Pacific slope of s.-e. Oaxaca in the drainage of río Tehuantepec (lat. 95°–96°20′W) and apparently disjunct in e.-centr. upland Chiapas (mun. La Trinitaria; mun. Tzimol), Mexico.—Fl. XI–II(–V).

Between its discovery in 1842 and the publication of Britton & Rose's revision in 1930 *Cassia liebmannii* was known only from the type-collection and the nature of its relationship to *Senna racemosa* (or *Cassia ekmaniana*) could not be evaluated. Thanks to the collections of Thomas MacDougall in the Isthmus region in the last decade we have a much clearer picture of it and find that it differs from typical *S. racemosa* chiefly in the pubescent ovary with slightly longer style and in the less prominent venulation of the leaflets. The newly described vars. *sororia, coalcomanica* and *moctezumae,* collectively different in the more numerous leaflets, differ among themselves in the same features, the first having the venulation and pistil of var. *liebmannii* and the other two the glabrous ovary and reticulate mature leaflets of var. *racemosa*. The morphological discontinuities between the five varieties are of about equal magnitude and would be insignificant unless supported by an emphatically vicariant pattern of dispersal.

197. Senna andrieuxii (Bentham) Irwin & Barneby, comb. nov. *Cassia andrieuxii* Bentham, Trans. Linn. Soc. London **27**: 548. 1870.—"Mexico [more precisely "in ditione Puebla"], *Andrieux,* [Pl. Mex.] n. 414, *Bates.*"—Holotypus, *Andrieux 414,* G! isotypi, FI, K (originally numbered

177) = NY Neg. *1419*, M, P, W!—*Gaumerocassia andrieuxii* (Bentham) Britton ex Britton & Rose, N. Amer. Fl. **23**(4): 252. 1930.

Cassia fulva T. S. Brandegee, Zoë **5**: 233. 1906.—"Collected at El Riego, near Tehuacan, Puebla. [*C. A. Purpus*] No. 1195."—Holotypus, collected VI.1905 (fl, old fr), UC! isotypi, GH, MO, NY!—Correctly equated with *C. andrieuxii* by Macbride, Contrib. Gray Herb. n. ser. **59**: 27. 1919.
Cassia andrieuxii sensu Standley, 1922, p. 410.

Shrubs and treelets 1–3.5 m, the leafless annotinous and older branchlets stiffly tortuous brown glabrate, the soft, densely leafy hornotinous ones with foliage and axes of inflorescence densely softly pubescent throughout with weak flexuous or simply spreading-incurved hairs up to 0.7–1.2 mm, the vesture (dry) strikingly bicolored, on lower face of lfts white- or gray-lanate, elsewhere golden-, brown- or rufous-pilosulous, the short subcorymbose racemes axillary or pseudoterminal to young branchlets, shortly or not exserted.

Stipules linear-subulate 1–4.5 mm, densely lanulose caducous.

Lvs (3–)4–14.5 cm; petiole 1–3.5 cm, at middle 0.6–1 mm diam, its very narrow ventral sulcus obscured by vesture; rachis 1.5–6(–7) cm; petiolar gland 0; lfts 3–5 pairs, moderately accrescent upward, the distal pair broadly obovate obtuse or emarginate 2–4.3 × 1.1–2.8 cm, 1.6–2 times as long as wide, inequilaterally rounded at base, the margin strongly revolute, the stout midrib with 5–7 pairs of secondary and ultimately an open mesh of tertiary venules all finely engraved on upper face, sharply prominulous beneath, the proximal lfts shorter and proportionately wider, sometimes orbicular.

Peduncles 1–4.5 cm; racemes closely, at first subcorymbosely ±20–35-fld, the several simultaneously expanded fls raised to or beyond level of the crowded nodding lanate-pubescent buds, the axis little elongating, up to 3(–5) cm long in fruit; bracts lance-elliptic cymbiform 2–7 × 1.5–2 mm, tardily deciduous; pedicels 12–23 mm; sepals moderately graduated ovate-obovate obtuse, the firmer outer ones 4–6 mm, the membranous-margined inner ones 6.5–8 mm; petals rich yellow, pilosulous dorsally near claw and along midrib, when dry brownish-yellow coarsely dark-venulose, the vexillary one, 2 lateral and 1 abaxial all broadly obovate emarginate similar in shape, the first sometimes a trifle shorter than the rest 8.5–11 mm, one abaxial heteromorphic, thickly short-clawed, the oblong-spatulate limb turned at ±90° to claw, the whole petal longest of all and 11.5–15 mm; androecium functionally 7-merous, the 3 staminodes 2.5–3.5 mm, their spatulate limb 0.5–0.8 mm wide, the 7 fertile stamens subhomomorphic in shape but dwindling in size very gradually toward abaxial side of fl, their filaments stout glabrous 1.6–2 mm, the anthers oblong often slightly enlarged upward, truncately rounded at both ends, 2.3–4.2 × 1.4–1.7 mm, scarcely or not incurved, thinly pilose, the minute beaks infraterminal scarcely 0.2 mm long dehiscent by parallel slits; ovary lanate; style glabrous subulate 2–3 mm, gently incurved and tapering distally, just below minute terminal stigmatic cavity 0.3–0.4 mm diam; ovules 14–18.

Pod broadly linear plano-compressed 7–13 × 1–1.3 cm, the valves when ripe papery fuscous densely rusty-pilosulous, the seminal locules ±5–8 mm long, as wide as the pod's cavity; seeds transverse, obovate-paddle-shaped, compressed parallel to the valves, 6–7.5 × 4–5.5 mm, the testa light brown dull smooth, ultimately crackled, the elliptic areole 1.5–2.2 × 0.6–0.7 mm.—Collections: 14.

Dry hillsides in matorral or disturbed oak-woodland, 1250–1850 m, commonly on limestone, locally plentiful on the sources of río Papaloapan in s. Puebla and adjoining Oaxaca, Mexico, thence extending s.-w. in mun. Huajuapan to the margin of the Balsas basin.—Fl. VI–X(–XI).

Senna andrieuxii appears to be a xeromorphic derivative of *S. racemosa* adapted to the severe climate of the Tehuacán Desert and cognate calcicolous thornforest of northern Oaxaca. Its compact known area of dispersal fits into the wide interrupted one of *S. racemosa* as neatly as a piece missing from a puzzle, and it seems likely to be, in reality, a member of the replacement series described under the latter. However that may be, the prevailingly small leaves, simplified down to five, four or three pairs of leaflets, numbers significantly less than found in vicariant varieties of *S. racemosa,* combine with the dense woolly indumentum, gray above and golden beneath, to establish a distinctive facies. With some reluctance therefore we maintain *S. andrieuxii* as specifically distinct.

198. Senna haughtii (Macbride) Irwin & Barneby, comb. nov. *Cassia haughtii* Macbride, Field Mus., Bot. **13**(Fl. Peru), part 3(1): 166. 1943.—"[Peru.] Piura: Saucecito, 40 miles north of Sullano, [*Oscar*] *Haught F10,* type. Cerro Viento Ridge [±15 km n.-w. of Saucecito], *Haught 62.*"—Holotypus (*F10*) and paratypus (*62*), F! the first = F Neg. *53988.*

Shrubs at anthesis 4–12 dm, the angulate lenticellate branchlets, lf-stalks, margin and dorsal midrib of lfts with all axes of inflorescence strigulose-pilosulous with very short incurved and subappressed mixed with random longer spreading hairs up to 0.4–0.8 mm, the pale dull olivaceous, firmly membranous foliage subconcolorous, the racemes axillary to distal lvs, not or little exserted.

Stipules erect firm linear-attenuate 2.5–5 × 0.2–0.5 mm, mostly persistent after lf-fall.

Major lvs 4.5–16 cm, some upper ones subtending brachyblasts bearing smaller lvs, the latter not further described; petiole including scarcely modified pulvinus 7–26 mm, 0.5–0.9 mm diam, narrowly sulcate ventrally; rachis 2.5–10 cm, its longer interfoliolar segments 4–15 mm; petiolar gland 0; pulvinules 0.6–1.7 mm; lfts 8–10 pairs, accrescent distally, the distal pair broadly oblong-oblanceolate or -obovate obtuse mucronulate 12–38 × 5–17 mm, at base inequilaterally rounded-subcordate, the margin plane or toward pulvinule incipiently revolute, the centric midrib immersed or almost so above, cariniform beneath, the 5–8 pairs of camptodrome secondary and few random connecting tertiary venules sharply prominulous beneath, only faintly so above.

Racemes ±15–35-fld, the axis including very short peduncle becoming 3–7.5 cm; bracts lanceolate 3–5 mm, early dry caducous; mature pedicels 18–32 mm; young fl-buds globose, puberulent at base, thence glabrate, the sepals (and petals) separating long before maturity; sepals subpetaloid graduated, the smallest outer one ovate 4.5–6 mm, the innermost oblong- or broad-obovate 7–10.5 × 4.5–5.5 mm; petals yellow drying dull yellow dark-veined, pilosulous dorsally especially along veins, 4 homomorphic except for size, broadly elliptic beyond slender claw 14–17 × 7–8 mm, one abaxial (opposed to divaricate pistil) more coarsely clawed and its elliptic blade incurved through nearly 90°, shielding the androecium, measured along midvein 19–23 × 7.5–9 mm; androecium glabrous, functionally 7-merous, the staminodes 1–1.3 mm wide, the fertile stamens isomorphic except for filaments abaxially a trifle longer, the filaments 1.8–3.3 mm, the oblong anthers obtusely truncate at both ends, obscurely sigmoidally arched, each theca 2-sulcate lengthwise and contracted into a very short lateral beak dehiscent by a vertical slit; ovary thinly pilosulous-ciliolate; style glabrous 2.5–3 mm, gently incurved distally and tapering to a conic apex, there 0.2–0.3 mm diam; ovules 10–12.

Pod erratically pendulous, the stipe 8–10 mm, the plano-compressed body 8–11.5 × 1.5–1.8 cm, bicarinate by the sutures, the almost flat greenish papery-membranous valves venulose toward the sutures, densely finely pilosulous overall; seeds (seen only half-ripe) compressed parallel to the valves, the presence or absence of areole to be demonstrated.—Collections: 5.

Habitat not recorded, but to be expected on dry brushy hillsides or quebradas ±400–950 m, known only from the arid Pacific Andean piedmont in n. Peru, between n.-w. Piura (Talará and Sullano) and w.-centr. Cajamarca (Santa Cruz and w. Chota), lat. 4°20′–6°40′S.—Fl. III–VI.

A senna readily recognized in its region of occurrence by a syndrome of highly asymmetric corolla, isomorphic, longitudinally 8-sulcate anthers, and papery plano-compressed 10–12-ovulate pod. It most closely resembles distantly allopatric Mexican *S. racemosa,* which differs primarily in the more stiffly papery pod containing 16–28 seeds. Sympatric *S. pistaciifolia* var. *picta,* with which *S. haughtii* was compared in the protologue, has a zygomorphic corolla, two abaxial anthers elevated on long filaments, and belongs in the remotely related sect. *Senna*. The only genuinely related senna encountered in the range of *S. haughtii* is *S. mollissima*. This has similar but somewhat larger flower, only 2–5 pairs of much ampler leaflets, more than 50 ovules, and a greatly elongate, narrower, firmly coriaceous pod thickened at the sutures. A curious feature of *S. haughtii,* duplicated in some species of *Cassia* sens. str., is the late appearance of pubescence on the pod. At anthesis the ovary is only thinly pilosulous or ciliolate along the sutures, but following fertilization the valves quickly acquire a dense fine pilosulous vesture that persists into maturity of the fruit.

199. Senna atomaria (Linnaeus) Irwin & Barneby, comb. nov. *Cassia atomaria* Linnaeus, Mantissa 68. 1767.—"*Habitat in* America. Dr. Jacquin."—Described from a plant growing but not flowering in the botanical garden at Upsala.—Holotypus, LINN *528/17* (lvs only, ex hort. upsal.)!—Correctly equated by Britton & Rose, 1930, p. 269, with *Cassia emarginata* sensu Benth. et auct. posterior., non *C. emarginata* Linn., quae = *Senna bicapsularis* nob.

Cassia arborescens Miller, Gard. Dict. ed. 8, *Cassia* no. 15. 1768.—"*Senna spuria tetraphylla arborea, siliquis compressis, angustis, longissimis, pendulis* Houst[on] MSS . . . sent me from La Vera Cruz, in New Spain, by the late Dr. Houston."—Holotypus, with annotations in hands of Houston and Miller, BM! = BH Neg. *5170* = NY Neg. *169*.—Correctly equated with *C. emarginata* sensu suo by Bentham, 1871, p. 548.

Cassia longisiliqua Linnaeus fil., Suppl. 230. 1781.—"*Habitat in* America."—Based entirely on *Cassia* [no. 3] *arborescens diffusa, siliquis longis compressis* P. Browne, Hist. Jamaica 223. 1756.—Holotypus, the protologue, which calls for a shrub 12–15 ft. tall, very common about Kingston, with long, compressed pod.—Mistakenly equated by DeCandolle, 1825, p. 497, with a plant collected by Bertero in Santo Domingo = our *Senna mexicana* var. *mexicana*; and reduced wrongly to *C. occidentalis* by Bentham, 1871, p. 532, to *Ditremexa occidentalis* by Britton & Rose, 1930, p. 256.

Cassia triflora Vahl, Ecl. Amer. **3**: 11. 1807.—"Habitat in insula St. Crucis [St. Croix]. *von Rohr*."—No typus seen, but the detailed protologue and origin decisive.—Overlooked by Bentham, 1871.

Cassia elliptica Humboldt, Bonpland & Kunth, Nov. Gen. & Sp. **6**(qu): 356. 1824.—"Crescit locis calidissimis, prope Cumanam [Venezuela], Turbaco [Colombia], et Campeche [Mexico]."—Lectoholotypus, Herb. Willd. *7942/1, 2,* labelled 'Cumaná, *Humboldt*,' B! paratypus, labelled '[Humboldt & Bonpland] n. *1466*. Turbaco,' P-HBK, this said by Kunth in the protologue to differ slightly from the plant from Cumaná principally described; no material from Campeche seen.—Correctly equated by Bentham, 1871, p. 548, with *C. emarginata* sensu suo.

Cassia grisea A. Richard in Sagra, Hist. fis., pol. y nat. Cuba **10** [Essai Fl. Cuba **1**]: 493. 1846.—"Crescit in *Vuelta de Abajo* [=Pinar del Río, w. Cuba]."—Holotypus, *R. de la Sagra s.n.*, P (hb. Richard.)!—Equated by Bentham, l.c. with *C. emarginata* sensu suo.

Cassia chrysophylla A. Richard in Sagra, op. cit. p. 500. 1846.—"Crescit in *Vuelta de Abajo*."—

Holotypus, *R. de la Sagra s.n.*, P (hb. Richard.)!—Equated with *C. emarginata* sensu suo by Bentham, l.c.

(?)*Cassia michoacanensis* Sessé & Mociño, Pl. Nov. Hisp. 61. 1888 & Fl. Mex. 101. 1893.—"Habitat in montibus inhospitalibus ab oppido Tepalcatepec, Coahuayanam versus interjectis [60 km w. of Apatzingán, Michoacán]."—No typus known, but the description suggests no other known species of the region.

Cassia elliptica Sessé & Mociño, Fl. Mex. 101. 1893.—"Habitat in montibus de Puerto Rico."—Holotypus, Hb. S. & M. *1132*, MA! = F Neg. *44434*; isotypi, FI (hb. Webb., misit Pavón), OXF!

Cassia emarginata var. *subunijuga* Robinson & Bartlett, Proc. Amer. Acad. Sci. **43**: 53. 1907.—"Gualan, Department of Zacapa, Guatemala, 15 January, 1905, C. C. Deam, no. 220 . . ."—Holotypus, GH! isotypi, MO, NY!—Reduced by Britton & Rose, 1930, p. 269 to *Isandrina emarginata*.

Isandrina maxonii Britton & Rose, N. Amer. Fl. **23**(5): 269. 1930.—"Nicaragua. Type from Tamagasta Peninsula, west of Managua, June 25, 1923, *Maxon, Harvey & Valentine 7313*."—Holotypus, NY! = NY Neg. *9390*.

Cassia planisiliqua sensu Lamarck, Encycl. Méth. **1**: 645. 1785.—"On trouve cette espèce dans l'île de Guadeloupe."—Non *C. planisiliqua* Linnaeus, 1753, quae = *Senna occidentalis*.

Cassia emarginata sensu Colladon, 1816, p. 101; DeCandolle, 1825, p. 499; Vogel, 1837, p. 33 (these all exclus. syn. Sloan.); sensu Bentham, 1871, p. 548, exclus. syn. tertio; Standley, 1922, p. 410; Standley & Steyermark, 1946, p. 114; Schery, 1951, p. 54; Little, Woodbury & Wadsworth, Trees Pto. Rico, second vol. 274, fig. 370. 1974; Isely, 1975, p. 82.—Non *C. emarginata* Linnaeus vera, quae = *Senna bicapsularis* (Linnaeus) Roxburgh.—*Isandrina emarginata* sensu Britton & Killip, 1936, p. 182; Britton & Rose, 1930, p. 269.

Cassia elliptica sensu Vogel, 1837, p. 33.

Cassia atomaria sensu auct. ann. 1871 usque, non Bentham, 1871, nec auct. posterior.

Xerophilous but not obviously xeromorphic, drought-deciduous, potentially arborescent shrubs or trees at anthesis (2–)3–17(–20) m, with 1–several gray or brownish trunks attaining 7–15(–?) cm diam and a rounded head of ill-scented foliage, the annotinous branchlets scaley and lenticellate, the hornotinous ones together with foliage and axes of inflorescence softly, often densely pilosulous or tomentulose with straight spreading, simply incurved or shorter, variously curly or even crimped hairs up to 0.2–0.6(–0.7) mm, the bicolored lfts dark olivaceous and dull or sublustrous above, pallid beneath, pubescent on both faces but more thinly and shortly so above, the lf-stalks and inflorescence nearly always charged with small thickened orange, erratically branched trichomes, the young vesture often lutescent fading whitish, the racemes arising either from a) leafless or depauperately leafy brachyblasts borne on otherwise leafless branches, or b) from brachyblasts lateral to branches leafy distally, or c) axils of ordinary foliage lvs, the inflorescence normally coinciding with renewed vigorous growth following rains, the foliage in actuality either syn- or hysteranthous, but mature lvs usually coeval with pods.

Stipules erect, subulate, triangular-subulate or rarely setiform 1.2–4(–5) × 0.3–1 mm, firm, early dry stramineous, often but not always persistent after fall of lf, either pilosulous or glabrous dorsally, ± thickened at base and externally convex, superficially nerveless.

Larger mature lvs (8.5–)10–24(–28.5) cm, those directly associated with fls or crowded on brachyblasts smaller in all dimensions (and not further described); petiole including firm discolored pulvinus (2–)2.5–6.5 cm, at middle 0.8–1.6 mm diam, terete except for 2 low thick ventral ridges; petiolar gland 0; rachis (0–)1.5–9(–11) cm, its longer interfoliolar segments 1–3(–3.4) cm; pulvinules (1.5–)2–5 mm; lfts (1–)2–5 pairs, accrescent distally, the distal pair broadly obovate to elliptic or less commonly ovate, when broadest beyond middle obtuse or emarginate at apex and either rounded or cuneate at the subequilateral base, when broadest near or below middle either obtuse or deltately acute mucronulate, the distal pair (2–)3–11(–13) × (1–)1.5–5.4(–6) cm, (1.4–)1.6–2.8(–3.2) times as long as wide, the margin strongly revolute, the midrib with 5–15 (commonly 6–11 + random intercalary) pairs of secondary camptodrome veins and a complex mesh of reticular venules all sharply prominulous (and often pallid) beneath, on upper face only faintly raised or mostly immersed, rarely engraved.

Racemes commonly (3–)5–15(–18) but occasionally (in Jamaica, Hispaniola, Venezuela) 20–55-fld, 2–several simultaneously expanded fls raised ± to or above level of succeeding buds, the slender, when long pendulous axis together with short peduncle becoming (1.5–)2–7 (exceptionally to 11–17) cm; bracts ovate- or lance-attenuate 1.5–3.5 mm, early caducous; young fl-buds globose, pilosulous, puberulent or glabrate, the sepals separating before true anthesis; pedicels at anthesis subfiliform, in fruit much thickened (13–)15–24(–28) mm; sepals greenish, fuscous or yellow and petaloid, well graduated, the small outermost ovate 2.5–4(–4.5) mm, the largest innermost one obovate, oblong-obovate or suborbicular (4.5–)5–7(–8) mm, all reflexed in late anthesis; petals rich orange-yellow drying bright or brownish yellow, delicately dark-veined, either glabrous or thinly pilosulous dorsally, the 3 adaxial and one abaxial (opposed to pistil) slenderly clawed and homomorphic except of different sizes, (7–)8–16(–17) mm long, the blades oblong-oblanceolate or -obovate, the vexillar petal usually shortest, either obtuse or retuse, the abaxial one longest, the fifth petal thick-clawed and abruptly incurved through ±90°, convex and folded over androecium, measured along falcate midrib (12–)13–23 mm; androecium glabrous or the anthers more commonly thinly pilosulous with erect stiff hairs, the staminodes 1.2–1.6 mm wide, emarginate at both ends, the filaments subequal or the abaxial ones a trifle longer, the 7 fertile anthers yellow, isomorphic except the abaxial ones sometimes slightly longer, all narrowly oblong obcompressed (2.8–)3–4.7(–5) × (1.1–)1.2–1.7(–1.8) mm, at apex bluntly truncate and incurved into 2 very short, latero-infraterminal separated beaks opening by a vertical slit, the whole anther sulcate lengthwise at the connective and carinate by the sutures, each theca obscurely sulcate proximally between suture and connective; ovary glabrous at anthesis, often permanently but sometimes becoming densely pilosulous following fertilization, the vesture persistent along the sutures and sometimes also the valves of the mature pod; style stout, gently incurved 0.9–1.8 mm, at conic tip 0.3–0.5 mm diam, the minute stigmatic cavity terminal glabrous; ovules 46–70.

Pod pendulous, the stipe 2–4(–5) mm, the body elongately linear straight or almost so, strongly compressed but thick-textured, potentially 2.2–3.5(–3.7) dm long but often following abortion of some or many proximal or random distal ovules shorter or irregularly strangulated (or both), between the greatly thickened sutures (8–)8.5–12(–14) mm wide and at them 2–3.5 mm thick, the cross section narrowly oblong or coarsely I-shaped, the firm subsucculent green valves becoming woody, dark brown or blackish, prominently venulose, their incrassate margins splitting transversely, the interseminal septa lignescent, the seed-cavities 3–5.5 mm long, as wide as the valves, the whole pod long-enduring on the tree, falling indehisced except by transverse fracture through the septa, the seeds released only by trampling or rot of valves; seeds transverse, obovoid or oblong-obovoid, accommodating themselves to the width of the cavity and thus moderately compressed parallel to the valves or rarely subterete, 4.2–5.3 × (2.2–)2.7–3.6 mm, the testa light reddish-brown, smooth but not highly lustrous, the elliptic or broadly oblong areole 1.2–2.4(–3) × (0.6–)0.8–1.7 mm.—Collections: 265.

Thorn-forest, chaparral, tropical deciduous and semi-deciduous woodland and brush-savanna, locally surviving in xeric microhabitats (outcrops, barranca-terraces, stream-banks, coral reefs) within mesic climax associations, mostly 1–600 m but attaining 1000–1150(–1600) m in s. Mexico (Michoacán, Oaxaca, Chiapas), Honduras and n. Venezuela (Maracaibo Basin, Cordillera Costanera), interruptedly widespread and locally abundant over much of seasonally arid lowland Mexico, Pacific Central America, West Indies and n. South America: Pacific lowlands and foothills of Mexico and Central America from s. Sonora (lat. ±28°N) and s. Baja California Sur to n.-w. Costa Rica (Guanacaste, lat. ±10°N), in Mexico extending inland to Balsas Depression (in Michoacán and México) and e. across the Continental Divide to the Papaloapan and Grijalva valleys in n. Oaxaca and and Chiapas, in Guatemala e. to Río Motagua; Gulf lowlands of e.-centr. Mexico in s. Tamaulipas, centr. Veracruz and immediately adjacent San Luis Potosí (lat. 19–23°30′N); Bahamas (Eleuthera); circum-Caribbean from Yucatan Peninsula through Greater Antilles to Puerto Rico (s.-centr. only) and, becoming rarer, around the Lesser Antilles to Guadeloupe, St. Vincent and Curaçao; arid coastal n. Colombia from Bolívar to Guajira Peninsula; n. Venezuela from periphery of Maracaibo Basin e. to Nueva Esparta, Sucre and Monagas, thence s. locally to the lower Orinoco and Caroní valleys in Bolívar (Cd. Bolívar, Parque Caroní).—Fl. in Mexico and Central America primarily (I–)II–V, in West Indies primarily II–VIII, in South America primarily III–X, but sparse flowering with leaves erratic throughout the year except in extreme desert conditions, the fruit long persisting, often contemporaneous with new fls.—*Vara* or *flor de San José, alcaparro, barba de jolote, lengua de perro, chile perro, palo de zorrilla, p. hediondo, mora hedionda, patita, vainillo, arguchoco* (Mexico and C. America); *chicharong, guncho* (Nicaragua); *xtuab, xtuabin* (Yucatan); *carbonera, cañafistola cimarrona, júpiter amarillo, guacamaya amarilla* (Cuba); *yellow candlewood* (Jamaica); *bois cabrite, palo de chivo, p. de burro, sopalpo extranjero* (Hispaniola); *velamuerto* (Pto. Rico); *chivato, chilmeto cuchillito, caraganito, jinnu* (Colombia); *mora, brusca* (*brucha*) *macho, brusquillo, carángano, platanillo* (*platonico*), *tarantán, cacaito, cañafistolín, mucutena extranjera* (Venezuela).

Due to seasonal dimorphism, *S. atomaria* is not easily visualized through descriptions. A tree flowering at winter's end from naked branches will be transformed in summer by a mantle of foliage sometimes mixed with erratic axillary racemes; while another may flower first from brachyblasts mixed with relatively small leaves destined not to attain their full amplitude until the pods begin to form. The species is most closely related to *S. mollissima,* which has the same long compressed many-seeded indehiscent pod but caducous stipules, all leaflets ovate-acuminate and all racemes axillary to coeval and ample leaves. In Pacific Mexico, where *S. mollissima* var. *glabrata* and *S. atomaria* are sympatric, the latter is easily distinguished by the pattern of inflorescence and the always 2–3 pairs of obovate (not 4–6 pairs of acuminate) leaflets; but in a wider view of these sibling species over their whole ranges these differential characters are fallible. Individual trees of *S. atomaria* in West Indies and Venezuela have many leaflets broadest below the middle and if not long-acuminate then distinctly deltate-acute at tip; others have many or all racemes pendulous and many-flowered from coeval leaf-axils, just like those of *C. mollissima,* although the individual flower is then smaller. Accompanying this small flower are generally shorter and slightly narrower anthers, and the seeds of *S. atomaria* are, so far known, a little smaller, as might be expected from the generally slightly narrower pod. Despite the fact that we have not found any one fully reliable diagnostic character, the identity of specimens is usually instantly clear and we have no difficulty in following tradition in the matter. A second close relative is the semiscandent *S. peralteana,* which see for comment.

Within *S. atomaria* there is evidence of incipient racial differentiation coincident with dispersal. Mexican and Central American *S. atomaria* except for the populations in Yucatan, have in mature leaves only (1–)2–3 pairs of relatively large leaflets, the distal pair up to about 6–11.5 cm long, with 8–15 pairs of secondary veins, whereas the average mature leaf in West Indies, Venezuela and Colombia consists of 3–5 pairs of leaflets up to about 3–6.5 cm long, with only 5–9 pairs of secondary veins. Exceptions are relatively few and scattered: the Puerto Rican populations, for example, have few large leaflets of the Mexican type, unlike any in Hispaniola or Guadeloupe, as have a very few from northern South America; moreover starved foliage from Mexico may resemble in size if not in number of leaflets fully expanded foliage from the Antilles. It is possible that the Puerto Rican plant, curiously localized in a small area around Coamo, Salinas and Cayey, is an old introduction from the continent; if this could be proved, the separation of geographic subspecies might become justified, in which case the varietal epithet *subunijuga* would be available for the large-leaved Mexican and Central American form. A peculiar sterile specimen from Haiti (St. Marc, l'Artibonite, *Leonard 2922,* NY), left unnamed by Britton, is probably abnormal *S. atomaria* but is anomalous in having nothing but bifoliolate leaves. It may be juvenile, but field-observation is obviously indicated.

The hard yellow wood of *S. atomaria* is gathered for fuel and formerly furnished a yellow dye. Animals feeding on the foetid foliage and firm fleshy pods are said to suffer loss of hair, as has been reported for the related *S. haughtii* in northern Peru. *Senna atomaria* often shares with the latter the curious development on the initially glabrous ovary of a fine pilosulous vesture that persists on the mature pod.

200. Senna mollissima (Willdenow) Irwin & Barneby, comb. nov. *Cassia mollissima* Humboldt & Bonpland ex Willdenow, Enum. pl. hort. Berol. 44. 1809.—Typus infra sub var. *mollissima* indicatur.

Amply leafy arborescent shrubs at anthesis 3–10 m, the main trunks attaining ±5–6 cm diam, the annotinous and older branches gray or pale brown lenticellate, the hornotinous branchlets and foliage pilosulous throughout with fine, simply or sinuously incurved or sometimes straighter spreading-ascending hairs up to 0.3–0.6(–0.7) mm, the dull olivaceous bicolored lfts pale and densely softly pubescent beneath, the vesture of their upper face thinner and shorter (the hairs sometimes basally dilated), of the emergent foliage often lutescent, the loose racemes axillary to and shorter than coeval lvs, pliantly pendulous.

Stipules erect firm, narrowly lanceolate or subulate 2.5–5.5 × 0.3–0.7 mm, early dry, deciduous before the lf.

Longer lvs when fully expanded (1–)1.5–4(–4.7) dm; petiole including wrinkled pulvinus (3–)4–13 cm, at middle 0.9–2 mm diam, subterete, obtusely 2-ribbed and very shallowly grooved ventrally; rachis (1.5–)2.5–21(–24) cm; petiolar gland 0; pulvinules (1.5–)3–6.5 mm; lfts of larger lvs 2–6 pairs accrescent distally, ovate, ovate-elliptic or elliptic short-acuminate 6–16 × (2.5–)3–7(–7.5) cm, 1.7–3 times as long as wide, at base subsymmetrically cuneate or rounded, the margin revolute, the midrib immersed or shallowly depressed above, cariniform beneath, the 9–12 pairs of slender camptodrome and random intercalary secondary veins with connecting tertiary and often loosely reticular venules all pallid and sharply finely prominulous beneath, faintly so or immersed above; shorter proximal pair of lfts often proportionately broader, ovate-acuminate.

Racemes 15–125-fld, the several fls fully or partly open often but not always surpassed by succeeding fl-buds; bracts ovate- or lance-acuminate or -caudate 2–3.3(–4) × 0.8–1.2 mm, firm but early dry and deciduous before anthesis; pedicels at anthesis very slender, flexuous and often refracted from the pendulous axis (heliotropic), much thickened in fruit, at and after full anthesis 12–23 mm; young fl-buds globose pilosulous, the early emergent inner sepals dorsally glabrous or glabrescent ciliolate; sepals thinly herbaceous greenish, well graduated, the outermost elliptic-obovate 4–5.5 mm, the innermost broadly obovate-suborbicular (5.5–)6–8 mm; petals glabrous yellow drying brownish-stramineous, coarsely brown-veined, highly heteromorphic, the 3 adaxial smallest, oblong-obovate or -oblanceolate obtuse 10–19 × 4–7 mm, one abaxial like the latter in outline but larger, 13–20 × (6–)8–11 mm, opposed to the oblique pistil, the other abaxial one coarsely short-clawed and incurved through nearly 90°, concave and incurved over the androecium, measured along the arcuate midrib 19.5–36 × 10–18.5 mm; androecium glabrous, functionally 7-merous, the 3 staminodes 1.2–1.7 mm wide, the fertile stamens almost isomorphic, parallel and contiguous, their stout filaments 1.5–3.2 mm, the anthers oblong or slightly dilated upward, obtusely rounded at both ends, nearly straight, in dorsal view (4–)4.5–6.6 × 2–2.6 mm, the connective on ventral side dilated distally so as to separate the thecae, these abruptly contracted into a divaricate beak 0.3–0.8(–1) mm, each dehiscent by a short vertical slit; ovary glabrous or very sparsely pilosulous; style often scarcely differentiated externally from the ovary, 2–3 mm long, gently incurved distally and abruptly contracted to the minute terminal stigmatic cavity, at base of the conical tip 0.5–0.9 mm diam; ovules 52–96.

Pod pendulous from pliantly pendulous raceme axis, the stipe 2–6 mm, the linear compressed but thick, tapelike, usually straight body 20–40 × (1–)1.2–1.5(–1.9) cm, the stout obtuse sutures a little broader than the cavity (the cross section hence dumbbell-shaped), the firmly coriaceous purplish- or livid-brown, ultimately nigrescent valves not or scarcely elevated over seeds, their margins thickened, internally pithy and at maturity splitting transversely at intervals of 1–2.5 mm, the seed-locules 4–6 mm long, not extending into the thickened margin, the ripe valves finely sharply venulose, not dehiscent lengthwise but fragile and easily broken along the transverse interseminal septa, the seeds apparently released only by weathering; seeds oblong-obovoid moderately compressed 5.8–7.2 × 3.5–4.2 mm, the smooth or remotely pitted, light reddish-brown testa dull, the glossy obtusely rhombic or oblong-elliptic areole 2.5–3.7 × 2–2.5 mm.

Our concept and definition of *S. mollissima* coincide with those developed in herbaria and in post-Benthamian literature under the name *Cassia atomaria*. Following Britton & Rose we equate genuine Linnaean *C. atomaria* with *C. emarginata* auct. (not Linn.) and consequently revive for *C. atomaria* in the sense of Bentham and most modern floras the next available epithet *mollissima*.

Senna mollissima is very closely related to *S. atomaria,* of which it has the pod and seeds, and differs principally from it only in the ampler, acuminate rather than obtuse or abruptly acute leaflets and larger, commonly more numerous flowers produced always from axils of coeval leaves, never from old wood. From a theoretical viewpoint these are weak differential characters. In practice, however, Mexican var. *glabrata,* the range of which lies within the western lobe of that of *S. atomaria,* has very much larger leaves and much longer racemes than its sympatric relative; while var. *mollissima* of northern Peru and southern Ecuador is at once allopatric and much larger-flowered. Between the Mexican and South American populations, which are similar in gross aspect and have both passed as

C. atomaria in recent literature, we find several small but significant differences which enable us to define in morphological terms two mutually remote geographic varieties.

Key to the Varieties of *S. mollissima*

1. Longer lvs 2–4.7 dm, their rachis (8–)12–24 cm, the lfts (3–)4–6 pairs; racemes (25–)30–125-fld, the axis becoming 1–4 dm; oblique abaxial petal, measured along its curved midrib, 19–24 mm; Pacific Mexico (Sinaloa to w. Oaxaca); s.-e. Guatemala. 200a. var. *glabrata* (p. 594).
1. Longer lvs (1.1–)1.5–2.5(–2.8) dm, their rachis (1.5–)2.5–12(–14) cm, the lfts 2–4(–5) pairs; racemes 15–35-fld, the axis becoming 5–18(–23) cm; oblique abaxial petal 26–36 mm; n.-w. Peru and s. Ecuador. 200b. var. *mollissima* (p. 595).

200a. Senna mollissima (Willdenow) var. **glabrata** (Bentham) Irwin & Barneby, comb. nov. *Cassia atomaria* var. *glabrata* Bentham, Trans. Linn. Soc. London **27**: 548. 1871.—"New Spain, Pavon in herb. Boiss."—Holotypus, "N[ueva] E[spaña]: *Cassia atomaria*. Hb. Pavon.," G (hb. Boiss.)!—Perhaps a typotypus of *C. nutans* following.

Cassia nutans Colladon, Hist. Casses 113, t. 4. 1816.—"*C[assia]* sp. nova. Moc[iño] et Sessé *pl. mex. ined. ic*. Hab. in Mexico."—Described from the original plate made from life in Mexico, copied at G for A. P. DeCandolle and recopied to illustrate the protologue of *C. nutans*.—Lectotypus, Colladon, op. cit. t. 4! presumed typotypi, Hb. S. & M. *1136* (=F Neg. *44230*) & *1165* (=F Neg. *44239, 44240*), MA!

Isandrina acapulcana Britton & Rose, N. Amer. Fl. **23**(5): 269. 1930.—"[Mexico:] Vicinity of Acapulco, Guerrero, October, 1894–March, 1895, *E. Palmer 478*."—Holotypus, NY! = NY Neg. *9389*; isotypus, K!

Isandrina xanthophylla Britton & Rose, N. Amer. Fl. **23**(5): 269. 1930.—"Casillas, Department of Santa Rosa, Guatemala, May, 1893, *Heyde & Lux* [in *J. Donnell Smith*] *4471*."—Holotypus, NY! = NY Neg. *9391*; isotypi, G, K, US!—*Cassia xanthophylla* (Britton & Rose) Lundell, Phytologia **1**: 215. 1937.

Cassia atomaria sensu Bentham, 1871, p. 548, ex parte, exclus. syn. Humboldt. & Colla. plantisque austro-americanis; Sessé & Mociño, Pl. Nov. Hisp. 59. 1893 & Fl. Mex. 101. 1894 ("Habitat in Apatzingani."), perhaps = *C. nutans* Collad.; Standley, 1922, p. 410.

Characters as given in key to varieties.—Collections: 41.

Open places and woodland borders in chaparral, tropical deciduous forest and mesic coastal forest, 4–650 m, sometimes persisting along fences and highways, foothills and coastal plain of Pacific Mexico from s. Sinaloa to w. and s.-centr. Oaxaca, in Guerrero and Michoacán sparingly inland to Balsas Depression and in Oaxaca ascending in Sa. Madre to 800 m; apparently disjunct at ±1300 m on the Pacific slope in s.-e. Guatemala (Casillas, dept. Sta. Rosa), but possibly only cultivated.—Fl. V–XII, intermittently later, the fruit ripening slowly and long persistent.—*Quiedondilla* (*hediondilla*), *mora hedionda* (Sinaloa); *vainillo, frijolillo* (Oaxaca); pods said to be edible when young.

Somewhat variable in outline of leaflets, the narrow extreme being equivalent to *Isandrina acapulcana,* not otherwise different. The var. *glabrata* is notable for the extremely large leaves and greatly elongated, pendulous racemes, the axis of which grows out rapidly, at an early stage elevating the unopened buds in a tapering spike far beyond the expanding and fully expanded flowers.

The typus of *Isandrina xanthophylla* consists of branchlets barely coming into anthesis, the only expanded flower being immature and hence atypically small for the species. The plant is unusual in var. *glabrata* for having only three pairs of leaflets, like Ecuadorian var. *mollissima,* but the inflorescence, already lengthened into a narrow spike, is typical of var. *glabrata*. In the protologue *I. xanthophylla* was distinguished from *I. atomaria* (=our *S. mollissima* var. *mollissima*) simply by the dilutely yellow vesture, a feature actually common in *S.*

mollissima when the foliage is young. Standley & Steyermark (1946, p. 132) have rightly questioned its specific status which deserves, nonetheless, reevaluation when mature flowering and fruiting material can be obtained and the native status in Guatemala established.

If our speculation is correct, that the *Cassia atomaria* of Sessé & Mociño's manuscripts is the same as that described from the *calque* as *C. nutans,* the type-locality of the latter is near Apatzingán, Michoacán, where the variety has been collected in modern times (*Leavenworth & Hoogstral 1314,* NY).

200b. Senna mollissima (Willdenow) Irwin & Barneby var. **mollissima**. *Cassia mollissima* Humboldt & Bonpland ex Willdenow, 1809, l.c., sens. str.—"*Habitat in* America *meridionali.*"—Holotypus, *Hb. Willd. 7947,* B! consists of 2 lvs of a plant cultivated in Berlin Botanical Garden from seed supplied by Humboldt, presumably from that, collected at Guayaquil in fruit, which Kunth described independently as *C. canescens,* q.v. infra.

Cassia canescens Humboldt, Bonpland & Kunth, Nov. Gen. & Sp. **6**(qu): 357. 1824.—"Crescit prope Guayaquil Quitensium."—Holotypus, labelled 'Guayaquil,' where collected by Humboldt & Bonpland in III.1803 (fl, fr), P-HBK!—Incorrectly referred by Bentham, 1871, p. 548 to *Cassia emarginata* (=our *S. atomaria*).

Cassia laeta Humboldt, Bonpland & Kunth, Nov. Gen. & Sp. **6**(qu): 340. 1824.—"Crescit prope Guayaquil, ad litora Maris Pacifici."—Holotypus, labelled "n. 3842, Guayaquil," P-HBK!—*Cathartocarpus laetus* (Humboldt, Bonpland & Kunth) G. Don, Gen. Hist. Dichl. Pl. **2**: 454. 1832.—Correctly equated by Macbride, 1943, p. 156, with *C. atomaria* auct.

Cassia siliquosa F. W. C. Areschoug, Kgl. Sv. Fregatten Eugenies Resa 141. 1910 (n.v.), reprinted by Fedde, Repert. Sp. Nov. **10**: 303. 1912.—"Ecuador: in insula Puna."—Holotypus, collected in 1852 by *N. J. Andersson* (n. *172*), S (n.v.); isotypus, K!

Cassia weberbaueri Macbride, Field Mus. Bot. **8**(2): 92. 1930.—"Peru: . . . southeast of Hacienda La Choza, Dept. of Tumbez, Feb. 28–March 3, 1927, *Weberbauer 7709.*"—Holotypus, F! isotypus, K!—Equated by Macbride himself, 1943, p. 156, with *C. atomaria* auct. austro-amer.

Cassia atomaria sensu Bentham, 1871, p. 548, ex parte, quoad pl. ecuador.

Cassia canescens sensu Vogel, 1837, p. 33.

Cassia laeta sensu Vogel, 1837, p. 45; Bentham, 1871, p. 548.

Cassia atomaria sensu Bentham, 1871, p. 548, minore ex parte, quoad pl. ecuador.; Macbride, 1943, p. 156.

Characters as given in key to varieties.—Collections: 32.

Arid tropical scrub forest and thickets along streams, sometimes on rocky river banks or stream-beds, 5–1250(–1800) m, locally common on the coastal plain and Andean piedmont in s. Ecuador (s.-ward from near 2°S both on the coast and along the Guayas estuary) and n.-w. Peru (Tumbes and Piura), thence locally e. across the Piura-Huancabamba divide to dry valleys of the Marañón basin in n. Cajamarca (Río Chamaya) and adjoining Amazonas (Río Utcubamba in prov. Bagua); collected once in the highlands of Bolivia (La Paz, *Rusby 1042,* NY), where presumably cultivated.—Fl. throughout the year, the fruit often contemporaneous with flowers.—*Machete de burro, machetilla* (Ecuador); *vainilla de montaña, chapra* (Peru).

The var. *mollissima* is superficially very similar to Mexican var. *glabrata* but has on the average shorter (mature) leaves, leaflets fewer by one or two pairs, and flowers at once fewer and larger. Because the raceme axis is relatively short, the expanded flowers are usually raised to the level of succeeding buds which, consequently, do not form the exserted, narrowly conical spike characteristic of var. *glabrata.*

The specimens of var. *mollissima* seen by Bentham prior to 1871 were scanty and ambiguous. The typus of *Cassia laeta* from Guayaquil consists of one large mature leaf and two detached flowering racemes; that of *C. canescens* from the same place consists of a young flowering branchlet and a detached pod. This pod is indistinguishable from that of *S. atomaria* and, we suppose, was the decisive factor in Bentham's misinterpretation of *C. canescens* as conspecific with the common Caribbean species. There is no modern evidence that genuine *S. atomaria* occurs in Ecuador or that there is more than one senna of its group native on the Pacific lowlands in equatorial latitudes. We are confident that Humboldt's two specimens represent aspects of one entity.

A collector in the Bosque Nacional de Tumbes (*D. R. Simpson 503,* NY) reports that browsing animals will eat the pods of var. *mollissima* but suffer loss of hair from this diet.

200bis. Senna sousana Irwin & Barneby, sp. nov., inter *S. atomariam* et *S. mollissimam* quasi intermedia, hanc foliolis ovato-acuminatis, illam racemis ineunte anthesi erectis abbreviatis subcorymbosis ipsisque floribus parvulis simulans, ab ea inflorescentiae forma florumque magnitudine, ab illa foliolorum ambitu racemisque omnibus e folii coetanei axillo nascentibus diversa, ab ambabus ecologia montano-temperata abstans.—MEXICO. Oaxaca, distr. Sola de Vega: Sola de Vega, 1300 m, 23.VI.1979 (fl), *M. Sousa et al. 10489.*—Holotypus, MEXU; isotypus, MO.

Drought-deciduous trees (3–)4–14 m tall, the stout annotinous branchlets, foliage and immersed panicle of racemes densely pilosulous and tomentulose with fine curly and incumbent golden hairs up to ±0.5–1 mm, the membrano-chartaceous lfts bicolored, brownish-olivaceous and relatively thinly pubescent above, densely so or at first pannose-tomentose beneath, the racemes arising singly or geminate from axils of relatively short synanthous lvs, the later lvs longer.

Stipules erect, subulate or lance-triangular 3–5 × 0.5–1.5 mm, firm, except from inhibited budlike brachyblasts early caducous.

Lvs accompanying fls 9–22 cm, the later ones 15–40 cm; petiole 3–12 cm × 0.5–2 mm, appearing terete, the shallow ventral sulcus concealed by vesture; rachis 1.5–18 cm, its interfoliolar segments (0.6–)1–5 cm; pulvinules 2–5 mm; lfts 2–5 pairs, accrescent distally, ovate abruptly deltate-acuminate varying to ovate-subelliptic, the distal pair (2–)3.5–13 × 2.5–5 cm, at base subsymmetrically rounded or broadly cuneate, the margin revolute, the midrib with (8–)10–15 pairs of camptodrome secondary and reticular venules all at first immersed but sometimes becoming prominulous above, sharply raised beneath.

Racemes ±10–30-fld, the several simultaneously expanded fls raised almost to the level of the crowded nodding subglobose, glabrate or pilosulous fl-buds, the erect axis at full anthesis 3.5–11 cm; bracts triangular-acuminate or lance-caudate 3–5.5 mm, early caducous; pedicels at anthesis 13–18 mm; sepals fuscous yellow-margined, strongly graduated, the outermost ovate 3.5–5 mm, the innermost broadly obovate concave 5.5–8 × 4.5–5.5 mm; petals yellow drying dark or brownish-yellow dark-veined, 4 subhomomorphic except for size, the 3 adaxial and 1 abaxial (opposed to pistil) broadly oblanceolate, the 3 similar ones ±9–11 × 4–5 mm, the abaxial one 12–16 × 5–6.5 mm, the fifth petal much larger and strongly oblique, its claw thickened, its blade both incurved through nearly 90° and folded inward over the androecium, measured along midvein ±15–17 mm; androecium of *S. atomaria,* the thinly pilose anthers 2.8–3.4 × 1.3–1.7 mm; ovary glabrous except for random hairs along ventral suture; style incurved 1–1.5 × 0.5–0.6 mm; ovules ±48.

Pod pendulous, the stout stipe ±2 mm, the thickly linear, laterally compressed body (18–)25–37 × 0.9–1.3 cm, in all respects like that of *S. atomaria* or *S. mollissima*.—Collections: 10.

Wooded mountainsides and stream banks in pine-oak or oak-alder forest, on limestone, 1300–1850 m, locally plentiful near the crest of Sa. Madre del Sur in distr. Sola de Vega and Ejutla, Oaxaca, Mexico.—Fl. II, IV–VIII.—*Vainillo.*

This handsomely floriferous but taxonomically ambiguous senna suggests a hybrid derivative of *S. atomaria* × *mollissima,* and at first was tentatively so interpreted by us. Its mature foliage is that of *S. mollissima,* but in form of inflorescence and in size of the individual flower it appears much closer to *S. atomaria*. The abundant and uniform material assembled by Dr. Mario Sousa and his students at MEXU, however, cannot be passed over so lightly, for it testifies eloquently to the existence of a self-productive entity locally abundant in a cool montane environment (1300 m upward) in which neither *S. atomaria* nor *S. mollissima* could be expected to flourish. Both of these two species are primarily or obligately at home in the tierra caliente, *S. mollissima* below 300 (one record at 800) m and *S. atomaria,* a little more accommodating in its requirements, ascending from sea level, in matorral and low deciduous woodland or thornforest, up to about 1100 m, beyond that contour only feebly, on south slopes or in arid interior basins. MacDougall, to be sure, collected *S. atomaria* somewhere near Sola de Vega (NY, *s.n.*) where it may possibly come into marginal contact with *S. sousana,* but *S. mollissima* cannot be expected at such elevations. Moreover *S. atomaria* and *S. mollissima* var. *glabrata* are sympatric along the Pacific coast between southern Sinaloa and the isthmus of Tehuantepec without giving rise anywhere to known hybrid forms. Some of Dr. Sousa's collections were provisionally named *Cassia xanthophylla* (B. & R.) Lundell, the type of which, an immature collection from Guatemala superficially similar because of the dense golden pubescence of the young foliage, is referred herein to *S. mollissima* var. *glabrata,* of which it has the precociously elongating and pendulous raceme-axis. Although distinguished from *S. atomaria* chiefly by leaf-form and habitat, *S. sousana* seems to have acquired a specifically independent identity.

We take pleasure in dedicating the species to Dr. Mario Sousa, Jefe del Herbario Nacional, Instituto de Biología, UNAM, equally well known for his energetic and fruitful exploration of the flora of Oaxaca as for his generosity in sharing the results with colleagues.

201. Senna peralteana (Humboldt, Bonpland & Kunth) Irwin & Barneby, comb. nov. *Cassia peralteana* Humboldt, Bonpland & Kunth, Nov. Gen. & Sp. **6**(qu): 356. 1824.—"*Crescit prope Campeche Mexicanorum . . . a celeb. Peralta cum Bonplandio communicata.*"—Holotypus, labelled simply "Campeche," P-HBK!—*Gaumerocassia peralteana* (Humboldt, Bonpland & Kunth) Britton & Rose, N. Amer. Fl. **23**(4): 253. 1930, quoad nom., vix aliter.

Cassia anisopetala Donnell Smith, Bot. Gaz. (Crawfordsville) **37**: 209. 1904.—"In planitiebus ad Cubilquitz [=Gubilguitz], Depart. Alta Verapaz, Guat[emala], alt. 350 m, Mart. 1902, *von Tuerckheim,* no. 8194 ex Pl. Guat. etc., quas ed. Donn[ell] Sm[ith]."—Holotypus, US? isotypi, GH, K, NY!—*Pseudocassia* (?) *anisopetala* (Donnell Smith) Britton & Rose, N. Amer. Fl. **23**(4): 231. 1930.

Pseudocassia petensis Britton & Rose, N. Amer. Fl. **23**(4): 231. 1930.—"Petén, Guatemala, April 8, 1922, *O. F. Cook & R. D. Martin 193.*"—Holotypus, US! clastotypus, NY!—*Cassia petensis* (Britton & Rose) Standley, Carnegie Inst. Wash. Publ. **61**(Bot. Maya Area: Misc. Papers **4**): 61. 1935.

Precociously flowering shrubs 1.5–5 m becoming in forest or at forest margins woody vines, the annotinous and older branches gray, erratically lenticellate, often but not always armed at the nodes with ligneous aculeiform stipule-bases (described below), the short leafy unarmed annotinous branchlets like the foliage and axes of racemes villosulous with weak spreading-incurved or contorted simple hairs to 0.25–0.55 mm, the lf-stalks at insertion of petiolules and the raceme-axes in addition charged with rufescent, irregularly branched trichomes, the simple vesture at first yellowish, gray in age, the foliage strongly bicolored, above ± lustrous dark green (brunnescent in age or drying), thinly pubescent when mature, beneath dull, densely villosulous, the contracted subcorymbose panicle of erect or pendulous racemes leafy-bracteate, not or scarcely exserted, its main axis continuing to grow out during maturation of the fruit.

Stipules subulate 1–2.5 mm, rather firm but deciduous, their base during the second year often but not invariably transformed into ligneous conic or spreading-decurved and prickle-like thorns (1.5–)3.5–6 mm seated on a ± dilated pediment.

Lvs (7–)10–21 cm; petiole slender including livid pulvinus (1.5–)2–4(–5) cm, at middle (0.55–)0.7–1.5 mm diam, narrowly shallow-sulcate ventrally; rachis (1–)2–8(–9.5) cm, its longer interfoliolar segments 1–2.6 cm; petiolar gland 0; pulvinules 2–4.5(–5) mm; lfts 2–5, of most mature lvs 3 or 4 pairs, accrescent upward, the distal pair broadly elliptic, elliptic-obovate or obovate obtuse apiculate or deltately acute-acuminulate (3–)4–8 × (1.5–)2–4 cm, 1.8–2.7 times as long as wide, at base subsymmetrically broadly cuneate (rounded), the margin revolute, the centric straight midrib shallowly canaliculate above, cariniform beneath, the 6–10 pairs of camptodrome secondary, the connecting tertiary and the open irregular reticular venules all finely prominulous above, more sharply so beneath; proximal pairs of lfts either like the distal pair only shorter, or at once shorter and broader, becoming obovate.

Racemes incurved-ascending from axis, loosely 7–35-fld, several fls reaching anthesis together and these in the earlier stages of the raceme standing well below the at first erect, later nodding, serially racemose fl-buds, the axis including short peduncle becoming (4–)5–18 cm; bracts lance-subulate 1.5–2.8 mm caducous; pedicels at full anthesis filiform 9–19 mm, bracteolate at or above middle, the caducous bracteoles subulate 0.4–1.2 mm; buds globose when very young, thinly pilosulous, but the often dorsally glabrous ciliolate inner sepals early emergent; sepals separating long before true anthesis, submembranous yellowish, brunnescent when dry, well graduated, the outermost 3–4 mm, the innermost concavely obovate-suborbicular 5–5.5 × 2.6–4 mm; petals yellow drying brown or brownish-yellow, except for sometimes remotely pilosulous claw glabrous, when fully expanded highly heteromorphic, 4 alike except for size, beyond claw oblong or oblong-obovate, the 3 adaxial 4.5–8 mm, one abaxial 9–13 mm, the fifth petal asymmetrically oblong-ovate or dimidiately ovate, strongly incurved and folded over the androecium, measured along the arcuate midrib 16–20 mm; androecium functionally 7-merous, the fertile stamens isomorphic, their glabrous filaments 1–2 mm, the oblong, almost straight, dorsoventrally compressed anthers 3–3.6 × 1.3–1.4 mm, sulcate lengthwise only at the connective, bluntly truncate at apex and turned through ±90° into the separated lateral very short beaks, these opening by vertical slit; ovary linear-falcate glabrous; style ±1.5 mm, gently incurved, conic at tip; ovules ±50–66.

Pod pendulous, the stipe 4–6 mm, the elongately linear body 20–36 × 1.2–1.6 cm, plano-compressed, ±2 mm thick at the strong sutures, the flat lignescent valves differentiated into a paler internally pithy margin running parallel to the sutures which becomes transversely cracked in ripening and an atrocastaneous face irregularly coarsely venulose, the interseminal septa very narrow or subobsolete, the seed-cavity ±5 mm long, extending only to the thickened valve-margin; seeds transverse, oblong-ellipsoid compressed parallel to the valves, ±5–6 × 3–3.5 mm, the lustrous smooth brownish-ochraceous testa marked on each broad face with an oblong-elliptic areole ±3.5–4 × 1.5 mm.—Collections: 33.

A vine of high forest or riverine forest, surviving as a diffuse shrub in thin savanna woodland or disturbed forest, reported from calcareous and serpentinized lateritic soils, 10–550 m, widespread over the Gulf lowlands and low hill-country from n.-w. Campeche (Laguna de Términos) to Quintana Roo, s. in Mexico to n. Chiapas, thence s. into n.-e. Guatemala (Petén, Alta Verapaz, Izabal) and Belize (Belize and Cayo districts).—Fl. mostly II–VI, the fruit maturing slowly and long persisting.—*Kanchinaik* (Yucatán); *mayahua* (Quintana Roo).

Senna peralteana is closely related to *S. atomaria,* differing constantly in the more prominently venulose upper face of leaflets, the minutely bracteolate pedicels unique in *Senna,* and the broader pod 12–16 (not 8.5–12, exceptionally 14) mm wide. The enlarged thornlike stipule-bases, either conic or curved backward like those of Amazonian *S. spinescens,* are a conspicuous feature of some specimens and lacking in others. The species is plastic in growth-form and one might expect in theory to find stipules adapted to climbing in the forest liana and perhaps absent from the juvenile shrubby state, but this is not the case; for example *Jones & Facey 3344* (NY) from a liana 30 m long is thornless through growth of at least three consecutive seasons, whereas the low-shrubby *Croat 23347* (NY), fruiting at 1.5 m, has short conic vulnerant thorns on annotinous branchlets. These thorns are not directly modified stipules but are formed by secondary thickening of the pediment to the real stipules, which are always small and caducous. The cause of apparently random development of thorns at some but not all nodes of a stem, in some but not all plants of *S. peralteana,* is a subject for field inquiry.

The typi of *Pseudocassia petensis* and *Cassia anisopetala* are flowering specimens and the two species were doubtless referred by Britton & Rose to genus *Pseudocassia,* the first without reservation, the second provisionally, because of similarities in the asymmetric perianth and isomorphic anthers. They differ slightly in outline of leaflets, but are certainly conspecific. The only fruiting specimen of *S. peralteana* known to Britton, *Lundell 1027* (NY, received several years after publication of the monograph in 1930), was annotated by Britton as an undescribed *Isandrina,* the affinity to *S. atomaria* being expressed in the pod. Bentham knew *Cassia peralteana* only from the typus, acquired by Bonpland from "celeber Peralta," whom we suppose to be the Mexican botanist Francisco Peralta y Guzmán, mentioned in the bibliography of I. K. Langman (1964, q.v.). The epithet was transferred by Britton to a form of our *Senna racemosa* and therefore to the genus *Gaumerocassia.*

Fvii. ser. EXCELSAE (Bentham) Irwin & Barneby

Senna sect. **Peiranisia** ser. **Excelsae** (Bentham) Irwin & Barneby, comb. nov. *Cassia* sect. *Chamaefistula* ser. *Excelsae* Bentham in Martius, Fl. Bras. **15**(2): 109. 1870 & Trans. Linn. Soc. London **27**: 528. 1871.—Sp. lectotypica: *C. excelsa* Schrader = **Senna spectabilis** var. **excelsa** (Schrader) Irwin & Barneby.

Pseudocassia Britton & Rose, N. Amer. Fl. **23**(4): 230. 1930, pro gen.—Sp. lectotypica: *Ps. spectabilis* (DeCandolle) Britton & Rose ≡ *Cassia spectabilis* DeCandolle ≡ **Senna spectabilis** (DeCandolle) Irwin & Barneby.

Corolla of *Peiranisia,* the odd petal standing aside from the androecium; sepals strongly graduated; anthers of 7 fertile stamens equilong and isomorphic except at apex, all gently sigmoid in lateral view and shortly beaked, the beak of 4 median and of the centric abaxial stamen lateral or subinfraterminal, that of 2 latero-abaxial ones porrect, each theca of all anthers dehiscent by its own subvertical slit; pod subterete, with woody corrugated mesocarp, the seed-locules extending between the sutures but suspended laterally between empty cavities; seeds 1-seriate transverse, turned broadside to the interseminal septa, areolate.—Trees; stipules narrow firm caducous; petiolar gland 0; lfts (8–)11–20 pairs; inflorescence thyrsiform-paniculate exserted; style short and stout, scarcely differentiated from ovary, the minute stigmatic cavity obliquely terminal; ovules ±80–120.—Species 1 (varietally differentiated), widespread in the Neotropics, often planted for ornament, occasionally cultivated and subspontaneous in the Paleotropics.

The slender rodlike pod of *Cassia spectabilis* and *C. excelsa* was unknown to Colladon, DeCandolle or Vogel. Bentham found it to be associated with a characteristic androecium, and was induced thereby to describe a series *Excelsae* of sect. *Chamaefistula,* in which a fruit that simulated that of genuine *Cassia* coincided with the exserted thyrsiform-paniculate inflorescence of the flat-podded *Paniculatae* (our ser. *Sapindifoliae*). The *Excelsae* remain a sharply defined group to which nothing has been added in modern times, but the pod resembles that of *Cassia* sens. str. only in external form, which must have been acquired independently from flat-podded antecedents. A clue to its origin is preserved in the androecium which finds its nearest match in that of ser. *Isandrae,* and it is not by accident that the *Isandrae* have a similar irregular corolla, similar foliage, and similarly glandless petioles. The pod of ser. *Excelsae* stands in relation to that of some *Isandrae* as the pod of ser. *Corymbosae* stands to that of ser. *Pachycarpae*; in each case the pod's girth is determined by volume of seeds within a given vessel. The very high number of ovules common to the pistils of *S.* (*Isandrae*) *emarginata* and *S.* (*Excelsae*) *spectabilis* are accommodated in the first case broadside to the valves of a flat pod that attains an exaggerated length and in the second are crowded together and turned broadside to the septa in a shorter, more nearly cylindric cavity. It is simply the capacity of the individual seed-locule which determines the attitude of the seed and consequently the compression of the pod itself. In this connection it may be recalled that Britton & Rose described from Guatemala two species of *Pseudocassia* other than *Ps. spectabilis,* although the fruit of neither was known. These have both turned out to be conspecific with *S.* (*Isandrae*) *peralteana,* but the general resemblance in floral morphology to *S. spectabilis,* which must have been the determining factor in their assignment to *Pseudocassia,* provides strong evidence of genuine affinity.

202. Senna spectabilis (DeCandolle) Irwin & Barneby, comb. nov. *Cassia spectabilis* DeCandolle, Cat. pl. horti bot. monspel. 90. 1813.—Typus infra sub var. *spectabili* indicatur.

Amply leafy soft-woody trees of rapid growth, precociously flowering as bushy shrubs of rounded outline, at anthesis 2–15(–20) m tall, the trunk rarely attaining 25 cm diam, the stout hornotinous branchlets brown or fuscous pallid-lenticellate, the young growth except for sometimes glabrate upper face of lfts and dorsal face of sepals pilosulous throughout with fine spreading-ascending or partly (largely) straighter ascending or subappressed hairs up to 0.2–0.7(–0.9) mm, the foliage bicolored, the membranous lfts lustrous green above, paler dull beneath, the vesture of the inflorescence golden or tawny, of the foliage (variable in density) usually gray but sometimes at first lutescent, the inflorescence an ample thyrsiform panicle of racemes commonly leafy-bracteate proximally, then well exserted from foliage.

Stipules caducous (lacking from many flowering and all fruiting spms), widely spreading and falcately incurved, the firm, narrowly linear-attenuate blades 4–10 × 0.2–0.9 mm.

Major lvs (below and at base of inflorescence) 1.7–4(–4.5) dm; petiole including firm livid pulvinus 11–35(–40) mm, at middle (0.8–)1–2.5 mm diam, rounded dorsally, shallowly open-sulcate ventrally; rachis (10–)12–36 cm, the longer interfoliolar segments 10–28 (–30) mm; pulvinules 1.2–2.7(–3) mm; petiolar gland 0; lfts (8–)11–19(–20) pairs, decrescent toward both ends of rachis but more so proximally, the lowest pair often much smaller than the rest and readily deciduous, the largest beyond middle of rachis, these varying from ovate or elliptic obtuse mucronulate to narrowly ovate- or lance-acuminate (the acumen itself either obtuse or acute), the blade (2.6–)3–9.5 × (1–)1.1–2.7(–3) cm, 1.9–4.3(–4.7) times as long as wide, at inequilateral base rounded or (when broad) shallowly cordate, the margin at least proximally revolute, the straight or gently incurved midrib, the (10–)12–22 pairs of camptodrome and intercalary secondary veins and the reticular venulation all sharply prominulous beneath, somewhat less so or rarely almost immersed above.

Racemes (6–)12–60-fld, the several simultaneously expanded or expanding fls raised to or beyond level of fl-buds, the axis including peduncle becoming (3–)4–20(–24) cm; bracts ovate acute or ovate- to lance-acuminate (1–)1.5–3.5(–4.5) mm, their broad base shielding the very young fl-bud but early caducous; pedicels at and after full anthesis (16–)20–32(–36) mm; fl-buds globose when young, usually puberulent at least proximally, sometimes glabrous, the sepals separating long before true anthesis of fl; sepals early reflexed concave, well graduated, the firm ovate outermost (2.5–)3–6 mm, the inner ones subpetaloid or marginally membranous, broadly obovate or suborbicular, the innermost (5.5–)6–10(–14) × 5–9.5 mm; petals rich yellow drying dull or brownish-yellow dark-veined, glabrous or rarely puberulent near claw, strongly heteromorphic, 4 similar in outline but of different lengths, broadly or narrowly obovate beyond the claw 13–36 mm, of these the vexillar one usually smallest and one abaxial (opposed to excentric pistil) longest, the fifth petal abaxial, coarsely thick-clawed, its blade turned at ±90° to the claw and folded over the androecium, measured along the curved midrib 22–38.5 mm; androecium functionally 7-merous, glabrous or exceptionally remotely pilosulous, the blades of 3 staminodes deeply cordate both at base and at apex (appearing 2-auriculate) 1.8–2.4 mm wide, the filaments all of equal length or slightly longer abaxially 1.5–3.5(–5) mm, the oblong pale yellow anthers of subequal length and except for beak isomorphic 4.5–7 × (1.6–)1.8–2.4(–2.6) mm, truncately rounded at both ends, straight or slightly decurved in proximal half, gently incurved distally, 8-sulcate lengthwise, the thecae of 4 median stamens and the centric abaxial one contracted into a very short divaricate lateral or shortly infraterminal beak separate to base and dehiscent by short vertical slit, those of 2 latero-abaxial stamens into scarcely longer but porrectly incurved, less clearly separated beaks; ovary glabrous or (locally in Colombia) thinly remotely pilosulous; ovules (72–)80–120, seldom all maturing into seeds; style scarcely differentiated from ovary externally, 1.5–2.5 mm, gently incurved distally and just below the conic tip (0.4–)0.5–0.7 mm diam, the minute terminal stigmatic cavity glabrous or ciliolate.

Pod pendulous, the stipe 5–7 mm, the narrowly linear turgid, straight or nearly straight body 16–30 × 0.85–1.2(–1.3) cm, a trifle flattened along the broad sutures and a trifle compressed laterally, the cross section oblong-elliptic 7–10(–11) mm wide, the valves consisting of a) a thin succulent, early papery exocarp, b) a deeply corrugated woody mesocarp intruded between seeds as a complete partition, and c) a membranous endocarp less capacious than the cavity and free from the interior walls of the mesocarp, the narrowly oblong 1-seriate seed-locules defined by septa and endocarp 2–3.5(–4) mm wide, extending from suture to suture but separated on either side from the mesocarp by an empty cavity often as wide as themselves, the mature exocarp nigrescent, little or deeply depressed into the reentrant corrugations of the underlying mesocarp, irregularly rugulose but not venulose, splitting transversely along the thickened sutural margins, the whole valve finally dehiscent along the ventral suture and narrowly gaping to allow egress to the seeds; seeds turned with broad faces to the septa, broadly (narrowly) obovate in outline 4.8–6.8 × (3.3–)3.5–5.5 mm, compressed but moderately plump, the usually light orange-brown, sometimes fuscous or brown-olivaceous, exceptionally ochraceous testa dull or sublustrous, the oblong-elliptic or subrhombic areole (1.8–)2.1–3.3 × 1–2(–2.3) mm.

When Bentham admitted to his sect. *Chamaefistula* ser. *Excelsae* an eastern Brazilian *Cassia excelsa* and a circum-Caribbean *C. spectabilis* he established a tradition which, because of the discrete ranges of the pair, was not critically evaluated or challenged subsequently. Bentham found differential characters in number and shape of leaflets that coincided with varying depth of corrugation in the pod-valves. The corrugation of the pod is determined not, as in many sennas, by expression of the valve wall over the swelling seed, but by the unyielding woody endocarp which is in all *Excelsae* equally strangulated between the interseminal septa but becomes apparent externally only as the papery exocarp accommodates itself to the underlying mesocarpic armature. The degree of accommodation is quite variable, depending at least in part on maturity of the fruit when dried and perhaps in part on intrinsic differences in thickness of the exocarpic tissue. What is certain is that pods equally deeply corrugated are known to accompany all types of leaflet and are not limited to any one segment of the full range of the *Excelsae*. In a provisional separation of *C. excelsa* from *C. spectabilis* Isely (1975, p. 126) abandoned this character, but supposed the pedicels and pods of *C. excelsa,* as defined primarily by relatively numerous obtuse leaflets, to be on the average longer and its flowers more ample than those of *C. spectabilis*. There is an element of truth here, but we have found so wide an overlap in all measurements (see comparative data in our descriptions following) and such complete agreement in form of perianth and androecium and in structure of the pod that we can no longer recognize two discrete species. The best that can be said of the Brazilian populations which we here treat as var. *excelsa* is that they generally have two or three more pairs of generally smaller and more obtuse leaflets than var. *spectabilis* and flowers often but not consistently larger. Some specimens from Argentina that have been segregated as *Cassia carnaval* have leaflets in size and number like those of var. *excelsa* but in outline like those of var. *spectabilis*.

Key to the Varieties of *S. spectabilis*

1. Largest lfts of fully expanded lvs (4.5–)5–9.5 × ±1.5–2.5 cm, in outline narrowly ovate- to lance-acuminate; longer interfoliolar segments of rachis mostly 14–28 mm; widespread n. of Amazon valley, extending s. along e. foothills of Andes to Argentina, thence e. into s.-e. Bolivia and Paraguay. 202a. var. *spectabilis* (p. 603).
1. Largest lfts of fully expanded lvs (2.6–)3–5.5 × ±1–2 cm, in outline oblong-elliptic, elliptic

or ovate, obtuse mucronulate; longest interfoliolar segments of rachis mostly 10–17 mm; e. Brazil (Piauí to Rio Grande do Norte, s.-e. Bahia, n.-centr. Minas Gerais and e.-centr. Goiás). 202b. var. *excelsa* (p. 604).

202a. Senna spectabilis (DeCandolle) Irwin & Barneby var. **spectabilis**. *Cassia spectabilis* DeCandolle, 1813, l.c., sens. str.—"Hab. ad Caracas [Venezuela]."—Described from plants cultivated at Montpelier.—Holotypus (3 lvs only), "cult. in h[ort]. m[eo] oct. 1809," G-DC! Spms at MPU, dated 14.XII.1817 and 23.XI.1831 are doubtless from the same source.—*Pseudocassia spectabilis* (DeCandolle) Britton & Rose, N. Amer. Fl. **23**(4): 230. 1930.

Cassia speciosa Humboldt, Bonpland & Kunth, Nov. Gen. & Sp. **6**(folio): 338. 1824.—"Crescit in ripa fluminis Magdalena, juxta oppidum Teneriffe [Colombia]."—Holotypus, so labelled, collected by Humboldt & Bonpland in V.18 (fl, fr), P-HBK!—*Cassia humboldtiana* DeCandolle, Prod. **2**: 489. 1825, a legitimate substitute (non *C. speciosa* Schrader, 1821). *Cathartocarpus speciosus* (DeCandolle) G. Don, Gen. Hist. Dichl. Pl. **2**: 453. 1832. *Cathartocarpus humboldtianus* (DeCandolle) Loudon, Hort. Brit. 167. 1832.—Correctly equated with *C. spectabilis* by Bentham, 1871, p. 529.

Cassia trinitatis Reichenbach ex DeCandolle, Prod. **2**: 489. 1825.—"[F. W.] Sieb[er] pl. exs. trin[itatis] n. 57 . . . in ins[ula] Trinitatis."—Holotypus, G-DC! isotypi, M, MO, W!—*Cathartocarpus trinitatis* (DeCandolle) G. Don, Gen. Hist. Dichl. Pl. **2**: 453. 1832; Loudon, Hort. Brit. 167. 1832.—Correctly equated with *C. spectabilis* by Bentham, 1871, p. 529.

Cassia edulis Posada, Adansonia **10**: 187. 1872.—"Croît a Médellin [Colombia] (1500 metres), où on lui donne le nom de *Canafistula macho*. Les enfants mange la pulpe et même l'épiderme du fruit."—No typus found at P, but the description and locality decisive, the vernacular name corroborative.—Overlooked by Britton & Killip, 1936.

Cassia totonaca Sessé & Mociño, Fl. Mex. 100. 1893.—Habitat in Oppido de Xoxocolco [±45 km s.-w. of Veracruz], ubi totonaci Lettaanaxanatl adpellant."—Lectoholotypus, labelled "*Cassia arborescens* N[obis]," *Hb. S. & M. 1149,* MA! = F Neg. *44235*.

Cassia excelsa var. *acutifolia* Hassler in Feddes Repert. Sp. Nov. **8**: 128. 1910.—"Paraguay: Caballero cué . . . flor. mens. Febr.; Fiebrig no. 4800 in herb. Hassler."—Holotypus, G! isotypus, E!

Cassia amazonica Ducke, Arch. Jard. Bot. Rio de Janeiro **3**: 112. 1922.—"Habitat . . . prope Montealegre [Pará, Brazil] . . . (Ereré, n. 16.132, florif., 3-5-1916; Igarapé das Pedras, n. 17.149, flor. et fruct., 28-7-1918), l[egit] *A. Ducke*."—Lectoholotypus, *MG 17149 = RB 11322,* RB! isotypi, NY, US!

Cassia carnaval Spegazzini, Revista Argentin. Bot. **1**(2–3): 82, fig. A(1–3). 1925.—"In apricis collium et convallium Provinciarum Salta et Jujuy sporadice non rara . . . [sub fig. cit.:] 'La Vina, valle Calchaqui, Salta, Febr. 1897."—No typus seen, but the protologue definitive.

Cassia spectabilis sensu Colladon, 1816, p. 115, t. 7; Vogel, 1837, p. 46; Bentham, 1871, p. 529; Fawcett & Rendle, 1920, p. 104; Standley, 1922, p. 407; Record, Trop. Woods **10**: 23. 1927; Macbride, 1943, p. 181; Schery, 1951, p. 49, fig. 120; Steyaert, 1952, p. 506; De Wit, 1955, p. 267; Worthington, Ceylon Trees 192 (photo). 1959; Little, Woodbury & Wadsworth, Trees Puerto Rico, second vol. 282, 283 (pl). 1974; Isely, 1975, p. 126; Steyermark & Huber, Fl. del Avila 560, t. 167 (optima!). 1978; Lasseigne, 1979, fig. 29 (seed).

Cassia trinitatis sensu Vogel, 1837, p. 46.

Pseudocassia spectabilis sensu Britton & Killip, 1936, p. 170.

Cassia carnaval sensu Burkart, 1952, p. 164; Boelcke, Darwiniana **7**(2): 303, t. IX, fig. b (seed); Dimitri & Rial Alberti, 1954, fig. 13 (optima!).

Arborescent in forest but commonly seen as a shrub or treelet of disturbed woodland and other secondary habitats; major lvs (2–)2.5–4(–5) dm, the petiole including pulvinus 1.5–3.5(–4) cm, the rachis (10–)14–36 cm; lfts (8–)10–16 pairs, the largest (4.5–)5–9.5 × (1.2–)1.5–2.7(–3) cm, the outline as given in key; outermost sepal (2.5–)3–6.5 mm, innermost (5.5–)6–10(–14) × 5–9.5 mm; isomorphic petals 13–27.5 mm; heteromorphic abaxial petal 22–34.5 mm.—Collections: 123.—Fig. 11 (androecium), 12 (pod, seed).

Moist and seasonally dry forest and woodland of various composition but now less common there than in secondary associations, often planted for ornament or as a living fence (for which purpose readily propagated by cuttings) and thence naturalized, the pre-Columbian range of dispersal consequently obscure, mostly below 850 m, but ascending to 1000 m in pine-forest in Honduras, to 1300 m in n. Venezuela and to 1800 m in the e. cordillera in Colombia, discontinuously widespread in parts of tropical N. and S. America: s.-e. Mexico (s.-e. San Luis Potosí, Veracruz and Tabasco to n. Oaxaca and Chiapas) s. to w. Panama, in most sites probably introduced; Greater Antilles (Cuba, Jamaica, Hispaniola and Puerto Rico), common only in Dominican Republic and probably nowhere truly autochthonous; native in Magdalena and Cauca valleys in Colombia, extending weakly to the Pacific slope in Valle; frequent and both native and adventive around the n. periphery of the Orinoco basin in Venezuela (Mérida to Delta-Amacuro and adjoining Bolívar); planted and escaped in Trinidad and Tobago; not recorded from the Guianas; very local and of unknown status on the e. slope of the Peruvian Andes in Junín and Cuzco; apparently again native along the Andean piedmont and in savanna-woodlands of Gran Chaco in s.-e. Bolivia, n. Argentina (Salta, Jujuy, Entre Ríos) and Paraguay, on Río Apa closely approaching the Brazilian border (to be sought in s.-w. Mato Grosso); adventive (?) in capoeira along the lower Amazon around Montealegre, Pará.—Fl. in N. America primarily (VII–)VIII–XII(–I), in S. America primarily I–IV, in equatorial latitudes throughout the year; pods maturing slowly, long persistent on branches of sometimes drought-deciduous trees.—*Canchín* (Tabasco); *flor de todos los Santos, f. de los Muertos* (Oaxaca); *candelilla* (Puebla); *frijolillo, candelillo* (Honduras); *bruscón* (Dominican Republic); *velero, cañafístola macho* (Colombia, Venezuela); *chiquichique, tarantán, vasito sucio* (Venezuela); *ramo* (Bolivia); *carnaval, yuto* (Argentina).

One of the handsomest ornamental sennas, recommended for rapid growth, ease of propagation and culture, and fragrance of the curiously asymmetric flowers arranged in bold thyrsiform panicles which may rise a meter above the crown of foliage. The Argentine and Paraguayan populations are not morphologically separable from Venezuelan and Colombian var. *spectabilis* but have been compared always with Brazilian var. *excelsa* and only for this reason maintained as an endemic *Cassia carnaval*. Ducke described *C. amazonica* as distinct from *C. spectabilis* in its somewhat fewer (8–12 pairs) of leaflets, but such numbers have been recorded repeatedly in material from the Orinoco basin and the Greater Antilles. It seems most likely, though we have no proof of it, that *S. spectabilis* was merely adventive and naturalized around Montealegre where it must have appeared to Ducke as notably distinct from all other Amazonian sennas. The flowers are said to poison bees in Salta, Argentina (Dimitri & Rial Alberti, l.c. sub *C. carnaval*).

202b. Senna spectabilis (DeCandolle) var. **excelsa** (Schrader) Irwin & Barneby, stat. nov. *Cassia excelsa* Schrader, Götting. Gelehrte Anz. **1**(72): 717. 1821.—"[. . . der Prinz Maximilian von Neuwied . . . in Brasilien gesammelten Gewächse . . .] no. 39."—Holotypus not found; isotypus (fragm + drawing), misit Schrader anno 1825, G-DC! = F Neg. *7010*.

Cassia fastigiata Nees von Esenbeck, Flora (Regensburg) **4**: 303. 1821 (21.V).—Described from a plant collected by Martius, no locality recorded.—No typus seen, but equated with *C. excelsa* by Nees himself, op. cit. p. 329. 1821 (28.V) and by Bentham, 1870, p. 109, 1871, p. 528; the name itself a posterior homonym of *C. fastigiata* Vahl (1794).

Cassia excelsa sensu Bentham, 1870, p. 109, t. 33 (optima); 1871, p. 528, exclus. syn. ult.

Shrubby and weakly arborescent, seldom above 7 m; major lvs 1.7–3.2 dm, the petiole including rachis 11–36 mm, the rachis (10–)12–26 cm; lfts (12–)13–19(–20) pairs, the largest (2.6–)3–5.5 × (1–)1.1–2(–2.4) cm, in outline as given in key; outermost sepal 3–4.5 mm, innermost (5.5–)6–9(–10) × 5–8 mm; isomorphic petals 18–36 mm; heteromorphic petal 29–38.5 mm.—Collections: 47.

Cerrado and caatinga, in s.-e. Bahia entering dense mata de cipó and wet coastal forest, surviving and flourishing in disturbed woodland, cocoa plantations and pasture thickets, sometimes planted as a living fence or as a dooryard ornamental shade-tree, widespread between ±350–900(–1100) m in interior e. tropical Brazil from Ceará to Rio Grande do Norte (there coming out near sea level to stabilized coastal dunes) s. to s.-e. Bahia, n.-centr. Minas Gerais, and the São Francisco-Tocantins divide in e.-centr. Goiás; planted in Belém, Pará (Museu Goeldi) and in coastal Ecuador (Manabí), to be expected elsewhere.—Fl. primarily I–V, occasional through the year.—*Canafistola, canafistola do boi, canafista.*

Numbered Exsiccatae of genus *Senna*

(Figures in parentheses refer to taxa as numbered in the text)

Abbott, W. L. 1056 (169b), 1674 (168a), 2246 (5)
Acosta Solís, M. 10626 (10b), 11370 (174), 12268 (14d)
Acuña, J. 12433 (135), 16738 (157), 17121 (199), 17464 (177j)
Agatha, Sister 455 (80)
Aguilar, J. I. 321 (177k), 1624 (26a), 1684 (177a)
Aguilar, R. G. 119 (95a)
Aguilar H., M. 127 (16)
Agostini, G. 1027 (199), 1134 (10a), 1139 (129a), 1722 (10a)
Alain, Bro.—see Liogier
Alain, J. Acuña & M. López Figueiras 5341 (4)
Alain, Bro. Clemente & Bro. Crisógono 1035 (4)
Albers, C. A. 43PH012 (84)
Albuquerque, B. W. de 168 (8b), 750 (8b), 819 (20), 877 (172), 1347 (3a), 1375 (20)
Albuquerque, B. W. de & J. Lima 245 (8b), 278 (8b)
Albert de Escobar, L. 780 (10b)
Alexander, E. A. 131 (177a), 139 (16), 167 (178a), 247 (199), 355 (196e), 535 (177q), 794 (178a), 1897 (72a)
Allard, H. A. 13204 (168a), 13385 (168a), 13389 (168a), 13830 (148), 13859 (168a), 14134 (168a), 15903 (71), 21732 (25a), 22296 (25a), 22475 (25a)
Allart, A. 199 (10a), 369 (29)
Allem, A. 884 (43)
Allem, A. & G. Vieira 2102 (66), 2279 (60)
Allen, C. 431 (13a), 747 (13a), 857 (27)
Allen, P. H. 327 (14a), 910 (14a), 1043 (27), 1119 (8a), 1180 (167), 2220 (14c), 2911 (167), 2917 (126k), 3333 (8b), 4151 (8a), 5231 (53a), 6930 (149), 7079 (177k), 7099 (16), 7114 (94a), 7282 (199)
Almeda, F. 642 (27), 4208 (199)
Almeida, J. 1488 (12)
Alston, A. H. G. 5440 (69), 6010 (171), 5860 (71), 6344 (129a), 6408 (129a)
Altson, R. A. 537 (161/Ia)
Alvarenga, M. 1 (44b)
Alvim, P. 3540 (44a)
Ancibor, E. & H. Schwabe 5797 (181)
Ancuash, E. 82 (17a), 289 (50), 1424 (147), 1469 (147)
Anderson, R. & S. Mori 47 (13a)
Anderson, W. R. 6214 (43), 6630 (56a), 6810 (140), 6817 (69), 6825 (60), 6932 (3f), 6961 (148), 7314 (126a), 7609 (43), 7656 (43), 7699 (69), 7826 (63b), 7840 (3f), 7851 (43), 8072 (43), 8300 (43), 8370 (56a), 8475 (43), 8494 (57), 8612 (43), 8671 (126a), 8894 (57), 8973 (42b), 9025 (43), 9047 (69), 9121 (3f), 9141 (71), 9146 (202b), 9148 (44b), 9207 (44b), 9344 (44b), 9459 (126a), 9539 (3f), 9682 (43), 9703 (60), 9936 (126a), 10007 (56a), 10013 (3a), 10088 (3f), 10172 (69), 10364 (43), 10438 (43), 10685 (58), 10781 (58), 11036 (3a), 11060 (140), 11193 (61), 11203 (43), 11242 (140), 11253 (3c), 11271 (63a), 11281 (126a), 11303 (3e), 11320 (60), 11329 (126a), 11405 (126a), 11408 (63a), 11522 (42b), 11569 (42b), 11581 (42b), 11597 (57), 11598 (57), 11655 (61), 11671 (124), 12017 (3a), 12033 (20), 34851 (60), 35152 (42b), 35701 (42a), 35737 (161/IIa), 35921 (57), 36121 (42a), 36158 (42b), 36439 (202b), 36877 (59a), 36879 (3d), 36924 (202b), 36993 (202b), 36998 (42d), 37062 (44b), 37100 (3a), 37141 (59a)
Anderson, W. R. & C. Anderson 4634 (193b), 568a (193a), 6070 (177c), 6135 (26a)
Anderson, W. R. & C. W. Laskowski 3745 (152), 4191 (149)
Anderson, W. R. & D. C. Sternberg 3008 (168b), 3095 (29), 3468 (29)
Andersson, N. J. 172 (200b), 237 (153c)
André, E. 180 (147), 215 (8a & 16), 697bis (27), 1811 (17b), 2308 (13a), 2744bis (177a)
Andrews, L. M. 3–10 (8a), 558 (129a), 564 (177a), 714 (71)
Andrieux, G. 177 (197), 414 (197), 416 (178a), 417 (138f), 418 (146), 419 (144), 421 (193a)
Angeli, C. 28 (42a), 40 (42a), 369 (12), 370 (12)

Anisits, J. D. 503 (63a), 2848 (66), 2660 (160), 2494 (157)
Apolinar-Angel, Bro. 751 (177a)
Apolinar-María, Bro. 525 (17b)
Appun, C. F. 363 (161/Ia)
Araquistain, M. 339 (71)
Araujo, D. 446 (45), 506 (165), 1431 (32), 1560 (42a)
Archer, W. A. 726 (126k), 1421 (147), 1479 (138g), 2310 (26a), 2478 (26a), 2619 (138g), 2707 (26a), 2953 (177a), 3007 (177a), 7649 (161/Ia), 7706 (161/Ia), 7667 (140), 7957 (36), 8268 (69), 8304 (20)
Arévalo y Lara, – 4 (200b)
Argüelles, E. 376 (144)
Aristeguieta, L. 1091 (177a), 1651 (17a), 2945 (8a), 3011 (8a), 4005 (8a), 4622 (159), 5468 (147), 5833 (8a), 5852 (161/IIb), 5868 (199), 5927 (170), 5981 (171), 5987 (199), 6822 (129a), 6856 (163), 7352 (26a), 7660 (29), 7662 (27), 7731 (10a), 91578 (8a)
Aristeguieta, L. & G. Agostini 6420 (69), 6422 (55)
Armour, A. V. 541 (28)
Arnodo-Boeders, F. 3615 (199)
Arreguín, M. L. 2534 (177a)
Arsène, Bro. 510 (116), 2964 (138a), 8762 (116 × 119), 8786 (119)
Artemio, Hno. 71 (177a)
Asplund, E. 8056 (116), 9029 (25a), 10215 (161/Ia), 13893 (174), 14194 (17a), 16507 (147), 16543 (129a), 18998 (116), 19232 (161/Ia), 19523 (26a), 20005 (25a), 20274 (154)
Assis, V. 97 (44a)
Assumpção, S. P. 3 (38)
Atchison, E. 67 (8a), 87 (202a)
Atencio, G. 5 (16)
Atwood, J. T. 356 (148), 2030b (68), 2031 (148), 2638 (177a), 2725 (68), 2815 (71), 3176 (16), 3467 (177a), 3556 (68)
Atwood, N. D. 2136 (82), 2223a (80)
Aubréville, A. 151 (20), 383 (26a)
Austin, D. F. 4188 (26a), 7025 (3/Ia), 7160 (36)
Aviles, S. 84 (14b)

Bacle, – 106 (128)
Badcock, W. J. 321 (68), 544 (110b), 545 (110b), 800 (63a)
Badillo, F. 2128 (10a)
Badillo, V. M. 3051 (10a), 3303 (129a), 3776 (169), 3824 (17a), 6271 (20), 7163 (27)
Baenitz, C. 1391 (95a)
Bailey, L. H. & E. Z. Bailey 137 (168a), 195 (69), 197 (177a), 205 (138g), 249 (168a), 377 (27), 389 (10a), 516 (177e), 544 (178a), 707 (29), 1118 (42a), 1315 (42a), 1614 (147), 1882 (199)
Baker, C. F. 4 (126k), 41 (147), 45 (46a), 125 (177a), 142 (147), 641 (177a), 687 (177a), 2523 (30), 3981 (126m), 4126 (138f)
Baker, R. E. D. & N. W. Simmonds 4892 (8a)
Balansa, B. 1399 (66), 1399a (66), 1404 (136), 1407 (63a), 1408 (63a), 1412 (160), 4414 (139), 4416 (3c)
Balegno, B. 1150 (118), 1235 (118)
Balls, E. K. 5319 (116)
Balslev, H. 1321 (174), 1787 (116)
Balslev, H. & E. Madsen 10606 (8b)
Baltmer, L. J. 1577 (16), 1622 (27)
Bamonte, A. G. 33 (55)
Bamps, P. & A. Robyns 63 (113a)
Bancroft, – 52 (29)
Bang, M. 24 (96), 464 (25b), 1090 (118), 1171 (117), 1345 (161/Ia), 1561 (46b), 1605 (138f), 1610 (3a), 1782 (117), 1985 (108), 2161 (63c), 2822 (202a), 2823 (202a), 2824 (126e), 2658 (126e)
Bangham, W. N. 603 (14b)
Barbados Bot. Sta. Herb. 559 (140)
Barbosa, V. P. 47 (42e), 219 (202b)
Barbosa da Silva, M. 9 (46a), 136 (36)
Barbour, P. J. 4361 (25a), 4528 (25a)
Barcia, L. 68 (40)
Barclay, G. W. 376 (153c), 868 (14d), 2318 (110f), 2610 (177a), 2716 (177a)
Barkley, F. A. 440 (177a), 3303 (90), 14121 (177a), 14132 (177a), 14166 (177a), 14348 (140), 17C269 (16), 17C337 (126k), 17C360 (177a), 17C430 (8b), 18S268 (177a), 18V253 (170), 19An479 (17b), 18C084 (8a), 18M031 (13b), 18M019 (17b), 18N097 (13a), 38C295 (10a)
Barneby, R. C. 2432 (87), 14459 (84), 17772 (196d), 17773 (152), 17777 (55), 17780 (72a), 17781 (129a), 17790 (116 × 119)
Barnes, C. R. & W. J. G. Land 292 (177a)
Barker, E. E. 1 (169a)
Barnhart, J. H. 2235b (69)
Baron, R. 334 (116 × 119), 5042 (116 × 119)
Barrios, – & V. González 102 (42g)
Barros, W. D. 164 (42a), 301 (120)
Barroso, G. et al. 575 (56a)
Bartlett, A. W. 8622 (9), 8758 (27)
Bartlett, H. H. 12017 (16), 12048 (126j), 12404 (16), 19717 (118)
Basilio Augusto, Bro. 823 (199), 1687 (69)
Bastos, N. C. 139 (46a)
Basurto, P. 686 (138f)
Beard, J. S. 208 (177a)
Beard, P. 1380 (8a), 1398 (129a)
Beck, S. G. 2699 (69), 2716 (140), 4192 (110c), 4259 (96), 4514 (202a), 5277 (126o), 5423 (157), 6155 (96), 6335 (126e), 6339 (202a), 6372 (138b), 6405 (138b), 6556 (26a), 6571 (69), 6782 (118), 6812 (122)
Backner, J. 767 (126m)
Beetle, A. A. 26274 (177a)
Bélanger, C. P. 90 (8a)
Belém, R. P. 29 (46a), 133 (3f), 1206 (42b), 1208 (44a), 1224 (161/IIa), 1498 (161/IIa), 1565 (44a), 1568 (44a), 1570 (44a), 1601 (137b), 1649 (44a), 1762 (147), 1800 (148), 1936 (43), 3566 (161/Ib), 3583 (44a), 3596 (42d), 3603 (161/IIa), 3634 (161/IIa), 3654 (43), 3658 (42d), 3769 (137b)
Belém, R. P. & J. M. Mendes 158 (148), 201 (164c), 334 (202b)
Belém, R. P. & G. Mendes Magalhães 223 (126a),

346 (42b), 551 (161/IIa), 564 (161/IIa), 664 (161/IIa), 846 (161/Ib), 852 (12), 966 (161/Ib), 1020 (126a), 1090 (42b), 1122 (161/Ib)
Belém, R. P. & R. S. Pinheiro 2091 (41), 2190 (12), 2201 (161/Ib), 2245 (44a), 2254 (26a), 2261 (12), 2266 (44a), 2354 (44a), 2385 (126a), 2448 (39a), 3291 (202b), 3372 (202b), 3407 (202b)
Belshaw, C. M. 3164 (126s), 3257 (140), 3258 (69), 3264 (161/Ia), 3504 (17a)
Benacchio, S. 3 (115), 9 (157), 20 (171), 37 (129a)
Benítez de Rojas, C. E. 223 (17a), 439 (10a), 522 (8a), 725 (10a), 870 (199), 877 (199), 1583 (199), 1588 (177a), 1645 (8a), 1830 (177a)
Bennett, – 677 (193b)
Benoist, R. 854 (140)
Benson, L. 9223 (138e)
Bequaert, J. 94 (146)
Berlandier, J. L. 840 (83b), 897 (71), 2206 (83b), 2234 (11), 2293 (11), 2294 (11), 2327 (71)
Berlin, B. 196 (17a), 340 (3a), 706 (25a), 731 (17a)
Bernardi, L. 196 (115), 1286 (17a), 1705 (26a), 2367 (13a), 3319 (202a)
Bernardi, A. L. 6101 (166), 6492 (27), 7355 (3a)
Bernoulli, C. & Cario 1128 (16)
Bertero, C. G. 148 (119), 149 (125), 875 (98d), 887 (131), 892 (116), 893 (109b), 894 (109b), 895 (106), 896 (106), 897 (125), 1420 (134a)
Berthoud-Coulon, M. 145 (69), 149 (140), 151 (26a), 155 (69)
Berti, L. M. 349 (161/Ia), 384 (161/Ia), 500 (161/Ia)
Bertoni, M. S. 514 (138a)
Bierner, M. & B. L. Turner 86 (76a)
Bigelow, J. M. 295 (73)
Bilimek, D. 144 (178a)
Biltmore Herb. 6955a (84)
Black, G. A. 1882 (36), 15154 (157), 15157 (126p), 16159 (138g), 16694 (26a), 16779 (147)
Blakley, G. A. B-1751 (90)
Blanchet, J. S. 43 (26a), 132bis (44a), 361 (18), 654 (44a), 844 (140), 1243 (26a), 1346 (44a), 1559 (176), 1592 (41), 1597 (161/Ib), 2069 (61), 2581 (164a), 2895 (59a), 3486 (44a), 3553 (202b), 3667 (42b), 3668 (18 and 42b) 7937 (44a)
Blanco, C. 286 (202a), 313 (8a), 345 (161/Ia), 375 (161/Ia), 570 (20)
Blanco, E. 89 (177a)
Blanco, F. 167 (69)
Blohm, H. 22 (95a)
Blum, J. L. 3525-14 (8a)
Blum, K. E. 381 (16), 598 (16), 1230 (16), 1312 (13a), 2050 (16)
Blum, K. E. & E. Tyson 2006a (8a)
Blumer, J. C. 1453 (138e), 1642 (138e), 1758 (73)
Bockerman, W. 166 (58), 167 (58)
Boeke, J. D. 106 (116), 1197 (25a)
Boldingh, I. 50B (129a), 2441 (159), 2843 (140), 5306 (141), 6215 (159), 7311 (129a)
Bolland, G. 2 (26a)
Bond, F. E. 45 (8a), 144 (148), 148 (126k)
Bondar, G. 3007 (12)
Bonpland, A. 1425 (14b)
Bósquez, J. 35 (16)
Botteri, M. 493 (72b), 783 (138f), 784 (126m)
Bourgeau, E. 1305 (116), 1536 (68), 2499 (119), 2816 (138f), 2875 (68), 2877 (72b), 3047 (126m)
Bourke-Borrowes, D. 71 (63a)
Bowie, J. & Cunningham 17 (40)
Boutin, F. & F. Brandt 2383 (92)
Boyan, R. 257 (20)
Bracamontes León, R. 276 (196b)
Brace, J. L. K. 332 (146), 1663 (131), 1718 (131), 1796 (131), 3535 (131), 3931 (134d), 4072 (134d), 4510 (140), 5283 (140), 6728 (131), 6927 (131), 7162 (146)
Brade, A. C. 6315 (70a), 10637 (12), 13408 (43), 15295 (39a), 15699 (44a), 16813 (39a), 17052 (105), 19776 (137d), 21204 (120)
BRADE Herb. 15712 (62), 33409 (202b), 61005 (42b), 128007 (164c)
Braga, P. I. S. 2211 (161/IIc)
Braga, R. 39 (161/IIa), 551 (121)
Brandegee, T. S. 172 (146)
Brantjis, R. B. M. 702414 (44a)
Brass, L. J. 14311 (80)
Braun, O. 3 (138f), 5 (140)
Bravo H., H. 3536 (196d)
Bray, W. L. 115 (84)
Breedlove, D. E. 1533 (177a), 7591 (16), 9086 (199), 9629 (149), 9839 (200bis), 10055 (119), 10107 (119), 10305 (140), 10369 (13a), 10444 (94c), 10445 (72b), 10596 (177k), 10697 (55), 10914 (72b), 11290 (72b), 11655 (119), 11690 (116), 11796 (71), 11802 (199), 11822 (173), 12048 (173), 12186 (173), 12186 (72b), 12425 (72b), 13515 (149), 13848 (70a), 14644 (119), 14737 (119), 14920 (16), 19884 (173), 20188 (149), 20199 (177a), 20223 (173), 20446 (177a), 20926 (199), 21775 (72b), 23283 (173), 23334 (177k), 23433 (147), 23729 (177a), 23797 (177a), 24127 (199), 26107 (13a), 26457 (161/III), 27253 (94c), 27746 (94c), 27929 (26a), 28039 (69), 28142 (13a), 28725 (72b), 29833 (177k), 29950 (173), 30321 (126k), 37204 (63b), 37326 (94c), 39001 (63b), 40092 (63b), 40177 (70), 42394 (177g, bis)
Breedlove, D. E. & P. H. Raven 8432 (177a), 8451 (178a), 13052 (72b), 13142 (177k), 13253 (71), 13272 (146), 13514 (71), 13534 (55), 13602 (202a), 13616 (138f)
Breedlove, D. E. & R. F. Thorne 17896 (177c), 20882 (140), 20912 (140), 30610 (178a)
Brenckle, J. F. 47-445 (68), 47/446 (138f)
Brenes, A. M. 4 (177o), 11 (138f), 26 (138f), 37 (16), 360a (138f), 374 (177o), 2724 (177o), 3738 (177o), 4417 (202a), 4611 (16), 4658 (69), 4584 (68), 5048 (68), 5220 (177o), 5488a (13a), 5841 (68), 5896 (13a), 5898 (173), 5923 (177o), 6202 (13a), 6339 (16), 6456 (69), 6606 (177a), 6815 (13a), 12651 (13a), 13233 (138f), 13264 (16), 13674 (13a), 14976 (68), 16121 (69), 16661 (173), 16974 (68), 17187 (13a), 17451 (16), 18657 (138f), 19155 (173), 21384 (16),

21638 (68), 22041 (68), 22901 (69), 23004 (69), 23058 (13a), 23058a (13a), 23115 (148), 23131 (177o), 23138 (173)
Breteler, F. J. 3206 (199), 3426 (13a), 3492 (138g), 4259 (13a), 4350 (199), 4723 (3a), 4755 (8b), 4892 (129a), 4912 (13a)
Bridges, T. 10 (96), 97 (99), 100 (109b), 1206 (99), 1295 (98c)
Bristan, N. 1283 (14b)
Britton, E. G. 6398 (141)
Britton, N. L. 103 (29), 140 (134e), 212 (134e), 761 (29), 1193 (131), 1237 (141), 1291 (29), 1647 (71), 1667 (199), 1897 (199), 2223 (30), 2718 (29), 2776 (129a), 3417 (29), 3791 (71), 3830 (29), 4122 (131), 5372 (134e), 7805 (129a), 8582 (138g), 9479 (141), 10055 (2), 13992 (170)
Britton, N. L. et al. 100 (27), 544 (30), 883 (27), 2157 (27), 2406 (27), 2754 (28), 5877 (28), 5988 (199), 6559 (28), 7504 (140)
Britton, N. L. & F. E. Boynton 6170 (138g)
Britton, N. L. & J. L. K. Brace 335 (134e), 657 (71), 712 (177j), 725 (140), 754 (134e)
Britton, N. L. & E. G. Britton 250 (28), 2180 (8a), 7256 (119), 7808 (28), 7898 (126m), 7899 (28), 8971 (28), 9007 (169a), 9335 (138g), 9795 (169a)
Britton, N. L., E. G. Britton & F. E. Boynton 8272 (169a)
Britton, N. L., E. G. Britton & M. Brown 5997 (169a)
Britton, N. L., E. G. Britton & S. Brown 5861 (169a), 6517 (169a)
Britton, N. L., E. G. Britton & J. F. Cowell 2320 (148)
Britton, N. L., E. G. Britton & T. E. Hazen 96 (8a)
Britton, N. L., E. G. Britton & D. Marble 2223 (169a)
Britton, N. L., E. G. Britton & J. A. Shafer 64 (141), 183 (140), 565 (141)
Britton, N. L. & W. E. Broadway 2821 (126k)
Britton, N. L. & J. F. Cowell 82 (159), 677 (140), 719 (69), 1276 (169a), 4081 (28), 10201 (131), 12589 (30), 12652 (199), 12664 (5), 12802 (133)
Britton, N. L., J. F. Cowell & S. Brown 4669 (169a)
Britton, N. L., J. F. Cowell & W. E. Hess 1872 (140)
Britton, N. L., J. F. Cowell & J. A. Shafer 12941 (170), 12980 (133)
Britton, N. L. & F. S. Earle 6536 (71), 7622 (131)
Britton, N. L., F. S. Earle & P. Wilson 5984 (199)
Britton, N. L. & W. C. Fishlock 995 (141), 1034 (169c), 1040 (129a)
Britton, N. L. & T. E. Hazen 1923 (147)
Britton, N. L. & N. Hollick 1721 (199)
Britton, N. L. & E. P. Killip 1450 (8b)
Britton, N. L. & D. Marble 400 (28), 402 (129a), 1265 (148), 2273 (199)
Britton, N. L. & C. F. Millspaugh 2259 (134e), 2557 (131), 2809 (134e), 2931 (177j), 2937 (134e), 5438 (177j), 5806 (177j), 5821 (134e), 5903 (141), 6182 (177j)
Britton, N. L. & J. A. Shafer 191 (129a), 783 (129a), 788 (28), 1661 (148), 1720 (28), 1807 (169a), 3089 (129a)
Britton, N. L. & W. M. Wheeler 83 (148), 99 (129a)
Britton, N. L. & P. Wilson 88 (199), 4906 (199), 5516 (30), 5546 (71), 14012 (30), 14314 (30), 14854 (30), 15251 (131)
Brizicky, G. K. & W. L. Stern 259 (134e)
Brizuela, A. 948 (118)
Brizuela, J. 201 (118)
Broadway, W. E. 16 (140), 38 (69), 159 (147), 160 (140), 165 (129a), 176 (8a), 373 (148), 475 (69), 525 (161/Ia), 588 (129a), 678 (36), 705 (129a), 914 (161/Ia), 1674 (177a), 2451 (8a), 2778 (161/Ia), 4301 (8a), 5241 (8a), 5575 (27), 5886 (27), 6332 (8a), 7904 (27), 9172 (27), 9795 (27)
Brooke, W. M. A. 5618 (180), 5641 (69), 6373 (95a)
Brown, G. G. & B. W. Brown 26 (129a)
Brown, M. 131 (27)
Brown, S. 332 (131), 334 (129a), 355 (141), 358 (199), 539 (140), 1765 (131)
Brown, S. & N. L. Britton 411 (129a), 1772 (140)
Brown, T. W. 23 (170)
Bruff, J. R. 1389 (201)
Bruijn, C. de 1172 (13a)
Brumbach, W. C. 8175 (126a), 9384 (134e)
Buch, W. 1918 (168a)
Bucher, Mrs. G. C. 78 (131), 217 (170)
Buchtien, L. P. 25 (95a)
Buchtien, O. 84 (96), 127 (96), 770 (68), 772 (138f), 773 (20), 774 (161/Ia), 799 (3a), 1779 (68), 1781 (46), 2262 (68), 4100 (63c), 8095 (126e), 8096 (138f)
Budowski, G. 64 (202a)
Bunting, G. S. 3108 (27), 3353 (17a), 3993 (140), 4004 (148)
Bunting, G. S. & L. Licht 176 (177a), 894 (147), 959 (177a)
Burbidge, R. B. 75/355 (177a)
Burch, D. 1135 (69), 1199 (8a), 1224 (16), 1308 (129a), 4230 (134d), 4758 (13a), 5886 (201)
Burchell, W. J. 889 (116), 2827 (39a), 2914 (12), 3813 (161/IIa), 3884 (120) 3964 (42b), 4627 (44a), 4980 (43), 6045 (60), 6237 (161/Ia), 7286 (46a), 7400 (175), 7432 (46a), 8749a (69), 8768 (60), 9342 (20), 9572 (36), 9926 (26a), 9932 (36), 9996 (20)
Burger, W. C. 4068 (177a), 4309 (148), 4832 (177a), 6538 (69), 6910 (13a), 7830 (199), 8121 (13a), 8445 (13a), 9189 (199)
Burger, W. C. & R. L. Liesner 6642 (138f), 6649A (16), 6552 (16), 6874 (68)
Burger, W. C. & G. Matta U. 4477 (48b)
Burger, W. C. & R. W. Pohl 7821 (16)
Burgot, – 1418 (69), 1419 (140), 1420 (138g)
Burkart, A. 5562 (110a), 7764 (110a), 14117 (181), 16502 (171), 16647 (138g), 17184 (55)

Busey, P. 297 (8a)
Bush, B. F. 25 (69), 197 (69)
Buza, J. 14 (115)

Caballero, B. de 14 (26a), 41 (68)
Cabrera, A. L. 2055 (128), 3151 (140), 11111 (181), 15189 (110a), 15025 (110a), 16254 (118)
Cabrera, A. L. & Muñoz 50 (83b)
Cabrera, E. 76 (27), 250 (27), 309 (71), 320 (16), 337 (16), 347 (71), 377 (16), 395 (27), 510 (199), 580 (27), 695 (27), 727 (27), 1177 (196a), 1199 (196a), 1663 (16), 1422 (196a), 1540 (196a), 1579 (196a), 1624 (27)
Cabrero, I. 3086 (118)
Caicedo, V. 6 (16)
Calderón, S. 206 (16), 207 (147), 318 (177a), 1103 (16), 1115 (202a), 1399 (177a), 1409 (149), 2485 (178a)
Calvárez, R. N. 1339 (118)
Calzada, J. I. 897 (140), 970 (63b), 1046 (202a), 1081 (161/III), 2023 (72b), 3859 (149)
Camargo, J. M. F. 123 (126a)
Camargo, O. 62595 (124)
Cameron, C. E. R. 30 (83b)
Camp, W. H. 888 (148), 1525 (8b), 1844 (116), 1923 (110g), 2380 (17a), 2454 (178a), 2527 (110g), 2422 (141), 2906 (95a), E-3965 (113b)
Campos Porto, J. 190 (12)
Campos Porto, P. 708 (161/IIa), 1278 (42e), 1762 (42b), 2453 (44b), 2458 (44a), 2467 (44b), 2476 (44b)
Campos Porto, S. 891 (120), 3382 (120)
Cañaveras, F. 172 (13a)
Cantino, P. 521 (122)
Carabia, J. P. 4075 (134c)
Carauta, J. P. P. 419 (124), 822 (42b), 1670 (124), 1826 (124), 2881 (126d)
Cárdenas, M. 1104 (3a), 2024 (21), 2274 (122), 3024 (103), 3636 (103), 4431 (122), 5857 (110b)
Cárdenas de Guevara, L. 620 (140), 626 (157), 727 (170), 945 (10a), 1094 (129a), 1208 (171), 1405 (171), 1412 (177a), 1512 (199), 1520 (171), 1523 (170), 1553 (199), 1605 (177a), 1686 (199), 1860 (177a), 1926 (177a), 2003 (177a), 2009 (199), 2025 (199), 2166 (13a)
Cardona, F. 33 (26a), 524 (26a), 609 (148), 2234 (8a), 2528 (3/Ia), 4002 (55), 4052 (202a), 4069 (199), 4076 (69), 4080 (171), 4086 (202a), 4139 (129a), 4143 (172), 4207 (27), 4217 (199), 4249 (115)
Carlson, M. C. 566 (177a), 1354 (178a), 2210 (178a), 3091 (11), 3649 (177a)
Carrasquilla, L. 385 (14b)
Carrick, A. 1166 (8a)
Carrillo, E. 47 (129b), 59 (174)
Carter, A. 2169 (199), 3605 (177a), 4621 (144), 5130 (144)
Carter, A. & R. Moran 5263 (144)
Carvalho, G. H. 11 (42e)
Case, P. et al. 163 (140)
Casaretto, G. 2941 (39b)
Castellanos, A. 3783 (202b), 24447 (120), 24733 (138a), 25016 (44a), 25096 (42b), 25308 (165), 25804 (71), 26003 (138f), 26061 (52), 26420 (161/Ib)
Castellanos, A. & A. P. Duarte 506 (46a)
Castellanos, A. & L. Duarte 505 (44b), 584 (147)
Castrejón, C. 19 (16)
Castrejón, L. A. 8 (16)
Castro, J. de 9 (136)
Castroviejo, S. & Ginés López 170 (157)
Cavalcante, P. 1408 (20), 1753 (3a), 2013 (20), 3004 (138g)
Cavalcante, P. & M. Silva 2899 (36)
Cazalet, P. C. D. & T. D. Pennington 7764 (161/Ia), 7801 (25a)
Cerbin, W. 116 (177j)
Cerrate, E. 3488 (110f)
Chagas, F. 9 (46a)
Chapline, W. R. 812 (73)
Chardon, C. E. 1234 (170)
Chase, A. 7649 (37), 7695 (164a), 9709 (104a), 715
Chase, V. H. 7154 (196b), 7177 (75), 7507 (71)
Chavelas P., J. 245 (202a)
Chaves, D. 363 (16)
Chávez, R. 438 (153b), 449 (153b)
Chazaro B., M. 396 (13a), 436 (161/III)
Chiang, F. et al. 7566B (87), 7633 (194), 7723 (194), 7895 (85), 7906 (193b), 7938 (76b), 7958 (193b), 8095 (193b), 8267B (76a), 8698 (193c), 8723 (73), 9000d (194), 9519 (82)
Chiang, F. & T. Delevoryas 837 (195), 843 (145)
Chinchay, J. 3648 (116), 3706 (95a)
Chindoy B., P. J. 39 (8b)
Choussy, F. 74 (202a)
Chumpitaz, E. 1810 (129b)
Cid, C. A. et al. 207 (3/Ia), 609 (161/Ia)
Clarke, O. F. 73 (13a)
Clarke-Essig-Pringle 1334-4 (177a)
Clausen, R. T. 6037 (71), 6045 (77), 6123 (69), 6188 (177a), 6189 (177a)
Claussen, P. 13 (61), 78 (39b), 87 (57), 89 (3f), 195b (39b), 224 (57), 630A (63b), 850 (161/IIa), 851 (126a), 862 (3f), 864 (42b), 867 (3f), 944a (39b)
Clemente, Bro. 154 (133), 195 (133), 4499 (135), 4826 (170)
Clewell, A. F. 3621 (27)
Cobra, L. Q. & R. P. Belém 150/62 (42b)
Cochrane, T. S. 8611 (13a), 8617 (161/III)
Coelho de Moraes, J. 718 (150), 880 (202b), 888 (44b), 2183 (126h)
Cohn, T. T31 (8d), T38 (84)
Coker, W. C. 9 (134e)
Colaris, W. J. J. 1070 (43)
Colchester, M. E. M. 2012 (140)
Coleman, J. R. 515 (176)
Collenette, C. L. 471 (16)
Collins, G. N. & C. B. Doyle 71 (11), 256 (13a)
Collins, G. N. & J. H. Kempton 22 (26a), 60 (177c), 66 (177c)
Colma, A. 6 (199), 64 (199), 83 (199), 98 (140), 169 (55), 215 (199), 303 (55)

Comber, H. F. 163 (181)
Combs, R. 92 (141), 243 (69), 244 (71), 504 (30), 617 (131), 673 (148)
Comissão Geográfica de Minas 1135 (44a), 1808 (124)
Commerson, P. 214 (126d)
Conrad, J. 2784 (13a)
Constance, L. 3138 (81)
Constantino, A. 132 (44a), 142 (44a)
Conzatti, C. 1499 (144), 1824 (144), 2042 (199), 3059 (199), 3450 (199), 4388 (177c), 4409 (11), 4584 (178a), 5297 (119)
Cook, M. T. 1 (119)
Cook, C. D. K. 113 (8a)
Cook, O. F. & G. N. Collins 697 (169a), 940 (28)
Cook, O. F. & C. B. Doyle 407 (8a)
Cook, O. F. & G. B. Gilbert 1504 (202a), 1747 (107)
Cook, O. F. & R. D. Martin 193 (201)
Cook, O. F. & R. D. Morris 18 (126j)
Cooke, C. D. K. 37 (27)
Coradin, L. 587 (69), 960 (177a), 1021 (8a), 1022 (20)
Cordeiro, M. dos R. 33 (3a), 109 (3a), 113 (38), 729 (3a), 880 (3a), 1318 (69)
Cordeiro, N. A. 1462 (140)
Correa, J. B. 37063 (118)
Correa, M. D. 1706 (27), 1813 (16 & 26a)
Correll, D. S. 10186 (69), 13634 (73), 27860 (82), 36770 (126m), 40388 (134e), 40389 (177j), 40393 (69), 40727 (177j), 40728 (134e), 40810 (177j), 40877 (177j), 41642 (168c), 42142 (134e), 42762 (131), 43310 (134d), 43648 (134e), 43836 (69), 43874 (177j), 44105 (177j), 44514 (126a), 44736 (131), 44808 (131), 45083 (134e), 45322 (69), 45837 (168c), 46037 (140), 46056 (177j), 46214 (71), 46249 (177j), 46380 (168c), 46467 (129), 46541 (134d), 47395 (140), 47482 (168c), 47535 (168c), 47964 (71), 48064 (134e), 48160 (177j), 48251 (141), 48429 (131), 48951 (71), 49122 (51), 49550 (131), 50198 (71), 50204 (69), 50564 (131), 50704 (199), 50758 (199), 51156 (71), 51185 (69)
Correll, D. S. & I. M. Johnston 20009 (193d), 20010 (82), 20040 (88), 20251 (83a)
Correll, D. S. & R. C. Rollins 23691 (73), 23952 (88)
Cory, V. L. 49823 (69), 49853 (69), 50679 (69), 50856 (69), 51855 (84), 53202 (73), 53449 (88)
Coues, E. & E. Palmer 171 (80)
Cowan, R. S. 1116 (8a), 1319 (27), 1985 (147), 2309 (81), 38459 (36), 39270 (20), 39390 (147)
Cowan, R. S. & R. Foster 1320 (8a)
Cowan, R. S. & T. H. Soderstrom 2015 (26a), 2133 (20)
Cowan, R. S. & J. J. Wurdack 31527 (3a), 31580 (140)
Cowell, J. F. 56 (140), 49099 (1)
Cowgill, H. B. 560 (140)
Cox, D. I. & E. A. Hubbard 22 (20), 101 (69), 148 (140)
Crawford, J. 576 (140), 604 (141), 697 (131)
Cremers, G. 4240 (35), 5002 (26a)
Crisman, G. E. & W. D. Willis 202 (11)
Croat, T. B. 482 (13a), 4151 (8a), 4263 (27), 4710 (27), 4842 (27), 4861 (27), 5169 (27), 5612 (27), 5946 (14b), 6012 (14b), 6371 (14b), 7117 (8a), 7257 (27), 7355 (8a), 7392 (8a), 7557 (17a), 7576 (17a), 7609 (25a), 7688 (27), 7779 (8a), 7944 (27), 11826 (14b), 11827 (27), 12012 (16), 12253 (14a), 12542 (16), 14414 (26a), 16885 (16), 17639 (17a), 17811 (3a), 17928 (69), 18021 (26a), 18233 (17a), 18346 (26a), 18610 (17a), 18660 (26a), 19064 (17a), 19111 (147), 19212 (3a), 19384 (26a), 20211 (161/Ia), 21243 (25a), 21799 (171), 23312 (71), 23347 (201), 23382 (68), 23480 (71), 23724 (71), 23975 (27), 24221 (27), 24493 (27), 24937 (27), 25446 (13a), 25759 (13a), 25920 (26a), 26534 (16), 26556 (13a), 27405 (14c), 32780 (16), 32814 (177a), 35370 (13a), 35757 (13a), 38087 (8a), 42819 (140), 42820 (71), 43191 (13a), 43552 (13a), 43680 (71), 43686 (71), 44211 (76a), 45444 (11), 45310 (138f), 45365 (26a), 45408 (148), 45494 (196d), 45738 (177e), 45769 (177a), 45970 (91c), 46218 (178a), 46224 (178a), 46272 (178a), 47490 (68), 47614 (94c), 47901 (68), 48198 (178a), 49055 (13a), 49498 (161/Ia)
Crockett, R. L. 87 (177h)
Croft, M. 106 (88)
Croizat, L. 345 (3a), 752 (147), 760 (8a), 920 (147), 923 (161/Ia), 931 (8a)
Cronquist, A. 4180 (69), 9911 (90)
Crosby, M. A. et al. 108 (199), 1033 (199)
Crossland, C. 441 (153c)
Cruckshanks, A. 139 (98c), 5153 (98c)
Crum, E. 1827 (90)
Cruxent, – 11 (27), 139 (129a)
Cruz, J. S. de la 963 (161/Ia), 1075 (140), 1088 (26a), 1317 (26a), 1745 (26a), 1760 (20), 1883 (161/Ia), 1897 (26a), 2058 (138g), 2090 (138g), 2229 (2), 2289 (26a), 2292 (26a), 2411 (26a), 3066 (148), 1340 (8a), 3462 (26a), 3697 (140), 3777 (148), 3910 (26a), 4208 (161/Ia), 4329 (140), 4353 (9), 4374 (161/Ia), 4493 (26a)
Cruz Cisneros, R. 2318 (178a), 2611 (145)
Cuatrecasas, J. 1140 (42f), 1922 (42f), 1360 (8b), 3550 (13b), 3551 (13b), 3991 (34), 4110 (16), 4232 (16), 4786 (8b), 7008 (8b), 7325 (8b), 7723 (8b), 7931 (126k), 9103 (17a), 10781 (17a), 10893 (8b), 11111 (17a), 11113b (8b), 11358 (25a), 12104 (10a), 12181 (102), 13169 (26a), 13934 (17b), 18697 (17b), 19479 (153a), 19741 (26a), 19868 (32), 22618 (17b), 22867 (17b), 24415 (16), 25419 (29), 27019 (116), 27021 (154), 27116 (8b), 27202 (8b), 27267 (25a)
Cuezzo, R. 932 (118)
Cuming, H. 341 (125), 708 (109b), 874 (98a), 1149 (63b), 1221 (200b)
Curial, O. 724 (104c)
Curran, H. M. 2 (161/IIa), 17 (177a), 64 (177a), 74 (44a), 145 (170), 191 (44a), 210 (177a), 212

(10a), 248 (44b), 348 (13a), 366 (161/IIa), 381 (199), 471 (10a), 534 (177a), 540 (55), 571 (171), 586 (10a), 590 (199), 658 (157), 767 (124), 897 (10a)
Curran, H. M. & M. Haman 740 (55), 779 (199)
Curtiss, A. H. 27 (146), 50 (177j), 187 (141), 202 (71), 240 (148), 603 (129a), 707 (69), 710* (88), 4259 (69), 5211 (69), 5630 (134e)
Cutler, H. C. 4730 (80)
Cutler, H. C. & M. Cárdenas 3888 (110c)

Dahlgren, B. E. 790 (42c), 902 (44b), 917 (42e), 918 (44b), 937 (43), 948 (175)
Dahlgren, B. E. & E. Sella 81 (161/Ia), 165 (20), 166 (26a), 199 (46a), 366 (20), 386 (20), 395 (20), 475 (20), 674 (20), 684 (20), 770 (20)
Damanu, E. 31 (161/IIa)
Damasio, L. 907 (39b)
Danforth, R. E. 47 (13a)
Daniel, Hno. 3879 (17b), 4749 (17b)
Danin, A. 76-34-9 (140), 76-35-2 (55)
D'Arcy, W. G. 6670 (16), 9295 (16), 9455 (16), 9508 (16), 10261 (13a), 10295 (8a), 10370 (177a), 10407 (55), 10473 (13a), 11142 (126k), 11145 (8a), 11971 (146), 11976 (71), 12030 (55), 13785 (108)
Daubenmire, R. 261 (177a), 268 (177a), 446 (177a), 485 (16)
Davidse, G. 1352 (69), 1384 (27), 3850 (157), 3927 (69), 3972 (1261), 4132 (8a), 4223 (55), 4602 (42g), 5034 (8a), 5038a (8a), 5326 (8b), 5799 (31), 10818 (3f), 10896 (161/IIa), 11243 (126a), 11443 (42b), 11559 (148), 12207 (43), 12892 (34), 13956 (10a), 14020 (1261), 14362 (10a), 14439 (53a), 14471 (148), 14752 (147), 14844 (10a), 15209 (3a), 15891 (10a), 16447 (161/Ia)
Davidse, G. & R. W. Pohl 1372 (16), 1510 (13a)
Davidson, A. 1342 (87)
Davidson, C. 3443 (25a), 3764 (111b), 3766 (122), 5339 (22), 9250 (25a), 9920 (17a)
Davidson, M. E. 854 (8a)
Davis, D. H. 49 (20)
Davis, P. H. 61079 (44b), 61120 (176), 61124 (150), 61136 (202b)
Davis, P. H. & P. Gibbs 2427 (175)
Dawson, E. Y. 14157 (43), 14364 (126a), 14392 (69), 14446 (56a), 14757 (140), 14999 (161/Ia)
Dawson, G. 633 (128), 3296 (181)
Day, F. H. 254 (141)
Deam, C. C. 220 (199), 221 (147), 291 (177a), 6273 (55), 6276 (199), 6289 (140), 6362 (177a)
Debeaux, G. 82 (10a), 97 (10a)
Debrot, E. 1 (10a)
Degener, O. 965 (131), 5025 (84), 18841 (131), 18893 (134e), 26471 (146)
Dehesa, M. P. 1600 (177c)
Delascio, F. & R. Jackson 4664 (10a), 4685 (171)
Delascio, F. & R. L. Liesner 7342 (140)
Delascio, F. & R. López 2750 (3a)
Delgado, E. 84 (8a), 90 (172)
Delgado S., A. 5 (68), 39 (153c), 142 (11), 161 (149), 575 (178a), 601 (178a), 618 (200bis), 661 (149), 663 (177a), 839 (94c), 88a (173), 965 (199)
Demaree, D. 5628 (69), 8613 (69), 9449 (69), 17937 (69), 21622 (69)
Deward, – 110 (161/Ia)
Diaz, C. 15 (8b), 206 (26a), 360 (140)
Diaz, D. 72 (147)
Diaz Luna, C. L. 2112 (149), 2949 (177c), 4154 (101), 5782 (91a), 7656 (91a)
Diaz S., C. & N. Jaramillo 108 (17a)
Dieckman, L. 279 (16)
Dieterle, J. V. A. 4019 (11), 4358 (55)
Diggs, G. M. 2124 (140)
Dillon, M. 1056 (95a)
Dixon, R. A. 476 (69)
Dodge, C. W. 6258 (177a), 16647 (16), 16823 (149)
Dodson, C. H. 2841 (17a)
Dodson, C. H. & L. B. Thien 515 (10b), 1298 (200b), 1789 (153c), 1924 (25a), 2147 (200b)
Dombey, J. B. 832 (129a), 836 (110f)
Dombrowski, L. T. 7237 (42b), 7371 (126a), 7430 (148)
Domínguez Peña, F. 130 (147), 131 (161/Ia)
Don, G. 69 (46a), 70 (36)
Donselaar, J. van 3257 (20)
Dorantes, J. 37 (177a), 417 (177a), 442 (71), 547 (177a), 841 (177a), 1240 (11), 1372 (177a), 3752 (177p), 5273 (71)
Doreste, E. 510 (69)
Dorrien Smith, Miss 65 (63b)
Dorsett, Shamel & Popenoe 225 (44a), 430b (164a)
Doyle, C. B. 53 (178a), 259 (161/III)
Dressler, R. L. 2321 (141), 4933 (25a)
Drouet, F. 1971 (69), 1983 (63b), 2433 (165), 2566 (148), 2691 (44b), 2893 (165), 3617 (177d), 3959 (177d)
Drushel, J. A. 11190 (84), 11242 (84)
Dryander, E. 71 (17b)
Duarte, A. P. 179 (126k), 1312 (44b), 1319 (44b), 1422 (42c), 3136 (161/IIb), 3686 (12), 5155 (43), 5799 (126d), 6141 (39a), 6672 (126a), 6703 (39a), 6712 (148), 6967 (20), 7229 (46a), 7256 (3a), 7711 (202b), 7718 (161/IIa), 7719 (175), 7721 (120), 8034 (126a), 8669 (120), 8698 (44a), 8793 (3f), 8860 (3b), 8975 (39a), 9010 (59c), 9015 (61), 9113 (57), 9324 (176), 9611 (3f), 9908 (3f), 10159 (60), 10424 (57), 10432 (56a), 10552 (176), 10566 (150), 10571 (164c), 11760 (43), 13967 (161/IIa), 13973 (156), 14180 (37), 14181 (44a)
Duarte, A. P. & A. Castellanos 292 (44a), 314 (42d), 351 (176), 367 (42c), 600 (140)
Duarte, A. P. & A. Mattos 8304/634 (43)
Duarte, A. P. & E. Pereira 9221/10131 (164c), 9306/10019 (43)
Duarte, L. & R. Santos 142 (148), 185 (43)
Ducatillon, C. & A. Gelly 70 (140)
Ducke, A. 271 (46a), 431 (26a), 432 (38), 482 (3a), 558 (147), 891 (20), 948 (165), 1014 (8b), 1502 (17a), 1503 (126p), 1504 (53a), 1561 (8b), 1947

(20), 2130 (8b), 2616 (46a), 16167 (20), 16729 (20), 17031 (20), 17285 (26a), 20331 (20)
Ducke, A. & J. G. Kuhlmann 2306 (42a)
Dudley, D. A. 1032 (177a)
Dufrenoy, J. 16 (177a)
Dugand, A. 172 (15), 278 (177a), 733 (177a), 928 (8a), 6260 (177a)
Dugand, A. & H. García Barriga 2300 (8a), 2706 (8a)
Dugand, A. & R. Jaramillo 390 (17b), 511 (177a), 2808 (177a), 3923 (126i), 4051 (177a)
Duke, J. A. 4130 (8a), 4180 (16), 5093 (14b), 5661 (14a), 5922 (16), 8545 (8a), 8923 (14b), 9135 (68), 9546 (8a), 9738 (140), 9804 (16), 10103 (8a), 10190 (16), 10249 (8a), 10312 (16), 10477 (16), 10633 (16), 10416 (140), 14405 (138g)
Dunlap, V. C. 21 (8a)
Dunn, D. B. 8741 (84), 17338 (178a), 17357 (71), 17388 (178a), 17547 (202a), 17629 (71), 18617 (178a), 18677 (196d), 18691 (138f), 18729 (178a), 18748 (146), 19080 (119), 19231 (73), 20150 (73), 12748 (68 and 69), 21870 (140), 23069 (193a)
Duque-Jaramillo, J. M. 2965 (154), 3116 (116), 3196 (202a)
Duran, V. 3223 (90)
Dusén, P. 33/66 (63a), 38 (46a), 113 (124), 188 (124), 1980 (12), 3562 (69), 3816 (137a), 4339 (44a), 7430 (161/IIb), 7805 (44a), 7920 (136), 8282 (126a), 9615 (43), 9659 (43), 9724 (43), 9816 (44a), 9973 (44a), 11442 (137a), 11710 (43), 14369 (161/IIa), 14863 (43), 18018 (43)
Duss, P. 824 (69), 828 (16), 829 (138g), 1035a (1), 1115 (129a), 1118 (161/Ia), 1119 (148), 1120 (140), 2620 (199), 2622 (141), 2626 (129a), 3048 (138g), 3224 (148), 3225 (140), 4468 (161/Ia)
Dutra, J. 1318 (124)
Dwyer, J. D. 519 (27), 562 (13a), 1531 (16), 1533 (16), 3086 (8a), 3638 (14b), 4698 (8a), 5555 (16), 6851 (16), 7140 (8a), 7355A (26a), 7356 (26a), 7963 (142), 8083 (26a), 8393 (177a), 8545 (16), 8626 (14c), 8679 (26a), 9649 (25a), 10398 (16), 10451 (16), 11008 (27), 11008a (27), 11203 (27), 11454 (140), 11853 (16), 11999 (26a), 12031 (16), 12089 (27), 12124 (68), 12514 (201)
Dwyer, J. D., T. Elias & R. Maxwell 19 (27), 519 (27)
Dwyer, J. D. & J. Kirkbride 7444 (16)
Dwyer, J. D. & R. Liesner 12339 (16)
Dziekanowski, C. 1868 (138e), 3189 (71), 3300 (193b), 3368 (55)

Earle, F. S. 24 (71), 29 (131), 221 (84), 550 (84), 555 (87)
Earle, F. S. & S. M. Tracy 137 (84)
Earle, W. H. 81 (87)
Ebinger, J. B. 898 (14b), 1068 (16)
Echevarría, J. A. 338 (16), 820 (16), 859 (16)
Edwards, J. B. 504 (177a), P-703 (202a), P-709 (16), 724 (177k)
Edwin, G. 3729 (177f)
Edwin, G. & J. Schunke V. 3911 (95a)
Eggers, H. F. A. 80 (28), 133 (28& 129a), 211 (148), 212 (129a), 213 (140), 315 (169a), 905 (28), 1084 (27), 2063 (131), 3360 (10a), 3947 (177j), 4043 (134e), 4535 (141), 5087 (131), 5349 (202b), 5364 (199), 5475 (69), 5518 (8a), 5704 (202a), 5945 (138g), 6886 (8a), 15483 (200b), 15527 (200b)
Eggleston, W. W. 16701 (84), 17438 (84), 17455 (84)
Egler, W. A. 848 (58), 1310 (3a), 47656 (26a)
Egler, W. A. & H. S. Irwin 45974 (46a), 46047 (161/Ia)
Eiten, G. (many with L. Eiten) 1615 (136), 1644 (3f), 1772 (43), 1853 (43), 1989 (126a), 3068 (44a), 3567 (69), 4145B (62), 4280 (71), 4286 (165), 4339 (165), 4486 (3d), 4490 (3d), 4514 (69), 4515 (71), 4550 (147), 4618 (59 affin), 4628 (140), 4629 (63b), 4631A (147), 4694 (147), 4769B (138g), 4769C (140), 4845 (3d), 4849 (70b), 4990 (42e), 5006 (71), 5008 (202b), 5012 (42d), 5364 (147), 5604 (136), 5647 (69 & 140), 5696 (43), 5937 (43), 6669 (111), 6990 (57), 7012 (57), 7039 (57), 8038 (161/IIa), 8330 (60), 9898 (43), 9980 (3f), 10118 (3a), 10145 (140), 10781 (165), 10850 (202b), 10886 (164b), 10897 (42a)
Ekman, E. L. 905 (168a), 1519 (131), 1692 (89), 1758 (138a), 1759 (69), 1892 (170), 2182 (134c), 3671 (134b), 4399 (5), 6010 (168a), 6193 (3a), 6567 (141), 6614 (134e), 7070 (134a), 7545 (30), 7782 (199), 8249 (133), 8402 (5), 9034 (159), 9169 (134a), 9846 (170), 10281 (129a, p.p.), 11113 (168a), 12253 (126k), 12537 (69), 12611 (138f), 13098 (199), 13314 (30), 15150 (4), 16038 (63b), 16144 (71), 18304 (157), 18738 (196a)
Elias, Bro. 126 (199), 334 (10a), 431 (8a), 693 (8a), 1058 (199), 1106 (8a), 1360 (177a)
Elias, T. S. 81 (199), 238 (116)
Elias de Paula, J. 218 (20)
Ellenburg, H. 308 (179)
Elliott, C. 28 (98d), 245 (99), 514 (98d)
Emmerich, C. 4 (150)
Emrick, G. M. 29 (177a), 101 (177a), 104 (177a), 243 (26a)
Englesing, F. C. 127 (148), 294 (13a)
Enríquez, A. 84 (87)
Epling, C. & Játiva 156 (129a)
Erlanson, C. O. 42 (8a), 195 (8a)
Ernst, W. R. 2250 (178a), 2288 (144), 2512 (177c)
Espinal T., S. 1960 (47)
Espinosa, R. 526 (138f), 588 (42h), 625 (174), 689 (174), 1294 (174), 2878 (154), 3020 (140), 3032 (8b)
Espinosa & Giacometti A17 (8a)
Eupunino, A. 2 (44a), 45 (162), 106 (3b), 235 (42b), 240 (12), 261 (12), 267 (44a), 312 (44a)
Evans, W. H. 9 (80)
Everett, C. P. 17618 (90)
Ewel, – 11 (199)

Eyerdam, W. J. 488 (159)
Eyerdam, W. J. & A. A. Beetle 22106 (110d), 22933 (180)

Fagerlind, F. & G. Wibom 96 (129a)
Fairchild, D. 2765 (8a)
Falcão, J. I. A. 1160 (26a)
Falcão, J. I. A., W. A. Egler & E. Pereira 1021 (44b)
Fallen, M. 672 (69)
Fanshawe, D. B. 2006 (9)
Fariñas, M. 340 (8b), 466 (3a)
Faris, J. A. 47 (168a), 437 (168a)
Fassett, N. C. 25156 (170)
Feddema, C. 14 (152), 323 (11), 461 (11), 751 (11), 1935 (147), 2599 (26a), 2709 (177a)
Feliciano, J. B. 10 (42a)
Fendler, A. 86 (27), 87 (8a), 88 (16), 322 (177a), 325 (63b), 326 (10a), 327 (10a), 328 (126k), 331 (202a), 332 (8a), 333 (27), 1850 (13a), 1857 (29), 2230 (171), 2239 (10a), 2596 (17a)
Fernald, M. L. & B. Long 13352 (69)
Fernández, A. 320 (10a), 342 (71), 422 (10a), 678 (8b), 702 (10a), 1022 (26a), 1064 (26a), 1159 (20), 1190 (69), 1234 (140), 1340 (69), 2126 (17a)
Fernández, A. & L. E. Mora 1220 (116), 1335 (202a)
Fernández, H. 7 (69)
Fernández Acosta, I. 8 (199)
Fernández Yeper, F. 398 (202a), 806 (140), 808 (202a)
Fernando, Hno. 74 (141)
Ferrari, G. 158 (69), 160 (10a), 317 (199), 380 (3a), 497 (115), 985 (172), 1003 (27)
Ferreira, A. 1189 (38)
Ferreira, J. 2315 (42b)
Ferreira, V. F. 171 (111), 698 (43), 794 (57), 1113 (44a)
Ferreyra, R. 243 (23), 524 (114), 761 (102), 1869 (25a), 2589 (110d), 3006 (110e), 3159 (198), 3195 (112), 3353 (147), 3364 (17a), 3883 (110f), 4415 (3a), 4512 (147), 4520 (17a), 4573 (69), 4634 (22), 4710 (17a), 4736 (22), 4772 (22), 4893 (26a), 4956 (3a), 5054 (3a), 5208 (129a), 6168 (110f), 6793 (25a), 7059 (174), 7077 (174), 7268 (110f), 7499 (95a), 7613 (129a), 7747 (126s), 7855 (26a), 7888 (161/Ia), 8396 (129b), 8426 (112), 8659 (110f), 9161a (174), 9312 (161/Ia), 9955 (95a), 10089 (22), 10109 (22), 10112 (147), 10179 (3a), 10336 (25a), 10763 (153c), 10897 (63b), 11337 (68), 11714 (42h), 12424 (174), 13368 (129b), 13422 (110f), 13478 (129b), 13482 (110f), 13644 (177f), 14179 (200b), 14195 (177f), 14338 (174), 14597 (174), 16391 (20), 16585 (110f), 17024 (3a), 17055 (110f), 17191 (126e), 17244 (69), 17266 (140), 17323 (26a), 17440 (26a), 17523 (22), 18012 (126e), 18297 (110f), 18340 (174)
Ferreyra, R. & C. Ancleto 15183 (155), 15194 (113a)
Ferreyra, R. & O. Tovar 7909 (22)
Ferrigni, N. R. 34E (3a)
Ferris, R. S. 2921 (73), 5072 (200a), 5357 (11), 5566 (177a), 5568 (177a), 5784 (26a), 7216 (90), 8587 (101), 8696 (81)
Ferris, R. S. & G. D. Duncan 2379 (87), 2500 (87), 2722 (84)
Ferris, R. S. & Y. Mexia 5085 (177a), 5260 (177a)
Fevereiro, P. C. 213 (37)
Fiebrig, K. 111 (63a), 1329 (63a, 63b and 66), 3007 (110a), 4064 (63a), 4164 (160), 4800 (202a)
Fischer, P. C. 218 (80)
Fisher, G. L. 35202 (119), 35501 (199), 35512 (55)
Fishlock, W. C. 191 (140), 282 (28), 325 (69)
Flores, R. S. 2 (16)
Flores M., D. 251 (119)
Florschütz, J. & P. A. 270 (161/Ia), 1059 (138g), 1158 (147), 1653 (26a)
Foldats, E. 340A (3a), 349A (138f)
Folsom, J. P. 1227 (26a), 2446A (26a), 6419 (16)
Fonseca, M. 2 (44a), 89 (37)
Forero González, E. 947 (147), 1907 (14a), 4856 (147), 5844 (146)
Forero González, E. & B. L. Wrigley 7080 (161/Ia)
Forest Dept. (British Guiana) 91 (161/Ia), 2205 (8a), 2210 (161/Ia), 3047 (140), 3328 (138g), 3486 (129a), 4736 (9), 6791 (8a), 7022 (147), 7728 (26a)
Fortín, G. M. 263 (157)
Fosberg, F. R. 3374 (80), 3920 (73), 21403 (13a), 22567 (174)
Foster, R. 712 (27), 1369 (14b), 2066 (148)
Frambach, J. 113 (12)
Frankie, G. W. 30a (13a), 30c (13a)
Fredholm, A. 3144 (140), 3324 (69)
Freire Allemão, F. 455 (44b), 460 (37), 473 (26a & 46a)
Friedrichstal, E. 118 (28), 819 (138f), 875 (177a)
Frödin, J. 171 (98d)
Froehner, C. 28 (25a)
Fróes, R. L. 1739 (161/Ia), 1745 (3a), 1798 (138g), 1814 (36), 11735 (161/Ia), 20276 (161/Ia), 26996 (126g), 31539 (27), 31908 (20)
Fromm, E. 1332 (61)
Frye, T. C. & E. M. 2557 (119), 2575 (200a)
Fryxell, P. 2261 (199), 3230 (202a)
Fuentes, E. 45 (17a)
Fuertes, M. 192 (199), 197 (168a), 252 (141), 1390b (148)
Funck, N. 369 (10a), 459 (10a)
Funk, V. A. 226 (69), 1842 (138f), 2348 (126k), 2376 (178a), 2531 (178a), 2534 (178a), 2536 (178a), 2542 (71), 2548 (177a), 2549 (140), 2586 (161/III), 2590 (71), 2592 (201), 2598 (126j), 2632 (177a), 2629 (68), 2633 (177a), 2648 (126k), 2655 (177a), 2678 (138f)

Galeotti, H. 3144 (119), 3145 (178a), 3183 (11), 3218 (78), 3227 (145), 3260 (126m), 3293 (11), 3312 (68), 3313 (119), 3372 (75), 3429 (13a & 16), 3465 (177q)

Galland, Y. 6 (137c)
Galvão, R. 5533 (12)
Gandara, J. M. & J. Dorantes 136 (140)
Garber, A. P. 61 (169a)
Garcia, P. 982 (118)
García-Barriga, H. 4962 (8b), 5135a (13b), 10882 (8b), 11836 (31), 12309 (31), 12309 (31), 12458 (17b), 13263 (170), 13848 (3a), 17614 (17b), 18540 (16), 20304 (6)
Gardner, G. 366 (161/IIa), 367 (161/IIa), 368 (42a), 990 (46a), 1463 (161/IIa), 1568 (37), 1571 (44b), 1575 (165), 2119 (165), 2120 (42e), 2123 (45), 2124 (62), 2548 (43), 2551 (157), 3690 (42b), 3691 (63b), 4532 (175), 5701 (137c)
Garibaldi, C. 114 (8a)
Garnier, Bro. A. 62 (26a), 153 (36), 222 (177a), 530 (177a), 3079 (177a), 4273 (177a)
Garroni, P. 79 (42g)
Gastony, G. C. 393 (140)
Gaudichaud-Beaupré, C. 115a (98d), 124 (200b), 258 (125), 867bis (126d), 878 (120), 879 (12), 886 (126d)
Gaumer, G. F. 354 (196a), 383 (199), 391 (177h), 519 (140), 757 (146), 1090 (146), 1330 (27), 1331 (177h), 1686 (196a), 1687 (196a), 1939 (146), 2315 (177h), 23245 (177h), 23261 (196a), 23408 (146), 23438 (124), 23490 (177h), 23611 (196a), 24066 (177h), 24106 (27), 24110 (196a), 24157 (27), 24157a (27), 24405 (27), 24407 (27), 24433 (177h)
Gay, C. 115 (98d), 139 (98d), 233 (125)
Gehriger, W. 12 (115), 445 (16)
Gehrt, A. 8025 (120)
Gehrt, G. 8292 (43)
Gentle, P. H. 69 (27), 75 (27), 191 (27), 324 (126j), 598 (27), 1012 (16), 1045 (68), 1060 (126j), 2493 (201), 3636 (27), 3715 (16), 4830 (71), 4995 (27), 5020 (202a)
Gentry, A. 512 (199), 1302 (13a), 1492b (14b), 1552 (16), 1805 (8a), 1843 (14c), 2113 (26a), 2934 (16), 4435 (26a), 5060 (27), 6596 (138f), 7412 (26a), 7619 (201), 8412 (27), 9200 (31), 9787 (17a), 10229 (171), 10491 (26a), 10664 (27), 10758 (8a), 11012 (171), 13372 (38), 14848 (199), 15008 (26a), 15219 (14b), 16287 (21), 16350 (69), 16406 (68), 17200 (14b), 18427 (69), 18641 (69), 18928 (69), 19618 (22), 19638 (161/Ia), 19646 (21), 20178 (177a), 21039 (8b), 21489 (44a), 22086 (161/Ia), 22203 (17a), 22607 (153b), 22724 (177f), 22779 (177f), 22820 (138f), 23220 (153b), 23333 (95a), 23782 (26b), 23864 (14b), 24368 (26b), 25768 (25a), 25779 (147), 28738 (25a), 28798 (177a)
Gentry, H. S. 295 (177d), 894M (177d), 967 (177c), 1035 (177c), 1171 (177c), 1295 (177d), 4952 (11), 4967 (177c), 5123 (63b), 5213 (11), 5247 (140), 6775 (126m), 16555 (8b), 17831 (177d), 19495 (26a), 19527 (11), 23317 (101)
Gerhart, G. A. 616 (28)
Germán, M. T. et al. 223 (200a), 955 (72b)
Giacometto, J. 50 (16)
Gianotti, E. 8391 (3f)
Gibbs, P. E. 3411 (161/IIb), 3428 (120), 4285 (57), 5109 (175), 5269 (56a), 5393 (126a)
Gibson, A. C. & L. C. 2169 (140)
Gifford, D. R. & S. Fonseca 318 (42c), 325 (202b)
Giler, M. A. 58 (110g)
Gillett, J. M. 16484 (147)
Gillis, W. T. 6954 (2), 6978 (2), 6979 (161/Ia), 7148 (169a), 7150 (126), 7160 (148), 7388 (131), 7468 (131), 7468A (177j), 7490 (202a), 7804 (140), 8029 (199), 1856 (140), 8370 (141), 8406 (199), 8544 (134e), 8588 (202a), 8732 (177j), 8807 (177j), 8915 (199), 8985 (69), 9056 (177a), 9515 (202a), 9735 (202a), 9811 (202a), 9999 (8a), 10269 (177a), 10325 (177e), 13033 (134d)
Gilly, C. L. 8 (196d), 15 (178a), 45 (138f), 98 (177a), 114 (11), 122 (177a)
Gilly, C. L. & R. F. Simpson 13 (178a)
Gilmartin, A. J. 193 (10b), 245 (14d), 400 (153c), 567 (200b), 654 (63b)
Ginés, Hno. 25 (13a), 32 (27), 804 (147), 2294 (172), 2858 (177a), 2859 (177a), 3479 (8a), 4283 (55), 4350 (27), 4367 (171), 4402 (172)
Glassman, S. F. 1960 (119)
Glaziou, A. 132 (120), 605 (124), 1034 (42a), 1459 (124), 2981 (12), 3721 (104a), 3942 (104a), 5848 (120), 6182 (61), 6869 (120), 7580 (40), 7597 (124), 8413 (162), 8416 (161/IIa), 8417 (111, affin), 8663 (136), 8664 (137c), 9413 (44a), 9750 (45), 10653 (39a), 10657 (63b), 10684 (1), 12611 (163), 12613 (43), 13743 (59c), 13746 (161/IIa), 13748 (44a & 136), 14628 (57), 14633 (63a), 15936 (120), 16756 (44a), 18208 (61), 19063 (126a), 19077 (57), 19079 (57)
Gleason, H. A. 13 (26a), 50 (26a), 155 (140), 390 (161/Ia), 538 (140), 776 (26a), 808 (20), 858 (140)
Glocker, E. F. von 165 (44a), 588 (26a)
Goés, O. C. & D. Constantino 5 (40), 89 (40), 52 (40), 131 (137c), 309 (124), 617 (40), 1087 (136), 1088 (12)
Goldman, E. A. 221 (140), 241 (177c), 284 (177d), 381 (140), 384 (177a), 452 (16), 522 (196a), 632 (177i), 939 (94c), 7160 (101)
Goll, G. P. 166 (140)
Gould, F. W. 2550 (80)
Gould, F. W. & H. S. Haskell 3208 (138e)
Gomes, F. & A. Mattos 390 (42a)
Gomes, J. C. 1074 (43), 1247 (62)
Gómez, M. P. 31 (118)
Gómez, R. 1036 (177k)
Gómez-Pompa, A. 1251 (47), 2557 (72b)
González, A. 25 (172), 556 (199), 1060 (129a), 1101 (170)
González, E. 11 (16)
González, Gloria 11 (16)
González Medrano, F. 7299bis (94f), 11571 (195), 11952bis (199)
González-Quintero, L. 905 (196b), 968 (196b), 1832 (138f), 3758 (94e)
Goodding, L. N. 289 (138e), 718 (80), 2391 (193c), 2431 (138e)
Goodland, R. 279 (69)

Goodman, G. J. 3061 (84)
Goodspeed, T. H. 15635 (102), 15827 (98b), 16250 (98a), 35036 (8b), 35083 (17a)
Gottsberger, I. S. 106 (43), 1020 (161/IIa)
Graham, E. H. 173 (161/Ia)
Graham, H. W. 16 (80)
Graham, V. 484 (27)
Graham, W. L. 1135 (199), 1187 (26a)
Grant, M. L. 9938 (13b), 10571 (8b)
Grant, V. 714 (177a), 911 (177a), 1111 (177a)
Granville, J. J. de 778 (35), 1637 (35), 2381 (35), 2679 (35), 2809 (140), B3725 (26a), B4218 (35), B4426 (20), B5129 (26a), B5241 (35), BC38 (129a)
Grassl, C. O. 21 (84)
Greenman, J. M. 119 (11), 225 (177a), 353 (196a), 5059 (16), 5261 (27), 5668 (147), 5679 (140), 5706 (177a), 5715 (177a)
Gregg, J. 1014 (26a)
Gregory, D. P. & G. Eiten 55 (116 × 119)
Gregory, L. E. 172 (28), 282 (199)
Grenand, – 1234 (35)
Grether, R. 1253 (91c), 2161 (177a)
Griffiths, A. 3883 (202a), 3897 (161/IIa), 4224 (116), 4270 (202a), 4607 (116), 4909 (161/IIa), 4990 (196b), 5311 (116)
Griffiths, D. 22 (80), 1621 (138e)
Groth, B. H. A. 57 (84)
Grubb, P. J. et al. 24 (25a), 161 (161/Ia), 1620 (25a)
Guedes, E. P. 58 (26a)
Guedes, T. N. 362 (165), 374 (202b), 424 (42c)
Guillemin, M. 404 (42a), 659 (12), 967 (104a)
Gutiérrez, M. M. 14 (16)
Gutierrez V., G. & R. E. Schultes 708 (53b), 717 (8b), 831 (53b), 910 (8b)
Guzmán, M. & D. Castro 73 (199)

Haas, H. de 31 (43), 292 (59 affin), 1138 (124)
Hackett, H. E. 172 (168c)
Hage, J. L. 96 (148), 137 (3b)
Hagee, – & Kerr 1059 (83b)
Hagen, C. & W. van 1060 (202a), 1263 (177k), 1341 (16)
Hahn, L. 267 (119), 339 (69), 439 (140), 1140 (140), 1166 (177a)
Hall, E. 167 (84)
Hall, J. S. & S. M. Bockus 7830 (71), 7908 (13a), 7960 (13a)
Hallé, F. 1007 (46a)
Hamblett, R. B. 216 (177a), 1110 (69)
Hammarlund, C. 13 (69), 490 (96)
Hammen, Th. van der 1066 (16)
Hammel, B. 4751 (8a), 5262 (138g), 5534 (8a)
Hamilton, S. H. 103 (141), 104 (141), 106 (69), 107 (71), 164 (199)
Hanbury-Tracy, J. 23 (126k)
Handro, O. 426 (161/IId)
Hansen, B. & J. 1515 (193a), 1536 (149)
Hanson, H. C. 574 (88)
Harbison, C. F. 27176 (101)
Harley, R. M. 15020 (52), 15022 (164c), 15152 (42c), 15217 (59d), 15370 (42c), 15373 (59b), 15483 (164a), 15661 (59b), 15994 (59a), 16144 (164a), 16151 (202b), 16170 (37), 16306 (42e), 16314 (42e), 16323 (150), 16481 (176), 16493 (42d), 16713 (59a), 16732 (164a), 16733 (164a), 16745 (37), 16822 (42c), 16886 (202b), 17203 (3b), 17516 (26a), 18308 (26a), 18574 (59d), 18625 (164a), 18626 (59d), 18906 (45), 18933 (164a), 19051 (202b), 19172 (202b), 19302 (59a), 19323 (42c), 19491 (42c), 19492 (59b), 19536 (42c), 19819 (59b), 19987 (166), 20013 (202b), 20202 (202b)
Harling, G. et al. 7361 (17a), 7799 (26a), 9475 (69), 9469 (170), 13616 (42h), 13806 (25a), 14057 (148), 14129 (174), 14155 (63b), 14269 (112)
Harmon, W. E. 1309 (26a), 1792 (177g), 2573 (55), 3303 (71), 3506 (55), 4430 (138f), 4489 (68), 4880 (177a)
Harmon, W. E. & J. D. Dwyer 2737 (201), 2875 (140), 3290 (55), 3552 (69)
Harmon, W. E. & J. A. Fuentes 1881 (147), 2138 (177a), 4718 (16), 4974 (147), 4988 (177a), 5029 (177a), 5132 (13a), 5236 (138f), 5933 (55), 6408 (140)
Harriman, N. A. 12318 (196b), 13692 (159), 13818 (141), 13860 (129a), 14683 (94c), 15952 (72b)
Harris, J. A. C. 1539 (199), 16362 (80)
Harris, W. 6722 (131), 9031 (168b), 9039 (8a), 9653 (131), 9927 (71), 9968 (71), 10045 (148), 11693 (134a), 12149 (134a)
Harris, W. & N. L. Britton 10616 (202a), 10719 (129a)
Harrison, S. G. 863 (20), 1041 (20), 1528 (129a), 1555 (161/Ia)
Harrison, S. G. & R. Persaud 1106 (140)
Hart, J. 538 (29), 612 (68), 1220 (8b), 1361 (161/Ia)
Hartshorn, G. S. 1516 (202a), 944 (13a)
Hartweg, T. 283 (144), 652 (10b), 729 (174), 960 (166)
Harvey, D. R. 5035 (16), 5163 (16), 5240 (27)
Hassler, E. 605 (69), 627 (138b), 1192 (140), 1385 (69), 1630 (69), 1659 (69), 1923 (148), 2797 (118), 3266 (63a), 3598 (138b), 3675 (138b), 3736 (63a), 3946 (126a), 4300 (160), 5071 (63a), 5312 (69), 5627 (63a), 5696 (140), 5815 (127), 5889 (3c), 5916 (127), 6013 (138b), 6487 (139), 6683 (3c), 7050 (136), 7121 (160), 7235 (157), 7240 (118), 7361 (126o), 7791 (126a), 8017 (3e), 8017a (3e), 8434 (63a), 8971 (127), 9309 (160), 9354 (66), 10504 (43), 10504a (43), 10767 (138b), 10953 (3e), 10980 (202a), 10981 (126a), 10993 (60), 11050 (44a), 11084 (136), 11208 (63a), 11238 (126a), 12111 (140), 12126 (126o), 12466 (148), 12539 (69), 12716 (138b), 12716a (138b)
Hatschbach, G. 1754 (104c), 3155 (63a), 5892 (104c), 5897 (44a), 6618 (44a), 7963 (44a), 11132 (3a), 13941 (43), 14099 (3a), 15913 (43), 16200 (39a), 17596 (161/IIb), 22152 (43), 23556 (60), 23602 (3c), 23678 (126a), 23773 (63a), 23840 (160), 23904 (104c), 23942 (137a), 24025 (39a), 24228 (63c), 24249 (60), 25095 (160), 25604 (121), 26029 (126a), 26051 (3c), 26138 (63a), 26189 (161/IIb), 26290 (137a), 26345 (124), 26434 (121), 27468 (43),

28109 (104c), 28662 (126a), 29411 (60), 29479 (157), 29558 (157), 31395 (42a), 31502 (3f), 31707 (59a), 31765 (161/IIa), 32043 (43), 32868 (121), 33094 (3c), 33770 (137a), 33811 (66), 34004 (126a), 34060 (3e), 34172 (43), 34421 (126a), 34889 (124), 35666 (137a), 35699 (161/IIb), 36721 (56a), 36916 (66), 37121 (63a), 37297 (121), 37731 (63a), 37741 (43), 37881 (120), 38030 (43), 38085 (161/IIa), 38088 (3a), 38122 (3f), 38158 (126a), 38271 (60), 38246 (69), 38356 (43), 38406 (3a), 38460 (126g), 38549 (69), 38551 (161/IIa), 38632 (126o), 39108 (71), 39396 (42d), 39435 (43), 39485 (59a), 39507 (42d), 39524 (164a), 39547 (52), 39920 (126a), 40113 (157), 40882 (59a), 41036 (71), 41244 (56a), 41272 (43), 41297 (59a), 41357 (3f), 41382 (44a), 41397 (69), 41491 (43), 41771 (161/IId), 41820 (121), 41921 (161/IIb), 41975 (161/IIa), 42011 (202b), 42040 (59a), 42118 (164a), 42141 (66), 42331 (45), 42336 (44b), 42467 (44a), 42483 (121), 42520 (121), 42686 (121), 42837 (59a), 42933 (42b)

Haught, O. F10 (198), 61 (200b), 62 (198), 129 (198), 270 (153c), 1450 (8b), 1600 (47), 3025 (10b), 3712 (8a), 3799 (13a), 3985 (8a), 4428 (177a), 4474 (8a), 4546 (14a), 5158 (177a), 6360 (202a), 6735 (147)

Havard, V. 14 (82), 151 (30)

Hawkes, J. G. 1917 (177l), 2088 (94a)

Hayden, S. M. V. 1015 (16)

Hayes, S. 41 (63b), 158 (140), 360 (8a), 403 (16), 531 (63b), 548 (14b), 572 (16), 694 (14a), 746 (129a), 880 (141)

Hazlett, D. 825 (16), 1331 (16), 2403 (11)

Heinrichs, E. 14 (116), 253 (25a)

Heithaus, E. R. 9 (177a), 424 (16)

Heller, A. A. 148 (140), 186 (69), 539 (28), 918 (148), 1666 (84), 4107 (69), 6068 (169a), 6115 (69)

Helper, I. W. 31 (161/Ia)

Hemming, J. H. 5 (58)

Henderson, N. C. 63-435 (84)

Henrickson, J. 2 (90), 2411 (177d), 6078b (194), 10309 (90), 11874 (76a), 12156 (76a), 12170b (194), 13567 (76a)

Heriberto, Bro. 38 (16), 163 (69), 181 (16), 232 (8a), 359 (199), 426 (161/IIb)

Heringer, E. P. et al. 41 (161/Ib), 99 (42e), 346 (42b), 405 (42e), 458 (42e), 602 (44b), 913 (148), 951 (42e), 1384 (126a), 1545 (140), 3011 (42b), 3150 (3f), 3203 (42b), 3350 (3/Ib), 3360 (43), 3491 (60), 3503 (175), 3683 (126a), 7107 (63b), 7136 (3f), 7497 (39b), 9665 (3f), 10321 (43), 11449 (56a), 11766 (161/IIa), 12791 (161/Ia), 13095 (60), 13251 (43), 14724 (161/Ia), 14742 (161/Ia), 17035 (43)

Heringer, E. P. & D. de A. Lima 623 (165)

Heringer, E. P. & Rizzini 7527 (60)

Heringer, E. P., Rizzini & Mattos 15894 (161/Ia)

Hermann, F. J. 10886 (147), 11207 (8b)

Hermann, H. A. van 165 (30), 193 (138f), 209 (126m), 259 (148), 267 (30), 602 (131), 635 (140), 636 (71), 688 (126m)

Hernández M., R. 546 (177a), 1348 (202a), 4495 (79)

Hernández X., E. 10026 (55), 16766 (199)

Herrera, A. 51 (16)

Herter, G. 368 (128)

Herzog, T. 1149 (180), 1562 (21)

Hespenhide, H. A. 1118 (29), 1604 (199)

Hess, L. 1251 (119), 1263 (178a)

Hess, W. E. 1425 (119), 1675 (140), 2907 (126n), 3079 (169a), 4008 (126n), 4640 (169a)

Heyde, A. W. & T. Lux 3712 (119), 3719 (11), 4152 (177n), 4158 (68), 4175 (177k), 4176 (94a), 6113 (177a)

Heywood, V. 30/76 (69)

Hieronymus, G. 77 (110a), 92 (130), 93 (130)

Higgins, J. E. 454 (8a)

Higgins, L. C. 2566 (177a), 4025 (80), 5458 (84), 6742 (84), 6777 (87), 6791 (73), 6992 (84), 7066 (84), 8637 (84), 8679 (84), 8703 (73), 8707 (87), 8985 (84), 9869 (82), 9989 (82), 9995 (73), 10266 (84), 11632 (80)

Hill, R. J. 1784 (116 × 119), 1785 (116), 1810 (177a), 1822 (71), 1823 (199), 1827 (193a), 1828 (140), 1829 (193a), 1830 (71), 1831 (71), 1832 (71), 1833 (140), 1838 (148), 1841 (140)

Hill, S. R. 1974 (71), 2174 (177j), 2343 (141), 2437 (134e), 2933 (69), 3004 (134e)

Hinckley, L. C. 251 (84), 328 (73), 815 (82), 1102 (193c), 1328 (87), 1510 (73), 1517 (82), 2052 (73), 2706 (84), 2779 (87), 2794 (87), 2798 (193c), 3004 (193c), 3035 (84), 3035 (84), 3302 (73), 3612 (88), 3852 (193c), 4554 (138e), 4608 (82), 4655 (88)

Hinton, G. B. 355 (177r), 1125 (72a), 1449 (71), 1517 (11), 2972 (177r), 3151 (196d), 4330 (11), 4816 (63b), 5582 (177r), 5660 (148), 5677 (140), 5742 (177r), 5889 (199), 5894 (196d), 6021 (69), 6144 (177r), 6216 (116), 6305 (193a), 6530 (72a), 6612 (193a), 6904 (148), 6961 (77), 7053 (11), 8278 (116 × 119), 8917 (177r), 9034 (177r), 9177 (55), 9359 (140), 9360 (69), 9585 (138f), 9945 (91b), 10177 (91b), 10237 (200a), 10842 (138f), 10876 (200a), 10887 (148), 10928 (177a), 10956 (148), 10983 (11), 11262 (177k), 11297 (178a), 11370 (140), 11422 (71), 11425 (149), 11430 (138f), 11439 (70a), 11479 (11), 11512 (177m), 11539 (177c), 11559 (63b), 11600 (140), 11617 (69), 11740 (200a), 11746 (63b), 11750 (71), 11796 (11), 11876 (72a), 12112 (55), 12199 (138f), 12352 (91a), 12487 (152), 12491 (72b), 12547 (26c), 12618 (177a), 12640 (196c), 12688 (177a), 12980 (68), 13008 (177a), 13546 (116), 13582 (72b), 13613 (199), 13628 (199), 13634 (177a), 13751 (199), 14119 (199), 14153 (177e), 14795 (91d), 14915 (177m), 15176 (140), 15436 (116 × 119), 15847 (177a), 15879 (199), 15995 (152), 16105 (196c), 16565 (194)

Hioram, Bro. 1229 (202a), 1903 (71), 2283 (148), 2541 (131), 4186 (202a)

Hitchcock, A. S. 16530 (147), 16546 (8a), 16643 (69), 16705 (69), 16715 (129a), 16717 (140), 17258 (20), 20023 (153c), 21272 (200b)

Hitchcock, C. L. 6100 (90)

Hladick, - 234 (8a)
Hodge, W. H. 598 (140)
Hoehne, F. C. 27 (126a), 232 (161/IIa), 1900 (43), 1987 (43), 4666 (161/IIb), 6612 (43), 20042 (70a), 28788 (42a)
Hoehne, W. 3957 (120)
Holdridge, L. R. 219 (202a), 327 (199), 827 (168a), 884 (148), 1169 (134b), 1184 (199), 1911 (202a), 1279 (132), 2284 (177k)
Hohn, T. 70 (28)
Hollermayer, P. A. 41 (109a)
Holmgren, A. H. 15499 (80)
Holmgren, N. H. 3568 (90)
Holmgren, N. H. & P. K. 7413 (84), 7573 (90), 7954 (80)
Holm-Nielsen, L. 2601 (24), 2608 (147), 2612 (24), 2690 (24), 2767 (69), 4886 (179), 6946 (23), 7250 (10b)
Holroyd, R. 4 (20)
Holt, E. G. & W. Gehriger 9 (199), 112 (140), 229 (147)
Holt, E. G. & E. R. Blake 432 (3a)
Holton, I. F. 17 (126k), 980 (17b)
Holway, E. W. D. 134 (177g), 182 (177g), 3675 (178a), 5390 (178a)
Hoover, W. S. 240 (201)
Horn, C. L. 300 (28)
Hostmann, F. W. 150 (26a), 327 (8a), 400 (140), 436 (69), 670 (8a)
Hostmann, F. W. & Kappler 327 (36), 728 (36), 829 (161/Ia)
Howard, R. A. 4799 (69), 4817 (141), 4987 (30), 5771 (199), 6201 (134c), 6524 (30), 8244 (202a), 8280 (199), 8297 (199), 8400 (199), 8634 (168a), 9220 (202a), 9940 (148), 10079 (131), 10240 (134e), 10698 (169a), 10738 (129a), 10832 (129a)
Howell, J. T. 9389 (153c)
Huashikat, V. 106 (25a), 367 (50)
Huber, J. 45 (46a), 678 (46a), 705 (138g), 1470 (3a), 4248 (26a), 4252 (20), 4513 (20)
Huber, O. 807 (69), 1038 (69), 1405 (3/Ia)
Hugh-Jones, D. 117 (26a)
Huidobro, A. 1798 (139), 3631 (128), 3667 (128)
Huidobro, R. 1863 (126o), 2002 (139), 2004 (126o), 2053 (140), 2059 (126o)
Humbert, H. 30632 (110c)
Humbles, J. 6143 (116)
Humboldt, F. W. H. A. & A. Bonpland 253 (10a), 615 (140), 1177 (126k), 1178 (147), 1180 (161/Ia), 1389 (8a), 1466 (199), 1630 (147), 1754 (126i), 2133 (153a), 3662 (177f), 3842 (200b), 3948 (77), 4138 (119), 4216 (144), 4256 (75)
Hunnewell, F. W. 18274 (146)
Hunt, D. R. 64 (16), 70 (27), 5464 (43), 5583 (3f), 5595 (3a), 6293 (161/Ia)
Hunt, D. R. & J. F. Ramos 5868A (66), 5986 (60), 6663 (42b)
Hunter, A. A. & P. H. Allen 236 (8a), 714 (68), 719 (16)
Hunziker, A. T. 18176 (122)
Hutchinson, P. C. 352 (98d), 1186 (23), 1506 (177f), 1665 (95a), 2785 (8a), 4190 (95a), 4192 (95a), 4421 (110e)
Hutchinson, P. C. & J. K. Wright 3435 (153b), 3602 (42h), 3882 (23), 4429 (153b), 5306 (95b), 6712 (129b), 6729 (179), 6730 (42h), 6773 (200b), 7154 (102), 7226 (110d)

Ibarrola, T. S. 190 (63a), 1482 (69), 2097 (67)
Idrobo, J. M. & A. Fernández 104 (202)
Idrobo, J. M. & R. E. Schultes 586 (3/Ia), 588 (8b)
Ijjasz, E. 104 (199), 157 (129a), 214 (27), 232 (170), 235 (13a), 295 (115)
Iltis, H. H. & C. M. 288 (25a), 526 (95a), 527 (116), 674 (193a), 829 (144), 1331 (144)
Iltis, H. H., C. M. Iltis & D. & V. Ugent 320 (126e)
INPA 1385 (26a), 3448 (26a), 3555 (26a), 5348 (26a), 6998 (26a), 7229 (26a), 7906 (26a), 11116 (69), 12733 (8b), 12981 (38), 37467 (8b), 37608 (8b), 37913 (26a), 38901 (8b), 50541 (38)
IPEAN 684 (126g)
Irwin, H. S. 14 (69), 355 (8a), 360 (20), 397 (169a), 503 (129a), 645 (8a), 657 (20), 1225 (74), 1226 (73), 1231 (82), 1234 (83a), 1236 (75), 1247 (138e), 1251 (140), 1272 (71), 1306 (138e), 1328 (138e), 1337 (72a), 1340 (116), 1342 (75), 1343 (94e), 1349 (196b), 1350 (138f), 1359 (199), 1394 (138e), 1395 (73), 1876 (140), 1901 (73), 2020 (147), 2024 (42a), 2106 (136), 2146 (69), 2196 (161/IIa), 2215 (126a), 2328 (42b), 2332 (3f), 2334 (44a), 2362 (43), 2374 (69), 2375 (42b), 2387 (3a), 2388 (57), 2389 (126a), 2397 (126a), 2402 (44a), 2403 (42a), 2404 (161/IIa), 2413 (3f), 2418 (42a), 2426 (57), 2436 (42a), 2476 (57), 2477 (56a), 2500 (3a), 2507 (3a), 2509 (42a), 2510 (57), 2511 (161/IIa), 2518 (126a), 2521 (63a), 2552 (3f), 2529 (43), 2530 (60), 2537 (63a), 2539 (63a), 2544 (43), 2545 (126a), 2585 (63a), 2586 (138g), 2615 (42b), 2625 (42a), 2628 (161/IIa), 2638 (42a), 2743 (104a), 2806 (104a), 37196a (177j), 37205 (38), 37254 (69), 37255 (140), 37256 (199), 37257 (168a), 37258 (168a), 37259 (69), 37260 (158), 37261 (199), 37264 (199), 37265 (141), 37266 (69), 37267 (202a), 37269 (199), 37270 (168a), 37271 (140), 37272 (168a), 37273 (199), 37274 (168a), 37275 (202a), 37277 (199), 37278 (168a), 37279 (141), 37280 (140), 37281 (199), 37303 (141), 37306 (199), 37309 (140), 37318 (71), 37319 (199)
Irwin, H. S. et al. 7846 (43), 8607 (69), 8114 (140), 8445 (126a), 9120 (43), 9141 (126a), 9160 (43), 9277 (56a), 9563 (140), 9705 (43), 9871 (43), 10302 (43), 10440 (3f), 10474 (43), 10505 (42b), 10555 (126a), 10826 (43), 10854 (3f), 10968 (42b), 11056 (126a), 11107 (43), 11247 (42b), 11612 (126a), 11827 (56a), 11879 (3f), 11960 (140), 12145 (126a), 12211 (43), 12463 (43), 12710 (56a), 13048 (126r), 13058 (42b), 13173 (148), 13394 (43), 13640 (17a), 13704 (56a), 13783 (42a), 13862 (126a), 14016 (3f), 14101 (60), 14168a (43), 14304 (126a), 14417 (43), 14418 (148), 14477 (69), 14487 (59a), 14606 (43), 14670 (3d), 14806 (59 affin), 14899

(59a), 14913 (43), 14940 (70b), 14948 (202b), 14977 (126a), 15076 (138g), 15092 (126a), 15119 (70b), 15134 (60), 15182 (60), 15221 (69), 15267 (63b), 15320 (69), 15377 (42b), 15413 (3f), 15455 (56a), 15474 (42b), 15491 (126a), 15611 (119), 15641 (69), 15654 (126a), 15683 (161/Ia), 15742 (161/Ia), 15865 (126a), 16038 (66), 16073 (3a), 16160 (60), 16197 (43), 16233 (3f), 16324 (126a), 16367 (60), 16515 (43), 16575 (43), 16649 (126a), 16710 (60), 16791 (3f), 16830 (60), 16947 (60), 16995 (66), 17462 (126a), 17611 (3f), 17634 (60), 17657 (43), 17669 (69), 17720 (66), 17737 (126a), 17904 (3f), 18102 (161/Ia), 18246 (42b), 18254 (42b), 18564 (43), 18807 (126a), 18866 (63b), 18940 (3f), 19086 (140), 19218 (43), 19518 (57), 19577 (43), 19922 (136), 20235 (57), 20373 (57), 20565 (42a), 20710 (42b), 20857 (3a), 20873 (3a), 20954 (57), 21236 (43), 21304 (148), 21325 (3f), 21333 (60), 21780 (69), 21969 (200a), 22013 (136), 22214 (42b), 22623 (56a), 22644 (42b), 22645 (57), 22646 (43), 22816 (57), 22980 (59a), 23003 (202b), 23211 (3f), 23410 (56a), 23499 (59a), 23659 (3f), 23735 (57), 23745 (59a), 23915 (3f), 23921 (3f), 24181 (148), 24288 (43), 24336 (59 affin), 24444 (3f), 24546 (126a), 24626 (66), 24654 (43), 24659 (56a), 25007 (126a), 25013 (3f), 25623 (43), 25785 (126a), 25941 (60), 25973 (43), 26110 (140), 26388 (3f), 26511 (42b), 26589 (126a), 26730 (43), 26792 (69), 26837 (126a), 26900 (42b), 26934 (3f), 27071 (3f), 27153 (63b), 27195 (3f), 27258 (57), 27338 (136), 27339 (140), 27344 (71), 27387 (44a), 27404 (202b), 27412 (42b), 27662 (43), 27663 (57), 27732 (42b), 27965 (42b), 28002 (57), 28074 (43), 28075 (42b), 28091 (56a), 28132 (136), 28170 (57), 28595 (56a), 28598 (56a), 28616 (43), 28629 (42b), 28431 (57), 28778 (126a), 28868 (161/IIa), 28878 (42a), 28986 (3a), 29111 (57), 29123 (137c), 29175 (42a), 29184 (136), 29186 (3a), 29371 (57), 29677 (57), 30202 (136), 30268 (42b), 30447 (57), 30520 (39b), 30712 (176), 30760 (43), 30831 (37), 30862 (59a), 30920 (43), 30981 (37), 30984 (59a), 31165 (164a), 31188 (202b), 31202 (164a), 31299 (140), 31462 (59a), 31547 (43), 31581 (202b), 31637 (70a), 31893 (43), 32075 (42b), 32396 (59a), 32448 (42c), 32566 (59a), 32586 (59a), 32632 (140), 32699 (56a), 32807 (43), 34370 (43), 34641 (56b), 34798 (126a), 34886 (140), 34887 (69), 34899 (3f), 35701 (42b), 36438 (43), 45940 (126p), 46711 (46a), 47157 (26a), 47345 (161/Ia), 47260 (36), 47940 (36), 48027 (36), 48092 (26a), 48114 (36), 48116 (147), 48257 (26a), 48263 (36), 48394 (26a), 48417 (36), 48619 (147), 48677 (26a), 48781 (148), 55383 (26a), 55402 (161/Ia), 55427 (36), 55734 (161/Ia), 55965 (140), 57603 (161/Ia)

Irwin, H. S. & T. Soderstrom 5008 (175), 5052 (46a), 5083 (43), 5499 (126a), 5504 (175), 5510 (69), 5511 (66), 5945 (148), 6006 (140), 6295 (60), 6722 (66), 6852 (69), 6857 (140), 7189 (69)

Isely, D. 9491 (90), 10855 (138e), 10894 (87), 10898 (87), 10903 (138e), 10908 (87), 10978 (134e)

Jack, J. G. 4688 (126m), 5224 (30), 5445 (30), 6175 (69), 6634 (202a), 6827 (30), 7721 (8a), 8071 (30), 8427 (1), 8448 (148), 8495 (141)

Jacquemin, H. 1420 (138g), 1735 (147), 1821 (35)

Jaeger, B. 36 (131)

Jaén, J. 2 (16)

Jaffuel, F. 3638 (106)

Jahn, A. 328 (10a), 368 (10a), 477 (17a), 666 (129a), 1171 (115), 1301 (10a), 1325 (8a)

Jameson, W. 214 (95a), 532 (200b), 599 (10b)

Jangoux, J. 255 (3a), 358 (140)

Jansen, J. 358 (140)

Janzen, D. H. 18 (177a), 19 (16), 20 (138f), 21 (69), 116 (119), 287 (199), 310 (199), 333 (140), 352 (69), 378 (69), 692 (126m), 1086 (199)

Jaramillo, J. 124 (25a)

Játiva, C. & C. Epling 623 (14d), 683 (14d), 941 (200b), 2162 (14d)

Jeffrey, C. & B. Trujillo 2372 (13a)

Jenman, G. S. 670 (161/Ia), 1176 (20), 1328 (161/Ia), 1337 (26a), 2158 (8a), 2160 (138g), 4204 (138g), 4233 (20), 4408 (140), 5242 (147), 5243 (26a), 5625 (129a), 5630 (69), 5845 (8a), 7175 (161/Ia), 7176 (161/Ia)

Jermy, G. 317 (73)

Jesus, J. A. de 464 (44a), 628 (3b)

Jesus, J. A. de & T. S. dos Santos 345/394 (18)

Jesus, J. O. de 36 (43)

Jiménez, A. 760 (13a)

Jiménez, F. 97 (193c)

Jiménez, J. de J. 445 (169b), 2221 (134b), 2348 (168a), 2788 (168a), 2808 (168a), 4643 (168a), 5534 (69), 5627 (168a), 5735 (202a), 6225 (71), 8111 (126k), 8254 (134a), 8458 (134a), 8391 (169a)

Jiménez, Otón 777 (177a)

Jiménez L., O. 763 (16)

Jiménez M., A. 534 (177a), 592 (199), 1112 (55), 596 (177a), 1175 (149), 1286 (16), 1523 (177a), 1770 (68), 2086 (10a), 2572 (68), 4020 (8a), 4016 (177o), 4167 (13a), 4173 (202a)

Jiménez Saa, H. 1241 (147), 1277 (13a), 1280 (126k), 1304 (26a)

Job, M. M. 626 (126o), 971 (139), 1024 (139)

Jobert, C. 737 (53a), 1088 (45)

Jobert, C. & C. A. W. Schwacke 31 (20), 211 (46a), 885 (36)

Johansen, H. 1 (11), 5 (154), 14 (11)

Johnson, C. D. 136/76 (69), 199/68 (140), 227/68 (193a), 253/68 (193a), 253/68 (72a), 366/73 (68), 421/73 (11), 646/79 (126k), 660/79 (200a), 1148 (140), 1177 (177a), 1180 (71), 1255 (126k), 1277 (71), 1297 (177a), 1309 (140), 1315 (200a), 1359 (11), 1393 (149), 1396 (26a),

1405 (68), 1551 (126m), 1560 (68), 1640 (146), 1650 (199), 1679 (146), 1701 (199), 1802 (27), 1809 (177h), 1824 (196a), 1826 (177h), 1929 (68), 1930 (16), 1935 (199), 1957 (177a), 1965 (55), 1973 (126k), 2011 (202a), 2042 (140), 2083 (199), 2145 (8a), 2193 (202a), 2290 (16)
Johnson, C. D. & R. Conway 277-78 (116)
Johnson, E. P. 33 (16)
Johnston, I. M. 3573 (81), 5166 (98b), 5238 (98b), 5454 (102), 5671 (98b), 5743 (98a), 5852 (100), 8094 (193c), 8776 (76a), 9082 (76a), 9309 (194)
Johnston, I. M. & C. H. Mueller 32c (193c), 77 (193c), 177 (73), 343 (194), 898 (73), 941 (82), 1340 (193c)
Johnston, J. R. 24 (199), 62 (8a), 700 (148), 1055 (177a)
Johnston, M. C. 2584A (194), 4580 (193b), 5388 (193b), 10883 (194), 11610a (88), 11905 (74), 12084F (194), 12211 (193d)
Joly, A. B. 1936 (57)
Jones, G. C. 3208 (140)
Jones, G. C. & L. Facey 3344 (201), 3373 (26a)
Jones, M. E. 178 (193a), 3864 (90), 3922 (80), 22482 (177a), 23018 (177a), 23027 (177a), 23059 (140), 23060 (126k), 25021 (80), 27196 (146), 27210 (177c), 28716 (138e)
Jones, P. 1046 (134e)
Jorge León, Bro. 104 (16)
Jorgensen, P. 3592 (67), 3632 (66), 3633 (69), 3634 (136), 4190 (126a), 4624 (160), 4827 (127)
José Eugenio, Fr. 606 (44b), 619 (46a), 620 (26a)
Junge, C. 2632 (109a)
Jurgensen, – 51 (11), 162 (200a), 705 (144), 724 (16)
Juzepczuk, S. 1374 (177a)

Kalbreyer, W. 695 (16), 696 (29), 1774 (47)
Kaltenborn, M. & A. 146 (95a)
Kappler, A. 1408 (20)
Karling, J. S. 17 (27)
Kayap, R. 98 (17a), 242 (25a)
Kegel, A. H. H. 66 (8a), 315 (138g), 809 (26a)
Keil, D. 1150 (80)
Kellerman, W. A. 4362 (177g), 5189 (177a), 5635 (177a), 6345 (177g), 6496 (177k), 7144 (147), 7284 (149), 7312 (16), 7592 (149), 7607 (149), 7789 (126k), 7920 (177g), 7954 (177a)
Kennedy, H. 194 (14d)
Kenoyer, L. A. C279 (149), 381 (14b), 1815 (144), 1892 (128), 2285 (79), 2333 (119)
Kenoyer, L. A. & Brum 3938 (11)
Kerber, E. 109 (68), 357 (177a)
Kibler, N. C. 104 (144), 121 (2)
Killip, E. P. 5369 (138f), 5421 (126k), 12055 (16), 13933 (30), 33052 (14d), 33268 (17b), 33590 (14b), 34343 (17b), 34363 (8b), 34470 (17b), 35565 (17b), 37079 (172), 41112 (129a)
Killip, E. P. & T. E. Hazen 8380 (69), 8381 (147), 9631 (166), 11061 (147)
Killip, E. P. & A. C. Smith 14027 (199), 14038 (69), 14040 (177a), 14071 (199), 14089 (69), 14195 (69), 14229 (140), 14319 (8a), 14481 (199), 14517 (8a), 14595 (147), 14621 (177a), 14782 (13a), 14919 (8b), 15180 (126k), 15403 (8b), 15405 (17b), 15415 (8b), 16101 (115), 16173 (202a), 16276 (16), 16400 (166), 16459 (42f), 16697 (115), 16828 (42f), 16886 (115), 18397 (115), 19224 (115), 19404 (166), 19415 (126k), 19447 (42f), 19447 (42f), 19518 (115), 19655 (154), 19807 (166), 20038 (126k), 20495 (115), 20750 (170), 20850 (27), 20950 (140), 21096 (199), 21102 (177a), 21886 (95a), 21983 (95a), 22939 (21), 23375 (138f), 23531 (25a), 23805 (68), 24264 (126e), 24967 (68), 26204 (3a), 26208 (161/Ia), 26291 (126e), 26299 (147), 26401 (161/Ia), 26965 (26a), 27098 (26a), 27126 (17a), 27154 (140), 27392 (51), 27934 (3a), 29557 (22), 29824 (53a), 30027 (26a), 30282 (46a), 30283 (36), 30493 (27), 30565 (36)
King, R. M. 151 (140), 313 (140), 370 (199), 544 (199), 772 (69), 773 (71), 887 (161/III), 1016 (69), 1287 (55), 1297 (199), 1348 (55), 1471 (71), 1512 (71), 1529 (177b), 1608 (199), 1641 (177b), 1789 (55), 2373 (177a), 2387 (177a), 2391 (177a), 2404 (177a), 2459 (178a), 2485 (178a), 2505 (178a), 2705 (126m), 2834 (72b), 2880 (199), 2915 (193a), 2969 (193a), 2979 (55), 3175 (94b), 3184 (116), 3268 (119), 3464 (141), 4048 (126m), 4060 (199), 4183 (177e), 4185 (177c), 4212 (196b), 4448 (196b), 5356 (13a)
King, R. M. & A. E. Guevara 6115 (8b), 6159 (17a)
King, R. M. & T. Soderstrom 4622 (138e), 4638 (152), 4918 (148), 4970 (200a)
Kirkbride, J. H., Jr. 248 (147), 288 (14b), 1029 (167), 1063 (26a), 1577 (140), 1724 (137c), 1997 (8a), 2487 (199), 2603 (199), 2614 (177a), 2623 (199), 2639 (8a), 2640 (15), 2646 (140), 2670 (69)
Kirkbride, J. H., Jr. & E. Lleras 2766 (3a), 2956 (43), 2959 (26a), 2970 (58), 2982 (69), 3001 (60)
Kirkbride, J. H. Jr. & J. A. Duke 1179 (14b), 1364 (138g)
Klug, G. 202 (26a), 279 (17a), 283 (140), 1014 (17a), 1029 (8b), 1107 (147), 1393 (69), 1622 (17a), 1634 (161/Ia), 1751 (26a), 2776 (22), 2799 (3a), 2862 (3a), 2917 (8b), 3000 (20), 3097 (161/Ia), 3598 (161/Ia), 3674 (126s), 4039 (3a)
Knechtel, – 736 (140)
Knight, D. H. 438 (200b)
Koch, S. D. & P. A. Fryxell 73158 (177a), 78191 (126m), 78192 (13a)
Koelz, W. N. 34175 (26a), 34182 (177e), 34243 (11)
Koeppen, R. & H. Iltis 1111 (55)
Koyama, T. 13743 (140)
Kramer, K. U. & W. H. A. Hekking 2686 (26a), 3178 (36)
Krapovickas, A. 1699 (180), 28693 (63a), 29073 (148), 19076 (126o), 29182 (126o), 30669

(118), 30806 (202a), 30814 (69), 30847 (180), 31250 (180), 31330 (63a), 31636 (161/IIa), 31742 (60), 31934 (66), 32125 (63a), 32140 (148), 32262 (43), 32370 (69), 32429 (63a), 32609 (148), 32708 (60), 32714 (3c), 32774 (63a), 32873 (60), 33081 (126a), 33298 (60), 34432 (3c), 34655 (69), 34687 (138g), 34983 (126e), 34998 (126e), 36133 (161/Ia), 36188 (3e), 36189 (60), 36239 (138b), 36304 (140)

Krause, E. H. L. (herb.) 11828 (147)

Krieger, P. L. 7307 (120), 7622 (120), 8403 (3/IIf), 9757 (57), 10148 (202b), 10490 (39a), 12007 (3/Ia), 12344 (3/Ia)

Kruckeberg, A. R. 4741 (73)

Krukoff, B. A. 1082 (140), 1087 (161/Ia), 1091 (69), 1111 (3a), 1570 (17a), 2059 (3d), 4510 (147), 4514 (26a), 4661 (161/Ia), 5093 (3a), 5292 (161/Ia), 5754 (147), 8282 (3a), 8974 (53a), 10003 (46b), 11242 (3a)

Kubitzki, K. 71/73 (42e), 71/78 (150), 71/83 (202b)

Kuhlmann, J. G. 51 (12), 163 (26a), 1116 (26a), 1424 (12), 1545 (8b), 1969 (26a), 2202 (12), 3242 (8a), 6015 (12), 6475 (3b), 17795 (64)

Kummrow, R. 41 (121), 230 (137a), 353 (137a), 479 (126a), 493 (136), 533 (136), 555 (126a), 799 (63a), 874 (137a), 975 (136), 1005 (121)

Kuntze, O. 85 (60), 186 (169a), 281 (140), 514 (169a), 619 (147), 622 (8a), 698 (27), 885 (138g), 1341 (10a)

Kurtz, F. 7093 (181), 7138 (99), 7666 (181), 8391 (110a), 9256 (118)

Labouriau, M. S. 15 (43), 34 (60), 168 (126a)

Lakela, O. 27381 (69), 29967 (126a), 30199 (134e)

Lamb, F. H. 356 (71), 368a (140), 517 (126k)

Lana, J. P. 707 (164b)

Landrum, L. R. 834 (125)

Lange, R. B. 1366 (126a)

Langenheim, J. 3285 (27), 3288 (8b)

Langlassé, E. 52 (177k), 113 (193a), 181 (200a), 448 (11), 479 (138f), 645 (26a)

Langman, I. 3440 (141), 3792 (72b)

Lanjouw, J. 71 (8a), 501 (36), 584 (8a), 1174 (36)

Lanjouw, J. & J. C. Lindeman 280 (161/Ia), 1432 (8a), 1950 (26a), 1975 (147) 3218 (147)

Lankester, C. H. 78 (63b)

Lanna Sobrinho, J. P. & A. Castellanos 1520/2626 (137b), 1550/26590 (12)

Lansing, O. E. 2085 (134e)

Lao, E. A. 99 (13a), 374 (13a)

Lara Rico, R. 2b (110a)

LaRivers, I. & N. P. Hancock 109 (90)

Lasseigne, A. 4307 (69), 4308 (138f), 4310 (202a), 4329 (126a), 4354 (43), 4370 (126a), 4395 (169a), 4424 (148), 4449 (126k), 4468 (170), 4872 (73), 4875 (126m), 4876 (71), 4877 (126m), 4878 (73), 4886 (116 × 119), 4887 (116), 4889 (116 × 119), 4892 (126m), 4895 (126m), 4898 (126m), 4901 (126m), 4907 (126m), 4908 (126m), 4909 (126m), 4917 (177a), 4924 (177a), 4925 (126m), 4931 (126m), 4933 (148), 4934 (126m), 4943 (202m), 4947 (126m), 4948 (177a), 4949 (178a), 4960 (177a), 4962 (55)

Lasser, T. 1316 (17a), 1441 (161/Ia), 1515 (20), 1528 (140), 1664 (3a)

Laughlin, R. M. 191 (199), 234 (147), 1105 (94c), 1208 (173), 1641 (71), 1676 (199), 1691 (140), 1834 (173), 1878 (72b), 1896 (94c), 1946 (177k), 2128 (177k), 2140 (173), 2281 (173), 2409 (72b), 2436 (173), 2543 (94c), 2560 (177k), 2586 (173), 2620 (149), 2678 (177k), 2696 (55), 2892 (147), 2957 (177k)

Lavastre, Bro. B. A. 305 (202a), 433 (168a), 553 (126k), 633 (126k), 824 (168a), 1152 (202a), 1209 (158), 1684 (202a), 1720 (199), 1763 (71), 1767 (131), 1961 (134a), 2117 (71), 2159 (168a)

Lawrence, A. E. 364 (17b)

Leavenworth, W. C. 493 (71), 586 (116 × 119)

Leavenworth, W. C. & H. Hoogstraal 1314 (200a), 1661 (55)

Lechler, W. 1704 (110d), 1932 (110f), 2435 (126e), 3244 (110d)

Ledingham, G. 4435 (130)

Ledoux, D. 2507 (73)

Ledoux & Torke 2185 (71)

Ledoux, Dunn & Wallace 2193 (199)

Le Gallo, C. 727 (129a), 791 (199)

Legname, P. R. & F. Vervoorst 62 (110a)

Lehmann, F. C. 271 (153c), BT372 (148), BT607 (138f), 822 (177a), 5360 (47), 5364 (138g), 6679 (17b), 7343 (8b), 7792 (177a), 7793 (153a), 8478 (13a)

Lehr, J. H. 1084 (80)

Lehto, E. 5685 (82), 12596 (90), 13190 (87), 15276 (87), 18625 (199), 19479 (199), 19503 (177d), 20292 (138e), 21563 (193c), 21574 (82), 21595 (73), 21749 (76b)

Leibold, – 3040 (107)

Leitão Filho, H. de H. 4 (169a), 5053 (164b)

Leite, E. 722 (124)

Lellinger, D. 1569 (202a)

Lemoine, M. 7834 (26a), 7835 (161/Ia)

Lems, K. 640309 (12)

Lent, R. W. 423 (148), 1329 (202a), 2569 (13a), 3135 (13a)

León, Bro. 419 (69), 668 (138f), 780 (129a), 1293 (71), 1538 (138f), 2798 (138f), 3733 (199), 4827 (131), 5439 (30), 5799 (30), 6016 (199), 6585 (131), 9571 (131), 10743 (131), 10950 (131), 11013 (170), 11037 (170), 11698 (138f), 12039 (199), 12402 (177j), 12529 (138f), 14707 (138f), 18658 (170)

León, Bro. & Bro. Hioram 4764 (71)

León, Bro. & Bros. Victorin & Clemente 20919 (135)

León, H. 477 (16)

León, J. 656 (27), 1224 (48b), 1740 (13a)

León, I. de 2 (16)

Leonard, E. C. 1386 (199), 2862 (141), 3200 (141), 3474 (168a), 3654 (131), 4264 (168a), 5081 (71), 5231 (159), 7148 (168a), 7181 (168a), 7193 (199), 7434 (168a), 7453 (141), 7700 (168a), 8025 (168a), 8799 (168a), 8891 (199),

8969 (168a), 9483 (148), 9488 (168a), 9752 (141), 9765 (168a), 9815 (168a), 9930 (168a), 9984 (168a), 11070 (168a), 11094 (199), 11121 (69), 11726 (168a), 12079 (134a), 12526 (141), 13704 (168a), 14261 (69), 14653 (199), 14659 (141), 14970 (131), 15088 (134a), 15101 (199), 15460 (168a), 15767 (199)

Lescure, – 275 (140), 517 (147), 625 (36), 652 (36), 707 (140), 717 (36)

Lewis, W. H. 120 (8a), 686 (16), 961 (138g), 3053 (8a), 3315 (16), 7233 (177j), 7450 (134e)

Lewton, F. L. 399 (16)

Lick, H. B-619 (58)

Liebmann, F. M. 21 (161/III), 32 (13a), 35 (11), 36 (11), 37 (16), 88 (178a), 160 (13a), 4988 (177a), 4991 (11), 5017 (196e), 5020 (178a)

Liesner, R. L. 19 (68), 3701 (148), 3841 (8b), 4036 (8b), 5770 (69), 5787 (8a), 6180 (161/Ia), 6464 (140), 6491 (69), 7359 (161/Ia), 7654 (10a), 7662 (129a), 7870 (13a), 9741 (17a)

Liesner, R. L. & J. D. Dwyer 1496 (16), 1518a (16), 1618 (16)

LIL 102325 (69)

Lima, A. 53/1365 (20), 53/1420 (26a), 58/2915 (52), 58/2924 (3/Ia), 58/2972 (43), 58/3115 (43), 60/3529 (37), 61/3651 (42e), 61/3736 (42a)

Lima, N. 126 (43), 153 (126a)

Lima, H. C. de 970 (43), 944 (43), 1080 (57), 54S (126d)

Linden, J. 735 (68), 1336 (42f), 1838 (131)

Lindberg, G. A. 396 (44a), 398 (136), 399 (70a), 725 (138f)

Lindeman, J. C. 4540 (43)

Lindeman, J. C. & J. H. Haas 990 (126a)

Lindheimer, F. J. 751 (84), III/381 (84)

Lindman, C. A. M. 2275 (157)

Lino, A. M. 27 (12)

Liogier, A. (Alain) 1025 (4), 9445 (119), 10190 (69), 10253 (69), 10316 (138g), 11585 (169b), 11782 (168a), 12044 (131), 12472 (168a), 13639 (168a), 13860 (134b), 14166 (131), 14221 (168a), 14368 (158), 14640 (168a), 14942 (168a), 14957 (168a), 15353 (131), 15470 (119), 15641 (141), 15707 (169a), 15715 (168a), 15746 (69), 16596 (168a), 16606 (168a), 16613 (168a), 16659 (168a), 16691 (202a), 16703 (134b), 16801 (168a), 16887 (168a), 16888 (134a), 16899 (168a), 16954 (168a), 16978 (134a), 17014 (71), 17035 (134b), 17491 (131), 17492 (69), 17498 (140), 17552 (158), 17560 (168a), 17568 (134a), 17640 (140), 17655 (134a), 17714 (168a), 17803 (126k), 17817 (168a), 17828 (134b), 18864 (69), 19310 (199), 19334 (119), 19475 (202a), 19688 (202a), 19855 (71), 20019 (69), 20161 (141), 20235 (131), 20236 (69), 20455 (168a), 20520 (168a), 20744 (148), 21172 (169b), 23295 (168a), 24791 (134a), 24967 (159), 25440 (134a), 25985 (159), 26070 (134a), 29837 (28), 30243 (71)

Lisboa, R. L. 13 (147), 14 (69), 15 (26a), 4749 (36)

Little, E. L. 6210 (14d), 6426 (200b), 6608 (200b), 6690 (200b), 7183 (8b), 7596 (147), 8334 (13b), 8371 (3a), 9221 (154), 9581 (17a), 21253 (154), 21748 (202a)

Little, E. L., A. Ortega & F. Vivar 702 (161/Ia)

Lloyd, F. E. 408 (140)

Lobo, M. G. A. 202 (3a)

Löfgren, A. 257 (62), 328 (62), 368 (63a), 548 (44b), 552 (43), 640 (138f), 669 (37), 936 (44b), 1006 (3f), 13212 (120)

Long, L. E. 19 (13a), 194 (13a), 214 (13a)

Long, R. W. & K. J. Norstog 3414 (131)

Looser, G. 2182 (99)

López M., A. & A. Sagastegui A. 3359 (42h), 4109 (177f), 5199 (198), 7979 (129b)

López Guillén, J. 112 (110d), 381 (110f)

López Figueiras, M. 208 (30), 8097 (17b)

López Figueiras, M. & H. Rodríguez 9058 (126k)

López-Sánchez, A. 1 (177r), 2 (177r)

Lorentz, P. G. 246 (126f), 250 (130), 658 (130), 739 (139), 893 (139)

Lorentz, P. G. & Hieronymus 189 (110a), 247 (126f), 1181 (63a)

Lourteig, A. 842 (181), 3025 (27)

Lüdenwaldt, H. 105 (44a), 444 (120), 4171 (136)

Luetzelburg, P. von 45 (202b), 109 (45), 272 (43), 345 (157), 738 (150), 999 (161/IIa), 1276 (43), 1534 (59 affin), 1915 (150), 5915 (157), 12265 (59b), 12725 (42a), 20066 (124), 21884 (26a), 25807 (46a), 26137 (42c), 26748 (44b), 26766 (165)

Lugo S., H. 557 (116), 1501 (17a), 1592 (8b), 1630 (8b), 2017 (25a), 2749 (8b), 2768 (25a), 3298 (8b), 3357 (17a), 3606 (8b), 3952 (148), 3988 (148), 4054 (161/Ia), 4132 (161/Ia), 4265 (148), 4323 (148), 4496 (25a), 4622 (8b), 5000 (8b), 5011 (8b), 5074 (8b)

Luna, A. 303 (141)

Luna Peixoto, A. 402 (3/Ib), 846 (43)

Lund, P. W. 180 (63b), 181 (136), 185 (43), 186 (57), 190 (66)

Lundell, C. L. 14 (27), 112 (16), 130 (27), 311 (27), 874 (16), 945 (140), 1027 (201), 1180 (16), 1361 (196a), 2543 (16), 2954 (27), 4369 (16), 4383 (201), 5345 (85), 8076 (199)

Luschnath, B. 161 (39a), 170 (44a), 805 (44a), 819 (44a), 824 (3b), 1471 (16)

Luteyn, J. L. 1471 (16), 4783 (147)

Luteyn, J. L. & M. Lebrón-Luteyn 6069 (13a), 6070 (138g), 6071 (126k), 6197 (115), 6711 (116), 7240 (126i), 7340 (69), 7344 (202a), 7609 (138g), 7610 (170), 7642 (126k), 7643 (170), 7682 (166)

Lutz, A. 52 (39a)

Lutz, F. E. 4040B (138e), 4035 (80)

Lyle Wynd, F. & C. H. Mueller 113 (73), 291 (84), 255 (84)

Lyonnet, E. 1182 (178a), 1236 (138f), 1713 (63b), 1718 (193a), 2253 (78)

Maas, P. J. M. 299 (69), 330 (140), 3418 (124)

Maas, P. J. M. & Westra 3917 (161/Ia), 4354 (20)

Maasola, J. 43 (16)

Macbride, J. F. 384 (174), 1598 (116), 1986 (116), 2750 (16), 2860 (119), 3194 (114), 3559 (114), 4186 (23), 4825 (155)
Macbride, J. F. & Featherstone 1183 (110e), 2526 (110e), 5477 (126e)
MacBryde, B. 314 (174), 422 (69), 594 (174), 623 (129b), 1326 (17a), 1504 (8b)
McConnell, – & J. J. Quelch 23 (8a)
McCorkle, J. S. C75 (68)
McDaniel, S. 1359 (69), 2614 (17a), 4990 (16), 5052 (82), 16124 (161/Ia), 16143 (17a), 17354 (161/Ia), 17610 (8b)
MacDaniels, L. H. 872 (11)
MacDougall, T. 44 (178a), 45 (146), 46 (178a), 324 (68), 534S (71), 535S (69), 567S (196e), 581S (178a), 5665 (178a), H-138 (149), H256 (178a), H369 (199), H415 (196e), H424 (200a)
Macedo, A. 169 (63a), 643 (43), 697 (44a), 1899 (175), 4485 (42b), 5306 (46a),
MacFadden, F. A. 13611 (90)
McFarlin, J. B. 5954 (69), 9619 (69)
Machado, O. 617 (12)
Machado de Campos, S. S. 206 (43)
McKee, H. S. 10647 (28), 10959 (78)
McPherson, G. D. 746 (55), 853 (138f)
Macrae, J. 304 (109b)
Macurdy, A. C. 1018 (25a)
McVaugh, R. 11845 (11), 20864 (11), 21334 (93), 22094 (177c), 22867 (177a), 25011 (177c), 25085 (177c), 25249 (177c), 25561 (11), 26319 (11)
McVaugh, R. & W. N. Koelz 529 (26a), 1163 (92), 1559 (177a), 1578 (196c), 1760 (177c)
Madison, M. T. 5158 (14d), 10440 (25a)
Magallanes, A. S. 30 (77), 32 (193a), 193 (193a), 197 (55), 198 (146), 205 (69)
Maguire, B. 10796 (80), 10890 (80), 11220 (80), 11522 (87), 11793 (80), 11915 (87), 12030 (80), 22867 (26a), 23383 (20), 23620 (26a), 23893 (147), 23893a (140), 24911 (161/Ia), 27310 (3a), 27320 (148), 29033 (3a), 29291 (8b), 29295 (148), 29765 (3a), 30815 (140), 31028 (53a), 31593 (3a), 32668 (27), 34749 (26a), 35031 (3a), 35812 (161/Ia), 35998 (8a), 36132 (34), 39428A (115), 40027 (3a), 40228 (161/Ia), 40312 (8a), 40347 (161/Ia), 40356 (140), 40465 (161/Ia), 40466 (161/Ia), 40508 (26a), 40831 (26a), 44086 (20), 44147 (3a), 44448 (110d), 44480 (110b), 44599 (104a), 45975 (8a), 46251 (3a), 46955 (9), 47090 (26a), 47956A (20), 49007 (43), 53916 (148), 53919 (20), 54145 (148), 56096 (3d), 56260 (43), 56525 (3a), 56593 (3a), 56810 (3a), 61359 (55), 61878 (177a)
Mahler, W. F. 5757 (85), 5934 (138f), 5936 (177a), 5959 (177a)
Makrinius, E. 689 (177c)
Malme, G. O. A. 682 (138a), 1026 (157), 1093 (29), 1253 (124), 1273 (138a), 1303 (138a), 1682 (66), 1682a (66)
Maltby, F. S. 49 (177a)
Malvárez, M. R. 813 (118)
Mandon, G. 751 (110c), 751bis (107), 1506 (95a)
Marcelino, L. 25 (16)
Marinho, L. R. 213 (17a)
Marsh, E. G. 449 (84), 733 (74), 782 (73), 1368 (74), 2149 (84), 2269 (73),
Marsh, E. M. 2060 (194)
Marshall, S. A. & D. A. Neill 6540 (148), 7070 (69)
Martin, R. T. 102 (26a), 1095a (26a), 1205 (26a), 1287 (134e), 1303 (8b)
Martin, R. T. & C. A. Lau-Cam 1037 (25a), 1255 (17a), 1256 (8b)
Martin, R. T. & T. Plowman 109 (126i)
Martinelli, G. 7 (40), 312 (57), 162 (137c), 171 (161/IId), 4536 (61), 5489 (59a), 6656 (59d), 6665 (44a)
Martínez-Calderón, G. M. 178 (140), 179 (69), 182 (140), 203 (161/III), 261 (202a), 320 (177p), 475 (16), 534 (147), 644 (13a), 658 (147), 675 (138f), 700 (138f), 831 (177p), 1017 (69), 1022 (140), 1144 (177a), 1187 (126m), 1311 (71), 1333 (63b), 1417 (13a), 1557 (126m), 1559 (69), 1576 (177a), 1635 (147), 1795 (126m), 1881 (138f), 2058 (69), 3019 (140), 3080 (161/III)
Martínez Ojeda, E. 202 (199)
Martins, H. F. 351 (61)
Martius, K. F. P. von 128 (39a), 129 (126d), 141 (161/IIa), 220 (60), 323 (3a), 348 (3f), 434 (161/IIa), 585 (126a), 715 (126a), 716 (44a and 57), 717 (39b), 718 (126a), 719 (44a), 727 (137c), 805 (44a), 862 (126a), 936 (42a), 1072 (161/IIa), 1073 (3e), 1074 (43), 1075 (42b), 2349 (150), 2524 (63b), 3183 (17a)
Maruñak, V. 106 (124)
Mason, H. L. 1762 (177a)
Mathews, A. 738 (110f), 740 (129a), 876 (110e), 1018 (119), 1589 (20), 1591 (22), 3276 (108)
Mathewson, K. 49 (177g)
Mathias, M. 49 (177g), 3902 (26a), 5035 (20 & 26a), 5556 (17a)
Mattos, A. 504 (42a)
Mattos, J. M. & N. F. 8568 (63a), 14481 (3/IIf), 16304 (43)
Mattos Silva, L. A. et al. 254 (59d), 272 (52), 316 (161/IIa), 329 (162), 334 (161/IIa), 343 (12), 350 (162), 389 (39a), 413 (138g), 418 (161/Ib)
Matuda, E. 29 (16), 68 (177a), 502 (13a), 1456 (177a), 1676 (119), 1708 (138f), 1911 (177k), 2108 (126k), 2266 (199), 2425 (116), 3035 (27), 3593 (13a), 4049 (177k), 16300 (177a), 16550 (16), 17081 (16), 17114 (177a), 17128 (16), 17207 (68), 17287 (147), 17310 (177a), 17337 (177a), 17412 (16), 18415 (138f), 18506 (16), 19687 (116), 26096 (193a), 27422 (178a), 28117 (116), 29266 (72a), 30593 (177r), 37408 (199)
Maxon, W. R. 2116 (199), 3984 (199), 4704 (27), 7254 (140), 10263 (29)
Maxon, W. R. & A. D. Harvey 6611 (16), 7313 (199)
Maxon, W. R. & R. Hay 3762 (177a)
Maxon, W. R. & E. P. Killip 380 (129a), 1662 (131)

Maxwell, R. H. 260 (11), 1970 (27)
Mearns, E. A. 229 (87), 593 (193c), 715 (138e)
Medrano, F. G. 8135 (85), 8140 (85)
Mélinon, – 25 (36), 252 (161/Ia), 302 (20), 435 (20)
Mell, C. D. 3 (16), 6 (199), 9 (161/Ia), 12 (161/Ia), 2040 (27), 2131 (55), 2134 (177b), 2197 (177a), 2207 (199)
Mello, F. 30 (8b)
Mello Barreto, M. 1423 (42a), 5875 (44a), 5877 (44a), 5880 (43), 5884 (43), 5886 (43), 5887 (43), 5892 (39b), 5910 (42b), 10283 (43), 10580 (44a)
Mello Filho, L. E. de 60 (120), 145 (3f), 242 (43)
Mendes Magalhães, G. 1451 (42b), 5113 (175), 5438 (120), 9656 (161/Ia), 10476 (43), 15207 (44a), 15263 (59c), 15290 (163), 15300 (44a), 15323 (44a), 14953 (40), 15603 (40), 17302 (40), 18351 (56a), 18918 (42a), 18960 (57), 18972 (175), 18978 (43), 18981 (43), 18982 (60), 18983 (43), 18987 (59a), 18991 (43), 18998 (60), 19129 (43), 19557 (12), 19599 (12), 19602 (161/Ib), 19615 (12), 19766 (161/Ib), 19900 (126a)
Mercano-Berti, L. 200 (26a)
Messer, E. 72/127 (178a)
Messing, H. M-113 (69)
Metcalfe, O. B. 941 (84), 1275 (87), 30013 (177a)
Metz, Sr. Mary 131 (84), 132 (84)
Meyer, F. G. 9703 (107)
Meyer, T. 3350 (138b), 3558 (126f), 3815 (118), 4673 (122), 5991 (67), 8384 (202a), 8743 (118), 9887 (138b), 9959 (118), 13076 (118)
Meyer, W. C. 118 (27)
Mexia, Y. 550 (161/IIa), 762 (11), 916 (177a), 923 (200a), 1026 (138f), 1274 (177c), 1523 (26a), 1675 (11 & 26a), 2662 (119), 4176 (42a), 4218 (161/IIa), 4398 (42a), 4405 (161/IIa), 4437 (136), 4441 (161/IIa), 4515 (161/IIa), 4559 (42a), 4671 (126a), 4701 (42a), 5533 (3f), 5619 (175), 5823 (57), 5950 (26a), 5950a (36), 5993 (147), 6074 (3a), 6214 (25a), 6472 (161/Ia), 6501 (8b), 6943 (17a), 6949 (25a), 7258 (8b), 7263 (8b), 7686 (154), 8953 (177e), 9000 (177k), 9285 (202a)
MG 330 (42a), 16338 (64), 16729 (20), 17285 (27), 31143 (202b), 31860 (126g)
Mick, G. & K. Roe 335 (140)
Mileski, E. 192 (53a), 262 (36)
Mille, A. 30 (147)
Miller, F. 196 (124)
Miller, G. S. 1899 (16), 1997 (16)
Miller, O. O. & J. R. Johnston 56 (71), 61 (177a), 62 (8a), 201 (199), 228 (159)
Millspaugh, C. F. 186 (177h), 236 (69), 503 (129a), 1481 (196a), 1509 (196a), 1658 (177h), 1690 (177h), 2285 (134e), 2490 (177j), 2497 (146), 2499 (134e), 9019 (129a), 9156 (134d)
Mimura, I. 27 (43), 58 (43), 236 (43), 320 (43), 381 (43)
Miranda, A. 110 (73)
Miranda, F. 1014 (199), 4171 (202a), 4660 (193a), 5491 (11), 5797 (173), 7189 (201), 7905 (173), 7982 (201), 8871 (161/III), 9110 (196c)
Mocquerys – 873 (13a)
Moldenke, H. N. 156 (69), 446a (129a), 488 (2), 515 (134e), 545a (134e), 1001 (69), 3662 (134e), 30677 (90)
Molina, J. A. 352 (16)
Molina R., A. 19 (27), 563 (13a), 659 (16a), 721 (11), 1086 (11 & 13a), 1926 (27), 1984 (27), 2463 (27), 2490 (27), 2715 (178a), 3494 (13a), 3631 (119), 8636 (199), 10143 (119), 10409 (147), 11016 (199), 11103 (71), 11896 (177a), 12066 (119), 12344 (199), 12879 (149), 12898 (140), 12903 (16), 12940 (138f), 13002 (149), 13010 (202a), 13060 (149), 13210 (199), 14125 (119), 14165 (119), 14210 (199), 14243 (177k), 14348 (199), 14682 (119), 14877 (27), 15144 (27), 15323 (177k), 15382 (177k), 15797 (16), 15992 (177g), 16553 (116), 16589 (94b), 16660 (177k), 17126 (68), 18118 (26a), 18182 (202a), 18734 (177k), 20192 (177a), 20540 (177a), 20791 (16), 21285 (116), 21421 (94b), 21448 (149), 21506 (177k), 21597 (147), 21884 (94a), 21940 (140), 22081 (119), 22265 (119), 22413 (140), 22753 (69), 22769 (55), 22901 (119), 23088 (149), 23176 (55), 24591 (149), 24616 (71), 24670 (68), 24684 (138f), 24699 (177a), 24745 (16), 24807 (177k), 25115 (126k), 25330 (16), 25674 (147), 25817 (138f), 25924 (147), 25968 (199), 26030 (199), 26220 (71), 27153 (177a)
Montaldo, P. 3133 (171), 3321 (170), 3331 (199), 3954 (177a), 3978 (171)
Monteiro da Costa, R. C. 219 (26a)
Montes, J. E. 1 (138a), 138 (126c), 711 (63a), 1982 (69), 14776 (63a), 14875 (138a), 14928 (140), 27495 (69), 27545 (140), 27546 (69)
Montes, J. S. 1134 (63a), 2924 (63a)
Moore, A. D. 1119 (71), 1827 (177a)
Moore, H. E., Jr. 2655 (196b), 5004 (11), 5763 (177a), 5239 (152)
Moore, H. E., Jr. & C. E. Wood 3673 (196b), 4212 (196b), 4478 (94e)
Moore, J. A. & J. Steyermark 3301 (87), 3416 (73), 3606 (84)
Moore, S. L. 36 (60), 907 (157), 919 (63b), 967 (148), 968 (69)
Mora, L. E. 2392 (177a), 2492 (177a)
Moraes Vasconcellos, J. de 340 (446)
Moran, R. 8197 (101), 8967 (81), 9426 (81), 9975 (177a), 9985 (26a), 10229 (101), 17060 (101), 28168 (81)
Moreira, A. X. 132 (104a)
Moreira Filho, H. 2 (42a), 94 (136), 95 (137a)
Morel, I. 452 (118), 2868 (118), 3505 (118)
Moreno, P. P. 38 (199), 291 (199), 330 (27), 343 (196a), 475 (146), 490 (199), 491 (196a), 697 (146), 801 (199)
Mori, S. A. 8096 (26a), 8251 (161/Ia), 9290 (140), 9428 (202b), 9432 (3a), 9468 (57), 9477 (202b), 9497 (52), 9589 (161/IIa), 9708 (61), 9729 (12), 9790 (3b), 9791 (39a), 9935 (59d), 9962 (42e), 9963 (202b), 9964 (126a), 9977 (12), 9999 (42a), 10043 (42c), 10062 (42c), 10088 (140), 10091 (161/Ib), 10403 (126a), 10576 (61),

10964 (44a), 11181 (52), 11773 (44a), 11783 (18), 11830 (164a), 11831 (164a), 11871 (39a), 12193 (42d), 12286 (59b), 12567 (59d), 13271 (164a), 13317 (59d), 13375 (59b), 13445 (71), 13446 (52), 13461 (59a), 13488 (44b), 13554 (43)
Mori, S. A. & J. Kallunki 1896 (16), 2516 (13a), 4030 (8a)
Mori, S. A. & G. T. Prance 9143 (53a)
Mori, S. A., G. T. Prance & S. R. Hill 19220 (147)
Moritz, J. W. K. 561 (171), 1788 (29)
Moroles Ruano, J. 1319 (177a)
Morong, T. 41 (138b), 82 (138b), 175 (69), 350 (126o), 426 (67), 613 (66), 643 (148), 665 (63a), 776 (126o), 1015 (118), 1095 (126o)
Morrow, C. F. 64 (28)
Morton, C. V. 4082 (30), 5090 (8a), 5579 (8a)
Morton, C. V. & E. Makrinius 2412 (177c), 2628 (177c), 2639 (177c)
Morton, G. H. 4788 (69)
Mosén, H. 1238 (136), 1240 (136), 1242 (138f), 1243 (70a), 1244 (66), 1964 (136), 3371 (161/IIa), 4098 (66), 4099 (124)
Mostacero L., J. 606 (116)
Mowbray, R. N. 6982 (8b), 6995 (8b), 17337 (17a), 69042 (17a), 69415 (17a), 70361 (17a), 70366 (7)
Moyne, Lord 9 (177a)
Mueller, C. H. 3631 (177d), 8033 (73), 8034 (82)
Mulford, A. I. 84 (87), 197 (87), 200 (73)
Muller, B. L. 58 (118)
Müller, F. 83 (140), 97 (140 & 141), 150 (72b), 221 (161/IIa), 455 (72b), 466 (138f), 545 (138f)
Munz, P. A. 1208 (138e), 17401 (90)
Munz, P. A. & D. Keck 4862 (80)
Mutis, J. C. 270 (8b), 2053 (8b), 2055 (17b), 2387 (27), 2408 (17b), 2846 (17b), 3597 (16), 4142 (17b), 4199 (17b), 4295 (16), 4725 (16)
Myers, J. G. 5690 (161/Ia)

Narváez Montes, M. & A. E. Salazar 38 (177c), 481 (11), 546 (177c), 710 (11)
Narváez S., E. 462 (68 & 71), 1255 (177a), 1947 (177a), 2526 (177a), 2526a (177a)
Nascimento, O. C. 24 (148), 581 (3a)
Nash, G. V. 82 (140), 649 (202a), 984 (199), 1005 (168a), 2230 (69)
Nash, G. V. & N. Taylor 1171 (168c), 1387 (168c), 1435 (71), 1516 (199), 1540 (71), 1603 (168a), 1644 (168a), 3837 (129a)
Nealley, G. C. 49 (80), 280 (138e)
Nee, M. 6967 (14a), 7165 (16), 7418 (16), 7450 (16), 7555 (16), 8223 (16), 8862 (8a), 8908 (149), 8996 (8a), 9064 (8a)
Nee, M. & S. A. Mori 3491 (177a), 3523 (68)
Neill, D. A. 567 (13a), 598 (55), 1089 (199), 1140 (177a), 1181 (149), 1704 (199), 2160 (140), 2305 (119), 2439 (71), 2649 (177a), 2678 (69), 2679 (140), 2722 (177a), 2723 (202a), 2791 (16), 2802 (138f), 2819 (177a), 2889 (147), 3046 (140), 3131 (177a)
Nelson, E. W. 3476 (72b), 3481a (94c), 4194 (177a), 4196 (177a), 4513 (193b), 4595 (83a)
Nelson, E. W. & E. A. Goldman 7208 (81), 7238 (144), 7341 (199), 7418 (146)
Nelson, G. 36 (177a), 42 (177a), 5954 (173)
Neves Armond, F. A. 45 (12)
Nichols, C. E. 1060 (177a), 1132 (71), 1470 (177a), 1843 (177a)
Niles, W. E. 940 (90)
Norris, D. H. 20260 (26a)
Northrop, J. & A. 14 (146), 55 (177j), 103 (134e), 105 (141), 123 (131), 422 (131)
Novaes, C. 223 (42b)
Novelo, A. 81 (126j)
Nowicke, J. W. 3601 (16)

Oberwinkler, F. 12551 (126k), 12726 (157), 13429 (115), 14234 (115)
Occhioni, P. 5482 (175), 5657 (43), 8092 (136), 8356 (12), 8360 (161/IIa), 8697 (104b), 8711 (111)
Ochoa G., S. 932 (196b)
O'Donnell, C. A. 2016 (181), 2041 (99), 2174 (99), 2493 (138b), 2495 (140), 2690 (140), 2934 (126f), 3110 (140), 3111 (69), 3204 (118)
Œllgard, B. & H. Balslev 9073 (8b)
Œrsted, A. S. 29 (16), 34 (28), 37 (149), 38 (149), 45 (16)
Oldeman, R. A. A. 1385 (26a), 3357 (129a), 4540 (148), B1249 (36), B1441A (20), B1591 (148), B1708 (26a), B2133 (35), B2218 (35), B3602 (147), B3984 (69), B4024 (69), B4083 (36), B4107 (26a), B4186 (26a), B4339 (35), B5241 (35)
Oldeman, R. A. A. & Arévalo 92 (17a)
Oldenburger, F. H. F. ON/257 (161/Ia)
Olea, D. 133 (130), 197 (110a)
Olivcira, E. 4288 (140), 4322 (20), 4553 (20), 4596 (3a), 4723 (20), 4745 (3a), 6129 (26a), 6137 (26a), 6295 (20), 6296 (20), 6513 (36), 6741 (3a)
Oliver, R. L. 909 (55), 1096 (73)
O'Neill, H. 8610 (27)
Opler, P. A. 440 (69), 1565 (16)
Orbigny, A. d' 893 (63c)
Orcutt, C. R. 653 (74)
Orozco, J. M. 13 (16), 172 (13a)
Ortega, J. G. 4026 (11), 4094 (177a), 4423 (177a & 177c), 4868 (11), 4871 (177a), 4992 (26a), 5599 (177a), 5834 (177a), 5915 (177a), 5916 (177a), 5975 (177a), 6070 (72b), 6172 (11), 6372 (177a), 6416 (200a), 7208 (200a), 6464 (71), 6496 (11), 6724 (177a), 6749 (177a), 7020 (177a), 7457 (11)
Ortiz, R. T. 394 (16), 1419 (126j)
Otero, J. I. 2 (129a), 4 (28)
Otto, E. 69 (140), 322 (140) 1108 (27)
Ouesset, – 25 (140), 131 (138b)
Ownbey, G. B. 1866 (83a), 1887 (177a)
Ownbey, M. & G. 2996 (90)

Pabst, G. F. J. 4078 (140), 4079 (43), 4178 (43), 4435 (42a), 5519 (161/IIa), 5537 (126a), 6815 (12), 6823 (42a), 6829 (63b), 6849 (136), 7109 (43), 7153 (124), 9064 (61), 9236 (120), 9416 (126b)

Pabst, G. F. J. & E. Pereira 3181/4016 (44a), 2382/3218 (43), 2653/3218 (43), 2653/3489 (39b), 4840/5165 (43), 6236/6409 (124), 6720/6894 (161/IIb)
Pachano, A. 101 (116)
Padilla, I. 1070 (8b)
Padillo, S. A. 407 (177a)
Paiva, F. 1009 (165)
Palacios-Balegno-Cuezzo 3804 (44a)
Palmer, E. 5 (71), 27 (140), 36 (193b), 44 (71), 52 (81), 72 (177c), 87 (140), 93 (177a), 116 (90), 119 (193c), 181 (177d), 190 (140), 201 (148), 210 (199), 232 (73), 260 (177e), 271 (83a), 273 (88), 274 (87), 276 (73), 281 (76a), 283 (140), 299 (199), 323 (199), 363 (199), 393 (177d), 455 (82), 461 (11), 478 (200a), 518 (193d), 538 (68), 599 (75), 632 (126k), 738 (69), 906 (71), 983 (200a), 1052 (11 & 126m), 1058 (71), 1161 (126k), 1350 (177c), 1423 (199), 1495 (177a), 1509 (138e), 1773 (126m), 1821 (71), 2132 (82), 11515 (84)
Palmer, E. J. 11515 (84), 11807 (88), 12408 (88), 12413 (84), 12969 (73), 24449 (73), 34015a (88), 34194 (87)
Palmer, W. 84 (69), 1289 (177c)
Palmer, W. & O. Bryant 87 (8a)
Paray, L. 137 (178a)
Parish, S. B. & W. F. 1285 (90), 1409 (90), 2981 (90)
Parry, C. C. 254a (82), 296 (193c)
Parry, C. C. & E. Palmer 204 (138e), 206 (85), 207 (75), 208 (193b), 210 (88)
Parker, K. 40-F (110a)
Parks, C. R. 265 (138f), 296 (69), 298 (177a)
Pastore, F. 2067 (130)
Paul, Bro. 177 (16)
Pavón Herb. 14/6 (138g), 14/9 (109a), 14/10 (95a), 14/14 (102), 14/15 (114), 14/16 (69), 14/17 (129a), 14/18 (10b), 14/21 (153c), 14/22 (119), 14/23 (114), 14/26 (200b), 14/27 (200b), 14/27 (153c), 14/30 (25a), 14/31 (147)
Peck, M. E. 36 (27), 263 (16)
Peckolt, T. 139 (42a), 218 (42a), 592 (42a)
Pedersen, T. M. 1719 (128), 1923 (139), 3280 (63a), 3700 (139), 3734 (63a), 8195 (139), 8599 (160), 9646 (63a), 10074 (126o), 10165 (126o), 11056 (3c), 12045 (43), 12082 (43), 12089 (3f), 12245 (43)
Peebles, R. H. 220 (90), 1297 (80), 5368 (87)
Peebles, R. H. & G. J. Harrison 2260 (138c)
Penland, C. W. 127 (25a)
Pennell, F. W. 1585 (13b), 1895 (42f), 2345 (116), 3278 (47), 3346 (138g), 3714 (8b), 3774 (27), 3816 (140), 3863 (69), 3908 (126k), 3938 (129a), 3974 (177a), 4052 (69), 4075 (129a), 4102 (27), 4666 (14a), 4723 (129a), 4726 (177a), 4727 (129a), 4749 (8a), 5638 (202a), 7137 (116), 10429 (84), 14229 (96), 14278 (110d), 14424 (97), 14748 (174), 16765 (73), 17569 (193b)
Pennell, F. W., E. P. Killip & T. E. Hazen 8522 (126k), 8613 (17b)
Pennell, F. W. & H. H. Rusby 42 (147)
Pennell, M. 200 (178a), 362 (55)
Pennell, M., Dunn & Dziekanowski 221 (199)
Pennington, T. D. 1481 (8b), 9141 (161/III)
Perazzo Barbosa, V. 22 (176), 133 (37), 139 (37), 150 (44b), 178 (176)
Peredo, G. 46 (69), 411 (140)
Pereira, C. 330 (60)
Pereira, E. 349 (66), 1036 (150), 1248 (12), 1249 (26a), 1416 (56a), 1445 (43), 1813 (3a), 2045 (43), 3227 (36), 3350 (26a), 3391 (161/Ia), 3410 (161/Ia), 3821 (42a), 4351 (120), 4475 (12), 4501 (161/IIa), 4535 (42a), 4537 (42a), 4943 (147), 5054 (46a), 5564 (44a), 7000 (12), 7261 (136), 7270 (43), 7292 (43), 7295 (3f), 7309 (60), 7547 (126a), 8686 (176), 8692 (138a), 8937 (136), 9053 (43), 9648 (60), 10037 (176)
Pereira, E. & A. P. Duarte 10037/9924 (176)
Pereira, E., W. Egler & G. Barroso 349 (63b)
Pereira, E. & G. F. J. Pabst 8534/7809 (138a), 8616/7891 (138a), 8631/7906 (63a), 8633/7908 (127), 9488/8377 (202b), 9528/8417 (42b), 9534/8423 (52), 9658/8547 (59d), 9666/8555 (42d), 9724/8613 (42e), 9752/8641 (164c)
Pérez, S. 46 (69), 47 (71), 53 (138e), 54 (140)
Pérez-Moreau, H. & B. Piccinini 3318 (181)
Periani, – 21 (104b), 64 (104b)
Persaud, R. 15 (8a), 55 (8a), 97 (26a), 194 (9)
Pflanz, – 947 (180)
Phelps, K. D. & C. B. Hitchcock 304 (161/Ia)
Philcox, D. 3979 (43)
Philcox, D. & A. Ferreira 3798 (140), 3838 (60), 3841 (3f), 3858 (140), 4408 (43), 4310 (43)
Philcox, D. & Onishi 4310 (43)
Philippi, R. A. 218 (109a)
Philipson, W. R. 784 (29)
Philipson, W. R. & J. M. Idrobo 1307 (8b), 1326 (63b), 1444 (13b)
Pickel, D. B. 83 (176), 1067 (176), 2485 (37), 2642 (126g), 3057 (44b), 3280 (26a), 3693 (69), 3711 (44b), 3863 (41), 3911 (42d)
Pierotti, S. A. 8 (138b), 35 (202a), 119 (138b), 131 (138b), 316 (140), 1221 (69), 1396 (118), 1397 (140), 4174 (118), 6521 (126f), 7082 (118)
Pimenta Velloso, H. 959 (44a)
Pinheiro, R. S. 32 (161/IIa), 42 (3b), 89 (42d), 90 (164a), 92 (26a), 102 (42b), 1059 (202b), 1106 (161/IIa), 1109 (42b), 1116 (148), 1155 (26a), 1189 (12), 1217 (138g), 1247 (161/Ib), 1445 (164c), 1800 (162), 2148 (126a)
Pinkava, D. J. 4212 (194), 12010 (90), 12209 (101), 12306 (101), 14844 (138e)
Pinkley, H. 154 (17a), 177 (8b), 341 (25a)
Pinkus, R. 94 (8a), 1159 (147)
Piper, C. V. 5131 (8a), 5132 (27), 5138 (27), 5141 (27), 5142 (16), 5357 (16)
Pipoly, J. J. 1623 (177a), 1625 (199)
Pires, J. M. 75/15 (161/Ia), 75/21 (147), 75/24 (140), 3863 (58), 4533 (161/Ia), 8190 (147), 9002 (126a), 9312 (161/Ia), 9313 (126a), 9713 (60), 9936 (161/Ia), 10006 (38), 10029 (3a), 12554 (3a), 15685 (17a), 16096 (3d), 16109 (46a), 50374 (161/Ia), 50835 (26a), 51046 (3a), 51500 (36), 51958 (147), 57895 (161/Ia), 57920a (43), 57949 (12), 58014 (43), 58033 (43), 58126 (70b), 58169 (138g)

Pires, J. M. & G. A. Black 1014 (26a), 1177 (38), 1533 (161/Ia), 1635a (46a), 2429 (3d), 2508 (138g), 2824 (3f), 10509 (20)

Pires, J. M. & P. Cavalcante 51981 (27), 52155 (3a), 52227 (27), 52280 (69), 52288 (161/Ia), 52453 (18g), 52532 (36), 52578 (161/Ia)

Pires, J. M. & N. T. Silva 4762 (36), 10864 (26a)

Pires, O. 111 (38)

Pittier, E. 14 (202a)

Pittier, H. F. 22 (10a), 150 (10a), 151 (10a), 156 (177a), 264 (10a), 465 (16), 1724 (27), 1763 (177a), 1890 (177n), 2009 (177k), 2096 (27), 1226 (69), 2312 (16), 2319 (8a), 2343 (177a), 2363 (149), 2367 (63b), 2550 (8a), 2555 (69), 2688 (27), 3345 (10a), 3942 (26a), 4253 (14b), 4405 (16), 4491 (48a), 4517 (16), 4746 (69), 5146 (8a), 5147 (26a), 5406 (167), 5407 (167), 5776 (177a), 5909 (177a), 6060 (171), 6579 (14a), 7411 (10a), 7426 (10a), 7434 (177a), 7501 (10a), 7560 (170), 7710 (177a), 7767 (10a), 7771 (10a), 8334 (13a), 8391 (171), 8678a (171), 9113 (27), 9134 (138g), 9382 (27), 9702 (202a), 9801 (10a), 9876 (10a), 9900 (10a), 10316 (199), 10402 (10a), 10489 (159), 10500 (199), 10529 (15), 10569 (167), 10705 (199), 10856 (27), 11239 (29), 11311 (171), 11335 (171), 11692 (171), 11943 (8a), 12023 (157), 12026 (170), 12028 (147), 12247 (10a), 12251 (10a), 12452 (10a), 12548 (126k), 12561 (10a), 12641 (129a), 12773 (27), 12890 (115), 13093 (42g), 13098 (172), 13159 (129a), 13167 (115), 13363 (177a), 13369 (10a), 13390 (10a), 13549 (69), 13551 (149), 14640 (42g), 14709 (8a), 14738 (177a), 15307 (171), 15391 (171), 15441 (17a), 15494 (202a), 15608 (10a)

Plowman, T. 1920 (8a), 2104 (17a), 3515 (129a), 4204 (8b), 4629 (95a), 4863 (21), 5482 (200b), 5947 (25a), 6390 (147), 6406 (8b), 6485 (26a), 6982 (20), 8294 (69), 8678 (3/Ia), 8735 (20), 8901 (140), 9867 (3/Ia)

Pohl, J. E. 120 (157), 350 (3f), 1114 (136), 1120 (136), 1338 (43), 1559 (56a), 1702 (66), 1724 (63b), 1777 (161/Ia), 1794 (3f), 1804 (46a), 1821 (60), 1852 (56b), 1904 (161/Ia), 5649 (120), 5651 (120), 5702 (126a)

Pohl, R. W. 11186 (138f), 11222 (202a)

Pollard, C. L. 291 (30)

Pollard, C. L. & G. N. Collins 72 (134e), 243 (134e)

Pollard, C. L. & W. Palmer 277 (71), 323 (126k), 336 (126k)

Pollard, G. R. M. 62 (27)

Poole, J. M. 1735 (140), 3078 (26a)

Poortmann, H. 122 (154), 351 (17a)

Pöppig, E. F. 129 (109b), 743 (99), 1950 (17a & 25a), 2429 (20), 2460 (3a), 2817 (69)

Porter, D. M. 4122 (27)

Prance, G. T. 2996 (3a), 3286 (3a), 3962 (3a), 3993 (64), 4088 (64), 4106 (161/Ia), 4127 (3a), 4272 (161/Ia), 4330 (27), 4544 (3a), 4896 (38), 5192 (38), 5209 (3a), 5254 (147), 5303 (126p), 5346 (26a), 5347 (161/Ia), 5395 (3a), 5451 (3a), 5502 (26a), 5582 (3a), 5759 (3a), 5801 (3a), 5851 (161/Ia), 5911 (161/Ia), 5989 (161/Ia), 6015 (3a), 6111 (161/Ia), 6208 (147), 6473 (3a), 6497 (46a), 6565 (3a), 6708 (38), 6931 (161/IIa), 6994 (3a), 7312 (3a), 7322 (3a), 9516 (69), 9748 (161/Ia), 9787 (20), 10005 (26a), 10335 (147), 10733 (126q), 10905 (17a), 11105 (26a), 11185 (26a), 12221 (20), 12283 (17a), 12521 (69), 12814 (3a), 12943 (161/Ia), 13018 (20), 13217 (126e), 13284 (161/Ia), 13394 (147), 13490 (126r), 13491 (8b), 13717 (38), 13761 (161/Ia), 13771 (3a), 13880 (38), 14112 (38), 14233 (3a), 14415 (3a), 14552 (3a), 14565 (26a), 14570 (3a), 15100 (147), 15181 (3a), 15962 (17a), 16663 (147), 16690 (147), 16797 (17a), 17248 (147), 17354 (147), 17388 (3a), 17421 (17a), 17499 (3a), 18688 (3a), 18703 (161/Ia), 19700 (147), 20902 (20), 21008 (147), 21170 (38), 21225 (8b), 22029 (38), 22163 (69), 22580 (38), 23292 (140), 23628 (17a), 24576 (38), 24901 (3a), 25889 (20)

Prance, G. T. & T. D. Pennington 1821 (148)

Prance, G. T., T. D. Pennington & J. M. Pires 1280 (3a)

Prance, G. T. & N. T. Silva 58516 (46a), 58547 (140), 58580 (3d), 58654 (148), 58774 (126g), 58946 (36), 59031 (126a)

Prescott, G. W. 306 (116), 391 (17a)

Prevost, M. F. 202 (35)

Prey, N. 32 (169a)

Prieto, F. 176 (110g)

Pringle, C. G. 45 (193c), 46 (73), 365 (73), 872 (193c), 2138 (138f), 2429 (177c), 2783 (76b), 3154 (75), 3731 (144), 3926 (126m), 4103 (71), 4408 (199), 4883 (178a), 5441 (11), 6248 (145), 6340 (55), 6773 (195), 7478 (193a), 8376 (116), 8485 (119), 8665 (77), 8672 (193a), 9392 (195), 9642 (193a), 9674 (77), 9728 (85), 9729 (138e), 9730 (140), 9732 (144), 9733 (71), 9734 (69), 10331 (149), 10383 (193a), 11401 (177c)

Prior, A. 1029 (131), 1036 (29)

Proaño, – 92 (110e), 152 (110e)

Proctor, G. R. 25593 (168b), 26742 (29), 32491 (177a), 33521 (168b)

Puga, B. 11 (16)

Purpus, C. A. 1 (177a), 11 (178a), 109 (126m), 1113 (88), 1185 (145), 1193 (195), 1195 (197), 1370 (85), 1739 (116), 1739a (116), 2342 (177a), 2342 (177k), 2343 (63b), 2344 (68), 2657 (144), 2658 (145), 2659 (144), 2660a (79), 2661 (94d), 3208 (82), 3217 (79), 3218 (145), 3219 (193a), 4831 (75), 5188 (193b), 5883 (13a), 6075 (146), 7153 (55), 7384 (11), 7482 (63b), 8720 (199), 8974 (177a), 9005 (177a), 9267 (11), 9268 (72b), 9368 (72b), 9382 (119), 9383 (126m), 9385 (177a), 9387 (11), 10033 (140), 10115 (11), 10479 (11), 10485 (177a), 10570 (177a), 10575 (69), 10576 (138f), 10577 (126m), 10579 (119), 10694 (145), 10786 (11), 10834 (161/III), 11162 (177a), 16118 (177a), 16244 (11)

Purseglove, J. W. 6332 (8a)

Pursell, R. A. 8010 (147), 8206 (69), 8210 (129a), 8370 (199), 8425 (129a), 8612 (199), 8633 (8a), 8669 (129a), 8694 (129a), 8733 (202a), 8915 (199)

Questel, A. 23 (69), 518 (69), 582 (69), 683 (129a)
Quirós Calvo, M. 605 (173), 851 (177a)

R 39932 (161/IIa), 49801 (46a), 65625 (12), 65758 (36), 65839 (42a), 65895 (12)
RB 310 (120), 1277 (64), 2565 (64), 3232 (33), 5132 (12), 5240 (42a), 105112 (42a), 11023 (37), 11073 (36), 11321 (36), 11323 (8b), 11326 (44b), 11332 (42a), 16132 (202a), 16357 (12), 16980 (20), 16982 (26a), 16983 (26a), 16986 (17a), 16989 (38), 16990 (38), 16993 (202a), 17149 (202a), 17164 (53a), 17191 (53a), 19084 (150), 20332 (38), 22997 (39b), 23308 (120), 23309 (120), 23310 (44a), 23311 (26a), 23900 (39b), 24106 (44a), 29068 (44a), 57236 (161/IIa), 57239 (44a), 67609 (44a), 75711 (41), 81502 (120), 90292 (37), 90547 (58), 92644 (42e), 93803 (44b), 105112 (42a), 111216 (120), 111747 (120), 112654 (40), 196702 (46a)
Rabelo, B. 179 (36), 587 (3a), 615 (20)
Raben, Count F. C. 365 (39a), 373 (136)
Rabia, T. R. 198 (147)
Ramalho, L. 42 (44b)
Rambo, B. 44007 (124), 49639 (124), 60320 (63a)
Ramia, M. 43 (10a), 628 (138g), 629 (126k)
Ramos, C. H. 318 (199)
Ramos B., A. J. 31 (157)
Ratter, J. A. 177 (60), 850 (60), 893 (60), 1302 (43), 1319 (3a), 1409 (3a), 4079 (60), 4437 (3/Ia)
Rauh, W. & G. Hirsch P-366 (95a)
Raven, P. H. 21504 (26a)
Raven, P. H. & D. Breedlove 19945 (173)
Ray, J. D. 5360 (69)
Raynal-Roques, – 19712 (26a)
Realino, Fr. M. 15 (69)
Reason, C. 32 (147)
Record, S. J. 19 (8a), 36 (177a), 46 (199), BH28 (27)
Redford, A. E. 29674 (69)
Reed, E. 368 (16)
Regnell, A. F. I/48 (124), I/66 (161/IIa), I/67 (161/IIa), I/69 (136), I/70 (44a), I/71 (42b), I/71bis (42b), I/72 (70a), II/76 (126a), II/3380 (63b), III/475 (124), III/477 (66), III/478 (63a)
Reichlen, – 45 (112), 46 (112)
Reineck, E. M. 202 (124)
Reiss, S. von 222 (69)
Reitz, R. & R. M. Klein 603 (126b), 1694 (3a), 1786 (126a), 2856 (42a), 3389 (124), 4111 (161/IIa), 5997 (161/IIb), 7872 (121), 17427 (161/IIb), 24731 (121)
Reko, B. P. 3410 (26a), 3699 (177m), 4446 (11), 4884 (11), 5056 (177e), 5102 (177e), 6043 (68)
Renson, C. 3 (140), 6 (140), 7 (16), 35 (126k), 208 (126k)
Rentería, E. 710 (138g), 764 (16)
Renvoise, S. A. 3246 (121), 3534 (180)
Reveal, J. L. & D. A. Atwood 3319 (82), 3357 (86)
Reverchon, J. 707 (84), 807 (88), 808 (73)
Revilla, J. 440 (8b), 504 (126q), 641 (17a), 998 (138f), 1907 (53a), 1999 (8b), 2299 (8b), 2434 (17a), 2526 (8b)
Ribeiro, B. G. S. 1068 (3a), 1131 (126g), 1322 (126g)
Ribeiro Rosa, F. 71 (43)
Richards, P. W. 6422 (60), 6803 (3a)
Ricksecker, A. E. 21 (141), 52 (129a), 204 (169a)
Riedel, L. 45 (40), 110 (12), 111 (42a), 112 (161/IIa), 115 (69), 132 (40), 158 (12), 168 (40), 184b (42b), 377 (57), 409 (138f), 533 (3b), 555 (57), 603 (160), 619 (63a), 703 (160), 705 (57), 721 (39a), 722 (137c), 838 (61), 1191 (124), 1255 (61), 1333 (43), 1918 (136), 1947 (43)
Riedel, L. & G. H. von Langsdorff 439 (137c)
Rimachi Y., M. 382 (26a), 2537 (147), 2872 (8b)
Rimbach, A. 179 (116), 260 (25a), 882 (174), 16908 (116)
Ripley, H. D. D. & R. C. Barneby 13252 (87), 13309 (75), 13398 (86), 13904 (86), 13954 (88)
Rivera, A. 175 (199)
Rizzini, C. T. & Mattos 1105 (164c)
Robbins, S. B. 5487 (69), 5488 (140), 6046 (71), 6047 (140)
Robertson, J. 28 (126j)
Rocha e Silva, I. 75 (137c)
Roden, J. van 111-1970 (69)
Rodrigo, A. P. 547 (118), 904 (140), 743 (128), 999 (139), 2439 (118)
Rodrigues, I. A. 81 (126d)
Rodrigues, W. 2078 (26a), 2673 (38), 2811 (38), 4569 (38), 8563 (26a), 9099 (38), 9107 (38), 9434 (58)
Rodrigues, W. & J. Chagas 1666 (161/Ia), 2453 (161/Ia), 5245 (161/Ia), 4747 (161/Ia)
Rodrigues, W. & D. Coelho 5879 (8b)
Rodríguez, C. 1318 (71), 1362 (71)
Rodriguez, D. 1284 (110a)
Rodríguez, H. 96 (199), 168 (157), 177 (13a), 218 (55), 343 (63b), 482 (199), 488 (177a), 536 (8a), 1974/3 (199)
Rodríguez, J. V. 938 (177a), 1088 (161/Ia), 1410 (16), 1537 (16), 1999 (177a), 2397 (177a), 3694 (16)
Roe, K. et al. 67 (193b), 525 (200a), 874 (71), 1002 (138f), 1125 (177k), 1309 (71), 1360 (161/III), 2160 (119), 2286 (138f), 2292 (138f), 2309 (126m), 2421 (73)
Rohr, J. P. B. von 70 (199)
Rohweder, O. 2932 (69), 2935 (69), 2968 (140)
Roig, J. T. 1835 (131), 2244 (30), 3896 (196a), 3899 (196a), 5542 (134e), 6572 (133)
Roig, R. 1 (48b)
Rojas, T. 4758 (136), 8658 (66), 10077 (66), 10999 (66), 12044 (118), 12569 (66), 12698 (3c), 13620 (63a)
Rombouts, H. E. 621 (26a), 857 (161/Ia)
Romero-Castañeda, R. 677 (177a), 990 (8a), 2128 (8a), 3467 (3a), 4389 (199), 6227 (140), 7256 (129a), 9097 (8a), 9181 (129a), 9268 (69), 9309 (8a), 9320 (177a), 9381 (147), 9709 (199), 9817 (161/IIb), 10438 (129a), 10459 (140), 10580 (177a), 10639 (199), 10793 (147)
Rosa, N. A. 410 (161/Ia), 994 (36), 1508 (27), 2023 (161/Ia), 2476 (69), 2489 (147), 2820 (26a)
Rosa, N. A. & O. Cardoso 2556 (161a)

Rosa, N. A. & M. R. Cordeiro 1746 (20), 1918 (126p)
Rosa, N. A. & M. R. Santos 2164 (161/Ia)
Rosário, C. da R. 8 (3a)
Rosário, C. S. 18 (3d vers. 3a)
Rosas R., M. 116 (126m), 146 (72b), 221 (119), 604 (119), 703 (68), 758 (126m), 947 (68), 1334 (69), 1342 (202a), 1346 (202a), 1349 (13a), 1389 (161/III)
Rose, J. N. 1222 (80), 1398 (140), 1582 (200a), 1401 (177a), 1720 (26a), 2219 (63b), 2303 (11), 2437 (79), 2448 (177c), 2913 (140), 11184 (79), 11275 (78), 11868 (80), 1659a (81), 16692 (81), 17831 (87), 17910 (87), 17950 (73), 19032 (95a), 19933 (176), 20029 (148), 21718 (10a), 21762 (202a), 21892 (177a), 22430 (174), 22461 (153c), 22467 (10b), 23505 (153a), 23586 (200b), 24134 (73)
Rose, J. N. & Fitch 3516 (129a), 3917 (129a)
Rose, J. N., Fitch & Russell 3255 (141), 3710 (141), 3795 (168a), 3795a (168a), 3816 (168a), 3828 (199), 3856 (168a), 4163 (141), 4309 (148)
Rose, J. N. & Hay 3666 (116)
Rose, J. N. & Hough 5889c (145), 9955 (145)
Rose, J. N. & Painter 6591 (71), 6897 (55), 9778 (79), 10002 (195)
Rose, J. N., Painter & Rose 8583 (193a), 8999 (85), 9399 (140), 9705 (138e), 9706 (75), 9743 (144)
Rose, J. N., Painter & Russell 9807 (193b)
Rose, J. N. & J. S. Rose 11231 (145), 11414 (195), 11433 (197)
Rose, J. N. & Russell 19653 (44a), 19856 (42d), 19951 (37), 20201 (124), 21424 (42e), 24110 (84)
Rose, J. N., Standley & Russell 12688 (177d), 12718 (177d), 12755 (71), 12890 (140), 12904 (199), 13190 (177d), 13484 (177d), 13532 (177d), 13699 (126m), 13712 (71), 13718 (140), 13858 (200a), 13900 (200a), 13948 (71), 14400 (138c), 14411 (11), 14490 (69), 14505 (199), 14506 (71), 14507 (140), 14527 (177a), 14527 (177a), 14716 (71), 14746 (177a), 14797 (199), 14807 (140), 14834 (126m), 14906 (140), 14983 (71)
Rose, L. S. 57102 (90)
Ross, G. B. 34-57 (16)
Roth, L. 1827 (44a), 1832 (43)
Rothéry, H. C. 107 (69)
Rothrock, J. T. 64 (29), 553 (177j), 555 (177j)
Rovirosa, J. N. 114 (27), 202 (13a), 267 (202a), 273 (140), 355 (147), 664 (68)
Rowell, C. M. 2280 (193a)
Rowlee, W. W. & H. E. 285 (8a), 285 (13a), 353 (13a)
Rowlee, W. W. & H. E. Stork 792 (140)
Rozynski, H. W. von 66a (73), 362 (193b)
Rubio, A. 123 (140)
Rudd, V. 1041 (94e)
Rugel, F. 68 (140 & 141), 69 (141), 70 (138f), 71 (71), 344 (131), 779 (69), 825 (131)
Ruiz Z., T. 922 (129a)
Runyon, R. 1337 (193b), 1338 (85), 1373 (138e), 2992 (83b)
Rusby, H. H. 41 (116), 112 (87), 821 (46a), 837 (25b), 980 (96), 981 (116), 982 (96), 983 (140), 984 (63c), 986 (161/Ia), 987 (68), 988 (126e), 989 (116), 1042 (200b), 2400 (46a)
Rusby, H. H. & F. W. Pennell 278 (148), 348 (177a), 639 (47), 850 (166), 856 (47)
Rusby, H. H. & R. W. Squires 13 (140), 19 (148), 24 (53a), 189 (126k), 319 (27)
Ruth, A. 14 (84), 292 (69), 1250 (88), 2207 (69)
Ruthsatz, B. 564 (96), 1014 (96)
Rzedowski, J. 14580 (152), 17818 (177c), 18858 (202a), 19295 (178a), 19594 (144), 19643 (200bis), 19697 (178a), 20972 (91c), 27069 (196d), 27621 (144), 27779 (196b), 29034 (119)

Saer d'Heguaert, J. 8 (172), 185 (199), 190 (172), 254 (172), 264 (10a), 282 (71), 290 (10a), 416 (171), 707 (171), 733 (10a), 761 (171), 3750 (174)
Sagástegui, A. 7925 (174), 8663 (200b), 8723 (174), 9169 (113a), 9332 (113a)
Sagot, P. 178 (69), 181 (36), 182 (26a), 477 (140), 803 (161/Ia), 1156 (20)
Sagra, R. de la 48 (140), 372 (141), 391 (71), 1825 (69)
Saint Hilaire, A. F. C. P. de 2727 (127), B/1832 (175), B1/1394 (12), B1/1566 (163), C/623 (175), C1/56 (137c), C1/224 (63a), C1/737 (175), C2/1303 (63a), C2/2550bis (89), C2/2748bis (63a), D/159 (137c)
St. John, H. 20512 (166), 20735 (13a)
Sakane, M. 541 (44a), 567 (44a)
Saldanha, J. de 5175 (42a)
Sampaio, A. 2852 (12), 4614 (44a), 6089 (126a), 6446 (43), 6615 (39b), 6949 (69), 6957 (136), 7115 (44a), 7393 (44a), 7895 (42a), 7895 (39a), 8189 (12), 8887 (12)
Sampaio, A. J. de 5557 (26a)
Sampaio, E. 6463 (136)
Samuels, J. A. 15 (8a), 300 (8a), 309 (148), 479 (148)
Sánchez Vega, I. 96 (112), 208 (95b), 847 (112)
Sandeman, C. 24 (95b), 325 (109a), 2095 (46a), 2136 (63b), 2144 (126c), 2149 (46a), 2526 (95a), 3633 (95a), 3757 (110d), 4035 (129a), 4615 (110e), 5588 (8b), 5904 (86)
Sandwith, N. Y. 674 (26a), 1286 (26a), 1497 (9), 1869 (8a)
Santaro, J. 508 (44a)
Santiago, A. F. 12 (43)
Santiago, S. A. 20 (27)
Santos & Castellanos 24094 (175), 24320 (175)
Santos, E. 1498 (26a), 2061 (126d)
Santos, E. & E. Fromm 1129/1092 (37)
Santos, E. & J. Sacco 1736/1969 (43), 1942/2203 (42c), 1945/2206 (176), 1968/2229 (176)
Santos, J. M. et al. 72 (3/Ia), 93 (140)
Santos, T. S. dos 687 (41), 933 (61), 1530 (161/Ia), 1587 (42e), 2242 (59c), 2246 (42d), 2277 (42b), 2536 (3a), 2552 (164c), 2704 (3b), 2911 (3b), 3525 (199)
Saravia T., C. 3525 (199)
Sargent, C. S. 22 (16)

Sargent, F. H. 70 (169a), 134 (28), 490 (28), 3193 (169a)
Sartorius, C. 39 (68), 40 (68), 67 (11)
Sarukhán K., J. 168 (16)
Sastre, C. 31 (20), 287 (36), 1434 (148), 1707 (3a), 2970 (3a), 3106 (8b), 3532 (3a)
Sauleda, R. 2657 (131), 2693 (199), 2713 (134e)
Saunders, S. G. E. 33 (114)
Scala, A. 247 (138a)
Schaffner, W. 624/831 (85), 625 (830), 829/626 (193b)
Schaller, G. 7 (126o), 191 (157)
Schiede, C. W. J. & Deppe 702 (146)
Schimpff, H. J. F. 347 (200b), 696 (25a), 1066 (63b), 1091 (10b), 1145 (10b), 1182 (200b)
Schinini, A. 8983 (43), 12370 (63a), 12548 (139), 12553 (63a), 12745 (69), 12996 (126o)
Schipp, W. A. 6 (26a), 6A (27), 824 (140), 847 (68), 1231 (126j), 1144 (147), 1329 (13a)
Schlim, L. 2 (170), 6 (16), 268 (63b), 291 (8a), 374 (126k)
Schmidt, L. J. 302 (3a)
Schnee, L. 1400 (157)
Schnell, R. 11183 (140)
Schomburgk, Robert 86 (27), 94 (8a), 522 (161/Ia), 614 (8a), 845 (20)
Schomburgk, Robert & Richard 683 (161/Ia), 775 (51a), 881/1523 (9)
Schott, A. 292 (88), 414 (177h), 982 (177h), 5646 (39a), 5654 (138d), 5655 (161/Ia)
Schott, – & Drewe 54 (83b)
Schubert, B. G. 1992 (11), 1999 (140), 2018 (177a), 2019 (69)
Schubert, B. G. & M. Sousa 1919 (119), 1991 (138e)
Schüch da Campanema, G. 85 (3f)
Schultes, R. E. 3387 (17a), 3751 (161/Ia), 5375 (8b), 5706a (8b), 17964 (17a), 18548 (3a)
Schultes, R. E. & I. Cabrera 14933a (8b), 15183 (53b), 15535 (53b), 15851 (53b), 17392 (8b), 19068 (17a)
Schulz, A. G. 8239 (118)
Schulz, C. L. 334 (139), 1356 (118)
Schumann, W. 162 (116 × 119)
Schunke, C. 62 (25a), 1438 (25a)
Schunke, J. M. 2 (161/Ia), 165 (8b), 180 (53a), 206 (22), 207 (17a), 208 (147)
Schunke V., J. 1610 (17a), 1867 (20), 2043 (126e), 2885 (25a), 3355 (126e), 3926 (3a), 4060 (161/Ia), 4531 (25a), 5812 (25a), 6080 (140), 6195 (17a), 6466 (25a), 6863 (26a), 8003 (25a)
Schwacke, C. A. W. 47 (161/IIa), 299 (26a), 6148 (42a), 6151 (37), 6202 (104a), 6740 (104a), 6835 (44a), III/320 (38)
Schwarz, E. 1745 (63a), 6606 (63a)
Schwarz, G. J. 235 (139), 584 (63a), 606 (138a), 679 (138a)
Schwindt, E. 3074 (138a), 3131 (63a)
Scribner, F. L. 8067 (119)
Seaton, H. E. 148 (141)
Seemann, B. 225 (16), 226 (63b), 483 (27), 529 (129a)
Segadas-Vianna, F. 309 (140), 3595 (126d)
Seibert, R. J. 1541 (13a), 1745 (140), 1757 (134a)
Seler, E. & C. 51 (146), 114 (193a), 2065 (149), 2680 (71), 3172 (94b)
Sellow, F. 796 (42a)
Semir, J. 4971 (57)
Semir, J. & M. Sazima 221 (126d)
Semple, A. 1 (177a)
Senn, H. A. 4299 (99), 4370 (99)
Service Eaux & Forêts (Guiane Fr.) 4032 (147)
Service Forestier (Guiane Fr.), 4003 (36), 4071 (161/Ia)
Sesmero, E. 332 (138a), 347 (138a)
Sessé & Mociño Hb. 1132 (144), 1133 (129a), 1135 (149), 1137 (11), 1140 (71), 1143 (119 & 199), 1148 (147), 1161 (126m), 1162 (146), 1163 (119), 1164 (71), 1167 (75), 1170 (68), 1171 (69), 1174 (26a), 1192 (169a), 1194 (177c), 1197 (149), 1199 (73)
Seymour, F. C. 956 (140), 1858 (68), 2343 (69), 2358 (71), 2419 (140), 2452 (148), 3197 (68), 4406 (140), 5425 (177a), 5601 (27), 5999 (16), 7909 (119), 12626 (140)
Shafer, J. A. 21 (129a), 80 (141), 136 (71), 152 (141), 162 (71), 216 (148), 322 (69), 356 (30), 393 (1), 411 (69), 414 (148), 437 (140), 540 (141), 690 (134e), 904 (134e), 907 (134e), 928 (71), 933 (141), 1038 (30), 1043 (202a), 1308 (170), 1398 (30), 1481 (177j), 1646 (126k), 1800 (199), 1845 (30), 2411 (140), 2458 (69), 2493 (129a & 177j), 2567 (177j), 2575 (148), 2599 (134e), 2767 (134e), 2909 (202a), 3370 (134c), 3565 (28), 3803 (134c), 4072 (135), 8270 (135), 8913 (126k), 10581 (126m), 11110 (131), 11667 (126m), 12364 (170), 12391 (170), 12404 (199)
Shapiro, G. 145 (140), 179 (161/III), 184 (202a), 314 (177c), 325 (147), 381 (202a), 388 (126m), 392 (16)
Sharp, A. J. 441164 (119), 441626 (116)
Sharp, A. J. & E. Hernández X. 50 (94f)
Shattuck, O. E. 362 (14b), 407 (27), 496 (27), 513 (14b), 732 (8a), 786 (27)
Shepherd, G. J. 4108 (3/IIe), 4113 (140), 7410 (69)
Shepherd, Mrs. R. S. 316 (129a)
Shimek, B. & C. L. Smith 4 (177a)
Shreve, F. 6095 (177d), 6139 (177d), 6378 (193c)
Sick, H. B240 (3f)
Sieber, F. W. 57 (202a)
Silva, A. 97 (20), 179 (161/Ia)
Silva, M. 598 (20), 647 (20), 991 (3a), 1061 (138g), 1064 (148), 1457 (3a), 1460 (20), 1503 (3a), 1518 (36), 1606 (36), 1696 (26a), 1703 (26a), 1818 (3a), 1821 (36), 1907 (64), 2194 (3a), 2746 (46a)
Silva, M. & R. Souza 2224 (138g), 2226 (46a), 2270 (26a), 2316 (46a), 2417 (36), 2774 (36), 2775 (3a), 4202 (3a), 4766 (38)
Silva, M. G. & R. Bahia 2948 (26a), 3051 (69), 3058 (161/Ia), 3106 (148), 3472 (126g), 3492 (20)
Silva, M. G. & A. Pinheiro 4159 (43), 4469 (126a)
Silva, N. T. 963 (27), 1187 (49), 4856 (126p), 59728 (20)
Silva, N. T. & M. R. Santos 4732 (3a)
Silva Manso, A. L. P. da 267 (43), 268 (42b)

Silveira, A. 886 (43)
Simmonds, N. W. 14903 (27)
Simpson, D. R. 503 (200b), 527 (200b), 735 (17a)
Simpson, J. H. 178 (134e)
Sintenis, P. 166 (28), 366 (28), 574 (169a), 1988 (199), 2138 (126n), 2520 (126n), 2688 (28), 2724 (126n), 2902 (169a), 3074 (126n), 3077 (28), 3163 (148), 3231 (169a), 4145 (119), 5365 (28)
Skinner, G. U. 18 (177n)
Skottsberg, C. & I. 711 (98d)
Skutch, A. F. 475 (8b), 761 (177k), 1002 (177k), 1497 (16), 1673 (177k), 1952 (177k), 1996 (126k), 2428 (126k), 2616 (16), 2655 (48b), 2786 (16), 2809 (202a), 2879 (69), 2980 (29), 3815 (126k), 4440 (25a), 4517 (8b), 5274 (53a), 5419 (53a)
Small, J. K. 148 (69), 3860 (134e), 3893 (134e), 4490 (69), 5061 (134e), 6683 (69), 12676 (2)
Small, J. K. & J. J. Carter 725 (134e), 2981 (134e), 3174 (134e), 3553 (134e), 8631 (134e), 8637 (134e)
Small, J. K. & C. A. Mosier 6015 (134e)
Small, J. K. & P. Wilson 1736 (134e)
Smith, A. 54 (177o), 427 (177o), 644 (177o), 984 (177o), 2724 (177o), 2759 (177o), 10011 (177o), 10121 (16)
Smith, A. C. 2159 (20 & 26a), 2227 (8a), 2358 (33), 2395 (8a), 2434 (161/Ia), 2439 (27), 2519 (3a), 3313 (20), 10147 (129a), 10328 (140)
Smith, C. C. 1302 (69)
Smith, C. E. Jr. 3259 (149), 3265 (16), 3398 (8a), 3564 (193a), 3771 (145), 4004 (195), 4056 (78), 4057 (145), 4339 (144), 4391 (199), 4470 (178a), 4522 (177a), 4874 (174)
Smith, C. L. 89 (13a), 130 (140), 342 (144), 346 (178a), 569 (178a), 572 (138a), 1411 (177a)
Smith, F. D. 261 (202a)
Smith, H. H. 20 (199), 31 (69), 70 (140), 71 (69), 248 (147), 249 (177a), 250 (15), 251 (15), 253 (16), 256 (129a), 257 (129a), 699 (126i), 700 (8a), D11 (159)
Smith, H. H. & G. W. 363 (148), 1051 (8a), 1052 (138g)
Smith, J. Donnell 125 (119), 1789 (68), 2288 (177a), 2289 (147), 2290 (177g), 2830 (177g), 4115 (16), 4471 (200a), 5188 (16), 5191 (177a), 6116 (149), 6485 (13a), 6486 (16), 7736 (16)
Smith, L. B. 1316 (120), 6351 (126d), 6900 (43), 6929 (43), 7081 (43)
Smith, L. B. & R. M. Klein 12002 (161/IIa), 14947 (137a)
Smith, L. B., R. M. Klein & G. Hatschbach 14812 (44a)
Smith, L. B. & R. Reitz 7932 (127), 9046 (121)
Smith, L. B., R. Reitz & R. M. Klein 12302 (126a)
Smith, S. G. & J. M. Idrobo 1444 (8b)
Sneidern, K. von 422 (17b), 600 (17b), 1159 (69), 1161 (126k), 1162 (147), 5567 (17b)
Snethlage, E. H. 165 (46a)
Sobrinho, J. S. 19 (42e), 236 (59b)
Soejarto, D. D. 562 (13b)
Sohner, L. 1212 (171)
Sohns, E. R. 1239 (85), 1338 (75)
Solanilla, O. 5 (16)
Solano, S. C. 204 (144), 311, (178a), 395 (146)
Solano P., F. 70 (173), 425 (13a)
Solís, O. 30 (177a)
Solís Magallanes, J. A. 885 (149), 1045 (149)
Solís R., F. 70 (173), 425 (13a)
Solomon, J. 3431 (26a)
Soto Núñez, J. C. 263 (77), 372 (152), 1358 (177e), 1877 (152)
Soukup, J. 4389 (23), 5405 (102), 5425 (95a)
Sousa, M. 137 (13a), 842 (16), 1274 (126m), 1369 (126m), 1552 (63b), 2594 (161/III), 2595 (140), 2596 (69), 2672 (144), 2687 (91c), 2693 (146), 2703 (196e), 2797 (13a), 4136 (196a), 4144 (71), 4253 (55), 4256 (199), 4275 (193a), 4349 (178a), 4350 (91c), 4599 (199), 4609 (116), 4623 (178a), 4625 (144), 5007 (69), 5025 (178a), 5039 (55), 5051 (71), 5131 (145), 5146 (178a), 5181 (68), 5182 (126k), 5187 (68), 5196 (177a), 5200 (149), 5202 (177a), 5207 (68), 5231 (200a), 5242 (68), 5266 (147), 5276 (200a), 5283 (140), 5310 (126k), 5330 (149), 5363 (177q), 5369 (91c), 5381 (178a), 5384 (199), 5435 (144), 5461 (72b), 5497 (149), 5508 (177a), 5510 (200a), 5511 (200a), 5519 (140), 5530 (200a), 5547 (200a), 5556 (147), 5576 (140), 5586 (177m), 5608 (119), 5627 (199), 5678 (144), 5681 (144), 5740 (71), 5754 (77), 5762 (77), 5778 (138f), 5785 (197), 5804 (193a), 5806 (77), 5810 (72b), 5818 (177s), 5864 (55), 5865 (71), 5889 (55), 5891 (13a), 5902 (140), 6028 (178a), 6054 (138f), 6101 (116), 6126 (197), 6136 (199), 6141 (193a), 6197 (195), 6202 (178a), 6250 (116), 6271 (138f), 6280 (178a), 6297 (177g, bis), 6381 (140), 6382 (69), 6402 (11), 6404 (177a), 6426 (177a), 6443 (149), 6479 (26a), 6509 (91c), 6514 (178a), 6515 (199), 6536 (146), 6538 (193a), 6541 (146), 6568 (55), 6585 (193a), 6589 (178a), 6601 (178a), 6607 (148), 6608 (199), 6653 (149), 6707 (94c), 6722 (149), 6857 (149), 6911 (193a), 6933 (197), 6953 (195), 6960 (179a), 6962 (197), 6978 (177s), 7002 (68), 7015 (68), 7034 (177g, bis), 7054 (149), 7101 (11), 7151 (149), 7159 (11), 7184 (200bis), 7209 (144), 7234 (145), 7264 (147), 7269 (161/III), 7272 (13a), 7287 (177p), 7340 (197), 7344 (178a), 7352 (140), 7361 (199), 7399 (199), 7440 (55), 7504 (193a), 7515 (193a), 7537 (193a), 7555 (91c), 7578 (200a), 7593 (200a), 7609 (177a), 7621 (200a), 7624 (140), 7630 (200a), 7660 (147), 7663 (11), 7695 (72b), 7729 (195), 7734 (79), 7745 (195), 7747 (145), 7748 (193a), 7751 (199), 7803 (144), 7827 (116), 7842 (146), 7911 (116), 7947 (161/III), 7971 (202a), 8063 (199), 8228 (178a), 8229 (144), 8251 (138f), 8252 (178a), 8254 (200bis), 8300 (200bis), 8346 (177g, bis), 8353 (200bis), 8401 (69), 8402 (71), 8408 (140), 8483 (68), 8568 (55), 8631 (69), 8705 (202a), 8770 (161/III), 8783 (202a), 8867 (178a), 8935

(178a), 8950 (116), 9149 (178a), 9212 (177p), 9260 (68), 9291 (196a), 9316 (199), 9434 (69), 9476 (71), 9481 (55), 9567 (199), 9573 (71), 9610 (199), 9670 (69), 9678 (146), 9689 (144), 9819 (178a), 9839 (200bis), 9872 (68), 9878 (126m), 9910 (149), 9916 (55), 9917 (177a), 9919 (140), 9996 (178a), 10007 (178a), 10028 (146), 10034 (178a), 10063 (178a), 10076 (146), 10086 (55), 10106 (178a), 10145 (178a), 10156 (177a), 10241 (16), 10261 (202a), 10364 (144), 10375 (79), 10378 (79), 10399 (193a), 10419 (119), 10434 (193a), 10478 (178a), 10484 (200bis), 10489 (200bis), 10511 (200bis), 10522 (200a), 10548 (119), 10569 (200a), 10570 (140), 10578 (140), 10604 (152), 10605 (119), 10619 (72b), 10621 (116), 10656 (144), 10833 (196a), 10844 (27), 10904 (177i), 10907 (196a), 11310 (177g, bis), 11316 (196e), 11341 (196e), 11412 (72b), 11545 (173), 11661 (149), 11958 (145)

SP 39689 (120), 41851 (120)

Spada, J. 217 (12), 229 (39a)

Sparre, B. & Vervoorst 1565 (127)

Spellenberg, R. 3324 (101), 3433 (84), 3457 (193c), 3847 (193c)

Spellman, D. 209 (13a), 438 (140), 1420 (68)

Spellman, D. & W. W. Newey 1811 (71), 1874 (201)

Sprague, T. A. 138 (8b)

Spruce, R. 238 (36), 255 (26a), 484 (36), 916 (36), 1110 (8b), 1182 (26a), 1184 (8b), 1272 (26a), 1558 (8b), 1701 (126p), 1775 (38), 2125 (161/Ia), 2539 (8b), 3925 (126s), 4147 (161/Ia), 4170 (138f), 4299 (68)

Stafford, D. K-40 (102), 276 (102)

Stahel & Gonggrijp 3076 (36)

Standley, P. C. 1100 (177a), 4075 (16), 7710 (177n), 8130 (177a), 8441 (177a), 8572 (177a), 9229 (177a), 9605 (177a), 10530 (177k), 12442 (70a), 13137 (13a), 13222 (71), 14897 (177a), 15427 (16), 15856 (178a), 16164 (177a), 16511 (177a), 16726 (16), 17754 (178b), 18056 (177a), 19430 (16), 19853 (68), 20184 (177a), 20340 (147), 20676 (149), 20725 (177a), 20781 (138f), 20920 (69), 20939 (71), 21001 (138f), 21009 (71), 21036 (177a), 21385 (178a), 21813 (16), 21940 (177a), 22132 (177a), 22576 (68), 23200 (178a), 23443 (177a), 23585 (138f), 24028 (126k), 24117 (177a), 24278 (177a), 24359 (177a), 24515 (16), 24695 (178a), 24953 (26a), 24993 (177a), 25315 (69), 25346 (16), 25532 (16), 25796 (16), 25807 (177a), 26057 (63b), 26447 (16), 26568 (16), 26664 (69), 26757 (8a), 26810 (16), 26844 (16), 27069 (16), 27118 (63b), 27269 (177a), 27711 (27), 27766 (16), 27986 (16), 28183 (16), 28492 (16), 28671 (8a), 28918 (63b), 28960 (16), 29152 (16), 29231 (27), 29303 (63b), 29317 (16), 29459 (8a), 29524 (8a), 29597 (16), 29704 (27), 29711 (68), 29793 (16), 30096 (16), 30188 (8a), 30490 (16), 30546 (8a), 30572 (27), 32086 (16), 32215 (173), 34036 (16), 36667 (13a), 41236 (16), 53060 (68), 54438 (8a), 54999 (177a), 58066 (16), 58387 (177k), 58748 (94b), 59197 (149), 59203 (71), 59228 (177g), 59402 (177k), 59647 (177a), 59892 (177g), 60317 (177g), 60488 (177a), 60634 (16), 60687 (177a), 61019 (94b), 61052 (94b), 61193 (177g), 61278 (177k), 61681 (177k), 62247 (16), 63301 (177g), 63367 (26a), 64209 (177a), 64500 (177g), 64845 (16), 66524 (16), 66603 (177a), 66616 (177a), 66629 (177a), 69884 (177l), 72010 (177a), 72604 (26a), 73069 (26a), 73142 (26a), 73626 (177a), 73889 (177a), 73926 (177a), 74024 (178b), 74091 (177a), 74530 (178b), 74700 (16), 74800 (178b), 74842 (16), 75085 (177a), 75975 (177a), 76133 (16), 77127 (177a), 77283 (177g), 77475 (177a), 77700 (177a), 77896 (16), 77909 (177a), 78133 (177a), 78292 (177a), 78404 (16), 78835 (177a), 78843 (177a), 78845 (177a), 78847 (177a), 78868 (177a), 79020 (16), 79064a (16), 79168 (16), 79207 (177a), 79586 (16), 79908 (177a), 79931 (177g), 80510 (177k), 80277 (177k), 80283 (177k), 80439 (177a), 80441 (177a), 81516 (177g), 87302 (16), 87546 (177a), 87570 (177a), 87576 (177a), 87664 (16), 87747 (16), 88145 (16), 88473 (177a), 88728 (16), 89021 (16), 89074 (177a), 91066 (177l), 91066 (177l)

Standley, P. C. & J. Chacón P. 5319 (177a), 5338 (177a), 5595 (177a)

Standley, P. C. & A. Molina A. 4662 (177a)

Standley, P. C. & E. Padilla V. 2770 (16), 2910 (16), 3016 (177k), 3736 (178a)

Standley, P. C. & S. A. Padillo 2445 (177a), 3073 (177a)

Standley, P. C. & J. Valerio 44266 (16), 44868 (149), 44872 (16a)

Standley, P. C. & L. O. Williams 94 (177a)

Stanford, L. R., Weatherford & Northcraft 229 (76a), 353 (76b), 268 (88), 500 (85), 823 (75), 823 (75), 829 (76a), 1094 (71), 1730 (193d)

Stearns, E. 91 (84)

Stehlé, H. 91 (84), 199 (129a), 222 (138g), 1316 (69), 1735 (141), 3473 (138g)

Steiger, T. L. 193 (84), 198 (84), 247 (82), 443 (73), 879 (84), 893 (87), 896 (87), 897 (73), 898 (87), 914 (84)

Stein, S. 2 (8a), 3 (27), 5 (8a), 6 (202a), 7 (161/Ia)

Steinbach, J. 5279 (148), 5354 (69), 5364 (140), 5399 (202a), 5567 (126e), 5792 (117), 6020 (110b), 6024 (117), 6441 (46a), 6939 (140), 7019 (69), 7270 (202a), 7560 (3a), 8630 (110b), 8694 (96), 8818 (122), 9688 (116)

Steinbach, R. 150 (118), 376 (69), 418 (140)

Stern, W. L. 447 (14a), 472 (14a), 1869 (8a)

Stern, W. L. & K. L. Chambers 292 (134e)

Stevens, F. L. 20 (10b)

Stevens, F. L. & W. E. Hess 2765 (28)

Stevens, G. W. 1070 (84), 2587 (199), 2716 (71), 2732 (199), 2881 (199), 3251 (16), 3541 (199), 3570 (140), 3827 (199), 3901 (55), 4257 (140), 4276 (199), 4422 (199), 4465 (69), 4565 (69), 4590 (71), 4669 (69), 4728 (16), 4762 (177a), 4811 (13a), 5132 (16), 5190 (71), 5209 (177a), 5411 (177a), 5435 (177a), 5468 (199), 5513

(177a), 5679 (199), 5732 (71), 5736 (140), 5867 (68), 6037 (68), 6046 (69), 6100 (177a), 6146 (16), 6158 (140), 6224 (177a), 6584 (126k), 6648 (177a), 7209 (16), 7237 (27), 7269 (148), 7942 (27), 9214 (119), 9347 (177a), 9456 (199), 9582 (13a), 9752 (199), 9805 (199), 9906 (71)
Stevenson, J. A. 66 (28), 205 (28), 2557 (169a)
Steward, W. C. 136 (36), 252 (3a), 259 (27), 260 (36), 261 (3a), 266 (161/Ia), 457 (3a), 500 (3a)
Stewart, R. M. 379 (193c), 458 (73), 542 (193c), 578 (73), 719 (193c), 1253 (76a), 1374 (73), 1693 (73), 1911 (73), 2718 (82), 2779 (82), 2838 (82), 2948 (82), 2969 (82)
Steyermark, J. A. 29020 (177a), 29496 (178b), 30362 (16), 30685 (177n), 31516 (177n), 32413 (177g), 32951 (177n), 33193 (177l), 35076 (177l), 38137 (26a), 38494 (26a), 38587 (16), 42183 (177a), 42738 (177n), 42926a (177k), 44394 (16), 45131 (27), 45193 (16), 46283 (140), 47147 (177k), 47969 (26a), 50992 (177k), 51017 (177k), 51313 (177k), 51638 (177k), 51645 (177k), 51934 (116), 52193 (11), 53739 (179), 54054 (10b), 55968 (27b), 60795 (26a), 60955 (27), 60964 (8a), 61225 (10a), 61249 (27), 74719 (3a), 74723 (26a), 75422 (26a), 77617 (177n), 80379 (177n), 80392 (177n), 86746 (161/IIb), 86851 (170), 87238 (161/Ia), 87346 (20), 88184 (27), 88206 (129a), 88235 (202a), 88437 (8a), 88915 (202a), 88918 (8a), 90506 (3a), 90854 (171), 94241 (199), 94519 (159), 95950 (17a), 96431 (199), 96854 (115), 97430 (10a), 97508 (8a), 97580 (29), 98839 (115), 98854 (138b), 98881 (126k), 99163 (170), 99360 (126m), 99818 (13a), 100006 (148), 100009 (69), 100012 (13a), 100032 (140), 100118 (140), 101424 (17a), 103404 (17a), 103673 (10a), 103674 (10a), 103677 (172), 104045 (8a), 104533 (26a), 104533 (8a), 104647 (27), 104941 (115), 106922 (8a), 106993 (17a), 107577 (69), 108811 (10a), 110058 (129a), 110060 (172), 111247 (8a), 11249 (20), 111323 (161/Ia), 111428 (170), 111432 (199), 112107 (202a), 113005 (8a), 113008 (3a), 113578 (138g), 113656 (147), 114614 (69), 116367 (27), 117896 (3a), 118506 (115), 118771 (170), 119542 (8a), 120110 (170), 120168 (177a), 120269 (8a), 120386 (148), 120822 (138g), 121004 (8a), 121024 (171), 121222 (171)
Steyermark, J. A. & P. H. Allen 17213 (8a)
Steyermark, J. A. & Nilsson 219 (26a), 710 (26a)
Steyermark, J. A. & J. J. Wurdack 40 (3a)
Stimson, W. R. 2945 (140), 5352 (14a)
Stone, D. F. 3174 (177a)
Stork, H. E. & O. B. Horton 9399 (129b), 10101 (153b), 10230 (110c), 10398 (153b), 10634 (126e), 10949 (114), 11391 (63b),
Strang, H. E. 372 (42b)
Strong, E. 1000 (61)
Stuchert, T. 13087 (138b), 18412 (130), 21118 (110a)
Studhalter, R. A. 1053 (84), 1061 (84)
Sucre, D. 87 (120), 148 (120), 302 (43), 637 (161/Ia), 2420 (12), 6685 (137c), 9010 (161/Ib), 9232 (165), 9316 (43), 9605 (39a)
Sucre, D. & P. Braga 2219/77 (137c), 2224/82 (136), 2387/230 (137c)
Sullivan, G. A. 468 (26a)
Swallen, J. 3711 (43), 4744 (150)
Swingle, C. 143 (110e)

Tamashiro, J. V. 35 (44a)
Tamayo, F. 273 (199), 291 (172), 315 (10a), 1106 (42g), 1255 (10a), 1711 (129a), 1817 (10a), 1877 (115), 2458 (115), 2717 (20), 2980 (26a), 3021 (20), 3336 (172), 3337 (172), 3793 (42g), 3797 (157), 4142 (55), 4352 (157)
Tanner, R. E. S. 1249 (202a)
Tate, G. H. H. 119 (161/Ia), 145 (161/Ia), 248 (8a), 260 (8a), 732 (138f), 1019 (138f)
Tate, R. 118 (44b)
Tavares, S. 186 (44b), 557 (46a), 671 (42e), 875 (37)
Taylor, A. A. 115 (69), 119 (140)
Taylor, J. & C. 6221 (75), 11699 (13a), 12486 (199), 12514 (71), 12664 (177h), 15777 (199), 16279 (84)
Taylor, M. 153 (76b)
Taylor, N. 46 (199), 77 (199), 370 (148), 430 (131), 445 (69)
Taylor, T. W. J. 48 (153c)
Teixeira, G. 2832 (176)
Tejera, E. 32 (13a)
Téllez, O. 9 (138f), 18 (178a), 22 (144), 61 (146), 97 (146), 105 (178a), 108 (193a), 121 (69), 122 (71), 127 (199), 128 (140), 132 (55), 181 (177a), 228 (71), 285 (193a), 631 (27), 636 (196a), 660 (199), 703 (16), 885 (201), 974 (201), 992 (201), 1028 (196a), 1055 (126k), 1077 (197), 1112 (177i), 1118 (148), 1224 (16), 1233 (196a), 1269 (201), 1309 (27), 1347 (140), 1362 (146), 1402 (196a), 1426 (27), 1456 (177h), 1492 (201), 1524 (27), 1573 (177i), 1624 (27), 1709 (16), 1752 (196a), 1756 (27), 1763 (201), 1772 (196a), 1783 (177i), 1861 (196a), 2009 (201), 2084 (201), 2161 (201), 2173 (201), 2357 (199), 2407 (71), 2650 (201), 2822 (71), 3314 (199), 3334 (27), 3376 (201), 3607 (71), 3655 (71), 4022 (195)
Templeton, B. 7452 (11), 7628 (26a)
Terborgh, J. 94 (168a), 103 (199), 221 (168a), 297 (140)
Tessmann, G. 3109 (126e), 3125 (3a), 3542 (26a), 3545 (147), 3554 (161/Ia), 3769 (3a), 4789 (7)
Thackery, F. A. 296 (90)
Tharp, B. A. 43-613 (84), 43Q007 (84), 49-1023 (84)
Tharp, B. A. & M. C. Johnston 5418888 (83b)
Thornber, J. J. 71 (138e), 132 (87), 196 (80)
Thorne, R. F. 43919 (90)
Thurber, G. 388 (193c), 828 (82)
Thurm, D. 139 (193a), 214 (161/III)
Thurn, E. im 1910 (161/Ia)
Tiehm, A. 3285 (90)

Tilforth, C. W. 570 (126m)
Tillett, S. S. & C. L. 45383 (8a), 45690 (161/Ia), 45778 (9), 44957 (9)
Tina, F. 2449 (8b)
Tinsley, J. D. 7 (84)
Toledo, V. M. 49 (161/III)
Tolstead, W. L. 7297 (84), 7482 (88)
Ton, A. S. 784 (116), 1076 (72b), 1230 (72b), 1391 (16), 1490 (94c), 1536 (72b), 1539 (68), 1613 (16), 1820 (72b), 1847 (94c), 1865 (177k), 2490 (119), 2552 (119), 2581 (173), 2700 (177g, bis), 2911 (173), 2981 (71), 3062 (202a), 3080 (16), 3088 (177a), 3143 (177a), 3147 (138f), 3148 (71), 3172 (149), 3203 (199), 3207 (199), 3235 (140), 3390 (149), 3392 (178a), 3393 (177a), 3423 (199), 3610 (147), 3893 (147)
Tonduz, A. 697 (119), 1484 (173), 1563 (13a), 3197 (173), 3942 (26a), 4741 (26a), 7007 (177o), 7213 (173), 8334 (13a), 8652 (140), 9611 (8a), 9779 (138f), 9824 (177o), 12714 (177o), 12833 (126k), 13546 (177a), 13551 (149), 13552 (138f), 13555 (16), 17867 (140)
Torke, K. et al. 113 (75), 120 (88), 135 (75), 369 (193a)
Toro, R. A. 7 (138f), 61 (16), 70 (138f), 266 (202a), 416 (16), 561 (119), 584 (8b), 748 (116), 839 (153a), 866 (153a)
Toroes, R. S. 3107 (8a)
Toumey, J. W. 31 (87)
Tovar, O. 2282 (68), 5133 (110d)
Tracey, Mrs. 302 (154)
Tracy, S. M. 565 (69), 6457 (69), 8006 (88)
Traill, J. W. H. 172 (161/Ia)
Travassos, O. 196 (44b)
Trejo, C. 1107 (177a)
Triana, J. 457 (13a), 4384 (17b), 4401 (126i), 4402 (116), 4404 (31), 4405 (166), 7385 (47)
Trinidad Hb. 11129 (26a)
Trinta, Z. A. & E. Fromm 491/1567 (42a), 550/1626 (161/IIa), 639/1713 (43), 801/1877 (126h)
Troetsch, M. 9 (16)
Troll, C. 128 (139), 181 (63a), 182 (180), 441 (196b), 619 (11), 1136 (118), 2273 (68), 2587 (46b)
Trujillo, B. 231 (126k), 542 (177a), 688 (199), 1159 (159), 1427 (199), 1566 (71), 1741 (10a), 1747 (69), 1778 (69), 1795 (140), 1827 (202a), 1830 (157), 1851 (170), 1857 (10a), 2000 (170), 2062 (140), 2077 (147), 2154 (157), 2340 (69), 2612 (172), 2674 (172), 3281 (147), 3481 (140), 3485 (69), 3759 (161/Ia), 3760 (26a), 3924 (140), 4467 (140), 4750 (199), 4948 (199), 5051 (172), 5054 (202a), 5057 (177a), 5194 (115), 5452 (177a), 5484 (147), 5488 (55), 5546 (172), 5546a (170), 5573 (169), 5826 (140), 5891 (69), 5904 (161/Ia), 5904a (161/Ia), 6064 (161/Ia), 6249 (199), 6369 (129a), 6403 (71), 6654 (71), 6831 (172), 6952 (172), 7108 (69), 7265 (147), 7442 (199), 7447 (199), 7469 (129a), 7906 (71), 7985 (138g), 8694 (69), 10409 (202a), 11560 (8a), 11887 (69), 11892 (71), 12030 (157), 12094 (69), 12178 (69), 12221 (148), 12258 (69), 12309 (69), 12924 (69), 13021 (140), 13143 (148), 14013 (27), 14212 (147), 14346 (140)
Tucker, J. M. 555 (178a), 929 (178b), 1131 (94a)
Tuerckheim, H. von 442 (68), 1091 (138f), 1422 (177l), 2366 (177a), 2504 (168a), 2618 (71), 2782 (148), 3344 (131), 3607 (168a), 4129 (16), 7847 (68), 8194 (201), 8512 (138f), II/1773 (119)
Tun Ortíz, R. 463 (71), 559 (196a), 1103 (196a), 1280 (201), 1281 (27), 1835 (196a), 2189 (13a), 2669 (201)
Turner, B. L. 4600 (73)
Tutin, T. G. 237 (20, 26a), 1062 (95a), 1137 (16),
Tyson, E. L. 1137 (16), 1350 (16), 1458 (14a), 1628 (6), 2097 (16), 3104 (8a), 3762 (27), 5084 (8a), 5129 (16), 5215 (138g), 5238 (69), 5711 (122), 5937 (6), 6425 (167)

Ugent, D. & V. 3809 (95a), 3856 (95a), 4090 (95a), 4486 (95a), 4641 (96), 4875 (118), 5191 (110a), 5195 (110b), 5252 (95a), 5396 (174), 5520 (174), 5550 (95a)
Ugent, D. & C. Vargas 4431 (110c)
Ule, E. 1016 (121), 2482 (39b), 3670 (104d), 3708 (104b), 5434 (26a), 6729 (63b), 7250 (164c), 7789 (27), 8142 (8a), 8144 (161/Ia), 8869 (38), 8870 (26a), 9443 (126e)
Underwood, L. M. & F. S. Earle 128 (199), 863 (141), 1619 (141)
Underwood, L. M. & R. F. Griggs 29 (140), 121 (69), 475 (199), 534 (169a), 705 (169a), 746 (119)
Uribe Uribe, L. 543 (126k), 786 (170), 885 (170), 1328 (13b), 1706 (31), 2193 (17b), 4895 (47)
Usteri, P. A. 60b (42a), 61b (136)
Utley, J. & K. 4907 (8a)

Valdes R., J. 480 (193c), 533 (82)
Váldez, J. 26 (69)
Valerio, M. 497 (177a)
Valeur, E. J. 412 (140), 587 (126k), 857 (199)
Valverde, F. M. 321 (147), 379 (129a), 507 (10b)
Varela, – 730 (118)
Vareschi, V. & E. Foldats 4543 (3a)
Vargas C., C. 940 (153b), 985 (123), 1717 (17a), 3374 (108), 6449 (21), 6958 (110c), 14572 (68), 15197 (25b), 15214 (21), 16075 (25b), 16303 (20), 18768 (21)
Vasconcelos, J. M. 295 (176), 367 (176), 511 (37)
Vásquez, M. 122 (177p)
Vaughan, J. 983 (94e), 1071 (94e)
Vauthier 135 (57), 150 (124)
Velarde, M. O. 121 (110f)
Velasco, R. 130 (202a), 223 (129a), 408 (138g)
Velásquez C., A. 7 (16)
Vélez, C. 22 (16)
Vélez, I. 903 (169a)
Vellozo, H. P. 185 (12)
Ventura A., F. 592 (116), 944 (199), 1150 (68), 1295 (13a), 2431 (119), 2534 (177a), 2596 (140), 2626 (11), 2697 (177a), 2699 (177a),

2707 (199), 2708 (199), 2907 (146), 3243 (116), 7144 (126m), 9036 (138f), 10576 (146), 11201 (199), 11892 (138f), 12138 (177a)
Venturi, S. 119 (118), 2660 (128), 2871 (130), 3706 (138b), 4269 (110a), 6002 (130), 6170 (130), 7681 (118), 8152 (110a), 8167 (138b), 8815 (130), II/73 (118)
Vera, B. 14 (129a)
Vera Santos, J. 2558 (13a), 3633 (177a)
Vergara, C. G. 10 (16)
Vianna, M. C. 61 (42a)
Victorin, Bro. & Bro. Clemente 23103 (135)
Vidal, J. I/576 (43), I/578 (42b), I/622 (42b), I/812 (43)
Vieira, M. G. et al. 1055 (3a)
Vilas, F. 48 (178a), 49a (68), 55 (178a), 251 (178a)
Villa Carrenzo, – et al. 1723 (122)
Vincelli, P. C. 324 (13a), 446 (199), 750 (13a)
Vinha, S. G. da 112 (26a)
Vinha, S. G. da & R. S. Pinheiro 18/166 (44a), 70/218 (44a)
Vogel, C. 109 (170), 503 (10a), 566 (140), 571 (63b), 572 (177a), 574 (10a), 1341 (147)

Wachenheim, – 97 (26a)
Wagner, R. J. 413 (126n)
Walker, J. W. 297 (29)
Walker, S. 74H32 (138e), 74H72 (193d), 75H06 (199)
Walker, T. 78H29 (193d)
Wallace, Dunn & Ellis 369 (55), 374 (77)
Wallich, N. 5310 (161/Ia)
Ward, D. B. 7790 (199), 7833 (140), 7925 (11)
Warming, E. 2 (177a), 3 (177a), 8 (27), 9 (8a), 17 (10a), 179 (69), 209 (44a), 212 (12), 222 (57), 227 (136), 228 (138c), 230 (126a), 231 (124), 233 (63b), 234 (138c), 239 (63b), 249 (70a), 259 (57), 296 (61), 285 (161/IIa), 708 (63b), 747 (42b)
Warner, R. H. 176 (16)
Warnock, B. H. W-44 (86)
Wasshausen, D. C. & F. Encarnación 530 (126e)
Waterfall, U. T. 4266 (84), 4824 (193c), 5225 (84), 5377 (73), 6277 (73), 6625 (73), 7579 (69), 7783 (73)
Watkins, A. 35 (169a)
Watkins, Lecloux & D. Dunn 583 (193c), 595 (193c)
Wawra von Fernsee, H. 13 (40), 70 (140), 2467 (199)
Weaver, R. E. 1757 (72b)
Weber, O. 15 (138e)
Weberbauer, A. 237 (116), 4277 (42h), 5368 (129b), 5756 (102), 6173 (177f), 6285 (179), 6574 (25a), 7608 (110c), 7709 (200b)
Webster, G. L. 12495 (119), 12793 (177c), 12960 (11), 16336 (196b), 20150 (193a)
Weddell, H. A. 16 (69), 2102 (60), 2329 (46a), 3065 (63a), 3363 (46a), 3363bis (46a), 4171 (103)
Wedel, H. von 93 (13a), 171 (13a), 517 (131), 613 (8a), 1205 (8a), 2626 (13a), 2730 (8a)
Weir, J. 51 (148)
Werdermann, E. 122 (98d), 139 (98a), 427 (100), 844 (98b), 917 (98d), 2041 (117), 2380 (46a)
Wessels Boer, J. G. 226 (36), 227 (26a)
West, R. C. 26 (116)
Westcott, R. C. 230 (177k)
Wetmore, R. H. & E. C. Abbe 54 (8a), 55A (27), 60 (8a), 75 (8a)
Wheeler, W. M., R. Rose & W. Beebe 6 (71)
Whiffin, T. 37 (140)
White, J. W. & R. W. Warner 99 (14b)
White, O. E. 975 (46a)
White, S. S. 3525 (73), 5199 (177g)
Whitmore, J. L. 67 (16)
Whitmore, T. C. 793 (148)
Whitton, B. A. 270 (9)
Wied-Neuwied, Prince Maximilian zu 43 (42c), 52 (39a)
Wiggins, I. L. 5381 (101), 5582 (199), 5588 (146), 6304 (177d), 6307 (177d), 6336 (177d), 6383 (177d), 6452 (177d), 7724 (101), 10949 (112), 11318 (101), 13300 (199), 18821 (129a)
Wiggins, I. L. & W. R. Ernst 652 (101)
Wiggins, I. L. & D. Porter 712 (153c)
Wiggins, I. L. & R. C. Rollins 319 (177d)
Wilbur, R. L. 1416 (152), 7580 (140), 9783 (16), 13350 (27), 20196 (55)
Wilhelminagebirgte Expeditie 111 (2)
Wilkinson, E. H. 99 (88)
Williams, Ll. 50 (17a), 89 (8b), 365 (17a), 384 (17a), 392 (25a), 532 (17a), 566 (8b), 687 (26a), 1290 (26a), 1537 (26a), 1762 (8b), 2070 (3a), 2112 (26a), 2122 (26a), 2143 (26a), 3798 (26a), 3929 (22), 4020 (26a), 4171 (26a), 5541 (22), 5828 (22), 6125 (22), 6237 (22), 6312 (25a), 7633 (20), 7907 (25a), 7965 (8b), 10058 (171), 10254 (17a), 10389 (10a), 10391a (10a), 10471 (29), 10599 (177a), 11162 (10a), 11222 (8a), 11473 (8a), 11491 (8a), 12167 (10a), 12256 (10a), 12315 (171), 12403 (177a), 12620 (8a), 13247 (34), 13502 (172), 14191 (17a), 14725 (161/Ia), 14825 (26a), 15104 (17a), 15142 (26a), 15168 (17a)
Williams Ll. & A. H. G. Alston 43 (177a), 365 (27)
Williams, L. O. 5209 (44a), 10944 (149), 12345 (138g), 12466 (126k), 15246 (16), 22276 (177g), 23682 (8a), 23742 (119), 23781 (177a), 25228 (177k), 26415 (177a), 26546 (202a), 26699 (138f), 27445 (138f), 27842 (13a), 27857 (119), 28474 (26a), 28974 (177o)
Williams, L. O. & V. Assis 5896 (44a), 6074 (43), 6488 (44a), 6489 (39b), 6733 (161/IIa), 6965 (39b)
Williams, L. O. & A. Molina A. 10580 (16), 10631 (177a), 13598 (177a), 13639 (177a), 14528 (16), 14698 (16), 14915 (177a), 15029 (177a), 15851 (16), 20214 (55), 26613 (16), 40561 (16), 41139 (177k), 41245 (177k), 42389 (177a), 42554 (16)
Williams, R. S. 15 (27), 135 (16), 137 (202a), 161 (138f), 226 (21), 537 (69), 540 (140), 560 (8a), 567 (167), 596 (27), 700 (3a), 717 (147), 1679 (96), 2344 (95a), 2345 (96), 2383 (108)

Wilson, C. L. 55 (27), 81 (27), 109 (8a)
Wilson, P. 80 (147), 243 (140), 353 (68), 647 (147), 1228 (30), 1327 (138f), 7187 (134e), 7326 (177j), 7351 (134e), 7370 (177j), 7593 (134d), 7639 (177j), 7739 (129a), 7817 (134e), 8408 (134e), 9478 (129a), 9494 (141), 9561 (126m), 11399 (71)
Wilson-Browne, G. 63 (8a), 137 (33), 168 (161/Ia), 364 (161/Ia), 367 (8a)
Windler, D. 1006 (199), 2832 (69), 2833 (69), 2871 (177a), 2872 (140), 2878 (71), 2935 (177a), 2993 (173), 3023 (177a)
Wingfield, P. 5147 (199), 5354 (199)
Wingfield, R. 6052 (55)
Winzerling, H. W. 102 (16)
Witherspoon, J. D. 8569 (26a)
Wolf, C. B. 7087 (90), 8532 (90), 11353 (73)
Wolf, C. B. & P. C. Everett 11396 (138e)
Wood, C. E. 237 (17a)
Woodson, R. E. 1766 (16)
Woodson, R. E. & R. W. Schery 1012 (14a)
Woodworth, R. H. & P. A. Vestal 339 (27), 413 (8a)
Woolston, A. L. 169 (138b), 466 (69), 494 (63c), 662 (63a), 944 (148), 1355 (63c), 1454 (63c), 1455 (63c), 1456 (63c), 1617 (63a)
Wooton, E. O. 93 (87)
Worth, C. R., J. L. Morrison & O. B. Horton 8949 (63b)
Woytkowski, F. 383 (21), 1262 (161/Ia), 5369 (20), 5499 (22), 5970 (25a), 6256 (8b), 6927 (153b), 7130 (17a), 7273 (8b), 7277 (22), 7352 (25a), 7522 (25a), 7904 (20), 8328 (25a), 8365 (22), 34371 (147), 34413 (3a), 35140 (22)
Wright, A. E. 117 (131), 194 (129a)
Wright, C. 148 (71), 150 (4), 1030 (87), 1031 (88), 1032 (138e), 1033 (193c), 1601 (133), 1602 (30), 1612 (30)
Wright, W. G. 1215 (177a)
Wullschlägel, H. R. 138 (138g), 819bis (29), 955 (69), 1418 (36)
Wunderlin, R. 731 (55)
Wurdack, J. J. 121 (69), 148 (202b), 1958 (17a), 2225 (25a), 2233 (3a), B-147 (44b)
Wurdack, J. J. & C. S. Adderley 43395 (3a), 43587 (26a)
Wurdack, J. J. & J. Monachino 39688 (202a), 39819 (3a), 39877 (161/Ia), 39904 (34), 39964 (1261), 40916 (161/Ia), 41022 (3a), 41179 (53a), 41261 (27)
Wurth, A. 61 (126o)

Ybarrola, T. 328 (128), 2560 (128), 2692 (128), 2757 (128)
Yuncker, T. G. 8082 (140), 8213 (8a), 17246 (29), 17415 (168b)

Zarucchi, J. L. 1242 (53b), 1571 (3a), 1828 (53b), 1876 (147), 2025 (3a), 2242 (3a), 2338 (116), 2352 (25a), 2503 (53b), 2805 (43)
Zehntner, M. 124 (10a), 2055 (42e)
Zehntner, M. & P. von Luetzelburg 5036 (202b)
Zelaya, H. 151 (147)
Zetek, – 5592 (16)
Zizumbo & Colunga 202 (197), 282 (71), 360 (199), 373 (140), 473 (178a)
Zola B., M. G. 81 (202a)
Zöllner, O. 1849 (98a), 2938 (98a), 3769 (110d), 5180 (109b), 5812 (99), 6025 (98c), 6028 (125), 7148 (98c), 7837 (99), 7998 (125), 7999 (109b), 8290 (98a)

CHAMAECRISTA [Breyne] Moench

Chamaecrista [Breyne] Moench, Meth. pl. hort. bot. Marburg. 272. 1794.—Generitypus infra sub sect. *Chamaecrista* indicatur.

Cassia sect. *Lasiorhegma* Vogel, Syn. Gen. Cass. 8. 1837, nom. illegit. (incorporating 3 prior names of same rank). *C.* subgen. *Lasiorhegma* Vogel ex Bentham in Martius, Fl. Bras. **15**(2): 129. 1870.

Cassia subgen. *Absus* (DeCandolle) Symon, Trans. Roy. Soc. S. Australia **90**: 77. 1966.

Chamaecrista sensu Greene, Pittonia **3**: 238–243. 1897; Britton, Bull. Torr. Club **44**: 3–12 & Pennell in op. cit. pp. 339–362; Britton & Rose, N. Amer. Fl. **23**(5): 270–299. 1930; Britton & Killip, Ann. N.Y. Acad. Sci. **35**: 182–187. 1936; Pittier, Cat. Fl. Venezol. **1**: 373–375. 1945, these, however, defined to accommodate only sects. *Chamaecrista, Caliciopsis* and *Xerocalyx*; sensu ampliat. Irwin & Barneby, Brittonia **28**(1): 28–36. 1976; in Polhill & Raven, 1981, p. 106.

Cassia subgen. *Lasiorhegma* sensu Irwin & Barneby, Mem. N.Y Bot. Gard. **30**. 2–3. 1977.

General characters of subtrib. Cassiinae, with these particular ones: pedicels 2-bracteolate near or above middle; pod elastically dehiscent, the valves coiling; androecium (erratically) actinomorphic, the 2 cycles of stamens often of different lengths but not bilaterally zygomorphic, the thecae of the anthers ciliolate along the lateral sutures; extrafloral nectaries (Fig. 36), when present, dish- or cup-shaped, rarely flat, secreting nectar from a concave (flat) surface; seed-funicle deltately dilated; seed-coat either smooth or pitted, exareolate.—Trees, shrubs, herbs, these perennial or monocarpic; roots bearing bacterial nodules; phyllotaxy spiral or distichous; inflorescence of 1–many-fld racemes, these mostly axillary from hornotinous stems, sometimes cauliflorous, sometimes appearing supra-axillary when peduncle adnate to stem; petals yellow, sometimes red-spotted near claw or suffused with orange-red, nearly always highly heteromorphic, the 2 abaxial ones (sometimes by resupination appearing adaxial) variously oblique and dissimilar between themselves, the larger one often shielding the androecium or interposed between it and the obliquely divergent (enantiostylous) pistil, the vexillar petal commonly interior in praefloration but sometimes exterior on one or both sides; androecium functionally (2–)5–10-merous, the basifixed anthers longer than their filament, either iso- or heteromorphic, each theca apically dehiscent by a pore or short slit; pod plano-compressed, very rarely winged along the sutures, the valves papery, leathery or subligneous; $2n = 14, 16$.—A genus of 6 sections and ±265 species, all sections and 239 species native to the Americas, represented in the Old World by the circumtropical monotypic sect. *Absus* and by numerous species of sect. *Chamaecrista*.

The syndrome of bracteolate pedicels, radial or irregular but never bilateral symmetry of the androecium, basifixed anthers ciliolate along the permanently closed lateral sutures, elastically dehiscent pod and dilated seed-funicle present in every *Chamaecrista* known to us establishes the genus on an impregnable morphological foundation. Bacterial root-nodules (Corby, 1974) and concave extrafloral nectaries (on leaf-stalk or raceme-axis), when present, are further characteristic, for they occur in no other Cassiinae. The checkered history of the genus and its modern interpretation as equivalent to Bentham's *Cassia* subgen. *Lasiorhegma* have recently been examined in detail (Irwin & Barneby, 1976), and a summary redefinition of it has appeared in our synopsis of the subtribe Cassiinae (Irwin & Barneby in Polhill & Raven, 1981).

Chamaecrista comprises six sectional groups of very unequal size, lately defined as sections of *Cassia* subgen. *Lasiorhegma* in the introduction to our monograph of sects. *Absus* and *Grimaldia* (Irwin & Barneby, 1977, pp. 2–3). The present synopsis, while comprehensive, deals with each section in varying depth of detail, as seemed appropriate in relation to previously published studies:

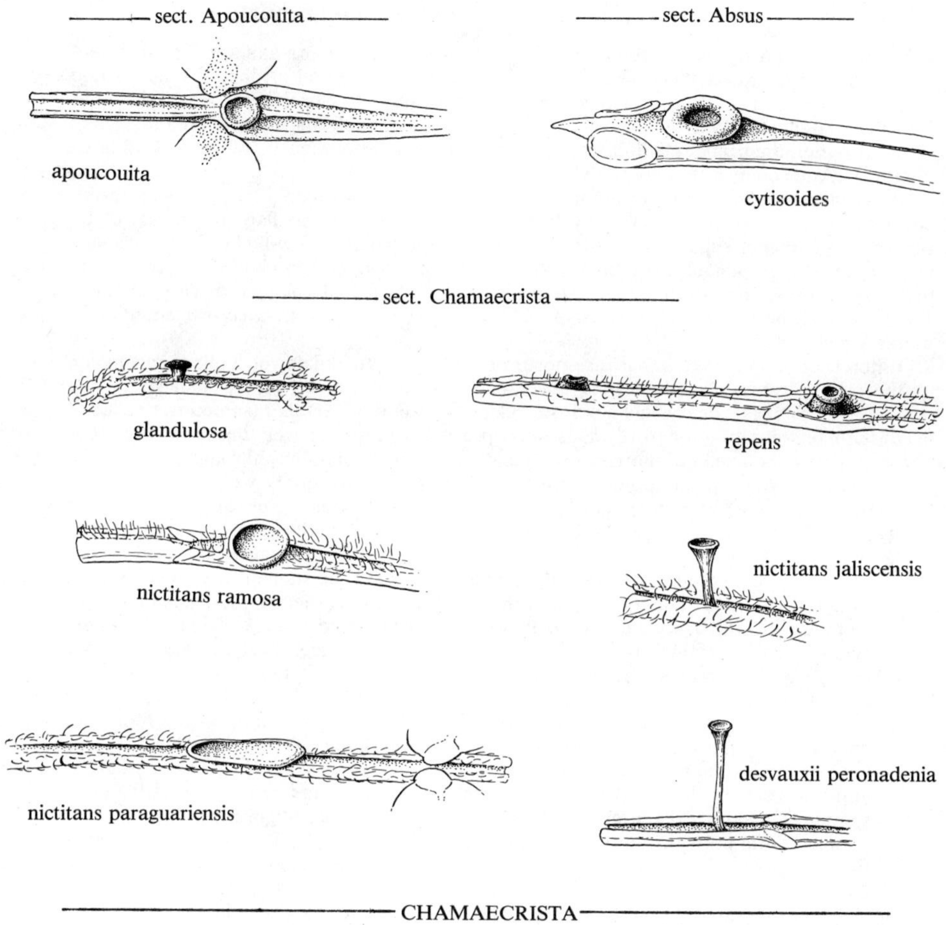

Fig. 36. Petiolar nectaries (glands) of genus *Chamaecrista*: *Ch.* (*Apoucouita*) *apoucouita* (Aubl.) I. & B.; *Ch.* (*Absus*) *cytisoides* (DC.) I. & B.; *Ch.* (*Chamaecrista*) *glandulosa* (Linn.) Greene, *repens* (Vog.) I. & B., *nictitans* (Linn.) Moench vars. *ramosa* (Vog.) I. & B., *jaliscensis* (Greenm.) I. & B. and *paraguariensis* (Chod. & Hassl.) I. & B.; *Ch.* (*Xerocalyx*) *desvauxii* (Collad.) var. *peronadenia* I. & B.

Sect. (A) *Apoucouita* was revised by Irwin & Rogers in 1967 and again, in light of new material, by Irwin & Barneby in 1977; a bare outline, accompanied by a key for identification, and a synoptical list of species are here presented.

Sect. (B) *Absus* and sect. (C) *Grimaldia* were revised by Irwin & Barneby in 1977, as sections of *Cassia*. A census of subsections, series, species and varieties, transferred as necessary to genus *Chamaecrista,* appears sufficient in this place.

Sect. (D) *Chamaecrista* (51 spp.) and (E) *Caliciopsis* (2 spp.) are subjected to detailed analysis and redefinition. Full descriptions, keys to the taxa and a list of numbered Exsiccatae are provided.

Sect. (F) *Xerocalyx,* subject of the senior author's doctoral dissertation (Irwin, 1964, the seminal paper in the series concluded herewith), has required extensive elaboration in order to accommodate materials collected on the Brazilian Planalto in recent years. New perspectives gained in fieldwork have imposed a radically revised taxonomy. New keys and descriptions for all taxa are given.

Key to the Sections of genus *Chamaecrista*

1. Inflorescences mostly cauliflorous, arising singly or in serial supra-axillary files from annotinous or older wood, but sometimes axillary to living lvs; pod at once pendulous and large (8–20 cm); testa of seeds finely minutely pitted overall. Shrubs and trees; glandular setae 0; scutellate or verruciform glands present on leaf-stalk and axes of inflorescence; stamens 10 homomorphic; phyllotaxy either spiral or distichous; spp. 17, of tropical America (Guianas, Amazonia, n.-e. and planaltine Brazil. A. sect. *Apoucouita* (p. 638).
1. Inflorescences arising from hornotinous branches, either terminal racemose, or terminal paniculate, or by reduction of axillary branchlets becoming axillary racemose, or by further shortening or suppression of the raceme-axis reduced to a few-fld fascicle or a single fl; pod very rarely pendulous, then smaller, exceptionally over 7 cm and if so erect; testa of seeds various, smooth or pitted, when pitted the pits usually aligned in vertical rows. Trees, shrubs, herbs, the latter (in sect. *Absus* rarely, in sect. *Chamaecrista* commonly) monocarpic.
   2. Inflorescence simple terminal racemose, or racemose-paniculate, or axillary racemose, the primary axis of each raceme developed and several-fld (if rarely reduced to one flower the pubescence viscid-setulose); glandular setae or setulae (sometimes reduced to their bulbous base or to resinous dots) present (except subsect. *Baseophyllum, Adenophyllum*) on axes of inflorescence, sepals, or ovary, commonly throughout.
      3. Perennial (except monotypic subsect. *Otophyllum*), commonly woody (suffruticose to arborescent) or functionally herbaceous from xylopodium; phyllotaxy of mature stems (except in very few shrubs and subshrubs) spiral; androecium 10-merous, concealed by a strongly heteromorphic, dimidiately ovate or falcately hemi-lanceolate, involute or convolute abaxial petal interposed between the stamens and the obliquely exserted pistil, the filaments all less than half as long as their anther, this dehiscent at anthesis of the fl; style terete, neither dilated nor hooked at apex, the minute stigma symmetrically terminal, poriform; no linguiform appendage between pairs of lfts and, if any gland on lf-stalk, these minute spiculiform (a large scutellate gland in subsect. *Baseophyllum*); spp. 166, tropical American. B. sect. *Absus* (p. 644).
      3. Monocarpic; phyllotaxy of mature stems distichous; androecium 2–7-merous, not concealed by an involute heteromorphic petal, the longer filaments at least half as long as their anther, this dehiscent before anthesis; style short (1.5–2 mm) stout, at anthesis scarcely longer than ovary, grooved ventrally and passing upward into a spatulate, distally dilated and incurved, obliquely truncate stigma; between each of 2 pairs of lfts a linguiform appendage, that between the distal pair narrower, sometimes setiform; fls small, 4.5–7 mm, at least facultatively cleistogamous; sp. 1, circumtropical. C. sect. *Grimaldia* (p. 664).
   2. Inflorescence composed of axillary racemes reduced either to a few-fld fascicle or to a solitary fl, the common peduncular axis often adnate to the stem-internode and the fl(s) then appearing supra-axillary; phyllotaxy of mature stem or branches distichous (but the 2 ranks of lvs themselves sometimes spiral); glandular setae 0.
      4. Venation of lfts pinnate (sometimes only from proximal side of forwardly displaced midrib; sepals subequal in length, parallel-veined only in small sect. *Caliciopsis*; fls not resupinate, the heteromorphic abaxial petal obliquely saucer-shaped, standing below and to one side of androecium; seeds in outline obovate to rhombic or trapezoid; $x = 8$.
         5. Sepals delicately reticulate-venulose or apparently veinless; spp. ±75, of these 51 American, both tropical and temperate, the rest in Africa, Asia, Australia, Oceania. D. sect. *Chamaecrista* (p. 664).
         5. Sepals parallel-veined, the veins prominulous and subcontiguous; spp. 2, of tropical and warm-temperate N. and S. America. E. sect. *Caliciopsis* (p. 857).
      4. Venation of lfts parallel; sepals at once strongly graduated in length and prominently parallel-veined, the veins subcontiguous; fls resupinate, the heteromorphic petal, really abaxial, becoming functionally adaxial, forming a shed-roof over the androecium; seeds in outline narrowly obovate or oblanceolate; $x = 7$; spp. 3, tropical American. F. sect. *Serocalyx* (p. 862).

A. Sectio APOUCOUITA (Bentham) Irwin & Barneby

Chamaecrista sect. **Apoucouita** (Bentham) Irwin & Barneby, Brittonia **31**(1): 155. 1979. *Cassia* sect. *Apoucouita* Bentham in Martius, Fl. Bras. **15**(2): 129. 1870.—Sp. typica: *C. apoucouita* Aublet ≡ **Chamaecrista apoucouita** (Aublet) Irwin & Barneby.

Cassia subgen. *Lasiorhegma* sect. *Apoucouita* sensu Bentham, 1871, pp. 577–578; Irwin & Rogers, Mem. N.Y. Bot. Gard. **16**: 71–118. 1967; Irwin & Barneby, Brittonia **29**(3): 277–290. 1977.

For description of the section and for full synonymy of each species see Irwin & Rogers, 1967 and Irwin & Barneby, 1977, ll. cc. We present here a key to the series, species and varieties of sect. *Apoucouita* which incorporates the known dispersal of each; a bare enumeration establishing the nomenclature in genus *Chamaecrista*; and minimal commentary as seemed appropriate.

Key to Taxa of sect. *Apoucouita*

1. Lfts of larger lvs at once small and numerous, 15–35 pairs and ±1–2 cm long, their blade simply pinnately veined (tertiary venulation fully immersed): ser. *Pteridophyllae*
  2. Phyllotaxy distichous, the lvs of annotinous branches arranged in a single plane; lvs subsessile, the petiole with pulvinus 3–6 mm; lfts of larger lvs 20–35 pairs, crowded along rachis at points ±3–5 mm apart; Amazonia and Guayana Highland; 1. *Ch. adiantifolia*:
    3. Pulvinules of lfts 0.6–1.1 mm, elevating the base of blade free from lf-stalk; apex of lfts obtuse; local on upper Rios Negro and Vaupés in n.-w. Amazonas, Brazil and Colombia, and on adjoining Río Guainia in s.-w. Amazonas, Venezuela. 1a. var. *adiantifolia* (p. 641).
    3. Pulvinules of lfts 0.2–0.4 mm, the base of blade standing sessile against the lf-stalk; apex of lfts abruptly emarginate; interruptedly widespread through middle and lower Amazon valley in Amazonas and Pará, Brazil, and around s. edge of Guayana Highland about Mt. Roraima, and on headwaters of the Orinoco in Guyana and in Amazonas, Venezuela. 1b. var. *petridophylla* (p. 641).
  2. Phyllotaxy spiral, the lvs scattered around the branchlets; lvs obviously petioled, the petiole with pulvinus 7–13 mm; lfts of larger lvs 13–19 pairs, inserted on rachis at points ±5–7 mm apart; coastal forest of Espírito Santo, Brazil, near 19°20′S, and valleys of rios Jequitinhonha, Mucuri and Doce in adjacent Minas Gerais. 2. *Ch. aspleniifolia* (p. 641).
1. Lfts of larger lvs commonly 2–10 pairs and their longer lfts 3+ cm long, in *Ch. aspidiifolia* 9–13 pairs and less than 2.5 cm long but then sharply reticulate-venulose: ser. *Apoucouita*.
  4. First gland of lf-stalk situated on petiole proper, distinctly below the proximal pair of pulvinules; lfts 2–3 pairs; local species of Guayana Highland and Orinoco-Amazon divide.
    5. Lfts asymmetrically ovate-elliptic short-acuminate, the midrib strongly curved; inflorescence from leafless annotinous branchlets; anthers villosulous overall; Rio Negro-Orinoco divide in Amazonas, Brazil and adjacent Terr. Amazonas, Venezuela. 3. *Ch. hymenaeifolia* (p. 641).
    5. Lfts subsymmetrically obovate-suborbicular, broadly obtuse or emarginate, the midrib straight; inflorescence from lf-axils of hornotinous branchlets; anthers ciliolate along sutures, otherwise glabrous; interior Guyana and adjacent Terr. do Roraima, Brazil. 4. *Ch. polystachya* (p. 641).
  4. First gland of lf-stalk situated exactly between or immediately below and contiguous to the proximal pair of pulvinules or, if shortly below this point, either the lfts more than 3 pairs, or the dispersal otherwise, or both.
    6. Inflorescences borne mostly on hornotinous branchlets, associated with living lvs; pulvinules of the 2–3 pairs of lfts 3–7 mm; fls small, the longer petals ±10–13 mm; coastal plain of Bahia and Espírito Santo, Brazil. 6. *Ch. duartei* (p. 642).
    6. Inflorescences cauliflorous, borne on leafless annotinous or older branchlets; if nearly or actually sympatric with the last then either the lfts more numerous and pulvinules shorter or, if the foliage similar, then the fls larger, the longer petals to 15–23 mm.
      7. Androecium glabrous; trunk of tree deeply sulcate lengthwise, appearing fluted; lfts 2–4 pairs, sessile or almost so; pedicels 15–30 mm; bicentric in range, of the Guianas and upper-middle Amazon valley in Brazil. 11. *Ch. apoucouita* (p. 642).
      7. Androecium villosulous or velutinous; trunk sulcate only in *Ch. scleroxylon* which, with foliage of the preceding, has pedicels only 5–10 mm.
        8. Pulvinule of lfts elongate, (3–)3.5–9 mm. Lfts 2–4 pairs.
          9. Sepals not over 6 mm, pubescent dorsally.

10. Petiole including pulvinus (8–)12+ mm; ovary puberulent; trees 8–20 m.
  11. Lfts either 2, 3 or 4 pairs, their blade rounded to subcordate at base; widespread in Amazonian Hylaea. 7. *Ch. negrensis*:
    12. Lfts 2–3 pairs, broadly ovate, short-acuminate (8–) 10–20 × 3.5–9 cm; Rio Negro and middle Amazon basins. 7a. var. *negrensis* (p. 642).
    12. Lfts 3–4 pairs, narrowly ovate, caudately long-acuminate 5.5–10 × 2–3.5 cm; middle Rio Solimões basin. 7b. var. *albuquerquei* (p. 642).
  11. Lfts exactly 2 pairs, their blade cuneate at base; lower Amazon valley and disjunctly on coastal plain of Bahia. 8. *Ch. bahiae* (p. 642).
10. Petiole subobsolete 4–5 mm, reduced to the pulvinus and length of proximal petiolar gland; ovary glabrous; small shrubs of white sand caatinga on sources of Rio Negro near Brazil-Colombia boundary. 10. *Ch. aiarana* (p. 642).

9. Sepals (6.5–)7–10 mm, glabrous dorsally; coastal plain of Bahia near 14°S. 9. *Ch. onusta* (p. 642).

8. Pulvinule of lfts short or subobsolete, not over 3(–3.5) mm. Lfts 2–10 pairs.
  13. Petiole subobsolete 2–2.5 mm, as long as petiolar gland; blade of lfts obliquely dilated below middle on distal side, the distal margin abruptly obtusangulate at base; Terr. do Roraima, Brazil. 5. *Ch. boyanii* (p. 642).
  13. Petiole 3.5+ mm, or lfts ± equilateral at base, or both; allopatric.
    14. Stipules obsolete or minute and caducous, not over 1 mm; widespread over e. and s.-e. Brazil, either shrubby or arborescent.
      15. Blade of lfts cordate at base on proximal or on both sides of pulvinule, consequently appearing sessile or subsessile against lf-stalk.
        16. Lfts 7–13 pairs and only 6–35 mm long.
          17. Pulvinules (viewed dorsally) ±1 mm; lfts 7–9 pairs, mostly oblong-elliptic, the longer ones 20–35 mm, at base on proximal side produced backward into a short blunt auricle; s.-e. Maranhão, near 6°30′S. 17. *Ch. subpeltata* (p. 642).
          17. Pulvinules 0.3–0.5 mm; lfts 9–13 pairs, obliquely ovate, angulately dilated on distal side, the longer ones 10–21 mm, at base on proximal side cordate; s.-e. Bahia, near 15°S. 16. *Ch. aspidiifolia* (p. 642).
        16. Lfts 2–5 pairs, the larger ones 4–10 cm long.
          18. Petiole deeply canaliculate; pedicels at anthesis 0.5–1 cm; fl small, the longer petals less than 1 cm; ovary glabrous; trees with fluted trunk of lower Amazonian Hylaea in Pará. 12. *Ch. scleroxylon* (p. 642).
          18. Petiole subterete; pedicels at anthesis 1–2 cm; fl larger, the longer petals 1.3–2.1 cm; ovary densely gray-strigulose; shrubs and treelets of seasonally arid formations in n.-e. Brazil; 14. *Ch. eitenorum*:
            19. Lfts of all lvs exactly 2 pairs, the rachis 1.4–2.5 cm; first gland of lf-stalk concave, shortly stoutly stipitate, 0.6–1.4 mm diam (not or scarcely wider than petiole), in profile squatly drum-shaped; pulvinules 1–1.5 mm; n.-e. Brazil (Maranhão to Rio Grande do Norte). 14a. var. *eitenorum* (p. 642).
            19. Lfts 3–4 pairs, the rachis (3–)4–8 cm; first gland discoid sessile 1.8–3 mm diam, wider than petiole; pulvinules (1.5–)2–3.5 mm; Chapada Diamantina in Bahia. 14b. var. *regana* (p. 642).
      15. Blade of lfts cuneate at base, usually on both sides, if rounded at base in some lfts then the pulvinule strongly developed.
        20. Lfts of all lvs exactly 2 pairs; trees 9–30 m sometimes flowering as treelets 3–10 m, of wet forest habitats in lower Amazonia and the coastal forest of Bahia. 8. *Ch. bahiae* (p. 642).

20. Lfts of all lvs or of at least some larger lvs of terminal branchlets at least 3 and sometimes up to 10 pairs; sympatric with the last but extending much further s. into the Brazilian Planalto and along the coast to n.-e. S. Paulo.
 21. First petiolar gland situated distinctly below the proximal pair of lfts; lfts of longer lvs 6–10 pairs; basins of lower rios Tapajós and Xingú in Pará. 13. *Ch. xinguensis* (p. 642).
 21. First petiolar gland situated between or immediately contiguous to the proximal pair of lfts; lfts of larger lvs commonly 3–5, rarely (and only locally in Bahia and Rio de Janeiro) up to 5–8 pairs; discontinuously dispersed over much of e. Brazil s.-ward from Maranhão; 15. *Ch. ensiformis*:
  22. Lfts of larger lvs (2–)3–5 pairs; range of the species.
   23. Annotinous branchlets and lf-blades thinly puberulent or glabrous; e. and s. of Maranhão, widespread. 15a. var. *ensiformis* (p. 642).
   23. Annotinous branchlets and often the lfts (at least beneath) ± densely pilosulous; Maranhão and immediately adjacent Pará and Goiás. 15b. var. *maranonica* (p. 642).
  22. Lfts of larger lvs 5–8 pairs; local and bicentric, in Paraíba valley of interior Rio de Janeiro and in Gongogi valley and vicinity of s. Bahia. 15c. var. *plurifoliolata* (p. 642).

14. Stipules subulate-setiform 1.5–8 mm, only tardily deciduous; shrubs 2–3 m, known only from middle Rio Pardo valley in s. Bahia (near 15°30′S). 18. *Ch. compitalis* (p. 644).

Conspectus of Taxa of sect. *Apoucouita*

i. ser. PTERIDOPHYLLAE (Irwin & Barneby) Irwin & Barneby

Chamaecrista sect. **Apoucouita** ser. **Pteridophyllae** (Irwin & Barneby) Irwin & Barneby, comb. nov. *Cassia* sect. *Apoucouita* ser. *Pteridophyllae* Irwin & Barneby, Brittonia **29**(3): 279. 1977.

1. Chamaecrista adiantifolia (Bentham) Irwin & Barneby, comb. nov. *Cassia adiantifolia* Spruce ex Bentham in Martius, Fl. Bras. **15**(2): 130. 1870.

1a. Chamaecrista adiantifolia (Bentham) Irwin & Barneby var. **adiantifolia**.

1b. Chamaecrista adiantifolia (Bentham) var. **pteridophylla** (Sandwith) Irwin & Barneby, comb. nov. *Cassia pteridophylla* Sandwith, Kew Bull. **1928**: 375. 1928. *C. adiantifolia* var. *pteridophylla* (Sandwith) Ducke, 1935.

2. Chamaecrista aspleniifolia (Irwin & Barneby) Irwin & Barneby, Brittonia **31**(1): 155. 1979.

ii. ser. APOUCOUITA (Irwin & Barneby) Irwin & Barneby

Chamaecrista sect. **Apoucouita** ser. **Apoucouita** (Irwin & Barneby) Irwin & Barneby, comb. nov. *Cassia* sect. *Apoucouita* ser. *Apoucouita* Irwin & Barneby, Brittonia **29**(3): 281. 1977.

3. Chamaecrista hymenaeifolia (Bentham) Irwin & Barneby, comb. nov. *Cassia hymenaeifolia* Bentham in Martius, Fl. Bras. **15**(2): 130. 1870.

4. Chamaecrista polystachya (Bentham) Irwin & Barneby, comb. nov. *Cassia polystachya* Bentham, J. Bot. (Hooker) **2**: 77. 1840.

5. **Chamaecrista boyanii** (Irwin & Barneby) Irwin & Barneby, comb. nov. *Cassia boyanii* Irwin & Barneby, Brittonia **29**(3): 281. 1977.

6. **Chamaecrista duartei** (Irwin) Irwin & Barneby, comb. nov. *Cassia duartei* Irwin, Mem. N.Y. Bot. Gard. **16**: 89. 1967.

7. **Chamaecrista negrensis** (Irwin) Irwin & Barneby, comb. nov. *Cassia negrensis* Irwin, Mem. N.Y. Bot. Gard. **16**: 101. 1967.—Fig. 14 (pod), 37 (androecium).

7a. **Chamaecrista negrensis** (Irwin) Irwin & Barneby var. **negrensis**.
7b. **Chamaecrista negrensis** (Irwin) var. **albuquerquei** (Irwin & Barneby) Irwin & Barneby, comb. nov. *Cassia negrensis* var. *albuquerquei* Irwin & Barneby, Brittonia **29**(3): 283. 1977.

8. **Chamaecrista bahiae** (Irwin) Irwin & Barneby, Brittonia **31**(4): 467. 1979.

9. **Chamaecrista onusta** Irwin & Barneby, Brittonia **31**(4): 464. 1979.—Fig. 38.

10. **Chamaecrista aiarana** (Irwin) Irwin & Barneby, comb. nov. *Cassia aiarana* Irwin, Mem. N.Y. Bot. Gard. **16**: 91. 1967.

11. **Chamaecrista apoucouita** (Aublet) Irwin & Barneby, comb. nov. *Cassia apoucouita* Aublet, Hist. pl. Guiane 379. 1775.—Fig. 36 (petiolar nectary).

12. **Chamaecrista scleroxylon** (Ducke) Irwin & Barneby, comb. nov. *Cassia scleroxylon* Ducke, Arch. Jard. bot. Rio de Janeiro **3**: 115. 1922.

13. **Chamaecrista xinguënsis** (Ducke) Irwin & Barneby, comb. nov. *Cassia xinguënsis* Ducke, Arch. Jard. bot. Rio de Janeiro **3**: 115. 1922.

14. **Chamaecrista eitenorum** (Irwin & Barneby) Irwin & Barneby, comb. nov. *Cassia eitenorum* Irwin & Barneby, Brittonia **29**(3): 285. 1977.

14a. **Chamaecrista eitenorum** (Irwin & Barneby) Irwin & Barneby var. **eitenorum**.
14b. **Chamaecrista eitenorum** (Irwin & Barneby) var. **regana** (Irwin & Barneby) Irwin & Barneby, comb. nov. *Cassia eitenorum* var. *regana* Irwin & Barneby, Brittonia **29**(3): 287. 1977.

15. **Chamaecrista ensiformis** (Vellozo) Irwin & Barneby, comb. nov. *Cassia ensiformis* Vellozo, Fl. Flum. 162. 1825 & Icones **4**: t. 79. 1835.

15a. **Chamaecrista ensiformis** (Vellozo) Irwin & Barneby var. **ensiformis**.
15b. **Chamaecrista ensiformis** (Vellozo) var. **maranonica** (Irwin) Irwin & Barneby, comb. nov. *Cassia maranonica* Irwin, Mem. N.Y. Bot. Gard. **16**: 105. 1967. *C. ensiformis* var. *maranonica* (Irwin) Irwin & Barneby, 1977.
15c. **Chamaecrista ensiformis** (Vellozo) var. **plurifoliolata** (Hoehne) Irwin & Barneby, comb. nov. *Cassia apoucouita* var. *plurifoliolata* Hoehne, Comm. Linh. Tel. Estrat. de Matto Grosso ao Amazonas **45**(5) Bot. **8**: 44. 1919.

16. **Chamaecrista aspidiifolia** Irwin & Barneby, Brittonia **31**(1): 149. 1979.

This species is here transferred from ser. *Pteridophyllae* to ser. *Apoucouita*, of which it has the reticulately venulose, even though atypically many and small leaflets. Closely related to the next.—Fig. 39.

17. **Chamaecrista subpeltata** (Rizzini) Irwin & Barneby, comb. nov. *Cassia subpeltata* Rizzini, Rodriguesia **28**(41): 173. 1976.

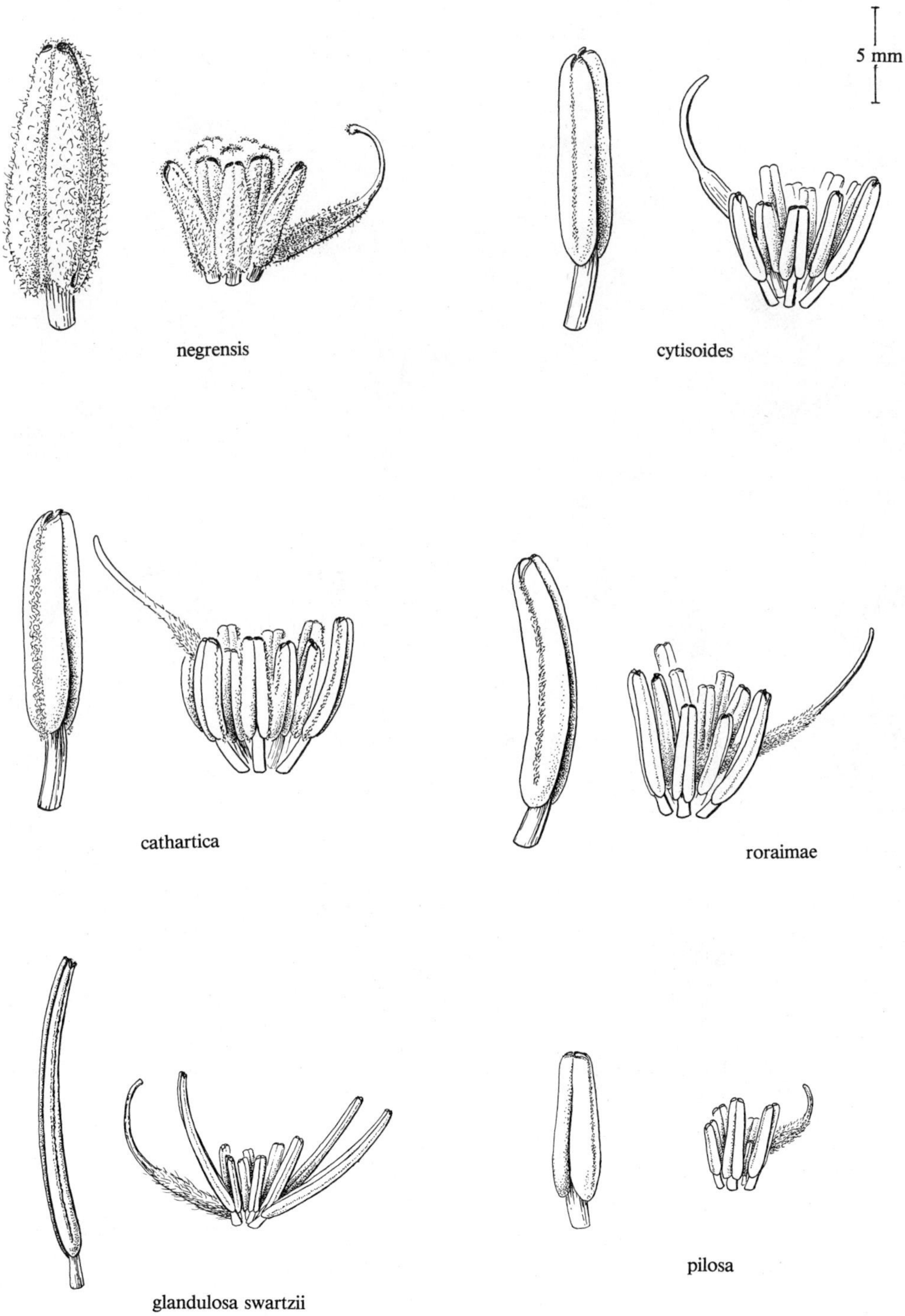

Fig. 37. Androceia of genus *Chamaecrista*. Sect. *Apoucouita*: *Ch. negrensis* (Irwin) I. & B. Sect. *Absus*: *Ch.* (*Baseophyllum*) *cytisoides* (DC.) I. & B.; *Ch.* (*Absus*) *cathartica* (Mart.) I. & B. Sect. *Chamaecrista*: *Ch.* (*Coriaceae*) *roraimae* (Benth.) I. & B.; *Ch.* (*Chamaecrista*) *glandulosa* (Linn.) var. *swartzii* (Wikstr.) I. & B. *Ch.* (*Prostratae*) *pilosa* (Linn.) Greene.

Fig. 38. *Chamaecrista* (*Apoucouita*) *onusta* I. & B. Flowering branch + leaf ×½; one sepal + petals ×1½; stamen ×5 (*Harley 18535*).

Known to us only from the type (*D. P. Lima 13325* in *RB 190622*), collected near S. João dos Patos in e.-centr. Maranhão.

18. Chamaecrista compitalis (Irwin & Barneby) Irwin & Barneby, comb. nov. *Cassia compitalis* Irwin & Barneby, Brittonia **29**(3): 289. 1977.

B. Sectio ABSUS (Colladon) Irwin & Barneby

Chamaecrista sect. **Absus** (Colladon) Irwin & Barneby, Brittonia **31**(1): 155. 1979. *Cassia* sect. *Absus* DeCandolle ex Colladon, 1816.—Sp. typica: *C. hispida* Colladon = **Chamaecrista hispidula** (Vahl) Irwin & Barneby.

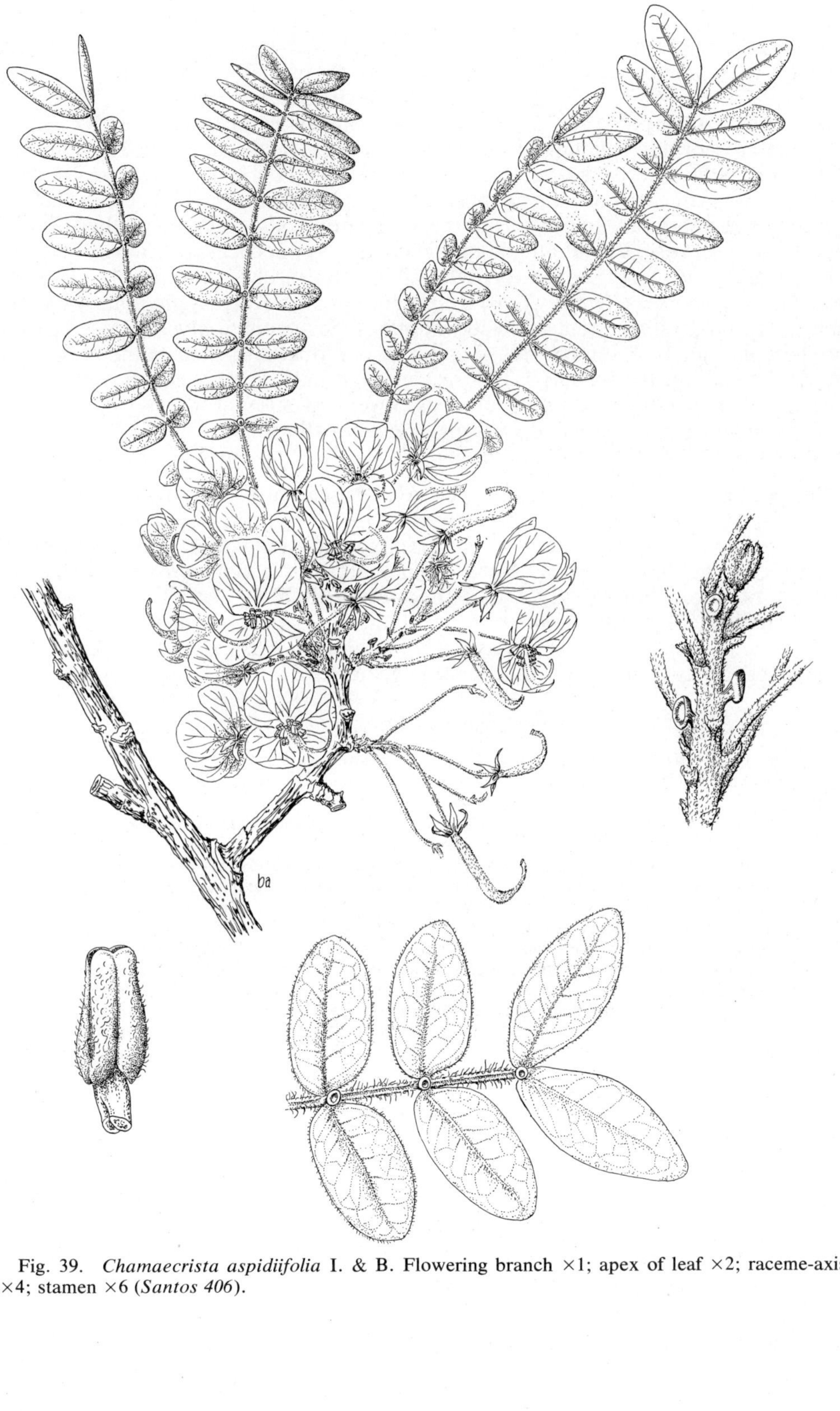

Fig. 39. *Chamaecrista aspidiifolia* I. & B. Flowering branch ×1; apex of leaf ×2; raceme-axis ×4; stamen ×6 (*Santos 406*).

Cassia sect. *Absus* Colladon emend. Irwin & Barneby, Mem. N.Y. Bot. Gard. **30**: 1–277. 1977.

Inflorescence from hornotinous leafy stems, either simple terminal racemose, or terminal paniculate-racemose, or by reduction of axillary branchlets axillary racemose, the primary axis of each raceme developed and several-fld, if (rarely) reduced to 1 fl the pubescence viscid-setulose; androecium 10-merous, concealed by a heteromorphic, dimidiately ovate or falcately hemi-lanceolate, involute or convolute petal interposed between it and the obliquely exserted style, the filaments all less than half as long as their anther, this ciliolate along its closed lateral sutures and dehiscent apically at anthesis of the fl; style terete, neither dilated nor hooked distally, the minute stigmatic cavity symmetrically terminal.—Perennial herbs, either from taproot or xylopodium, or shrubs, these rarely subarborescent, one (subsect. *Otophyllum*) frutescent but monocarpic; phyllotaxy mostly spiral, always so if the stem herbaceous; pubescence commonly composed partly of glandular setae or setulae, these sometimes reduced to their bulbous base or to a resinous dot, the foliage or inflorescence (or both) in consequence ± viscid, but the specialized excretory trichomes absent in subsect. *Baseophyllum* and *Adenophyllum*; petiolar glands absent except in the 4 spp. of subsect. *Adenophyllum, Baseophyllum* and *Otophyllum*; $x = 14$.—Spp. 167, Neotropical.

For a full descriptive account of sect. *Absus,* see our recent monograph (1977, sub *Cassia,* supra cit.). From this we reproduce here a) a key to subsections; and b) a bare synoptic enumeration of the serial, specific and varietal taxa transferred from *Cassia* to *Chamaecrista*. Note that no heterotypic synonymy is repeated in the enumeration. For brevity's sake the present authors' names are contracted to IB; references to our monograph in Mem. N.Y. Bot. Gard. vol. 30 to "IB, 1977, p. . . ."; and the generic names *Cassia* and *Chamaecrista* to initial *C.* and *Ch.* respectively. One newly discovered species is described ad calcem.

Key to Subsections of sect. *Absus*

1. Petiolar glands present, scutellate depressed sessile, situated between pairs of lfts, below the first pair of lfts, or (when petiole suppressed) between the proximal (or only) pair of lfts; glandular setae 0, but the pod sometimes glutinous.
   2. Shrubs, or subshrubs from a xylopodium; lfts 1–4 pairs, all normally foliaceous; peduncles exactly axillary; petals drying yellow or orange.
      3. Lfts plane-margined, strongly asymmetric at base, palmately 3–7-nerved from the pulvinule; axis of inflorescence charged with scutellate glands (as in sect. *Apoucouita*); e. Brazil, 2 spp. Ba. subsect. *Baseophyllum*.
      3. Lfts revolute-margined, subsymmetric at base, 3-nerved from pulvinule; axis of inflorescence glandless; e. Cuba, 1 sp. Bb. subsect. *Adenophyllum*.
   2. Monocarpic herbs (tall, sometimes of long duration and softly fruticose in age); lfts 10–20 pairs, the proximal 1–3 pairs modified into minute sessile deltate-reniform blades resembling floral bracts or stipules; peduncles adnate to the stem-internode, the fls appearing supra-axillary (as in sect. *Chamaecrista* ser. *Chamaecrista*); petals drying whitish; e. Brazil, 1 sp. Bc. subsect. *Otophyllum*.
1. Petiolar glands 0; glandular setae present, at least on inflorescence or on ovary (pod) or, if setae (setulae) lacking, the foliage and inflorescence glandular-verruculose or resin-dotted, viscid when fresh; widespread through tropical N. and S. America, feebly penetrating the warm subtropics, 166 spp. Bd. subsect. *Absus*.

Synopsis of *Chamaecrista* sect. *Absus*

Ba. subsect. BASEOPHYLLUM

Ch. sect. **Absus** subsect. **Baseophyllum** (Colladon) IB, comb. nov. *C.* sect. *Absus* subsect. *Baseophyllum* (Collad.) IB, 1977, p. 9.

1. **Ch. cytisoides** (Colladon) IB, comb. nov. *C. cytisoides* DeCandolle ex Colladon, Hist. Cass. 116. 1816.—Fig. 14 (pod), 36 (petiolar nectary), 37 (androecium).

1a. var. **cytisoides**.
1b. var. **micrantha** (IB) IB, comb. nov. *C. cytisoides* var. *micrantha* IB, 1977, p. 12.
1c. var. **confertiformis** (IB) IB, comb. nov. *C. cytisoides* var. *confertiformis* IB, 1977, p. 12.
1d. var. **brachystachya** (Bentham) IB, comb. nov. *C. cytisoides* var. *brachystachya* (Bentham) IB, 1977, p. 13 ≡ *C. brachystachya* Bentham, 1840.
1e. var. **blancheti** (Bentham) IB, comb. nov. *C. cytisoides* var. *blancheti* (Bentham) IB, 1977, p. 13 ≡ *C. blancheti* Bentham, 1840.
1f. var. **unijuga** (Bentham) IB, comb. nov. *C. brachystachya* var. *unijuga* Bentham in Martius, Fl. Bras. **15**(1): 139. 1870.
1g. var. **decora** (IB) IB, comb. nov. *C. cytisoides* var. *decora* IB, 1977, p. 14.

2. **Ch. coriacea** (Bentham) IB, comb. nov. *C. coriacea* Bongard ex Bentham, J. Bot. (Hooker) **2**: 78. 1840.

Bb. subsect. ADENOPHYLLUM

Ch. sect. **Absus** subsect. **Adenophyllum** (IB) IB, comb. nov. *C.* sect. *Absus* subsect. *Adenophyllum* IB, 1977, p. 16.

3. **Ch. bucherae** (Moldenke) IB, comb. nov. *Peiranisia bucherae* Moldenke, Bull. Torr. Club **68**: 675. 1941.

Bc. subsect. OTOPHYLLUM

Ch. sect. **Absus** subsect. **Otophyllum** (IB) IB, comb. nov. *C.* sect. *Absus* subsect. *Otophyllum* IB, 1977, p. 17.

4. **Ch. debilis** (Vogel) IB, comb. nov. *C. debilis* Vogel, Syn. Gen. Cass. 53. 1837.

Bd. subsect. ABSUS

Ch. sect. **Absus** subsect. **Absus** (Colladon) IB, comb. nov. *C.* sect. *Absus* Colladon emend. IB, 1977, p. 18.

i. ser. CATHARTICAE

Ch. sect. **Absus** ser **Catharticae** (IB) IB, comb. nov. *C.* ser. *Catharticae* IB, 1977, p. 27.

5. **Ch. cathartica** (Martius) IB, comb. nov. *C. cathartica* Martius in Spix & Martius, Reise Brasil. **1**: 548. 1823.

5a. var. **cathartica**—Fig. 37 (androecium).
5b. var. **paucijuga** (IB) IB, comb. nov. *C. cathartica* var. *paucijuga* IB, 1977, p. 30.

6. **Ch. catharticoides** (IB) IB, comb. nov. *C. catharticoides* IB, 1977, p. 31.

7. **Ch. microsenna** (IB) IB, comb. nov. *C. microsenna* IB, 1977, p. 32.

ii. ser. TRACHYCARPAE

Ch. sect. **Absus** ser. **Trachycarpae** (IB) IB, comb. nov. *C.* ser. *Trachycarpae* IB, 1977, p. 33.

8. Ch. trachycarpa (Vogel) IB, comb. nov. *C. trachycarpa* Vogel, Syn. Gen. Cass. 52. 1837.

8a. var. **trachycarpa.**
8b. var. **acutifolia** (Bentham) IB, comb. nov. *C. trachycarpa* var. *acutifolia* Bentham, Linnaea **22**: 528. 1849.

9. Ch. cavalcantina (IB) IB, comb. nov. *C. cavalcantina* IB, 1977, p. 35.

10. Ch. venatoria (IB) IB, comb. nov. *C. venatoria* IB, 1977, p. 37.

iii. ser. OLIGOSPERMAE

Ch. sect. **Absus** ser. **Oligospermae** (IB) IB, comb. nov. *C.* ser. *Oligospermae* IB, 1977, p. 38.

11. Ch. oligosperma (Bentham) IB, comb. nov. *C. oligosperma* Martius ex Bentham in Martius, Fl. Bras. **15**(2): 147. 1870.

iv. ser. ATROGLANDULOSAE

Ch. sect. **Absus** ser. **Atroglandulosae** (IB) IB, comb. nov. *C.* ser. *Atroglandulosae* IB, 1977, p. 39.

12. Ch. atroglandulosa (Taubert) IB, comb. nov. *C. atroglandulosa* Taubert, Feddes Repert. **20**: 128. 1924.

v. ser. INCANAE

Ch. sect. **Absus** ser. **Incanae** (IB) IB, comb. nov. *C.* ser. *Incanae* IB, 1977, p. 40.

13. Ch. incana (Vogel) IB, comb. nov. *C. incana* Vogel, Syn. Gen. Cass. 52. 1837.

vi. ser. NIGRICANTES

Ch. sect. **Absus** ser. **Nigricantes** (Bentham) IB, Brittonia **31**(1): 155. 1979. *C.* ser. *Nigricantes* Bentham in Martius, Fl. Bras. **15**(2): 89, 150. 1870.

14. Ch. pilicarpa (Harms) IB, comb. nov. *C. pilicarpa* Glaziou ex Harms, Feddes Repert. **20**: 135. 1924.

15. Ch. gilliesii (Harms) IB, comb. nov. *C. gilliesii* Harms, Feddes Repert. **20**: 133. 1924.

16. Ch. sophoroides (Bentham) IB, comb. nov. *C. sophoroides* Martius ex Bentham in Martius, Fl. Bras. **15**(2): 152. 1870.

17. Ch. urophyllidia (IB) IB, comb. nov. *C. urophyllidia* IB, 1977, p. 46.

18. Ch. calixtana (IB) IB, comb. nov. *C. calixtana* IB, 1977, p. 47.

19. Ch. tephrosiifolia (Bentham) IB, Brittonia **31**(1): 155. 1979. *C. tephrosiifolia* Bentham in Martius, Fl. Bras. **15**(2): 153. 1870.

20. Ch. philippi (IB) IB, Brittonia **31**(1): 155. 1979. *C. philippi* IB, 1977, p. 48.

21. Ch. glischrodes IB, Brittonia **31**(1): 151. 1979.—Fig. 40.

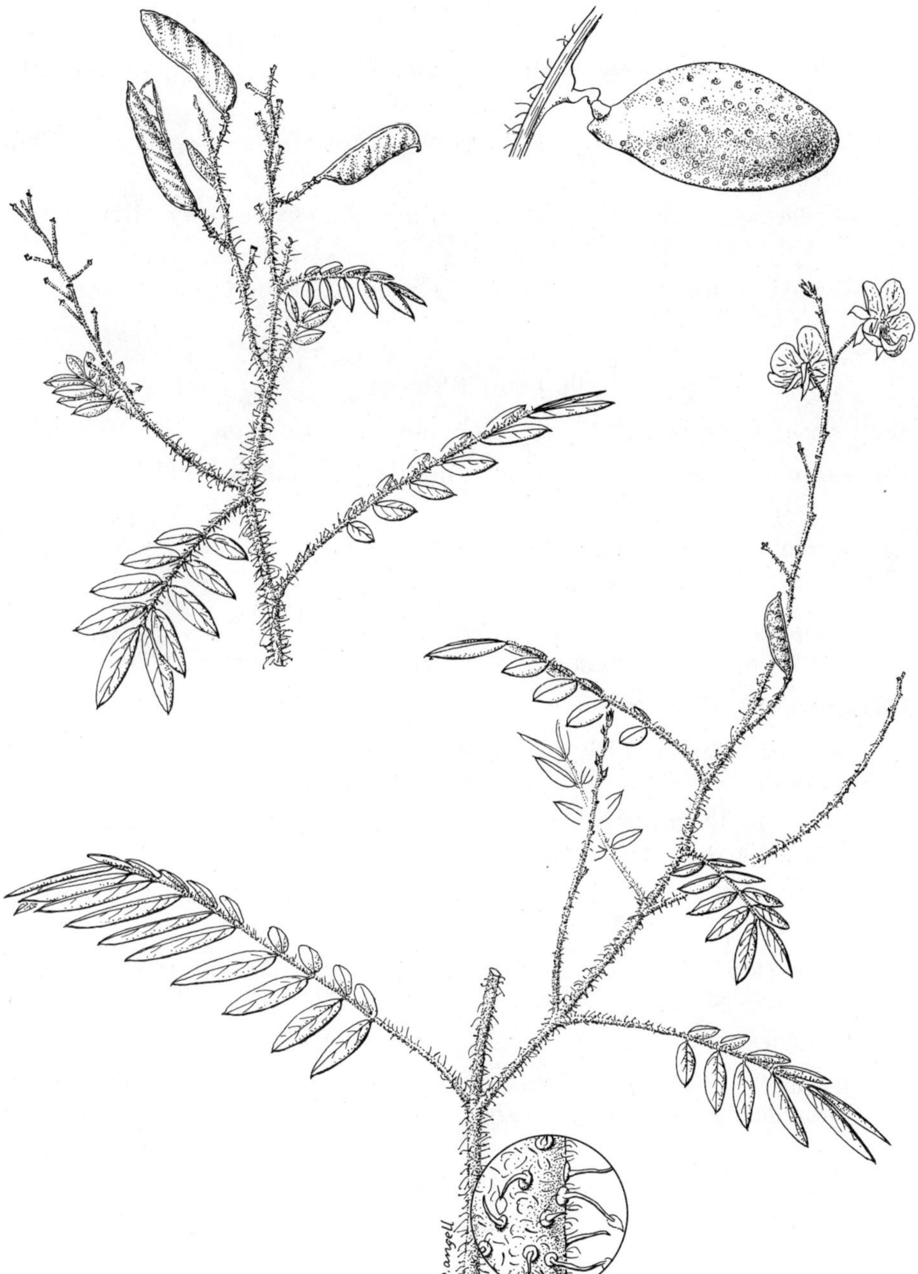

Fig. 40. *Chamaecrista* (*Absus*) *glischrodes* I. & B. Flowering branchlet ×1 (*Semir & Giulietti 5020*). Fruiting branchlet ×1; seed (*Joly et al. 2547*).

22. Ch. itambana (Bentham) IB, comb. nov. *C. itambana* Martius ex Bentham in Martius, Fl. Bras. **15**(2): 151. 1870.

23. Ch. machaeriifolia (Bentham) IB, comb. nov. *C. machaeriifolia* Bentham in Martius, Fl. Bras. **15**(2): 151. 1870.

vii. ser. BRACTEOLATAE

Ch. sect. **Absus** ser. **Bracteolatae** (IB) IB, comb. nov. *C.* ser. *Bracteolatae* IB, 1977, p. 51.

24. Ch. bracteolata (Vogel) IB, comb. nov. *C. bracteolata* Vogel, Syn. Gen. Cass. 52. 1837.

25. Ch. phyllostachya (Bentham) IB, comb. nov. *C. phyllostachya* Bentham in Martius, Fl. Bras. **15**(2): 150. 1870.

26. Ch. glaziovii (Harms) IB, comb. nov. *C. glaziovii* Taubert ex Harms, Feddes Repert. **20**: 135. 1924.

viii. ser. URSINAE

Ch. sect. **Absus** ser. **Ursinae** (IB) IB, comb. nov. *C.* ser. *Ursinae* IB, 1977, p. 54.

27. Ch. exsudans (Bentham) IB, comb. nov. *C. exsudans* Bentham, J. Bot. (Hooker) **2**: 80. 1840.

28. Ch. adamantina (IB) IB, comb. nov. *C. adamantina* IB, 1977, p. 57.

29. Ch. virginis (IB) IB, comb. nov. *C. virginis* IB, 1977, p. 57.

30. Ch. xanthadena (Bentham) IB, comb. nov. *C. xanthadena* (Bentham) IB, 1977, p. 58 ≡ *C. exsudans* var. *xanthadena* Bentham, 1870.

31. Ch. ixodes (IB) IB, comb. nov. *C. ixodes* IB, 1977, p. 59.

32. Ch. aurivilla (Bentham) IB, comb. nov. *C. aurivilla* Martius ex Bentham in Martius, Fl. Bras. **15**(2): 152. 1870.

33. Ch. ursina (Bentham) IB, comb. nov. *C. ursina* Martius ex Bentham in Martius, Fl. Bras. **15**(2): 151. 1870.

34. Ch. fuscescens (Bentham) IB, comb. nov. *C. fuscescens* Bentham in Martius, Fl. Bras. **15**(2): 153. 1870.

35. Ch. astrochiton (IB) IB, comb. nov. *C. astrochiton* IB, 1977, p. 63.—Fig. 14 (pod, seed).

36. Ch. centiflora (IB) IB, comb. nov. *C. centiflora* IB, 1977, p. 63.

37. Ch. leucopilis (Harms) IB, comb. nov. *C. leucopilis* Glaziou ex Harms, Feddes Repert. **20**: 136. 1924.

ix. ser. PINIFOLIAE

Ch. sect. **Absus** ser. **Pinifoliae** (IB) IB, comb. nov. *C.* ser. *Pinifoliae* IB, 1977, p. 66.

38. Ch. paniculata (Bentham) IB, comb. nov. *C. paniculata* Bentham in Martius, Fl. Bras. **15**(2): 153. 1870.

x. ser. SETOSAE

Ch. sect. **Absus** ser. **Setosae** (IB) IB, comb. nov. *C.* ser. *Setosae* IB, 1977, p. 68.

39. Ch. setosa (Vogel) IB, comb. nov. *C. setosa* Vogel, Syn. Gen. Cass. 51. 1837.

39a. var. **setosa**.

39b. var. **paucivenia** (IB) IB, comb. nov. *C. setosa* var. *paucivenia* IB, 1977, p. 70.

39c. var. **detonsa** (Bentham) IB, comb. nov. *C. setosa* var. *detonsa* Bentham in Martius, Fl. Bras. **15**(2): 71. 1970.

39d. var. **subsetosa** (Malme) IB, comb. nov. *C. setosa* var. *subsetosa* Malme, Ark. f. Bot. 23A, **13**: 62. 1930.

40. Ch. amphibola (IB) IB, comb. nov. *C. amphibola* IB, 1977, p. 72.

41. Ch. ochrosperma (IB) IB, comb. nov. *C. ochrosperma* IB, 1977, p. 73.

42. Ch. orenocensis (Bentham) IB, comb. nov. *C. orenocensis* Spruce ex Bentham in Martius, Fl. Bras. **15**(2): 150. 1870.

43. Ch. obtecta (Bentham) IB, comb. nov. *C. obtecta* Bentham in Martius, Fl. Bras. **15**(2): 136. 1870.

44. Ch. campicola (Harms) IB, comb. nov. *C. campicola* Harms, Feddes Repert. **20**: 126. 1924.

45. Ch. scabra (Bentham) IB, comb. nov. *C. scabra* Pohl ex Bentham in Martius, Fl. Bras. **15**(2): 135. 1870.

46. Ch. auris-zerdae (IB) IB, comb. nov. *C. auris-zerdae* IB, 1977, p. 80.

xi. ser. GEMINATAE

Ch. sect. **Absus** ser. **Geminatae** (IB) IB, Brittonia **31**(1): 155. 1979.

47. Ch. didyma IB, Brittonia **31**(1): 153. 1979.—Fig. 41.

48. Ch. geminata (Bentham) IB, Brittonia **31**(1): 155. 1979.

xii. ser. ANDROMEDEAE

Ch. sect. **Absus** ser. **Andromedeae** (IB) IB, comb. nov. *C.* ser. *Andromedeae* IB, 1977, p. 83.

49. Ch. andromedea (Bentham) IB, comb. nov. *C. andromedea* Bentham in Martius, Fl. Bras. **15**(2): 135. 1870.

xiii. ser. LOMATOPODAE

Ch. sect. **Absus** ser. **Lomatopodae** (IB) IB, comb. nov. *C.* ser. *Lomatopodae* IB, 1977, p. 85.

50. Ch. lomatopoda (Bentham) IB, comb. nov. *C. lomatopoda* Bentham in Martius, Fl. Bras. **15**(2): 146. 1870.

xiv. ser. SECUNDAE

Ch. sect. **Absus** ser. **Secundae** (IB) IB, comb. nov. *C.* ser. *Secundae* IB, 1977, p. 86.

51. Ch. secunda (Bentham) IB, comb. nov. *C. secunda* Bentham in Martius, Fl. Bras. **15**(2): 148. 1870.

xv. ser. MICROPHYLLAE

Ch. sect. **Absus** ser. **Microphyllae** (Bentham) IB, comb. nov. *C.* ser. *Microphyllae* Bentham in Martius, Fl. Bras. **15**(2): 146. 1870.

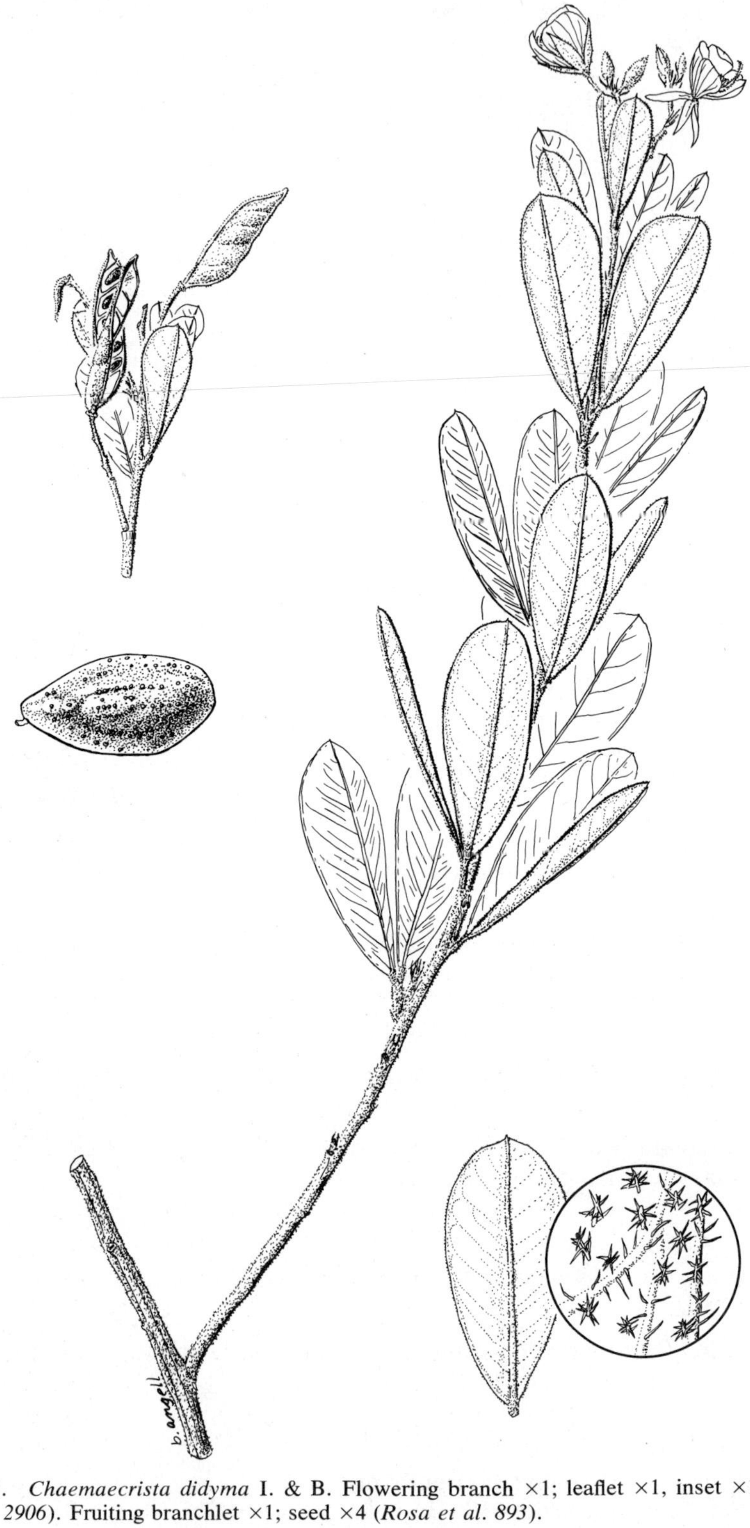

Fig. 41. *Chaemaecrista didyma* I. & B. Flowering branch ×1; leaflet ×1, inset ×12 (*Kirkbride & Lleras 2906*). Fruiting branchlet ×1; seed ×4 (*Rosa et al. 893*).

52. Ch. dalbergiifolia (Bentham) IB, comb. nov. *C. dalbergiifolia* Bentham in Martius, Fl. Bras. **15**(2): 149. 1870.

53. Ch. nuda (IB) IB, comb. nov. *C. nuda* IB, 1977, p. 91.

54. Ch. dumalis (Hoehne) IB, comb. nov. *C. dumalis* Hoehne, Comm. Linhas Estrat. Matto Grosso ao Amazonas **74**(5): = Botanica **12** (Contrib. Conh. Legum. Rondônia): 14. 1922.

55. Ch. huntii (IB) IB, comb. nov. *C. huntii* IB, 1977, p. 92.

55a. var. **huntii**.
55b. var. **correntina** (IB) IB, comb. nov. *C. huntii* var. *correntina* IB, 1977, p. 95.

56. Ch. sincorana (Harms) IB, comb. nov. *C. sincorana* Harms in Engler, Jahrb. **42**: 210. 1919.

57. Ch. isidorea (Bentham) IB, comb. nov. *C. isidorea* Bentham in Martius, Fl. Bras. **15**(2): 154. 1870.

58. Ch. imbricans (IB) IB, comb. nov. *C. imbricans* IB, 1977, p. 97.

59. Ch. foederalis (IB) IB, comb. nov. *C. foederalis* IB, 1977, p. 97.

60. Ch. neesiana (Bentham) IB, comb. nov. *C. neesiana* Martius ex Bentham in Martius, Fl. Bras. **15**(2): 148. 1870.

60a. var. **neesiana**.
60b. var. **laxiracemosa** (Harms) IB, comb. nov. *C. neesiana* var. *laxiracemosa* (Harms) IB, 1977, p. 101 ≡ *C. laxiracemosa* Harms, 1924.
60c. var. **subnitida** (Taubert) IB, comb. nov. *C. neesiana* var. *subnitida* (Taubert) IB, 1977, p. 102 ≡ *C. subnitida* Taubert, 1924.
60d. var. **goyazensis** (Taubert) IB, comb. nov. *C. neesiana* var. *goyazensis* (Taubert) IB, 1977, p. 102 ≡ *C. goyazensis* Taubert, 1896.

61. Ch. adenophora (Harms) IB, comb. nov. *C. adenophora* Harms, Feddes Repert. **20**: 127. 1924.

62. Ch. cristalinae (IB) IB, comb. nov. *C. cristalinae* IB, 1977, p. 103.

63. Ch. pohliana (Bentham) IB, comb. nov. *C. pohliana* Bentham, J. Bot. (Hooker) **2**: 80. 1840.

64. Ch. decrescens (Bentham) IB, comb. nov. *C. decrescens* Bentham, J. Bot. (Hooker) **2**: 80. 1840.

65. Ch. elachistophylla (Harms) IB, comb. nov. *C. elachistophylla* Harms, Feddes Repert. **20**: 132. 1924.

66. Ch. subdecrescens (IB) IB, comb. nov. *C. subdecrescens* IB, 1977, p. 109.

67. Ch. polymorpha (Harms) IB, comb. nov. *C. polymorpha* Glaziou ex Harms, Feddes Repert. **20**: 129. 1924.

68. Ch. harmsiana IB, nom. nov. *C. microphylla* Glaziou ex Harms, Feddes Repert. **20**: 128. 1924. Non *C. microphylla* Willd., Sp. Pl. **2**: 529. 1799.

69. Ch. psoraleopsis (IB) IB, comb. nov. *C. psoraleopsis* IB, 1977, p. 111.

xvi. ser. GRACILLIMAE

Ch. sect. **Absus** ser. **Gracillimae** (IB) IB, comb. nov. *C.* ser. *Gracillimae* IB, 1977, p. 112.

70. Ch. benthami (Ghesquière) IB, comb. nov. *C. benthami* Ghesquière, Rev. Zool. & Bot. Afr. **26**: 148. 1935 ≡ *C. gracillima* Bentham, 1870, non Welwitsch, 1858.

xvii. ser. RIGIDULAE

Ch. sect. **Absus** ser. **Rigidulae** (Bentham) IB, comb. nov. *C.* ser. *Rigidulae* Bentham in Martius, Fl. Bras. **15**(2): 142. 1870.

71. Ch. decumbens (Bentham) IB, comb. nov. *C. decumbens* Bentham, J. Bot. (Hooker) **2**: 79. 1840.

72. Ch. benthamiana (Harms) IB, comb. nov. *C. benthamiana* Harms, Feddes Repert. **20**: 128. 1924.

73. Ch. polita (IB) IB, comb. nov. *C. polita* IB, 1977, p. 118.

74. Ch. strictula (IB) IB, comb. nov. *C. strictula* IB, 1977, p. 120.

75. Ch. planaltoana (Harms) IB, comb. nov. *C. planaltoana* Harms, Feddes Repert. **20**: 127. 1924.

76. Ch. altoana (IB) IB, comb. nov. *C. altoana* IB, 1977, p. 122.

77. Ch. nummulariifolia (Bentham) IB, comb. nov. *C. nummulariifolia* Bentham in Martius, Fl. Bras. **15**(2): 144. 1870.

78. Ch. multipennis (IB) IB, comb. nov. *C. multipennis* IB, 1977, p. 125.

79. Ch. cipoana (IB) IB, comb. nov. *C. cipoana* IB, 1977, p. 127 ≡ *C. sipoana* IB ex Gray Herb. Index 285. 1978, lapsu.

80. Ch. ciliolata (Bentham) IB, comb. nov. *C. ciliolata* Bentham in Martius, Fl. Bras. **15**(2): 146. 1870.

80a. var. **ciliolata.**
80b. var. **pulchella** (IB) IB, comb. nov. *C. ciliolata* var. *pulchella* IB, 1977, p. 130.
80c. var. **caprina** (IB) IB, comb. nov. *C. ciliolata* var. *caprina* IB, 1977, p. 130.

81. Ch. nanodes (IB) IB, comb. nov. *C. nanodes* IB, 1977, p. 131.

82. Ch. brachyrachis (Harms) IB, comb. nov. *C. brachyrachis* Harms, Feddes Repert. **20**: 129. 1924.

83. Ch. densifolia (Bentham) IB, comb. nov. *C. densifolia* Bentham, J. Bot. (Hooker) **2**: 80. 1840.

84. Ch. feliciana (IB) IB, comb. nov. *C. feliciana* IB, 1977, p. 135 ≡ *C. densifolia* var. *felipensis* Bentham, 1870.

85. Ch. macedoi (IB) IB, comb. nov. *C. macedoi* IB, 1977, p. 136.

86. Ch. chaetostegia (IB) IB, comb. nov. *C. chaetostegia* IB, 1977, p. 137.

86a. var. **chaetostegia.**
86b. var. **obolaria** (IB) IB, comb. nov. *C. chaetostegia* var. *obolaria* IB, 1977, p. 139.

87. Ch. mollicaulis (Harms) IB, comb. nov. *C. mollicaulis* Harms, Feddes Repert. **20**: 131. 1924.

88. Ch. brachyblepharis (Harms) IB, comb. nov. *C. brachyblepharis* Harms, Feddes Repert. **20**: 130. 1924.

89. Ch. filicifolia (Bentham) IB, comb. nov. *C. filicifolia* Bentham in Martius, Fl. Bras. **15**(2): 148. 1870.

90. Ch. glaucofilix (IB) IB, comb. nov. *C. glaucofilix* IB, 1977, p. 143.

91. Ch. azulana (IB) IB, comb. nov. *C. azulana* IB, 1977, p. 144.

92. Ch. dawsonii (Cowan) IB, comb. nov. *C. dawsonii* Cowan, Los Angeles Co. Mus., Contrib. Sci. **13**: 15. 1957.

93. Ch. gymnothyrsa (IB) IB, comb. nov. *C. gymnothyrsa* IB, 1977, p. 146.

xviii. ser. PANICULATAE

Ch. sect. **Absus** ser. **Paniculatae** (Bentham) IB, comb. nov. *C.* ser. *Paniculatae* Bentham in Martius, Fl. Bras. **15**(2): 140. 1870.

94. Ch. rigidifolia (Bentham) IB, comb. nov. *C. rigidifolia* Bentham in Martius, Fl. Bras. **15**(2): 142. 1870.

- **94a.** var. **rigidifolia**.
- **94b.** var. **veadeirana** (IB) IB, comb. nov. *C. rigidifolia* var. *veadeirana* IB, 1977, p. 150.

95. Ch. clausseni (Bentham) IB, comb. nov. *C. clausseni* Bentham, J. Bot. (Hooker) **2**: 79. 1840.

- **95a.** var. **clausseni**.
- **95b.** var. **cyclophylla** (IB) IB, comb. nov. *C. clausseni* var. *cyclophylla* IB, 1977, p. 153.
- **95c.** var. **megacycla** (IB) IB, comb. nov. *C. clausseni* var. *megacycla* IB, 1977, p. 153.

96. Ch. lundii (Bentham) IB, comb. nov. *C. lundii* Bentham in Martius, Fl. Bras. **15**(2): 143. 1870.

97. Ch. pachyclada (Harms) IB, comb. nov. *C. pachyclada* Harms, Feddes Repert. **20**: 126. 1924.

98. Ch. orbiculata (Bentham) IB, comb. nov. *C. orbiculata* Bentham, J. Bot. (Hooker) **2**: 79. 1840.

- **98a.** var. **orbiculata**.
- **98b.** var. **ustulata** (IB) IB, comb. nov. *C. orbiculata* var. *ustulata* IB, 1977, p. 159.
- **98c.** var. **trichothyrsus** (Harms) IB, comb. nov. *C. trichothyrsus* Harms, Feddes Repert. **20**: 125. 1924 ≡ *C. orbiculata* var. *trichothyrsus* (Harms) IB, 1977, p. 159.
- **98d.** var. **cercidifolia** (IB) IB, comb. nov. *C. orbiculata* var. *cercidifolia* IB, 1977, p. 160.

99. Ch. celiae (IB) IB, comb. nov. *C. celiae* IB, 1977, p. 160.

xix. ser. ADENOPHYLLAE

Ch. sect. **Absus** ser. **Adenophyllae** (IB) IB, Brittonia **31**(1): 155. 1979. *C.* ser. *Adenophyllae* IB, 1977, p. 162.

100. Ch. adenophylla (Taubert) IB, Brittonia **31**(1): 155. 1979. *C. adenophylla* Taubert, Feddes Repert. **20**: 124. 1924.

101. Ch. itabiritoana (IB) IB, comb. nov. *C. itabiritoana* IB, 1977, p. 163.

102. Ch. chrysosepala (IB) IB, Brittonia **31**(4): 467. 1979. *C. chrysosepala* IB, 1977, p. 164.

103. Ch. hatschbachii IB, Brittonia **31**(4): 466. 1979.—Fig. 42.

xx. ser. LUCIDAE

Ch. sect. **Absus** ser. **Lucidae** (Bentham) IB, comb. nov. *C.* ser. *Lucidae* Bentham in Martius, Fl. Bras. **15**(2): 146. 1870.

104. Ch. amambaya (IB) IB, comb. nov. *C. amambaya* IB, 1977, p. 167 ≡ *C. cathartica* subsp. *floribunda* Hassler, Feddes Repert. **8**: 130. 1910.

105. Ch. seticrenata (IB) IB, comb. nov. *C. seticrenata* IB, 1977, p. 168.

106. Ch. caiapo (IB) IB, comb. nov. *C. caiapo* IB, 1977, p. 169.

107. Ch. lamprosperma (Bentham) IB, comb. nov. *C. lamprosperma* Martius ex Bentham in Martius, Fl. Bras. **15**(2): 169. 1870.

108. Ch. labouriaeae (IB) IB, comb. nov. *C. labouriaeae* IB, 1977, p. 171.

xxi. ser. HASSLERANAE

Ch. sect. **Absus** ser. **Hassleranae** (IB) IB, comb. nov. *C.* ser. *Hassleranae* IB, 1977, p. 172.

109. Ch. hassleri (IB) IB, comb. nov. *C. hassleri* IB, 1977, p. 172.

xxii. ser. HEDYSAROIDES

Ch. sect. **Absus** ser. **Hedysaroides** (IB) IB, comb. nov. *C.* ser. *Hedysaroides* IB, 1977, p. 174.

110. Ch. hedysaroides (Vogel) IB, comb. nov. *C. hedysaroides* Vogel, Syn. Gen. Cass. 52: 1837.

xxiii. ser. OCHNACEAE

Ch. sect. **Absus** ser. **Ochnaceae** (IB) IB, comb. nov. *C.* ser. *Ochnaceae* IB, 1977, p. 176.

111. Ch. ochnacea (Vogel) IB, comb. nov. *C. ochnacea* Vogel, Syn. Gen. Cass. 51. 1837.

111a. var. **purpurascens** (Bentham) IB, comb. nov. *C. ochnacea* var. *purpurascens* Bentham in Martius, Fl. Bras. **15**(2): 142. 1870.
111b. var. **ochnacea.**
111c. var. **speluncae** (IB) IB, comb. nov. *C. ochnacea* var. *speluncae* IB, 1977, p. 180.
111d. var. **mollis** (IB) IB, comb. nov. *C. ochnacea* var. *mollis* IB, 1977, p. 180.
111e. var. **latifolia** (Bentham) IB, comb. nov. *C. ochnacea* var. *latifolia* Bentham in Martius, Fl. Bras. **15**(2): 142. 1870.

Fig. 42. *Chamaecrista hatschbachii* I. & B. Flowering branch ×1 (*Hatschbach 41707*).

112. Ch. desertorum (Bentham) IB, comb. nov. *C. desertorum* Martius ex Bentham in Martius, Fl. Bras. **15**(2): 133. 1870.

113. Ch. geraldi (IB) IB, comb. nov. *C. geraldi* IB, 1977, p. 183.

114. Ch. vauthieri (Bentham) IB, comb. nov. *C. vauthieri* Bentham in Martius, Fl. Bras. **15**(2): 138. 1870.

115. Ch. crenulata (Bentham) IB, comb. nov. *C. crenulata* Bentham, J. Bot. (Hooker) **2**: 80. 1840.

116. Ch. punctata (Vogel) IB, comb. nov. *C. punctata* Vogel, Syn. Gen. Cass. 51. 1837.

117. Ch. bifoliola (Harms) IB, comb. nov. *C. bifoliola* Glaziou ex Harms, Feddes Repert. **20**: 125. 1924.

118. Ch. cotinifolia (G. Don) IB, comb. nov. *C. cotinifolia* G. Don, Gen. Hist. Dichl. Pl. **2**: 445. 1832.

118a. var. **cotinifolia.**
118b. var. **leptodictya** (IB) IB, comb. nov. *C. cotinifolia* var. *leptodictya* IB, 1977, p. 190.
118c. var. **percoriacea** (IB) IB, comb. nov. *C. cotinifolia* var. *percoriacea* IB, 1977, p. 191.
118d. var. **glaberrima** (IB) IB, comb. nov. *C. cotinifolia* var. *glaberrima* Hoehne ex IB, 1977, p. 191.

119. Ch. lavradiiflora (Harms) IB, comb. nov. *C. lavradiiflora* Harms, Feddes Repert. **20**: 124. 1924.

xxiv. ser. CONFERTAE

Ch. sect. **Absus** ser. **Confertae** (IB) IB, comb. nov. *C.* ser. *Confertae* IB, 1977, p. 193.

120. Ch. conferta (Bentham) IB, comb. nov. *C. conferta* Bentham in Martius, Fl. Bras. **15**(2): 142. 1870.

120a. var. **conferta**.
120b. var. **simulans** (IB) IB, comb. nov. *C. conferta* var. *simulans* IB, 1977, p. 196.
120c. var. **machrisiana** (Cowan) IB, comb. nov. *C. machrisiana* Cowan, Los Angeles Co. Mus., Contrib. Sci. **13**: 12. 1957 ≡ *C. conferta* var. *machrisiana* (Cowan) IB, 1977.
120d. var. **virgata** (IB) IB, comb. nov. *C. conferta* var. *virgata* IB, 1977, p. 197.

121. Ch. crommyotricha (Harms) IB, comb. nov. *C. crommyotricha* Harms, Feddes Repert. **20**: 126. 1924.

122. Ch. caespitosa (Bentham) IB, comb. nov. *C. caespitosa* Bentham in Martius, Fl. Bras. **15**(2): 136. 1870.

xxv. ser. GLUTINOSAE

Ch. sect. **Absus** ser. **Glutinosae** (IB) IB, comb. nov. *C.* ser. *Glutinosae* IB, 1977, p. 201.

123. Ch. dentata (Vogel) IB, comb. nov. *C. dentata* Vogel, Syn. Gen. Cass. 49. 1837.

124. Ch. fragilis (IB) IB, comb. nov. *C. fragilis* IB, 1977, p. 203.

125. Ch. stillifera (IB) IB, comb. nov. *C. stillifera* IB, 1977, p. 204.

126. Ch. catiarae (IB) IB, comb. nov. *C. catiarae* IB, 1977, p. 205. *C. catarae* IB ex Gray Card Index 285. 1978, lapsu.

127. Ch. punctulifera (Harms) IB, comb. nov. *C. punctulifera* Harms, Notizbl. Bot. Gart. Berlin **8**(80): 714. 1924.

128. Ch. lentiscifolia (Bentham) IB, comb. nov. *C. lentiscifolia* Bentham in Martius, Fl. Bras. **15**(2): 144. 1870.

129. Ch. myrophenges (IB) IB, comb. nov. *C. myrophenges* IB, 1977, p. 207.

130. Ch. semaphora (IB) IB, comb. nov. *C. semaphora* IB, 1977, p. 209.

131. Ch. caracensis (IB) IB, comb. nov. *C. caracensis* Taubert ex IB, 1977, p. 210.

132. Ch. kirkbridei (IB) IB, comb. nov. *C. kirkbridei* IB, 1977, p. 212.

133. Ch. echinocarpa (Bentham) IB, comb. nov. *C. echinocarpa* Bentham in Martius, Fl. Bras. **15**(2): 136. 1870.

xxvi. ser. INCURVATAE

Ch. sect. **Absus** ser. **Incurvatae** (IB) IB, Brittonia **31**(4): 467. 1979. *C.* ser. *Incurvatae* IB, 1977, p. 214.

134. Ch. planifolia (IB) IB, comb. nov. *C. planifolia* IB, 1977, p. 215.

135. Ch. lavradioides (Bentham) IB, comb. nov. *C. lavradioides* Bentham in Martius, Fl. Bras. **15**(2): 137. 1870.

136. Ch. linearifolia (G. Don) IB, comb. nov. *C. linearifolia* G. Don, Gen. Hist. Dichl. Pl. **2**: 446. 1832.

136a. var. **linearifolia**.

136b. var. **bradei** (Harms) IB, comb. nov. *C. bradei* Harms, Notizbl. Bot. Gart. Berlin **8**(80): 714. 1924 ≡ *C. linearifolia* var. *bradei* (Harms) IB, 1977.

137. Ch. incurvata (Bentham) IB, Brittonia **31**(4): 468. 1979. *C. incurvata* Bentham in Martius, Fl. Bras. **15**(2): 137. 1870.

137a. var. **incurvata**.

137b. var. **zanclodes** (IB) IB, comb. nov. *C. incurvata* var. *zanclodes* IB, 1977, p. 221.

xxvii. ser. SPINULOSAE

Ch. sect. **Absus** ser. **Spinulosae** (IB) IB, comb. nov. *C.* ser. *Spinulosae* IB, 1977, p. 221.

138. Ch. spinulosa (IB) IB, comb. nov. *C. spinulosa* IB, 1977, p. 222.

xxviii. ser. UNIJUGAE

Ch. sect. **Absus** ser. **Unijugae** (Bentham) IB, comb. nov. *C.* ser. *Unijugae* Bentham in Martius, Fl. Bras. **15**(2): 134. 1870.

139. Ch. monticola (Bentham) IB, comb. nov. *C. monticola* Martius ex Bentham in Martius, Fl. Bras. **15**(2): 134. 1870.

140. Ch. catapodia (IB) IB, comb. nov. *C. catapodia* IB, 1977, p. 226.

xxix. ser. STRICTIFOLIAE

Ch. sect. **Absus** ser. **Strictifoliae** (IB) IB, comb. nov. *C.* ser. *Strictifoliae* IB, 1977, p. 227.

141. Ch. strictifolia (Bentham) IB, comb. nov. *C. strictifolia* Bentham in Martius, Fl. Bras. **15**(2): 227. 1870.

xxx. ser. ERICIFOLIAE

Ch. sect. **Absus** ser. **Ericifoliae** (IB) IB, comb. nov. *C.* ser. *Ericifoliae* IB, 1977, p. 229.

142. Ch. ericifolia (Bentham) IB, comb. nov. *C. ericifolia* Bentham in Martius, Fl. Bras. **15**(2): 145. 1870.

xxxi. ser. ABSOIDEAE

Ch. sect. **Absus** ser. **Absoideae** (Bentham) IB, comb. nov. *C.* ser. *Absoideae* Bentham in Martius, Fl. Bras. **15**(2): 131. 1870.

143. Ch. multiseta (Bentham) IB, comb. nov. *C. multiseta* Bentham in Martius, Fl. Bras. **15**(2): 141. 1870.

144. Ch. egleri (IB) IB, comb. nov. *C. egleri* IB, 1977, p. 235.

145. Ch. longicuspis (Bentham) IB, comb. nov. *C. longicuspis* Bentham in Martius, Fl. Bras. **15**(2): 150. 1870.

146. Ch. paraunana (IB) IB, comb. nov. *C. paraunana* IB, 1977, p. 238.

147. Ch. barbata (Nees & Martius) IB, comb. nov. *C. barbata* Nees & Martius, Nov. Act. Acad. Caes. Leop.-Carol. Nat. Cur. **12**: 32. 1823.

148. Ch. rugosula (Bentham) IB, comb. nov. *C. desertorum* var. *rugosula* Bentham in Martius, Fl. Bras. **15**(2): 134. 1870 ≡ *C. rugosula* (Bentham) IB, 1977.

149. Ch. belemii (IB) IB, comb. nov. *C. belemii* IB, 1977, p. 241.

149a. var. **belemii.**
149b. var. **paludicola** (IB) IB, comb. nov. *C. belemii* var. *paludicola* IB, 1977, p. 243.

150. Ch. salvatoris (IB) IB, comb. nov. *C. salvatoris* IB, 1977, p. 244.

151. Ch. acosmifolia (Bentham) IB, comb. nov. *C. acosmifolia* Martius ex Bentham in Martius, Fl. Bras. **15**(2): 133. 1870.

151a. var. **acosmifolia.**
151b. var. **oropedii** (IB) IB, comb. nov. *C. acosmifolia* var. *oropedii* IB, 1977, p. 247.
151c. var. **euryloba** (IB) IB, comb. nov. *C. acosmifolia* var. *euryloba* IB, 1977, p. 247.

152. Ch. andersonii (IB) IB, comb. nov. *C. andersonii* IB, 1977, p. 248.

153. Ch. juruenensis (Hoehne) IB, comb. nov. *C. juruenensis* Hoehne, Comm. Linhas Estrat. Matto Grosso ao Amazonas **74**(5): = Botanica **12**(Contrib. Conh. Legum. Rondônia): 13. 1922.

154. Ch. brevicalyx (Bentham) IB, comb. nov. *C. brevicalyx* Bentham in Martius, Fl. Bras. **15**(2): 134. 1870.

154a. var. **brevicalyx.**
154b. var. **elliptica** (IB) IB, comb. nov. *C. brevicalyx* var. *elliptica* IB, 1977, p. 252.

155. Ch. zygophylloides (Taubert) IB, comb. nov. *C. zygophylloides* Taubert, Flora (Regensburg) **75**: 79. 1892.

155a. var. **zygophylloides.**

155b. var. **colligans** (IB) IB, comb. nov. *C. zygophylloides* var. *colligans* IB, 1977, p. 255.

155c. var. **deamii** (Britton & Rose) IB, comb. nov. *Grimaldia deamii* Britton & Rose, N. Amer. Fl. **23**(5): 300. 1930 ≡ *C. zygophylloides* var. *deamii* (Britton & Rose) IB, 1977.

155d. var. **caribaea** (IB) IB, comb. nov. *C. zygophylloides* var. *caribaea* IB, 1977, p. 256.

156. Ch. souzana (IB) IB, comb. nov. *C. souzana* IB, 1977, p. 256.

157. Ch. jacobinea (Bentham) IB, comb. nov. *C. jacobinea* Bentham in Martius, Fl. Bras. **15**(2): 132. 1870.

158. Ch. chapadae (IB) IB, comb. nov. *C. chapadae* IB, 1977, p. 258.

159. Ch. viscosa (Humboldt, Bonpland & Kunth) IB, comb. nov. *C. viscosa* Humboldt, Bonpland & Kunth, Nov. Gen. & Sp. **6**(qu): 360. 1823.

159a. var. **viscosa.**

159b. var. **paraguayensis** (Chodat & Hassler) IB, comb. nov. *C. viscosa* var. *paraguayensis* Chodat & Hassler, Bull. Herb. Boiss. II, **4**: 825. 1904.

159c. var. **major** (Bentham) IB, comb. nov. *C. viscosa* var. *major* Bentham in Martius, Fl. Bras. **15**(2): 132. 1870.

160. Ch. campestris IB, nom. nov. *C. camporum* Bentham, J. Bot. (Hooker) **2**: 79. 1840. Non *Ch. camporum* Greene, 1903.

161. Ch. hispidula (Vahl) IB, comb. nov. *C. hispidula* Vahl, Ecl. Amer. **3**: 10. 1807.—N.B.: the earlier taxonomic synonym *C. tetraphylla* Martyn, 1797, cannot be transferred to *Chamaecrista,* the epithet being preempted by *Ch. tetraphylla* (Desvaux) Britton & Killip, 1936.

162. Ch. amiciella (IB) IB, comb. nov. *C. amiciella* IB, 1977, p. 267.

163. Ch. carobinha (IB) IB, comb. nov. *C. carobinha* IB, 1977, p. 270.

164. Ch. punctulata (Hooker & Arnott) IB, comb. nov. *C. punctulata* Hooker & Arnott, Bot. Beechey Voy. 420. 1840.

165. Ch. roncadorensis (IB) IB, comb. nov. *C. roncadorensis* IB, 1977, p. 272.

166. Ch. fagonioides (Vogel) IB, comb. nov. *C. fagonioides* Vogel, Syn. Gen. Cass. 50. 1837.

166a. var. **fagonioides.**

166b. var. **macrocalyx** (IB) IB, comb. nov. *C. fagonioides* var. *macrocalyx* IB, 1977, p. 276.

Species Nova post num. 14 inserenda

Chamaecrista fodinarum Irwin & Barneby, sp. nov., juxta *Ch.* (ser. *Nigricantes*) *pilicarpam* (Harms) Irwin & Barneby, habitu foliisque similem inserenda, ab ea imprimis ovario (deinde legumine) verruculoso interdumque puberulo setis autem elongatis destituto diversa.—BRAZIL. Minas Gerais, mun. de Grão Mogol: campo rupestre, 1050 m, Serra do Pipiri, 22.III.1981, *G. Hatschbach 42886*.—Holotypus, MBM; isotypi, NY + 4 distributuri.—Mun. Grão Mogol: em mata baixa próximo ao Rio Jequitinonha, estrada ao Pôrto Mandacaru, 11.V.1979, *H. C. de Lima 1013* in RB *200227*; paratypi, MBM, NY, RB (not seen).

Densely branched shrubs and treelets at anthesis 1–4 m, with nigrescent lenticellate older branches, the young stems and foliage microscopically, sometimes only residually puberulent but charged throughout with secretory glands, the stems and inflorescence glutinous-verruculose, the firmly chartaceous lfts bicolored, above lustrous olivaceous glandless, beneath pallid and resin-dotted or glandular-punctate, the elongate, loosely fld racemes far exserted from foliage.

Stipules subulate 0.5–0.8 mm, deciduous.

Lvs 5–15 cm; petiole 1.5–3 cm, at middle 0.8–1.3 mm diam, the rachis (1–)1.5–7 cm firmly green-margined and narrowly deeply sulcate ventrally, the margins slightly enlarged upward and abruptly contracted at insertion of pulvinules; lfts 2–5 pairs, spreading-ascending face upward from rachis on pulvinules 1.5–2.5 mm, slightly accrescent distally, all in outline obovate or oblong-obovate emarginate, the distal pair 3–4 × 1.2–2 cm, ±1.5–2.2 times as long as wide, at asymmetric base subcordate on proximal and cuneate on distal side, the margin revolute, the midrib immersed above, cariniform beneath, the 5–8 pairs of major secondary camptodrome nerves prominulous on both faces but more sharply so beneath, the weak intercalary secondary and the tertiary venules scarcely perceptible above, faint beneath.

Racemes erect many-fld, the nodding buds in praefloration racemosely exserted above currently expanded fls, the primary axis including short peduncle becoming 12–21 cm; bracts subulate involute 2–2.5 mm, caducous; pedicels ±12–15 mm, 2-bracteolate beyond middle; buds ellipsoid or oblong-ellipsoid obtuse, densely verruculose; sepals elliptic-oblanceolate 8.8–10 × 3–3.5 mm; petals yellow glabrous, strongly heteromorphic, 4 obovate or oblanceolate, 3 or them larger up to 11–14 mm, 1 smaller, the cucullus involutely boomerang-shaped 12–16 mm, crenulate along the outer margin; filaments 0.1–1 mm; anthers 4–5 mm, the connective muticous; ovary either glabrous or minutely puberulent, but in either case densely glandular-verruculose; style glabrous or puberulent proximally, 8–10 mm long.

Pod unknown.—Collection: 1.—Fig. 43.

Campo rupestre and brush-woodland along rivers, ±(?) 800–1050 m, apparently local, known only from the municipio of Grão Mogol, near 16°30′S in n. Sa. do Espinhaço, n.-centr. Minas Gerais.—Fl. III–V.

Chamaecrista fodinarum bears a strong habital likeness to *Ch.* (ser. *Nigricantes*) *pilicarpa,* a species known only from its type-locality in Sa. de Caraça, distant about 400 km southward along Sa. do Espinhaço, and is doubtless closely related to it. Both have broad firm, bicolored and marginally revolute leaflets lustrous green above and pallid gland-dotted beneath, together with precociously elongating many-flowered racemes rising well above the leaves. The southern species is known only in the mature fruiting condition, whereas our material of *Ch. fodinarum* has flowers only, precluding a full comparison. The striking feature to which *Ch. pilicarpa* owes its epithet is the armament of stout setae on the pod, setae of a type that in sect. *Absus* are readily evident on the young ovary. There are no such setae on the ovary of *Ch. fodinarum,* of which the pod will surely turn out to be glabrous or almost so and glandular-verruculose.

Our two collections of *Ch. fodinarum* are not exactly alike. The type, from campo rupestre at 1050 m, is described by Hatschbach as a compactly branched shrublet a meter tall; all its leaves have only two or rarely three pairs of leaflets. The paratype from low woodland along Rio Jequitinonha, at elevation not recorded but probably near 800 m, was taken from treelets 3–4 meters tall, and has 3–5 pairs of leaflets. The general aspect of the bijugous type suggested affinity either to *Ch.* (*Glutinosae*) *punctulifera,* which is endemic to Pico de Almas in Bahia, and its long-exserted racemes of nodding buds pointed ambiguously to

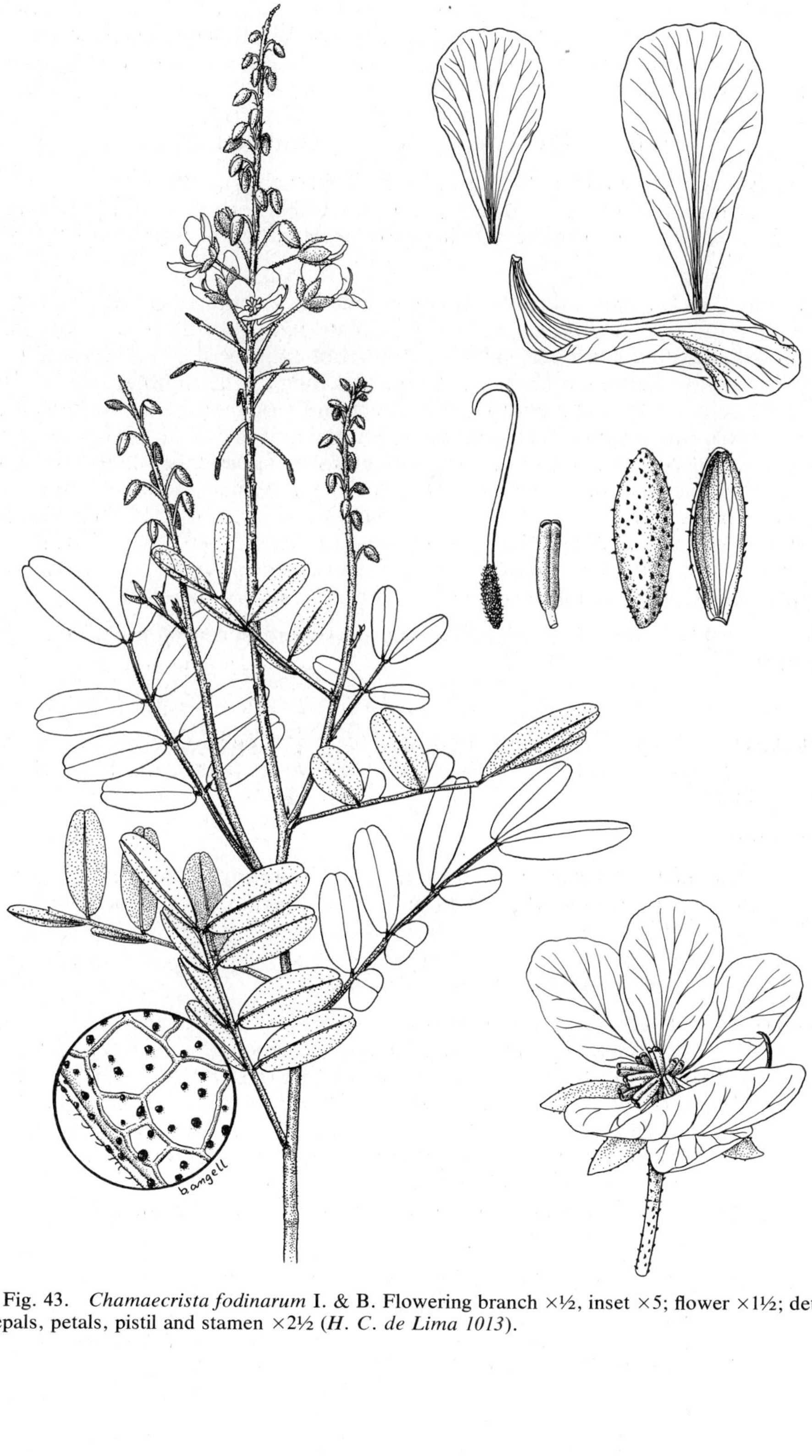

Fig. 43. *Chamaecrista fodinarum* I. & B. Flowering branch ×½, inset ×5; flower ×1½; detached sepals, petals, pistil and stamen ×2½ (*H. C. de Lima 1013*).

ser. *Ochnaceae,* a group of wide dispersal in Sa. do Espinhaço. However the ser. *Glutinosae* and *Ochnaceae* are both characterized by dorsally undifferentiated leaflets with plane margins, a criterion by which *Ch. fodinarum* and *Ch. pilicarpa* are handily excluded.

C. Sectio GRIMALDIA (Schrank) Irwin & Barneby

Chamaecrista sect. **Grimaldia** (Schrank) Irwin & Barneby, comb. nov. *Grimaldia* Schrank, Bot. Zeit. (Regensb.) **4**: 184. 1805, pro gen. ≡ *Cassia* sect. *Grimaldia* (Schrank) Irwin & Barneby, Mem. N.Y. Bot. Gard. **30**: 227. 1977.

Resembling sect. *Absus* subsect. *Absus* in glandular-setulose pubescence, and in particular the ser. *Absoideae* in the 4-foliolate lvs, but different as follows: phyllotaxy distichous; androecium 2–7-merous, not concealed by an involute heteromorphic petal, the longer filaments at least as long as their anther, this dehiscent before separation of the petals; style short and stout, grooved ventrally and passing upward into a spatuliform, distally enlarged and incurved, obliquely truncate stigma; between each pair of lfts a linguiform appendage, the distal one smaller, sometimes setiform.—Monocarpic herbs sometimes enduring into a second season and softly woody at base; fls small, 4.5–7 mm, at least facultatively cleistogamous.—Sp. 1, circumtropical, in the Americas 2 vars., the typical *Ch. absus* var. *absus,* with 5(–7)-androus fls and extending to Africa, s.-e. Asia and n. Australia, and the endemic var. *meonandra,* with (2–)3–4-androus fls.

A full description and synonymy has been presented in our monograph (Irwin & Barneby, 1977, l.c. sub *Cassia*). The transfer to *Chamaecrista* is here proposed:

1. Chamaecrista absus (Linnaeus) Irwin & Barneby, comb. nov. *Cassia absus* Linnaeus, Sp. Pl. 537. 1753 ≡ *Grimaldia absus* (Linnaeus) Link, Handbuch **2**: 141. 1831.

1a. var. **absus**.

1b. var. **meonandra** (Irwin & Barneby) Irwin & Barneby, comb. nov. *Cassia absus* var. *meonandra* Irwin & Barneby, Mem. N.Y. Bot. Gard. **30**: 282. 1977.

D. Sectio CHAMAECRISTA

Chamaecrista Moench sect. **Chamaecrista**. *Chamaecrista* Moench, 1794, sens. str. Sp. typica: **Ch. nictitans** (Linnaeus) Moench ≡ *Cassia nictitans* Linnaeus; the lectotypus proposed by Britton & Rose, 1930, p. 270, *Cassia chamaecrista* Linnaeus, is rejected because decandrous, thus at variance with the protologue which explicitly separates the genus from *Cassia* because of its pentandrous androecium.—*Cassia* sect. *Chamaecrista* (Moench) DeCandolle ex Colladon, Hist. Cass. 24, 118. 1816, accuratius definivit Irwin, Mem. N.Y. Bot. Gard. **12**(1): 44, by implication. 1964.

Cassia **Chamaecristae* (statu haud designato) Linnaeus, Sp. Pl. 379. 1753.

Cassia sect. *Lasiorhegma* C§ *calyce membranaceo* (statu haud designato) Vogel, Syn. Gen. Cass. 56. 1837.

Cassia subgen. *Lasiorhegma* (Vogel) Bentham sect. *Chamaecrista* subsect. *Leiocalyx* Bentham in Martius, Fl. Bras. **15**(2): 161. 1870.

Cassia subsect. *Leiocalyx* sensu Bentham, 1871, pp. 569–582.
Chamaecrista sensu Britton & Rose, 1930, pp. 270–299, exclus. ser. I, II (spp. 1–3).
Cassia sect. *Chamaecrista* sensu Irwin & Barneby, Mem. N.Y. Bot. Gard. **30**: 3, in clave. 1977.

Inflorescence composed of 1–few-fld racemes, the peduncle short or obsolete, exactly axillary or adnate to stem, the fls then appearing supra-axillary; fls enantiostylous; sepals subequilong, delicately net-veined or almost veinless; corolla irregular, 1 abaxial petal (often termed the *cucullus*) heteromorphic, opposed to the oblique pistil, ± shielding the anthers; androecium 5–10-merous, the 5 antesepalous stamens larger but commonly unequal among themselves, the 5 antepetalous ones smaller, or sterile, or lacking; seeds obovoid, rhomboid or trapezoid, the testa serially pitted; $x = 8$.—Shrubs (exceptionally subarborescent), subshrubs (some from a xylopodium) and herbs, these either perennial or monocarpic, the latter potentially weedy; phyllotaxy distichous, but the 2 ranks of lvs sometimes secundly spiral; pubescence commonly duplex, of shorter incurved and longer setiform (but never secretory) trichomes, but either sort predominant or sometimes absent; petiolar gland(s) present in all but ser. *Bauhinianae* (2 spp. with 2-foliolate or largely suppressed lvs), sessile or stipitate; lfts 1–many pairs.—Spp. (subject to controversial taxonomic interpretations) ±75, circumtropical and thence weakly invading the n. and s. temperate zones, primarily diversified in Brazil s. of the Amazonian forests, in Central and Caribbean North America, and tropical Africa, reaching South Africa, Madagascar, Japan, n. Australia and (adventively) the Hawaiian Is., 51 herein recognized from the New World.

The fifty-one American species of sect. *Chamaecrista* recognized herein are arranged in six series, so far as possible in what is thought to illustrate a sequence of evolutionary specialization. The *Coriaceae,* which are connected with sect. *Absus* by way of its subsect. *Baseophyllum,* are a basically woody group (the trunk sometimes condensed into a xylopodium), with relatively large longistyled flowers borne on strictly axillary peduncles. Of the twenty species all but one occurs in the planaltine provinces of eastern Brazil and all but two are endemic there, several narrowly so. The less numerous *Flexuosae* are essentially like the *Coriaceae* but are scarcely fruticose and have peculiar angulate leaf-stalks and stems and a characteristic venulation of the leaflets. All four are native to eastern Brazil but two extend discontinuously far abroad, one to Colombia, the other to Mexico, Texas and Argentina. The frankly herbaceous *Prostratae* have flowers axillary like the preceding but mostly small and brachystylous. Like the foregoing they all occur in eastern Brazil, but all but two extend thence into neighboring countries or far into North America. The *Bauhinianae* are like the *Prostratae* except that the petioles are glandless and the leaves are reduced to one pair of relatively broad leaflets, or in one of the two species are absent except at juvenile nodes, their function being assumed by a thatch of leaflike stipules. One of this pair is endemic to the Brazilian Planalto, the other widespread and weedy almost throughout the American tropics. The ser. *Greggianae* and *Chamaecrista* differ collectively from the remainder in their adnate peduncle and consequently supra-axillary flowers. The first consists of just one (but pluriracial) Mexican species, a xeromorphic shrub further peculiar in the section for its obtuse sepals and consequently obovoid flower-buds, reminiscent of compatriot species of *Senna.* The genuine chamaecristas form an immense and taxonomically perplexing set of often intergradient forms widespread throughout the tropics and into the temperate zones of both hemispheres. In the Americas there are a few truly shrubby, even subarborescent species, but the majority, like those of the Old World, are herbs of rapid growth and short duration, several obligately monocarpic. Almost all of them are excessively polymorphic in growth-habit, pubescence, flower-size and style-length, this sometimes at once short, dilated apically and hooked backward to receive pollen from the associated anthers prior to opening of the flower. The drift toward autogamy in sect. *Chamaecrista* appears to have taken place independently in several lines of inheritance, accompanied in individual species of ser. *Flexuosae, Bauhinianae, Prostratae* and *Chamaecrista* by decrease or total loss of the antepetalous whorl of stamens. Similarly a displacement of the midrib toward the distal margin of the leaflets, which finally become truly dimidiate, has occurred independently, it would appear, in African and American chamaecristas not immediately related, even though all would be accommodated in our ser. *Chamaecrista* as defined below. In the belief that *Cassia* ser. *Dimidiatae* Bentham (1870, p. 581), based on this single feature, is not phyletically natural, we do not recognize its existence in the New World.

Morphology and the known facts of dispersal suggest that the primary differentiation of sect. *Chamaecrista* occurred in upland savanna country in tropical eastern South America. From this nucleus independent evolutionary lines, following roughly parallel courses towards monocarpism and autogamy, have expanded into cognate environments in North America, Africa and Asia, culminating in eastern Asia and United States in short-lived, tiny-flowered annual forms which are adapted to survive winter in the form of seeds. Narrow endemism and confinement to a particular habitat is encountered commonly in the relatively primitive woody *Coriaceae* and in the calciphile desert-dwelling *Ch. greggii*, species relatively easy to define and classify. Most other members of sect. *Chamaecrista* thrive in disturbed or unstable habitats where a short immaturity and a short lifetime of continuous and abundant seed-production are advantageous. These conditions, which promote rapid conquest of territory and explosive evolutionary radiation, are reflected in the unsettled and still controversial taxonomic history of the group.

Key to Series of sect. *Chamaecrista* Represented in the Americas

1. Sepals ovate or lanceolate, acute or acuminate, the fl-buds in consequence ovoid-acuminate or -acute.
  2. Peduncles exactly axillary, the axis not adnate to the stem-internode beyond its point of origin.
    3. Lfts 1–many pairs, but if all lvs 2-foliolate then the petiole furnished with a gland; perfect lvs developed at all nodes.
      4. Lfts with midrib and one equally strong primary nerve adjacent on its proximal side produced to apical anastomosis, together enclosing a veinless areole that occupies the center of the blade; stems often abruptly flexuous. ii. ser. *Flexuosae* (p. 695).
      4. Lfts variably venose, but the primary vein adjacent to the midrib on its proximal side not produced to apical anastomosis and not enclosing a veinless areole; stems seldom flexuous.
        5. Shrubs (some arborescent) or subshrubs from xylopodium, of the Brazilian Planalto, 1 extending to the Guayana Highland and 1 endemic on Bahama Is. and adjacent Haiti; large-fld, the longest petal (11–)12–28 mm; petiolar gland (except in W. Indian *Ch. caribaea*) sessile or almost so. i. ser. *Coriaceae* (p. 667).
        5. Herbs (one, both shrubby and Brazilian, is keyed with the preceding and not further allowed for herein), with mostly diffuse stems radiating from the rootcrown or a short caudex, neither fruticose nor with xylopodium; mostly small-fld, with long petal 5–10 mm, but up to 20 mm in some forms of *Ch. serpens*, there coinciding with slenderly stipitate petiolar gland; petiolar gland either stipitate or (rarely) 0. iii. ser. *Prostratae* (p. 706).
    3. Lfts 1 pair in every lf; petiolar gland 0; foliage developed in one sp. only at a few early nodes, these lvs deciduous, the flowering branches clothed in imbricated stipules. iv. ser. *Bauhinianae* (p. 727).

2. Peduncles appearing supra-axillary, adnate to the stem-internode through (0.5–)1+ mm beyond point of origin. vi. ser. *Chamaecrista* (p. 738).

1. Sepals obtuse, the fl-buds in consequence subglobose or broadly obtusely ovoid. Xeromorphic microphyllous shrubs of arid e. Mexico and s. Texas. v. ser. *Greggianae* (p. 735).

i. ser. CORIACEAE (Bentham) Irwin & Barneby

Chamaecrista sect. **Chamaecrista** ser. **Coriaceae** (Bentham) Irwin & Barneby, comb. nov. *Cassia* sect. *Chamaecrista* ser. *Coriaceae* Bentham in Martius, Fl. Bras. **15**(2): 165. 1870.—Sp. lectotypica: *C. choriophylla* Vogel ≡ **Chamaecrista choriophylla** (Vogel) Irwin & Barneby.

Cassia sect. *Chamaecrista* ser. *Subcoriaceae* Bentham in Martius, Fl. Bras. **15**(2): 171. 1870, ex parte, quoad typum.—Sp. lectotypica: *C. mucronata* Sprengel ≡ **Chamaecrista mucronata** (Sprengel) Irwin & Barneby.

Cassia sect. *Chamaecrista* ser. 4. *Coriaceae* sensu Bentham, 1871, p. 572.

Shrubs (some subarborescent) or subshrubs from a xylopodium; petiolar gland always present, except in West Indian *Ch. caribaea* sessile or almost so; lfts (1–many pairs) mostly thick-textured, variably venulose, but the midrib and one adjacent primary nerve not apically anastomosing as in ser. *Flexuosae*; peduncles exactly axillary; sepals acute or acuminate; fls relatively large, the long petal (11–)12–28 mm.—Spp. 20, primarily of the Brazilian Planalto, there elaborately diversified in the highlands of Serra do Espinhaço, 1 extending to the Guayana Highland, 1 endemic on Bahama Is. and n.-w. Hispaniola.

Key to the Species of ser. *Coriaceae*

1. Lvs 4–many-foliolate.
   2. Lfts 10–27 pairs, their blades strongly asymmetrical, often subdimidiate; stipules never ovate and cordate-amplexicaul.
      3. Stipules 4–9 × 1–1.5 mm, persistent even after fall of lvs along the naked trunks; midrib of lfts submarginal; centr. and s. Sa. do Espinhaço in centr. and s. Minas Gerais. 2. *Ch. olesiphylla* (p. 670).
      3. Stipules 5–12 × 2–3.5 mm, deciduous before fall of the associated lf; midrib of lfts strongly displaced but intramarginal; Chapada Diamantina of Bahia and extreme n. Minas Gerais, disjunct on Guayana Highland. 1. *Ch. roraimae* (p. 669).
   2. Lfts 2–16 pairs, the blades usually not strongly asymmetrical and if so then not over 8 pairs, if more than 10 pairs the stipules ovate-amplexicaul.
      4. Stipules ovate to suborbicular, in all but *Ch. cinerascens* basally cordate-amplexicaul on one or both sides, and in *Ch. cinerascens* deciduous before the associated lf.
         5. Blade of stipules simply ovate, not basally cordate-amplexicaul, early deciduous. 4. *Ch. cinerascens* (p. 672).
         5. Blade of stipules at base cordate-amplexicaul on one or both sides, persistent.
            6. Stipules at first antrorsely imbricated but early reflexed against the stem, the blade ample and 17–35-nerved.
               7. Lfts 9–16 pairs. 5. *Ch. potentilla* (p. 673).
               7. Lfts 2–9 pairs.
                  8. Blade of stipules 7–25 × 5–25 mm, like the lfts ovate or oblong-ovate obtuse and only minutely mucronulate. 7. *Ch. rotundata* (p. 675).
                  8. Blade of stipules 6–9.5 × 2.5–5.5 mm, ovate and like the lfts abruptly acuminulate. 6. *Ch. distichoclada* (p. 674).
            6. Stipules at first and permanently erect, the blade 1–25-nerved.
               9. Lfts (4–)5–9 pairs; stems 1.5–5 dm, ascending and erect from xylopodium; mostly w. of Rio S. Francisco in w. Minas Gerais and s.-e. Goiás, rarely e. to n.-centr. Minas Gerais but the plant then entirely glabrous. 8. *Ch. cardiostegia* (p. 678).
               9. Lfts 2–4 pairs; stems fruticose to 9–12 dm; stems at least thinly pubescent. 9. *Ch. anceps* (p. 679).
      4. Stipules lance-acuminate, subulate or deltate, either less than 2.5 mm wide, or only 1–5-nerved, or both, in no case basally cordate or semi-cordate.
         10. Plants of Bahama Is. and n.-w. Hispaniola. 10. *Ch. caribaea* (p. 680).

10. Plants of e. Brazil.
    11. Petiolar gland urceolate slender-stipitate 0.2–0.45 mm diam, the stipe 0.5–1.6 mm long, the whole in profile taller than diam of head; n.-e. Brazil; cf. 25. *Ch.* (ser. *Prostratae*) *tenuisepala* (p. 707).
    11. Petiolar gland sessile or subsessile, in profile shorter than diam of head; Brazilian Planalto in Bahia, Minas Gerais and Goiás.
        12. Lfts obliquely ovate 15–30 × (8–)10–20 mm, tipped with a vulnerant spinule 1.7–2.7 mm; very local in n.-centr. Minas Gerais. 11. *Ch. aristata* (p. 683).
        12. Lfts various in shape and size, but only weakly aristulate or mucronulate, the excurrent projection less than 1.5 mm and never vulnerant.
            13. Lfts 4–10 pairs, at least 5 pairs in some larger lvs.
                14. Stipules relatively large, 5–15 mm long, 5–9-nerved; lfts strongly asymmetric, the midrib dividing the blade ±1:2. 3. *Ch. venulosa* (p. 671).
                14. Stipules 1.5–7 mm long, 1–5-nerved, if over 4.5 mm long only 1–3-nerved; lfts subsymmetrical or sometimes retrorsely falcate but the midrib centric or subcentric.
                    15. Lvs numerous, small and crowded, mostly 1–2, rarely up to 2.7 cm, the lfts 6–16 × 1.5–4 mm; low bushy plants, seldom over 3 dm, but sometimes developing long prostrate woody trunks. 14. *Ch. tragacanthoides* (p. 686).
                    15. Lvs larger, 2–6.5 cm, the lfts 10–26 × 2.5–14 mm; variable in habit, but if only 3 dm tall the larger lfts over 4 mm wide.
                        16. Lfts 5–10 pairs, oblong, 2.5–7 mm wide, densely papillate beneath; pod ±3–4 cm, 5–9-ovulate. 15. *Ch. papillata* (p. 688).
                        16. Lfts 5–6 pairs, obovate-oblanceolate, 4–14 mm wide, not papillate beneath; pod 5.5–7 cm, 10–14-ovulate. 12. *Ch. mucronata* (p. 684).
            13. Lfts 2–4 pairs, of larger lvs at least 3 pairs.
                17. Pedicels 3–5 cm; stems herbaceous erect from xylopodium 1.5–4 dm, relatively few- and remotely lvd; s.-centr. Goiás. 17. *Ch. burchellii* (p. 690).
                17. Pedicels 1.5–2.5 cm; stems mostly fruticose and taller but, if as short, or from a xylopodium, then densely leafy distally; Sa. do Espinhaço in Minas Gerais and s. Chapada Diamantina in s. Bahia.
                    18. Lvs 1–2 cm, the crowded lfts narrow, 6–12(–16) × 1.5–4 mm. 14. *Ch. tragacanthoides* (p. 686).
                    18. Lvs 2–4 cm, the lfts either longer or 4–14 mm wide or both.
                        19. Lfts of all lvs, including those subtending fls, 3–4 pairs, their blade mostly obovate-oblanceolate, obtuse or mucronulate; widespread. 12. *Ch. mucronata* (p. 684).
                        19. Lfts of cauline lvs 2 and (fewer) 3 pairs, or of lvs subtending a fl only 1 pair, the blades of all narrowly oblance-elliptic subacute; rare (environs of Grão Mogol and Diamantina).
                            20. Lfts 14–30 × 3.5–8.5 mm, 5–7-nerved from base, thence reticulately venulose; suffruticose from xylopodium; Diamantina. 13. *Ch. multinervia* (p. 685).
                            20. Lfts 10–14 × 2.2–3.8 mm, 4–5-nerved from base, faintly penninerved from midrib; spindly treelets with pendulous flowering branchlets; Grão Mogol and vicinity. 16. *Ch. ulmea* (p. 689).

1. Lvs all exactly 2-foliolate.
    21. Lfts glaucous, oblance-elliptic, mostly 3–8.5 × 0.8–1.6 cm, at base subsymmetrically broad-cuneate on both sides; petioles 4–9 mm; fruticose with naked trunks to 1.5 m; stipules deltate ±1 mm; local in Sa. do Cipó. 20. *Ch. lagotois* (p. 694).
    21. Lfts olivaceous, either smaller or proportionately broader than the preceding, at base cordate on proximal side, cuneate on distal one; suffruticose from xylopodium and stems 1.5–6 dm, leafless at base only; stipules various, but if less than 2.5 mm the associated petiole subobsolete or the larger lfts obliquely obovate; relatively wide-

spread, but where sympatric with the last the stipules very much larger and cordate-amplexicaul.

22. Blade of lfts, at least in larger lvs, broadly elliptic-oblanceolate to obovate, well over 1 cm wide; stipules rarely less than 2.5 mm or less than 5-nerved but, if so, then the petiole with pulvinus together over 2.5 mm; centr. and s. Sa. do Espinhaço. 18. *Ch. choriophylla* (p. 691).

22. Blade of lfts narrowly oblong- or oblance-elliptic up to 15–35 × 3–7.5 mm; stipules deltate 1–1.5 mm, obscurely 3-nerved; petiole and pulvinus together 1–2.5 mm, the petiole proper almost 0; centr. Sa. do Espinhaço near Diamantina. 19. *Ch. simplifacta* (p. 694).

1. Chamaecrista roraimae (Bentham) Gleason, Bull. Torr. Club **56**: 395. 1929. *Cassia roraimae* Bentham, Trans. Linn. Soc. London **27**: 574. 1871.—"British Guiana: Roraima, *Appun,* n. 1231; and probably in the same mountains, *Rob. Schomburgk,* 2nd coll. n. 582; *Rich. Schomburgk* n. *840.*"—Lectoholotypus (Pittier, Cat. Fl. Venezol. 374. 1945), *Appun 1231,* K (hb. Hook.)! = NY Neg. *1533*; paratypi, *Rob. Schomburgk 582,* FI, G = F Neg. *28013,* K, & *Rich. Schomburgk 840,* K, LE!

Shrubs up to 3.5 m (sometimes flowering precociously as herbs only 1.5–2 dm), with knotty trunks to 2 cm diam, slender, soon leafless branches, and densely leafy annotinous branchlets, the stems, lf-stalks and inflorescence villosulous and setulose with short incurved and (especially the lf-stalks) longer straight ascending often yellowish setules up to 0.4–1(–1.3) mm, the foliage ± bicolored, the stiffly chartaceous lfts drying brown, darker above, ciliolate, but the faces varying from glabrous to finely villosulous.

Stipules erect, broadly lance-acuminate, shortly above the subsymmetrically rounded base 2–3.5 mm wide, the blades either glabrous except ciliolate margins or densely pubescent dorsally, all deciduous before the lf.

Lvs spreading-ascending, 3–10.5 cm, short-petiolate, the expanded blade lance- or oblong-elliptic in outline; lf-stalk 2.5–9.5 cm, narrowly margined ventrally, the sulcus almost or quite closed; petiole including wrinkled pulvinule 3–5 mm; petiolar gland immediately below first pair of lfts, bucket- or squatly peg-shaped, (0.5–)0.6–1 mm diam, not or little wider than the head of the thick short stalk, 0.5–1.1 mm tall; 1–several smaller stalked glands between some distal pairs of lfts; lfts of major cauline lvs 14–27 (on occasional lvs of axillary branchlets fewer) pairs, diminished at base and toward apex of lf-stalk, separated by segments of lf-stalk slightly > or < than their greatest width, in outline narrowly lance- or linear-oblong, 3.5–17 × 1.2–3.6 mm, distally subfalcate and abruptly contracted into a cusp 0.3–1.2 mm, at base deeply angulate-cordate on the narrow adaxial side, more shallowly cordate on the other, from base 3–5-nerved, the midrib strongly displaced but intramarginal, the venation prominulous both sides or commonly immersed above.

Peduncles axillary solitary 2–5 mm, or sometimes serially geminate, the distal one then briefly adnate to stem, one or both 1–2-fld; bracts and bracteoles resembling stipules but only 3.5–7.5 mm; pedicels ascending, 10–25 mm; buds ovoid-acuminate, minutely pilosulous and often also ± yellow-setulose; sepals greenish or brownish where exposed in bud, with broad petaloid margins, lance- or ovate-acuminate, 12–16(–17) mm; petals yellow, 4 broadly oblanceolate to obovate-cuneate, 20–28 mm, the fifth falcately dimidiate, slightly longer or shorter than the rest; stamens 10, the anthers 5.5–9.5 mm; gynoecium divaricate; ovary densely strigulose or villosulous; style linear-cylindric, puberulent to middle or glabrous, 10–14 mm; ovules 12–15.

Pod spreading-ascending, 5–6.5 × 0.8–0.9 cm, the valves reddish-brown, minutely pilosulous or yellow-villosulous; seeds unknown.—Collections: 25.—Fig. 37 (androecium).

Savanna-margins and rock-outcrops (n.-ward) or cerrado and margin of gallery-forest (s.-ward), local in two widely disjunct areas: plateaux at 1000–1200 m about the s.-e. margins of the Guayana Highland in s.-w. Guyana and adjoining Venezuela (Kamarang trail; río Cuquenan) and Brazil (Mt. Roraima; Sa. Tepequem); and near 1000 cm on Chapada Diamantina and n. Serra do Espinhaço from centr. Bahia (Sa. Agua de Rega) s. to the headwaters of Rio Pardo in extreme n. Minas Gerais.—Fl. VIII–II (n.-ward), XII–II (s.-ward).

An exceptionally handsome chamaecrista, resembling *Ch. olesiphylla* in the large flower but differing in the more erect stature, deciduous stipules, and longer leaves with pliant footstalks and less crowded leaflets. The populations in southeastern Brazil have dorsally pubescent stipules, leaflets at least puberulent dorsally and often on both faces, and densely strigulose ovary, whereas those of the Guayana Highland, separated by a distance of 2900 km, have dorsally glabrate stipules, mostly glabrous (rarely puberulent) leaflets, and yellow villosulous ovary. In other respects they are closely similar. Bentham described the petiolar gland of the northern population as sessile, urceolate or scutellate; and we can find no discontinuity in this respect between the northern and southern lobes of the species range.

2. Chamaecrista olesiphylla (Vogel) Irwin & Barneby, comb. nov. *Cassia olesiphylla* Vogel, Syn. Gen. Cass. & Linnaea **11**: 714, descr. ampliat. 1837.—"Hbt. in Brasilia: Sellow leg. ad Itambé."—Holotypus, †B = F Neg. *1718*; neoholotypus, former isotypus, K (hb. Hook.)! = IPA Neg. *1122* = NY Neg. *1535*; isotypi, LE, P!

Cassia olesiphylla sensu Bentham, 1870, p. 169; 1871, p. 574.

Diffuse or prostrate shrubs up to 0.6 m tall but up to 4–5 times as wide, with stiff coarse blackish trunks angulate by prominent lf-spurs and through second year clothed with persistent stipules, the branchlets of current season densely leafy, their internodes 2–11(–14) mm, mostly as long or shorter than stipules, except for glabrate faces of the lfts puberulent throughout with forwardly incurved and some spreading hairs up to 0.2–0.6 mm, the crowded subcoriaceous leaflets concolorous.

Stipules erect, firm but not pungent, lanceolate, 4–9 mm, at base subsymmetrically decurrent and ±1–1.5 mm wide, 3–5-nerved from base.

Lvs stiffly, sometimes widely ascending, 2–7.5 cm, short-petiolate, the expanded blade broadly lanceolate; lf-stalk 1.5–7 cm, tapering distally, narrowly 2-winged ventrally, the sulcus closed; petiole 2.5–6 mm; petiolar gland 1, round or elliptic, 1–2.2 mm diam, sessile or almost so (sometimes 1–2 smaller glands above middle of rachis); lfts 10–25, of lvs associated with flowers at least 15 pairs, slightly diminished upward, separated by segments of rachis shorter than the width of the blades, in outline dimidiately lance-oblong, subfalcately incurved distally, mucronulate, 5–13 × 1.2–2.5 mm, at base cordate proximally, the blades 4–6(–7)-nerved from base, the midrib marginal or submarginal, the venation usually prominulous at first on both faces, becoming in age immersed above or on both faces.

Peduncles exactly axillary, ±1 mm long, concealed by stipules, 1-fld; bract and bracteoles resembling stipules, 4–5.5 mm; pedicels 1.8–3 cm; buds ovoid-acuminate, thinly softly pilosulous; sepals ovoid-acuminate, 11–15 mm, the parts exposed in bud reddish or greenish, the rest petaloid; petals yellow, 4 similar obovate-cuneate 13.5–17 mm, the cucullus strongly oblique, 16–22 mm; stamens 10, the anthers yellow, narrowly lanceolate, 5–6.6 mm; androecium abruptly divaricate; ovary densely gray-pilosulous; style linear-cylindric 7.5–9 mm; ovules 7–9.

Pod linear-oblong, commonly a little arched forward, 30–55 × 5–5.5 mm, the valves brown turning black, finely incurved-pilosulous; seeds not seen.—Collections: 21.

Rock ledges and barren stony slopes and hilltops, 1250–2500 m, uncommon along crest and uppermost slopes of Sa. do Espinhaço from Diamantina s. to Sa. do Cipó in lat. 18°10′–19°S, and apparently disjunct near S. João del Rei, lat. 21°10′S, in centr. and s.-centr. Minas Gerais.—Fl. X–III.

3. Chamaecrista venulosa (Bentham) Irwin & Barneby, comb. nov. *Cassia venulosa* Bentham in Martius, Fl. Bras. **15**(2): 170. 1870.—"Habitat in prov. Minas Geraës pascuis ad Vão de Paranan: Martius; locis umbrosis Serra da Lapa: Riedel."—Lectoholotypus, *Riedel 559,* K (hb. Benth. ex herb. Acad. Petrop.)! = NY Neg. *1536*; isotypi, LE! and renumbered *1005,* LE, US! paratypus, *Martius s.n.*, IX.1818 (fr), M!= F Neg. *6263*.—*C. venulosa* sensu Bentham, 1871, p. 574.

Cassia luetzelburgii Harms in Notizbl. K. Bot. Gart. Berlin-Dahlem **8**(80): 715. 1924.—"Brasilien: Bahia, Carrasco-Gebiet, Almas, 1700 m (Ph. von Luetzelburg n. 166 u. 257; 1913)."—Lectoholotypus, *Luetzelburg 257,* †B; neotypus, M! isotypus, NY!
Cassia venulosa sensu Bentham, 1871, p. 574.

Precociously flowering herbs becoming suffruticose, variable in habit, the primary stem erect, its branches often diffuse or humifuse, at anthesis 1.5–15 dm, except for the glabrous faces of the serrulately ciliolate lfts pilosulous with shorter incurved and longer, straighter spreading hairs to 0.7–1.3(–12.6) mm, the long, often yellowish hairs sometimes almost or quite lacking below the pedicels or calyx, the stems equally but usually not densely leafy, the straight internodes 8–20 mm, as long or longer than stipules.

Stipules erect, lance-acuminate from an obscurely cordate base, 5–13 (–15) × 0.7–1.5 mm, submembranous, from base 5–9-nerved, glabrate facially, ciliolate, feebly persistent.

Lvs ascending, straight or outwardly arcuate, 1.5–5(–6) cm, shortly petiolate, the expanded blade ovate in outline; lf-stalk 1–3.5(–4.2) cm, narrowly margined ventrally, the petiole 2–4(–5) mm, the rachis slightly dilated under the pulvinules; petiolar gland (sometimes 2) near middle of petiole or immediately above pulvinus, round or nearly so, 0.35–0.7 mm diam, squatly short-stalked or subsessile, 0.3–0.4(–0.5) mm tall; lfts 4–8 pairs, slightly diminished upward, obliquely oblong, lance-oblong, or oblong-elliptic, (5–)7–16(–21) × 1.5–4(–5.2) mm, at asymmetric base shallowly cordate, at abruptly acuminulate apex contracted into a cusp 0.4–0.8 mm, from base on both faces prominently 5–9-nerved, the excentric, penninerved midrib displaced to ± the distal ⅓ of blade, straight or slightly incurved distally, finely penninerved.

Peduncles exactly axillary, ±1 mm, concealed by stipules, 1-fld; bracts and bracteoles lance- or narrowly ovate-acuminate, (2–)2.5–5 mm; pedicels widely ascending, 2–3.3 cm; buds ovoid-acuminate, pilose; sepals membranous, greenish where exposed in bud but with wide petaloid margins, ovate- or lance-acuminate, 7.5–10.5 × 2.4–4.4 mm; petals yellow, the 4 similar obovate-cuneate, 9–13 mm, the strongly oblique landing-petal 12–14.5 mm; stamens 10, the anthers yellow 4.7–9 mm, the 3 longer ones 7–9 mm, all minutely puberulent along the sutures; ovary densely white-pilose; style glabrous linear 4.3–7 mm; ovules 8–12.

Pod linear-oblong, (2.5–)3–4 × 0.6–0.75(–0.9) cm, the green, softly pilose valves becoming dark brown or blackish; seeds 3.5–4 mm long, the testa black, dull, lineolate-pitted.—Collections: 18.

Open grassy, sometimes disturbed sites in cerrado, in seasonally inundated campo (brejo), and in quartz sand along streams and about rock outcrops, 1100–1320 m, along the crest and higher e. and w. slopes of Sa. do Espinhaço from Belo Horizonte n. to Sa. de Curumataí in lat. 17°50′–20°S; disjunct, on carrasco at 1700 m, near the s. end of Chapada Diamantina in s. Bahia (Pico de Almas) near 13°35′S; and (if data correct) found once by Martius on the São Francisco-Paraná divide in the n.-w. corner of Minas Gerais near lat. 15°S, the occurrence w. of Rio São Francisco requiring confirmation.—Fl. I–V(?VIII).

Although from a locality remote from the main range of *Ch. venulosa* in Minas Gerais, the cited isotypus of *Cassia luetzelburgii* matches closely some collections from the Diamantina region. Harms compared *C. luetzelburgii* with the small-flowered *Ch. serpens* and *Ch. trichopoda,* from which it is, of course, amply distinct.

4. Chamaecrista cinerascens (Vogel) Irwin & Barneby, comb. nov. *Cassia cinerascens* Vogel, Syn. Gen. Cass. 60 & Linnaea **11**: 711, descr. ampliat. 1837.—"In Brasilia: Sellow leg. in Serra do San Antônio."—Holotypus, †B = F Neg. *1673*! no isotypus found.

Cassia rotundata var. *angustifolia* Bentham in Martius, Fl. Bras. **15**(2): 167. 1870.—". . . in prov. Minas Geraës, e gr. Serra da Lapa et ad Caxoeira do Campo: Riedel."—Holotypus, *Riedel 558* from Sa. da Lapa, K! isotypus, *Riedel 558,* LE!
Cassia cinerascens sensu Bentham, 1870, p. 167, tantum quoad typum; 1871, p. 573.

Slender erect shrubs 0.6–2 m with pliant leafless trunks and horizontal or pendulous young leafy branchlets, the young stems, lf-stalks and pedicels softly pilosulous with fine, forwardly incurved whitish hairs up to 0.3–0.9(–1.2) mm, the veiny foliaceous stipules and forwardly imbricated concolorous, sometimes glaucescent lfts glabrous or sparsely erratically ciliolate, some lvs at tip of branchlets often inhibited and represented by loosely imbricated stipules.

Stipules erect appressed ovate from broadly rounded but not or very obscurely cordate base, (5–)7–11 × 3–6 mm, at apex acute and callous-mucronulate, prominently ±9–15-nerved from pulvinule, deciduous before the associated lf, longer than the first pair of lfts.

Lvs stiffly ascending or arched outward, (1.5–)1.8–3.7 cm; lf-stalk (1–)1.3–3 cm, narrowly blunt-margined, openly shallow-grooved; petiole 2–5.5 mm, 0.5–0.8 mm diam; petiolar gland immediately below first pair, sessile or almost so, suborbicular, 0.45–0.65 mm diam, ±0.2–0.3 mm tall; lfts (5–)6–9(–10) pairs, of subuniform size or slightly decrescent at either end of rachis, subsymmetrically oblong or ovate-oblong from obliquely semi-cordate base, (3.5–) 4.5–10.5 × (1.2–)1.5–3 mm, at apex abruptly acuminulate and callous mucronulate, from pulvinule ±7–9-nerved, the midrib straight subcentric, the venation of adult lfts prominulous beneath, immersed or nearly so above.

Peduncles exactly axillary, 0.5–1.5 mm, 1-fld; bracts and bracteoles papery, broadly ovate acute from clasping base, many-nerved, 3–5.5 mm, the bracteoles inserted near middle of pedicel, deciduous; pedicels (8–)10–18 mm; buds ovoid-acuminate glabrous; sepals subpetaloid, yellowish or red, ovate- to broadly lance-acuminate, 12–17 mm; petals (little known) yellow, the 4 similar ones obovate-cuneate ±20–23 mm; stamens 10, the larger anthers (probably yellow) 6–7 mm, their sutures minutely barbellate; ovary shaggy-pilosulous; style 8–9.5 mm, linear-cylindric; ovules 12–14.

Pod erect, linear-oblong, 6–7 × 0.7–0.85 cm, the castaneous valves finely softly gray-pilosulous; seeds (few seen) ±4.5 × 3 mm, the testa dull brown, pitted.—Collections: 19.

Stony campo and about outcrops, on sandstone and quartzite, 1100–1600 m, local, known only from the crest and higher e. slope of s. Sa. do Espinhaço between the head of R. São Antônio and Sa. de Caraça (18°45′–20°10′S) in s.-centr. Minas Gerais.—Fl. IV–VIII.

5. **Chamaecrista potentilla** (Bentham) Irwin & Barneby, comb. nov. *Cassia potentilla* Martius ex Bentham in Martius, Fl. Bras. **15**(2): 168. 1870.—Typus infra sub var. *potentilla* indicatur.

Shrubs 1–2 m with many slender ascending naked trunks forking upward into a rounded crown of horizontal or ultimately pendulous, obscurely ancipital annotinous branchlets clad in imbricated or at least crowded stipules and abruptly divaricate lvs of narrow outline, except for the faces of most of the many small crowded ciliolate lfts softly pilosulous throughout with fine spreading-incurved whitish hairs up to (0.3–)0.4–0.6 mm, the branchlets and axils of the early deflexed stipules densely so, the foliage dull olive-green, concolorous.

Stipules at first erect, appressed to stem, antrorsely imbricated but early reflexed, ovate-acuminate from the oblique cordate-auriculate base, 6–12 × (2.5–)3–6 mm, at apex weakly caudate-mucronulate, from base flabellately 9–13-nerved, finely pilosulous on both faces but more densely so ventrally.

Lvs 1.5–4.5 cm, in expanded outline narrowly oblong or lance-oblong; lf-stalk stiff straight ±1.2–4 cm, narrowly margined, the petiole 1.5–2.5 mm; petiolar gland immediately below first pair, urceolate subsessile, orbicular to broadly elliptic, 0.4–0.7 mm diam, 0.2–0.3 mm tall; lfts 9–16 (in seedlings to 18) pairs crowded along rachis and when closed antrorsely imbricated, of subequal size except the slightly smaller lowest and uppermost ones, in outline oblong to ovate-oblong from the obliquely shallow-cordate base, 2.5–11 × 1–4 mm, at apex abruptly contracted into a soft, erect or oblique mucro 0.2–0.5 mm, the firmly chartaceous blades palmately (5–)7–9-nerved from pulvinule, when adult prominently venulose only dorsally, the faces of all but the first pair glabrous, these commonly puberulent beneath.

Peduncles exactly axillary, 0.5–2 mm, 1–2-fld; bracts and bracteoles papery, broadly lance-acuminate, the latter 4–5.5 mm, caducous; pedicels ascending, 11–25 mm; buds narrowly ovoid-acuminate, pilosulous; sepals where exposed in bud firm, often reddish, pilosulous dorsally, where interior in bud yellowish subpetaloid, lance-acuminate to ovate acute, ±12–13.5 mm; petals yellow, the four similar ones obovate to oblong-cuneate, 13.5–16 mm, the dimidiate one broader and longer than the rest, up to 14.5–19 mm; stamens 10, the anthers yellow, minutely barbellate along sutures, the longer ones 6.5–7 mm; ovary densely white-pilosulous; style linear-cylindric, glabrous beyond the base, 7–9 mm; ovules 8–11.

A species clearly related to *Ch. distichoclada* and through its var. *interstes* to *Ch. rotundata,* these together with *Ch. cinerascens* forming a closely knit group similar in growth habit and virtually identical in form and size of the flower and fruit. *Chamaecrista distichoclada* has similar ovate-cordate sharply acuminate stipules, but *Ch. potentilla* differs from it, as from the other species mentioned, in the numerous (9–16, not 3–9) pairs of very small leaflets and (so far as known) also in the denser pubescence of the stems which extends out over both faces of the stipules, especially over the adaxial one.

Typical *Ch. potentilla* has been collected only along the crest and higher slopes of middle Sa. do Espinhaço between Diamantina and Sêrro, occupying a known range of scarcely 50 km diameter. A population of the species differing very slightly in the ampler foliage and longer pedicels has been discovered recently near the southern margin of the Planalto in southwestern Minas, at a point in the upper basin of Rio Grande over 400 km distant from Sêrro. It is of great interest to note that this aberrant and allopatric form of *Ch. potentilla* is associated in its isolated station with an endemic variety of *Ch. ochnacea* Vog., a species of sect. *Absus* otherwise endemic to Sa. do Espinhaço, though extending further south in that range than typical *Ch. potentilla*. Evidently there is near Alpinópolis in southwestern Minas a relict element of a flora once more widely dispersed over the Planalto that is now virtually confined to the Sa. do Espinhaço. In the case of both cassias, isolation has either permitted or induced enough morphological differentiation to deserve taxonomic status.

Key to the Varieties of *Ch. potentilla*

1. Lvs. relatively narrow, the lfts 2.5–7.5 × 1–2.5 mm; pedicels relatively short, 11–18 mm; Sa. do Espinhaço in lat. 18°10′–18°40′S. 5a. var. *potentilla* (p. 674).
1. Lvs wider and ampler, the lfts 4–11 × 2.2–4 mm; pedicels 17–25 mm; s.-w. Minas Gerais, near 20°50′S, 46°20′W. 5b. var. *specuum* (p. 674).

5a. Chamaecrista potentilla (Bentham) Irwin & Barneby var. **potentilla.** *Cassia potentilla* Martius ex Bentham, l.c., sens. str.—". . . in Brasilia, loco accuratius non indicato: Martius; in saxosis districtus Adamantium, prov. Minas Geraes: Gardner n. 4529; ad urbem Diamantina: Vauthier n. 154 . . ."—Holotypus, *Martius s.n.*, M! = F Neg. *6524* = NY Neg. *9203*; paratypi, *Vauthier 154,* G! *Gardner 4529,* BM, K!

Characters as given in key.—Collections: 15.

Quartzite and sandstone outcrops and in white sands along brooks or on seasonally moist stony campo, 790–1260 m, local but forming colonies along the crest and higher e. slope of centr. Sa. do Espinhaço (lat. 18°10′–18°40′S), centr. Minas Gerais.—Fl. V–VIII.

5b. Chamaecrista potentilla (Bentham) var. **specuum** Irwin & Barneby, var. nov., a var. *potentilla* foliis amplioribus, foliolis latioribus (2.2–4 nec 1–2.5 mm latis) pedicellisque longioribus (17–25 nec 11–18 mm) insuper patria procul aliena absimilis.—BRAZIL. Minas Gerais: Furnas, mun. Alpinópolis, 25.VII.1972 (fl), *Luiz Emygdio, L. Duarte, H. Becker & R. Silva Santos 3605.*—Holotypus, R; isotypus, NY!

Characters as given in key.—Collection: 1.

Habitat not recorded, but to be expected on outcrops or stony campo, known only from the type-locality on the highland between the Sapucaí and Pardo forks of Rio Grande, near 20°50′S, 46°20′W, in s.-w. Minas Gerais.—Fl. VII–VIII.

6. Chamaecrista distichoclada (Bentham) Irwin & Barneby, comb. nov. *Cassia distichoclada* Martius ex Bentham in Martius, Fl. Bras. **15**(2): 168. 1870.—". . . in altis montibus prope Ouro Preto prov. Minas Geraës: Martius."—Holotypus, *Martius s.n.,* labelled "in altis montibus ad Va. Rica [=Ouro Preto], V.1818.," M!= F Neg. *6234.*

Cassia distichoclada sensu Bentham, 1871, p. 573.

Habit of the lower, more compactly branched or many-stemmed forms of *Ch. rotundata,* ±1–1.5 m, the obscurely ancipital branchlets densely pilosulous with spreading gray hairs to 0.45–0.6 mm, the stipules and lfts often but not always hispidulous-ciliolate.

Stipules at first erect and forwardly imbricated, early reflexed, obliquely ovate-acuminate from a deeply cordate-auriculate base, (4–)6–9.5 × (2–)2.5–5.5 mm, callous-mucronulate, the firmly chartaceous blades (15–)17–23-nerved from base, persistent beyond fall of associated lf.

Lvs abruptly divaricate, 0.8–2.5(–3) cm; lf-stalk 5–22 mm, petiole (1–)1.5–3 mm; petiolar gland sessile or almost so, suborbicular 0.25–0.5 mm diam, 0.15–0.3 mm tall; lfts 3–5 pairs, forwardly imbricated, narrowly ovate to lance-oblong from a semicordate base, 2.5–8(–10) × 1.5–3 mm, abruptly acuminulate and obliquely callous-mucronulate, (5–)7–9-nerved from pulvinule, the midrib slightly decurved distally, scarcely stouter than the lateral veins, the venation of adult, chartaceous sublustrous minutely papillate blades prominulous only beneath.

Inflorescence with individual fl and pod of *Ch. rotundata* or *Ch. potentilla*; seeds (little known) ±3.5–3.8 × 2.5 mm, the testa dull brown, lineolate-pitted.—Collections: 4.

Sand-pockets of sandstone or quartzite outcrops near 1200–1300 m, very local, known in modern times only from centr. Sa. do Espinhaço between S. João da Chapada and Diamantina (lat. 18°05′–18°15′S) in centr. Minas Gerais, the type supposedly from near Ouro Preto (lat. 20°) in s. Sa. do Espinhaço where not since seen.—Fl. III–V.

7. Chamaecrista rotundata (Vogel) Irwin & Barneby, comb. nov. *Cassia rotundata* Vogel, Syn. Gen. Cass. & Linnaea **11**: 709, descr. ampliat. 1837.—Holotypus infra sub var. *rotundata* indicatur.

Commonly shrubs with 1–several, gracefully ascending slender naked trunks 1–3 m branching distally into fan-shaped, horizontal or finally pendulous, ancipitally compressed, densely leafy annotinous branchlets clad in ample foliaceous, at first forwardly but early retrorsely imbricated stipules and usually short, distichously spreading lvs, rarely arborescent to 4 m with trunk near base up to 1 dm diam, the young branchlets (especially along the angles) lf-stalks (ventrally) and pedicels pilosulous with fine spreading-incurved white hairs to 0.25–0.6 mm, the faces and margins of stipules and lfts usually glabrous, but the young stipules sometimes puberulent near base and the chartaceous concolorous lfts (especially of young plants) sometimes erratically ciliolate, the lvs of some terminal shoots often inhibited and seemingly replaced by imbricated stipules.

Stipules broadly ovate or suborbicular from asymmetrically 2-auriculate, cordate-amplexicaul base, (7–)9–25 × 5–25 mm, at apex broadly rounded to broadly deltate (but not acuminate), the margin membranous, sharp-edged, the chartaceous blades always larger than the lfts of the associated lf, palmately 17–35-nerved from base, when young commonly yellowish or anthocyanic, green when adult, persistent after fall of lf.

Lvs divaricate or incurved-ascending, in expanded outline ovate-oblong, 1.2–7(–10) cm; lf-stalk 1–6(–9) cm, petiole 3–11(–13) mm; petiolar gland below first pair, suborbicular subsessile 0.4–1.3 mm diam, 0.2–0.5 mm tall; lfts 3–7 pairs, slightly decrescent distally, ovate or ovate-oblong, 4.5–18(–24) × 3–14(–19) mm, either cordate or semi-cordate at base, at apex minutely obliquely callous-mucronulate.

Peduncles exactly axillary ±1 mm, 1–2-fld; bracts and bracteoles papery, ovate from clasping base, several-nerved, the bracts 2–4 mm, the deciduous bracteoles 3–5.5 mm; pedicels ascending 13–29 mm; buds ovoid acute-apiculate, glabrous; sepals subpetaloid, the outer either yellowish or reddish, ovate-acuminate 11–16 mm; petals yellow, 4 obovate-cuneate somewhat unequal, up to 16–18 mm, the falcately dimidiate fifth one up to ±25 mm; anthers yellow, the larger ones 6.5–7.5 mm (minutely barbellate along sutures); ovary densely white-pilosulous; ovules 9–12.

Pod erect or incurved-ascending, linear-oblong, 5–6.5 × 0.6–1.0 cm, the reddish-castaneous valves and stout sutures gray-pilosulous; seeds not seen.

A shrubby, rarely arborescent chamaecrista of characteristic fountain-like outline, the naked trunks gracefully ascending and passing upward into horizontally divergent, finally pendulous fans of leafy branchlets clothed when young with a thatch of forwardly imbricated, often colored stipules, which become green and reflexed in age. The leaves, composed of 3–7 pairs of leaflets always a little smaller and narrower than the subtending stipules, are when well developed early deciduous and liable to sudden reduction in size or atrophy toward the tips of some or all branchlets. Thus both in adult stems, following anthesis, and in the young terminal twigs preceding anthesis, the photosynthetic function is transferred from the leaf proper to the proportionately much dilated, long persisting stipules.

As here defined, *Ch. rotundata* is polymorphic and clearly embraces several distinct races which, while maintaining a common habit of growth and virtually identical flowers and fruits, differ in the number and size of leaflets and in the size and outline of the stipules. Our concept of the species differs from that of Bentham (1870, p. 167) in that we transfer the var. *angustifolia,* which has narrowly ovate stipules deciduous before the leaf, to the closely related *Ch. cinerascens*; and introduce a form unknown to Bentham, our var. *interstes,* which is only precariously excluded from the closely related *Ch. distichoclada* by the deltate, not acuminate stipule-tips and forms a transition between these two species.

Key to the Varieties of *Ch. rotundata*

1. Stipules at tip broadly and evenly rounded and amply ovate-orbicular, 12–25 × 9–25, flabellately 27–35-nerved from point of attachment; centr. Sa. do Espinhaço within a radius of ±30 km of Diamantina and s. to Itacolomí, Minas Gerais (18–19°S). 7a. var. *grandistipula* (p. 676).
1. Stipules either broadly low-deltate at tip or rounded as in the preceding, but if the latter then smaller, 7–14(–15) × 5–12 mm and in any case only 17–25-nerved from point of attachment; if sympatric with the last then the stipule-tips deltate.
   2. Blade of stipules suborbicular, rounded at apex as in var. *grandistipula* but smaller; s. Sa. do Espinhaço near lat. 21°S. 7b. var. *rotundata* (p. 677).
   2. Blade of stipules ovate, at apex broadly low-deltate; centr. Sa. do Espinhaço, within the range of var. *grandistipula*. 7c. var. *interstes* (p. 677).

7a. Chamaecrista rotundata (Vogel) Irwin & Barneby var. **grandistipula** (Vogel) Irwin & Barneby, comb. nov. *Cassia rotundata* var. *grandistipula* (Vogel) Bentham in Martius, Fl. Bras. **15**(2): 167. 1870. *C. grandistipula* Vogel, Syn. Gen. Cass. 58 & Linnaea **11**: 710, descr. ampliat. 1837.—"In Brasilia: Sellow! leg. in Serra do Galheiro."—Holotypus, presumably †B; no isotypus found, but the description decisive.

Stipules 12–25 × 9–25 mm, 27–35-nerved from point of attachment; lfts 4–7 pairs, ovate from scarcely oblique cordate base, 6–18(–24) × 4–14(–19) mm.—Collections: 22.

Outcrops, stony campo and thin gray quartzite sands overlying shelving rock, 900–1300 m, local but forming colonies, known only from middle Sa. do Espinhaço within a circumference of ±30 km of Diamantina (S. João da Chapada to Gouveia, lat. 18°–18°30′S) and Sa. do Itacolomí (19°S) in centr. Minas Gerais.—Fl. ±IX–XI, possibly earlier.

In its extreme form var. *grandistipula* is a striking plant, notable for the ample, broadly rounded overlapping stipules bordered when young with red or yellow, and is clearly different in its coarse foliage both from allopatric var. *rotundata* and from the nearly or quite sympatric var. *interstes.*

7b. Chamaecrista rotundata (Vogel) Irwin & Barneby, var. **rotundata**. *Cassia rotundata* Vogel, 1837, ll. cc., sensu str.—"In Brasilia: Sellow leg."—Holotypus, †B = F Neg. *1741*! neoholotypus, former isotypus, K (hb. Hook.)! = IPA Neg. *1112* = NY Neg. *1529*; isotypi, G, LE (2 sheets), P!

Stipules (7–)9–13(–15) × (6–)7–12 mm, ±25-nerved from base; lfts 4–5 pairs, ovate or oblong from semi-cordate base, 6–13 × 3–6 mm.—Collections: 12.

Habitat little known, but to be expected on outcrops or in shallow rocky soils near 1000 m, known only from vicinity of S. João del Rei (Sas. de Lenheiro and Tiradentes) in s. Minas Gerais.—Fl. VI–?.

7c. Chamaecrista rotundata (Vogel) Irwin & Barneby var. **interstes** Irwin & Barneby, var. nov., a var. *grandistipula* iisdem fere locis obvia stipulis minoribus 7–14 (nec 12–25) mm longis apice deltatis (nec latissime rotundatis) ex ipso basi ±17–23 (nec 27–35)-nerviis abstans.—BRAZIL. Minas Gerais: cerrado, 790 m, 25 km by road n.-e. of Diamantina, 2 km w. of Rio Jequití, 9.IV.1973 (fl), *W. R. Anderson 8376.*—Holotypus, UB; isotypi, F, K, M, NY, RB, US.

Habit of var. *rotundata,* but blades of stipules and lfts smaller, the stipules as described in diagnosis, the lfts commonly 3–4, rarely up to 7 pairs, ovate to ovate-oblong, 4.5–12 × 2.5–5 mm.—Collections: 9.

Cerrado and stony campo, commonly on white or gray quartzite sands, 790–1200 m, known only from the headwaters of R. Jequetinonha between Diamantina and Mendanha, centr. Minas Gerais.—Fl. IV–VI.

In form of the stipules, narrowly membranous-margined as in other forms of *Ch. rotundata* but incipiently acute and smaller than those of sympatric var. *grandistipula,* this newly described variety suggests morphological passage between its species and the closely related, also sympatric *Ch. distichoclada.* By shifting emphasis from the small vegetative difference to the shared habit, flower and fruit, essentially identical throughout the group, a case could be developed for considering these species, as well as *Ch. potentilla* and perhaps *Ch. cinerascens,* as deviant expressions of one major type, differing one from the next in shape of stipules and in size and number of leaflets. However, despite an essential sympatry within a radius of some 30 kilometers of Diamantina, and no known ecological separation, we have no evidence that the two varieties of *Ch. rotundata, Ch. distichoclada* and *Ch. potentilla* are connected by a clinal series of intermediate forms; they emerge, on the contrary, as small, weakly but exactly differentiated entities. Furthermore the difference between the extreme forms, *Ch. potentilla* on the one hand, with its many tiny leaflets and small, firm-margined, acuminate stipules, and *Ch. rotundata* var. *grandistipula* on the other, with few ample leaflets accompanied by suborbicular, membranous-margined stipules, is altogether too great to be compressed within a convenient specific concept.

8. Chamaecrista cardiostegia Irwin & Barneby, sp. nov., *Ch. cinerascenti* et *Ch. rotundatae* affinis, illi quacum confudit cl. Bentham. verosimiliter propior, a prima caulibus maturis supra xylopodium herbaceis (nec fruticosis), stipulis alte cordatis persistentibus, glandula petiolari majuscula (0.7–1.5 nec 0.45–0.65 mm diam), foliolis subduplo majoribus, pedicellis longioribus (14–31 nec 8–18 mm) legumineque breviori (35–45 nec 60–70 mm), a *Ch. rotundata* quoad staturam *Ch. cinerascentem* simulanti imprimis stipulis semper erectis (nec mox reflexis) facilius separanda.—BRAZIL. Minas Gerais: Tejuco to Veraba legítima [=n.-e. of present Uberaba, near 48°W], 13.IX.1827 (fl, fr), *W. J. Burchell 5756*.—Holotypus, K! – IPA Neg. *1116* = NY Neg. *9188*; isotypi, LE, P!

Subshrubs with several or many erect, simple or weakly branched stems (1.5–)2–5 dm, arising directly from a massive xylopodium, at anthesis leafless and glabrate in lower ⅓, densely leafy distally, the young stems, lf-stalks, pedicels and pods puberulent with weak incurved hairs up to 0.1–0.2 mm, the stiffly chartaceous lfts and ample cordate stipules glabrous on both faces, sometimes remotely erratically ciliolate, the whole plant exceptionally glabrous.

Stipules erect, appressed to stem, broadly asymmetrically ovate from the cordately amplexicaul base, (6–)8–15(–19) × 5–12 mm, the basal lobes rounded, at apex abruptly acuminulate-cuspidate, the stiffly chartaceous blades flabellately ±15–25-nerved from pulvinule, tardily deciduous with associated lf.

Lvs stiffly ascending and often gently arched outward, 2.5–7.5 cm, the expanded blade narrowly oblong, the short petiole concealed by stipules; lf-stalk 2–6 cm; petiole 3.5–8 mm, (0.5–)0.6–1 mm diam; petiolar gland immediately below first pair, suborbicular scutellate sessile or nearly so, 0.7–1.5 mm diam, 0.2–0.5 mm tall, sometimes a second smaller gland near pulvinus and rarely others below upper pairs; lfts when closed forwardly imbricated along rachis, (4–)5–9 pairs, obliquely ovate-oblong from shallowly cordate base, (7–)9–20 × 3–10 mm, at apex abruptly acuminate and sharply spinulose-cuspidate, ±7–11-nerved from pulvinule, the subcentric costa straight or gently arched outward, the venation subequally prominulous on both faces.

Peduncles from several or many distal axils, exactly axillary, solitary or sometimes paired, 1–7 mm, 1–4(–6)-fld; bracts and deciduous bracteoles alike broadly ovate-acuminate from clasping base, papery, several-nerved, 2.5–5.5 mm; pedicels ascending 14–31 mm; buds ovoid-acuminate, glabrous or thinly puberulent at base; sepals subpetaloid, ovate- or broadly lance-acuminate, 9.5–12.5 mm; petals yellow, the 4 similar ones oblong- to obovate-cuneate 12–17 mm, the dimidiate fifth slightly longer; stamens 10, the larger anthers 5.5–6.5 mm; ovary densely whitish-puberulent (exceptionally glabrous); style 5.5–7.5 mm, filiform, gently incurved distally; ovules 5–10.

Pod erect, linear-oblong, ±35–45 × 5–6.5 mm, the valves usually puberulent, exceptionally glabrous; ripe seeds unknown, probably ±3.5 mm.—Collections: 14.

Cerrado, especially about sandstone or crystalline outcrops, sometimes in red sandy clays or arenitic sands, apparently local, known in typical form from the Paranaíba-Grande basin in the Triângulo of Minas Gerais and adjacent s.-e. Goiás (Cristalina s. to Uberaba, e. to Catiara and Patrocínio) and apparently, in completely glabrous form, from the e. side of the R. São Francisco valley in n.-centr. Minas Gerais (Agua Bôa, near 17°S).—Fl. VIII–X.

A well marked species, related to the tall shrubby *Ch. rotundata* and *Ch. cinerascens* endemic to the heights of Sa. do Espinhaço and associated mountain ranges in central and southern Minas, and while resembling these in the dilated stipules, stiff venulose leaflets and large flowers, very different in the functionally herbaceous stems rising yearly from a xylopodium, the growth-form common in campo habitats subject to intermittent fires. Apparently not uncommon, at least formerly, in the Triângulo Mineiro and adjoining southeastern Goiás, *Ch. cardiostegia* was collected first by St. Hilaire in 1819, by Burchell in 1827, and by Riedel in 1834; it only escaped description because Bentham (1870, p. 168) confused it, though not without comment on the broader stipules and short pod, with the then poorly known *Ch. cinerascens*. From this it differs not only in the habit of growth already referred to and in the stipular character noticed by Bentham, but also in the larger petiolar glands, larger leaflets, longer pedicels and shorter pod. The Tejuco of the type-locality, misinterpreted by Bentham as the town of Tejuco in central Minas Gerais now known as Diamantina, is actually a small left affluent of Rio Paranaíba rising a short distance northwestward from Uberaba, 500 km west of Diamantina. The related cassia with cordate stipules common around Diamantina, *Ch. rotundata,* is immediately distinguished by its tall fruticose arching stems and in particular by the early reflexing stipules.

9. Chamaecrista anceps (Bentham) Irwin & Barneby, comb. nov. *Cassia anceps* Bentham in Martius, Fl. Bras. **15**(2): 166. 1870.—". . . in rupibus prope Diamantina prov. Minas Geraës: Riedel, Vauthier n. 155."—Lectoholotypus, *Vauthier 155* ipso manu Bentham. determinatus, G! = F Neg. *27994*; isotypus, P! paratypi, *Riedel 483,* XII.1824, LE, US! = *Riedel 1239,* LE!

Cassia anceps sensu Bentham, 1871, p. 573.

Shrubs reportedly 9–12 dm, the stout purplish, densely leafy annotinous branchlets ancipitally angulate by a narrow wing descending from each petiole and by a rib descending from each stipule, the lf-stalks ventrally together with margins of stipules and of some lfts and axes of inflorescence thinly pilosulous or strigulose with ascending-subappressed hairs to 0.4–0.7 mm, the dull olivaceous foliage concolorous, the chartaceous lfts glabrous on both faces, strongly venulose.

Stipules ascending subappressed to stem, broadly asymmetrically ovate-acuminate 5–12 × 2.5–7.5 mm, at base cordate on both sides but more deeply so and obtusely auriculate on that opposed to petiole, the firm blades from base flabellately 11–21-nerved, glabrous on both faces, deciduous only with the associated lf.

Lvs stiffly ascending 2–5 cm, the expanded blade oblong in outline; petiole with poorly differentiated pulvinus 3–5 mm, at middle 0.8–1.2 mm diam, narrowly winged ventrally, the sulcus forced open by the gland but closed distally; petiolar gland 1 sessile patelliform, situated immediately below first pair of lfts, in outline orbicular or broadly elliptic 1–2 mm diam, coarsely margined; rachis 7–18 mm;

lfts (2–)3–4 pairs, narrowly oblong-elliptic and abruptly acuminate to broadly ovate-elliptic obtuse acuminulate, the larger ones 15–25 × 5–12 mm, at base varying from subequally rounded on both sides to asymmetrically shallow-cordate on the broader one, the plane blades from base coarsely flabellately 6–12-nerved, the subcentric midrib and major primary veins few-forked distally, the secondary venulation weakly anastomosing but not forming a regular mesh.

Peduncles exactly axillary 0.5–2 mm, 1–2-fld; pedicels 11–18 mm, bracteolate 3.5–7.5 mm below calyx; bracteoles like bracts submembranous ovate or broadly lanceolate several-nerved 2–3.5 mm, tardily deciduous; buds ovoid-acuminate pilosulous; sepals dull reddish with petaloid margins, ovate-acuminate 11–13.5 mm; petals (little known) yellow, 4 broadly obovate beyond a short claw, slightly unequal, up to 16–19 mm, the dimidiate abaxial petal scarcely as long; ovary densely white-strigulose; style linear, gently incurved, 6–7.5 mm; ovules 10–14.

Pod (35–)50–60 × 6.5–8 mm, the purplish-nigrescent valves densely strigulose; seeds (few seen) ±5 mm, brown, dull, deeply lineolate-pitted.—Collections: 2.

Outcrops, probably near 1200 m, apparently highly localized on the crest of Sa. do Espinhaço in the vicinity of Diamantina, centr. Minas Gerais.—Fl. XII–I.

Little can be added to Bentham's account of this evidently rare chamaecrista which has not been rediscovered in modern times. In general aspect *Ch. anceps* suggests a gigas form of *Ch. distichoclada,* having leaflets of the same shape and number, but these up to 15–25, not 2.5–10 mm long. It differs from nearly or actually sympatric *Ch. rotundata* in the permanently erect, cordate-acuminate (not early deflexed and obtuse) stipules, and from allopatric *Ch. cardiostegia* in habit and in the fewer (2–4, not 5–9) pairs of leaflets. The two known collections of *Ch. anceps* differ in shape of leaflets, those of Vauthier's collection being narrowly oblong-elliptic, those of Riedel's more broadly ovate-oblong. Extrapolating from what is known about the extremely localized dispersal of the *Lasiorhegma* chamaecristas of Sa. do Espinhaço, we surmise that these represent populations confined to mutually isolated quartzite outcrops emergent from the grassfields of the serra crest.

10. Chamaecrista caribaea (Northrop) Britton, Bull. Torr. Club **44**: 8. 1917. *Cassia caribaea* Northrop, Mem. Torr. Club **12**: 39. 1902.—Typus infra sub var. *caribaea* indicatur.

Erect shrubs of ultimately bushy, round- or flat-topped outline, when crowded few-stemmed and spindly, at anthesis 4–15 dm, with slender pallid or fuscous furrowed trunks stiffly branched distally, the annotinous branchlets and lf-stalks minutely strigulose with appressed or subappressed straight or incurved hairs up to 0.1–0.4 mm and sometimes in addition pilosulous with spreading ones up to 0.5–1(–1.4) mm, exceptionally glabrate, the pedicels sometimes thinly pilose, the stiffly chartaceous lustrous concolorous olivaceous lfts brunnescent on drying, glabrous or minutely ciliolate at base.

Stipules erect appressed, subsymmetrically triangular-subulate or lance-attenuate (1–)1.5–6.5 × 0.4–1.2 mm, from base (3–)4–8-nerved, in age often nigrescent, persistent.

Major lvs up to 1.7–8 cm, petioled or subsessile; pulvinus swollen, wrinkled when dry, 0.6–4 mm; petiole proper 0.4–9 mm, at middle 0.5–0.9 mm diam, narrowly shallow-sulcate ventrally; petiolar gland 0–1, when present usually much above middle of petiole proper, discoid or cupulate, stipitate but sometimes only shortly so, in profile commonly tack- or broadly trumpet-shaped, sometimes squatly drum-shaped or obconic, (0.4–)0.6–1.5 × (0.4–)0.45–1.4(–1.5) mm, in profile (0.7–)0.4 mm shorter up to 0.9 mm longer than diam of head; similar but smaller glands below each or below the last pair of lfts; lfts of major lvs 2–10 pairs, elliptic, oblong-, oblance-, lance- or obovate-elliptic, obtuse-acuminulate, acute, obtuse, or sometimes obovate-emarginate, all mucronulate, up to (12–)13–31 × 4–19 mm, at base broadly rounded to cordate on proximal and cuneate to subcordate on distal side, from base 4–9-nerved by the subcentric midrib with on the proximal side 2–6, on the distal side 0–3 short incurved-ascending primary veins, the midrib thence penninerved with 4–8 major secondary veins with or without 1–3 intercalaries, these faint or sometimes almost as strong as the major ones, the lft-blade then multistriate, the venulation above immersed or bluntly prominulous, beneath either faintly or sharply prominulous.

Peduncles exactly axillary 0–5 mm, 1–2-fld; pedicels 15–55 mm, bracteolate 2.5–20 mm below calyx; buds ovoid-acuminate glabrous or thinly pilosulous; sepals ovate-acuminate up to 8.5–11 mm; petals yellow heteromorphic, 2–3 smaller obovate- or oblong-obovate-cuneate, 1–2 larger obovate- to flabellate-cuneate, 1 of these abaxial up to (12.5–)13–16 mm, the oblique cucullus a little shorter; androecium 10-merous, the anthers yellow or reddish, the 2–3 longest up to 5.5–7.3 mm, ovary strigulose, sometimes thinly so; style linear 4.5–6 mm, incurved distally; ovules 8–13.

Pod linear-oblong or linear 36–60 × 5–6.5 mm, the ripe valves atrocastaneous, dark purplish, or pinkish-brown, thinly or remotely strigulose, or glabrate except along the sutures; seeds 3.2–5.3 mm, the testa dull brown, faintly pitted, reticulately crackled.

A truly shrubby chamaecrista, notable for its glossy glabrous foliage and relatively large flowers, *Ch. caribaea* is everywhere sympatric with some form of the polymorphic *Ch. lineata* var. *lineata,* which differs in its dull, usually glaucescent and commonly at least dorsally pubescent leaflets combined with sessile petiolar glands and at least shortly adnate, supra-axillary peduncles. Our concept of *Ch. caribaea* readily accommodates four Brittonian chamaecristas but these admittedly are not identical, and we have been able to sort the material into three geographic categories: a well differentiated var. *caribaea,* with subsessile leaves lacking a petiolar gland and 2–4 pairs of relatively small, simply venulose leaflets, endemic to the central Bahamas; and the very closely related vars. *lucayana* and *inaguensis,* collectively different from var. *caribaea* in the well-developed glanduliferous petiole and larger, more elaborately venulose leaflets, distinct from each other in leaflet-number and a geographic gap in the Bahamas corresponding with the latitudinal interval 22–23°N. A similar taxonomic solution was independently worked out by Gillis (Phytologia **35**(2): 90–91. 1976), who, however, did not recognize named varieties of *C. caribaea* sens. lat.

Key to the Varieties of *Ch. caribaea*

1. Lvs subsessile, the major ones up to 1.7–4.2 cm, their proper petiole (excluding pulvinus) 0.4–2.2(–2.7) mm; petiolar gland 0 (but stipitate glands on rachis); lfts 2–4 pairs up to 14–22 × 4–8 mm, relatively simply nerved, with only 1 primary vein from base on distal side of midrib and from this 5–8 pairs of major with or without faint intercalary secondaries, these not connected by tertiary venules; pedicels 15–26(–30) mm; centr. Bahamas (Andros, New Providence, Cat). 10a. var. *caribaea* (p. 682).
1. Lvs petioled, the major ones up to 3–8 cm, their proper petiole (excluding pulvinus) (2–)2.5–9 mm; petiolar gland present, as well as glands on rachis; lfts 2–10 pairs, but if 2–5 pairs then broader than in var. *caribaea* and in any case more elaborately nerved, with 2–3 primary veins from distal side of midrib and from this both major and 1–3 fainter or almost as strong intercalary secondaries, these ± irregularly anastomosing or connected by tertiary venules; pedicels (20–)25–55 mm; widespread from n. Bahamas s. to Caicos Is. and n.-w. Haiti, sympatric with the last only on Cat I. in centr. Bahamas.

2. Lfts of major lvs 5–9(–10) pairs, 13–23 × 6–10.5(–12) mm; centr. and n. Bahamas (Rum Cay and Exuma Chain n.-ward). 10b. var. *lucayana* (p. 682).
2. Lfts of major lvs 2–4(–5) pairs, up to (16–)20–31 × 9–19 mm; extreme s. Bahamas (Inagua), Turk and Caicos archipelago and n.-w. Hispaniola (Tortue I.). 10c. var. *inaguensis* (p. 682).

10a. Chamaecrista caribaea (Northrop) Britton var. **caribaea**. *Cassia caribaea* Northrop, l.c., pl. 6. 1902, sens. str.—". . . Fresh Creek, Andros [Island, Bahamas], June 10 [1890, *John I. & Alice R. Northrop*] (638)."—Holotypus, NY! clastotypus (fragm), US!

Chamaecrista caribaea sensu Britton & Rose, 1930, p. 282.

Annotinous branchlets and lf-stalks minutely strigulose with hairs up to 0.1–0.2(–0.4) mm; lvs as described in key, the lfts subsymmetrically elliptic, oblong- or oblance-elliptic; pod 36–45 mm.—Collections: 11.

In coppice, or in shallow earth pockets on coral reefs or outcrops, mostly near sea-level, known only from the centr. Bahamas in lat. 24–25°10′N (Andros, New Providence, Cat Is.)—Fl. I–III, VI–IX, perhaps irregularly through the year.

10b. Chamaecrista caribaea (Northrop) Britton var. **lucayana** (Britton) Irwin & Barneby, stat. nov. *Cassia lucayana* Britton, Bull. N.Y. Bot. Gard. **4**: 138. 1906.—"Cay north of Wide Opening, Exuma Chain (*Britton & Millspaugh 2774* . . .)."—Holotypus, collected 18.II.1905 (fr), NY! isotypus, US! paratypi, *Britton & Millspaugh 2841, 3041,* NY!—*Chamaecrista lucayana* (Britton) Britton in Bull. Torr. Club **44**: 8. 1917.

Chamaecrista lucayana sensu Britton & Rose, 1930, p. 282.

Minutely strigulose as var. *caribaea* or sometimes in addition pilosulous with spreading-ascending hairs up to 0.5–1(–1.4) mm, or (Grand Bahama) the branchlets and lf-stalks glabrate; major lvs (4–)4.5–8 cm, the lfts 5–10 pairs, broadly lance- to ovate- or oblong-elliptic, rarely obovate-emarginate; gland (below) middle of petiole usually tack- or trumpet-shaped, but the stipe sometimes coarsely dilated and then squatly drum-shaped or obconic, in profile (0.5–)0.6–1.3 × (0.6–)0.7–1.4(–1.5) mm, (0.7–)0.4 mm shorter to 0.3 mm longer than diam of head; pod 43–60 × 5–6.5 mm.—Collections: 9.

Habitats of var. *caribaea,* of apparently bicentric dispersal in n. and centr. Bahamas: Grand Bahama (±26°40′N); Exuma Chain, Cat and Conception Is., Rum Cay (23°20′–24°30′N).—Fl. XI–II, VII, perhaps at intervals through the year.

The var. *lucayana* is variable in pubescence, the disjunct populations on Grand Bahama being nearly glabrous, those in the central Bahamas either strigulose or pilose. Variation in the petiolar gland, as described, appears quite random and without taxonomic significance. On Cat Island, the only island where two varieties of *Ch. caribaea* are known to occur, var. *lucayana* is sharply distinct from sympatric var. *caribaea.*

The epithet *lucayana* refers to the long-extinct tribe of Arawak Indians found living on the Bahamas by the first European visitors.

10c. Chamaecrista caribaea (Northrop) Britton var. **inaguensis** (Britton) Irwin & Barneby, stat. nov. *Cassia inaguensis* Britton, Bull. N.Y. Bot. Gard. **3**: 443. 1905.—"Inagua (*Nash & Taylor 910,* type; *1261*)."—Holotypus, collected by G. V. Nash & N. Taylor, 11.X.1904 (fl), NY! paratypus (*1261*), NY!—*Chamaecrista inaguensis* (Britton) Britton in Bull. Torr. Club **44**: 8. 1917.

Cassia tortuensis Urban & Ekman ex Urban in Arkiv Bot. **21A**(5): 91. 1927. "[Haiti:] Insula la Tortue prope Pointe Petit-Bois ad orientem versus in arenosis litoralibus, m. Majo defl. et fr.: n. (*Ekman*) 4162."—Holotypus, †B; isotypi, NY (2 sheets), US!—*Chamaecrista tortuensis* (Urban) Britton ex Britton & Rose in N. Amer. Fl. **23**(5): 282. 1930.
Chamaecrista inaguensis sensu Britton & Rose, 1930, p. 282.

Minutely strigulose; major lvs up to 3–7(–7.5) cm, the lfts 2–4(–5) pairs, varying from elliptic subacute or lance- or oblance-elliptic to obovate and obtuse to emarginate; petiolar gland tack- or trumpet-shaped, (0.4–)0.6–1.5 × (0.4–)0.45–0.8(–0.9) mm, as long as to 0.9 mm longer than diam of head; pod of var. *lucayana*.—Collections: 15.

Coppice, in either sandy or rocky soils, sometimes on exposed rock, maritime or submaritime, extreme s. Bahamas (Inagua I.); Turk and Caicos Is. (Grand Turk, N., Middle and S. Caicos, Ambergris and Pine cays); and n.-w. Hispaniola (Ile de la Tortue).—Fl. X–III, VI, perhaps irregularly through the year.

Variable in width and outline of the leaflets, but almost constantly distinguished from var. *lucayana* by leaflet number, these two to four pairs in all but one specimen (*Proctor & Gillis 34033,* NY) from North Caicos, of which one branchlet has five pairs. Plants from South Caicos (*Millspaugh 9303,* GH, NY) and Ambergris Cay (*P. Wilson 7687,* GH, NY) with only two pairs of broadly obovate leaflets are essentially identical to the type-series of *Cassia tortuensis* from northwest tip of Haiti.

11. Chamaecrista aristata (Bentham) Irwin & Barneby, comb. nov. *Cassia aristata* Bentham in Martius, Fl. Bras. **15**(2): 170. 1870.—"Habitat in Brasilia, loco accuratius non indicato: Martius."—Holotypus, annotated by Bentham, *Martius s.n.,* M! = F Neg. *6226*.

Cassia aristata sensu Bentham, 1871, p. 574.

Shrubs or treelets up to 1.5–3 m with blackish trunks remotely & awkwardly branching upward, the ribbed annotinous branchlets, lf-stalks, and pedicels finely pilosulous with soft erect hairs to 0.2–0.4 mm, the coriaceous spine-tipped dull olivaceous or when young sublustrous, concolorous, coarsely venulose lfts either similarly pilosulous only near pulvinule dorsally or sometimes overall on both faces, the inflorescence often, by loss of some uppermost lvs, becoming pseudopaniculate.

Stipules widely spreading or reflexed, lanceolate to narrowly ovate-acuminate, (1.5–)2.5–7 mm, at insertion 1–1.8 mm wide, prominently 3–5-nerved, deciduous with or shortly before the lf.

Lvs stiffly ascending (1.5–)2–7 cm, heteromorphic, those associated with fls commonly shorter and simpler than those below, all very shortly petiolate or subsessile, the expanded blade broadly ovate to oblong in outline; petiole including wrinkled pulvinus 1.5–3.5 mm, shallowly open-grooved; glands situated immediately below each pair of lfts, sessile, round or broadly elliptic, shallowly concave, thick-rimmed, 0.45–1.1 mm diam; rachis 0.5–4.7 cm, narrowly thick-margined and sulcate ventrally, the interfoliolar segments scarcely widened upward; lfts of some lower lvs 4–7, of upper lvs mostly 2–3 pairs, ascending face upward on insensitive broad but very short pulvinule, the blades obliquely ovate (1.5–)1.7–3 × (0.8–)1–2 cm, abruptly acuminate and tipped with a stout stramineous or brownish vulnerant spinule 1.7–2.7 mm, at base cordate on proximal and more shallowly so or merely rounded on distal side, the margins thickened, crenulate, the stout subcentric midrib closely penninerved, the inner pair of lateral nerves almost as thick, incurved-ascending through ±⅔–¾ the blade's length to anastomosis with the marginal nerve, the outer laterals weaker and shorter, the secondary and close tertiary venulation bluntly prominulous on both faces.

Peduncles axillary to short upper lvs or arising from between pairs of stipules at a series of leafless nodes, mostly 3–7-fld, up to 14 mm long; pedicels flexuously ascending ±2–4 cm, bracteolate immediately beneath calyx; bracteoles subulate, 1.5–3 mm; buds slenderly ovoid-acuminate, puberulent at base; sepals subpetaloid, ovate-elliptic acute 10.5–13 mm; petals yellow, 4 obovate to broadly oblanceolate, shortly or obscurely clawed, of subequal length, 17–18 × 8.5–10 mm, the cucullus much longer, dimidiately ovate-reniform, ±25 × 12 mm; androecium 10-merous, not seen complete; ovary gray-strigulose; ovules 7–9.

Pod erect, linear or linear-oblanceolate in profile, slightly curved or straight, 3.5–5 × 0.6–0.8 cm, the castaneous coriaceous valves thinly minutely strigulose; seeds unknown.—Collections: 3.

Sandstone outcrops, 600–950 m, very local in the n. Sa. do Espinhaço of n.-centr. Minas Gerais, known precisely only from Sa. do Grão Mogol (near 16°30′S), where perhaps endemic.—Fl. VII–VIII.

12. Chamaecrista mucronata (C. Sprengel) Irwin & Barneby, comb. nov. *Cassia mucronata* C. Sprengel, Syst. veg. ed. 16, **2**: 341. 1825.—"Brazil."—Holotypus, labelled "Brasilia: Sellow" and ticketed by Sprengel "*Cassia mucronata*. Brasil. Otto," B (hb. Spreng.) = F Neg. *1712*! Vogel, 1837, p. 58, expanded the data to "In Brasilia: Sellow leg. pr. Villa Rica, ad Itambe aliisque locis."—Presumed isotypus, M!

Cassia arrojadoana Harms, Notizbl. Bot. Gart. Berlin **8**(80): 715. 1924.—"Brasilien: Bahia, Rio de Contas, casa de Pedra, Campo (Ph. von Luetzelburg n. 8. 1913)."—Holotypus, †B; neotypus, *Luetzelburg s.n.*, coll. VII.1913 (fl), M!

Cassia mucronata sensu Bentham, 1870, p. 171; 1871, p. 575.

Slender prostrate or obliquely ascending shrubs at anthesis 0.5–1.5 m, with brown, ± lenticellate twiggy trunks and densely leafy, divergent or subpendulous hornotinous branchlets, pilosulous with widely ascending or divaricate gray hairs up to 0.4–0.7 mm, the stems and lf-stalks densely, the concolorous foliage more thinly so, the lfts sometimes glabrous above and glabrescent beneath, always ciliolate.

Stipules erect firm lanceolate to triangular, 1.5–4.5 mm, at symmetric base 0.4–1 mm wide, (1–)3–5-nerved, persistent but becoming dry and fragile.

Lvs (1.5–)2–4(–4.5) cm, very shortly petioled, the expanded blade ovate in outline; lf-stalk 1–3 cm, the petiole (1.5–)2–5 mm, the rachis narrowly wing-margined, the wing-segments not dilated upward; petiolar glands 1–4, inserted immediately below pairs of lfts, the lowest round or nearly so, (0.5–)0.6–1.3 mm diam, squatly short-stalked or sessile, the obconic stalk or pediment 0.3–0.7 mm tall; lfts 2–5(–6) pairs, slightly diminished upward, subsymmetrically elliptic-obovate to elliptic- or oblong-oblanceolate, 10–26 × 4–12(–14) mm, semicordate at oblique base, at obtuse or emarginate apex contracted into a stiff but non-vulnerant erect or retrocurved mucro (0.2–)0.4–1.2 mm, the chartaceous blades subequally venulose on both faces or more prominently so beneath, 4–6-nerved from pulvinule by the straight, slightly displaced midrib with 2–4 primary veins on its proximal and 1 on the distal side, the midrib penniveined with 2–5 major and sometimes weaker intercalary secondaries.

Peduncles exactly axillary, 0.5–2.5 mm, 1–2-fld; bracts and bracteoles resembling stipules, slightly smaller; pedicels ascending (1.2–)1.4–2.4 cm; buds ovoid-acuminate, finely pilosulous or glabrescent; sepals subpetaloid, elliptic- or ovate-acuminate, 11–14 mm; petals yellow, the 4 similar ones obovate- or oblong-cuneate up to 14.5–20 mm, the dimidiate fifth one 20–23 mm; stamens 10, the anthers yellow, the long ones 5–7.5 mm; ovary appressed-pilosulous; style linear 7–8.5 mm, glabrate beyond middle; ovules 10–14.

Pod erect, linear-oblong, 55–70 × 7–8.5 mm, the castaneous nigrescent valves finely puberulent; seeds 4–4.8 × 2.6–3.8 mm, the testa dull rufous-brown or blackish, deeply lineolate-pitted.—Collections: 34.

Rocky hills and outcrops, mostly on schistose or sideritic ("canga") soils, locally plentiful at 1150–1600 m on the crest and e. slope of Sa. do Espinhaço from Itambé do Mato Dentro s. to Congonhas do Campo and Itacolomí near Ouro Preto (±19°30′–20°30′S) in s.-centr. Minas Gerais, e. (disjunctly?) to w.-centr. Espírito Santo (Reserva de Linhares); distantly disjunct on the headwaters of Rio Paraguaçú (? and Contas) in Chapada Diamantina (±12°30′–13°30′S), s.-centr. Bahia; a report from Maranhão (Bentham, 1871, p. 575) was based on specimens of *Ch. pascuorum* collected by Martius at Rio Itapicurú.—Fl. VII–IV.

13. Chamaecrista multinervia (Bentham) Irwin & Barneby, comb. nov. *Cassia multinervia* Martius ex Bentham in Martius, Fl. Bras. **15**(2): 166. 1870.—"Habitat in campis altis prope Diamantina prov. Minas Geraës: Martius."—Holotypus, labelled "in campis altis Prov. Minarum prope Tejuco, Majo 1818," M! = NY Neg. *9205*; isotypi, one labelled "in inter. Serro Frio prope Tejuco, 1818" = F Neg. *6248*, the other without locality data, both M!

Cassia multinervia sensu Bentham, 1871, p. 572.

Frutescent of unknown stature, certainly 4+ dm and probably like kindred *Ch. choriophylla* and *Ch. simplifacta* erect-ascending from a xylopodium, glabrous up to the strigulose ovary, the dark brown stems angled by stout ribs decurrent from each petiole and stipule, paniculately branched and densely leafy distally, the leathery foliage dull olivaceous concolorous, the large flowers borne in distal axils.

Stipules erect, lanceolate or triangular 2–3.5 mm, the firm blades 3–5-nerved, persistent.

Lvs ascending at narrow angles, (1.5–)2–4 cm; petiole with short wrinkled pulvinus 2.5–5.5 mm, at middle 0.5–0.8 mm diam, coarsely margined, the groove opened by the gland; rachis 0–10 mm; glands subsessile immediately below each pair of pulvinules, orbicular, shallowly urceolate 0.4–0.8 mm diam, coarsely thick-margined, substipitate and in profile squatly drum-shaped; lfts of most cauline lvs 2(–3) pairs, of lvs subtending most peduncles only 1 pair, narrowly ascending, turned half face to face, on wrinkled pulvinule broader than long, when more than 1 pair accrescent upward, in outline subsymmetrically or intro-falcately narrowly oblong- or oblance-elliptic, abruptly acuminulate or obtuse and coarsely mucronate, (12–)14–30 × 3.5–8.5 mm, at oblique base deeply cordate on proximal and broadly cuneate on distal side, from base coarsely 5–7-nerved by the subcentric midrib with on its proximal side 2–4, on its distal one 2 major primary veins, the outer quickly marginal or nearly so, the inner one on the proximal side produced to or almost to apex of blade, the midrib beyond base closely alternipinnate with narrowly ascending, irregularly forking and reticulately anastomosing secondaries, the venulation prominent on both faces, but more sharply so beneath.

Peduncles exactly axillary 0.5–2 mm, 1–2-fld; pedicels 15–22 mm, bracteolate 1.5–5.5 mm below calyx; bracteoles lanceolate 1.5–4 mm, persistent; buds ovoid acute glabrous; sepals (probably red when fresh) ovate-acuminate up to 13–14.5 mm; petals yellow, 4 obovate or broadly oblanceolate unequal, the longest ± as long as sepals, the falcate fifth one longer; androecium 10-merous, the long anthers ±6.5 mm; ovary white-strigulose; style 6.5–7 mm, gently incurved and slightly dilated distally; ovules ±10–13.

Pod narrowly oblong tapering abruptly at both ends, straight or curved outward, 40–50 × 6–6.5 mm, the purplish-castaneous valves minutely puberulent; seeds unknown.—Collection: 1.

Habitat not recorded but to be sought on campo sujo about quartzite or sandstone outcrops near 1000–1200 m, known only from the vicinity of Diamantina, centr. Minas Gerais.—Fl. V.

A rare species, apparently related to bifoliolate *Ch. choriophylla, Ch. simplifacta* and *Ch. lagotois,* resembling the last two in the elliptic-oblanceolate outline of the leaflets but these, in all but a few leaves subtending the flowers, 2(–3) pairs. More data are needed to settle the question as to whether *C. multinervia* is specifically distinct from *C. mucronata,* very similar, even though placed by Bentham in different series, except in number of leaflets, as indicated in our key. The equally rare and sympatric *C. anceps,* resembling *C. multinervia* in shape of leaflets but these 3–4 pairs in all leaves, is now readily distinguished by the basally cordate stipules and sparse but evident pubescence of the leaf-stalks, stipule-margins and flower-buds. A feature common to all members of this group of species endemic to the heights of central and southern Sa. do Espinhaço is the strongly angled stem, the angles formed by stout ribs decurrent from each petiole and from the midrib of each stipule.

14. Chamaecrista tragacanthoides (Bentham) Irwin & Barneby, comb. nov. *Cassia tragacanthoides* Martius ex Bentham in Martius, Fl. Bras. **15**(2): 166. 1870.—Typus infra sub var. *tragacanthoide* indicatur.

Subshrubs with many procumbent and ascending, at length freely branching stems, at anthesis 1.5–3.5 dm, arising after fire directly from a xylopodium but if undisturbed from elongating prostrate tortuous trunks and at length forming matted thickets up to 1 m diam, the densely leafy annotinous branchlets, lf-stalks and at least margins of lfts strigulose or puberulent with appressed, incumbent or incurved hairs up to 0.2–0.4 mm sometimes mixed with spreading setules up to 0.5–0.8 mm, the whole vesture extremely variable in density, sometimes extending as a dense silky appressed indument over both faces of the lfts, sometimes sparse and lfts glabrous on both faces, then concolorous dull pallid-papillate.

Stipules stiffly erect, linear-lanceolate or triangular-subulate, (1.5–)3–7 × 0.4–1 mm, faintly 3-nerved, persistent.

Lvs arcuately spreading 1–2.7 cm, ovate in expanded outline; petiole with poorly differentiated pulvinus 1.5–4 mm, at middle 0.4–0.6 mm diam, narrowly margined, the groove opened by the gland, closed distally; gland 1, situated near or above middle of petiole proper, sessile or subsessile, in outline round or elliptic 0.3–1 mm diam, when sessile urceolate-patelliform, when stoutly short-stipitate becoming drum-shaped, in profile much shorter than or up to 0.2 mm longer than diameter of the head; rachis 3–11(–18) mm; lfts 3–7(–8, commonly 3–5) pairs, slightly decrescent upward, in outline subsymmetrically or retrofalcately linear-oblong, narrowly oblong-oblanceolate, or (especially the distal pair) obliquely obovate, obtuse and stoutly mucronate or abruptly apiculate, the larger of median lvs up to (5–)6–16 × 1.5–4 mm, at base cordate on proximal and cuneately rounded on distal side, from base 2–4-nerved by the straight or outwardly curved, centric or moderately displaced midrib with, on its proximal side, 1–3 weaker primaries, the inner (or only) one of these incurved-ascending beyond middle of blade, the outer ones much shorter, the midrib from near middle penniveined with 2–3(–4) pairs of secondary venules, tertiary venulation faint or immersed, primary venation usually prominent beneath, immersed or nearly so above, sometimes concealed by vesture.

Peduncles exactly axillary subobsolete 1(–2)-fld; pedicels 9–30 mm, bracteolate 3–6 mm below calyx; bracteoles ovate or lanceolate 1.2–4 mm; buds plumply ovoid-acuminate puberulent; sepals ovate- or lance-acuminate, often reddish, up to 7.5–10 mm; petals yellow, 4 obovate to broadly oblanceolate up to 11–15 mm, the falcately dimidiate one slightly longer; androecium 10-merous, the long anthers 4.8–6 mm; ovary white-strigulose; style linear, gently incurved, (4.5–)5–7.5 mm; ovules 6–8(–11).

Pod ascending ±25–40(–45) × 4.5–5.5(–6) mm, the reddish-brown valves puberulent or finely pilosulous; seeds (scarcely known) ±3 mm, the testa dark brown sublustrous, finely pitted.

A species well characterized by the low bushy habit of growth, densely leafy branchlets, and relatively few pairs of thick-textured leaflets. We can find in the now fairly copious material no difference other than the dense silky vesture to separate *Cassia tragacanthoides* sens. str. from glabrescent *C. pachyphylla*, which are, moreover, sympatric (if not actually directly associated in any known station) around Diamantina, in the same environment of the campos rasos near the Serra crest. The difference in pubescence is, in this particular case, sharply marked, the leaflets being either densely pubescent on both faces or green and minutely ciliolate, permitting the retention of *C. pachyphylla* in varietal rank.

Key to the Varieties of *Ch. tragacanthoides*

1. Foliage densely silky-velutinous throughout with appressed sublustrous hairs, the branchlets similarly pubescent and often in addition shortly setulose. 14a. var. *tragacanthoides* (p. 687).
1. Foliage green, the leaflets glabrous above, glabrous or almost so beneath, minutely ciliolate, the branchlets puberulent, not setulose. 14b. var. *rasa* (p. 688).

14a. Chamaecrista tragacanthoides (Bentham) Irwin & Barneby var. **tragacanthoides**. *Cassia tragacanthoides* Martius ex Bentham, 1870, l.c., t. XLV, fig. 1, sens. str.—"Habitat in campis desertis inter Formigas et Contendas prov. Minas Geraës: Martius."—Holotypus, collected late VII or early VIII, 1818, M! = NY Neg. *9210*.

Densely silky-pubescent throughout with very short appressed and often some longer stiffly spreading hairs to 0.5–0.8 mm, the foliage silvery or greenish-cinereous.

Lvs 1–2 cm; petiole 1.5–2.5 mm; gland 0.3–0.7 mm diam, commonly drum-shaped and 0.3–0.7 mm tall; rachis 3–10 mm; lfts 3–6 pairs.—Collections: 9.

Stony campo, on sandstone and quartzite, 1100–1300 m, local but locally plentiful, known only from the crest of Sa. do Espinhaço between Montes Claros and Diamantina, lat. ±16°45′–18°15′S, n.-centr. Minas Gerais.—Fl. X–VIII, probably most freely in the dry season.

14b. Chamaecrista tragacanthoides (Bentham) Irwin & Barneby var. **rasa** Irwin & Barneby, nom. & stat. nov. *Cassia pachyphylla* Martius ex Bentham in Martius, Fl. Bras. **15**(2): 166. 1870.—"Habitat in altis Serra de Itambé, prov. Minas Geraës: Martius; in Serra dos Cristaës [in reality near Montes Claros, n.-centr. Minas Gerais]: Pohl."—Lectoholotypus, dated V.1818, M (3 sheets)! = NY Neg. *9209*; paratypi, *Pohl 3135* from Fazenda Vieira do Mato, VIII.20 (fl), M! and *s.n.*, K!

Young branchlets, lf-stalks and margins of lfts puberulent with incumbent or incurved ascending hairs up to 0.2–0.4 mm, the lfts green, densely papillate, glabrous above or on both faces.

Lvs 1.5–2.7 cm; petiole 2–4 mm; gland sessile or nearly so, commonly urceolate-patelliform, sometimes drum-shaped, (0.3–)0.4–1 mm diam; rachis 4–11(–18) mm; lfts 3–7(–8, mostly 3–4) pairs.—Collections: 8.

Stony campo and thin cerrado near 1100–1200 m, apparently local, best known from the vicinity of Francisco Sá and Montes Claros in n. Sa. do Espinhaço near lat. 16°30–55′S, and collected once near Diamantina, lat. 18°15′S, in n.-centr. Minas Gerais, the holotypus supposedly collected on Sa. Itambé do Mato Dentro at 19°20′S, the locality requiring confirmation.—Fl. VIII–V, probably most freely in the dry season.

15. Chamaecrista papillata Irwin & Barneby, sp. nov. *Ch. mucronatae* proxima sed foliolis plerumque 5–10 (nec 3–6)-jugis oblongis nec obovatis inferne elevatius venulosis ac densissime papillatis necnon legumine subdimidio minori ±3–4 (nec 5.5–7) cm longo, 5–9 (nec 10–14)-ovulato certe distincta.—BRASIL. Minas Gerais: frequent on sandstone, 1200 m, summit of Sa. do Cipó, km. 111–120, road from Hotel Chapeu do Sol, 6.VIII.1960 (fl), *B. Maguire, G. Mendes Magalhães & C. K. Maguire 49019*.—Holotypus, NY.

Procumbent or prostrate subshrubs with several radiating simple or few-branched stems in age flexuous and woody at base, at anthesis 3–7 dm, softly pilosulous throughout or almost so with fine horizontal or widely ascending gray hairs to 0.3–0.7 mm, the relatively ample concolorous lvs equally pubescent on both faces, rarely glabrate above.

Stipules stiffly erect, narrowly lanceolate, 2–7.5 mm, 1–3-nerved, at symmetric base 0.3–1 mm wide, persistent.

Lvs (2–)2.5–6.5 cm subsessile or shortly petioled, the expanded blade broadly ovate-oblong in outline; lf-stalk (1.6–)2.5–5.2 cm; petiole stout 2.5–6(–7) mm, 0.7–1 mm diam; rachis narrowly winged, the wing-segments scarcely dilated upward; petiolar gland below first pair of lfts elliptic urceolate subsessile, 1.2–3.2 mm diam, not over 0.5 mm tall; smaller glands often present below each subsequent pair; lfts 5–10-pairs, slightly decrescent distally, oblong or oblong-oblanceolate 10–23 × 2.5–7(–8) mm, asymmetrically cordate at base, at deltate to truncate-emarginate apex contracted into a stiff but not vulnerant cusp 0.4–1.6 mm, from base ±3–5-nerved by the straight, subcentric midrib with usually 2 strong nerves on the adaxial side and 1(–2) weaker ones on the abaxial, these with the few secondary venules beneath strongly elevated and cordlike, above faintly raised, the dorsal face between the venules densely and prominently papillate.

Peduncles exactly axillary, 1–2-fld, 1–3 mm; bracts and bracteoles resembling stipules, slightly smaller; pedicels (13–)17–28 mm; buds plumply ovoid-acuminate, finely pilosulous; sepals subpetaloid, the outer often reddish, ovate- or broadly lance-acuminate, 10–13 mm; petals yellow, sometimes reddish at the claw, four similar obovate- to oblong-cuneate, up to 15–19 mm, the dimidiately oblong-oblanceolate fifth one slightly longer; stamens 10, the longer anthers 5.5–7 mm, minutely crisp-pilosulous along the sutures; ovary densely whitish-pilosulous; style linear-cylindric, 6.5–9 mm, puberulent proximally, beyond middle glabrous incurved; ovules 5–9.

Pod (scarcely known ripe) apparently ±3–4 × 0.6–0.7 cm, the valves finely pilosulous.—Collections: 24.

Stony campo and about outcrops, 1050–1300 m, in red sandy clays or sideritic gravels, apparently local, known from scattered stations on the crest and higher w. slope of Sa. do Espinhaço between Gouveia and vicinity of Ouro Branco (18°30′–20°30′S), s.-centr. Minas Gerais.—Fl. VIII–X and sporadically to II.

Specimens of *Ch. papillata* have been referred either to *Ch. repens* or *Ch. drepanophylla,* both of which are distinguished by supra-axillary peduncles. The close affinity is to the sympatric *Ch. mucronata* which differs in the fewer (mostly 2–5, not 5–10) pairs of broader leaflets less prominently venulose beneath and lacking the dense granular papillae on the dorsal face. Most of the material available has been collected between October and January, at a season of sparse sporadic or infertile bloom, and we still lack good fruiting examples of the species. The pod appears however to be substantially smaller and fewer-ovulate than that of the related *Ch. mucronata.*

16. Chamaecrista ulmea Irwin & Barneby, sp. nov., *Ch. multinerviae* proxime affinis sed ab ea statura arborescenti, caulibus teretibus (nec angulatim costatis) foliolisque angustis 2.2–3.7 (nec 3.5–8.5) mm latis simpliciter pauciveniis (nec argute reticulatis) diversa.—BRAZIL. Minas Gerais: ±15 km w. of Grão Mogol, 19.II.1969 (fr), *Irwin et al. 23535.*—Holotypus, UB; isotypi, F, LE, K, NY, P, R, S, US.—Corrego do Pasto no alto da Sa. Geral, mun. Grão Mogol, 21.X.1978 (fl), *Hatschbach 41627.*—Paratypus, NY.

Spindly treelets of vaselike outline, at anthesis 1–3 m, with few slender, narrowly ascending trunks, the terminal subterete atrocastaneous branches pendulous, glabrous throughout, the 4–6-foliolate lvs stiffly ascending, the plane thick-textured lfts dull olivaceous concolorous.

Stipules erect, narrowly triangular or lance-subulate (1.5–)2–3.5 mm, the firm blades 3–5-nerved, deciduous with or after the lf.

Lvs (13–)15–23 mm; petiole with slender wrinkled pulvinus 2.7–3.5 mm, the former coarsely narrow-margined, 0.4–0.8 mm diam, the groove opened by the gland; gland near middle of petiole proper, urceolate sessile 0.6–0.9 mm diam thick-margined, in profile squatly tub-shaped and 0.4–0.6 mm tall; rachis 1.2–5 mm; lfts 2, or in a few lvs 3, pairs, the distal pair slightly larger, all narrowly oblanceolate, abruptly acuminulate-mucronulate, 10–19 × 2.2–3.8 mm, at base cordate on proximal and rounded on distal side, from base 4–5-nerved by the slightly excentric costa with on its proximal side 1–2 very short outer nerves and 1 strong inner one produced beyond middle of blade and on its distal side 1 very short nerve, the costa thence faintly penniveined with 4–5 pairs of secondaries, the venulation either immersed or bluntly prominulous above, bluntly prominulous beneath.

Racemes exactly axillary 1-fld, the axis almost 0; pedicels obliquely ascending toward the meridian, appearing refracted from pendulous branchlets, ±17 mm long, bracteolate near apex; fl-buds ovoid-acuminate glabrous; sepals reddish-purple ovate-acuminate 12–14.5 × 6 mm, finely several-nerved; petals yellow, 4 obovate 17–19 × 7.5–9 mm, the falcate cucullus ±24 × 10 mm, its inner margin crenulate-dentate; androecium 10-merous, the subequal anthers brownish ±7 × 0.9 mm; ovary minutely strigulose; ovules 12–14; style linear-attenuate 8 × 0.6 mm glabrous.

Pod 6–7.5 × 0.5–0.6 cm, the sublustrous atrocastaneous valves glabrate; seeds not seen.—Collections: 2.

Rocky creek banks in cerrado near 950 m, known only from the sources of Rio Itacambiruçu, on the e. slope of Sa. do Espinhaço in mun. Grão Mogol, n.-centr. Minas Gerais.—Fl. X–XI.

A remarkable chamaecrista, notable for the slenderly arborescent habit, the narrowly ascending primary trunk supporting a crown of pendulous, thinly leafy branchlets, the whole resembling in outline a diminutive *Ulmus americana* L., whence the epithet. It is closely related to *Ch. multinervia,* which appears, however, to be a subshrub from a xylopodium and therefore quite different in habit. It resembles *Ch. multinervia* in number of leaflets, but these are smaller, narrower, more simply nerved, and borne on smoothly terete, not angulately ribbed branchlets.

17. Chamaecrista burchellii (Bentham) Irwin & Barneby, comb. nov. *Cassia burchellii* Bentham in Martius, Fl. Bras. **15**(2): 165. 1870.—"Habitat inter Goyaz et Cavalcante [ex num. (Phytologia **14**: 502) = Guardamôr to Riberão de S. João, s.-w. of Niquelândia]: Burchell; prope S. José prov. S. Paulo [doubtless = S. José de Tocantins = modern Niquelândia, Goiás]: Pohl."—Lectoholotypus, *Burchell 7582,* K! = NY Neg. *1527*; paratypus, *Pohl 2126,* K!

Functionally herbaceous subshrubs with few, few-lvd and few-fld, at anthesis basally leafless, ribbed stems 1.5–4 dm ascending or stiffly erect from a knotty xylopodium, appearing glabrous but the stems distally, stipules and bracts, ventral ridges of lf-stalk, and ovary all minutely puberulent with incurved hairs, the leathery veiny foliage concolorous, usually quite glabrous, but some lfts sometimes sparsely long-ciliate.

Stipules erect appressed, deltate, triangular, or subulate, 1.5–5 mm, coarsely 3–5-nerved, persistent.

Lvs narrowly ascending (4–)6–8.5 cm, stoutly short-petiolate; lf-stalk (2–)2.5–5 cm, obtusely coarsely 2-marginate, the groove open, shallow, the petiole 5–12 mm, the rachis strongly dilated under the pulvinules; petiolar gland solitary near or above middle of petiole, elliptic sessile, 1.3–3.4 mm diam (sometimes a smaller gland high on rachis); lfts 2–4 pairs, rising vertically from rachis, turned half face to face, in outline narrowly oblong-obovate or -oblanceolate, very obtuse but sharply mucronulate, (1.7–)2–4.8 × 0.8–1.8(–2) cm, at base deeply cordate on proximal side, rounded to cuneate on the other, the coriaceous plane blades dull olivaceous on both faces, from base coarsely 5-nerved, the outer pair short, early anastomosing with a stout marginal nerve, the inner pair produced to or nearly to apex, these and the slightly excentric midrib subequally prominulous above and beneath, penniveined and thence intricately reticulate, the ultimate areoles much <1 mm diam.

Peduncles axillary 1–2 mm, solitary or sometimes geminate, 1–2-fld; bracts and bracteoles resembling stipules but only ±1–2 mm; pedicels narrowly stiffly ascending (2.5–)3–5 cm; buds plumply ovoid-acuminate, glabrous; sepals submembranous, petaloid-margined, ovate- to lance-elliptic, abruptly acute, 12–14 mm;

petals yellow, 4 narrowly oblong- or obovate-cuneate, unequal, 15.5–21 mm, the fifth falcately dimidiate slightly longer; stamens 10, the anthers 6–9 mm, the sutures minutely puberulent; gynoecium divaricate; ovary puberulent; ovules 7–9; style linear-cylindric, 7–9 mm.

Pod erect, narrowly oblong or oblong-oblanceolate, 4.5–5.5 × 0.7–0.8 cm, the purplish-brown valves sublustrous, minutely puberulent; seeds not seen.—Collections: 4.—Fig. 44.

Gravelly hillsides in cerrado, near 1000 m, very local on the upper forks of Rio Tocantins between 14° and 15°S in centr. Goiás.—Fl. XII–I.

18. Chamaecrista choriophylla (Vogel) Irwin & Barneby, comb. nov. *Cassia choriophylla* Vogel, Syn. Gen. Cass. 56 & Linnaea **11**: 706, descr. ampliat. 1837.—Typus infra sub var. *choriophylla* indicatur.

Coarsely leafy subshrubs 1.5–5 dm, with several or many simple or distally few-branched stems erect or incurved-ascending from a xylopodium, glabrous up to the minutely puberulent ovary and sometimes vestigially strigulose stipules and lf-stalks, floriferous from many axils above the middle, the internodes terete or angulate or alate by sharp-edged ribs decurrent from each stipule and petiole, the ample coriaceous 2-foliolate lvs olivaceous concolorous, sometimes glaucescent when fresh.

Stipules ascending, variable (according to variety) in size and outline, 1.5–14 mm, at base 1–8 mm wide, 3–21-nerved, the blades firm, subpungently mucronulate.

Lvs stiffly ascending toward the meridian, (2–)2.5–7.5 cm, stoutly short-petioled, the petiole rounded dorsally, narrowly thick-margined ventro-laterally, widely shallowly sulcate ventrally, dilated at apex; petiolar gland adjacent to or between the pulvinules, variable according to variety, sessile or stoutly short-stipitate, 1–2.7 mm diam; lfts 1 pair, narrowly ascending, turned half face to face, in outline obliquely elliptic, oblong-elliptic, or elliptic-obovate, obtuse or subacuminate, pungently mucronulate, (1.7–)2.2–7 × (0.45–)0.7–4 cm, at asymmetric base broadly rounded-subcordate on proximal side, broadly cuneate on the other, the coriaceous blades subequally venose on both faces, mostly 5-nerved from base, the midrib excentric, only 1–2 primary lateral nerves produced nearly or quite to apex, the outer ones early anastomosing with a strong corneous marginal nerve, the midrib pinnately many-branched, thence more finely reticulate-venulose.

Peduncles axillary, solitary or geminate, 0.5–2 mm, the axis 1–3-fld, 0–6 mm; pedicels ascending, strongly compressed and ribbed, bracteolate near or above middle, sometimes near apex; bracts and bracteoles resembling stipules but smaller, persistent; buds ovoid-acute glabrous; sepals elliptic-ovate acute 10.5–19 mm; petals (much eaten, poorly known) yellow, 4 (unequal) obovate-cuneate to 2–2.8 cm, the dimidiate fifth one as long or longer, falcate; stamens 10, the anthers 4–8 mm, minutely puberulent along the sutures; gynoecium divaricate; ovary strigulose; ovules 9–15; style linear-cylindric, incurved distally, 7–15 mm.

Pod erect, 5–8 × 0.6–1.15 cm, straight or curved outward, the valves reddish-brown nigrescent, minutely puberulent; seeds (little known) quadrate or oblong, the testa atrocastaneous, densely lineolate-pitted.

Key to the Varieties of *Ch. choriophylla*

1. Stems subterete, faintly obtusely ribbed; stipules narrowly triangular, 1.5–2.5 mm, 3-nerved; petiolar gland sessile, elliptic, longer than wide, 2–2.7 mm; pedicels (±3 cm) bracteolate

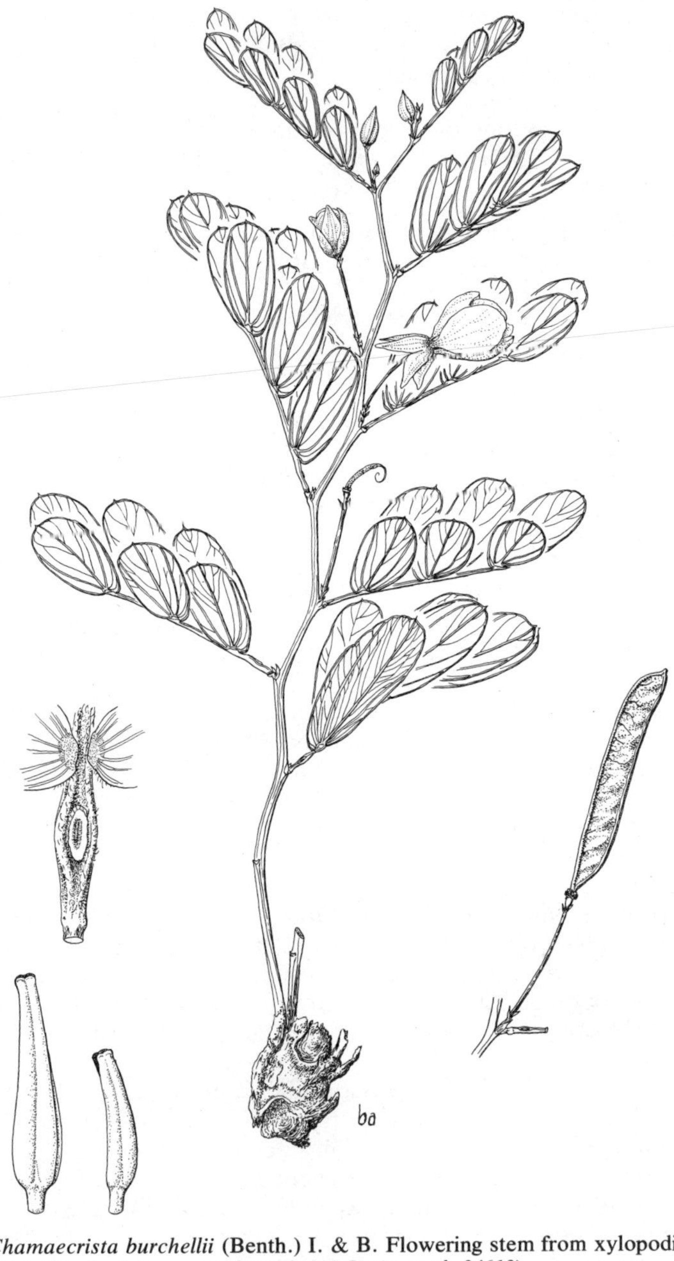

Fig. 44. *Chamaecrista burchellii* (Benth.) I. & B. Flowering stem from xylopodium ×½; petiole, ventral view ×4; detached stamens ×3; pod ×½ *Irwin et al. 34613*).

0.5–3 mm below calyx; pod probably narrow as in var. *rossicorum* but lfts broad as in var. *latifolia*; w. slope of Sa. do Espinhaço near 18°30′S. 18a. var. *choriophylla* (p. 693).

1. Stems sharply 4-angulate or -alate by sharp-edged ribs descending from each stipule and petiole; stipules 2.5–14 mm, 5–21-nerved; petiolar gland raised on a short stout pediment thicker than high, in outline round to deltate-reniform and wider than long, 1–2 mm diam; pedicels variable in length but bracteolate well below calyx.
   2. Stems 4-angulate, scarcely winged; stipules 2.5–5 mm, at base not over 2 mm wide,

5–7(–9)-nerved, at insertion not contracted or cordate; lfts 2–4.5 × 0.6–2 cm; pedicels 0.8–2 cm; anthers 4–6 mm; pod 6–7.5 mm wide; Sa. do Espinhaço around Diamantina, ±18°–18°30′S. 18b. var. *rossicorum* (p. 693).
2. Stems tetrapterous; stipules 6.5–14 mm, at base 4.5–8 mm wide, from base 13–21-nerved, at insertion cordate-amplexicaul; lfts (3–)3.5–7 × 1.3–4 cm; pedicels 3–4 cm; anthers 6.5–8 mm; pod 8–11.5 mm wide; crest and summits east of crest of Sa. do Espinhaço (Pico da Itambé near 18°30′S and Sa. do Cipó near 19°15′S). 18c. var. *latifolia* (p. 693).

18a. Chamaecrista choriophylla (Vogel) Irwin & Barneby var. **choriophylla**. *Cassia choriophylla* Vogel, 1837, ll. cc., sens. str.—"In Brasilia [acc. the original label, †B = F Neg. *1672*: Sa. do Galheiro, Minas Gerais]: Sellow leg."—Holotypus, †B; neoholotypus, former isotypus, K! = IPA Neg. *1108* = NY Neg. *1526*; isotypus (fragm), P (hb. Glaziov.)!

Cassia choriophylla sensu Bentham, 1870, p. 165; 1871, p. 572, exclus. syn.

Characters as given in key; sepals ±11–12 mm.—Collections: 4.

Outcrops, near 1200–1300 m, apparently rare, known only from the high w. slope of Sa. do Espinhaço in mun. Gouveia (Sa. do Galheiro; Rodovia BR 259, 40–50 km s.-w. of Diamantina); centr. Minas Gerais.—Fl. IX–X.

18b. Chamaecrista choriophylla (Vogel) var. **rossicorum** Irwin & Barneby, var. nov., a var. *choriophylla* caule acute 4-angulo, stipulis latioribus 5–9 (nec 3)-nerviis, glandula petiolari reniformi-rotunda substipitata minori, pedicellisque longioribus (±3 nec 0.8–2 cm) abstans.—BRAZIL. Minas Gerais: rocky cerrado, 1300 m, 15 km n.-e. of Diamantina on rd to Mendanha, 26.I.69 (fr), *Irwin et al. 22621*.—Holotypus, UB; isotypi, C, F, GH, K, RB, S, UC, US.

Cassia gonoclada Bentham, Trans. Linn. Soc. **27**: 572, pro syn. 1871; non Bentham in Martius, 1870, p. 169.—A name applied in herb. to *Riedel 562,* LE!

Characters as given in key and diagnosis; sepals 10.5–16 mm; seeds ±4.3 mm.—Collections: 14.

Rocky or sandy campo and about outcrops, 1150–1300 m, local, known only from the crest and higher slopes of Sa. do Espinhaço within a radius of 40 km from Diamantina, centr. Minas Gerais.—Fl. VIII–XI.

This, the best known form of *Ch. choriophylla,* was collected first by Riedel and provisionally named from his specimens in Petrograd by Bentham *Cassia gonoclada,* an unpublished combination later used by Bentham himself for a quite different, multifoliolate chamaecrista. It differs from typical var. *choriophylla* in the sharply tetragonal stems, larger stipules, smaller petiolar gland elevated on a stout but very short stalk, and short pedicels bracteolate well below the calyx. The angled stem and petiolar gland are features common to var. *rossicorum* and var. *latifolia,* which differs markedly however in the again larger, many-nerved, cordate-amplexicaul stipules, broader and longer leaflets, broader pod, and larger seeds.

18c. Chamaecrista choriophylla (Vogel) Irwin & Barneby var. **latifolia** (Bentham) Irwin & Barneby, comb. nov. *Cassia chloriophylla* var. *latifolia* Bentham in Martius, Fl. Bras. **15**(2): 165. 1870.—". . . in prov. Minas Geraës: Claussen."—Lectoholotypus, *Claussen 208,* P! isotypi, G, LE!

Characters as given in key; sepals 15–19 mm; seeds 5.5–6 mm.—Collections: 15.

High rocky and sandy campo or about outcrops, 1250–1300 m, local, on and near the crest of Sa. do Espinhaço from (?Sa. de Caraça) Sa. do Cipó n. to mun. Gouveia, and at 1700 m near summit of Pico da Itambé, centr. Minas Gerais.—Fl. VIII–III, perhaps irregularly through the year.

Readily recognized by the broad, multistriate, clasping stipules, ample leaflets, and very large flowers. On the road between Diamantina and Gouveia Irwin et al. collected under one number typical var. *latifolia* growing with var. *rossicorum*; the dissonant elements are instantly separable, and there are no intermediates as yet. The vigor and coarse foliage, reminiscent of *Vicia faba* L., are suggestive of a polyploid derivative of var. *rossicorum*. With the exception mentioned, the ranges of the two varieties appear geographically disjunct.

The cited isotypus at LE, received from BR, gives the locality as Serra de Caraça, but the label is not an original from Claussen and the data require confirmation.

19. Chamaecrista simplifacta Irwin & Barneby, sp. nov., inter *Ch. multinerviam*, quoad foliolorum ambitum ac venationem praesimilem, et *Ch. choriophyllae* var. sympatricam *rossicorum* aequaliter 2-foliolatam quasi intermedia, ab illa primo intuitu foliorum numero, ab hac imprimis stipulis parvis 1–1.5 (nec 2.5–5) mm longis obscure 3 (nec 5–7)-nerviis foliolisque anguste oblong-ellipticis 3–7.5 (nec 6–20) mm latis recedens.—BRAZIL. Minas Gerais: campo pedregoso, Cons. Mata, mun. Diamantina, 12.VIII.72 (fl) *G. Hatschbach 30215*.—Holotypus, MBM; isotypus, NY.

Closely resembling *Ch. multinervia* in habit, and in texture, venation and outline of lfts, glabrous below the ovary, the stems erect from a xylopodium up to ±5–6 dm.

Stipules deltate 1–1.5 mm, obscurely 3-nerved.

Lvs below the fertile axils 1.5–3.7 cm, thence shorter, the inflorescence becoming pseudo-racemose; petiole with pulvinus 1–2.5 mm, the petiole proper sometimes almost suppressed; rachis 0; lfts 1 pair narrowly oblong- or oblance-elliptic, those of larger lvs 15–35 × 3–7.5 mm.

Peduncles 1–3-fld; pedicels 7–13 mm; bracteoles ovate-deltate 0.7–1.2 mm; sepals ± 9–11 mm, all shorter than the petals.—Collections: 3.

Stony campo near 1200 m, apparently very local, known only from the heights of centr. Sa. do Espinhaço in município do Diamantina (Biribiry; Cons. Mata), centr. Minas Gerais.—Fl. III, VIII, probably most freely during the dry season.

The description of *Ch. simplifacta* is based on three collections, the type, a fragment of *Glaziou 19103* (P) of which the greater part consists of *Ch. choriophylla* var. *latifolia* and a sterile stem (with attached xylopodium) in the Damazio herbarium (R-*38647*) without locality data. This rather fragmentary and unsatisfactory material is, however, very uniform in detail and impossible to refer to any described chamaecrista. It represents a species obviously related to *Ch. multinervia* and *Ch. choriophylla,* resembling the first in the narrow outline and venation of the leaflets and the second in their number, a single pair to each leaf. Features unique to *C. simplifacta* in the context of its close kindred are the very short or essentially obsolete petiole and the abrupt reduction in size of leaves at floriferous nodes, whereby the successive 2–3-flowered peduncles come to form a sort of bracteate false raceme.

20. Chamaecrista lagotois Irwin & Barneby, sp. nov., *Ch. multinerviae* propiusque *Ch. simplifactae* affinis, ab ea foliis omnibus 2-foliolatis glandulaque petiolari magna arcte sessili, ab ambabus foliolis magnis glaucescentibus, plerisque 3–8.5 (nec 1.5–3.5) cm longis, ulterius a *Ch. simplifacta* statura fruticosa elatiori, petiolo cum pulvino 4–9 (nec 1–2.5) mm longo, glandula petiolari medium petiolum versus (nec inter ipsos foliolorum pulvinulos) collocata, pedicellisque longioribus 16–22 (nec 7–13) mm longis satis ut videtur distincta.—BRAZIL. Minas Gerais: w. slope of Sa. do Cipó, near km 105 (±130 km n. of Belo Horizonte), 1150 m, 17.11.68 (diu deflorata), *Irwin, Maxwell & Wasshausen 20356.*—Holotypus, UB; isotypi, F, K, LE, M, NY, R, US.

Stiffly erect few-branched glabrous shrubs to 1.5 m, with early leafless blackish trunks knotty from prominent leaf-scars, the angulately ribbed annotinous branchlets densely clad in stiffly erect coriaceous, narrowly 2-foliolate, concolorous olivaceous-glaucescent lvs.

Stipules erect appressed, deltate ±1 mm, thick-textured, obscurely 1–3-nerved.

Lvs stiffly ascending at narrow angles (3–)4–9 cm, the stout petiole sessile 1.4–2.2 mm; lfts 1 pair, stiffly ascending in the form of a narrow V from tip of petiole, turned half face to face, in outline oblanceolate or narrowly oblong-elliptic, obtuse to shortly acuminate, pungently mucronulate, (2–)3–8.5 × (0.5–)0.8–1.6 cm, at base subsymmetrically narrowed into the petiole, scarcely differentiated dorsoventrally, 3-nerved from base, the outer pair of nerves early anastomosing with the thickened, cordlike marginal one, the tapering midrib prominulous on both faces, giving rise to many crowded, narrowly ascending secondaries, thence finely reticulate, the ultimate areoles much <1 mm diam.

Peduncles axillary, 1 mm or less, mostly 2-fld; pedicels (in fruit) stout 1.6–2.2 cm, bracteolate near apex; bracts and bracteoles resembling stipules, slightly smaller, subpersistent; fls and pods unknown.—Collections: 2.

Steep rocky slopes, 1150–1200 m, very local, known only from the crest of Serra do Cipó near 19°10–15′S in Serra do Espinhaço, s.-centr. Minas Gerais.—Fl. probably VI or thereabouts, collected sterile with persistent pedicels but no fruits I–II.

A strikingly peculiar xeromorphic chamaecrista, notable for the single pair of coriaceous glaucescent leaflets which stand erect from the tip of the short petiole in the attitude of hare's ears at alert. It is closely related to *Ch. multinervia* and particularly to *Ch. simplifacta,* both locally endemic further north on the crest of Sa. do Espinhaço around Diamantina, and similar in outline and venation of the leaflets; but it differs from both in the large size of the leaflets and their glaucous coloration. From *Ch. multinervia* it differs further in the constantly bifoliolate leaves and large discoid-patelliform (not drum-shaped) gland, and from *Ch. simplifacta,* like it in leaflet number, in the developed petiole charged at middle with the gland, and in the longer pedicels. The poorly known *Ch. simplifacta* probably differs from *C. lagotois* in stature and habit, being a lowly suffrutex with stems arising annually from a xylopodium, and foliage strikingly reduced upward, whereas *C. lagotois* is known in a fruticose state only, with naked trunks of several years' duration attaining 1.5 m in length and annotinous branchlets equably leafy to the tip. The variable and less closely allied *Ch. choriophylla,* represented on Sa. do Cipó by var. *latifolia,* is immediately distinguished where sympatric by the broad green leaflets and foliaceous stipules.

ii. ser. FLEXUOSAE Irwin & Barneby

Chamaecrista sect. **Chamaecrista** ser. **Flexuosae** Irwin & Barneby, ser. nov., a ser. *Coriaceis* inter quas a Benthamio dispositae praesertim foliolorum venatione peculiari, costa cum nervo primario de latere latiori adjacenti in folioli ipsum mucronem apicalem anastomosanti, cito segregandae, ulterius caulibus abrupte flexuosis 4-angulatis, glandulis arcte vel fere sessilibus, foliolorumque rachi de latere producto notabiles.—Sp. typica: **Ch. flexuosa** (Linnaeus) Greene.

Chamaecrista Flexuosae Britton & Rose, 1930, p. 271, in clave, sine descr.

Herbs and subshrubs, the stems 4-angulate, when herbaceous arising either from a taproot or a xylopodium, the secondary roots sometimes tuberous; lf-rachis narrowly winged between lfts and ridged behind them; petiolar gland sessile or subsessile, in profile shorter than diam of head; lfts 3–65 pairs, their midrib and the primary vein next to it on the lft's broader side produced to apex of blade and there confluent into an excurrent mucro or spinule, the two nerves enclosing a veinless areole; peduncles exactly axillary; sepals ovate acute or lance-acuminate; longest petal (9–)10–19 mm.—Spp. 4, all present on the Brazilian Planalto, 2 endemic to it, 1 disjunct in Colombia, and 1 widely dispersed to s.-e. coastal Brazil, n.-e. Argentina, Paraguay, e. Bolivia, s.-w. Mexico, s. Texas and Cuba.

The series *Flexuosae* is well defined by the syndrome of quadrangular stem, margined leaf-rachis, venation of leaflets (as described above) and exactly axillary peduncles.

Key to the Species of ser. *Flexuosae*

1. Lfts of adult upper lvs 3–8 pairs, abruptly contracted at tip into an obliquely divergent, vulnerant spinule 0.8–1.3 mm; local in coastal and interior Bahia. Stipules amply cordate; androecium 10-merous. 22. *Ch. swainsoni* (p. 701).
1. Lfts of adult lvs 9–65 (in Bahia at least 25) pairs, obtusely mucronulate or minutely aristulate, the cusp when present either weak, or much less than 0.5 mm, or both.
   2. Lfts of middle and upper cauline lvs 9–65 pairs, crowded along the rachis at points 1–2(–2.5) mm apart, if less than 20 pairs in larger lvs then these leaves only 1.5–4.5 cm long. Androecium 10-merous and blades of stipules broader than the proximal adjacent lfts, at base cordate on one or both sides; s. Texas and Cuba to n. Argentina, where sympatric with the next the lfts over 25 pairs in larger lvs. 21. *Ch. flexuosa* (p. 696).
   2. Lfts of middle and upper cauline lvs 9–25 pairs, inserted on rachis at points up to 2.5–5.5 mm apart, the longer lvs mostly 4–10 cm. Androecium either 5- or 10-merous; stipules various, either as above or much narrower; s.-e. Brazil and immediately adjoining Paraguay and Argentina.
      3. Androecium 10-merous; stipules broader than the immediately adjacent lfts, semicordate-amplexicaul, (5–)7–15 × (2–)2.5–7 mm; s. Mato Grosso to Paraguay and n. Argentina. 23. *C. gonoclada* (p. 702).
      3. Androecium 5-merous; stipules not or hardly broader than the adjacent lfts, ovate or lanceolate, not cordate-amplexicaul, 1.5–8 × 0.7–2.5(–3) mm; centr. Mato Grosso e. through Goiás into Distrito Federal; n.-e. Colombia. 24. *C. parvistipula* (p. 704).

21. Chamaecrista flexuosa (Linnaeus) Greene, Pittonia **4**: 27. 1899. *Cassia flexuosa* Linnaeus, Sp. Pl. 379. 1753.—Typus infra sub var. *flexuosa* indicatur.

Precociously flowering but often ultimately suffruticose or rarely truly frutescent herbs variable in habit, stature, and pubescence, with 2–several humifuse, procumbent, assurgent, or erect, always abruptly flexuous and bluntly 4-angular stems at anthesis (1–)1.5–12, rarely 30 dm, arising from the sometimes dilated crown of the blackish woody root or (in age) from a distinct lignescent caudex, except for the always strigulose ovary nearly (quite) glabrous or variably pubescent with short, forwardly incurved hairs, pilosulous with fine spreading-incurved hairs, or hispid with fine septate setae to 1–2.5 mm, either type of vesture predominant or either 0, the stems and foliage in consequence green or grayish, the small firm crowded, often very numerous, concolorous, dorsally venulose lfts sensitive and in most dried specimens closely imbricated along the lf-stalk.

Stipules ± heteromorphic, the lowest small, ovate, not further noticed, the median and upper stiffly ascending, obliquely lance- to dimidiately or subsymmetrically ovate-acuminate, apiculate-mucronate or stiffly aristulate, (3–)3.5–18(–28) × 1.5–8(–12) mm, at base broadly semi-cordate-amplexicaul or, when broad, cordate on both sides, the stiffly chartaceous persistent blades coarsely (7–)8–15-nerved from pulvinule, on either or both sides either glabrous or puberulent, the margins smooth, scabrous-ciliolate, or erratically setose.

Lvs stiffly widely ascending or slightly recurved, short-petiolate, heteromorphic, the lowest relatively short and simple (not further described), the median and upper most commonly 4–11(–14, in var. *texana* only 1.5–4.5) cm, the expanded blade varying from narrowly ovate to linear-lanceolate or almost linear-oblong, the lowest and uppermost lfts commonly smallest; petiole (1–)1.5–5.5(–10) mm, openly shallow-sulcate; petiolar glands 1–4(–5), shallowly cup- or saucer-shaped, circular or rarely elliptic in outline, thick-rimmed, pallid or livid, the largest of them 0.2–0.9(–1) mm diam, sessile or raised on a thick cylindric or distally dilated stipe mostly not over 0.3, rarely to 0.6–0.7 mm, in profile commonly tub- or drum-shaped, rarely tack-shaped; rachis not over 7 mm shorter than lf, rounded dorsally, sharply winged ventrally, the wing emarginate at insertion of each pair of lfts, hence crenate in profile, and also on each side, behind insertion of lfts, again narrowly winged or ridged; lfts (variable both within and between vars.) 9–60(–65) pairs inserted 1–2(–2.5) mm apart along rachis, linear-oblong, -oblanceolate, or narrowly oblong-elliptic, straight or slightly falcate distally, (1–)1.5–11(–13) × (0.4–)0.5–2.1(–2.5, in Colombia exceptionally 3) mm, at apex abruptly deltate acute or subacute, callous-mucronulate or rarely aristulate, at base on proximal side angulately semi-cordate-auriculate, on distal side narrowly cuneate, from pulvinule coarsely venose by the strongly displaced or submarginal midrib and (on broad side) always by 1 equally strong lateral nerve produced to anastomosis at tip of lft, these both either simple or weakly forked distally, relatively broad lfts also with 1–2(–3) shorter exterior nerves ascending ¼–¾ length of the coriaceous or firmly chartaceous blade, the venation always coarsely prominent beneath, fainter above.

Peduncles exactly axillary 0–1.5 mm, 1–2(–3)-fld; pedicels ascending, straight or incurved under the pod, 11–42 mm, glabrous or puberulent, exceptionally setulose, bracteolate shortly below calyx; bracteoles narrowly ovate or subulate, 1–3-nerved, 0.8–1.7 mm; buds ovoid-acuminate, strigulose or minutely pilosulous; sepals lance-acuminate to ovate acute, the outer often red or red-tinged, the inner hyaline-margined, the longer to 5–13 mm; petals pale or bright yellow, wilting by mid-day and fading during the first day to orange or pinkish-brown, 4 similar, obovate-cuneate or obovate narrowed to a short distinct claw, up to (10.5–)11.5–18(–19) mm, the cucullus ± as long but broader, dimidiately or very obliquely ovate, incurved distally, commonly enfolding the gynoecium; stamens 10, the inner series shorter, sometimes infertile, the anthers yellow, the adaxial pair of the larger outer whorl incurved-connivent, the 3 abaxial straight or almost so, to 4.5–8.5(–9) mm; style usually thinly puberulent to middle, linear-cylindric and obscurely clavate toward apex, incurved or hooked distally, 4–9 mm; ovary gray-strigulose; ovules 8–22.

Pod ascending, linear-oblong, straight or slightly curved, 2.5–7 × 0.35–0.5 cm, the castaneous or purplish-brown, finally nigrescent valves minutely puberulent; seeds obtusely quadrate, 3–3.9 mm, the testa pinkish- or grayish-ochraceous, dull or scarcely lustrous, sprinkled or lineolate with small shallow fuscous pits, sometimes faintly fuscous-lined along truncate or retuse distal keel.

Our concept of the weedy and polymorphic *Ch. flexuosa* is here expanded a little beyond the traditional limits to admit plants differing from what has been considered typical in their relatively short adult leaves composed of fewer pairs of otherwise identically similar leaflets. Such leaves occur in all individuals of the species, but in the races called typical only low on the stems or branchlets, quickly giving way to longer, more elaborately pinnate ones distally. With this modification *Ch. flexuosa* incorporates without strain the North American *Cassia texana* and *C. itzana* which are, however, still maintained collectively as a well-marked variety.

As redefined, *Ch. flexuosa* is distinguished from all other American chamaecristas, except the closely related *Ch. parvistipula,* by form of the leaf-rachis and venation of the leaflets, further from all those with exactly axillary peduncles and pale seeds by the stiff, abruptly flexuous stems and high number of leaflets, and from *Ch. parvistipula* and *Ch. gonoclada,* which see for emphasis, in the form of the root and the closely spaced leaflets. The characteristic rachis of *Ch. parvistipula* and *Ch. flexuosa* is rounded on the abaxial side, narrowly winged or ridged laterally, and more emphatically winged ventrally, the wider interfoliolar wing, which corresponds to the herbaceous margins of the ventral sulcus closely appressed or partly fused together, being indented between each pair of pulvinules, in consequence crenulate in profile. Between this and the lateral ridges lie a series of little oblong depressions running parallel with each interfoliolar segment and of the same length. Venation of the leaflet varies with breadth of blade in the number of exterior venules arising from the pulvinule, but is uniform in one striking feature: two equally strong veins arising from the pulvinule and attaining the tip of the leaflet blade, where they anastomose into the mucro, that on the blade's narrow distal side being the strongly displaced or submarginal midrib, the other the first lateral nerve on the blade's broader side. This basic venulation pattern serves by itself to distinguish the group of *Ch. flexuosa* and *Ch. parvistipula* from all American chamaecristas at all similar in general facies.

Key to the Varieties of *Ch. flexuosa*

1. Major cauline lvs, especially those subtending fls, 4–14 cm, with up to (20–)25–65 pairs of lfts; pedicels always far surpassed by the lf-stalk; pod (3.5–)4–7 cm, 10–22-seeded; s. Mexico and Cuba to n. Argentina. 21a. var. *flexuosa* (p. 698).
1. Major cauline lvs 1.5–4.5 cm, with 9–20(–25) pairs of lfts; pedicels either longer or only a trifle shorter than the lf-stalk; pod 2.5–4.7 cm, 8–12-seeded; s. Texas to s. Mexico, where confluent with the preceding. 21b. var. *texana* (p. 700).

21a. Chamaecrista flexuosa (Linnaeus) Greene var. **flexuosa**. *Cassia flexuosa* Linnaeus, 1753, l.c., sens. strict.—"*Habitat in Brasilia.*"—Lectotypus (Greene, 1899, l.c.): *Chamaecrista pavonis brasiliana siliqua singulari* Breynius, Exot. Pl. Cent. Prima t. 23. 1678!—"Brasiliae campos ac colles floriferos diligens, mihi ope *Excellentissimi Domini Adolphii Vorstii p. m.* communicata."

Cassia arenaria Humboldt, Bonpland & Kunth, Nov. Gen. & Sp. **6**(folio): 370. 1824.—"Crescit in arenosis, umbrosis, prope Maypures [on Río Orinoco in Vichada, Colombia]."—Holotypus, *Humboldt & Bonpland 883,* P-HBK! isotypi, B (*Hb. Willd. 8005*), P (hb. Bonpland.)!—Reduced to *C. flexuosa* by Bentham, 1871, p. 574.

Cassia flexuosa var. *pubescens* Bentham in Martius, Fl. Bras. **15**(2): 169. 1870.—". . . prope Oeiras, prov. Piauhy: Gardner n. 2131."—Holotypus, collected IV.1829, K! = IPA Neg. *1120*; isotypi, G, NY (hb. Meisner.)!

Cassia flexuosa var. *cuyabensis* Pilger in Engler, Jahrb. **30**(2): 158. 1901.—"Mattogrosso: in Regenrinnen auf Steinboden bei Cuyabá ([*Pilger*] n. 285) . . . Marz 1899."—Holotypus, †B; isotypus (fragm), US!

Chamaecrista amplistipulata Rose, Contrib. U.S. Nat. Herb. **12**(7): 267. 1909.—"Collected by E. W. Nelson near Santa Efigenia, Oaxaca, July 18, 1894 (no. 2850)."—Holotypus, US! isotypus (fragm), NY!—Reduced to *C. flexuosa* by Standley, Contrib. U.S. Nat. Herb. **23**: 404. 1922.

Cassia picachensis Brandegee, Univ. Calif. Publ. Bot. **6**: 180. 1915.—"Collected on Cerro de Picacho, Oaxaca, [*C. A. Purpus*] No. 7201."—Holotypus, UC; isotypi, GH, NY, US!

Cassia flexuosa sensu Colladon, 1816, p. 126; Vogel, 1837, p. 68; Bentham, 1870, p. 169; 1871, p. 574. *Chamaecrista flexuosa* sensu Britton & Rose, 1930, p. 278.

Root variable in thickness, sometimes slender and monocarpic, sometimes (especially in savanna) developing a dilated crown or xylopodium, but not tuberously thickened or giving rise to subterranean caudex; stems prostrate or erect (1–)1.5–12(–30), commonly 2–7 dm; pubescence highly variable, of lvs commonly scanty, sometimes 0, less often densely gray-pilosulous; lvs 4–11(–14) cm, the expanded blades mostly narrowly lance- to linear-oblong; petiolar glands 1–4(–5), 0.2–0.9(–1) mm diam; lfts of adult lvs 20–65, of some upper lvs at least 25 pairs, (1–)1.5–11(–13) × (0.4–)0.5–2.1(–3) mm; pedicels 11–28(–31) mm; sepals mostly 7–13 mm; petals either pale or bright yellow, up to (10.5–)11.5–18(–19) mm.—Collections: 323.

Open sandy or grassy places, native in seasonally arid grassland (savanna, llano, campo), maritime sands, open pine-forest, and on rock-outcrops, becoming weedy and locally abundant in disturbed woodland, pasture, waysides, lakeshores and seasonally inundated river-banks or flood-plains, mostly below 450 m but ascending to 700 m in Guayana Highland and to 1300 m on the Brazilian Planalto in Goiás and Minas Gerais, discontinuously widespread from s. Mexico (s.-w. México and the Pacific slope of Guerrero to Veracruz and Tabasco) and w. Cuba (Pinar del Río and Isla de Pinos) to n. Colombia (Magdalena), Venezuela (Maracaibo Basin and middle and lower Orinoco valleys, there extending to the Colombian shore in Vichada) and periphery of Guayana Highland in Venezuela (Bolívar, Amazonas), n. Brazil (Terr. do Roraima) and interior Guyana; lower Amazon valley in Brazil, downstream from mouth of Rio Tapajós (Pará); interruptedly on sandy shores and restinga of the Atlantic coast from Surinam to s.-e. Brazil (Rio Grande do Sul); and widely dispersed in caatinga, cerrado, campo, and capoeira over much of the Brazilian Planalto, from Maranhão, Rio Grande do Norte and Paraíba to São Paulo and n.-e. and s. Mato Grosso, thence feebly into the upper Madeira basin in Terr. Rondônia, Brazil, e. Bolivia (Sta. Cruz), Paraguay, and Argentina (Corrientes).—Fl. almost throughout the year, but most prolifically toward the end of the wet season and into the months following its end.—*Tepeguajito* (Oaxaca); *canela de Macarico* (Roraima); *espadim de Mata* (R. Grande do Norte); *estrella matutina* (Bolivia).

The synonymy listed under var. *flexuosa,* much of it already established in the literature, is of two sorts, based on: exactly typical plants (Colombian *C. arenaria*; Mexican *C. amplistipulata*) encountered in places remote from the original Brazilian one, this not exactly known but surmised to be one of the early Atlantic coastal settlements; or minor variants, almost certainly of independent origin, that differ in more or less densely pilosulous stems and foliage, the Brazilian vars. *pubescens* and *cuyabensis,* the Mexican *C. picachensis.* Other notable but unnamed variants not involving pubescence characters are a few examples (e.g. *Lundell 6990* from Belize, GH) with conspicuously stipitate petiolar glands, and a notable population from La Jagua, Magdalena, Colombia (*Haught 3575,* NY, US) remarkable for its immense foliaceous stipules and correspondingly ample leaflets. The majority of populations from the Brazilian Planalto are characterized by very numerous and very small, firm leaflets coupled with relatively large flowers that open pale yellow (as in *Ch. parvistipula*) and fade pinkish-brown; but golden yellow petals that fade orange-brown have been noted from this region and appear to be commoner in the maritime habitats (as in restinga near Rio de Janeiro!) and in Central America. An unusual erect, fruticose form with stems up to 3 m tall (*Irwin et al. 31262*) encountered once in Sa. Agua de Rega in interior highland Bahia is not otherwise different from sympatric herbaceous or lowly suffruticose ones. A variant completely glabrous, even to the ovary, is known from Venezuela only (*Pittier 14283,* US).

21b. Chamaecrista flexuosa (Linnaeus) Greene var. **texana** (Buckley) Irwin & Barneby, stat. nov. *Cassia texana* Buckley, Proc. Acad. Nat. Sci. Philad. **1861**. 452. 1861.—"Sandy soil, Bastrop County [Texas]."—Holotypus, PH! isotypus (fragm ex. hb. Durand.), GH!—*Chamaecrista texana* (Buckley) Pennell, Bull. Torr. Club **44**: 344. 1917.

Cassia itzana Lundell, Bull. Torr. Club **69**: 390. 1942.—"MEXICO–YUCATAN: Progreso . . . on low sand dunes, July 23, 1938, *C. L. Lundell & Amelia A. Lundell 8055* . . ."—Holotypus, MICH! isotypi, A, US (3 sheets)! paratypi, *Lundell & Lundell 7959, 8127 & Steere 3100,* all MICH!

Chamaecrista texana sensu Britton & Rose, 1930, p. 284.

Root (at least sometimes in Texas) tuberously fusiform, often deeply penetrating, its crown buried and giving rise to subterranean woody (rootlike) caudex-branches; stems humifuse, 1–5 dm; pubescence less variable than in var. *flexuosa,* mostly of incurved hairs <0.5 mm, occasionally on stipule-margins, upper stems and pedicels mixed with weak setae up to 0.6–1 mm, the lfts ciliolate, but either glabrous or pilosulous dorsally; lvs 1.5–4.5 cm, the expanded blade ovate or lance-ovate in outline; petiolar gland 1(–2), 0.15–0.35 mm diam, often concealed in the petiolar sulcus but sometimes elevated on a thick stalk as long as its diam; lfts of adult lvs 9–21(–25) pairs, 1–8.5 mm; pedicels 2–4.2 cm; sepals red-brown 5–7.5 mm, often proportionately wider than in var. *flexuosa*; petals bright yellow, up to 10.5–16 mm.—Collections: 35.

Sandy plains, sand flats behind barrier beaches, and dunes, sometimes weedy on roadsides and in old fields, 1–500 m, locally plentiful on the coastal plain of s. Texas (s.-ward from Maverick, Karnes and Bastrop cos) and Tamaulipas above Tropic of Cancer; disjunct on the n. coast of Yucatán (Progreso and vicinity); and along the Pacific coast of s. Mexico from Colima to Isthmus of Tehuantepec, occasionally inland to the floor of the Balsas Depression in Guerrero, here passing imperceptibly into var. *flexuosa*.—Fl. II–V n.-ward, VI–III s.-ward.

The first collection of var. *texana,* the only one known to Bentham (*Berlandier 2427*), was provisionally referred by him (1871, p. 578) to *Cassia procumbens,* but noted as different in several features. In modern times it has been maintained as a distinct species, and no allusion has been made to the multiple similarities in habit, stipules, leaf and flower to the widespread tropical *Ch. flexuosa*. Britton & Rose (1930, p. 278, 284) placed *Chamaecrista flexuosa* and *Ch. texana* in separate series, the former in monotypic *Flexuosae,* the latter in *Deeringianae,* groups supposedly, but to us imperceptibly, different in texture of the leaflets. As known, and it is there best known, from the Gulf Coastal Plain of southern Texas and Tamaulipas, *Cassia texana* does at first sight appear quite distinct, compared with average Venezuelan or Brazilian *Ch. flexuosa,* in its humifuse stems, short leaves, relatively few leaflets, and long pedicels that raise the flowers beyond the subtending leaf. Furthermore the average calyx is shorter, the average pod shorter and fewer-ovulate. This idealized picture of abrupt contrasts fades into indistinctness when we consider those southern Mexican populations, traditionally referred to *C. flexuosa,* in which short leaves composed of 21 or fewer pairs of leaflets are combined with pedicels either longer or shorter than themselves and resemble in every single feature (even if not in all simultaneously) some individual Texan plants. The early leaves of tropical, long-leaved *C. flexuosa* are essentially identical in length and leaflet-number with the adult leaves of largely extratropical *C. texana,* in which it appears that a juvenile condition persists the length of the stem, or, if one prefers, the ontogenetic sequence of shorter to longer leaves upward along the stem fails to be carried out. Such an occasional failure would handily explain the random appearance in Paraguay of plants (*Archer 4742,* US) with no more than 16–19 pairs of leaflets.

On the Brazilian Planalto the petals of typical var. *flexuosa* commonly open pale yellow and by mid-day fade pinkish-brown, as in the related and sympatric *C. parvistipula*; whereas in typical var. *texana* they open golden yellow and fade orange-brown. But golden yellow flowers have been seen on the restinga of the Guanabara coast, and noted by collectors in inland Bahia and elsewhere. In Texas the root-crown is sometimes (*Irwin 1209*) rigid and deeply buried, fusiformly dilated, and gives rise to subterranean caudex-branches, each of which appears from superficial viewpoint to give rise to an independent plant. In tropical *C. flexuosa* the root-crown is superficial, and the root itself, varying from subfiliform (and short-lived) to a centimeter or more in diameter, sometimes dilated upward into a knotty platform or xylopodium, is apparently never tuberously enlarged. Too few collections show enough of the root-system to provide systematic information on this point, but the collected root of at least some Texan populations (e.g. *Cory 55285,* US; *M. C. Johnston 5353,* NY) show lateral rootlets arising from what appears to be an ordinary taproot without any indication of superimposed caudical structure.

The population on the coast of Yucatan, described as *C. itzana,* was compared by Lundell with the "clearly allied" *C. jalapensis,* a species in our opinion entirely different in venation of the leaflets and supra-axillary pedicels. These plants do differ somewhat from Texan var. *texana* in slightly broader leaflets, looser pubescence, and shorter pedicels; but in no such fundamental feature as would lead us to believe in the existence of a specifically distinct type. Some dwarfed *Ch. flexuosa* from very dry sites in the Balsas Depression in Guerrero are intermediate in every sense between ideal vars. *flexuosa* and *texana* and are here arbitrarily disposed of according to the highest number of leaflets on the longest leaf.

22. Chamaecrista swainsoni (Bentham) Irwin & Barneby, comb. nov. *Cassia swainsoni* Bentham in Martius, Fl. Bras. **15**(2): 167. 1870.—"Habitat in Brasilia, loco accuratius non indicato: Swainson . . ."—Holotypus, K (hb. Hook.)! = NY Neg. *1531.*

Cassia swainsoni sensu Bentham, 1871, p. 573.

Suffruticose herbs with several coarse, obtusely 4-angular, stiffly abruptly flexuous stems 1.5–5(–?) dm obliquely ascending or diffusely radiating from the crown of a woody taproot, densely hispidulous throughout with stiff erect trichomes 0.2–0.5 mm mixed on some stipule- or lft-margins and some pedicels with few random or many crowded weak setae 0.5–1.6 mm, the foliage dull olivaceous concolorous, the firmly chartaceous, obliquely spine-tipped lfts coarsely venulose on both faces.

Stipules at first erect, in age spreading or reflexed, asymmetrically cordate-acuminate 7–16 mm, stiffly spinulose at apex, at base on both sides but unequally auriculate-amplexicaul 5–11 mm wide, from base 9–17-nerved, the major veins often penninerved, the blades becoming papery, marcescent.

Lvs ascending often slightly incurved, shortly petioled, the expanded blade broadly oblong-ovate or elliptic-oblong in outline; petiole coarse, 4–9 mm, in profile 0.6–1.1 mm diam, openly grooved ventrally; gland situated near middle of petiole, subsessile, cup-shaped, ±0.2–0.6 mm diam, often concealed in the sulcus; rachis 7–25 mm, faintly ridged laterally, margined and narrowly grooved ventrally, the ventral wings of each interfoliolar segment dilated upward; lfts 3–8 pairs, inserted on rachis ±2–3.5 mm apart, in outline obliquely oblong to broadly lance-oblong-elliptic, (5.5–)7.5–25 × (1.5–)2–7 mm, at apex abruptly contracted into an outwardly divergent vulnerant spinule 0.8–1.3 mm, at base on proximal side deeply cordate-auriculate, on distal side narrowly cuneate, from base coarsely 3–5-nerved, the strongly displaced midrib and first lateral vein on broad side of blade equally strong and anastomosing into the spinule, the 2–3 exterior on the broad side weaker and shorter, but the innermost of these sometimes almost attaining tip of lft.

Peduncles exactly axillary 0.5–2 mm, 1(–2)-fld; pedicels 20–26 mm, at anthesis ascending, in fruit widely ascending or refracted, bracteolate well above middle; bracteoles lance-acuminate 2–3 mm, 3–5-nerved; buds slenderly ovoid-acuminate, puberulent; sepals ovoid-acuminate 10–11 mm; petals yellow, highly irregular, 2 oblanceolate ±12 × 5 mm, 2 obovate-cuneate only slightly longer but nearly twice as wide, the cucullus ±14–15 mm and slightly wider than long, reniform-obovate, not strongly asymmetric, apparently incurved over the androecium; stamens 10, the inner whorl slightly smaller, the 3 adaxial incurved-connivent, the anther of the largest of the 3 abaxial ones lance-oblong ±5 mm; ovary densely strigulose; style linear-cylindric, gently incurved distally, 4.5–6 mm; ovules 9–16.

Pod erect or ascending, linear-oblong, straight, the greenish or castaneous, minutely puberulent valves becoming stiffly chartaceous; seeds (scarcely known) ±4 mm, the testa ochraceous dull, minutely fuscous-pitted.—Collections: 6.—Fig. 45.

Caatinga thickets and open sandy places, 20–1000 m, apparently rare, known from only two precise stations in centr. and e.-centr. Bahia, one on the coastal plain ±30 km n.-w. of Salvador, the other in Sa. do Tombador w. of Morro do Chapeu.—Fl. VIII–II.

Doubtless because of the relatively few leaflets combined with broad foliaceous stipules, Bentham associated *Ch. swainsoni* with *Ch. cinerascens* and *Ch. rotundata*. By shift of emphasis from these characters to the venation of the leaflets and the structure of the leaf-stalk we discern a closer relationship to *Ch. flexuosa*, from which *Ch. swainsoni* differs in the fewer but by compensation larger leaflets. The stiff zigzag stems and stiffly chartaceous leaflets abruptly contracted into a spinulose mucro are distinctive and unmistakable.

23. Chamaecrista gonoclada (Bentham) Irwin & Barneby, comb. nov. *Cassia gonoclada* Bentham in Martius, Fl. Bras. **15**(2): 169. 1870.—Habitat ad Rio Pardo, prov. S. Paulo [actually s.-e. Mato Grosso]: Riedel . . .''—Lectoholotypus, *Riedel 468*, collected IX.1826, LE (2 sheets)! = NY Neg. *1532* = IPA Neg. *1119*; paratypus, *Riedel 177* labelled ''S. Paulo,'' LE (mounted with one sheet of no. *468*)!

Cassia gonoclada sensu Bentham, 1871, p. 574.

Functionally herbaceous from a xylopodium, the incurved-ascending, simple or weakly branched, strongly ribbed, distally flexuous stems at anthesis 2.5–4.5

Fig. 45. *Chamaecrista swainsoni* (Benth.) I. & B. Habit ×1; stipules + lower leaflets ×3; detached stamens ×4; seed ×3 (*Irwin et al. 32631*).

dm, appearing glabrous but the young stem, lf-stalks, axes of inflorescence, ovary, and sometimes the lft-margins minutely strigulose with appressed or incumbent hairs up to 0.15–0.4 mm, the foliage concolorous, the lfts when dry dull brownish-olivaceous, conspicuously pallid-veined dorsally.

Stipules stiffly ascending, very obliquely ovate-acuminate, those subtending major lvs (5–)7–15 × (2–)2.5–7 mm, at base deeply cordate on the broader amplexicaul side, rounded to cuneate on the other, the firmly chartaceous blades from base strongly 9–14-nerved, glabrous on both faces, only minutely if at all ciliolate.

Lvs mostly narrowly ascending, heteromorphic, the lowest and sometimes some distal ones relatively short and simple, the middle and most upper (as described hereafter) 5.5–10 cm, short-petiolate, the expanded blade narrowly lance-oblong-elliptic in outline; petiole with pulvinus 2.5–8 mm, below the gland 0.55–1.2 mm diam, narrowly wing-margined, the sulcus forced open by the gland but beyond this, and throughout the rachis, closed, the rachis-segments thus winged ventrally and also faintly ridged laterally; gland solitary near middle of petiole sessile, suborbicular-patelliform 0.7–1.7(–2) mm diam; lfts (9–)12–25 pairs inserted on rachis at intervals up to 3–3.5 mm, the firmly chartaceous blades linear-oblong or narrowly oblong-elliptic, symmetrically or obliquely acute mucronulate, up to 9–20 × (1.4–)2–4.4(–5) mm, at base cordate or cordately angulate on the proximal side, cuneate on the distal one, from pulvinus coarsely 4–7(–8)-nerved by the strongly excentric midrib with, on its narrow side, 1 submarginal nerve produced nearly to apex, and on the broad side 1 strong interior nerve anastomosing with the midrib into the terminal mucro together with 1–4(–5) exterior ones progressively shorter, these all raised on both faces but more sharply so beneath, the secondary pinnate venulation fainter, not forming a closed mesh.

Peduncles exactly axillary 0–1 mm, 1–3-fld; pedicels strictly ascending (1.3–)1.5–4 cm, bracteolate 2–9.5 mm below calyx; bracteoles membranous ovate or lance-acuminate 1–2.8 mm, persistent; buds ovoid-acuminate, minutely strigulose at base only or to middle; sepals brownish, pallidly membranous-margined, lance- or ovate-acuminate 9–12 mm; petals (of *Ch. flexuosa*) the plane ones up to 12–15.5 × 7–9 mm, the cucullus scarcely or not longer but obliquely dilated; androecium 10-merous, the long anthers up to 6–7.5 mm; ovary strigulose; style 5–7.5 × 0.2–0.5 mm, not dilated at tip; ovules 10–14.

Pod (little known) erect, ±45–55 × 5–5.5 mm, the reddish-brown valves minutely sparsely strigulose; seeds unknown.—Collections: 7.

Campo and cerrado, probably not above 400 m, local, best known from Sa. Maracajú in s.-w Mato Grosso and adjoining Paraguay, from Camapuã to the sources of río Apa (19°30′–22°30′S), collected once in the upper Paraguay valley near Amolar, Mato Grosso (18°S) and once (fugitive from upstream?) on the banks of the lower Paraná (near 27°20′S) in extreme n.-w. Corrientes, Argentina.—Fl. X–V.

A precariously characterized species, intermediate between the already very closely related *Ch. flexuosa* and *Ch. parvistipula,* having as it were the cordate stipules and decamerous androecium of the former combined with the relatively few, larger and more distant pairs of leaflets of the latter. The leaflets of *Ch. gonoclada* tend to be a trifle broader and the petiolar glands larger than in its immediate kindred, but no discontinuity has been found in the ranges of variation. It seems reasonable to suppose that *Ch. gonoclada* arose from the richly variable matrix of *Ch. flexuosa* and in turn gave rise, with reduction of stipules and androecium, to the yet more specialized *Ch. parvistipula,* now vicariant in range to the northeastward. The latter and *Ch. gonoclada,* endemic to the cerrado (or llano) grasslands, are clearly more specialized ecologically than the widespread and adaptable *Ch. flexuosa.*

24. Chamaecrista parvistipula (Bentham) Irwin & Barneby, comb. nov. *Cassia parvistipula* Bentham in Martius, Fl. Bras. **15**(2): 170. 1870.—". . . in campis prov. Minas Geraës: Weddell; in siccis graminosis Serra da Chapada: Riedel; in Serra da Natividade prov. Goyaz: Gardner n. 3690., Burchell."—Lectoholotypus, *Gardner 3689,* collected II.1840, K (hb. Benth.)! = IPA Neg. *1125* = NY Neg. *1538*; isotypi, †B = F Neg. *1726,* G, NY, P! paratypi, *Riedel 481,* K! = *992,* LE (3 sheets), US! *Weddell 1915b,* P!

Cassia parvistipula sensu Bentham, 1871, p. 574.

Suffrutescent perennial herbs of long duration with ultimately many incurved-ascending and erect, in age leafless and sublignescent, simple or freely branching, quadrangular, ± abruptly flexuous stems (1–)1.5–5(–6) dm arising from the crown of an early enlarged, finally carrot-like taproot up to 2 cm diam, appearing glabrous but the young stems, the ventral margin of lf-rachis, the margins of stipules and lfts (or some of these) usually minutely pilosulous with incurved hairs to 0.2 mm, but nowhere setose, the concolorous lvs olivaceous dull or sublustrous, when fresh often subglaucescent, the usually few, pale yellow nodding fls immersed in foliage.

Stipules erect, ovate, lanceolate acute-mucronulate, or broadly lance-acuminate, 1.2–8(–10) × 0.7–2.5(–3) mm, at base obscurely contracted or, when relatively wide, incipiently semi-cordate, the blades firmly chartaceous, 3–9-nerved from base, persistent.

Lvs strictly ascending, straight or slightly incurved, (2–)3–8(–9) cm, shortly petiolate, the expanded blade lance-elliptic in outline; petiole (2.5–)3–6(–7) mm, openly shallow-grooved ventrally; gland 1, situated immediately below first pair of lfts, sessile or almost so, shallowly saucer-shaped, circular or broadly elliptic in outline, pallidly thick-rimmed, 0.4–1(–1.1) mm diam; rachis (1.2–)2–6.5(–7) cm, laterally and ventrally ridged as in *Ch. flexuosa* but the interfoliolar segments longer and less deeply notched at ends; lfts (6–)9–18(–22, of some lvs at least 10) pairs, inserted on rachis (2–)2.5–4.5 mm apart, in outline linear, linear-oblanceolate, or narrowly oblong-elliptic, straight or slightly falcate, (3–)5–15(some at least 6) × 0.7–2.3(–2.5) mm, at apex obtuse and bluntly mucronulate or deltately acuminulate, at base angulately auriculate on proximal side, the coriaceous lfts coarsely venulose beneath, smooth or faintly so above, the venation of 2 subparallel, apically anastomosing and 1–2 shorter exterior primary veins exactly as in *Ch. flexuosa*.

Peduncles precisely axillary, 0–2 mm, 1–2(–3)-fld; pedicels at anthesis slender 6–21(–25) mm, in fruit thickened, either erect or divaricate, bracteolate above middle; bracteoles ovate obtuse or lance-acuminate, 1–2.5 mm, persistent; buds ovoid-acuminate, minutely strigulose at least to middle; sepals reddish or the inner subpetaloid, ovate- or lance-acuminate, 8–11.5 mm; petals newly expanded pale or cream yellow, by noon of first day in sun fading pinkish or pale brownish, 4 obovate-cuneate or spatulate, subequal, up to 9–14.5 mm, the fifth dimidiately ovate, no longer but much broader; stamens 5, the 2 adaxial shorter and connivently incurved, the 3 abaxial straight, the longest anther up to 5–8.5 mm; ovary gray-strigulose; ovules 7–12.

Pod erect, linear, straight or slightly decurved, 3.5–5 × 0.45–0.5 cm, the reddish-castaneous or purplish, ultimately nigrescent valves minutely strigulose, the persistent stylobase stout straight 1.5–3 mm; seeds quadrately oblong, 4–5.8 mm, the pale ochraceous testa dull, profusely lineolate with minute fuscous or castaneous pits.—Collections: 35.

Campo and cerrado, mostly in sandy soils, resistant to fire but apparently never weedy, locally plentiful at 680–1200 m over both slopes of the watershed between the Tocantins and Xingú to the n. and the Paraguay and Paraná to the s. in centr. and s. Goiás and Distrito Federal (Sa. da Natividade s. to Contraforte Geral in lat. 12°–17°S), w. into n.-e. and centr. Mato Grosso (Sa. do Roncador and Cuiabá) and extreme s.-e. Rondônia (Vilhena); reported from Minas Gerais by Bentham (1870, l.c.) but not since seen there; disjunctly, near 220 m, on llanos of n.-e. Colombia (Vichada).—Fl. X–V, but freely only from I onward.

In hindsight the epithet *parvistipula,* although originally descriptive, proves to have been ill-chosen, the stipules now being known to vary from ovate and narrower than the stem to broadly lance-acuminate and shallowly semicordate-amplexicaul, when they are hardly different from those of *Ch. flexuosa*. The species is, however, readily distinguished from the latter, its closest ally, by the more distantly spaced lfts which are at once fewer and longer than those of sympatric var. *flexuosa,* in the stout fusiform taproot developed even in young seedlings (cf. *Anderson 11325,* NY), and decisively in the pentandrous androecium.

iii. ser. PROSTRATAE (Bentham) Irwin & Barneby

Chamaecrista sect. **Chamaecrista** ser. **Prostratae** (Bentham) Irwin & Barneby, comb. nov. *Cassia* sect. *Chamaecrista* ser. *Prostratae* Bentham in Martius, Fl. Bras. **15**(2): 162. 1870, char. emend.—Sp. lectotypica: *C. serpens* Linnaeus ≡ **Chamaecrista serpens** (Linnaeus) Greene.

Ophiocaulon Rafinesque, Sylva Tell. 129. 1838, pro gen.—Generitypus: *O. serpens* (Linnaeus) Rafinesque ≡ **Chamaecrista serpens** (Linnaeus) Greene.

Senna subgen. *Tagera* Rafinesque, Sylva Tell. 129. 1838, saltem ex parte.—Subgeneritypus: "*Tagera filiformis*" Rafinesque, l.c. = **Chamaecrista kunthiana** (Chamisso & Schlechtendal) Irwin & Barneby.

Cassia sect. *Chamaecrista* ser. *Paucijugae* Bentham in Martius, Fl. Bras. **15**(2): 164. 1870, max. ex parte.—Sp. lectotypica: *C. cordistipula* Martius ≡ **Chamaecrista cordistipula** (Martius) Irwin & Barneby.

Cassia sect. *Chamaecrista* ser. 2. *Prostratae* sensu Bentham, 1871, p. 570, exclus. *C. rotundifolia*.

Cassia sect. *Chamaecrista* ser. 3. *Paucijugae* sensu Bentham, 1871, p. 571, exclus. spp. africanis duabusque ultimis americanis.

Chamaecrista Tagerae & *Pilosae* Britton & Rose, N. Amer. Fl. **23**(5): 270, in clave, sine descr.

Mostly diffuse or prostrate taprooted herbs, some suffrutescent in age, one (*Ch. tenuisepala*) frutescent erect; petiolar gland always present, slenderly stipitate; lfts in most 2–9, in one (*Ch. trichopoda*) up to 22(–26) pairs; peduncles exactly axillary; sepals acute or acuminate; fls mostly small, the longest petal 3.5–11, in 2 spp. up to 17–20 mm.—Spp. 7, widespread throughout the American tropics, extending to n. Argentina and (adventively) to s. Florida, all represented in e. Brazil and 2 confined to that region.

This series is precariously distinguished in principle from ser. *Coriaceae*, similar in the strictly axillary peduncles, by the coincidence of diffuse herbaceous stems, slenderly stipitate petiolar glands and small flowers (petals 3.5–10 mm). However, some forms of *Ch. serpens* are large-flowered, and *Ch. tenuisepala* is both large-flowered and suffrutescent, thus different from the *Coriaceae* only in the slenderly stipitate gland. The series is composed of those elements of Bentham's *Cassia* ser. *Prostratae* and *Paucijugae* remaining after the transfer to ser. *Chamaecrista* of the African and American species with supra-axillary peduncles.

Key to the Species of ser. *Prostratae*

1. Stipules symmetrically to obliquely lanceolate or ovate, at base cordately auriculate, if at all, only on side opposed to the associated petiole.
   2. Blade of stipules cordate at base on neither side. Fl variable in size, the longest petal 5–20 mm, but where less than 8 mm the androecium 10-merous, the anthers all present even though those of the inner whorl smaller; style less than 3 mm only in 10-androus *Ch. serpens* var. *serpens*.
      3. Erect shrubs or shrublets of arid n.-e. Brazil, with stiff, eventually lignescent trunks clothed in furrowed bark and erectly virgate hornotinous branchlets, the whole at anthesis (3–)5–10 dm; lfts 3–5 pairs; seeds dark-castaneous or sooty. 25. *Ch. tenuisepala* (p. 707).
      3. Diffuse herbs, sometimes suffruticulose in age but not so in n.-e. Brazil, the stems in any case procumbent or weakly assurgent, if basally sublignescent in age the lfts mostly 5–9(–12) pairs; seeds usually pale brown, dark-spotted, with dark transverse line across distal end; widespread in N. and S. America. 26. *Ch. serpens* (p. 708).
   2. Blade of stipules cordately auriculate on side opposed to petiole. Fl always small, the longest petal 3.5–7 mm; androecium 5–10-merous, but only 3–5 anthers perfect; style 1–2.2 mm.
      4. Lfts 11–26 pairs; pedicels 10–26 mm. 27. *Ch. trichopoda* (p. 716).
      4. Lfts commonly 2–5 pairs, up to 11 pairs only in uncommon Brazilian var. *luxurians*; pedicels (20–)25–50 mm. 28. *Ch. pilosa* (p. 718).

1. Stipules subsymmetrically cordate-amplexicaul, at base auriculate on both sides. Lfts 3–6 pairs; androecium 3–9-merous, the inner whorl reduced or 0.
   5. Fl relatively large and longistylous, the longest petal 8.5–10 mm, the style 5.5–9.5 mm. 29. *Ch. cordistipula* (p. 721).
   5. Fl small, the longest petal 3.5–5 mm, the style 1–2 mm.
      6. Lfts (3–)4–5, of most adult lvs exactly 5 pairs; pedicels 2.5–6.5 mm. 30. *Ch. supplex* (p. 722).
      6. Lfts of all lvs exactly 3 pairs; pedicels 5–30 mm. 31. *Ch. kunthiana* (p. 724).

25. Chamaecrista tenuisepala (Bentham) Irwin & Barneby, comb. nov. *Cassia tenuisepala* Bentham in Martius, Fl. Bras. **15**(2): 164. 1870.—"Habitat in campis alpestribus vel in pascuis convallium prope Ouro Preto et in deserto ad Formigas, prov. Minas Geraës: Martius; in collibus saxosis prope Oeiras prov. Piauhy, frequens: Gardner n. 2125 . . ."—Lectoholotypus, *Gardner 2125,* collected in fl and fr in III.1839, K (hb. Hook.)! = NY Neg. *1525*; isotypi, B = F Neg. *1756,* G, GH, K, NY (hb. Meissn.), P!—The paratypi, *Martius s.n.* (M!), supposedly from Formigas, VIII.1818 and Vila Rica, IV.1818, all at the same stage of anthesis and probably representing one gathering, are very likely mislabelled and possibly collected by Martius at or near Oeiras, Piauí in V.1919.

Cassia tenuisepala sensu Bentham, 1871, p. 571, exclus. loc. paraguayensi.

Small shrubs and subshrubs with gnarled woody roots, slender but stiff, fuscous, longitudinally furrowed trunks and erect virgate, often distantly and sparsely leafy annotinous branchlets, at anthesis (3–)5–10 dm, except for often ventrally glabrous lfts thinly pilosulous nearly throughout with incurved hairs up to 0.25–0.6 mm and the stems, lf-stalks, stipule margins and pedicels (or some of these) in addition often finely pilose with spreading or loosely incurved setae up to 0.8–1.2 mm, the foliage dull olivaceous concolorous, the leaflets varying from glabrous ciliolate to pilosulous on lower or both faces.

Stipules erect appressed, narrowly ovate- or lance-acuminate, (2.5–)3–6.5 mm, at oblique base 0.7–1.7 mm wide, from base coarsely 5–9-nerved, the blades either glabrous or puberulent dorsally.

Lvs spreading-ascending 10–26 mm, short-petioled, the expanded blade ovate or flabellate in outline; petiole including small pulvinus 1.5–3(–3.5) mm, coarsely thick-margined, above middle 0.5–0.8 mm diam, the margin like that of each segment of rachis dilated upward; petiolar gland solitary, situated immediately below first pair of lfts, urceolate slender-stipitate, the circular red or atropurpureous head 0.2–0.45 mm diam, the stipe 0.5–1.6 mm; rachis (1–)2.5–8(–10) mm, shorter than the longer lfts; lfts (2–)3–5 pairs, in outline linear-oblong to narrowly oblong-elliptic or (especially the distal pair) narrowly obliquely obovate, obtuse mucronulate or aristulate, (5–)7–14 × 2–5.5 mm, at base cordate on proximal and cuneate to rounded on narrow side, from base (4–)5–7-nerved by the straight or gently retrofalcate, moderately excentric midrib dividing the blade 1:(1.5–)2(–2.4) with, on narrow side 1–2, on broad side (3–)4–5 incurved-ascending primary veins, the midrib thence penniveined with (2–)3–5 pairs of major and sometimes a few faint intercalary minor secondary venules, the venation prominulous on both faces, sometimes more sharply so beneath.

Peduncles obsolete 1-fld; pedicels appearing axillary, at anthesis (7–)13–23 mm, in fruit either ascending or divaricate (11–)15–26 mm, bracteolate 1.5–5.5 mm below calyx; bracteoles submembranous lanceolate 1.2–2.7 mm; buds ovoid-acute, sparsely setose; sepals submembranous pallid or orange-tinged, ovate and lance-elliptic acute 5.5–10.5 × 2–4.3 mm; petals yellow, sometimes with red flare at claw, 4 subhomomorphic except of different lengths, either oblanceolate or cuneate-obovate, narrowed to a slender claw, 7–14 × 3.5–6.5 mm, the cucullus conspicuously longer than the rest, reniform-falcate, 11–17 × 6–9.5 mm; stamens 10, the long anthers 5.5–7.7 mm; ovary minutely strigulose; style linear, straight or almost so, 6.5–10 × 0.3–0.35 mm; ovules 9–12.

Pod narrowly oblong, 30–42 × 5–7 mm, the stramineous or reddish-brown valves minutely puberulent; seeds (3.1–)3.4–4 mm, the testa dark chestnut- or sooty-brown, smooth but scarcely lustrous, faintly shallow-pitted.—Collections: 11.

Dry hilltops, stony knolls in chapada, caatinga thickets, open stony places in gallery forest, and on sandstone outcrops along streams, ±300–500 m, apparently local, known only from n.-e. Brazil, from the upper Parnaíba in s. Maranhão e. within ±7–9°S latitude through central Piauí to interior Paraíba and w. Pernambuco.—Fl. I–V.

Bentham (1871, p. 571) placed *Cassia tenuisepala* in his heterogeneous ser. *Paucijugae,* associating it directly with *C. wrightii,* which is herbaceous and North American, and *C. pascuorum* which, though similarly frutescent and Brazilian, differs importantly in the supra-axillary flowers. We presently interpret *C. wrightii* as a variety of *Chamaecrista serpens,* and believe that it is to this species, sensu lato, that *Ch. tenuisepala* is mostly nearly allied. It has essentially the leaves, petiolar glands, and flowers of *Ch. serpens* var. *grandiflora* from which the fruticose life-form alone seems to distinguish it. The difference is a pronounced one, however, for the stems of *C. serpens* die back annually to the root-crown or divisions of a caudex, and never develop a trunk or persistent woody framework of branches clothed in vertically splitting bark. We have noted however that some North American forms of *Ch. serpens* acquire, especially in desert or subdesert climate, a true multicipital lignescent caudex, and we suspect that *C. tenuisepala* represents a parallel, but more emphatic or advanced stage of the same trend, achieved under the stress of the seasonally dry climate of northeastern Brazil.

26. Chamaecrista serpens (Linnaeus) Greene, Pittonia **4**: 29. 1899. *Cassia serpens* Linnaeus, Syst. ed. 10, 1018. 1759.—Typus infra sub var. *serpenti* indicatur.

Slender herbs with several or many pliantly humifuse (exceptionally assurgent), simple or sparingly branched stems radiating from the crown of a blackish taproot or ultimately (especially in arid climate) a weakly suffruticulose caudex, commonly both puberulent with short incurved hairs and hispid with at least a few, sometimes with many fine divaricate septate setae up to (0.5–)0.8–1.4(–1.8) mm, the incurved hairs often aligned in vertical strips along stem and pedicel, the latter often concentrated on the petiole, but sometimes extending to stipule- and lft-margins, the stems, and dorsal face of lfts (or to some of these), but sometimes wholly lacking, the pod often likewise both puberulent and pilose with yet longer, ± flexuous filamentous hairs, but sometimes only puberulent, the usually thin-textured lfts either concolorous dull olivaceous or olivaceous above and cinnamon-brown beneath.

Stipules erect, subappressed to stem, narrowly lanceolate to narrowly ovate-acuminate or -caudate, (1.5–)2–7 × (0.2–)0.3–1.2 mm, the submembranous, persistent blades 3–7(–9)-nerved from pulvinule, at base slightly oblique or (var. *oaxacana*) incipiently semicordate.

Lvs divaricate or ascending, commonly displayed horizontal to the meridian (7–)10–35(–42) mm, short-petiolate, the expanded blade ovate in outline; petiole (including wrinkled pulvinus) 1.5–4 mm, narrowly sulcate ventrally, in profile 0.2–0.8 mm wide; gland situated near middle of petiole proper, discoid or cupulate, usually reddish, 0.2–0.5 mm diam, elevated on a stipe 0.2–1.1(–1.4) mm, sometimes accompanied by a second smaller petiolar gland, and exceptionally by 1–2 yet smaller on rachis; rachis (3–)5–26(–30) mm; lfts 3–9(–12), of some adult lvs at least 4 pairs, the middle usually slightly largest, in outline narrowly oblong to oblong-oblanceolate, dimidiately narrow-obovate, or obliquely obovate, obtuse or deltately acute, mucronulate or aristate by excurrent costa, 3–12(–14) × 0.4–6.3 mm, at base auriculate on proximal side, straight or slightly falcate, the submembranous or less often firmly chartaceous blades commonly glabrous on both faces, sometimes setose beneath, the margin glabrous, minutely ciliolate, or erratically setose, the displaced midrib with, on its broader side, 2–3(–4) primary veins palmately ascending from pulvinule and on the narrow side 3–4 weak secondary pinnate venules all finely or (adult) coarsely sharply prominulous on both faces, sometimes when young subimmersed.

Peduncles axillary, up to 1(–1.5) mm, 1(–2)-fld; pedicels at anthesis ascending toward meridian, subfiliform or very slender, in fruit a little thickened and commonly divaricate or deflexed and geniculate under the pod, 1.2–4(–4.5) cm, glabrous or puberulent; bracteoles above middle of pedicel, lanceolate to narrowly ovate, 0.8–2.5 mm; buds ovoid-acuminate, glabrous or thinly setose; sepals membranous, pallid or greenish seldom tinged with red or yellow, the longer ones lanceolate or ovate-acuminate, 3–12(–13.5) mm; petals yellow, at anthesis expanded into a campanulate but highly asymmetrical perianth, the 3 adaxial ones smallest, oblanceolate to narrowly obovate, a little shorter to a little longer than the sepals, the 2 abaxial longer, one of them resembling the 3 adaxial in outline, the other broadly falcate-obovate, 5–19(–20) mm, its distal margin incurved but not concealing the androecium; stamens 10, the 5 oppositisepalous larger than the reduced inner whorl, the yellow anthers all subsessile, the longest 2.5–9.5 mm; style varying in length with petals and stamens, 2–10 mm; ovary often both strigulose and pilose with weak ascending hairs, but the latter sometimes few or 0; ovules 3–10(–12).

Pod linear-oblong, short-stipitate, 14–40(–45) × 3–5 mm, the pale green or reddish valves ripening stramineous, castaneous, or ultimately dark brown, at once incurved-puberulent and pilose with fine flexuous hairs to 1.5–3 mm, or the latter wanting; seeds compressed-pyriform or prismatic, 2.3–3(–3.3) mm, the testa dull, pale or dark brown, lineolate with crateriform pits usually appearing as small dark, pale-centered rings, mucilaginous when wetted.

In the foregoing description of *Ch. serpens* we have extended the concept of Bentham (1870, p. 162) so far as to admit two or three more pairs of leaflets to the average adult leaf and a corolla slightly larger than that of the already relatively large-flowered var. *grandiflora.* As redefined, the species accommodates at one extreme the small-flowered Jamaican plant to which the epithet was originally applied and on the other a series of forms progressively larger-flowered that culminates in the Mexican entities that have passed recently as *Cassia wrightii, C. palmeri* and segregates from these. In habit *Ch. serpens* resembles the related and partly sympatric *Ch. pilosa,* a usually more coarsely leafy and, in the common range, more hispid plant that is, however, most reliably distinguished by the pentandrous androecium. The also related *Ch. trichopoda,* which where sympatric with var. *serpens* is similar in flower size, will be readily recognized by its many (11+ not 3–8) pairs of leaflets.

Conspicuous variation within *Ch. serpens* follows two independent modes, one trichotic, the other floral. The double vesture of short incurved and long septate setose hairs characteristic of the group is present at least potentially in all forms of the species, but the proportions of the two sorts of hair, the relative and absolute density and length of each, and the extent to which setae extend outward from the petiole, where most prevalent, to the stem, stipules, leaflets and sepals are too often subject to erratic modification to serve as taxonomic characters. It is otherwise with flower-size, which appears generally to have strong correlation with geographic dispersal patterns, and is furthermore linked to some degree with minor traits of foliage and even pubescence. The large-flowered varieties appear to have arisen independently in North and South America, and have acquired the greatest substance near and outside the Tropic lines in southern Paraguay and adjoining Argentina (var. *grandiflora*), and in northwest Mexico (var. *wrightii*). Each is tied in to the matrix of small-flowered tropical forms by sympatric or vicariant types of intermediate flower-size which they resemble more closely in minor characters of foliage than they do races of comparable flower-size in the opposing hemisphere.

A unicate specimen from Yhú in southeastern Paraguay (*Hassler 9531,* G) strongly suggests a hybrid *Ch. serpens* × *rotundifolia,* the 3–4 pairs of leaflets being fewer and broader than elsewhere in *Ch. serpens,* but obviously different from the single broad pair of *Ch. rotundifolia*. The intermediate flower size might have been derived from combination of the typical variety of either species with the var. *grandiflora* of the other.

Key to the Varieties of *Ch. serpens*

1. Fl small, the asymmetric long petal 5–7 mm, the longer anthers to 2.5–4 mm, the style 2–3.6(–4) mm; s.-centr. Mexico (Michoacán, Guerrero, México, Oaxaca) and w. Cuba s. to Argentina. Lfts of larger lvs 4–7(–8) pairs, but these either glabrous or thinly setose. 26a. var. *serpens* (p. 711).
1. Fl larger, the asymmetric petal 9–19(–20) mm, the longer anthers to 4–9.5 mm, the style (4–)4.5–10 mm; discontinuously dispersed from s. Arizona to n. Argentina.
    2. Lfts of larger lvs 3–7(–8) pairs; s. Mexico (Puebla, Oaxaca); Central America (Honduras and Nicaragua); South America (middle Orinoco valley to Argentina, localized and discontinuous).
        3. Lfts of relatively narrow outline, the widest of a plant mostly less than 2.5 mm broad. Petiolar gland usually 1; either Mexico or S. America.
            4. Venulation of lfts coarsely sharply prominulous; stipules 1.5–2 mm wide, incipiently cordate basally on side further from petiole; s. Puebla and n. Oaxaca, Mexico, 1400–1700 m. 26f. var. *oaxacana* (p. 715).
            4. Venulation of lfts scarcely prominulous; stipules less than 1.5 mm wide, not cordate at base; S. America, lowland. 26b. var. *grandiflora* (p. 712).
        3. Lfts of broader outline, the widest of larger lvs (2.5–)2.7–6 mm wide. Petiolar gland 1–several; either lowland Oaxaca, or highland Central America.
            5. Glands on all or at least 2 distal interfoliolar segments of rachis; larger lfts 3.5–12 × 2.5–4.5 mm, more than twice as long as wide, delicately venulose; highland s.-w. Honduras and adj. Nicaragua, 600–1300 m. 26c. var. *mensarum* (p. 713).
            5. Gland solitary; larger lfts relatively very broad, 4–9 × 3–6 mm, only 1.5–2 times as long as wide, coarsely sharply venulose; s. Oaxaca below 200 m. 26g. var. *isthmogenes* (p. 715).
    2. Lfts of larger lvs 7–9(–12) pairs; Pacific slope of Sa. Madre Occidental from s. Arizona to Jalisco and e. through Sa. Madre del Sur to Guerrero; Chiapas.
        6. Fls of medium size, the long petal 10–12 mm, the long anthers to 4.5–5.5 mm, the style ±4 mm; local within the range of the next (Durango to s. México and Guerrero). 26d. var. *delicata* (p. 713).
        6. Fls large, the long petal 13–19(–20) mm, the long anthers to 6–9.5 mm, the style 6–10 mm; Arizona to Chiapas. 26e. var. *wrightii* (p. 714).

26a. Chamaecrista serpens (Linnaeus) Greene var. **serpens**. *Cassia serpens* Linnaeus, l.c. 1759 & Sp. Pl. ed. 2, 541 (descr. ampliat.). 1762, sens. strict.—"*Habitat in* Jamaica . . . Brown. Jam. 225 [=*Cassia* no. 15, P. Browne, Nat. Hist. Jam. 225. 1756: ". . . grows everywhere on the pastures of Jamaica, and creeps among the grasses."]."—Holotypus, acquired by Linnaeus from Patrick Brown in 1758, LINN *528/24*!—*Ophiocaulon serpens* (Linnaeus) Rafinesque, Sylva Tell. 129. 1838.

Cassia flexuosa Miller, Dict. ed. 8, *Cassia* no. 16. 1768.—"This grows naturally in Jamaica."—Based partly on *Senna occidentalis, foliis herbae mimosae, siliqua singulari etc.*, Sloane, Hist. Jam. **2**: 51. 1725.—". . . in sandy places of the *Savanna,* near the town of *St. Jago de la Vega.*"—Holotypus, labelled "Jamaica, Houston, 1730," BM! Non *C. flexuosa* Linnaeus, 1753.

Cassia prostrata Humboldt, Bonpland & Kunth, Nov. Gen. & Sp. **6**(folio): 365. 1824.—"Crescit juxta fluvium Orinoci."—Holotypus, P-HBK! isotypi, P (*hb. Bonpland.*)! = NY Neg. *9183,* B (*hb. Willd. 7953*)!

Stems usually very slender, glabrous, puberulent, or setose; lvs mostly 1–3 cm, with up to 5–7(–8) pairs of lfts, these varying from glabrous to thinly, rarely densely setose dorsally; fl small, the cucullus 5–7 mm, the longer anthers 2.5–4 mm, the style 2–3.6(–4) mm; pod usually both puberulent and weakly long-pilose, mostly 1.4–3.2 cm, 5–9-seeded.—Collections: 134.

Savannas, campos, roadsides and waste places, becoming weedy in parks, fallow fields and lawns, mostly on light or sandy soils, near sea level up to 800 m on the Brazilian Planalto, 1200 m in Central America, 1300 m in n. Venezuela, and 1600 m in Colombia, widely but discontinuously dispersed through the Neotropics: around the w. and s. periphery of the Caribbean (Jamaica, w. Cuba, Isthmus of Tehuantepec in Mexico, Belize, Honduras, Guatemala, Nicaragua, Colombia, and Venezuela n. of the Orinoco R.) extending w. to the Pacific slope in Nicaragua and n.-w. to coastal Guerrero, Mexico, in Colombia s. up the Magdalena Valley to Huila; disjunct on the margins of Guayana Highland in interior Guyana and adjacent Terr. do Roraima, Brazil; in s. hemisphere reappearing on Rio Tocantins near 4°30′–8°S in Pará and Maranhão; in Rio Grande do Norte; in s. Pernambuco and w. and coastal Bahia; over the s. Planalto from w. Minas Gerais to n. São Paulo and s. Mato Grosso; not recorded from Goiás, Brazil or from Paraguay but again locally frequent in Corrientes, Argentina; introduced in peninsular Florida (Hillsboro Co.); and in s. Mexico (Michoacán and s.-w. México) represented by forms intermediate in foliage to var. *delicata*.—Fl. nearly throughout the year, except where inhibited by seasonal drought.

The range of var. *serpens* is, on present evidence, puzzlingly discontinuous. The plants are, however, inconspicuous at best and easily passed over, and the herbarium record is probably not an accurate reflection of the full dispersal. On the other hand, the ability of *Ch. serpens* to thrive in disturbed places may well have encouraged its spread by artificial means.

Variation in pubescence is in this case clearly correlated with dispersal. In North America, and in South America s.-ward to Bahia, the leaflets are glabrous except for minutely ciliolate margins; but from Minas Gerais s.-ward setae become more abundant on the stems and petioles and extend outward, few or many, to the margins and sometimes to the dorsal face of some or all leaflets.

A notable collection from the lower Orinoco valley in Venezuela (Cd. Bolívar, *Holt & Gehriger 53,* US) suggests a hybrid between var. *serpens* and *Ch. pilosa* or possibly a distinct entity coincidentally sharing some aspects of the latter, for neither parent is known from the immediate region. This plant has the hispid stem and few (2–3) pairs of leaflets expected in *Ch. pilosa;* but the floral morphology, including ample oblique cucullus and decandrous androecium, are typical of var. *serpens*.

The protologue of *Cassia serpens* is somewhat ambiguous, Linnaeus having described seven pairs of leaflets and a pentandrous androecium. The latter character agrees with *Ch. pilosa,* also native to Jamaica, but not to the species traditionally known as *C. serpens,* which has always a full set of ten stamens. On the other hand the leaflets are only 4–5 pairs in Jamaican *Ch. pilosa*. Both species are represented in the Linnaean herbarium by specimens acquired from Patrick Browne, and by simply accepting these as the types the difficulties inherent in the description of *Ch. serpens* can be passed over.

26b. Chamaecrista serpens (Linnaeus) Greene var. **grandiflora** (Bentham) Irwin & Barneby, comb. nov. *Cassia serpens* Linnaeus var. *grandiflora* Bentham in Martius, Fl. Bras. **15**(2): 163. 1870.—". . . loco haud indicato Brasiliae: Weddell."—Lectoholotypus, *Weddell 2286,* from "rives de l'Araguay," P! presumed isotypi, *Weddell s.n.,* G(3 sheets)!

Cassia tenella Humboldt, Bonpland & Kunth, Nov. Gen. & Sp. **6**(folio): 365. 1824.—"Crescit in ripa fluminis Orinoci, prope San Borja [in present Edo. Bolívar at 6°N lat.]"—Holotypus, labelled "n. 1182. San Borja Orinocensium," P-HBK! isotypi, P (hb. Bonpland.) = F Neg. *39839*; B (*hb. Willd. 7954*)! *Chamaecrista tenella* (Humboldt, Bonpland & Kunth) Pittier, Cat. Fl. Venezol. 375. 1945.

Habit, stature, foliage, pod of var. *serpens,* but flower larger, the cucullus (8–)9–13(–16) mm, the longer anthers up to 5–8 mm, the style (4–)5–6.5(–7) mm.—Collections: 41.

River banks, sandy floodplains, savanna (n.-ward) and campo (s.-ward), mostly below 600 m, known from 3 widely separated areas in S. America: banks of the Orinoco about the common boundaries of Apure, Amazonas, and Bolívar, Venezuela (to be expected in n.-e. Vichada, Colombia); Araguaia and Tocantins valleys in Pará and centr. Goiás, Brazil; middle and lower Paraguay and Paraná valleys in s.-w. Mato Grosso, Brazil, s.-centr. Paraguay and adjoining Argentina (Formosa; about the Misiones-Corrientes boundary); ?n.-centr. Bolivia (Mamoré valley; cf. discussion infra).—Fl. at low latitudes nearly throughout the year, in Paraguay and Argentina mostly X–III, sometimes later.

The three disjunct major populations which we refer to var. *grandiflora* are essentially alike in corolla-size, and differ only minutely in other characters. The plants of the Orinoco valley (*C. tenella* H.B.K.) tend to have, at maturity, very long stems, up to 5–10 dm, and relatively long petioles (2.5–4 mm) incurved-puberulent but lacking setae. In the context of copious material now available from the banks of the middle Orinoco the typus of *C. tenella* is peculiar only for its diminutive stature, the stems about 12 cm long from a knotty root-crown. It is evidently a depauperate individual, and led Bentham (1871, p. 571) to maintain the species in his ser. *Paucijugae,* next to *C. schlimii,* artificially separated from its real affinity within ser. *Prostratae*. The short-stipitate glands, now known to occur in typical *C. serpens,* may well have exerted an influence on the misinterpretation. The plants of Argentina and Paraguay are more compact, with stems 1–4 dm and short petioles (1.2–2.5, rarely 3 mm) setose as var. *serpens*. The two known collections from Goiás are intermediate in stature and petiole-length. North of Capricorn the leaflets are uniformly glabrous; in Argentina and Paraguay either glabrous or thinly setose, the latter prevailing in Argentina and the former around Asunción. A fragmentary specimen (*O. Braun 2,* NY) from the lower Mamoré valley in dept. Beni, n.-centr. Bolivia, appears to represent a broad-leaflet form of var. *grandiflora* but lacks expanded flowers or fruits and is inadequate for analysis.

The discontinuous range of var. *grandiflora* lies almost within that of var. *serpens* but they appear nowhere to overlap, the former indeed fitting neatly into gaps in the dispersal pattern of the latter. Perhaps var. *grandiflora* consists not of a phylogenetically pure strain but is a composite of parallel variants, at present morphologically inseparable, that have arisen independently from var. *serpens* and replace it locally.

26c. Chamaecrista serpens (Linnaeus) Greene var. **mensarum** (Molina) Irwin & Barneby, stat. nov. *Cassia mensarum* Molina, Ceiba **1**: 255. 1951.—"HONDURAS: . . . Agua Amarilla, Dept. Morazán, alt. 1300 m, Oct 11, 1946, *Williams & Molina 10579* . . ."—Holotypus, olim in herb. Esc. Agric. nunc in US translatus! isotypi, A, F, GH, MEXU, MO! paratypi, *Standley 12814, 23246* etc., F, US!

Stems and foliage varying from densely incurved-puberulent to subglabrous, exceptionally also weakly setose, the diffusely trailing stems suffruticulose in age; lvs mostly 1–3.3 cm, with (3–)4–5(–6) pairs of narrowly oblong-obovate to obliquely obovate lfts 3.5–12 × (1.3–)2.5–4.3 mm; fls of var. *grandiflora*; pod puberulent, not weakly setose.—Collections: 17.

Open rocky hillsides, stony meadows, glades in pine-oak forest, sometimes trailing over roadside banks, locally abundant at 610–1300 m in the highlands of s.-w. Honduras (Morazán, El Paraíso) and extreme n.-w. Nicaragua (Volcán Viejo, Chinandega).—Fl. (VIII–)IX–XII.

Described as a relative of *C. serpens,* but different in its larger flower and pod and in the presence of glands between most or all pairs of leaflets in addition to that on the petiole. The flower is however, in effect, identical with that of var. *grandiflora,* and occasional glands are found on the leaf-stalk in this and other forms of *Ch. serpens*. The relatively wide leaflets, which vary around an obliquely obovate rather than narrowly oblong mean, contribute a recognizably distinct facies to var. *mensarum,* but we can find no solid basis for separating it from *C. serpens* at specific level. The leaflets are substantially broader and always fewer than in the races of var. *wrightii* geographically nearest in Chiapas, Mexico.

26d. Chamaecrista serpens (Linnaeus) Greene var. **delicata** (Britton & Rose) Irwin & Barneby, stat. nov. *Ch. delicata* Rose ex Britton & Rose, N. Amer. Fl. **23**(5): 277. 1930.—"Cerro Pedregosa [sic], Ocote (Michoacan or Guerrero [s. Guerrero]), 550 m. altitude, October 29, 1898, *E. Langlassé 544*."—Holotypus, US! isotypi, GH, K, NY (fragm), P (2 sheets)!

Habit and foliage of var. *wrightii,* but relatively small-fld; stems both incurved-pilosulous and weakly long-hispid, but lfts glabrous; cucullus ±10–11 mm; longest anthers 4.5–5.5 mm; style ±4 mm.—Collections: 4.

Grassy openings in the oak-belt, ±500–1460 m, apparently local within the range of var. *wrightii,* known only from Sa. Tres Picos, Durango, from distr. Temascaltepec, México, and from the type station in s. foothills of Sa. Madre del Sur (cf. McVaugh, Candollea **13**: 203. 1951).—Fl. IX–I.

While easily detached from var. *wrightii* by the technical character of its abruptly smaller flower and shorter style, var. *delicata* may be no more than a set of individual, independently derived, minor variants of the common Sierra Madrean variety of *Ch. serpens*. It differs from most of the Mexican populations of var. *serpens* in larger flowers combined with the foliage of var. *wrightii,* but as already mentioned under var. *serpens,* this foliage is encountered rarely, in Michoacán and México, with the yet smaller flower of the typical variety. These last, with var. *delicata,* forge indissoluble links between the largest- and smallest-flowered expressions of the *Ch. serpens* complex.

26e. Chamaecrista serpens (Linnaeus) Greene var. **wrightii** (Gray) Irwin & Barneby, stat. nov. *Cassia wrightii* Gray, Pl. Wright. **2**: 50. 1853.—"Hillside, on the Sonoita, near Deserted Rancho, Sonora [now Santa Cruz Co., Arizona]; Sept. (1034)."—Holotypus, *Wright 1034,* GH! isotypi, K, NY, US!—*Chamaecrista wrightii* (Gray) Wooton & Standley, Contrib. U.S. Nat. Herb. **19**: 335. 1915; Pennell, 1917, p. 342 (as comb. nov.).

Cassia palmeri S. Watson, Proc. Amer. Acad. **22**: 408. 1887.—"Rio Blanco [near Guadalajara, Jalisco]; June [1886]. (29.)."—Holotypus, GH! isotypi, G, LE, NY (2 sheets), US (2 sheets, one labelled "Barranca near Guadalajara")!—*Chamaecrista palmeri* (S. Watson) Greene, Pittonia **3**: 242. 1897; Britton & Rose, 1930, p. 242.

Chamaecrista mazatlensis Rose ex Britton & Rose, N. Amer. Fl. **23**(5): 283. 1930.—"Vicinity of Mazatlan, Sinaloa, March 31, 1910, *Rose, Standley & Russell 13824.*"—Holotypus, US! isotypus, NY!

Chamaecrista ortegae Britton ex Britton & Rose, l.c. 1910.—"Capule, Mazatlán, Sinaloa, 200 m. altitude, 1925, *J. G. Ortega 6088.*—Holotypus, US! clastotypus + photo, NY!

Chamaecrista mexiae Britton ex Britton & Rose, op. cit. 284. 1930.—"Between Yxtlan and Barranca del Oro, 1480 m. altitude, Nayarit, September 27, 1926, *Ynes Mexia 790.*"—Holotypus, US! isotypi, A, CAS, GH, NY!

Chamaecrista chiapensis Britton & Rose, l.c. 1930.—"Rocky soil, Monserrate, Chiapas, May, 1925, *Purpus 10592.*"—Holotypus, NY! isotypus, US!—*Cassia monserratensis* Lundell, Phytologia **1**: 215. 1937 (non *C. chiapensis* Standley, 1919).

Chamaecrista alamosensis Britton & Rose. l.c. 1930.—"Sierra de Alamos, Sonora, March 18, 1910, *Rose, Standley & Russell 13046.*"—Holotypus, US! isotypus, NY!

Cassia wrightii sensu Bentham, 1871, p. 571; Pollard, 1894, p. 217; Kearney & Peebles, 1951, p. 406. *Chamaecrista wrightii* sensu Britton & Rose, 1930, p. 286.

Stems mostly (1–)1.5–4 dm, in desert environment weakly suffrutescent basally; vesture highly variable, the stems and petioles thinly incurved-pilosulous or thinly to very densely hispid with erect or subretrorse septate setae; larger lvs 2–3.5(–4) cm; petiole (including pulvinus) 2–4 mm; gland 1(–2, and rarely 1 or more on rachis), 0.2–0.5 mm diam, the stipe (0.2–)0.3–1 mm; lfts of larger lvs (6–)7–10(–12) pairs, up to 5–10(–12) × 1.5–2.5(–3) mm, either olivaceous both sides or cinnamon-brown dorsally; fls large, the sepals up to 8–12(–13.5) × 3–5.5 mm, the cucullus 13–19(–20) mm long; longer anthers 6–9.5 mm; style 6–10 mm; pod mostly 2.5–3.5(–4) mm, 6–11(–12)-ovulate.—Collections: 53.

Dry hillsides, often in grassy places, extending from the coastal plain upward through short-tree forest into the pine-oak belt, 10–2100 m, locally plentiful along the Pacific slope of Sa. Madre Occidental from s. Arizona (Santa Cruz Co.) to centr. Jalisco and s.-e. locally and interruptedly to Sa. Madre del Sur in Guerrero, s.-w. México, and highland Chiapas, unknown as yet from Michoacán, Morelos, or Oaxaca.—Fl. (VI–)VII–I, exceptionally later.

Like var. *serpens* itself, the large-flowered var. *wrightii* varies greatly in density and proportions of the short incurved and long setose vesture and to considerable degree likewise, but independently, in the diameter of the petiolar gland and development of its stipe. The pubescence varies in a clinal mode, setae being absent or almost so (*C. wrightii* sens. str.) from Arizona south into Sinaloa (that is, within and around the periphery of the Sonoran Desert) and thence southward become increasingly long and many (*C. palmeri* sens. str.), culminating in a densely hirsutulous minor variant known from a few collections from s.-w. Durango (*Rose 2253, 3482*). Many collections of var. *wrightii* are characterized by a bright cinnamon-brown discoloration of the lower leaf-surface, but this sometimes appears erratically on only some leaves, or on only some plants of a population, and has no taxonomic value. Two of the three known collections from Sa. Madre del Sur differ slightly either in abnormally assurgent branches (*Hinton 10536*) or unusually many (–12) pairs of leaflets, but seem to represent no more than minor variants. The Chiapas populations, described by Britton & Rose as a species supposedly distinguished by the combination of strongly veined leaflets with submarginal costa, a short-stipitate gland, and relatively short pedicels, really do have prominently and closely venulose and narrow leaflets, but are otherwise matched at all points by distantly allopatric forms of var. *wrightii*. The types of *Ch. alamosensis* and *Ch. mazatlensis* were both collected in March, that is at the height of the dry season, and have lost their larger leaves in consequence; no structural differences from sympatric var. *wrightii* collected in the vigor of new growth can be detected. Despite the fact that Britton referred *Ch. mexiae* and *Ch. palmeri* to different groups, supposedly separable by leaf-texture, we can find no substantial difference between the types, collected at points in the río Santiago valley about 100 km apart in Jalisco and adjoining Nayarit respectively.

26f. Chamaecrista serpens (Linnaeus) Greene var. **oaxacana** Irwin & Barneby, var. nov., ab affini var. *grandiflora,* quoad florum magnitudinem comparabili, stipulis latioribus de latere a petiolo semoto basi subcordatis necnon foliolorum lamina grosse venulosa distincta.—MEXICO. Oaxaca, distr. Huajuapan: 6 km s.-w. of Huajuapan de León, 1400 m, 18.VIII.1976 (fl, fr), *M. Sousa 5767.*—Holotypus (2 sheets), MEXU; isotypus, NY.

Stems 1–2.5 dm from suffruticulose caudex; stipules (2–)3–5 × 1.5–2 mm, shallowly cordate on side further from petiole; lvs 12–20 mm; lfts 5–7 pairs, the larger ones obliquely oblong-elliptic 5–7 × 1.5–2.5 mm, 2.8–4.2 times as long as wide, sharply coarsely venulose on both faces; fls relatively large, the longest petal 12–14 mm, the longer anthers 4.5–6.5 mm; style 4–6.5 mm.—Collections: 7.

Open places in low deciduous brush-woodland, on red lateritic or calcareous soils, 1400–1750 m, narrowly localized in seasonally arid hill-country of n.-w. Oaxaca (distr. Huajuapan) and immediately adjacent Puebla (mun. Chila), Mexico.—Fl. VI–VIII.

This variety and the next following, both highly localized in southern Mexico, seem to have arisen as locally modified races of the more widely dispersed var. *wrightii,* from which they differ collectively in the fewer pairs of leaflets. In leaflet number they resemble var. *grandiflora,* but differ in the coarsely prominent venulation and remotely allopatric distribution. The relatively broad stipules of var. *oaxacana,* which may be distinctly cordate at base on the side further from the associated petiole, and the proportionately narrow leaflets distinguish it from lowland maritime var. *isthmogenes*.

26g. Chamaecrista serpens (Linnaeus) Greene var. **isthmogenes** Irwin & Barneby, var. nov., a var. *grandiflora* et var. *oaxacana* necnon ab aliis speciei varietatibus grandifloris imprimis foliolis late obovatis, majoribus ±1.5–2-plo tantum longioribus quam latioribus, diversa, adspectu ideo idiosyncratica patriaque peculiari recognoscenda.—MEXICO. Oaxaca, distr. Tehuantepec: 2 km w. of Salina Cruz, 10 m, 19.IX.78 (fl), *M. Sousa 9499.*—Holotypus, MEXU.

Stems prostrate from suffruticulose caudex, 1–2 dm, together with foliage either pilosulous and setose or only thinly setose; stipules 2.5–6 × 0.4–1(–1.4) mm; lvs 10–22 mm; lfts 3–5(–6) pairs, the larger ones obliquely obovate very obtuse 4–9 × 3–6.3 mm, 1.5–2 times as long as wide, sharply coarsely venulose on both faces; fls relatively large, the long petal 11–15 mm, the longer anthers 4–6 mm; style 5–7 mm.—Collections: 6.

Rocky hillsides overlooking the Pacific Ocean, in open places in low deciduous brush-woodland, 5–140 m, known only from the coastal hills 1–5 km w.-ward from Salina Cruz in distr. Tehuantepec and from one station ±160 km distant along the immediate coast of distr. Pochutla (Barra de Sta. Elena, a glabrate form), Oaxaca.—Fl. IX–XII.

The var. *isthmogenes* differs from all other forms of *Ch. serpens* in its broad veiny leaflets which endow it with a strongly individual facies. Near Salina Cruz it is a member of a xeromorphic coastal scrub flora which includes several locally endemic Leguminosae, among them *Mimosa sousae* Grether, *Acacia deamii* Standl. and *Dalea sousae* Barneby.

27. Chamaecrista trichopoda (Bentham) Britton & Rose ex Britton & Killip, Ann. N.Y. Acad. Sci. **35**: 185. 1936. *Cassia trichopoda* Bentham in Martius, Fl. Bras. **15**(2): 163. 1870.—". . . in campis prov. Minas Geraës et S. Paulo: Weddell, Burchell, Regnell coll. III. n. 487., Warming; in prov. Goyaz prope Curralinho: Pohl; ad Porto Imperial ad flum. Tocantins et prope Goyaz: Burchell . . ."—Lectoholotypus, *Pohl 1395,* collected probably in III.1819, at Curralinho, Goiás, K (hb. Benth.)! = NY Neg. *1522*; isotypus, NY! paratypi, *Weddell 1860, Burchell 6774, 8679, Regnell III/487,* all K!

Cassia trichopoda sensu Bentham, 1871, p. 570.

Slender, precociously flowering perennial herbs of short duration, in early anthesis appearing annual and perhaps occasionally monocarpic, with (1–)several–many, in open places humifuse or procumbently radiating, in competitive situations assurgent, sarmentose or rarely erect stems from blackish roots, becoming (1–)1.5–11 dm, incurved-puberulent or -pilosulous nearly throughout with short incurved hairs and, especially the stems, stipule-margins and lf-stalks also setose or setulose with coarser spreading-incurved, sometimes yellowish trichomes up to 0.6–2.2 mm, the pedicels pilose with extremely fine flexuous lustrous setae up to 1.2–2.7(–3.2) mm, the olivaceous foliage concolorous, the sensitive lfts glabrous above, often thinly pilosulous especially along midrib beneath.

Stipules erect, thinly herbaceous, persistent, heteromorphic, the lowest, subtending small early lvs, narrowly ovate ±1.5–4 mm, 3–7-nerved, those subtending adult lvs and fls longer, obliquely lance-acuminate, sometimes falcately arched outward, 5–14 mm, at semiamplexicaul, shallowly semi-cordate base 0.8–2.8(–3.8) mm wide, cuspidate at apex, from base palmately (5–)7–15-nerved, commonly glabrous dorsally but setulose-ciliate.

Lvs ascending, very shortly petioled, the earliest much shorter and simpler than the rest, not further noticed, the fully developed ones, especially those associated with fls, (1.5–)2–4.8(–6.5) cm, the expanded blades lance-ovate in outline, the lfts slightly decrescent both up- and downward from ± the 4–6th pair; petiole 1.5–3 mm, narrowly margined and like the rachis shallowly open-sulcate; gland 1(–2) situated near middle of petiole proper, saucer- or cup-shaped 0.35–0.8(–0.9) mm diam, reddish or pallid, elevated on a stout, linear or upwardly thickened stipe 0.4–1 mm; rachis mostly 1.7–4.3(–5.5) cm; lfts of largest lvs to 11–22(–26) pairs, crowded (±1 mm apart) along rachis, in outline obliquely linear-oblong or incipiently linear-oblanceolate, straight or subfalcate, the largest of any plant (4–)5–9.5(–11.5) × 0.8–2 mm, at apex abruptly rounded and aristulate by the excurrent costa, at base angulately shallow-cordate on proximal side, the blades submembranous, the highly excentric midrib weakly penninerved with, on the broad side, 3–5 short accessory nerves from the pulvinule, the venation prominulous beneath, faintly so above.

Peduncles from almost all axils above immediate base of stem, subobsolete to 4.5 mm, exactly axillary or adnate to stem through 0.5–2.5 mm, 1–3-fld; pedicels at anthesis filiform, in fruit wiry, usually divaricate or refracted and abruptly incurved under the pod, (9–)11–22(–26) mm, shorter than lf, bracteolate immediately below calyx; bracteoles linear-lanceolate, 0.8–1.6 mm; buds ovoid-acuminate, at least thinly setulose; sepals submembranous, greenish or reddish-tinged, lance- or narrowly ovate-acuminate, (3.6–)4–5.2 mm, widely spreading or refracted at full anthesis; petals yellow, never widely opening and fading by noon of first day, highly heteromorphic: vexillary and 1 alar petal small, obovate, 2.6–4.2 mm, usually shorter than sepals; 1 alar and 1 carinal petal obovate but larger, the carinal conduplicate and enclosing the divaricate gynoecium and its opposed large anther; the cucullus very obliquely obovate, 4.2–6.5 mm, usually as long as or a little longer than sepals, rarely shorter; stamens usually 10, heteromorphic, the inner whorl and 2–3 of the outer whorl depauperate or sterile, with relatively long filament and small oblong anther, one or more of the inner whorl sometimes lacking, the 3 abaxial of the outer whorl, or only the central abaxial and that next to it but opposed to the cucullus, normally developed, with very short filament and linear-lanceolate anther 2.4–3.6 mm; ovary gray-strigulose; style linear, 1.1–1.6(–1.8) mm; ovules 7–16.

Pod linear-oblong, (13–)16–36 × 2.5–3.6 mm, the green or violaceous valves turning castaneous and ultimately blackish, minutely strigulose or incurved-pilosulous; seeds quadrate or quadrately oblong in outline, 1.7–2.4(–2.7) mm, the dull, ochraceous or pale brown testa sublustrously pitted, the transparent outermost coat in age peeling in scurfy flakes, mucilaginous when wetted.—Collections: 78.

Campo, cerrado, and becoming weedy along highways, in pastures, and in cut-over forest, usually a diffuse herb of open sunny places but becoming erect in scrub or in partial shade of gallery-margin, 250–1150 m (on Chapada dos Veadeiros up to 1300) m, common and widespread over the Brazilian Planalto in centr. and s. Goiás, s.-w. Minas Gerais and adjacent São Paulo, and n.-e. Mato Grosso, in Minas extending e. to s. Sa. do Espinhaço, n., becoming rarer, up the Tocantins and Araguaia valleys just into s.-w. Pará and s. Maranhão; widely disjunct, possibly introduced, in the Andean foothills of n.-w. Bolivia (near Tumpasa, La Paz), in n. Venezuela (D.F.) and n. Colombia (foothills of Sa. Nevada de Sta. Marta); reportedly collected by Aublet in French Guiana (Bentham, 1870) but not known from there in modern times.—Fl. in Brazil mostly X–V, but moisture permitting almost throughout the year.

A modest, weedy chamaecrista, resembling in habit *Ch. serpens* and *Ch. pilosa,* but readily distinguished from the latter and from the former's typical form, each similarly small-flowered, by the many (to 11–26) pairs of leaflets in all mature leaves. The androecia of these three species show three stages of reduction; that of *Ch. serpens* decandrous, with the inner whorl of five anthers smaller than the outer, yet fertile; *Ch. trichopoda* usually decandrous but with only two or three of the abaxial members of the outer whorl perfect, the rest (1–2 sometimes completely suppressed) short and sterile; and *Ch. pilosa* pentandrous. The strongest anther of *Ch. trichopoda* is always that one immediately exterior to the excentrically displaced gynoecium, and together with it is enclosed by the conduplicate larger wing-petal. It seems likely that the flower is self-fertilized for the tips of the style and the associated anther lie close together immediately before anthesis and the ovary has often begun to expand by the time the perianth opens.

The species appears less variable than most chamaecristas of similar range and habitat. The slenderest plants have generally the fewest pairs of leaflets and relatively short, few-ovulate pods, but we cannot separate these taxonomically on any but arbitrary grounds.

28. Chamaecrista pilosa (Linnaeus) Greene, Pittonia **4**: 28. 1899. *Cassia pilosa* Linnaeus, Syst. ed. 10, 1017. 1759.—Typus infra sub var. *pilosa* indicatur.

Precociously flowering perennial herbs at length weakly suffrutescent but of short duration, with 3–several simple or distally much-branched stems (1–)1.5–13 dm procumbently radiating, assurgent, or less often suberect from the crown of a deeply penetrating blackish taproot, the stems, margins of stipules and sometimes of lfts pilose with fine divaricate lustrous septate setae up to (1–)1.5–4.5(–5) mm and the stems often in addition, either all around or in lines, puberulent with short incurved hairs, these exceptionally extending to the dorsal face of the usually ample, thin-textured ciliolate lfts, the foliage pale green concolorous, the stramineous wiry pedicels smooth and glabrous.

Stipules erect, appressed to stem, submembranous, disregarding some small narrower ones at base of stems all obliquely lance- to semi-ovate, -acuminate or -caudate, stiffly mucronulate or weakly aristate, (5–)6–16 mm, at semicordate base 1.5–4 mm wide, palmately 7–13-nerved from pulvinule, the nerves all sharply prominulous beneath, the margins ciliolate and often also ± setose.

Lvs (1–)1.5–4.5(–5) cm, shortly petioled, the larger cauline mostly at least 2 (but on basal shoots regenerating after browsing or late on some vigorous stems only 1–2) cm, all shortly petiolate, the expanded blade commonly ovate (in a rare var. lanceolate) in outline, the lfts of subequal size or commonly shorter and progressively more sharply pitched forward distally, the last pair strictly ascending from tip of rachis and often broadest; petiole 2–6 mm, narrowly margined and shallowly open-sulcate; gland (sometimes 2 in var. *luxurians*) situated either below or above middle of petiole proper, shallowly cup-shaped, usually thin-rimmed, greenish, livid or reddish, 0.1–0.25 mm diam, elevated on a very slender stipe 0.1–2.7 (commonly 0.1–0.35) mm, in var. *pilosa* often either deciduous or obsolete; lfts mostly 2–5, in var. *luxurians* to 11 pairs, oblong-oblanceolate or the distal pair semi-obovate, all obtuse but abruptly apiculate and mucronulate, 7–24 × 2–8(–9.5) mm, at base broadly obtusely auriculate-semi-cordate on proximal, cuneate on distal side, the moderately excentric penninerved midrib with, on broad side of blade, 2–4(–5) incurved ascending primary veins all sharply prominulous on dorsal, finely so on ventral face.

Peduncles subobsolete to 2.5 mm, exactly axillary or adnate to stem through 0.5–1.5(–2.5) mm, commonly 1–2-, rarely 3–6-fld; pedicels ascending, straight or almost so, at anthesis filiform, in fruit wiry stramineous (2–)2.5–5 cm, the longest at any one node usually as long as or surpassing the lf, bracteolate far above middle; bracteoles membranous lance-subulate 1–3-nerved 0.9–3.3 mm; buds slenderly ovoid-acuminate, at least thinly setose; sepals membranous, greenish-yellow or reddish, ovate- or narrowly lance-acuminate, 3.4–5.7(–7) mm; petals yellow, early fading and discolored (pinkish-orange), 4 obovate-cuneate to oblong or elliptic-oblanceolate, ± unequal (2 commonly shorter), the longest up to 3.5–6.5(–7) mm and either a little longer or shorter than the sepals, 1 lateral one folded to enclose the divaricate gynoecium and one large anther, the cucullus scarcely differentiated, obliquely obovate (4–)4.5–7 mm; stamens of outer whorl perfect, their lanceolate yellow or reddish anthers up to 3.2–4.5 mm, the inner whorl represented by 2–3 adaxial, greatly reduced or rudimentary members, sometimes obsolete; ovary densely gray-strigulose; style cylindric, stoutish, incurved, (1–)1.3–2.2 mm; ovules (6–)8–17.

Pod linear in outline, straight or commonly a little decurved, (1.7–)2.3–4.7 × 0.3–0.4 cm, the pale green valves becoming stramineous and finally nigrescent, finely incurved-pilosulous, never setose, the persistent style-base short stout, apically dilated, a trifle declined; seeds quadrately oblong-pyriform, 2–2.7(–3) mm, the testa pale ochraceous or brownish with fuscous transverse line at truncate distal end, dull rugulose, faintly pitted, mucilaginous when wetted.

A chamaecrista of weedy aspect and often weedy habits, notable for the small size of the long-pedicellate flower relative to the usually ample, thin-textured foliage, for the lustrously hispid stems, and a flower which superficially resembles that of the partly sympatric *Ch. serpens* var. *serpens* but is in reality different in the always incomplete, if partly present wholly staminodal inner whorl of stamens, the scarcely differentiated cucullus, and the usually shorter style. The var. *serpens* differs further in its lanceolate, not semi-cordate stipules, and very commonly but not quite reliably in shorter (mostly under not over 1.5 mm) setiform vesture of the stems, on the average smaller and slightly more numerous (5–8, not 3–5) pairs of leaflets, and shorter, commonly 5–9- not 8–17-seeded pod clothed in part with long weak setae, not simply pilosulous. However, starveling individual plants of *Ch. pilosa,* and some maimed ones regenerating after cutting, resemble var. *serpens* in the small size (but not number) of leaflets and short, few-seeded pod; these retain the floral morphology intact and usually some distinctly semicordate stipules. A particular, possibly intermediate form from Venezuela is mentioned further under var. *serpens*. The relationship of *Ch. pilosa* to the largely allopatric *Ch. trichopoda,* always easily distinguished from var. *pilosa* by the many (mostly 11–22) pairs of leaflets, is discussed under *Ch. pilosa* var. *luxurians* below.

In its extensive but interrupted range north of the Equator *Ch. pilosa* is, disregarding the minor variation already mentioned, essentially monomorphic and easily distinguished. Remotely distant in southeastern Brazil the species is represented by a variety instantly recognized by leaflet number and great development of the petiolar gland.

Key to the Varieties of *Ch. pilosa*

1. Lfts 2–5 pairs, often relatively large, the longest of the major lvs mostly 12+ mm, the expanded lf-blade ovate in outline; petiolar gland either obsolete or minute, elevated on a stipe 0.1–0.35(–0.5) mm; widespread circum-Caribbean, to lower Orinoco valley in Venezuela, the Pacific slope in central America, and introduced in Florida. 28a. var. *pilosa* (p. 719).
1. Lfts (5–)6–11 pairs, relatively short, 7–12 mm, the expanded blade in consequence lanceolate in outline; petiolar gland (often 2) long-stipitate, the stipe 0.8–2.7 mm; n.-e. Brazil. 28b. var. *luxurians* (p. 720).

28a. Chamaecrista pilosa (Linnaeus) Greene var. **pilosa**. *Cassia pilosa* Linnaeus 1759, l.c. & Sp. Pl. ed. 2, 540, descr. ampliat. 1762, sens. str.—". . . Brown. jam. 224 [=*Cassia* no. 12, P. Browne, Civ. Nat. Hist. Jamaica

224. 1756—". . . grows in many parts of the island . . ."]. *Habitat in* Jamaica."—Holotypus, identified as *C. pilosa* by Solander, LINN *528/18*!—*Disterepta pilosa* (Linnaeus) Rafinesque, Sylv. Tell. 126. 1838.

Cassia emarginata Miller, Gard. Dict. ed. 8, *Cassia* no. 13. 1768.—"This grows plentifully in Jamaica."—Holotypus, presumably a plant cultivated at Chelsea from seed received from Houston, BM! = NY Neg. *112*.—*C. milleri* Colladon, Hist. Casses 132. 1816, non *C. emarginata* Linnaeus, 1753.—Equated with *C. pilosa* by Bentham, 1871, p. 570.

Chamaecrista macropoda Standley in Contrib. Nat. Herb. **17**: 431. 1914.—"Type [US] no. 258959, collected on the Cerro Redondo, Department of Santa Rosa, Guatemala, . . . in September, 1893, by Heyde and Lux (J. D. Smith, no. 6133)."—Holotypus, US! isotypi, GH, US (hb. Donn. Sm.)!—*Cassia macropoda* (Standley) Standley, Field Mus., Bot. **4**: 213. 1929.—Equated with *Chamaecrista pilosa* by Britton & Rose, 1930, p. 277.

Cassia pilosa sensu Colladon, 1816, p. 122, t. 20, fig. A; Bentham, 1871, p. 570.

Lvs (1–)1.5–4.5(–5) cm; lfts 2–5, in larger lvs at least 3 and commonly 4–5 pairs, oblong-oblanceolate or the distal pair semi-obovate, 7–24 × 2.5–9(–9.5) mm, the larger ones of any plant at least (10–)12 × 3.5 mm; petiolar gland 0–1, the slender, fragile stipe 0.1–0.35(–0.5) mm.—Collections: 76.—Fig. 37 (androecium).

Savannas, dunes, waste places, waysides, sometimes a weed in orchards or sandy pastures, mostly 0–500 m, but ascending in Jamaica to 750, in Guatemala to 1350 m, interruptedly circum-Caribbean: w. Cuba (Matanzas, La Habana, Pinar del Río, Isla de Pinos), Jamaica, Hispaniola (rare, Dajabón in Rep. Dominicana), n. Colombia (Magdalena, La Guajira), n. Venezuela (both slopes of Cordillera Costanera from Carabobo to Guárico, weedy at Pto. Ordaz on lower Orinoco) and Trinidad; very local and perhaps only a casual weed of cultivated land in Central America on Pacific slope of Mexico (Chiapas), Guatemala (Santa Rosa) and El Salvador, and in Guatemala once found on the Usamacinta slope (El Quiché); weedy in s. peninsular Florida (Palm Beach & Highland cos.).

28b. Chamaecrista pilosa Linnaeus var. **luxurians** (Bentham) Irwin & Barneby, comb. nov. *Cassia trichopoda* var. *luxurians* Bentham in Martius, Fl. Bras. **15**(2): 163. 1870.—". . . ad Porto Imperial cum forma typica: Burchell."—Lectoholotypus, *Burchell 8747,* K (hb. Hook.)! = IPA Neg. *1102* = NY Neg. *9187*; isotypi, GH, LE! paratypi (*Burchell 8664*), GH, K, LE, P!

Lvs (2–)2.5–5.3 cm; lfts (5–)6–12 pairs, oblong-oblanceolate 7–15 × 2–4.5 mm; petiolar glands often 2, the larger elevated on slender stipe 0.8–2.7 mm.—Collections: 6.

Fields and waste places, known from scattered stations in e. Brazil: on middle Tocantins in n.-centr. Goiás (Pôrto Nacional) and adjacent Maranhão (Carolina); Ceará (Fortaleza); and on e. slope of Espigão Mestre in s.-w. Bahia.—Fl. II–V.

While clearly distinct in the form of the leaves and long-stipitate petiolar gland, var. *luxurians* preserves precisely the habit, the lustrous setose cauline vesture, the long glabrous pedicels, the almost regular small perianth with reduced androecium, and the long, many-seeded pod of *Ch. pilosa*. With more collections, its rank may require reappraisal, but its relationship can hardly be in question. The prostrate chamaecrista to be expected in or near its range is *Ch. trichopoda,* similar to *Ch. pilosa* in general habit except usually more slender, and different from var. *luxurians* in the yet more numerous (11–22, not 6–12) pairs of leaflets, in the shorter (mostly 11–22, not 25–50 mm), finely pilose pedicels, and in a perianth equally small but notably dissimilar in the obliquely enlarged cucullus and an androecium of 10 stamens, usually 3 large and functional and the rest sterile, as opposed to five exterior functional with 0–3 interior staminodal members.

29. Chamaecrista cordistipula (Martius) Irwin & Barneby, comb. nov. *Cassia cordistipula* Martius, Flora (Regensburg) **21**, Beibl. II(4): 59. 1838.—"[Herb. Mart.] 223. [Mato Grosso:] in inundatis et paludosis prope Cujabá . . . [*Silva Manso*] No. 219."—Holotypus, *Herb. Mart. 223,* M! = F Neg. *6229* = NY Neg. *9204*; isotypi, FI (hb. Webb.), G, GH, K = IPA Neg. *1105* = NY Neg. *1524,* LE, MO, P!

Cassia schlimii Bentham, Trans. Linn. Soc. London **27**: 571. 1871.—"Columbia, near Ocaña, *Schlim,* n. 281."—Holotypus, K (hb. Hook.)! = NY Neg. *1565*; isotypi, BM, BR, G = F Neg. *28019,* MPU, NY (hb. Meisn.), P, US (fragm)!—*Chamaecrista schlimii* (Bentham) Britton & Rose ex Britton & Killip, Ann. N.Y. Acad. Sci. **35**: 184. 1936.
Cassia cordistipula sensu Bentham, 1870, p. 164; 1871, p. 571.
Cassia psammophila Bentham, 1871, l.c., pro syn.—Spm. authent.: *Riedel 808,* A, GH, LE, US!
Cassia serpens var. *grandiflora* sensu Hoehne, Comm. Linh. Tel. Estrat. de Matto Grosso ao Amazonas 45(5) Bot. **8**: 49, t. 145. 1919; non Benth.

Slender, humifusely matted perennial herbs with several simple or randomly branching pliant stems 1–6 dm radiating from the crown of a deeply penetrating blackish taproot and (in age) a weakly developed caudex, sometimes adventitiously rooting toward the base, very variably pubescent with short incumbent hairs and longer spreading or widely spreading-incurved fine septate setae to 0.4–1.2 mm, either type of hair sometimes predominating or subobsolete, but some setae usually persisting on stipule- and lft-margins and sepals, the foliage olivaceous or when dry brunnescent, concolorous, the lfts either glabrous or thinly setose beneath, glabrous above.

Stipules erect persistent, subsymmetrically cordate-acuminate or -caudate from amplexicaul base, (2–)2.5–7 × (1–)1.5–3(–3.7) mm, prominently 7–13-nerved from base, the nerves incurved distally into a strong marginal one, the blades between nerves submembranous, in age subhyaline.

Lvs obliquely ascending toward the meridian or appressed to ground, 1–2.3 cm, petioled, the expanded blade broadly ovate in outline; petiole 2–5.5 mm, rather coarsely but narrowly margined, shallowly open-sulcate; gland situated immediately below first pair, shallowly cup-shaped, reddish, 0.2–0.35 mm diam, elevated on a filiform or narrowly claviform stipe 0.2–0.45 mm; rachis 5–12 mm; lfts 3–6, in some lvs at least 4 pairs, the first and last pairs smallest, the first divaricate from rachis and the rest pitched forward at increasingly narrow angles, the last ascending from tip of rachis, all in outline obliquely narrowly oblong to oblanceolate or semi-obovate, mostly a little curved forward, 3–11 × 1–3.3(–4) mm, at apex abruptly deltate-acute or -acuminulate, minutely mucronulate, at base semicordate on proximal side, the blades submembranous, the moderately excentric midrib few-branched distally, accompanied from pulvinule on the broader side by 1 strong and 1–3 weaker and shorter exterior veins, on the narrower side often by 1 weak submarginal one, the venation prominent beneath, less so above.

Peduncles exactly axillary, to 1 mm, 1-fld; pedicels (3–)6–19 mm, bracteolate near or well above middle, at anthesis very slender, in fruit firmer and often sigmoid; bracteoles narrowly ovate-acuminate to lance-caudate, (1.5–)2.5–6 mm, striate-nerved; buds slenderly ovoid-acuminate, pilosulous; sepals at anthesis widely spreading or deflexed, membranous, pallid or reddish, ovate or broadly lanceolate acute, to 7.7–9.5 mm; petals yellow, widely expanding in anthesis, 4 oblanceolate-spatulate, narrowed to a slender claw, unequal, 2 of them often smaller, the longest 8.5–10 mm, the cucullus dimidiately ovate 10–12 mm; stamens 5–9, the outer whorl complete but the 3 abaxial (yellow) anthers much larger, to 6.5–7.5 mm, the inner whorl either obsolete or of up to 4 members, but these sterile, the anthers shortly oblong; ovary densely white-strigose or -pilose; style filiform (5.5–)6–9.5 mm; ovules (3–)4–9.

Pod (little known) linear-oblong, 8–19 × 3–3.5 mm, the castaneous valves either minutely strigulose or loosely finely pilosulous; seeds (few seen) exactly of *Ch. supplex* except sometimes larger, to 2.8 mm.—Collections: 27.—Fig. 46.

Campo and savanna, mostly below 350 m, bicentric: Guajira Peninsula and lower Magdalena valley in n. Colombia (La Guajira, Magdalena, Bolívar, Norte de Santander) e. into Maracaibo Basin in Zulia, Venezuela; and in the middle Paraná-upper Paraguay basin in s.-e. Brazil (centr. and s. Mato Grosso, extreme s.-w. Goiás and adjacent Triângulo of w. Minas Gerais) and e.-centr. Paraguay.—Fl. VII–XI in Colombia, II–IV, VII–X s.-ward.

In the form of the stipules and humifuse habit of growth *Ch. cordistipula* resembles *Ch. rotundifolia,* obviously distinct in its bifoliolate leaves, and *Ch. supplex,* which has similar foliage but is on the average a more delicate, less enduring plant, distinct in its very much smaller flower (cucullus 3.5–5, not 10–12 mm long) with proportionately diminished anthers and short style (1.5–2 not 5.5–9 mm long). A range of variation in flower-size quite equivalent to that encountered in *Ch. cordistipula* and *Ch. supplex* collectively is admitted within the single species *Ch. serpens.* The situations are not identical, however, the extremes in *Ch. serpens* being linked by a series of intermediates whereas *Ch. supplex* and *Ch. cordistipula* are separated, so far as we now know, by a relatively immense discontinuity.

We find no grounds for maintaining *Cassia schlimii* as even varietally distinguishable from *Ch. cordistipula.* When Bentham described the Colombian species he possessed for comparison with the type-collection only two examples of *Ch. cordistipula* from Mato Grosso, both collected at Cuiabá, more than 3000 km southeast of the lower Magdalena valley. With increase of material from both areas the supposed differential characters of broader leaflets with "peculiar venation" attributed to *C. schlimii* by Bentham have proved ineffective. The venation is a function of broader leaflets, which provide space for some major secondary venules arising from the midrib to fork dichotomously more than once before attaining the leaflet margin. Relatively broad leaflets are neither confined to the northern lobe of the specific range, nor are they constant therein. Pubescence is about equally variable in Colombia and Brazil-Paraguay.

30. Chamaecrista supplex (Bentham) Britton & Rose ex Britton & Killip, Ann. N.Y. Acad. Sci. **35**: 185. 1936. *Cassia supplex* Martius ex Bentham in Martius, Fl. Bras. **15**(2): 164. 1870.—". . . ad praedia Angico et Alegre prov. Pernambuco: Martius; prope Utinga prov. Bahia: Blanchet n. 2747.; in collibus siccis circa Oeiras, prov. Piauhy: Gardner n. 2129.; prope S. João et Porto Imperial prov. Goyaz: Burchell."—Lectotypus (Britton & Rose, 1936, p. 185, by implication): *Martius Obs. 2433,* M (3 sheets, all ticketed by Bentham)! = NY Neg. *9208*; paratypi, *Blanchet 2747,* BM, G, K, LE, P! *Gardner 2129,* G, GH, K, NY, US! Burchell *8483,* K! *8548,* K, NY! *8554,* GH, K, LE, NY, US!

Cassia supplex sensu Bentham, 1871, p. 571.

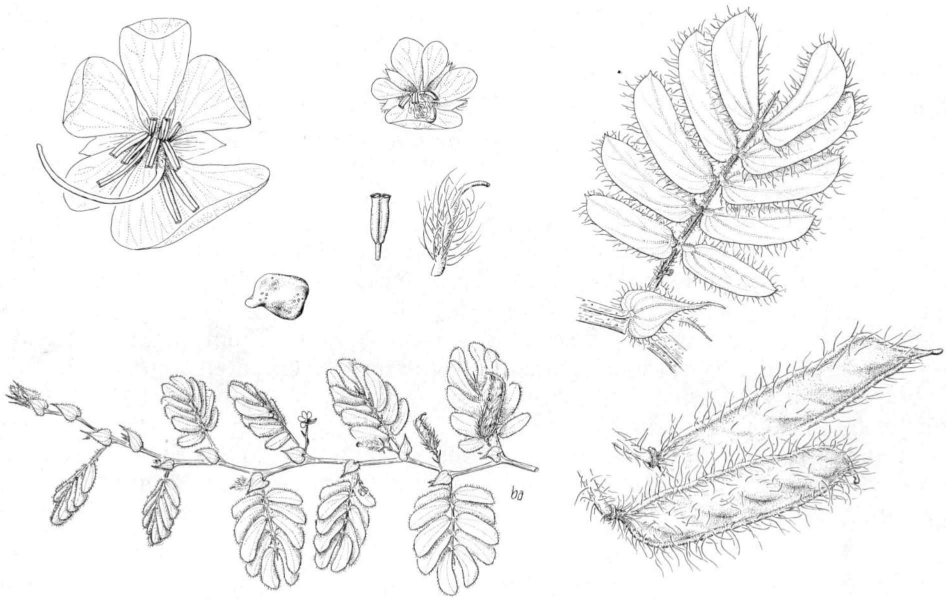

Fig. 46. *Chamaecrista supplex* (Benth.) I. & B. Flowering stem ×1; leaf + stipules (top right) ×3; flower (top center) + 1; stamen and pistil ×6; pod (bottom right) ×3 (*Hatschbach 39520*). *Ch. cordistipula* (Mart.) I. & B. Flower (top left) ×3; pod (middle right) ×3; seed ×4 (*Hatschbach 29518*).

Very slender, sometimes diminutive, humifusely matted herbs of short duration, precociously flowering and then appearing annual but in age ± lignescent at base, with 3–several simple or few-branched, pliantly flexuous, densely leafy stems 0.4–3.5 dm radiating from the root-crown, the stems except for some faint lines of short incurved hairs commonly glabrous, but the stipule margins, lf-stalks, and dorsal face and margins of lfts ± densely pilose or pilosulous with fine white spreading and straight or widely incurved-ascending septate setae to 0.6–1.3 mm, the foliage olivaceous or brownish, concolorous, the tiny axillary short-pedicelled fls often concealed by lvs.

Stipules erect, appressed to stem, persistent, bluntly deltate-ovate-caudate or -acuminately aristate from an asymmetrically cordate-amplexicaul base, 2.4–5 × 1.3–2.6 mm, from base prominently (5–)7–9(–11)-nerved, the lateral nerves incurved distally into a marginal one, the blades submembranous between veins.

Lvs obliquely ascending toward the meridian, petioled, 9–22 mm, the expanded blade ovate in outline; petiole 2.5–4 mm, narrowly margined and open-sulcate; gland shallowly cup-shaped reddish, 0.2–0.35 mm diam, elevated on a slender stipe 0.25–0.5 mm; rachis 3.5–10 mm; lfts (3–)4–5, of most adult lvs exactly 5 pairs, when expanded the first pair divaricate from rachis and the succeeding ones pitched forward at progressively sharp angles, the terminal pair ascending from rachis side-by-side, all in outline obliquely oblong to dimidiately obovate, a trifle forwardly incurved, obtuse but abruptly mucronulate, at base semicordate on proximal side, the blades firmly membranous, the weakly penniveined, moderately excentric costa accompanied from pulvinule on broad side by 2–3 and often on narrow one by 1 incurved-ascending veins, these all sharply prominulous beneath, less so above.

Peduncles exactly axillary, 0–1 mm, 1–2-fld; pedicels 2.5–6.5 mm, bracteolate near and above middle, at anthesis ascending, in fruit firm, sigmoidally divaricate; buds narrowly ovoid acute, pilosulous; sepals membranous, pale green or reddish-tinged, at anthesis widely spreading to deflexed, narrowly ovate to lance-elliptic acute, 3.1–4.1 mm; petals yellow early fading orange or purplish-orange, sub-marcescent, 4 narrowly oblanceolate to spatulate-obovate, narrowed downward into a slender claw, unequal, 2 smaller up to 1.7–3.3 mm, 2 larger up to 2.8–4 mm, the cucullus dimidiately obovate or broadly ear-shaped, 3.5–4.9 mm, not over 1 mm longer than sepals; stamens 3–5, at most 3 fertile and the 2 accessory, if present, minute and sterile, the 2–3 fertile with linear-oblong or shortly oblong anthers to 1.2–1.9 mm, the largest that directly opposed to the divaricate gynoecium; ovary densely white-pilose; ovules 2–5; style 1.5–2 mm, incurved to bring the stigma almost to quite in contact with the large contiguous anther.

Pod narrowly or, when very short, broadly oblong in outline, 4–13 × 2–3 mm, tipped by the short, incurved style-base, the castaneous valves densely pilose with fine flexuous white setae to 1–2.6 mm; seeds quadrate or prismatically compressed-pyriform, 1.2–2.3 mm, the testa pale pinkish-ochraceous with faint black line along the distal keel, sublustrous, rugulose, faintly pitted, mucilaginous when wetted.—Collections: 31.—Fig. 46.

Campo, open places in cerrado and capoeira, and caatinga, sometimes weedy along highways, usually in thin, especially sandy soils, (20–)150–950 m, local, scattered over e. Brazil from middle Tocantins valley in s.-w. Maranhão e. to Ceará and the lower S. Francisco valley in w. Pernambuco and adjoining Bahia, s. through e. Goiás and w. Bahia to Distrito Federal; e.-centr. Bahia (Ibicaraí); doubtfully recorded from s. Minas Gerais (*Glaziou 12614,* P); disjunct on dry campo of the lower Amazon valley near Montealegre, Pará; and on the Atlantic coast of s.-e. Rio de Janeiro (Paratimirím); reported (Britton & Killip, 1936, l.c.) erroneously from Colombia.—Fl. II–VI.

The smallest of the prostrate Brazilian chamaecristas, *Ch. supplex* is readily distinguished from sympatric small-flowered forms of *Ch. serpens* var. *serpens* by the cordate stipules, broader more obtuse leaflets, short pedicels and an androecium reduced to at most five members, only two or three of them bearing fertile anthers. In habit and in details of foliage and stipules *Ch. supplex* resembles, and in consequence has been confused with, *Ch. cordistipula,* which is, however, easily distinguished by its large, probably out-crossing, long-styled flower. Bentham drew attention to the uncinately recurved style-base that tips the small pod of *Ch. supplex* and contrasted it with the straightish style-base of *Ch. cordistipula*. The difference in this tiny organ has become blurred by modern collections.

31. Chamaecrista kunthiana (Schlechtendal & Chamisso) Irwin & Barneby, comb. nov. *Cassia kunthiana* Schlechtendal & Chamisso, Linnaea **5**: 598. 1830.—"[*Schiede & Deppe*] 708 . . . In collibus prope Hacienda de la Laguna [near Jalapa, Veracruz]."—Holotypus, †B; isotypus NY (hb. Meissner., commun. Schlechtend.)!

Cassia humilis Willdenow ex Steudel, Nom. ed. 2, **1**: 305. 1840, pro syn.—Taken from *herb. Willd. 7941,* B!

Cassia kunthiana sensu Vogel, 1837, p. 57.

Cassia tagera sensu Bentham, 1870, p. 162; 1871, p. 570; auct. plur. recentior. *Chamaecrista tagera* sensu Britton & Rose, 1930, p. 276. Non *Cassia tagera* Linnaeus, nom. confus.; cf. Irwin & Barneby, Brittonia **28**(4): 436. 1976.

Perennial herbs from blackish, white-tipped, often richly nodular roots, the many slender pliant, freely branching stems humifusely radiating from the root-crown or in age from a shortly branched, weakly suffruticulose caudex, forming dense mats of small foliage (1–)2–11 dm diam, either subglabrous below the always basally hispidulous calyx and puberulent ovary to thinly pilosulous and setose with fine spreading septate hairs to 0.7–1.6 mm, the pilosulous vesture of incurved short hairs often reduced to a minute puberulence along the ventral side of the petioles, the pedicels, or the stem (in lines), the setae confined mostly to the dorsal side of petiole, the margins of stipules and lfts, rarely extending to the lower face of lfts, the herbage green concolorous, the small flowers raised from the mat of foliage on filiform pedicels, the perianth little expanded even at sunrise, closing by noon.

Stipules erect, appressed to stem, persistent, broadly lance- or ovate-acuminate from symmetrically cordate-auriculate base, (2–)3–10(–11) × (0.7–)1–4.5 mm, the blades membranous becoming hyaline, prominently (7–)9–15-nerved from base, the nerves all passing distally into a strong marginal one, the central nerve produced as a spinule, the faces usually glabrous, the margin either glabrous or thinly setose.

Lvs 5–17(–20) mm, the expanded blade 6-foliolate subcircular or broadly flabellate and longer than the abbreviated rachis; petiole including wrinkled pulvinus (2–)2.5–6 mm, at middle 0.3–0.5 mm diam, openly grooved ventrally; gland situated near or above middle of petiole, urceolate, the head 0.2–0.4 mm diam, the slender stalk 0.25–0.6 mm, at least as long as the head's diam; rachis (1–)1.5–3 mm, dilated under each lft; lfts 3 pairs, by day displayed face upward, the proximal pair directed backward, the median pair horizontally divaricate, the distal (and widest) pair forward, in sleep (or wilted, as in most spms) all forwardly imbricated, in outline dimidiately obovate, obtuse or deltately acuminate, mucronulate by the excurrent, strongly displaced midrib, at base cuneate on the distal and shallowly auriculate on the proximal side, the distal pair 3–10(–12) × (1.5–)2–5(–6) mm, the rest usually at once a little shorter and narrower, the proximal pair sometimes dimidiately oblanceolate, the wider ones palmately 2- the narrower 1-nerved from base on proximal side of midrib, these primary veins with few pinnate venules prominulous on both faces but beneath more sharply so and usually pallid.

Peduncles exactly axillary, up to 1 mm, 1(–2)-fld; pedicels at anthesis filiform, incurved under calyx, firmer in fruit, (3–)5–30(–34) mm; bracteoles linear-lanceolate or subulate, 1.3–3.3 mm; buds ovoid-acuminate, horizontally spreading, glabrous beyond the usually hispidulous base; sepals green or brownish-red, ovate- or broadly lance-acuminate, the longest 3–4.2 mm; petals yellow, at full anthesis forming a subglobose perianth (never widely opening), 4 obovate-cuneate of somewhat unequal length, the 2 adaxial slightly larger, up to 3.5–5 mm, the fifth very obliquely obovate, broader than the rest but scarcely longer, incurved over the androecium and gynoecium; stamens 5–7, heteromorphic, 2–3 adaxial with short oblong anthers borne on a slender filament ± as along as themselves, 1 abaxial, borne on a short stout filament, 2–3 variably degenerate or fully sterile, sometimes reduced to a linear rudiment; gynoecium only slightly askew; ovary appressed-strigulose or spreading-hispidulous; style linear, 1–1.9 mm, incurved distally; ovules 1–4(–5).

Pod horizontal or declined, minutely stipitate, the body oblong (when 1-seeded ovate) in outline, (4–)7–12(–15) × 3–4.5(–5) mm, the brown, ultimately nigrescent valves finely strigulose or pilosulous; seeds irregularly obtusely rhomboid,

2.2–3.1(–3.6) mm, the testa light brown or ochraceous dull, faintly lustrous-pitted, with faint fuscous transverse line at distal end, filamentous when moist.—Collections: 163.

Savannas, llanos, campos, glades in pineforest (Centr. Amer.) and cerrado (Brazil), often in sandy or disturbed soils, an abundant and successful weed of waysides and fallow fields, and of pastures where resistant to browsing and forming part of the sod, mostly below 600 m but ascending in Mexico (Veracruz) to 1000, in Honduras to 1200, in Panama to 1500, and in the Brazilian Plateau (Goiás) to 1150 m, discontinuously dispersed to the n. and s. of the Amazonian Hylaea: s. Mexico (Guerrero to Veracruz and Chiapas, e. into Yucatan Peninsula) and w. Cuba (Isla de Pinos) s. through Central America to Panama; middle Orinoco valley, upstream from Parmana, in Guárico and Bolívar, Venezuela and adjoining Vichada, Colombia, to headwaters of Rio Negro in Amazonas, Brazil; Brazilian Planalto from Tocantins valley in n. Goiás and adjoining Maranhão s. through Goiás, extreme w. Bahia and Distrito Federal to the Triângulo of w. Minas Gerais, thence w. through s. Mato Grosso into Santa Cruz, Bolivia; sometimes cultivated, as at Belém (Pará) from seed collected on upper Rio Negro.—Fl. both n. and s. of the equator primarily at end of the rainy season and onward, but in lowlands irregularly throughout the year.—*Golondrina* (Ven.).

A species readily recognized by its prostrate, mat-forming habit of growth and especially by the small, sensitive, 6-foliolate leaves. The six leaflets, densely set along a very short rachis, form when expanded a blade of broadly fan-shaped or almost circular outline, face upward to the meridian, but in sleep or when wilted (as in most herbarium specimens) fold forwardly together, face to face, each pair clasping that next above it. The small fleeting flower, which generally fades by noon of the day it opens, is well suited to autogamous fertilization, the reproductive organs of both sexes being enclosed together within the sac of the larger incurved fifth petal. As would be expected, a fertile pod is produced by virtually every flower.

The variable vesture of *Ch. kunthiana* is of the double type common to many chamaecristas, composed of short incurved hairs and fine spreading septate setae. In Mexico, Cuba, and Central America southward to Belize and El Salvador the average plant appears almost glabrous, setae being absent or nearly so, and the stem, ventral margins of the petiole and pedicels are alone thinly puberulent. From Nicaragua southward and in the Orinoco valley setae appear regularly on the dorsal keel of the petiole, or along margins of some upper stipules and leaflets, or on some of these, occasionally extending to the dorsal face of the leaflets. North of the Amazon the ovary and pod are strigulose with subappressed hairs, but in southeastern Brazil, where the vesture of the stems and foliage is essentially that of southern Central America, the ovary is densely hispidulous and the pod that ripens from it is loosely pilose. The correlation between vesture and dispersal is, however, imperfect, two forms occurring occasionally together and the few collections known from Bolivia being of the glabrate type. A slight variation in flower-size appears to occur sporadically through the range of the species, which is in other respects variable only in stature and amplitude of leaflets, subject to influence of habitat, age and season.

In Panama (La Jagua, prov. Panamá) Dr. J. A. Duke observed leaf-cutter ants of the genus *Atta* carrying away leaves and especially flowers of *Ch. kunthiana,* presumably for use in their subterranean nurseries.

iv. ser. BAUHINIANAE (Colladon) Irwin & Barneby

Chamaecrista sect. **Chamaecrista** ser. **Bauhinianae** (Colladon) Irwin & Barneby, comb. nov. *Cassia* sect. *Chamaecrista* ser. *Bauhinianae* Colladon, Hist. Casses 119. 1816, sens. valde restr.—Sp. lectotypica: *C. rotundifolia* Pers. ≡ **Chamaecrista rotundifolia** (Persoon) Greene.

Cassia sect. *Chamaecrista* ser. *Subaphyllae* Bentham in Martius, Fl. Bras. **15**(2): 162. 1870.—Sp. typica (unica): *C. basifolia* Vogel ≡ **Chamaecrista basifolia** (Vogel) Irwin & Barneby.
Cassia sect. *Chamaecrista* ser. *Bauhinianae* sensu DeCandolle, Prod. **2**: 501. 1825, char. ("foliolis 2–3-jugis") manifeste fallaci necnon spp. compluribus exceptis.
Cassia sect. *Chamaecrista* ser. 1. *Subaphyllae* sensu Bentham, 1871, p. 569.
Chamaecrista Rotundifoliae Britton & Rose, N. Amer. Fl. **23**(5): 270, in clave, sine descr. 1930.

Precociously flowering monocarpic, biennial or short-lived perennial herbs, the stems exceptionally soft-woody in age; petiolar gland 0; lfts 1 pair, but most lvs of adult *Ch. basifolia* suppressed, their function assumed by forwardly imbricated stipules; peduncles either axillary or obscurely adnate to stem (supra-axillary), or commonly obsolete; sepals ovate- or lance-acuminate; petals 3–17 mm; stamens commonly 5, sometimes ±1–3 barren vestiges of an upper whorl.—Spp. 2, one ubiquitous in lowland Neotropics, extending both n. and s. (often weedy) into warm temperate zones, one endemic to the Brazilian Planalto.

Although Bentham referred our two *Bauhinianae* to different series, they seem closely akin in their cordate stipules, short-petioled, amply bifoliolate leaves lacking the petiolar gland otherwise characteristic of sect. *Chamaecrista,* and functionally pentamerous androecia. Although very different in adult habit, which recalls that of some compatriot Melastomataceae, *Ch. basifolia* is indistinguishable in the leafy seedling stage from *Ch. rotundifolia*.

Key to the Species of ser. *Bauhinianae*

1. Lvs present at all nodes; stems commonly prostrate, sarmentose when supported, exceptionally erect; widespread in N. and S. America. 32. *Ch. rotundifolia* (p. 727).
1. Lvs present usually at no more than ±8–12 nodes subsequent to the cotyledons, a few sometimes developed low on secondary branches, these all commonly deciduous by anthesis, the stems distally clothed in a file of imbricated stipules; stems characteristically erect, at first simply virgate, in age awkwardly ascending-branched; Brazilian Planalto, extending just into Paraguay. 33. *Ch. basifolia* (p. 733).

32. Chamaecrista rotundifolia (Persoon) Greene, Pittonia **4**: 31. 1899. *Cassia rotundifolia* Persoon, Syn. Pl. **1**: 456. 1805.—Typus infra sub var. *rotundifolia* indicatur.

Herbs from woody-fibrous blackish roots, the wiry stems at first and often permanently radially humifuse, at anthesis 1–7 dm, sometimes at first simple erect, more rarely (especially in competitive microhabitat) assurgent, in age basally suffrutescent, exceptionally bushy-branched or sarmentose and up to 1 m, highly variable in amplitude of foliage and stipules, in flower-size, and in vesture, this commonly duplex, composed of shorter incurved and longer straightish spreading septate hairs up to 1–3.5 mm, but either type of hair potentially predominant or absent, the olivaceous concolorous 2-foliolate lvs either pilosulous or glabrous on the faces, ciliolate.

Stipules erect appressed to stem, commonly accrescent upward, broadly lance- to deltate- or ovate-acuminate or -caudate, (2–)3–22(–26) mm, shorter than or a little exceeding the internodes, at apex aristulate, at subsymmetrically cordate-amplexicaul base (0.5–)1.2–9(–11) mm wide, the basal angles rounded, the green submembranous blades palmately 7–21-nerved from pulvinule, the faces either pilosulous or glabrous, the margins coarsely ciliate to minutely ciliolate.

Lvs ascending, (3–)7–45(–50) mm, shortly petioled; petiole (including pulvinus) 2–8(–10) mm, at middle openly shallow-grooved and (0.25–)0.3–0.7 mm diam; petiolular gland 0; lfts 1 pair, by day ascending, face upward from tip of petiole, folded face to face at night, in outline obliquely or dimidiately obovate to broadly oblanceolate, at apex broadly rounded, truncate, or openly and shallowly emarginate, often minutely mucronulate by the excurrent, excentric midrib, at base cordate on proximal and cuneate on distal side, the plane membranous or chartaceous blades prominulously venulose on both faces by the penninerved midrib with (on its proximal side) 4–5 nerves palmately incurved-ascending from the pulvinule, the tertiary venulation likewise prominulous on both faces, forming an open reticulum.

Peduncles either obsolete or if developed adnate to stem through 1–3 mm, 1–3-fld, the raceme-axis almost 0, the pedicels (if more than 1) closely contiguous, the distal one commonly shorter than the one below; pedicels subfiliform becoming stiff and wiry, straight or almost so, ascending toward the meridian, (1.2–)1.5–4.5(–6) cm, varying from a little shorter than to 3 times as long as the subtending lf, bracteolate well above middle; bracteoles lanceolate or lance-acuminate, 1–4 mm; buds horizontal or somewhat nodding, slenderly ovoid-acuminate, pilosulous at least below middle; sepals either greenish or reddish-brown, narrowly ovate-acuminate to lance-elliptic-caudate, the longest 3.2–13 mm; petals yellow, drying pinkish- or orange-brown, at full anthesis campanulately ascending, obovate- to oblong-obovate-cuneate, obtuse or truncate, slightly differentiated in size but neither of the abaxial petals strongly oblique, the longest ± as long up to ±2 mm longer than the longest sepal; androecium commonly 5-merous but often with 1–3 accessory (adaxial) staminodia, one of the 5 fertile yellow anthers sometimes dwarfed and the staminodes sometimes developed into short (perhaps infertile) stamens, the longer of the 5 oppositisepalous anthers (1.7–)2–9(–11.5) mm; ovary gray- or yellowish-strigulose-pilosulous; style linear-cylindric, incurved distally, (0.5–)0.7–6(–8.5) mm, the stigma raised ± to level of longest anther; ovules (4–)6–18(–21), commonly ±8–15.

Pod linear-oblong, straight or slightly curved forward, (1.5–)2–5.5(–6) × 0.3–0.5(–0.6) cm, the purplish-green or brown, finally castaneous or nigrescent valves finely pilosulous; seeds obtusely oblong-prismatic from an obconic base, 2.2–3.2(–3.4) mm long, truncate-emarginate distally and there marked on each face with a dark line, the testa otherwise pale-brown or ochraceous, smooth or almost so, dull or sublustrous, emitting mucilage-threads when wetted, the pits low-crateriform or immersed and appearing as pallid dots.

In the context of its section *Ch. rotundifolia* is instantly recognized by the shortly petioled bifoliolate leaves borne at every node of the stem. In herbaria it is occasionally confused with *Ch.* (*Xerocalyx*) *diphylla,* similar in number and sometimes in form of the leaflets but different in the venation of the leaflets, in presence of a petiolar gland, in the strongly graduated, multistriate calyx-lobes, and in the decandrous androecium. As defined at present, the species is composed of numerous geographic races, differing among themselves in stature, pubescence and especially in size of the perianth together with length of style and of the five fertile anthers. Thus the stems, most commonly slender and humifuse or weakly assurgent, may become coarse, and in age bushy-branched and suffruticose. Vesture varies in a mode characteristic of sect. *Chamaecrista,* consisting of shorter incurved and longer spreading, in this case septate hairs variously proportioned on stem, face of stipules, and elsewhere, either type of trichome being in some cases predominant and in others lacking. Stature and vesture vary independently of each other and pubescence independently of either, while the leaf-blades and the stipules, either together or separately, vary from small and firm to amply foliaceous and of thinner texture. And further superimposed on this complex set of inconstant features the perianth and androecium vary in size to an extraordinary degree, the expanded corolla from 6 mm to 3 cm in diameter, the longest of the four or five large oppositisepalous anthers from about 2 to 11 mm long. Over the whole range of the species, and exclusively north of Honduras, *C. rotundifolia* is represented by small-flowered races, with petals 3–7 mm, anthers up to 4 mm, and style up to about 2 mm long. In these forms it seems likely, because the incurved style of newly opening flowers is raised just to the level of the anthers' terminal pores and the stigma is often seen to be receptive and already coated in pollen, that the flower is autogamous, and the extremely regular and successful production of fruit, even when isolated under cultivation, supports this assumption. Autogamy combined with the ability to colonize disturbed and ruderal habitats (especially those proliferating in the path of man) would encourage not only the fixation within expanding populations of particular selections of potential phenetic characters but in addition promotes a disorderly dispersal of these over immense distances. It is precisely such a pattern of partly random and partly logical distribution that emerges from analysis of the very extensive material now available for study.

The populations of *Ch. rotundifolia* now present in the West Indies and Florida, perhaps aboriginally native only in Cuba, are almost without exception extremely small-flowered (petals 3–5 mm), small-leaved (leaflets mostly less than 2 cm), and short-hairy (septate hairs seldom over 1 mm). Similar plants predominate in the Orinoco valley in Venezuela, in Costa Rica and neighboring Panama, and in eastern Brazil (Maranhão; about Rio de Janeiro, where possibly only a weed of long standing). Along the Pacific coast of Mexico the common form (*C. fabaginifolia* H.B.K.) has slightly larger flowers, at least in normally rainy seasons ample foliage (some leaflets well over 2 cm), and stems pilose with horizontal hairs up to 2–3.5 mm. But occasionally in this same range, we do not know whether truly native or whether, as seems very possible about the port of Mazatlan, brought in by man, small-flowered, small-leaved and short-hairy types have been collected. On the Brazilian plateau, where the species is both common and prolific in minor variants, the predominant forms closely parallel those of Mexico, except only that the pod tends to be narrower (mostly 3–4 rather than 4–5 mm wide). Northward from the Amazon valley the pedicels of *Ch. rotundifolia* seem to be uniformly glabrous, but over a rather extensive area in southeastern Brazil (São Paulo and Paraná) they are commonly, if not quite constantly puberulent. Puberulent and glabrous pedicels in one small colony of plants (e.g. *Eiten 2919,* NY, pubescent; US, glabrous) and the reappearance of glabrous ones in Argentina and Paraguay diminish the significance of this character. In summary we find that while certain combinations of phenetic characters tend to become emphasized in particular segments of the range of small-flowered *Ch. rotundifolia,* no particular combination is anywhere exclusively present or, when carefully observed, truly monomorphic. The degree to which natural dispersal patterns have been deranged and modified by human agency is impossible to estimate with any accuracy, but we can be sure that it contributed to the contemporary mingling of what may well have been in origin (and perhaps remain) genetically distinct forms.

Within small segments of the range of small-flowered *Ch. rotundifolia,* as well as slightly beyond its known range, occurs a most interesting set of larger-flowered forms, with petals mostly 8–17 mm and style 2.5–8.5 mm long, which collectively incorporate, however, the whole spectrum of vegetative, trichotic, and habital variation encountered in the smaller-flowered types. The large-flowered plants can be sorted into four geographic groups characterized by the syndromes of: a) erect, ultimately suffrutescent habit, large stipules, copious long septate hairs on stems, and glabrous pedicels (S. America north of the Amazon, Pacific Ecuador and Bay of Honduras); b) like a, but commonly procumbent stems and small stipules (Atlantic slope of s.-e. Brazil); c) procumbent habit, leaflets and stipules of moderate size, few and short septate hairs, and puberulent pedicels (s.-e. Bolivia); and d) as the last but glabrous pedicels (Paraguay basin). Up to the flower each of these four variants can be matched for foliage, habit, and pubescence with some particular small-flowered one. It seems possible that these large-flowered races have arisen independently of each other from small-flowered antecedents themselves not exactly uniform. Within their highly discontinuous collective range, although this lies almost entirely within that of small-flowered types, each is ordinarily the one and only representative of the species; they cannot be interpreted as merely large-flowered or luxuriant individuals, except possibly for some few from southern Mato Grosso and adjacent Paraguay and Argentina (syndrome d) which in their relatively long glabrous pedicels suspiciously resemble sympatric or subsympatric small-flowered plants.

While recognizing two large-flowered varieties of *C. rotundifolia,* Bentham (l.c.) admitted that his var. *bauhiniifolia* (=our syndrome a, and including *C. filipes*) differed scarcely at all from luxuriant typical states of var. *rotundifolia* except in the slightly larger flower, whereas his var. *grandiflora* (=our syndrome b) appeared better characterized by the combination of large, long-anthered flowers with smaller stipules. We now have evidence that long anthers (and associated long style) occur remotely disjunct in the upper Paraguay basin in southwestern Brazil and adjoining countries, associated there with a variety of foliage and pubescence types. The flower in which elongate anthers (8–11.5 mm) coincide with long style (5–8.5 mm), this perhaps more widely divaricate from the androecium than is possible in the small autogamous flower of var. *rotundifolia,* is theoretically interesting because its configuration suggests the probability of outcrossing. Moreover the less reduced androecium (with sometimes more than 5 fertile anthers) that coincides with this type of flower, might point to a primitive condition and therefore an origin for the species on the Brazilian Plateau. However this may be, the racial situation in *C. rotundifolia* will not be understood until experimental biosystematic methods can be brought to bear on its study. The maintenance of Bentham's two large-flowered varieties would require, at the present level of our knowledge, at least equivalent taxonomic recognition for several parallel large-flowered types and, in addition, a subdivision of the small-flowered ones at equivalent stages of differentiation. The pragmatic but, we believe, only provisional solution of the problem is to recognize no more than two admittedly variable and probably genetically heterogeneous categories characterized by size and proportions of the flower alone.

Key to the Varieties of *Ch. rotundifolia*

1. Fl relatively small, the largest petal 3–6.5(–7) mm; largest of the (4–)5 fertile anthers (1.7–) 2–4.4 mm; style (0.5–)0.8–1.8(–2) mm; range almost of the species (Florida and Sinaloa to Argentina). 32a. var. *rotundifolia* (p. 731).
1. Fl larger, the longest petal 7–17 mm; largest of the 5–7 fertile anthers 5–11.5 mm; style (2–) 2.5–7(–8.5) mm; more local and unknown from n. of Honduras. 32b. var. *grandiflora* (p. 732).

32a. Chamaecrista rotundifolia (Persoon) Greene var. **rotundifolia**. *Cassia rotundifolia* Persoon, 1805, l.c., sens. strict.—"Hab. in Amer. austr. (Herb. Juss.)."—Holotypus, collected at Rio de Janeiro by Dombey, P-JUSS (no. *14531*)!—*C. bifoliolata* var. [C.] *rotundifolia* (Persoon) Desvaux, Ann. Sci. Nat. I, **9**: 439. 1826, comb. illegit.

Cassia bifoliolata DeCandolle ex Colladon, Hist. Cass. 120, t. 9, fig. B + 19. 1816.—Origin and collector not recorded.—Holotypus, collected in Brazil, probably at Rio de Janeiro, by Commerson, G-DC! isotypus, *Commerson 888*, FI (hb. Webb.)!—*Chamaecrista bifoliolata* (DeCandolle) Greene, Pittonia **4**: 31. 1899.

Cassia pentandra Raddi, [Quar. Piant. Nuov. 20] in Mem. Acad. Moden. **18**: 399. 1820.—". . . dappertutto nei contorni di Rio-Janeiro . . ."—Holotypus, *Raddi s.n.*, FI! isotypus, *Raddi s.n.*, G (hb. Moricand.)!—*C. bifoliolata* var. *pentandra* (Raddi) Desvaux, Ann. Sci. Nat. I, **9**: 438. 1826.

Cassia fabaginifolia Humboldt, Bonpland & Kunth, Nov. Gen. & Sp. **6**(folio): 363. 1824.—"Crescit in monte ignivomo Jorullo, alt. 630 hex. (Regno Mexicano [Michoacán])."—Holotypus, P-HBK! isotypi, P (*hb. Bonpland.*), B (*hb. Willd. 7919*)!

Cassia monophylla Vellozo, Fl. Flum. 166. 1825 & Atlas **4**: t. 61. 1835.—"Habitat pascuis gramineis sub-urbanis [=outskirts of Rio de Janeiro]."—Lectotypus, the cited plate!

Cassia pentandria Larrañaga, Escritos de don D. A. Larrañaga **1**: 408 ("*pentardria*"). 1922.—Collected at Rio de Janeiro, Brazil (not Uruguay), no typus known to survive; referred by Irwin, Mem. N.Y. Bot. Gard. **12**: 52. 1964, to *C. diphylla*, but the type-locality and the epithet strongly suggestive of *C. rotundifolia*.—Non *C. pentandra* Raddi, 1820 (vide supra).

Cassia tenuivenosa Jones, Extracts from Contrib. West. Bot. **18**: 41. 1933, nom. provis. Cf. Morton in Contrib. U.S. Nat. Herb. **29**: 100. 1945.—Spm. authentica: *Jones 27192*, GH, NY, POM, US!

Cassia rotundifolia sensu Bentham, 1870, p. 161; 1871, p. 570. *Chamaecrista rotundifolia* sensu Britton & Rose, 1930, p. 277.

Stems commonly humifuse, sometimes assurgent, only rarely suffrutescent; pubescence variable, the pedicels commonly glabrous, but in s.-e. Brazil often puberulent, the stems at once long-pilose and incurved-pilosulous, but sometimes either one type of hair predominating or one absent; stipules mostly less than 13 mm; flowers small (as given in key); fertile anthers commonly 5, one rarely much smaller than the rest, and commonly 1–3 subfiliform staminodia.—Collections: 206.—Fig. 47.

Savannas, campos, openings in light woodland, and an abundant colonizer of waste places, roadsides, and disturbed forest, a persistent weed of lawns and pastures resistant to cutting and grazing, entering many treeless and brushy plant-communities of the Neotropical zone and extending n. in w. Mexico to 27° and in s.-e. United States (as an immigrant) to 28°N, and s. (mostly weedy beyond 30°) in Uruguay and adjacent Argentina to 34°S, characteristic of tropical lowlands but ascending in Mexico (Nayarit, Jalisco, Oaxaca) and central America (Costa Rica, Panama) to ±1100(–1550) m, in n. Venezuela to 800 m, and on the Brazilian Planalto to 1300(–1600) m, locally abundant but of notably discontinuous range: Pacific slope of Mexico from centr. Sinaloa to Isthmus of Tehuantepec in Oaxaca, inland to the middle Lerma-Santiago valley and (uncommon) the Balsas Depression; Costa Rica (Alajuela and San José) and adjacent Panama (Chiriquí), centr. and w. Cuba; rare and perhaps only introduced on Jamaica and Puerto Rico (unknown as yet from Hispaniola); introduced but persisting in centr. Florida; in n. South America widespread over the savannas and llanos of the Orinoco-Apure-Meta basin in Venezuela and n.-e. Colombia (Boyacá and Meta) and occasional in the Cordillera Marítima of Venezuela (Distrito Federal and Carabobo) and upper Magdalena Valley in Colombia (Huila); absent from the Amazon basin proper (except as a weed on upper Rio Branco near Bôa Vista, Terr. Roraima, Brazil) and the Guayana Highland; again widespread over s.-e. Brazil from the

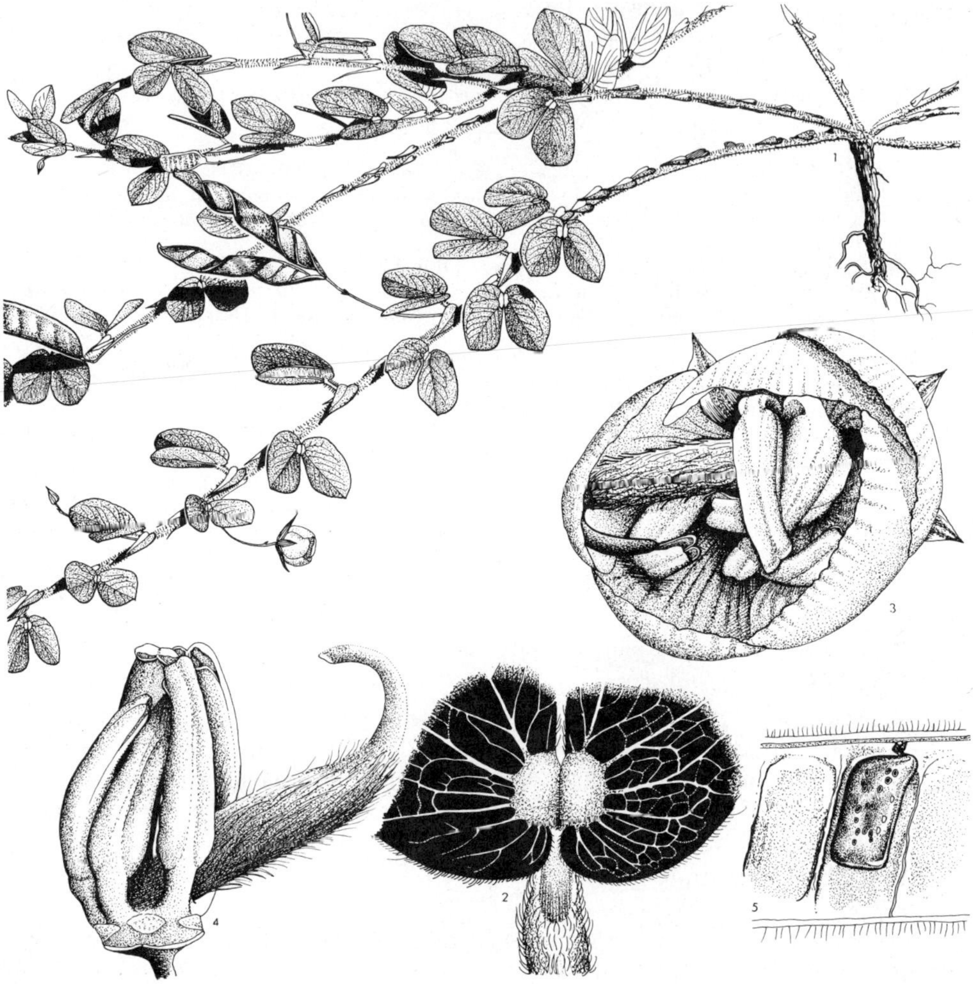

Fig. 47. *Chamaecrista rotundifolia* (Pers.) Greene var. *rotundifolia*. 1) Habit ×½; 2) leaflet pulvinules ×5; 3) flower ×10; 4) androecium + pistil ×10; 5) seed ×5 (Drawn from life by W. S. Moye III).

upper Araguaia, middle Tocantins, and middle Parnaíba valleys in n.-e. Mato Grosso, n. Goiás, and Maranhão s. and s.-w. to Guanabara, Paraná, and s. Mato Grosso, almost ubiquitous throughout the Planalto but on the Atlantic coast only (and perhaps only weedy) in Rio de Janeiro and Guanabara, thence s. through Paraguay to n. Argentina (Formosa, Corrientes, where weedy s.-ward, Misiones) and adjoining Bolivia and Uruguay.—Fl. and fr. in the N. Hemisphere mostly VII–III and in S. Hemisphere XI–VI, but granted warmth and water indefinitely through the year.—*Pega-pega chiquita* (Ven.).

32b. Chamaecrista rotundifolia (Persoon) Greene var. **grandiflora** (Bentham) Irwin & Barneby, comb. nov. *Cassia rotundifolia* var. *grandiflora* Bentham in Martius, Fl. Bras. **15**(2): 162. 1870.—". . . in arenam prostrata ad Caravellos prov. Bahiae: Princeps Maximilianus Neovidensis; in campis arenosis prov. Bahiae et ad Vão do Paranan prov. Goyaz: Martius; ad Tamandua prov. Bahiae: Blanchet n. 3844."—Lectoholotypus, *Blanchet 3844,* collected at Tamanduá, between Jacobina and S. Tomé, BM! isotypi, G, GH, P! paratypus, *M. von Neuwied s.n.* e. Caravellas, GOETT!

Cassia bauhiniifolia Kunth, Mimoses 123, t. 37. 1819 ("*bauhiniaefolia*").—"Crescit in ripa arenosa fluminis Orinoci, prope Carichana [edo. Bolívar, Venezuela] . . ."—Holotypus, P-HBK (no. *823*)! isotypi, B (*hb. Willd. 7920*), P!—*C. rotundifolia* var. *bauhiniifolia* (Kunth) Bentham in Martius, Fl. Bras. **15**(2): 162. 1870. *Chamaecrista bauhiniifolia* (Kunth) Gleason, Bull. Torr. Club **56**: 395. 1929; Britton & Rose, Ann. N.Y. Acad. Sci. **35**: 184. 1936.

Cassia filipes Bentham in J. Bot. (Hooker) **2**: 81. 1840.—"Savannahs, about Fort St. Joaquin [Terr. Roraima, Brazil], Schomburgk, n. 787."—Holotypus, K! isotypi, GH, NY, US!—Reduced to the preceding by Bentham, 1870, p. 162.

Cassia rotundifolia var. *grandiflora* Micheli in Engler, Bot. Jahrb. **16**(Beibl. 37): 7. 1892, nom. nud.—"Columbia; in arenosis prope Neira, in prov. Cauca, 400 m. ([*Lehmann*] n. 2373).:—Spm. anthent.: *Lehmann 2373,* from "Savanen neben Wegen um Neiva, 3.I.1883" [dept. Tolima or Huila], US!

Cassia rotundifolia var. *longepedicellata* Chodat & Hassler, Bull. Herb. Boiss. **4**: 826. 1904, nom. nud.—"[Paraguay] . . . in sabulosis rivuli Arroyo Primero (Apa), Febr., [*Hassler*] n. 8462."—Spm. authent.: *Hassler 8462,* NY!

Cassia bifoliola "Persoon" ex Knuth, Fedde Repert. Sp. Nov., Beih. **43**: 374. 1927.—An incorrect name derived from a herbarium label (*Pittier 10592*).

Stems commonly (n. of the Amazon R. almost always) assurgent to suberect, but s.-ward either procumbent or assurgent; stipules 7–26 mm, the size not correlated with habit; pubescence variable, as in var. *rotundifolia*; flowers relatively large (as given in key); fertile anthers 5–7, the 1–2 antepetalous smaller, and often 1–3 staminodia.—Collections: 36.

Campos, llanos, and disturbed, commonly sandy soils, sometimes on riverbanks or seashores, mostly 5–500, in Bahia up to 1100 m, of highly discontinuous range within that of var. *rotundifolia,* but apparently not or seldom associated with it: Golfo de Honduras (I. de Roatan); upper Magdalena Valley in Colombia (Tolima and Huila); Pacific lowlands of Ecuador (Guayas); Maracaibo Basin in Zulia, Venezuela; Guayana Highland in Guyana, Venezuela (Bolívar) and Brazil (Terr. Roraima); and disjunctly s. of the Amazon R. in interior and coastal Bahia, coastal Espírito Santo; and about the common boundary of Mato Grosso, Brazil with Bolivia (Sta. Cruz), Paraguay, and Argentina (Corrientes).—Fl. in Brazil s.-ward mostly XII–IV(–VIII), n.-ward irregularly through the year.

33. Chamaecrista basifolia (Vogel) Irwin & Barneby, comb. nov. *Cassia basifolia* Vogel, Syn. Gen. Cass. 56 & Linnaea **11**: 707, descr. ampliat. 1837.—"In Brasilia: Sellow leg."—Holotypus, †B; no isotypus known to survive, but the protologue decisive.

Cassia basifolia var. *aristulata* Hassler in Feddes Repert. **8**: 129. 1910.—"Paraguay: Sierra de Amambay in campis arenosis pr. Estrella, flor. et fruct. mens. Jan.: Hassler no. 10038, leg. Rojas."—Holotypus, G! isotypi, BM, GH, K!

Cassia basifolia sensu Bentham, 1870, p. 161, t. XLIV; 1871, p. 569.

Erect, when young simple virgate, later awkwardly few-branched, in age basally suffrutescent herbs of 1–2 seasons' duration, at anthesis (1.5–)2.5–10(–15) dm, except at the lowest nodes leafless, the stems clothed in loosely imbricated or shortly discrete pairs of appressed, amply cordate-amplexicaul venulose stipules, the few early developed lvs 2-foliolate, the rest reduced to a seta concealed between pairs of stipules, the stems and olivaceous concolorous foliage varying from pilose nearly throughout with fine spreading septate hairs to 0.8–2(–2.7) mm to glabrous except for a few short incurved hairs on stems or minutely ciliolate stipules, the sepals, ovary and pod consistently pilosulous.

Stipules subsymmetrically ovate- or deltate-acuminate from deeply cordate base, (6–)9–21 × (3.5–)4–12(–14) mm, the basal angles rounded, the apex sharply mucronate or spinulose, the plane blades stiffly chartaceous, palmately many-nerved from base, the lowest of the primary stem and of some branchlets narrower and few-veined.

Lvs developed at ±8–12 nodes following the cotyledons, early shed, 11–28 mm, shortly petioled; petiole openly shallow-sulcate ventrally, 2.5–5.5 mm, at middle 0.3–0.5 mm diam; petiolar gland 0; lfts 1 pair, ascending from tip of petiole turned half face to face, in outline dimidiately obovate (8–)10–21(–24) × (4–)6–12(–14) mm, at apex obtuse to truncate and commonly also acuminulate or mucronulate, at base semi-cordate, the submembranous blades ±4–6-nerved from pulvinule by the strongly displaced, closely penni-nerved midrib with 3–5 major veins on its proximal side, the venation equally prominulous on both faces.

Peduncles exactly axillary, to 1 mm, 1-fld but often several-bracteate; pedicels widely ascending, at anthesis subfiliform thickened in fruit, (7–)9–21 mm, bracteolate near or above middle; bracteoles lanceolate or narrowly ovate, several nerved, 2–3.5 mm; buds slenderly ovoid-acuminate, loosely pilosulous at base or throughout; sepals dull red or reddish-brown, lance-acuminate to narrowly ovate-acute, 6–9.5 mm; petals yellow fading pinkish-brown, at anthesis campanulately ascending, not or shortly exserted beyond the longest sepal, obovate-cuneate to oblong-obovate, 7.5–10 mm, the abaxial pair sometimes a trifle larger than the 3 adaxial but neither oblique nor differentiated in texture; stamens 5, oppositisepalous, the often reddish anthers narrowly lance-acuminate, 4.5–6.6 mm, their sutures mealy-puberulent; style cylindroid, glabrous, 2.5–3 mm, at early anthesis about as long as the longest anther; ovary densely pilosulous; ovules (7–)8–11(–12).

Pod ascending, (2.5–)3–4.5 × 0.6–0.7 cm, the castaneous nigrescent valves transversely venulose, pilosulous; seeds obliquely oblong, 4.1–4.7 mm, the testa pale brown, dull, microscopically papillate, mucilaginous when wetted, the lines of pits visible externally as faintly paler dots, not excavated.—Collections: 79.

Virgin campo and cerrado, persisting and often flourishing in disturbed or burnt campo and along highways, occurring both in abundant colonies and as scattered individuals, 850–1400 m, widespread over the Planalto from the upper R. São Francisco valley in Minas Gerais (Montes Claros s. to Lagôa Santa), w. in scattered stations over the highlands of s. Goiás to the Araguaia-Xingú divide in e. Mato Grosso; apparently isolated on the upper Rios Pardo and Tieté in s.-e. São Paulo (Mogi-Mirim; Itú).—Flowering primarily IX–III, but occasional through the year.

A species instantly recognized by the dorsiventrally flattened, ribbonlike files of cordate stipules loosely imbricated along the stems and interrupted here and there by a few small axillary flowers. In the adult plant stipules assume the photosynthetic function of the foliage but normal foliaceous leaf-blades are regularly produced from about ten nodes of the primary stems and sometimes again at a few nodes low on the lateral branchlets. Upward all leaves are reduced to the terminal seta of the suppressed petiole, each concealed by a pair of appressed stipules. In the form of the bifoliolate perfect leaves and of the pentamerous almost regular corolla, *Ch. basifolia* greatly resembles the more erect forms of *Ch. rotundifolia*, and is presumably derived from it or an immediate common precursor.

There is some small variation in flower-size, perhaps partly seasonal and in any case without geographical correlation. Vesture of the plant varies below the always pilosulous flower-buds and ovary from relatively long and copious, composed of slender, widely spreading, multiseptate hairs scattered along the stems and over faces and margins of the stipule-blades, to almost none, when reduced to a few short cilia on the stipule margins and a line of short incurved hairs along one face of the stem-internodes. The range of the most hairy form extends from central Minas Gerais into east-central and southeastern Goiás; that of the most glabrous from east-central Mato Grosso south through southwestern Goiás and the Minas Triângulo to São Paulo. However a broad band of intermediate forms and of occasional heterogeneous populations has been traced from Chapada dos Veadeiros south into the Federal District, where glabrous and pubescent phases are found close together. While none of the glabrous form is yet known from the São Francisco basin and none of the hairy one from Mato Grosso the intergradation between the extremes is so gradual and so frequent that their taxonomic recognition should be avoided.

v. ser. GREGGIANAE Irwin & Barneby

Chamaecrista sect. **Chamaecrista** ser. **Greggianae** Irwin & Barneby, ser. nov., sepalis obtusis ideo alabastris globosis (nec ovoideis acutis vel acuminatis) inter *Chamaecristas* omnes tam neo- quam paleogeas unicae, ulterius habitu xeromorphico microphyllo fruticoso insignes.—Sp. unica: **Ch. greggii** (A. Gray) Pollard ex A. Heller.

Stiff microphyllous desert shrubs; petiolar gland present; lfts 2–6 pairs, small and firm; peduncles axillary or (rarely) shortly supra-axillary; sepals obovate-suborbicular obtuse; petals up to 9–16 mm; androecium 10-merous.—Spp. 1, of arid upland e. Mexico and s. Texas.

34. Chamaecrista greggii (A. Gray) Pollard ex A. Heller, Cat. N. Amer. Pl. ed. 2, 5. 1900. *Cassia greggii* A. Gray, Pl. Wright. **1**: 59, in nota. 1852.—Typus infra sub var. *greggii* indicatur.

Erect or diffuse, tortuously stiff-branched, microphyllous xeromorphic shrubs and treelets 5–20(–35) dm, with smooth or longitudinally fissured gray-barked trunks and annotinous branches, the hornotinous branchlets, lf-stalks and axes of inflorescence puberulent with appressed straight or curly-incumbent hairs up to 0.1–0.25(–0.3) mm, the branchlets canescently so, the firmly chartaceous olivaceous concolorous dull or sublustrous lfts either glabrous minutely ciliolate, or thinly puberulent beneath.

Stipules erect, subsymmetrically lanceolate or triangular-subulate, 1–3(–4) mm, the firm blades brunnescent persistent.

Lvs ascending and spreading, distant along young branchlets or sometimes subfasciculate on lateral brachyblasts, 7–30 mm, the expanded blade ovate-oblong in outline; petiole with short pulvinus (1.5–)2–5.5(–7) mm, at middle 0.4–0.6 mm diam, narrowly margined and openly shallow-sulcate; gland 1, situated near or well above middle of petiole proper, sometimes adjacent to first pair of pulvinules, urceolate short-stipitate or subsessile 0.15–0.45 mm diam, in profile varying from shortly tack- to drum-shaped or squatly obconic, sometimes subpatelliform, 0.2–0.6 mm tall, 0.3 mm taller to 0.15 mm shorter than diameter of head, rarely additional small glands on rachis distally; lfts of major lvs 2–5(–6) pairs, narrowly oblong or oblanceolate to oblong-elliptic or obovate, obtuse or emarginate, mucronulate, the longer ones up to 5–17 × 2–5 mm, at strongly asymmetric base cordate on proximal and subcordate to cuneate on distal side, from base 2–6-nerved by the centric straight costa with on proximal side 1–3, on distal one 0–1 primary veins, these all short or the innermost on the proximal side produced nearly to blade's apex, the costa from near or from far below middle penniveined, with 2–6 pairs of secondary venules, these sometimes connected by tertiary venulation into an irregular mesh, the whole venation usually prominulous both sides, always so beneath.

Peduncles 0.5–5(–9) mm, commonly exactly axillary, sometimes adnate to stem through 0.5–2(–6) mm, 1(–2)-fld; pedicels at anthesis (9–)12–30(–35) mm, bracteolate 4–14 mm below calyx, in fruit a little thickened but hardly longer; bracteoles ovate or subulate 1–2.5 mm; buds broadly ovoid or globose, obtuse; sepals yellowish subpetaloid, elliptic to elliptic-obovate or suborbicular up to 5–8(–9) mm; petals yellow, 4 unequally obovate to broadly oblanceolate beyond the cuneate claw, up to (8–)9–14.5(–16) × 4.5–8.5 mm, the dimidiate fifth one incurved over androecium; stamens 10, the two cycles not much differentiated in size, the longest anthers 3.4–5.3 mm; ovary strigulose, at least along the sutures; style linear, gently incurved, 2.8–5 mm; ovules (5–)6–9.

Pod (25–)30–46(–50) × (5–)6–7.5 mm, the castaneous or purplish-castaneous valves obliquely venulose; seeds (few seen) 4–5 mm, the black, dull or sublustrous testa faintly pitted.

As defined herein, *Ch. greggii* is easily distinguished from all other North American chamaecristas by the syndrome of xeromorphic fruticose growth-style, few firm leaflets, and obtuse flower-buds. We are unable to see in *Chamaecrista greggii, Ch. macdougaliana* and *Ch. potosina* as treated by Britton & Rose (1930) more than one major specific type, in segments of its full range somewhat modified as to venation and number of leaflets, but everywhere maintaining a characteristic facies along with essentially invariable flowers and fruits.

Over most of the range of *Ch. greggii* the peduncles emerge directly from the leaf axils, suggesting the possibility of affinity to the Brazilian species of ser. *Coriaceae* collectively so characterized; but in some populations of var. *greggii* (e.g. *Palmer 283*) they are distinctly adnate to the stem and supra-axillary as in ser. *Chamaecrista*. The species is likely to be confused only with the partly sympatric *Senna wislizeni,* superficially similar in growth-form and obtuse sepals but, as befits a member of genus *Senna,* totally different in the racemose inflorescence, bractless pedicels, glabrous anther-sacs, inertly dehiscent pod, and filiform seed-funicle.

Key to the Varieties of *Ch. greggii*

1. Lfts relatively numerous and small, in larger lvs 3–5(–6) pairs and up to 5–11 mm; venation various.
    2. Inner primary vein on proximal side of midrib short, not attaining middle of lft-blade, usually not or scarcely longer than the lowest pair of secondaries; tertiary venulation well developed, connecting the secondary pinnate venules into an irregular but sharply defined mesh; ovary strigulose overall, the faces of the pod puberulent; n.-e. Mexico, n.-ward from s. Tamaulipas to n.-e. Coahuila and adjacent s. Texas. 34a. var. *greggii* (p. 737).
    2. Inner primary vein on proximal side of midrib produced up to or beyond middle of lft-blade; tertiary venulation 0 or very obscure, not connecting the secondaries into a mesh; ovary strigulose only along the sutures, the faces of the pod glabrous; local in the Tehuacán desert, s.-e. Puebla and adjacent Oaxaca. 34c. var. *macdougaliana* (p. 737).
1. Lfts usually 2, rarely up to 4 pairs, in any case longer, those of larger lvs up to 11–17 mm; primary vein adjacent to midrib on proximal side produced well beyond middle of lft-blade, sometimes to its apex; tertiary venulation sometimes present, but fainter than in vicariant

var. *greggii,* sharper than in disjunctly allopatric var. *macdougaliana*; ovary of the latter, the faces of the pod glabrous; local in s.-w. Tamaulipas and s.-e. San Luis Potosí.
34b. var. *potosina* (p. 737).

34a. Chamaecrista greggii (A. Gray) Pollard ex A. Heller var. **greggii.** *Cassia greggii* A. Gray, 1852, l.c., sens. str.—"Northern Mexico, near Rinconada, Cerralvo, and Monterey, *Dr. Gregg*; North of Monterey, *Dr. Wislizenus*; Aug."—Lectoholotypus (Britton & Rose, 1930, p. 283, by implication, ratified by Isely, 1975, p. 203): *Gregg s.n.,* 25.V.1847, from w. of Rinconada, GH! isotypus, NY (hb. Torr.)! paratypus, *Gregg 823* from Cerralvo, MO!

Cassia greggii sensu Bentham, 1871, p. 575. *Chamaecrista greggii* sensu Britton & Rose, 1930, p. 282.

Lvs (0.7–)1–3 cm; lfts (2–)3–5(–6) pairs, the longer of a given plant up to 5–11(–14) × 2–4.2 mm; venation as given in key.—Collections: 42.

Open or brushy hillsides and thickets at 300–1250 m, often associated with *Larrea* or *Cordia,* locally plentiful along the e. foothills and first bench of Sa. Madre Oriental from Victoria and Soto la Marina in s.-centr. Tamaulipas n.-w., always on limestone or caliche soils, through centr. Nuevo León and e. Coahuila as far as Músquiz and Allende, and disjunctly to the s. Texas Coastal Plain in Jim Wells and Live Oak counties.—Fl. (III–)V–X(–XI).

34b. Chamaecrista greggii (A. Gray) Pollard ex A. Heller var. **potosina** (Britton & Rose) Irwin & Barneby, Phytologia **44**(7): 500. 1979. *Chamaecrista potosina* Britton & Rose, N. Amer. Fl. **23**(5): 283. 1930.—"Minas de San Rafael, San Luis Potosi, May, 1911, *Purpus 4832.*"—Holotypus, so labelled, NY! isotypi, sub eodem num. ex "Buena Vista, XI.1910," G, GH, US!—*Cassia potosina* (Britton & Rose) Standley, Field Mus., Bot. **22**: 78. 1940.

Lvs up to 1.2–2.3 cm; lfts commonly (1–)2, rarely 2–4 pairs, those of longer lvs oblanceolate and up to 11–17 × 2.5–5 mm; venulation as given in key.—Collections: 9.

In desert scrub on limestone or perhaps sometimes on gypsum, 620–2500 m, localized in intermontane valleys of Sa. Madre Oriental and plains w. of the crest in s.-w. Tamaulipas and e. San Luis Potosí, from Miquihuana and Jaumave s. to Río Verde.—Fl. V–XII.

34c. Chamaecrista greggii (A. Gray) Pollard ex A. Heller var. **macdougaliana** (Rose) Irwin & Barneby, stat. nov. *Cassia macdougaliana* Rose, Contrib. U.S. Nat. Herb. **12**: 267. 1909.—"Collected by J. N. Rose in company with Dr. D. T. MacDougal near Tehuacán, Puebla, September 1, 1906 (*no. 11253,* type) . . ."—Holotypus, US! isotypi, NY (2 sheets)!—*Chamaecrista macdougaliana* (Rose) Britton & Rose, N. Amer. Fl. **23**(5): 283. 1930.

Lvs up to 8–16 mm; larger lfts up to 5–7.5 × 2–3.7 mm; venulation as given in key.—Collections: 13.

Stony calcareous desert hillsides, ±1200–1800 m, local in the Tehuacán Desert of s. Puebla and immediately adjoining Oaxaca.—Fl. VII–IX.

vi. ser. CHAMAECRISTA

Chamaecrista Moench sect. **Chamaecrista** ser. **Chamaecrista**. *Chamaecrista* Moench, Meth. pl. hort. bot. Marburg. 272. 1794, sens. strict.—Generitypus: **Ch. nictitans** (Linnaeus) Moench.—*Cassia* sect. *Chamaecrista* (Moench) DeCandolle ex Colladon, 1816, l.c., sens. str.

Cassia sect. *Chamaecrista* (Moench) ser. *Mimosoideae* DeCandolle ex Colladon, Hist. Casses 121. 1816.—Sp. lectotypica: *C. mimosoides* Linnaeus ≡ **Chamaecrista mimosoides** (Linnaeus) Greene.

Cassia sect. *Chamaecrista* (Moench) ser. *Sesbanianae* DeCandolle ex Colladon, Hist. Casses 124. 1816.—Sp. lectotypica: *C. chamaecrista* Linnaeus = **Chamaecrista fasciculata** (Michaux) Greene.

Hepteireca Rafinesque, Sylva Tellur. 126. 1838, pro gen.—Generitypus: *H. glandulosa* (Linnaeus) Rafinesque ≡ **Chamaecrista glandulosa** (Linnaeus) Greene.

Dialanthera Rafinesque, Sylva Tellur. 127. 1838, pro gen.—Generitypus: *D. glandulosa* (Linnaeus) Rafinesque ≡ **Chamaecrista glandulosa** (Linnaeus) Greene.

Xamacrista Rafinesque, Sylva Tellur. 127. 1838, pro gen.—Generitypus: *X. triflora* Rafinesque ≡ *Cassia chamaecrista* Linnaeus = **Chamaecrista fasciculata** (Michaux) Greene.

Nictitella Rafinesque, Sylva Tellur. 128. 1838, pro gen.—Generitypus: *N. amena* Rafinesque ≡ **Chamaecrista nictitans** (Linnaeus) Moench.

Cassia sect. *Chamaecrista* ser. "*Chamaecristae verae*" Bentham in Martius, Fl. Bras. **15**(2): 172. 1870.—Sp. typica: *C. chamaecrista* Linnaeus = **Chamaecrista fasciculata** (Michaux) Greene.

Cassia sect. *Chamaecrista* ser. *Dimidiatae* Bentham, Trans. Linn. Soc. London **27**: 581. 1871.—Sp. lectotypica: *C. tristicula* Humboldt, Bonpland & Kunth ≡ **Chamaecrista glandulosa** var. **tristicula** (H.B.K.) Irwin & Barneby.

Chamaecrista Tristiculae, Lineatae, Chiapenses, Deeringianae, Glandulosae, Yucatanae, Pygmaeae, Patellariae, Fasciculatae & Tabascoënses Britton & Rose, N. Amer. Fl. **23**(5): 270–273, in clave, sine descr. 1830.

Mostly herbs, perennial or monocarpic, the latter often weedy, some softly woody in age, a few frutescent or exceptionally arborescent; petiolar gland(s) always present but highly variable in form and size, sessile or stipitate; lfts (1–)2–45 pairs, the midrib centric or variably displaced toward distal margin, the blade eventually becoming dimidiate; peduncles adnate to stem through (0.5–)1+ mm, the 1–few-fld racemes appearing supra-axillary; sepals acute or acuminate; fls variable in size, either dolicho- or brachystylous.—Spp. numerous, the exact number subject to taxonomic standards as yet not evenhandedly applied throughout the series, in range circumtropical and extending into temperate South Africa, n.-e. Asia, South America and into arid s.-w. and forested n.-e. United States, 17 here recognized from the Americas but some of these aggregations of taxa alternatively evaluable as a plexus of microspecies, with us ubiquitous throughout the lowland tropics except where excluded by climax forest, most numerous and diverse in e. Brazil, Central America and West Indies; highly diversified, often in parallel ways, in tropical Africa, less so in Asia, n. Australia, Madagascar.

Our ser. *Chamaecrista* is equivalent to Bentham's ser. *Verae* and *Dimidiatae* collectively, and approximately to ser. *Mimosoideae* of earlier classifications. We have found it impractical to segregate the *Dimidiatae,* marked solely by a more emphatic distal displacement of the midrib of the leaflets, for this is variable within some taxa (notably in *Ch. nictitans* subsp. *patellaria*) and may well have developed independently in several lines of inheritance. The series is only precariously separable from others by the apparently supra-axillary origin of the inflorescence due to concrescence of peduncle and stem through a space of 1 mm or more. Shortly supra-axillary peduncles are encountered occasionally in *Ch.* (ser. *Prostratae*) *pilosa* and *trichopoda* also, and we have been obliged to introduce these species into the keys following.

The taxonomy of ser. *Chamaecrista* is preeminently and notoriously difficult, alike in the American and African tropics (cf. Steyaert, 1952, pp. 514–531; Brenan, 1967, pp. 81–103). On both these continents and to less marked degree in tropical Asia the group is in evolutionary ferment, a process accelerated, no doubt, by disturbance of the primitive ecology and consequent increase of suitably weedy or unstable second-growth bush-savanna habitats in which the chamaecristas thrive. Great ranges of variation have developed in the duplex vesture of short incurved and longer setiform hairs, in form of the petiolar glands, in number and venation of the leaflets, in flower-size and, often correlated with this, in length of style and consequently in mode of pollination, but syndromes of character-states at once correlated with dispersal and discrete from others seem hardly to exist. In spite of an enormously enriched data base we are hardly better off than was Bentham a century ago (1871, p. 576) when he wrote that the "*Chamaecristae verae* seem to pass so gradually one into the other, that the delimitation of species . . . must be admitted to be very unsatisfactory." The formulation of specific concepts remains in many cases arbitrary, a pragmatic method of dealing with a bewildering plexus of which the genetic structure is yet to be systematically investigated. The rationale of our proposed classification is propounded in greater detail under the individual species following.

An inherent instability of particular phenetic characters within the taxon and concurrent parallelism between taxa believed to be not closely related are formidable obstacles to an efficient dichotomous key to all American members of ser. *Chamaecrista*. We have taken a practical approach, in the interests of simplification, by dividing the territory covered into three and presenting a key for each division: i (immediately following), continental North and equatorial South America to the south limit of the Amazonian Hylaea; ii (p. 740), South America southward from the preceding; and iii (p. 741) the West Indies.

Key (i) to Major Taxa of ser. *Chamaecrista* in the Andean and Hylaean Provinces of South America, in all South America n. of the Equator, and Continental North America

1. Midrib of lfts centric or moderately displaced, dividing the blade 1:1–2.25.
    2. Lfts of larger lvs 10 pairs or more.
        3. Monocarpic, large-fld dolichostylous, and extratropical, the longest petal 10–23 mm, the style (3.5–)4–7 mm, the androecium always 10-merous; endemic to e., e-.centr. and s.-e. United States. 50. *Ch. fasciculata* (p. 802).
        3. Either monocarpic or perennial, but if at once monocarpic and sympatric with the last the fl small brachystylous, the long abaxial petal 3.5–7(–8) mm, the style 0.8–1.5 mm, and the androecium 4–8-merous.
            4. Stems arising from shallowly buried, oblique, rhizomelike root; style 6–8 mm; local in s.-e. United States. 49. *Ch. deeringiana* (p. 801).
            4. Stems arising from vertical or obliquely ramified roots; allopatric if style over 3 mm; 2 multiracial megaspecies, morphologically confluent and widely sympatric.
                5. Perennial, suffruticose, fruticose or subarborescent; fl relatively long-pedicellate and large dolichostylous, the pedicels 7–30 mm, the longest petal (8–)8.5–23 mm, the always linear style (3–)3.4–9 mm, but, if pedicels less than 10 mm, or petals less than 10 mm, or style less than 3.5 mm, the plant decisively frutescent. 47. *Ch. glandulosa* (p. 780).
                5. Monocarpic, sometimes in age softly lignescent basally but not genuinely perennant; fl variable in size, commonly (except in subsp. *disadena*) short-pedicellate and (except in var. *disadena* and some forms of var. *paraguariensis*) small brachystylous, the pedicels mostly 1–10 mm, the longest petal mostly 3.5–10 mm, the style mostly 0.8–3 mm (or if up to 3.5 mm then distally dilated), but where these measurements are surpassed the plant decisively monocarpic, often monopodial virgate. 51. *Ch. nictitans* (p. 811).

2. Lfts of larger lvs 2–9 pairs.

6. Lfts 2–5 pairs; pedicels (2–)2.5–5 cm, as long as or longer than the associated lf; fl small brachystylous, the longest petal 3.5–6.5 mm, the style 1–2.2 mm; cf. widespread weedy 28. *Ch.* (*Prostratae*) *pilosa* (p. 718).

6. Lfts 5–9 pairs; pedicels shorter, or shorter than associated lf, or both; fl larger dolichostylous; highly localized taxa, not weedy.

7. Densely pilosulous herbaceous perennial herbs of s. peninsular Florida; var. b. *keyensis* of 40. *Ch. lineata* (p. 752).

7. If herbaceous perennial then not densely pilosulous, and in any case remotely allopatric on Mexican Plateau or in Andes of Colombia and Peru.

8. Stems decumbent or weakly assurgent from root-crown, fully herbaceous and not over 3.5 dm; local either in n. Colombia or on Mexican Plateau; vars. k. *andreana* and l. *parralensis* of 47. *Ch. glandulosa* (p. 780).

8. Stems erect shrubby 4 dm and over; local on e. slope of Andes in n.-centr. Peru. 48. *Ch. pedemontana* (p. 799).

1. Midrib of lfts strongly displaced abaxially or marginal, dividing the blade 1:1–2.25+.

9. Fl small brachystylous, the longest petal 3.5–6.5 mm, the style 0.8–1.8 mm.

10. Petiolar gland slenderly stipitate, in profile pin- or tack-shaped; pedicels 10–26 mm; style filiform; slender, at least potentially perennant herbs, mostly prostrate; cf. 27. *Ch.* (*Prostratae*) *trichopoda* (p. 716).

10. Petiolar gland sessile or almost so, patelli- or scutiform; pedicels 0.5–6 mm; style apically dilated; coarsely leafy, commonly erect or ascending, monocarpic herbs; var. b. *ramosa* of 51. *Ch. nictitans* (p. 811).

9. Fl larger dolichostylous, the longest petal 8–19 mm, the style 3–6 mm.

11. Stipules (5–)7–17 × (0.7–)1.2–3 mm; major cauline lvs 3–9(–12) cm; lfts (14–)18–45 pairs; discontinuously dispersed from Mexico to Amazonian Brazil, but on the Gulf slope in Mexico only above 800 m (if led hither by plants of Cordillera Oriental in Colombia cf. discussion of 47g. *Ch. glandulosa* var. *tristicula,* p. 793). 43. *Ch. rufa* (p. 765).

11. Stipules (2–)3–7 × 0.7–1.7 mm; major cauline lvs up to 1.5–3.5(–4) cm; lfts 10–17(–19) pairs; immediate coast and adjacent coastal plain of extreme s. Texas and e. Mexico s.-ward to Campeche, disjunctly on shore of Gulf of California in Sinaloa and in coastal Jalisco and Michoacán. 44. *Ch. chamaecristoides* (p. 769).

Key (ii) to Major Taxa of ser. *Chamaecrista* in Brazil s. of the Amazonian Hylaea, Lowland e. Bolivia, Argentina, Paraguay and Uruguay

1. Lfts of larger adult lvs 11–many pairs.

2. Midrib of lfts highly excentric, dividing the blade 1:2.25+.

3. Fls at once very small and long-pedicellate, the filiform pedicels 10–26 mm, the longest petal 4–6.5 mm; gland slender-stipitate, in profile tack- or pin-shaped; cf. 27. *Ch.* (*Prostratae*) *trichopoda* (p. 716).

3. Fls larger, or if as small then short-pedicellate, the pedicels not over 6 mm.

4. Monocarpic, small-fld brachystylous, the longest petal 3.5–5.5 mm, the distally dilated style less than 2 mm; var. b. *ramosa* of 51. *Ch. nictitans* (p. 811).

4. Perennial or fruticose, large-fld dolichostylous, the longest petal 10–19 mm, the linear style 3.5–7 mm.

5. Petiolar gland sessile patelliform; n. Argentina and feebly into Paraguay and e. Bolivia. 36. *Ch. venturiana* (p. 746).

5. Petiolar gland elevated on a mounded (often puberulent) pediment or columnar stipe as wide or wider than itself; e. Brazil, from e.-centr. Goiás and n. Rio de Janeiro n. to Maranhão; var. b. *multijuga* of 35. *Ch. repens* (p. 742).

2. Midrib of lfts centric or moderately displaced, dividing the blade 1:1–2.2.

6. Petiolar gland slenderly stipitate, in profile tack- or pin-shaped, taller than diam of head; III. subsp. *disadena* of 51. *Ch. nictitans* (p. 811).

6. Petiolar gland sessile or subsessile, shorter than diam of head.

7. Shrubs of the Atlantic slope in Rio de Janeiro and Guanabara; pedicels 13–35 mm; var. m. *brasiliensis* of 47. *Ch. glandulosa* (p. 798).

7. Herbs, perennial or monocarpic; pedicels not over 10 mm; interior Brazil to Argentina and Bolivia.

8. Erect, monocarpic, usually virgate and monopodial; midrib of lfts subcentric.
9. Lfts of larger lvs 10–12 pairs; subsp. I. *brachypoda* of 51. *Ch. nictitans* (p. 812).
9. Lfts of larger lvs 20–38 pairs; var. IIa. *paraguariensis* of 51. *Ch. nictitans* (p. 815).
8. Diffuse perennial, with several stems radiating from the root-crown; midrib of lfts moderately displaced. Lfts of larger lvs 12–18 pairs. 36. *Ch. venturiana* (p. 746).
1. Lfts of larger adult lvs 3–10 pairs.
10. Pedicels (2–)2.5–5 cm, the longer one at a node usually as long as the associated lf; fl small, the style 1–2.2 mm; gland small, slender-stipitate, or sometimes 0; cf. 28. *Ch.* (*Prostratae*) *pilosa* (p. 718).
10. Pedicels 0.4–2.3 cm, much shorter than associated lf, or fl much larger, or both, and if pedicels over 2 cm the plant both fruticose and large-fld.
11. Fls relatively large dolichostylous, the longest petal (8–)11–21 mm, the linear style at least 3.5 mm (if petals less than 11 mm, then the petiolar gland sessile).
12. Petiolar gland sessile or very shortly stipitate, sometimes raised on a mounded pediment or squat column but no longer than diam of head.
13. Erect shrubs of the Atlantic slope in Rio de Janeiro and Guanabara, at anthesis 8–15 dm; pedicels 13–35; var. m. *brasiliensis* of 47. *Ch. glandulosa* (p. 780).
13. Diffuse perennial or erect monocarpic herbs, mostly of interior planaltine Brazil; pedicels 5–23 mm.
14. Perennial, the stems diffusely radiating from crown of a woody root; gland raised on a puberulent mounded pediment as broad or broader than itself; var. a. *repens* of 35. *Ch. repens* (p. 742).
14. Monocarpic, virgately erect; gland sessile or short-stipitate, the stipe narrower than the head; subsp. I. *brachypoda* of 51. *Ch. nictitans* (p. 811).
12. Petiolar gland stipitate, in profile mostly peg- or trumpet-shaped and longer than diam of head (if slightly shorter, the stipe at once puberulent and much narrower than head); Bahia to Maranhão and Pernambuco.
15. Relatively tall, often monopodial 2–8 dm; stipules 4.5–10 mm; lvs 2.5–7 cm, the larger with 4–9 pairs of lfts up to ±1–2 cm; sepals 10–16 mm; pod 4–7.5 cm long, 14–28-ovulate; widespread in range indicated. 37. *Ch. pascuorum* (p. 747).
15. Dwarf shrublet 1–2 dm; stipules 1.6–2.2 mm; lvs 6–17 mm, the larger ones with 3(–4) pairs of lfts up to 2.5–7 mm; sepals 7–9.5 mm; pod ±2.5–3.5 cm long, 8–10-ovulate; local in w.-centr. Bahia. 38. *Ch. cuprea* (p. 749).
11. Fls small brachystylous, the longest petal 8–10 mm, the style 1.5–3.5 mm; gland slender-stipitate, in profile tack- or trumpet-shaped, longer than diam of head; s.-e. Brazil (s.-ward from São Paulo) and Paraguay. 39. *Ch. vestita* (p. 750).

Key (iii) to Major Taxa of ser. *Chamaecrista* in the West Indies

1. Midrib of lfts either centric or only moderately displaced, dividing the blade 1:1–2.2.
2. Lfts (mostly 5–8, rarely 14–16 pairs) stiffly chartaceous, callous-margined, the thick marginal nerve as strong or almost as strong as the prominent midrib; Hispaniola (rare), Puerto Rico, Virgin Is. 42. *Ch. portoricensis* (p. 763).
2. Lfts membranous or subchartaceous, but not callous-margined, the marginal nerve, when differentiated as such, filiform.
3. Lfts 2–5 pairs; pedicels (2–)2.5–5 cm, the longer ones as long or longer than the associated lf. Prostrate herbs of Cuba and Jamaica; cf. 28. *Ch.* (*Prostratae*) *pilosa* (p. 718).
3. Lfts over 5 pairs; pedicels either absolutely shorter, or much shorter than the lf, or both.
4. Plant sometimes precociously flowering the first season but potentially and when adult suffruticose, fruticose or subarborescent; fls relatively long-pedicelled, or relatively large, or dolichostylous, or some of these together, the pedicels (6–)7–25 mm, the longest petal (8.5–)9–19 mm, the always linear style 3–6 mm.
5. Lfts of larger lvs at least (8–)9–23 pairs; widespread except Bahamas and Cuba. 47. *Ch. glandulosa* (p. 780).
5. Lfts of larger lvs (1–)2–8 pairs, very occasionally in e. Cuba and Bahamas up

to 8–12 pairs; sympatric with the last in Jamaica, Hispaniola, Puerto Rico, Virgin Is and some Windward Is. e. to Guadeloupe, replacing it in Cuba and Bahamas.

6. Glands both on petiole and on each segment of lf-rachis, slenderly stipitate, in profile tack- or peg-shaped and mostly as long or longer than diam of head; lfts obovate-emarginate or obcordate, glabrous glaucous; known only from St. Barthélemy, St. Martin, and Guadeloupe. 41. *Ch. obcordata* (p. 762).

6. Glands either stipitate or sessile, but usually absent from lf-rachis; lfts, when coinciding with stipitate petiolar gland, not as above in all respects; Puerto Rico w.-ward. 40. *Ch. lineata* (p. 752).

4. Plant essentially monocarpic, sometimes softly woody in age; pedicels 1–8 mm; longest petal mostly less than 8, rarely to 10 mm and style 0.8–2 mm long. 51. *Ch. nictitans* (p. 811).

1. Midrib of lfts strongly displaced or submarginal, dividing the blade 1:2.5+.

7. Coarse monocarpic herbs with very small, short-pedicelled, brachystylous fls, the pedicels 2–5.5 mm, the longest abaxial petal 3.5–5.5 mm, the style 0.8–1.6 mm, distally dilated and obliquely truncate at apex; widespread and weedy almost throughout Cuba and Jamaica; Guadeloupe; var. IIb. *ramosa* of 51. *Ch. nictitans* (p. 811).

7. Perennial, mostly slender herbs, with usually longer pedicels, often larger fls, and always linear styles; pedicels and fls sometimes no longer than the last in w. Cuban *Ch. pygmaea,* but the style then linear.

8. Lfts 5–12 mm; pedicels 13–35 mm; petals up to 8.5–12.5 mm; style 2.6–4.6 mm; extreme e. Cuba (Oriente) and Hispaniola. 45. *Ch. pedicellaris* (p. 763).

8. Lfts 2.5–5.5(–6) mm; pedicels 3–20 mm; petals up to 4.5–8(–9) mm; style 1–2 mm; w. Cuba (w. Matanzas, Pinar del Río, Isla de Pinos). 46. *Ch. pygmaea* (p. 776).

35. Chamaecrista repens (Vogel) Irwin & Barneby, comb. nov. *Cassia repens* Vogel, Syn. Gen. Cass. 60 & Linnaea **11**: 712, descr. ampliat. 1837.—Typus infra sub var. *repenti* indicatur.

Perennial herbs with 2–several simple or few-branched, amply leafy stems 1.5–11(–14) dm arising from the crown of a woody taproot or from divisions of a sometimes subterraneously branched caudex, fully herbaceous or sometimes suffrutescent in age, variably puberulent or pilosulous with short incurved or short spreading, filiform or bulbous-based hairs and sometimes in addition longer spreading setae, the longer trichomes up to 0.4–1.4 mm, the foliage bright green, olivaceous or gray-olivaceous concolorous, the lfts varying from glabrous on both or only the upper face and ciliolate or rarely setose-ciliate to densely gray-pilosulous throughout, the lvs of mature plants commonly reduced at several shortened distal nodes and the inflorescence becoming pseudo-racemose.

Stipules erect persistent, narrowly lanceolate or narrowly ovate-acuminate, 2.5–8(–10.5) mm, at ± oblique base 0.7–1.6 mm wide, obscurely rounded to subcordate on broader side, from base prominently 5–8(–10)-nerved, the blades either glabrous or pilosulous dorsally, nearly always ciliolate.

Lvs at middle of flowering stems up to 2.5–10(–11) cm, those at the lower and uppermost nodes shorter, all shortly stout-petiolate, the expanded blade ovate to lanceolate or lance-elliptic in outline, the distal and often the first pairs of lfts smaller; petiole including pulvinus 3–8.5(–10) mm, at middle rather broadly margined and (0.45–)0.5–1.2(–1.5) mm diam, the marginal flanges commonly forced apart by the massive gland so as to form, in dorsal view, a pronounced nodule; gland almost always solitary, situated near middle or at top of petiole proper, in outline round, deeply concave, coarsely smooth-margined, (0.4–)0.5–1 mm diam, elevated on a broad mounded, commonly pilosulous pediment (dilated stipe), the whole in profile 0.5–0.9 mm tall, at base 0.65–1.5(–1.6) mm diam, at least a trifle broader than high, but the stipe becoming in some populations of var. *multijuga* cylindric and the whole gland drum-shaped in profile; lfts of adult median lvs (4–)

5–19(–20) pairs, relatively well-separated along rachis, in outline varying from linear-oblong to oblong-elliptic or -obovate, straight or falcate, obtuse to subemarginate or exceptionally subacute, abruptly mucronulate or spinulose, the largest (10–)12–23(–27) × (2–)2.5–8 mm, at strongly asymmetric base cordate or cordately angulate on the broad side, from base prominently 6–9(–11)-nerved by the subcentric to strongly excentric midrib with on the broad side 4–6(–7) and on the narrow one (0–)1–2(–3) veins, the costa thence penniveined, with 4–7(–8) pairs of major secondary and often as many almost as strong intercalated venules, the venation prominulous on both faces but often more sharply so beneath.

Peduncles with raceme-axis (2–)3–13(–15) mm, the peduncle adnate throughout or free up to 3 mm, the axis 0–3.5 mm, 1–4(–5)-fld; pedicels ascending, at anthesis slender 4–14 mm, in fruit thickened and 6–23 mm (or at some distal nodes shorter), bracteolate 1–5 mm below calyx; bracteoles ovate or lanceolate 1.3–3 mm; buds narrowly ovate-acuminate, thinly pilosulous and rarely also randomly setose; sepals commonly brownish or reddish-tinged, lance-acuminate or the inner ovate-acute, 8.5–13.5 mm; petals yellow fading orange, 3 adaxial oblanceolate to narrowly elliptic-obovate, (7.5–)8–15 × (2.5–)3.5–8 mm, 1 abaxial flabellate to broadly obovate, (10–)11–19 × 7–12 mm, the cucullus semi-obovate, obliquely reniform or falciform 11–20.5 mm; stamens 10, the long anthers 7–12 mm; ovary densely gray strigulose; style linear, gently incurved, (2.5–)3.5–7 × 0.2–0.4 mm; ovules (7–)10–22.

Pod erect or stiffly ascending, linear, slightly arcuate, (25–)35–70 × (3–)3.5–4.7 mm, the reddish-brown or purplish-castaneous, ultimately nigrescent valves minutely puberulent, shortly thinly pilosulous, or rarely remotely setulose; seeds (little known) 2.6–3.3 mm, the dull fulvous or sooty testa blackish-punctate.—Fig. 2 (petiolar nectary).

As here redefined in more comprehensive terms, the inappropriately named *Ch. repens* differs from related South American chamaecristas with large, supra-axillary flowers chiefly in its curious and characteristic (even though in fine detail variable; see note under var. *multijuga*) petiolar gland, which is so stoutly stipitate as to appear perched on a mounded or rarely columnar, usually puberulent pediment at least as wide and ordinarily wider than itself. Variation in number and venation of the leaflets supports the recognition of two geographically disjunct varieties, a southern, largely extra-tropical and extra-Brazilian var. *repens* with relatively few, subsymmetric leaflets, the range of which is centered on the lower Paraguay-Paraná basin; and an endemic Brazilian var. *multijuga,* widespread around the São Francisco basin from western Minas Gerais northeastward to inland Maranhão and Ceará, different in the more numerous, variably but always decidedly oblique leaflets, often but not consistently coinciding with a more erect and more robust habit of growth. Apart from the gland, the detached leaf of var. *repens* closely resembles that of *Ch. nictitans* var. *brachypoda,* vicariant northward on the Brazilian Planalto, but this is readily distinguished in the field by its monopodial monocarpic wandlike life-form and, when closely observed, by the larger, multistriate stipules and sessile patelliform or turbinate gland. Granted the same reservations about the petiolar gland, the leaf of var. *multijuga* is almost that of the distantly allopatric Argentine *Ch. venturiana* which resembles the dwarf and diffuse var. *repens* in habit, and is only somewhat arbitrarily excluded from *Ch. repens* sens. lat., a point discussed in the foreword to ser. *Chamaecristae.*

Variation in composition and density of vesture, everywhere rife in sect. *Chamaecrista,* is especially marked in *Ch. repens,* where it occurs both within and between populations and equally in each variety, the leaflets changing imperceptibly from bright green, glabrous and sharply striate-veined on both faces to gray-pilosulous and only obscurely (cryptically) venulose, or both pilose and striate. A feature of the species as a whole, seen however only in mature flowering and fruiting individuals, is the often sudden reduction in size of the upper leaves, exposing successive fascicles of flowers or pods in a sort of interrupted virgate raceme. The pinnate venulation arising from the midrib of the leaflets varies in number and emphasis of venules intercalary to the always raised 4–7 pairs of major secondaries; and in var. *multijuga* in the development of one pronounced primary vein arising next to the midrib on the leaflet's broad side and produced nearly or quite to the blade's apex. This last pattern characterizes the relatively narrow and falcate leaflets of the typus of *C. drepanophylla,* but occurs sometimes on the same plant with broader leaflets normally veined.

The small sample seen by Bentham (1870, 1871) provided only a defective and distorted view of the real *Cassia repens* and *C. drepanophylla,* taxa which emerge in modern experience as so closely related as to be no longer separable at the specific level. Bentham referred the two species to different series in sect. *Chamaecrista, C. drepanophylla* to ser. *Coriaceae*, from which it is handily excluded by the supra-axillary inflorescences, and *C. repens,* because of its relatively few leaflets, to ser. *Subcoriaceae*. An obscure *C. brachypoda* var. *multijuga,* based on a collection simultaneously (and correctly) associated by Bentham with *C. drepanophylla,* provides our varietal epithet.

Key to Varieties of *Ch. repens*

1. Larger lvs near middle of stems 2.5–6.5(–8) cm, with (4–)5–9(–10) pairs of lfts; midrib of lfts moderately or little excentric, dividing the blade 1:1–2; largely extratropical n.-e. Argentina, Uruguay, adjacent Brazil, and n. in Paraguay to ±22°S. 35a. var. *repens* (p. 744).
1. Larger lvs (4–)5–10(–11) cm, with up to (10–)11–19(–20) pairs of lfts; midrib strongly excentric, dividing the blade 1:2+–4.5; e. Brazil, from near 17°S in w. Minas Gerais to s. Ceará and Maranhão. 35b. var. *multijuga* (p. 745).

35a. Chamaecrista repens (Vogel) Irwin & Barneby var. **repens**. *Cassia repens* Vogel, 1837, ll. cc., sens. str.—"Hbt. In Brasilia. Sellow! leg."—Holotypus, †B = F Neg. *1739*! neoholotypus, former isotypus, K (hb. Hook.)! = NY Neg. *1539*; isotypi, FI (hb. Webb.), G (2 sheets), LE (2 sheets), P!

Cassia apaënsis Chodat & Hassler in Bull. Herb. Boiss. II, **4**: 827. 1904. "[Paraguay:] in campo San Rafael in regione cursus superioris fluminis Apa, Oct., [*Hassler*] n. 7713."—Holotypus, G! isotypi, K, NY, P!

Stems 1.5–4.5(–5) dm, procumbent or assurgent; lvs at middle of stems up to 2.5–6.5(–8) cm; lfts (4–)5–9(–11) pairs, varying from linear-oblong to oblong-elliptic or -obovate, straight, (2–)2.5–8 mm wide, the midrib little or moderately excentric, the pinnate secondary venulation either composed of 4–7(–8) major pairs or of these and as many intercalated ones nearly as strong.—Collections: 51.

Campo, matorral, becoming weedy along highways, paths, stream banks and in fallow pastures, mostly below 600 m, widespread in the basins of the lower-middle Paraguay, lower Paraná, and Uruguay rivers s.-ward to 32°S in Misiones, Corrientes and Entre Ríos, Argentina and adjoining Uruguay, n. in Paraguay to Río Apa and Sa. Amambay (±22°S) and recorded once from s.-e. Bolivia (prov. Chiquitas, near 18°S); extending rarely through Rio Grande do Sul, Brazil, e. to the Atlantic lowlands of Rio Jacuí; to be expected in far w. Santa Catarina and s. Mato Grosso.—Fl. X–III(–IV).

35b. Chamaecrista repens (Vogel) var. **multijuga** (Bentham) Irwin & Barneby, comb. nov. *Cassia brachypoda* (?) var. *multijuga* Bentham in Martius, Fl. Bras. **15**(2): 172. 1870.—". . . in prov. Bahia: Blanchet n. 3711."—Lectoholotypus, P (2 sheets)! isotypi, †B = F Neg. *21811*! G (labelled "Jacobina")! K (paratypus of next following)! LE!

Cassia drepanophylla Bentham in Martius, Fl. Bras. **15**(2): 170. 1870.—"Habitat in Serra de Araripé prov. Ceará: Gardner n. 1572., et ut videtur in prov. Bahia: Blanchet n. 3711 . . ." Lectoholotypus, *Gardner 1572,* collected X.1838 (fl), K (hb. Hook.)! = IPA Neg. *1124* = NY Neg. *1537.*

Cassia subtriflora Martius ex Bentham in Martius, Fl. Bras. **15**(2): 176. 1870.—". . . in campis petrosis Serra das tres Barras prov. Minas Geraës: Martius; in pascuis ad S. Gonçalo d'Amarante prov. Piauhy: Martius; in silvis umbrosis arenosis prope Oeiras ejusdem provinciae: Gardner n. 2127.; in prov. Bahia: Blanchet 3711.; . . . Serra de Araripe prov. Ceará, frequens: Gardner n. 1573.; in silvarum marginibus campisque ad Manáos prov. do Amazonas: Martius."—Lectoholotypus, *Martius Obs. 1268* (sub nom. ined. "*C. erythrocalyx*"), labelled "in campis petrosis Serra das 3 Barras et prope Tejuco, V.1818," M! = NY Neg. *9207*; paratypi, *Gardner 2127,* K (hb. Benth. = IPA Neg. *1133,* hb. Hook.), P! *Gardner 1573,* K (hb. Benth., hb. Hook. = IPA Neg. *1134*), P! *Blanchet 3711,* cited in 2 preceding paragraphs!—The Martius collections from Piauí and Amazonas = *Ch. nictitans* var. *ramosa.*

Cassia drepanophylla sensu Bentham, 1871, p. 574.

Cassia subtriflora sensu Bentham, 1871, p. 580, ex parte, exclus. loc. piauiens. & amazon.

Stems 3–11(–14) dm, sometimes as low and diffuse as in var. *repens* but often taller and more erect, sometimes weakly suffrutescent in age; lvs at middle of stems up to (4–)5–10(–11) cm; lfts (10–)11–19(–20) pairs mostly linear-oblong to oblong-elliptic, straight or falcate, (2–)2.5–4.5(–5) mm wide, the midrib usually strongly excentric, the secondary venulation variable as described for var. *repens,* but the first primary vein arising from the pulvinule on broad side of midrib sometimes as strong as midrib itself and produced almost or quite to blade's apex.—Collections: 56.

Grassy and brushy places in campo, cerrado and caatinga, mostly 450–1200 m, frequent on both slopes of Chapada Diamantina and Sa. do Espinhaço in upland inland Bahia and n. and e. Minas Gerais, extending w. to Espigão Mestre and to the headwaters of Rio Paraná in e.-centr. Goiás, thence s. through Sa. do Rio Preto to headwaters of Rio Paranaíba near 17° on the Goiás-Minas boundary, n. interruptedly to the s.-w. corner of Ceará (Sa. do Araripé) and the middle Parnaiba valley near 7°S in e.-centr. Maranhão, on the Atlantic slope s. from s.-e. Bahia through far e. Minas Gerais to the Doce valley; collected once by St. Hilaire on the Atlantic coast near Ibatapoana in extreme n.-e. Rio de Janeiro and (perhaps as a waif) by Glaziou (no. *12620*) within the city of Rio de Janeiro, Guanabara.—Fl. (VII–)IX–III.

The material of var. *multijuga* seen by Bentham was disposed of in three ways. As noted in the synonymy above, *Blanchet 3711* was simultaneously referred to *Cassia drepanophylla, C. subtriflora* and *C. brachypoda.* Among Martius's material at München we found collections of var. *multijuga* identified by Bentham as *C. brachypoda* (in campis prov. Goyaz) and as *C. subtriflora* (Sa. das Tres Barras). The latter, associated with an unpublished name of Martius, was cited in the protologue of *C. subtriflora* and is herein designated as lectotypus of the same. While it seems probable that the epithet *subtriflora* was intended by Martius to apply to his specimens of *Ch. nictitans* var. *ramosa,* these cannot be associated with the species actually described as *C. subtriflora* by Bentham who twice (1870, 1871, l.c.) emphasized the large flowers as the diagnostic character.

Reference has been made in the description of the species to a variation in the petiolar gland in which the stipe becomes squatly cylindric and not or scarcely wider than the head, and sometimes simultaneously glabrous. Such glands characterize most, but not all populations along the Atlantic slope from southeastern Bahia to Guanabara. The old collection from coastal Rio de Janeiro has, however, the mounded pubescent stipe prevalent inland in the São Francisco valley. We find no correlated differences in foliage or flower.

36. Chamaecrista venturiana Irwin & Barneby, sp. nov., *Ch. repenti* Vog. proxime affinis ejusque var. *repentem* habitu simulans, sed glandula petiolari sessili patelliformi (nec stipite verruciformi dilatato elevata) ulterius a var. *repenti* foliolorum 12–18-jugorum usque costa magis excentrica absimilis. A *Ch. nictitanti* var. *paraguariensi,* quoad glandulam comparabili, radice lignoso pluricauli perenni (nec annua) caulibusque procumbentibus (nec solitario erecto virgato) longius distare videtur.—ARGENTINA. Prov. Tucumán, dept. Cruz Alta: Campo Alegre, 416 m, XI.1921 (fl, fr), *S. Venturi 1471.*—Holotypus, NY: isotypi GH, US.

Perennial herbs with several simple or few-branched stems 1.5–4 dm procumbently or assurgently radiating directly from the crown or from the caudical branchlets of a blackish woody root, except for the always glabrous upper face of lfts puberulent almost throughout with short appressed or incurved hairs and the lf-stalks and rarely the lower face and margins of lfts sometimes in addition pilosulous with loosely spreading setules up to 0.6–1.1 mm, the foliage dull olivaceous concolorous.

Stipules erect appressed or stiffly ascending, linear- or narrowly ovate-acuminate 5–11 mm, at slightly oblique base 1–1.7 mm wide, on broader side rounded to shallowly subcordate, from base (7–)9–15-nerved, glabrous dorsally, ciliolate.

Adult lvs near and beyond middle of stem 2.5–8 cm, shortly petioled, the expanded blade narrowly lanceolate in outline; petiole 3–6 mm, at middle 0.4–0.8 mm diam, narrowly thick-margined; glands 1–2, the lower (or single) one situated near middle of petiole proper, sessile patelliform (0.5–)0.7–1.5 mm diam, when over 1 mm diam elliptic in outline, longer than wide; lfts of adult lvs up to 12–18 pairs, in outline narrowly oblong-elliptic, linear-oblong or lance-elliptic, straight or falcate, commonly acuminulate but sometimes acute, sharply mucronulate or spinulose, 5.5–18 × 1.2–3(–3.5) mm, at oblique base cordate on broad and cuneate on narrow side, from base 5–8-nerved by the straight or curved midrib with on its broad side 4–5 and on narrow side 0–2 primary nerves, the excentric midrib dividing the blade 1:(1.5–)2–3.5(–4), penniveined with 3–5 pairs of major secondary nerves and sometimes some intercalated minor ones, the venation prominulous on both faces but more sharply beneath.

Peduncles and raceme-axis (3–)5.5–14 mm, adnate almost to first bract, 1–2-fld; pedicels at anthesis very slender ascending 4–8.5 mm, bracteolate 1–3.5 below calyx, in fruit thickened erect 6.5–9.5 mm; bracteoles ovate-acuminate 1.5–3 mm; buds slenderly ovoid-acuminate, appressed-puberulent; petals yellow, 3 adaxial oblanceolate (or 1 broader), 7.5–12 × 3–5 mm, abaxial either oblanceolate or flabellate, 9–16 × 4.5–12 mm, the cucullus falcately semiobovate 10.5–13 × 6–7.5 mm; ovary gray-strigulose; style linear, gently incurved, (3.5–)4–5.2 × 0.2–0.25 mm; ovules 11–16.

Pod erect, linear 30–50 × 3–3.5 mm, the valves reddish or purplish, finally nigrescent, minutely puberulent; seeds (little known) 2.6–3.2 mm, the testa fuscous above the paler base, dull, sooty-pitted.—Collections: 17.

Seasonally dry grassland (campo, pajonal), stony hillsides and mountain meadows, mostly 420–2600, at its n. limit up to 3000 m, along the e. slope and foothills of the Andes from the upper Pilcomayo valley (near 20°30′S) in s.-e. Bolivia s. to Tucumán and Sa. de Córdoba (near 31°S) in n.-w. Argentina, extending e.-ward, perhaps disjunctly, into the Gran Chaco of w.-centr. Paraguay near 23°20′S.—Fl. XI–II, in the higher mountains into III.

A woody perennial root, diffuse herbaceous stems, a sessile, shallowly dish-shaped petiolar gland, and larger leaves with up to (10–)12–18 pairs of asymmetric, often falcate leaflets are characters all easily matched severally in other American chamaecristas, but occur all together only in *Ch. venturiana.* We think that the species is most closely allied to *Ch. repens,* having altogether the facies of var. *repens* except for narrower and somewhat more numerous leaflets, the foliage being in fact almost that of *Ch. repens* var. *multijuga,* remotely disjunct in eastern Brazil; but the petiolar gland is entirely different. This patelliform gland is essentially that of *Ch. nictitans* var. *paraguariensis* or subsp. *brachypoda,* but these are virgately erect monocarpic plants, differing greatly in appearance and life-form. The only chamaecrista known to share the subandean range of *Ch. venturiana* is the weedy annual *Ch. nictitans* var. *jaliscensis,* separable at a glance by the small brachystylous flower.

Of theoretical interest is the close resemblance in almost all gross phenetic characters between *Ch. venturiana,* endemic to the Argentine-Bolivian Andes and Gran Chaco, and *Cassia wallichiana* DC., which occupies a comparable set of habitats along the foothills of the Himalayas in India and China. The latter species appears separable from *Ch. venturiana* by the relatively long pedicels, a difference that must seem ludicrously inadequate to sustain specific distinctions. We suspect, however, that we have here a case of independent genesis from the protean matrix of *Chamaecrista* of two parallel but not identical forms adapted to similar climatic circumstances.

37. Chamaecrista pascuorum (Bentham) Irwin & Barneby, comb. nov. *Cassia pascuorum* Martius ex Bentham in Martius, Fl. Bras. **15**(2): 165. 1870.—"Habitat in pascuis prope Joazeira [=Joàzeiro] et in interioribus campis et Taboleiro serrado prope Caiteté seu Villanova do Principe prov. Bahia: Martius."—Lectoholotypus, *Martius Obs. 2372,* collected V.1819 (fl, fr), M! = IPA Neg. *1381* = NY Neg. *9206*; paratypi, *Martius s.n.* from Caiteté, presumably collected X.1818 (fl, fr), M (2 sheets)! = F Neg. *6252* = IPA Neg. *1380.*

Cassia aristulifera Harms in Engler, Bot. Jahrb. **42**: 209. 1908.—"Bahia: Sandcampo der Serra do São Ignácio (Ule n. 2 Ba—Febr. 1907)."—Holotypus, †B = F Neg. *1649*! no isotypus found.

Cassia pascuorum sensu Bentham, 1871, p. 572.

Precociously maturing, at first herbaceous but quickly suffrutescent, usually monopodial herbs with one erect or obliquely ascending, simple or either from base or only distally few-branched stem at anthesis (1–)2–8 dm, the bark of old stems sometimes splitting and striped lengthwise, the young stems, lf-stalks, and often one or both faces of the dorsally pallid-venulose lfts puberulent with incumbent hairs up to 0.3–0.6 mm and commonly in addition hispid(ulous) with stiffer, spreading-incurved (or the pedicels finely pilose with weak flexuous) setae up to 0.8–2.2 mm, the absolute and relative density of the two types of vesture highly unstable (sometimes the longer, sometimes both, entirely wanting), the foliage concolorous olivaceous, submembranous or less often subcoriaceous and then either green or glaucous.

Stipules erect lance-acuminate or -attenuate from oblique or semi-subcordate base 4.5–10 × 0.8–1.7 mm, subspinulose at apex, from base 8–13-nerved, the blades nearly always ciliolate but either glabrous or pubescent dorsally, persistent.

Lvs spreading-ascending, disregarding some early short and simple ones (1.5–) 2.5–7 cm, the expanded blade ovate or ovate-oblong in outline; petiole with well differentiated but little swollen pulvinus (2.5–)3–6 mm, at middle 0.4–0.7 mm diam, the shallow groove open; petiolar gland 1(–2), situated near base of petiole proper, urceolate or subdiscoid orbicular (0.3–)0.4–1.2(–1.4) mm diam, coarsely or slenderly stipitate, the stipe either linear or claviform, commonly hispidulous but sometimes glabrous, the whole in profile urceolate, trumpet- or less commonly tack-shaped (0.4–)0.6–2 mm tall, 0.2 mm shorter to 1 mm longer than diam of head; lfts (3–)4–9(–11) pairs, oblong, oblong-elliptic, -oblanceolate or (especially the distal pair) obliquely obovate, obtuse, emarginate or abruptly acuminulate, the larger ones up to (7–)11–20(–23) × (2.5–)4–9.5 mm, stiffly or weakly aristate (cusp 0.6–3 mm) at base cordate or cordately angulate on the proximal side and rounded to broadly cuneate on the distal one, the membranous or chartaceous blades from base 5–7(–8)-nerved by the moderately displaced midrib with, on its proximal side 4–5(–6), on its distal one 0–1(–2) primary veins, the strong inner one attaining ± middle of blade, the midrib from near base penniveined with 3–7(–8) pairs of major sometimes separated by faint intercalary minor secondaries, these simple or V-forked distally, tertiary venulation essentially obsolete, the major veins always strongly prominulous (and pallid) beneath, sometimes equally, sometimes faintly so above.

Peduncles adnate to stem through (1–)2–10 mm, the free axis 0–4.5 mm, 1–3-fld; pedicels 8–23 mm, early divaricate and abruptly geniculate 1–4 mm below calyx, where bracteolate; bracteoles lanceolate or lance-attenuate (2–)2.5–8.5 mm; buds slenderly ovoid-acuminate, varying from densely pilose to thinly puberulent, exceptionally glabrous; sepals yellowish or reddish, lance-acuminate, the longer (9.5–)10–16 mm; petals bright or greenish-yellow fading red or orange, very unequal, the 4 subsymmetric ones broadly oblanceolate to obovate beyond the cuneate claw, the larger up to 12–18(–21) mm, the falcately semiobovate fifth usually a trifle longer, curled over the 10-androus androecium; long anthers (usually red) up to 8–10.5 mm; ovary densely white-pilosulous or strigulose (glabrescent); style linear, gently incurved at tip, 5.5–9 mm; ovules 16–28.

Pod narrowly linear, (35–)40–75 × 2.5–4 mm, the reddish- or purple-castaneous, finally nigrescent valves puberulent and usually also weakly pilose; seeds 2.8–3.5 mm, the testa dull ochraceous often with dark transverse line across distal end, the sparse pits concolorous or brown.—Collections: 25.

Cerrado and taboleiro, commonly on sandy soils and in disturbed or secondary brush-woodland, most abundant at 250–600 m in the middle and lower São Francisco valley in w. and n.-e. Bahia, downstream from the mouth of Rio Carinhanha, e. of the highlands s. to middle Paraguaçú valley, and feebly n. to w. Pernambuco and to Rio Balsas and the headwaters of Rio Itapicurú in s. Maranhão; also in caatinga on sandstone caprock of Chapada Diamantina at 950–1200 m in centr. Bahia, from Sas. do Curral Feio, São Ignácio, and do Tombador s. to Pico de

Almas and middle Contas valley (Poções).—Fl. I–VII(–IX), perhaps intermittently through the year.

Within its range of dispersal *Ch. pascuorum* is readily recognized by the combination of pilose- (exceptionally glabrous) stipitate petiolar gland, mostly 4–9 pairs of leaflets prominently pallid-venulose beneath, long-acuminate flower-buds of narrow outline, large flowers, long narrow pods turned toward vertical by an abrupt bend near top of the pedicel, and ochraceous seeds. The foliage is reminiscent of some forms of *Ch. nictitans brachypoda,* but this differs (except irrelevantly, in populations distantly allopatric and habitally dissimilar) in a sessile gland, mostly shorter, non-geniculate pedicels, and black seeds. The large-flowered chamaecrista commonest in the range of *Ch. pascuorum* is *Ch. repens* var. *multijuga,* obviously different in the more numerous, more asymmetric leaflets and especially in the petiolar gland, raised on a mounded stipe wider than itself. The diagnostic characters of the more closely related *Ch. vestita* of southern Brazil and adjoining lands and of sympatric *Ch. cuprea* are emphasized under the two next following. The position assigned by Bentham (1871, p. 572) to *Cassia pascuorum,* between sympatric *C. tenuisepala,* which has exactly axillary peduncles, and the West Indian *C. grammica* (=*Chamaecrista lineata* var. *brachyloba* of this revision), seems arbitrary and fails to express natural affinity.

As here defined, *Ch. pascuorum* is highly variable in details of pubescence, diameter and height of petiolar glands, and texture of the foliage. Pilose and merely puberulent variants are about equally common. Both types were encountered by Martius but were not recognized as such by Bentham, whose misidentification (1870, p. 171) of one of the latter sort collected on Rio Itapicurú by Martius (M) as *Ch. mucronata* is the sole basis for extending the range of that species into Maranhão. Martius found in the same place a single individual (M) glabrous up to the minutely puberulent pod. In the lowland valleys of Bahia and Maranhão the foliage of *Ch. pascuorum* is membranous and delicately venulose. On Chapada Diamantina, above 900 m, a series of minor variants differ collectively in the stiffer, on the average smaller and very coarsely venulose leaflets, one of which was described as *Cassia aristulifera* Harms. In this latter series one collection from Sa. do Tombador (*Irwin et al. 30666*) is notable for a glaucous bloom on the leaflets, a feature occasionally encountered also in the kindred *Ch. vestita.*

38. Chamaecrista cuprea Irwin & Barneby, sp. nov., notulis multis cum affini *Ch. pascuorum* congrua sed ab ea statura pumila, stipulis parvis 1.6–2.2 (nec 4.5–10) mm longis ±5–7 (nec 8–13)-nerviis, foliis abbreviatis minus quam 2 (nec 2.5–7) cm longis, foliolis plerumque 3 (nec 4–9)-jugis, necnon floribus leguminibusque minoribus, his ±2.5–3.5 (nec 4–7.5) cm longis 8–10 (nec 14–28)-ovulatis diversa.—BRAZIL. Bahia: Lagôa Itaparica, 10 km w. of the São Inácio–Xique-Xique road at turning 13.1 km n. of São Inácio, 42°46′W, 11°01′S, 26.II.1977 (fl, fr), *R. M. Harley 19123.*—Holotypus, RB; isotypi, CEPEC, K, NY, UFP, US.

Dwarf microphyllous subshrubs 1–2 dm, with gnarled contorted blackish caudex and stiff, sparsely leafy incurved-ascending, simple or few-branched annotinous branchlets, the young stems, lf-stalks, sepals and ovary puberulent with fine appressed pallid hairs up to 0.1–0.2 mm, the foliage bicolored, the carnosulous lfts dark blue-green glabrous above, glaucescent but either glabrous or puberulent beneath.

Stipules erect, ovate-, triangular- or lance-acuminulate 1.6–2.2 × 0.7–1.2 mm, the often castaneous or purplish blades coarsely 5–7-nerved from asymmetric base, persistent.

Lvs 6–17 mm, the expanded blade ovate in outline; petiole including wrinkled pulvinus 1.5–3.5 mm; rachis (1–)2–6(–7.5) mm; gland inserted immediately below proximal pair of lfts stipitate, in profile slenderly peg-shaped 0.3–0.6 mm tall, at apex scarcely dilated, the concave head 0.2–0.25 mm diam; pulvinules 0.2–0.4 mm; lfts 2–4, in most lvs exactly 3 pairs, of subequal size or the penultimate pair slightly longer and broader than the rest, all in outline obliquely oblong-obovate or oblong, obtuse mucronulate or obscurely acuminulate, the larger pairs 2.5–7 × 1–4.5 mm, ±1.5–2.7 times as long as wide, all at base cordate on proximal and rounded on distal side, the blade 3–4-nerved from pulvinus, the moderately displaced midrib giving rise to (2–)3–4(–5) pairs of secondary veins, these all immersed above, bluntly prominulous beneath.

Peduncles adnate to stem through (1–)1.5–6 mm, 1–2-fld, the axis not over 1 mm; pedicels 5–11 mm, bracteolate 1–3 mm below calyx, incurved-ascending in fruit; fl-buds slenderly ovoid-acuminate puberulent; sepals narrowly ovate or lanceolate acute, delicately nerved from base, greenish-yellow or red-tinged, 7–9.5 mm; petals when fresh "dark reddish-orange with yellow margin," when dry livid pallid-edged, highly heteromorphic, the vexillary one and one adjacent petal smallest, semi-ovate beyond short claw, ±8–11 mm, one lateral and one abaxial petal larger, subsymmetrically obovate-flabellate 12–14 mm, the boomerang-shaped cucullus a little longer, 14–16 mm, incurved over androecium; androecium 10-merous 2-cyclic, the filaments of 5 outer stamens 1–1.4 mm, of 5 inner ±0.5 mm, the slenderly lanceolate anthers of outer cycle 5–6.5 mm, of inner cycle ±4 mm; style glabrous almost straight 5.5–7 × 0.25–0.3 mm; ovules 8–10.

Pod ascending linear 2.6–3.3 × 0.35–0.4 cm, the greenish valves red-margined; seeds compressed-rhomboid ±3.3 × 2 mm, the testa fuscous, sparsely pitted.—Collections: 1.—Fig. 48.

Open sandy places ±300–400 m, known only from the type-locality near 11°S in w.-centr. Bahia.—Fl. I–III.

A singularly attractive chamaecrista, notable for the wiry fruticulose habit, neat glaucescent foliage and proportionately large flowers that open rich orange margined with yellow and do not simply acquire (like many kindred) an orange flush as they fade. Striking as the flower color may be, it is not reliably diagnostic, for we have random instances of orange petals (cf. *Mori 9513,* CEPEC, NY) in populations of the closely related and sympatric *Ch. pascuorum.* The latter, widely dispersed through highland central Bahia and very likely the immediate progenitor of *Ch. cuprea,* is highly variable in detail but is consistently of coarser growth, with larger leaves and stipules, coarser petiolar glands, more numerous and longer leaflets, and especially a longer pluriovulate pod. It seems possible that *Ch. cuprea,* as known to us from only one population, represents merely an extreme and emphatic variant of *Ch. pascuorum* fated to be reabsorbed into the main genetic current of that species; but for the present the number and quality of differences, emphasized in the key and implicit in the descriptions, leave no alternative to treating it as an independent species.

39. Chamaecrista vestita (Vogel) Irwin & Barneby, comb. nov. *Cassia vestita* Vogel, Syn. Gen. Cass. & Linnaea **11**: 711, descr. ampliat. 1837.—"Hbt. In Brasilia: Sellow leg."—Holotypus, †B = F Neg. *1663*; no isotypus found and none seen by Bentham (1871, p. 575, in nota sub. *C. brachypoda*).

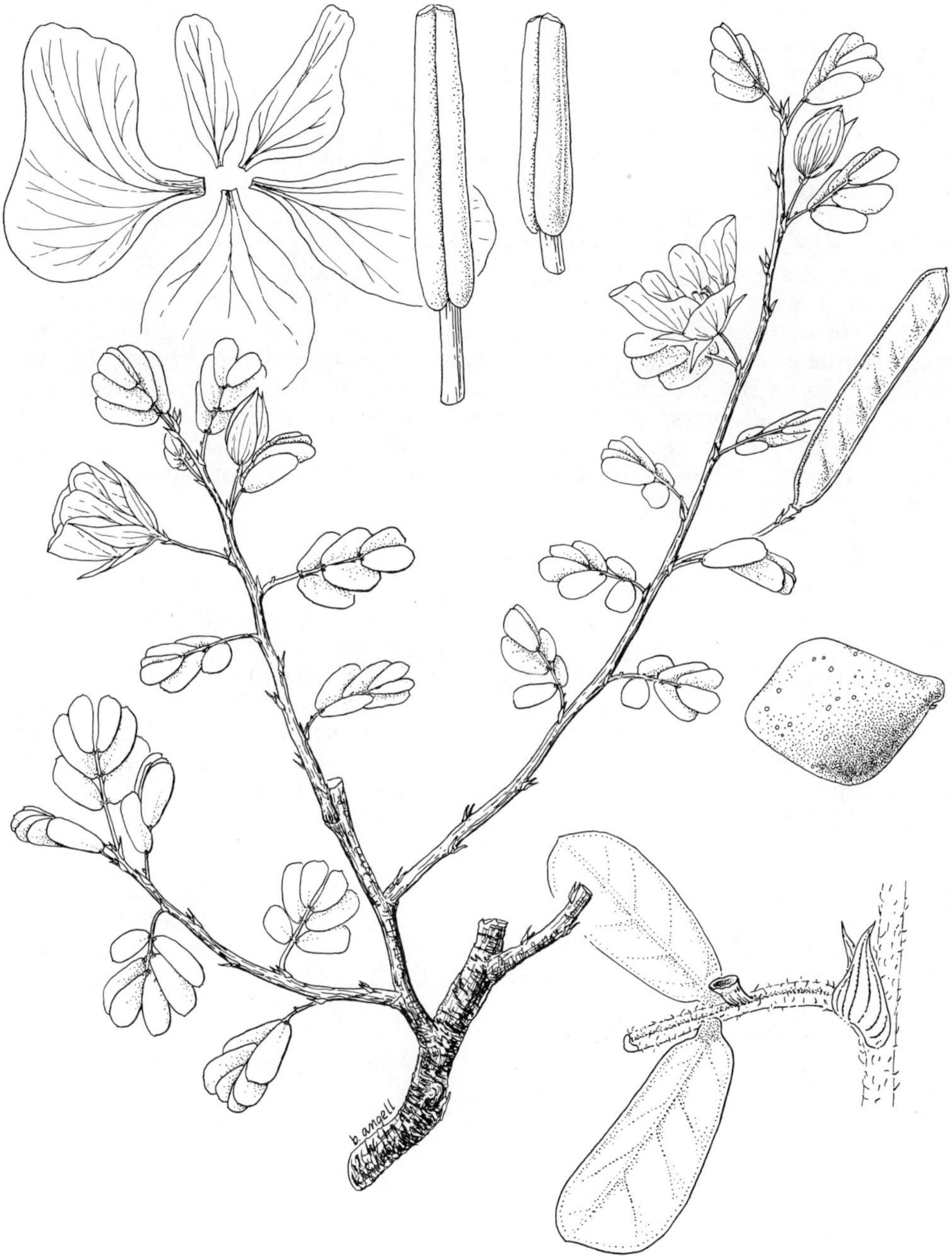

Fig. 48. *Chamaecrista cuprea*. I. & B. Habit ×1; leaf + stipules ×6; detached petals ×2; 2 stamens ×6; seed ×6 (*Harley 19123*).

Perennial herbs with 1–several simple or few branched stems 1.5–5 dm erect or ascending from the crown of a slender black root or from the divisions of a short caudex, except for the commonly glabrous upper face of the always ciliate lfts pilose throughout with fine divaricate setae up to 1.4–2.8 mm and the stems (sometimes only in lines) and pedicels in addition incurved-puberulent, the foliage dull olivaceous concolorous or the lfts sometimes glaucescently pallid-papillate beneath.

Stipules stiffly erect, obliquely lance-acuminate, (5–)6–13(–15) × 0.8–2.5 mm, from base prominently (5–)7–11-nerved, glabrous dorsally, ciliate.

Adult lvs 1.7–4.5(–5) cm, short-petiolate, the expanded blade ovate in outline; petiole including pulvinus 2.5–5 mm, at middle narrowly margined and 0.35–0.6 mm diam; petiolar gland 1(–2), situated close above pulvinus, 0.4–0.8 mm diam, stipitate, in profile tack- or trumpet-shaped, (0.5–)0.6–1.2 mm tall, always a little longer than diam of head; lfts 3–7(–8) pairs, the distal slightly smaller and proportionately broader, obliquely oblong or oblong-obovate obtuse (5–)6–14 × 2–4.5 mm, abruptly stiffly mucronate by midrib excurrent (0.4–)0.8–2 mm, at base deeply cordate or angulate-cordate on proximal and rounded to broadly cuneate on distal side, from base prominently 6–8-nerved by the moderately excentric midrib with on its broad side 4–5, on narrow one 0–2 lateral veins, the costa thence penniveined, with 4–7 pairs of secondary venules incurved-ascending to margin of blade, the venation prominulous on both faces but more sharply so beneath.

Peduncles adnate through (1–)2–6 mm, 1–2-fld; pedicels ascending, at anthesis 7–12 mm, in fruit 9–14 mm, bracteolate 1–2.5 mm below calyx; bracteoles narrowly lance-acuminate 3–5 mm; buds ovoid-acuminate, pilose; sepals lanceolate and ovate, 6.7–9 mm; petals pale yellow, 3 adaxial oblanceolate 7–9 mm, one abaxial broadly oblanceolate 8.5–10 × 4.4–5.5, the ample cucullus 7–9 mm, obliquely obovate, nearly as wide as long, incurved over the 10-merous androecium; long anthers 4–5 mm; ovary pilose, the style rather stout but not dilated distally, gently incurved, 1.5–3.4 × 0.3–0.4 mm; ovules 14–22.

Pod linear, slightly arcuate, 33–55 × 3–3.8 mm, the castaneous or purplish-brown valves thinly setose; seeds (scarcely known) ±3 mm, the testa dull pale brown with dark line at distal end.—Collections: 20.

Campo, near 600–800 m, apparently uncommon and scattered in s.-e. Brazil and s.-centr. and n.-e. Paraguay: extratropical, on the forks of middle and upper Iguaçú, Tibagi and Tieté rivers in n.-e., e. and s.-centr. Paraná, adjoining n.-e. Santa Catarina, and s.-e. São Paulo, w. to the Paraguai Valley in Paraguay, lat. 23°50′–26°10′S; and disjunctly, just north of the Tropic line on Río Apa, in n.-e. Paraguay.—Fl. I–III.

A rare and seldom collected chamaecrista, *Ch. vestita* might be visualized as a small *Ch. nictitans* var. *pilosa* with fewer (3–7, not 12–24) pairs of more strongly venulose, stiffly mucronate or subspinulose leaflets. Bentham, who knew it only from the original description, guessed (1871, p. 575) that it might be the same as his own *Ch. brachypoda,* but the petiolar gland, as he remarked, is quite different. Its one really near relative is *Ch. pascuorum,* disjunctly allopatric in northeastern tropical Brazil, which has essentially the same foliage and (except for its often hispidulous stipe) the same petiolar gland, long narrow pod, and ochraceous seed, but differs in the very much larger longistylous flower.

40. Chamaecrista lineata (Swartz) Greene, Pittonia **4**: 31. 1899. *Cassia lineata* Swartz, Prod. 66. 1788.—Typus infra sub var. *lineata* indicatur.

Precociously flowering herbs and shrubs, variable in habit and in pubescence as described under each variety, at anthesis (1.5–)2.5–12 dm, the foliage either concolorous or (commonly) in age glaucescent beneath, dull or exceptionally glossy.

Stipules erect, triangular-acuminate to lanceolate from subsymmetric or oblique and minutely semi-auriculate base (1–)1.5–9.5 × 0.6–2 mm, from base 5–11(–14)-nerved, the green, purplish or castaneous blades early papery brownish, either glabrous or puberulent dorsally, persistent.

Major lvs 1–5(–5.5, in var. *jamaicensis* –7) cm, the expanded blade (when more than 2-foliolate) ovate, suborbicular-flabellate, or broadly lance-oblong in outline; petiole with usually discolored, wrinkled pulvinus 1–5.5(–6.5) mm, the petiole proper (0.3–)0.4–0.8 mm diam, either longer or shorter than pulvinus; gland 1, situated at or above middle of petiole proper, discoid, shallowly concave, or shallowly convex and dimpled, (0.2–)0.25–1.2(–1.8) mm diam, either sessile or stipitate, described further under each var.; lfts of major lvs (1–)2–8, rarely up to 8–11 pairs, very variable in size and outline but when 5 pairs or more mostly oblong-oblanceolate or -elliptic and when 1–5 pairs relatively large or broad, mostly obovate, obovate-elliptic or obovate-cuneate in outline, obtuse, openly emarginate, or abruptly acuminulate, mucronulate or cuspidate, the largest (6–) 7–21(–24) × 2–8.5(–10) mm, at base broadly rounded or cordate on adaxial, narrowly rounded to cuneate on distal side, from base 4–7-nerved by the subcentric or slightly displaced midrib with on the adaxial side 2–4, on the abaxial side (0–) 1–2 primary veins, the innermost on the adaxial side produced beyond middle, sometimes to apex of blade, the midrib thence penniveined with, on the abaxial side 2–9(–10) major with or without finer intercalary secondary venules, on the adaxial side usually slightly fewer, the venulation always prominulous beneath but varying in emphasis, on the upper face either prominulous or immersed.

Peduncles with raceme axis 0.5–6(–10) mm, exactly axillary or adnate through 0.5–3.5(–8) mm, 1–3(–4)-fld; pedicels (8–)9–26(–32) mm, the more vigorous usually at least 10 mm, bracteolate 1–6 mm below calyx; buds ovoid-acuminate, strigulose, pilosulous, or glabrate; sepals lance- or ovate-acuminate up to 5.5–10.2(–11) mm; petals yellow, 3 adaxial smaller, broadly oblanceolate to obovate-cuneate, 1 abaxial larger, broadly flabellate or suborbicular-cuneate up to (8–)9–16 mm, the dimidiately obovate cucullus usually a trifle shorter, rarely longer; androecium 10-merous, 7(–8) anthers shorter yellow or sometimes red-tipped, (2–) 3 longer usually red, sometimes yellow, the longest up to 5.5–9 mm; ovary densely strigulose or pilosulous; style (2.5–)3–5(–6) mm, in some fls apparently abortive; ovules 6–16.

Pod linear, (18–)23–50(–54) × 3.5–5 mm, the usually reddish-brown or purplish, ultimately nigrescent valves either strigulose or pilosulous; seeds 2.5–3.8(–4.3) mm, commonly dull grayish-brown with fuscous pitting, sometimes dull brown overall.

Comprehensively delimited so as to include the four varieties described below and within these several insignificant variants of highly localized dispersal, *Ch. lineata* emerges as that major unit in a replacement series that assumes the role of *Ch. glandulosa* south from the Florida Keys through the length of the Bahamas, Cuba, and Grand Cayman to the northwest tip of Haiti, becoming marginally sympatric with var. *glandulosa* only in Jamaica, and with *Ch. glandulosa* var. *swartzii* in a narrow eastward extension along the south coasts of Hispaniola and Puerto Rico. It has precisely the flower of *Ch. glandulosa,* sens. lat., if sometimes a trifle smaller, and differs from the Caribbean forms of that species by the ordinarily fewer (1–7, rarely to 11) pairs of leaflets combined with a commonly sessile or subsessile petiolar gland. The specific distinctions are, however, precarious and arbitrary, for the gland of some otherwise typical var. *lineata,* in Jamaica, Cuba, and Haiti may be elevated on a short stalk and that of var.

brachyloba is characteristically so; while the leaflets rise in number to ten or exceptionally eleven pairs in isolated populations of var. *lineata* in northern Bahamas (New Providence) and eastern Cuba (Oriente). But within *Ch. lineata* sens. lat. stalked glands coincide only with leaflets fewer and broader than any Caribbean *Ch. glandulosa,* while leaflets over eight pairs are associated only with sessile glands. We have found no single differential character to separate all forms of *C. lineata* from all forms of *C. glandulosa,* and regard both as part and parcel of one turbulently evolving macrospecies. As maintained here, *Ch. lineata* is a geographic reality but in the rank of species a taxonomic fiction.

Key to the Varieties of *Ch. lineata*

1. Longest petal the plane abaxial one paired with the cucullus; West Indies and s. Florida, widespread.
    2. Petiolar gland stipitate, in profile taller than diam of head, and this gland coincident with up to (4–)5–8 pairs of lfts; s. coasts of Puerto Rico, Hispaniola, and extreme e. Cuba (Sa. Maestra, Oriente); s.-e. Jamaica; Isla Providencia in s.-w. Caribbean.
        3. Large lfts of larger lvs up to 6–12 mm, seldom glabrous on both faces; range as given, except for Jamaica and Providencia. 40c. var. *brachyloba* (p. 757).
        3. Large lfts of larger lvs up to 14–27 mm, glabrous on both faces; Jamaica (local in St. Andrew Pa.) and Providencia I., Colombia. 40e. var. *jamaicensis* (p. 757).
    2. Petiolar gland ordinarily sessile or subsessile, in profile shorter than diam of head, but if stipitate then coincident with fewer lfts; s. peninsular Florida, Bahamas, Cuba, n.-w. Haiti, Jamaica, Grand Cayman, approaching the last only in far e. Cuba and Jamaica.
        4. Weakly suffrutescent diffuse herbs of the Florida Keys, pilosulous throughout. 40b. var. *keyensis* (p. 756).
        4. Fruticose when adult, obviously so if pilosulous throughout, if precociously flowering as an herb then stiffly erect; West Indian.
            5. Lfts 1–10 pairs, but if only 1–3 pairs in larger lvs then the blades broadly oblanceolate to obovate, at apex below the mucro either broadly rounded, or truncate, or emarginate; widespread. 40a. var. *lineata* (p. 754).
            5. Lfts of larger lvs 2–3 pairs, acuminulate and stoutly mucronate; local about Guantánamo Bay, extreme s.-e. Cuba. 40d. var. *brevipila* (p. 758).
1. Longest petal the cucullus; s. Mexico (s.-e. Oaxaca and s.-w. Chiapas). Lfts of larger lvs up to 6–8 pairs, pilosulous on both faces; petiolar gland very small but shortly stipitate, a trifle longer than diam of head. 40f. var. *pinoi* (p. 761).

40a. Chamaecrista lineata (Swartz) Greene var. **lineata**. *Cassia lineata* Swartz, 1788, l.c. & Fl. Ind. Occ. **2**(1): 726, descr. ampliat. 1798.—"Jamaica" or (1798) ". . . ruderata et culta Jamaicae."—No typus seen, but the protologue, as interpreted by Bentham, 1871, p. 572, decisive.

Cassia cuneata DeCandolle ex Colladon, Hist. Casses 121. 1816.—"Hab. in America calidiore?"—Holotypus, s. loc., G-DC! = F Neg. *7004.—Chamaecrista cuneata* (Colladon) Greene, Pittonia **4**: 31. 1899 (but unqualified attribution to South America mistaken).

Cassia niqueroensis Urban & Ekman ex Urban, Feddes Repert. **22**: 355. 1926.—"Hab. in Cuba prov. Oriente in peninsula de Cabo Cruz prope Niquero in rupibus litoralibus, m. Jan. fl. et fr. Ekman no. 16173."—Holotypus, †B; isotypi, collected 16.I.1923, NY, US!—*Chamaecrista niqueroensis* (Urban & Ekman) Britton ex Britton & Rose, N. Amer. Fl. **23**(5): 279. 1930.

Chamaecrista clarensis Britton ex Britton & Rose, N. Amer. Fl. **23**(5): 278. 1930.—"Province of Santa Clara, Cuba . . . Punta Colorado, Cienfuegos Bay, March 18, 1910, *Britton & Wilson 5648.*"—Holotypus, NY! isotypus, US!—*Cassia clarensis* (Britton & Rose) R. A. Howard, Jour. Arn. Arb. **28**: 126. 1947.

Cassia cuneata sensu Bentham, 1871, p. 575.
Chamaecrista lineata sensu Britton & Rose, 1930, p. 279.

Shrubs mostly 5–12 dm, in exposed places freely branched and ultimately of bushy or globose outline, in thickets sometimes of spindly stature and up to 1.5–2 m, often flowering the first season as erect, few-branched or simply virgate, weakly suffrutescent herbs, the main trunk reaching 2 cm diam but mostly less, the annotinous leafy branchlets and lf-stalks strigulose or pilosulous with all straight appressed, all straight ascending-spreading, rarely all incumbent, whitish hairs to 0.15–0.4 mm, exceptionally mixed with few ascending setules up to 0.4–0.7 mm, the foliage usually bicolored, the firmly chartaceous lfts above dull olivaceous brunnescent, beneath commonly but not always glaucescent, strigulose-pilosulous like the stems on both faces, or on the lower only, or less often merely ciliolate, only exceptionally quite glabrous; stipules (1–)1.5–5(–6.5) × 0.6–1.2 (–1.4) mm; major lvs to (1.2–)1.5–5(–5.5) cm; gland (0.25–)0.35–1.2(–1.8) mm diam, usually sessile or almost so and either sunk within the ventral groove of petiole or resting at level of the wing-margins, locally (Jamaica; e. Cuba; n.-w. Haiti) shortly and stoutly stipitate when in profile squatly tack-shaped and up to 0.4–0.7 mm tall, exceptionally a trifle longer than diam of head; rarely similar glands on 1–3 segments of rachis; lfts of major lvs (1–)2–7, locally in Bahamas (Great Bahama, New Providence, Andros), Jamaica, e. Cuba (Oriente) up to 8–11 pairs, very variable in size and outline but when relatively many (5+ pairs) mostly oblong-oblanceolate or -elliptic and when fewer (1–5 pairs) obovate, obovate-elliptic or -cuneate, obtuse mucronulate, obscurely acuminulate mucronulate, or widely open-emarginate, the larger up to (7.5–)10–21(–24) × (2.5–)3–8.5(–10) mm, the secondary major venules from abaxial side of midrib (3–)4–9(–10), with or without finer intercalary ones, the venulation always prominulous beneath but variable in strength, usually finer so above; peduncles adnate through 0–3.5 mm; sepals up to (6–)7–10(–11) mm; long petals (8–)9–16 mm; ovules (8–)9–16; pod 25–50 × 3.4–5 mm.—Collections: 155.

Dunes, sand flats behind barrier beach, soil pockets in limestone (coral) pavement, limestone rock ledges, and extending inland in coppice thickets, savanna and palm-savanna, becoming locally dominant and sometimes especially vigorous and large-leaved in waste ground or along highways, over most of its range maritime and submaritime below 5 m, ascending inland in Jamaica and eastern Cuba (Sa. de Nipe) to ±700 m, in centr. and e. Cuba locally abundant on ferruginous and serpentine outcrops, widespread through the Bahamas s.-ward from Bimini, Grand Bahama and Abaco to the Caicos Is. and n.-w. corner of Hispaniola (Ile de la Tortue, St. Louis du Nord, Port Margot); s.-w., centr. and e. Cuba (Isla de Pinos; Matanzas, Las Villas, Camaguey, Oriente, but locally replaced in Sa. Maestra and vicinity by vars. *brachyloba* and *brevipila*); w. and n. coastal Jamaica; Grand Cayman.—Fl. interruptedly through the year, barren only during drought.

The length of our varietal description reflects the great latitude of variation encountered in var. *lineata,* variation that independently involves growth-habit, length and diameter of petiolar gland, vesture of stem and foliage, and number, outline and amplitude of leaflets. The patterns of variation seem to arise both from genetic differences, little variable within a population or within a given geographic limit, and from external influences of microenvironment and season; but the interplay of these two sets of forces is cryptic and will remain so until recourse can be had to controlled experimental plantings. Nevertheless our analysis of the variable features has brought out some interesting facts:

a) It is a myth, propagated by Britton & Rose's key to *Chamaecrista* ser. *Lineatae* (1930, p. 271) that the foliage of *Ch. lineata,* even as defined in the form of specimens at NY annotated in Britton's hand, is "glabrous or nearly so." In fact truly glabrous leaflets are exceptional, unknown from Jamaica (the type-region), whereas leaflets strigulose beneath or on both faces are the rule, and leaflets pilosulous with patent hairs on one of both faces occur at random points in the Bahama Chain (Grand Bahama, *Correll 44763*; Exuma, *Correll 40725*; San Salvador, *Gillis 8818*; Acklin, *Brace 4290,* all NY), but not, so far as known, beyond.

b) The petiolar gland is in the Bahamas uniformly sessile or almost so, commonly sunk in the ventral groove of the petiole or seeming in profile view to rest snugly on its narrow ventral flanges; but in either situation varying greatly in size, and in color from red to brown, yellowish, or green. In Jamaica, both on the seashore and inland on the Santa Cruz Mountains, as also in eastern but not in central Cuba or Isla de Pinos, and in coastal Haiti (Nord-Ouest) the gland is elevated on a distinct stalk, and becomes essentially like that of var. *brachyloba,* even while still associated with the distinctive glaucous foliage of Bahaman var. *lineata*.

c) Leaflet number, described here in terms of the highest number found in given plant's largest leaves, appears to be relatively stable within major populations and even over considerable segments of the range of var. *lineata*; but a survey of the whole range shows random inconsistencies and distantly allopatric duplications of phenetically identical types. In the Bahamas northward from New Providence a maximum of 6–8 pairs prevails; southward a maximum of 2–5 pairs, the numbers commonest in central Cuba (Matanzas, Las Villas, Camaguey), and never exceeded in Isla de Pinos, Grand Cayman, or Haiti. But Jamaica and the Caicos yield records of 3–7 and 5–7 pairs respectively, eastern Cuba, both maritime and interior, 3–8 pairs. Extremes of 9–11 pairs, which must be considered quite exceptional, occur on New Providence (*Curtiss 180,* A, G, GH, NY, US; *Britton & Brace 397,* NY) and on Sa. de Nipe in eastern Cuba (*Shafer 8414,* NY). Because number and shape of leaflets are to some degree correlated, a loss in number being compensated for by increase in amplitude, the forms with 1–3-jugate and 5–8-jugate leaves appear strikingly different, but are connected by forms, just as numerous and without clear geographic grouping, of intermediate number. We find no relation whatever between leaflet number and pubescence.

Of the synonyms listed above, *Cassia cuneata* remains somewhat doubtful, due to the fragmentary nature and unknown origin of the holotypus. The name has never hitherto been associated with anything more substantial than the type-specimen itself, and was supposed by Bentham to differ from *C. lineata* in having 4–6, not 2–4 pairs of leaflets, a character now known to be valueless and even at the time irrelevant, for Swartz's original *C. lineata* from Jamaica had five pairs. The typus of *Ch. clarensis* represents the minor variant of *C. lineata* in which the 2–3-jugate leaflets are puberulent on the upper as well as the lower face. That of *C. niqueroensis* represents the extreme reduction of the leaves to a single or (in the isotypus at NY) in some few leaves 2 pairs of leaflets. Approximate topotypes (*León 16335,* GH, US) have many, or (US) mostly bijugate leaflets and are essentially identical with the races of var. *lineata* found distantly allopatric on the coast of Haiti. This extreme form was collected first somewhere in Cuba by Ramón de la Sagra and annotated by Richard (P) as an undescribed species; but his manuscript name seems to have remained no more than that.

40b. Chamaecrista lineata (Swartz) Greene var. **keyensis** (Pennell) Irwin & Barneby, comb. nov. *Ch. keyensis* Pennell, Bull. Torr. Club **44**: 344. 1917.—". . . Big Pine Key, Monroe County, Florida, collected in flower May 2, 1917, *F. W. Pennell 9533* in the herbarium of the New York Botanical Garden."—Holotypus not found at NY in 1975–7; isotypus, US!—*Cassia keyensis* (Pennell) Macbride, Contrib. Gray Herb., n. ser. **59**: 24. 1919.

Chamaecrista keyensis sensu Britton & Rose, 1930, p. 278. *Cassia keyensis* sensu Isely, 1976, p. 103, map 42, p. 204 (exclus. "holotype, *Small, Mosier, Small 4454*" lapsu cit.).

Stems flexuously decumbent or weakly assurgent from the root-crown or shortly forked caudex, at anthesis (1–)1.5–7(–8) dm, the young branchlets, lf-stalks and both faces of lfts pilosulous with widely divergent hairs to 0.4–0.65 mm; stipules 3–9.5 × 0.7–2 mm; major lvs up to 1.7–3.5(–4) cm; gland 0.3–0.6 mm diam, sessile or almost so but not sunk in the ventral groove, in profile always a trifle wider than long, if visibly stipitate squatly obconical; lfts of major lvs (5–) 6–9 pairs, mostly oblong-oblanceolate, obtuse mucronulate, up to 7–12 × 2–4.5 (–5) mm, veined as in var. *lineata*; peduncle adnate through 1.5–8 mm; pedicels in fruit spreading-incurved; sepals up to 9–10 mm; long petal 11–15 mm; ovules 10–13; pod 33–45 × 4.5–5 mm, pilosulous.—Collections: 16.

Pineland, in thin soil overlying coral limestone, below 10 m, known only from the outer Keys (No Name, Big Pine, Ramrod), Monroe County, Florida.—Fl. IV–VI(–VIII).

Cassia keyensis, clearly different from any other chamaecrista native to Florida, was first identified (Pollard, 1894, p. 217) as *C. grammica* (=our var. *brachyloba* next following), often superficially similar in the pilosulous vesture but readily distinguished by the emphatically fruticose, not diffuse subherbaceous growth-form and the stipitate, not sessile petiolar gland. It escaped comparison with the immediately vicariant Bahaman var. *lineata* because that has long, but wrongly, been supposed to have glabrous or merely strigulose-puberulent foliage; yet it has the same sessile gland, the same venation of the leaflets, and it often flowers precociously as a scarcely suffruticose herb. Isely (1975, p. 103) considered *C. keyensis* an "undoubted derivative of *C. grammica* . . . which could . . . with reasonable justification be considered a northern subherbaceous variant of it." Granted the substitution of the epithet *lineata* for *grammica,* we readily accept this evaluation. Apart from the distinctive growth-habit, var. *keyensis* tends to have larger stipules than var. *lineata,* and peduncles further adnate to the stem.

The earliest collection of var. *keyensis* known to us was made on Pine Key in mid-XIX century by J. L. Blodgett (NY).

40c. Chamaecrista lineata (Swartz) Greene var. **brachyloba** (Grisebach) Irwin & Barneby, comb. nov. *Cassia lineata* var. *brachyloba* Grisebach, Mem. Amer. Acad. II, **8**(1): 179. 1891.—"([Cuba orientalis, *C. Wright*] 146)."—Holotypus, collected 29.IX.1859 (fl, fr) at Farallones, on s. coast of Oriente at 77°12′W, GOETT! isotypi, GH! = NY Neg. *9198,* LE!—Equated with the next following by Bentham, 1871, p. 572.

Cassia grammica K. Sprengel, Neue Entd. **3**: 55. 1822.—". . . ad oram maritimam Cubae et Hispaniolae."—No holotypus known to survive, none in hb. Willd. (B); presumed isotypi: "*Cassia grammica* Spreng. in litt., St. Domingo, *Bertero,*" G-DC! and "St. Domingue," FI (hb. Webb. via hb. Desf.)!—*Chamaecrista grammica* (K. Sprengel) Pollard, Field Mus. Bot. **2**: 47. 1900.

Cassia portoricensis var. (β) *granulata* Urban, Symb. Antill. **1**: 318. 1899.—". . . in Portorico . . . prope Salinas de Cabo-Rojo . . . Febr. fl. et fr., prope Guanica in fruticetis montis El Maniel et in litore maris ad salinas, Jan., Febr. flor.: Sintenis 631, 3450 (form. interm.), 3704."—Lectotypus, *Sintenis 631*: holotypus, †B; isotypi, GH, LE, NY, US, W!—*Chamaecrista portoricensis* var. *granulata* (Urban) Cook & Collins, Contrib. U.S. Nat. Herb. **8**: 113. 1903. *Ch. granulata* (Urban) Britton, Ann. Missouri Bot. Gard. **2**: 41. 1915. *Cassia granulata* (Urban) Macbride, Contrib. Gray Herb., n. ser. **59**: 26. 1919.

Cassia grammica sensu Bentham, 1871, p. 572; Pollard, 1894, p. 217, excl. pl. florid. *Chamaecrista grammica* sensu Britton & Rose, 1930, p. 279.

Chamaecrista granulata sensu Britton & Rose, 1930, p. 280.

Frutescent when adult but either erect or diffuse, to 2.5–12 dm, with dark brown or gray trunk bushily branched into a rounded crown or depressed thicket of stiff branchlets, the stems, lf-stalks and commonly both faces of lfts puberulent or pilosulous with all short incurved, mixed incurved and straighter ascending, all ascending, or all subhorizontally spreading hairs to 0.15–0.5(–0.6) mm, the vesture of branchlets often yellowish, the foliage dull olivaceous concolorous, the lfts rarely glabrate above, exceptionally glabrous ciliolate, the faces often granular-papillate; stipules (1.5–)2.5–5.5 × 0.7–1.2 mm, 5–10-nerved; major lvs to (1.2–)1.5–2.7(–3) cm; gland 0.25–0.5 mm diam, either stoutly or slenderly stipitate, exceptionally subsessile, in profile tack-, pin-, or shortly trumpet-shaped, (0.3–)0.4–1.2 mm tall, almost always 0.1–0.9 mm taller than diam of head; lfts of major lvs (4–)5–7(–8) pairs, oblong or oblanceolate (the short uppermost sometimes semiobovate) mostly obtuse or subemarginate mucronulate, rarely subapiculate, to (6–)7–12 × 2–3.7(–4) mm, the midrib with 3–6(–7) pairs of secondary venules immersed or faintly prominulous on upper face, faintly or sharply, sometimes (especially in Cuba) coarsely prominulous on lower; pedicels 9–17 mm; sepals to 5.5–8 mm; long petal up to 10–13 mm; ovules 6–11; pod (18–)23–32 × 3.5–5 mm.—Collections: 30.

Dunes, coastal scrubland, and in humus pockets of reefs and rock ledges within reach of sea-spray, mostly below 10 m, locally abundant along the s. shores of Puerto Rico and Dominican Republic between Cayo Muertos (66°30′W) and Bahía de Ocoa (70°45′W), unknown from but to be expected along the s. shore of Haiti, reappearing on the s. coast of Oriente Province, Cuba, along foot of Sa. Maestra w. to Farallones (77°15′W), and ascending in the Sierra to pinelands at ±650 m.—Fl. intermittently throughout the year.

If one can accept such moderate variation in pubescence as, while always theoretically if not always in practice unacceptable to Britton, must be expected to occur in all taxa of supra-axillary *Chamaecrista,* var. *brachyloba* becomes neatly definable as an offshoot of *Ch. lineata* in which the syndrome of a stipitate petiolar gland and small leaves composed of some 5–7 pairs of leaflets up to 6–12 mm long has become stabilized in an independent range of dispersal. Except for the gland, some plants of var. *brachyloba* are virtually identical with some small-leaved Bahaman var. *lineata*; and we have already emphasized the presence in some races of the latter, races characterized, however, by leaflets either larger, or fewer, or both, of glands at least incipiently elevated on a stalk. Separation of var. *brachyloba* as a distinct species is therefore no longer defensible.

In taking up the epithet *brachyloba* for what has been known as *Ch. grammica* if pubescent or *Ch. granulata* if glabrate, we have to admit that the collection on which the name rests, *Wright 146* from the south coast of Oriente Province, is taxonomically inconvenient. The type-locality of var. *brachyloba* is at the furthest westerly record for its kind, and as so often happens at the periphery of a taxon's range is not representative of its sort, having in fact a very shortly stipitate gland, but this combined with the leaflets expected of var. *brachyloba.* We feel confident, nevertheless, that it is conspecific in the narrowest sense with *León 10331* (NY) from slightly further east along Sierra Maestra, and in this the gland is as long-stalked as in any var. *brachyloba* from Hispaniola. In adopting the equation *C. grammica* = *C. lineata* var. *brachyloba,* we follow a tradition established by Bentham in 1871.

40d. Chamaecrista lineata (Swartz) Greene var. **brevipila** (Urban) Irwin & Barneby, comb. nov. *Cassia arduinervis* var. *brevipila* Urban, Symb.

Antill. **9**: 444. 1928.—"Prov. Oriente prope Guantánamo ad Caimanera . . . [*Ekman*] n. 15772 . . ."—Holotypus, †B; isotypus, collected 24.XI.22, NY!—*Cassia brevipila* (Urban) Alain, Candollea **17**: 110. 1960, comb. illegit.

Cassia arduinervis Urban in Feddes Repert. **15**: 310. 1918.—"Hab. in Cuba: Wright no. 2377."—Holotypus, †B; isotypi, collected (data at GH) in "pebbly bed of Yateras river, 2.VII (fl)," G, GH, P!—*Chamaecrista arduinervis* (Urban) Britton ex Britton & Rose, N. Amer. Fl. **23**(5): 280. 1930.

Chamaecrista hioramii Britton ex Britton & Rose, N. Amer. Fl. **23**(5): 278. 1930.—"Southern Oriente, Cuba. Type from Boca de Guantanamo, June 23, 1921, *Brother Hioram 4868*."—Holotypus, NY! isotypi, NY, US!—*Cassia hioramii* (Britton & Rose) León, Contrib. Ocas. Mus. Hist. Nat. Colegio de la Salle **9**: 11. 1950.

Like var. *lineata* in everything but the slightly modified foliage, variable in pubescence, the branchlets either incurved-puberulent or pilosulous with hairs to 0.15–0.5 mm, the lfts either pilosulous on both faces, or puberulent beneath, or glabrous ciliolate, or fully glabrous, in the latter case lustrous; major lvs 1–2 cm, the petiole with pulvinus 1–2.5 mm; gland subsessile 0.25–0.4 mm diam; lfts 2–3 pairs, obliquely oblanceolate to subdimidiately obovate, at apex triangular- or deltate-acuminulate, stoutly cuspidate, to 9–13 × 2.5–5 mm, the venation beneath very coarse and prominent, the secondary nerves from midrib 2–4(–5) pairs.—Collections: 4.

Limestone rock ledges and gravelly streambeds below 100 m, known only from immediate environs of Bahia Guantánamo (Caimanera; mouth of Río Guantánamo; Yateritas), s.-e. Oriente, Cuba.—Fl. VI, XII, perhaps at intervals through the year.

The type-collection of *C. arduinervis,* which Bentham (1871, p. 588) referred without comment to *C. lineata,* is remarkable in that context for its small, thick-textured, lustrous leaflets, acuminulate at apex, coarsely and relatively simply venulose, and entirely glabrous. It is precisely matched by part of topotypic *Ekman 10223* (A, NY, US) which, however, contains some plants with facially hairless but ciliolate leaflets. In the types of var. *brevipila* and *Ch. hioramii,* considered synonymous by Britton and united by Liogier under the technically superfluous name *C. brevipila,* the pubescence extends over both faces of the leaflets, which become in consequence less glossy, although unchanged in size or venulation. These four gatherings differ collectively from *C. lineata* var. *lineata* as this occurs, under several forms, in eastern Cuba, by the short (1–2 cm) leaves and small, abruptly acuminulate and coarsely cuspidate leaflets. The very closely related var. *brachyloba,* as represented in nearby Sa. Maestra, has almost as small but more numerous (5–7-jugate) leaflets of thinner texture commonly accompanied by a strongly stipitate gland. In Cuban var. *brachyloba* the pod, so far as known, is about 2.5 cm long and 6–7-ovulate. We have seen no pod of var. *brevipila,* but this was described by Urban as 3.5 cm long, presumably therefore pluriovulate. The supportive differential characters of pubescence, leaflet-number, flower-size and breadth of pod supposed by Urban to separate *C. arduinervis* from *C. lineata* have proved illusory.

40e. Chamaecrista lineata (Swartz) Greene var. **jamaicensis** (Britton) Irwin & Barneby, stat. nov. *Ch. jamaicensis* Britton, Bull. Torr. Club **42**: 515. 1915.—". . . near the southern coast of Jamaica. Type from the south slope of Long Mountain (*Britton 811*)."—Holotypus, collected 22–23.IX.1906 (fl jun), NY!—*Cassia jamaicensis* (Britton) Adams, Phytologia **20**(5): 311. 1970.

Chamaecrista jamaicensis sensu Britton & Rose, 1930, p. 280.

Fruticose erect or in exposed sites diffuse, up to 5–13 dm, the young branchlets and lf-stalks strigulose-pilosulous with straight appressed or narrowly ascending, sometimes mostly incumbent hairs up to 0.2–0.7 mm, the foliage glabrous except sometimes for minutely ciliolate lft-margin and -midrib beneath, the lfts dull or sublustrous olivaceous concolorous; stipules 2.5–5(–6) × 0.6–1.2(–1.4) mm, 5–7-nerved; major lvs up to 3.5–7 cm; petiolar gland 0.35–0.5 mm diam, stoutly stipitate, in profile peg- or pin-shaped 0.7–1.5 mm tall, 0.35–1 mm longer than diam of head; similar but smaller glands often but not always on all or some distal rachis-segments; lfts of major lvs 4–7(–8) pairs, oblong, oblong-oblanceolate or -elliptic, obtuse mucronulate, up to 14–23(–27) × 5–8.5(–9) mm, striate-venulose on both faces, the secondary veins sometimes but not always connected by tertiary venules; pedicels 12–21 mm; sepals up to 7–10 mm; long petals up to 8.5–11 mm; ovules (7–)8–10; pod 32–42 × 5–6 mm.—Collections: 6.

Open rocky slopes, 100–300 m, very local in two widely separated localities: s.-e. Jamaica (Long Mt. and Cane River gorge, St. Andrew Pa.); Isla de Providencia, Colombia, in the s.-w. Caribbean at 13°21′N, 81°23′W.—Fl. in Jamaica VI–VII, IX–II, on Providencia III–V, but perhaps in both stations irregularly through the year.

The var. *jamaicensis* seems to have been first mentioned specifically by Bentham (1871, p. 575) as one of three forms of West Indian *Cassia polyadena,* that form in which glabrous leaves are combined with glands at once "remarkably stipitate and rather numerous"; but the material then available in London (*Purdie, March,* BM) had already been referred to *C. polyadena* by Grisebach (1864, p. 210). Britton's description of *Ch. jamaicensis* aimed at correcting this error which, nevertheless, was perpetuated by Fawcett & Rendle (1920, p. 113). Bentham (l.c.) had separated his poorly characterized *C. polyadena* from his *C. virgata* (=*Ch. glandulosa* var. *swartzii* of this account) by the fewer and larger leaflets; and by more numerous oncs from *C. lineata,* supposed to have only 2–4 pairs. Thus it came about that *C. jamaicensis* long escaped direct comparison with its very close relative *C. lineata,* now known to vary considerably more than Bentham could forsee in size, number and pubescence of the leaflets. Adams (1972, p. 323 in key, p. 327) maintained *C. jamaicensis* as specifically distinct from *C. lineata* because of its glabrous, on average larger leaflets combined with stipitate glands. The pubescence is indeed significant in Jamaica, but not so in Bahaman *Ch. lineata,* and for leaf-size there is nothing to choose between *Harris 12120* from Long Mountain (NY, US = glabrous-leaved *jamaicensis* with long glands) and *Harris 6952* from Malvern in western Jamaica (US = pubescent *lineata* with sessile gland). The stipitate glands are, in the large view, the only decisive differential character, and ineffective in excluding *Ch. jamaicensis* from our comprehensive concept of *Ch. lineata.* Adams (l.c.) suggests that *C. jamaicensis* may have arisen as a neospecies from a cross between *C. lineata* and *C. glandulosa.* One would expect that the progeny of two pubescent parents to be pubescent, which is not the case; and although the largely calcifuge *Ch. glandulosa* is plentiful about Kingston, we have no evidence that mostly calciphile *Ch. lineata* is present in southeastern Jamaica. Furthermore the hybrid hypothesis would hardly account for the remotely disjunct population of var. *jamaicensis* on Providence Island, distant over 700 km off to the southwest of Kingston. We interpret var. *jamaicensis* as one more product of racial fission at the same level of differentiation as *Ch. lineata* var. *brevipila* or var. *pinoi.*

40f. Chamaecrista lineata (Swartz) Greene var. **pinoi** (Britton & Rose) Irwin & Barneby, stat. nov. *Ch. pinoi* Britton & Rose, N. Amer. Fl. **23**(5): 283. 1930.—"Between Jalisco and Tuxtla Gutierrez, Chiapas, 1923, *Purpus 9384*."—Holotypus, collected VIII (fl), US! isotypus, NY!—*Cassia pinoi* (Britton & Rose) Lundell, Phytologia **1**: 215. 1937.

Frutescent, ultimately fruticose, apparently erect, at anthesis 3–9(–?) dm, the slender, sometimes flexuously zigzag trunks brown or longitudinally striped, the hornotinous branchlets and foliage pilosulous throughout with fine incumbent, spreading-incurved, and (especially lf-stalks and pedicels) some few longer spreading hairs to 0.3–0.8 mm, the dull, pallidly olivaceous concolorous lfts pilosulous on both faces but sometimes more thinly so above; stipules 3–7 × 0.7–1.1 mm, 5–7-nerved; major lvs up to 2–4 cm; gland 0.2–0.3 diam, minutely stipitate, in profile squatly tack-shaped 0.25–0.4 mm tall, mostly a trifle (±0.1) mm longer than diam of head; lfts of major lvs up to 6–8 pairs, oblong-obovate or the short distal pair obliquely obovate, all very obtuse mucronulate, up to 6.5–11 × 3–4.5 mm, the slightly displaced midrib giving rise on each side to 2–4 major with fainter intercalary secondaries, the venation subequally prominulous on both faces; peduncles axillary 1-fld; pedicels 13–20 mm; sepals up to 7.5–9 mm; long adaxial petal 10.5–12.5 mm, the cucullus exceptionally ample and up to 14–15 mm; ovules 8–9; pod (little known) ±30–35 × 4.5 mm, incurved-pilosulous and sometimes in addition remotely pilose.—Collections: 3.

Habitat not recorded, but to be expected in xeromorphic, drought-deciduous scrub-forest, ±70–600 m, local, known only from three stations on the Pacific slope of Istmo de Tehuántepec in s.-e. Oaxaca (Cerro S. Pedro; Ixtepec) and in adjoining s.-w. Chiapas (Jalisco).—Fl. VII–X (and probably later).

Britton & Rose presented *Ch. pinoi* as kindred to *Ch. greggii,* gravely different in the leaflet venation, obtuse sepals and subglobose flower buds, and separated these collectively from *Ch. lineata* and its West Indian relatives in geographical terms alone. They described *Ch. pinoi* as a treelet 3.5–4 m tall, but there are no data directly associated with the type-collection to indicate that such was the case. The arborescent character is derived from a letter (NY, copy) written on February 24, 1925 in the village of Jalisco, Chiapas, in which Carl Anton Purpus describes to J. N. Rose his return to the original locality of *Ch. pinoi,* discovered in August two years previously. Here Purpus describes the plant as a small tree with "flowers hanging down on the stem," but in February only few of these present with old leaves, pods, and ripe seeds. These were said to have been collected, but there is no mention of fruit in the protologue and we have found no Purpus collection at US or elsewhere that we can identify as such. An arborescent habit is certainly not impossible in *Chamaecrista,* but the two collections of *Ch. pinoi* known to us from Oaxaca, both close matches for the holotypus in details of foliage, gland and pubescence, are noted by the collectors as only one or three feet tall. The treelet habit is therefore not characteristic, and the idea may have been introduced by some unexplained error on the part of Purpus. Disregarding this false, or if not false inconstant character of stature, we find scarcely any tangible quality of habit or foliage by which *Ch. pinoi* can be separated from loosely pubescent *Ch. lineata,* even though we have found no particular West Indian specimen that matches the Mexican material at all points. The peduncle of *Ch. pinoi* is exactly axillary, but so is that of some Bahaman var. *lineata.* Only the proportions of the perianth seem a little different, the longest petal in var. *lineata* being ordinarily the adaxial one paired with the cucullus, whereas in *Ch. pinoi* it is the cucullus itself. It is on this precarious morphological basis, reinforced by disjunct dispersal, that we provisionally maintain var. *pinoi* at varietal rank.

The person commemorated by the epithet, which Purpus at one place in the cited letter spells "*del Pinoi,*" is unknown to us.

41. Chamaecrista obcordata (Wikstrom) Britton, Bull. Torr. Club **44**: 6. 1917. *Cassia obcordata* Swartz ex Wikstrom, Kongl. Vetensk.-Akad. Handl. **1825**: 429. 1826.—"*C. obcordata* Swartzii Herb . . . in insula S:ti Bartholomaei: Forsström."—Holotypus, S! = NY Neg. *9233*; isotypi, S (5 sheets), LE!

Chamaecrista obcordata sensu Britton & Rose, 1930, p. 279.

Erect or diffuse shrubs and subshrubs up to 2–6(–?) dm, with fuscous or brownish woody stems, the annotinous branchlets and lf-stalks minutely, sometimes only remotely strigulose-pilosulous with incumbent or subappressed straight hairs to 0.15–0.3 mm, the firm concolorous pallidly olivaceous lfts glaucescent glabrous.

Stipules erect appressed, deltate, triangular-acuminate or broadly subulate 1–2 mm, from base faintly 4–6-nerved, persistent.

Major lvs up to 1.5–4.5(–6) cm; petiole including moderately swollen pulvinus 2–7 mm, at middle 0.4–0.6 mm diam, very narrowly winged; petiolar gland 1 above middle of petiole proper or close under first pair of lfts, round discoid or shallowly cupuliform 0.2–0.45 mm diam slenderly stipitate, in profile tack- or peg-shaped, 0.1 mm shorter up to 0.5 mm longer than diam of head; similar glands on each segment of rachis; lfts of major lvs 3–7(–9) pairs, oblong-obovate or obovate emarginate to obcordate up to 5–15(–18) × 3–7(–9) mm, at base shallowly semi-cordate to rounded on proximal, cuneate on distal side, from base 5–7-nerved by the dorsally prominulous centric midrib with on proximal side 2–4 and on distal side 1–2 shortly ascending primary veins, the midrib thence penni-nerved with (3–)4–9 major and (or without) fainter intercalary pairs of secondaries, the venulation prominulous on both faces, sometimes more sharply so beneath.

Peduncles 1.5–9 mm, adnate to stem through 1–8 mm, 1(–2)-fld; pedicels 12–25 mm; buds ovoid-acuminate, puberulent at base, glabrous distally; fls scarcely known, apparently not different from those of *Ch. lineata,* the sepals 9–10 mm; ovary pilosulous; ovules 6–12.

Pod ±30–45 × 4.5–6 mm, remotely pilosulous; seeds (little known) ±4–4.3 mm, dull brown, faintly pitted.—Collections: 9.

Cliffs and rocky slopes along and near the seashore, apparently calciphile, probably not above 100 m, local in the Leeward Islands of St. Martin, St. Barthélémy, and Guadeloupe.—Fl. X–II.

An ambiguous entity, poised morphologically and geographically between *Ch. lineata* var. *lineata* and *Ch. glandulosa* var. *swartzii,* having the few firm glaucescent leaflets of the former, but the elevated glands (on both petiole and rachis) and the subsymmetrically pinnate venulation of the broad-leaflet form of var. *swartzii* that has been segregated as *Cassia polyadena* DC. The precisely typical form of *Ch. obcordata,* known only from the two small islands of St. Barthélémy and St. Martin, has a quality of distinctness due to the small size of the leaves (1–3 cm) and their small (to 5–9 mm), heart-shaped leaflets; but we cannot detach from these some more amply leafy collections from Guadeloupe (e.g. *Questel 5169,* US; *Sastre & Jérémie 1969,* NY) which closely approach, except in their fewer leaflets, some ecotypes of var. *swartzii* from the Windward Islands and Barbados. Bentham (1871, p. 572) referred *Ch. obcordata* to *C. lineata*; but this arrangement obliterates the last tenuous distinction between the latter species, sens. lat., and *C. glandulosa*. We suspect that field observation may show *Ch. obcordata* to be little more than a local paucifoliolate glaucescent variant of *C. glandulosa* var. *swartzii*; we maintain it provisionally as a taxonomic convenience.

A peculiar plant from Guadeloupe (*Fournet 1760,* A), in which most leaves are reduced to a simple blade consisting of one leaflet subsessile on the stem (some, however, accompanied by a smaller deformed twin), is interpreted as abnormal *Ch. obcordata*. The suppression of leaf-stalk and consequently of petiolar gland is an unexpected anomaly. Texture and coloring of the individual leaflets and form of the flower are consistent with the Guadeloupe form of *Ch. obcordata*.

42. Chamaecrista portoricensis (Urban) Cook & Collins, Contrib. U.S. Nat. Herb. **8**: 113. 1903. *Cassia portoricensis* Urban, Symb. Antill. **1**: 317. 1903.—Typus infra sub var. *portoricensi* indicatur.

Suffruticose from a blackish root and pluricipital caudex, sometimes eventually fruticose, the stems at anthesis simple or weakly branched and decumbent, or assurgent and bushy-branching (1–)1.5–6(–7.5) dm, the young leafy stems and lf-stalks puberulent or strigulose with short incurved or short appressed, sometimes mixed with or replaced by longer incurved-ascending or exceptionally straight horizontal hairs to 0.15–0.5(–0.7) mm, the young pubescence either gray or lutescent, the stiffly chartaceous, callous-margined lfts commonly glabrous ciliolate and lustrous, but sometimes beneath or on both faces pilosulous and opaque.

Stipules erect appressed triangular- or lance-acuminate from subsymmetrical or obliquely dilated and semi-cordate base, (1.5–)2.5–8 × 0.6–1.3(–1.5) mm, from base prominently 6–11(–13)-nerved, the green or purplish, finally brunnescent blades dorsally glabrous or puberulent, persistent.

Major lvs mostly 1.5–4(–4.5), exceptionally up to 4–6 cm, in expanded outline ovate or lance-ovate; petiole with ± dilated and wrinkled pulvinus 1.5–3.5(–5) mm, the petiole at middle 0.35–0.6 mm diam, narrowly margined ventrally; petiolar gland 1(–4) near middle of petiole proper, round concave or discoid and dimpled 0.2–0.7(–0.9) mm diam, either stipitate or subsessile, the stipe very variable in length and thickness, the whole in profile commonly tack-, trumpet- or club-shaped, when relatively short becoming drum-shaped or squatly obconic, (0.2–)0.3–1.4 mm tall, (0.3–)0.1 mm shorter to 0.8(–0.9) mm longer than diam of head; lfts of major lvs mostly (4–)5–8(–10), exceptionally to 12–16 pairs, oblong, oblong-elliptic or the shorter distal ones obliquely obovate, obtusely rounded, obtuse-acuminulate or openly emarginate at apex, stoutly mucronulate, the larger ones up to (6–)7–14 × (2.2–)2.5–4.5(–5) mm, at base cordate or broadly rounded on proximal side, cuneate or narrowly rounded on the distal one, from base (3–)4–5(–6)-nerved by the moderately displaced midrib with on its proximal side (0–)1 primary nerves, the strongest inner proximal one ascending half or less than half the length of blade, the dorsally cariniform midrib thence penniveined with ±5–8 pairs of major with, between each successive one, 1–3 fainter intercalary secondaries, the secondaries sometimes all of almost equal strength and the lft then appearing multi-striate, the major ones or all confluent into a stout raised marginal vein as or almost as strong as the midrib.

Peduncles 1.5–7(–12) mm, adnate to stem through 1–4.5(–10) mm, 1–2(–3)-fld; pedicels (10–)11–22 mm, bracteolate 1–7 mm below calyx; buds ovoid-acuminate, thinly puberulent or pilosulous; sepals lance- and ovate-acuminate up to (6–)6.5–9.5 mm; petals yellow, like those of *Ch. lineata,* the abaxial one suborbicular-flabellate (8–)8.5–14 mm; androecium 10-merous, the 2–3 long anthers either red or yellow, up to 5–6.4 mm; ovary gray-strigulose; style linear 3–4.2 mm; ovules (7–)8–12.

Pod linear (21–)25–44(–50) × 3.5–7 mm, the nigrescent valves minutely sparsely strigulose or pilosulous; seeds (little known) 3.4–5 mm, the testa dull ochraceous-brown, dark-pitted.

The stiff and, when glabrous, varnished, callous-margined leaflets of *Ch. portoricensis,* which recall on a miniature scale those of the Brazilian *Ch. lagotois,* strike the eye forcefully as different from those of all other Caribbean chamaecristas, but there seems to be no correlated diagnostic character by which the species can be separated from the closely related *Ch. lineata.* Especially in its erect fruticose calciphile aspect mentioned further below, *Ch. portoricensis* sometimes becomes only tenuously distinct from immediately vicariant races of *Ch. lineata* var. *brachyloba* found in coastal southwestern Puerto Rico, one of which was actually described by Urban as *Cassia portoricensis* var. *granulata.* The material from Vieques and the Virgin Islands identified by Britton, because of its pilosulous foliage, as *Ch. grammica* (a taxonomic synonym of var. *brachyloba*), but here transferred to *Ch. portoricensis* because of the prominently callous leaflet margins, is really intermediate between the two and furnishes a strong argument for taxonomic subordination of the latter. Nevertheless the bulk of the material of *Ch. portoricensis* is so easily and decisively identifiable that we find it convenient to maintain the taxon at specific rank.

As delimited by the leaflet venation, *Ch. portoricensis* is extremely heterogeneous internally as regards stature, pubescence, and petiolar glands. Between Mayaguez and Guanica, around the southwestern corner of Puerto Rico, and in scattered disjunct stations on Hispaniola, the plants are generally confined to serpentine and are essentially herbaceous, the diffuse or prostrate stems dying back annually to the root-crown or to a shortly ramified suffrutescent caudex. Eastward from Guayanilla (the type-locality), the adult plant is fruticose, like *Ch. lineata* var. *brachyloba,* and the habitats so far as recorded are calcareous. Within each of these minor groupings we find an equal range of variation in the glands and in pubescence, tiny subsessile and long-stipitate glands appearing close together in otherwise similar plants and dull pilosulous and lustrous glabrous leaflets occurring sometimes together in a single collection (cf. *Heller 2615,* GH, US). The holotypus of *Ch. tuerckheimii* is herbaceous, pilosulous, with glands only ±0.2–0.3 mm diam and 0.2 mm tall; but part of an isotypus (G) is, with the same gland, merely puberulent, and we have from Sa. de Ocoa also (*Liogier 20631,* NY) an example of this strigulose pubescence combined with glands up to 0.5 mm diam and 1.3 mm tall.

Standing apart from all mentioned hitherto is a major population found on serpentine outcrops along the northeast coast of Hispaniola which seems decisively different in the longer leaves and more numerous leaflets, its general habit suggesting *Ch. pedicellaris,* but the leaflets callous-margined and centrically costate as in typical *Ch. portoricensis.* We describe this below as var. *atlantica.*

Key to the Varieties of *Ch. portoricensis*

1. Longest lvs 1.5–4(–4.5) cm, their lfts (4–)5–8(–10) pairs; s.-w. Dominican Republic (Azua) to Virgin Is. and Puerto Rico. 42a. var. *portoricensis* (p. 765).
1. Longest lvs 4–6 cm, their lfts 12–16 pairs; n.-e. Dominican Republic (Espaillat). 42b. var. *atlantica* (p. 765).

42a. Chamaecrista portoricensis (Urban) Cook & Collins var. **portoricensis**. *Cassia portoricensis* Urban, 1903, l.c., sens. str.—". . . Portorico var. α prope Guayanilla in montibus calcareis ad Peñón, Aug. flor., prope Mayaguez: Krug n. 295, Sintenis n. 4926, 4926b."—Lectoholotypus, *Sintenis 4926,* collected near Guayanilla 25.VIII.1886 (fr), †B; neoholotypus, former isotypus, GH!—*C. portoricensis* var. (α) *callosa* Urban, l.c., nom. superfl.

Chamaecrista tuerckheimii Britton, Bull. Torr. Club **44**: 8. 1917. "Near Maniel de Ocoa, Santo Domingo, in fields, 300 m. alt. (*Tuerckheim 3680*)."—Holotypus, NY! isotypus, G!—*Cassia tuerckheimii* (Britton) Macbride, Contrib. Gray Herb., n. ser. **59**: 26. 1919; Urban, Symb. Antill. **8**: 271, pro comb. nov. 1920.
Chamaecrista portoricensis sensu Britton & Rose, 1930, p. 281.

Lvs as described in key; pod mostly 3.5–5, rarely 6 mm wide.—Collections: 24.

Dry hills, stony barrens, thickets, savannas, on either limestone or serpentine, near sea level up to ±450 m, locally abundant in the dry hill-country of s. and w. Puerto Rico, e., becoming maritime, through Vieques I. to the Virgin Is. (St. Thomas; St. Jan), and disjunctly to Hispaniola on Sa. de Ocoa, prov. Azua, s.-w. Dominican Republic; to be expected in s. Haiti.—Fl. throughout the year, perhaps most freely X–IV.

42b. Chamaecrista portoricensis (Urban) Cook & Collins var. **atlantica** Irwin & Barneby, var. nov., a var. *portoricensi* foliis longioribus, majoribus (juxta flores) 4–6 (nec 1.5–4.5) cm longis, foliolisque suis magis numerosis 12–16 (nec 5–8, raro –10)-jugis diversa.—DOMINICAN REPUBLIC. Espaillat: locally common on serpentine roadside embankment ±10 m alt., ±7.5 km e. of Gaspar Hernández on road to Río San Juan, 22.II.1979 (fl, fr), *H. S. Irwin & A. L. Wolff 37316*.—Holotypus, NY; isotypi, F, GH, K, US, USD.

Lvs as described in key; pod 5.5–7 mm wide.—Collections: 2.

Serpentine outcrops ±10–100 m, known only from the vicinity of the type-locality near the Atlantic coast of Dominican Republic in Espaillat.—Fl. IX–III certainly, very likely whenever active growth permits.

First discovered, in fruit only, by A. Liogier in 1969.

43. Chamaecrista rufa (Martens & Galeotti) Britton & Rose, N. Amer. Fl. **23**(5): 274. 1930. *Cassia rufa* Martens & Galeotti, Bull. Acad. Bruxelles **10**(2): 306. 1843.—Typus infra sub var. *rufa* indicatur.

Precociously flowering but early suffrutescent, in age sometimes truly fruticose, at anthesis (1.5–)4.5–20 dm, either diffusely branching from near or shortly above the root-crown or with 1–few erect, feebly forking wandlike trunks or branches, the annotinous branchlets and lf-stalks variably pubescent with short incurved hairs and longer, narrowly ascending, widely incurved-ascending, or horizontal, pallid or faintly lutescent setae up to (0.5–)0.6–1.5(–1.7) mm, the density and relative proportions of the 2 types of vesture highly unstable, either sometimes predominant or either wanting, the foliage dull green subconcolorous, the narrow, usually crowded lfts varying from puberulent on both faces and setulose-ciliolate to setose dorsally and ciliolate, or glabrous on both faces and then either ciliolate or not.

Stipules erect, broadly lance-acuminate to narrowly lance-attenuate (5–)7–17 ×

(0.7–)1.2–2.6(–3) mm, from asymmetrically rounded or semi-subcordate base prominently (7–)9–17(–19)-nerved, the pale green or anthocyanic, dorsally glabrous or puberulent blades persistent.

Main cauline lvs up to 3–9(–12) cm, the expanded blade narrowly lance-elliptic in outline, the late-season foliage sometimes much reduced distally; petiole including swollen pulvinus (1.5–)2.5–6(–8) mm, near middle 0.4–0.75(–0.85) mm diam; gland 1(–2) situated near or well above middle of petiole proper, sometimes close to proximal pair of pulvinules, shallowly urceolate or discoid, round or elliptic (0.4–)0.5–1(–1.2) mm diam, stipitate or subsessile, the stipe usually dilated upward, sometimes stoutly columnar, the whole in profile varying from drum-shaped to squatly obconical, trumpet- or tack-shaped, 0.4–1(–1.2) mm long, 0.4 mm shorter up to 0.5 mm longer than diam of head; lfts of major lvs up to 18–45 pairs, inserted 0.8–3 mm apart along rachis, in outline obliquely or falcately linear-oblong or -lanceolate acute or subacute aristate, the longer of larger lvs up to 6–15 × 0.8–2.2(–2.6) mm, at base cordate or abruptly angulate on proximal side, cuneate on distal one, the midrib strongly displaced or marginal, the distal segment of blade 0–0.3(–0.5) mm wide, the secondary venulation of the broad side consisting of (2–)3–10 major venules with (usually) ± as many fainter (in var. *polyphlebia* equally strong) intercalary ones, the venulation usually finely, sometimes coarsely prominulous on both faces.

Peduncles with raceme-axis 3–20 mm, adnate through (1–)2–16 mm, (1–)2–4(–5)-fld; longer pedicels 7–17(–23) mm; buds ovoid-acuminate, thinly pilose or puberulent; sepals submembranous, brownish or lutescent, ovate- to lance-elliptic-acuminate, up to 5.5–13.5 mm; petals and androecium of *Ch. glandulosa,* the long abaxial petal up to 8–19 mm; long anthers (red or yellow) to (5.5–)6–10.5 mm; ovary pilose or strigulose; style linear 2–6 mm; ovules 12–18.

Pod linear in outline, usually slightly curved outward, (30–)35–60 × (3.5–)4–5.5 mm, the purplish-castaneous valves either incurved-puberulent or spreading pilose; seeds 2.8–3.7(–4) mm, the testa dull reddish-ochraceous concolorous, or the same but fuscous-pitted, or fuscous overall and then sometimes sublustrous.

We distinguish *Ch. rufa* from habitally similar *Ch. glandulosa* sens. lat. and from *Ch. nictitans* var. *disadena,* which can have pedicels and flowers in the same ranges of variation, solely by the dimidiate or at least very highly asymmetric leaflets which are at the same time either narrower, or more numerous, or both. The displacement of the midrib, which in var. *rufa* often and in var. *polyphlebia* constantly runs exactly parallel and contiguous to the abaxial margin, is unfortunately not a constant quantity, the abaxial segment of the blade, although always very narrow, becoming sometimes up to 0.3 mm wide in var. *rufa* and even 0.5 mm wide in var. *exsul,* the costa then dividing the blade into proportions up to 1:7 and 1:4.5 respectively. But such leaflets remain either narrower or more numerous than in sympatric or immediately vicariant expressions of kindred taxa. Such North American forms of *Ch. rufa,* the var. *rufa* of this revision, as were known to Bentham (1871, l.c.) were subordinated by him to the Colombian *C. tristicula,* our *Ch. glandulosa* var. *tristicula,* but we have found it possible to distinguish them by the constantly greater asymmetry of the leaflet blades:

1. Distal side of blade 0–0.2(–0.3) mm wide, at most $^{1}/_{7}$ as wide as the proximal one (Mexico to Costa Rica). *Ch. rufa* var. *rufa.*
1. Distal side of blade 0.3–0.7(–1) mm wide, ($^{1}/_{6}$–)$^{1}/_{5}$–$^{1}/_{3}$ as wide as the proximal one (Cundinamarca, Colombia). *Ch. glandulosa* var. *tristicula.*

In the var. *exsul,* remotely allopatric in southwestern Brazilian Amazonia, where the displacement of the midrib fails to provide constant distinction from

Ch. glandulosa tristicula, the leaflets of larger leaves are much more numerous, up to 32–45, not 14–22 pairs.

It must be mentioned that the foliage of *Ch. rufa* finds a close match in some forms of the polymorphic *Ch. nictitans* var. *ramosa,* but this is easily distinguished at anthesis by its small brachystylous flowers, in fruit by the short pedicels, and usually at all seasons by the character of the sessile, shallowly dish-shaped or discoid petiolar glands. Differential characters of the West Indian chamaecristas with dimidiate leaflets are emphasized elsewhere.

Key to the Varieties of *Ch. rufa*

1. Fls relatively large, the longer sepals (8–)9–13.5 mm, the long abaxial petal (10–)11–17(–19) mm; number and secondary venulation of lfts various; N. Hemisphere (Colombia to Mexico).
   2. Lfts of larger lvs mostly 18–32, exceptionally to 39 pairs, inserted 1.2–2.5 mm apart along rachis; costa marginal or intramarginal, the distal side of blade 0–0.3 mm wide; secondary venulation of proximal side of lfts consisting of (2–)3–5(–6) major veins with 0–4 weaker intercalary ones, these all separated by intervals much wider than themselves. Mexico and Central America (s. to Honduras and Costa Rica). 43a. var. *rufa* (p. 767).
   2. Lfts of larger lvs 30–44 pairs, inserted 0.8–1.5 mm apart along rachis; costa exactly marginal; secondary venulation consisting of 10–20 subcontiguous venules of nearly equal strength; e. slope of Cord. Oriental in Boyacá and Cundinamarca, Colombia (±4–5° N). 43c. var. *polyphlebia* (p. 769).
1. Fls small, the longer sepals 5.5–7.5 mm, the long abaxial petal 8–9.5 mm; lfts up to 32–45 pairs, the costa intramarginal, the distal side of blade 0.2–0.5 mm wide, the secondary venulation of proximal side consisting of 3–9 major with some faint intercalary venules, these separated by intervals much wider than themselves; local in the Madeira-Aripuanã basin in Rondônia and n.-w. Mato Grosso, Brazil (±9–10°S). 43b. var. *exsul* (p. 768).

43a. Chamaecrista rufa (Martens & Galeotti) Britton & Rose var. **rufa**. *Cassia rufa* Martens & Galeotti, 1843, l.c., sens. strict.—"3311 . . . dans les petits bois de la colonie de Mirador [Veracruz, Mexico], à 3,000 pieds."—Holotypus, *Galeotti 3311,* BR (4 sheets)! isotypi, G, GH, K, LE, NY, P = F Neg. *39835,* US!

Chamaecrista jalapensis Britton ex Britton & Rose, N. Amer. Fl. **23**(5): 274. 1930.—"Veracruz. Type from sandy fields, Jalapa, May 10, 1900, *Pringle 8337.*"—Holotypus, NY! isotypi, G, GH, K, US!—*Cassia jalapensis* (Britton & Rose) Lundell, Bull. Torr. Club **69**: 391. 1942.

Cassia tristicula sensu Bentham, 1871, p. 582, ex parte, quoad pl. centr.-amer. et Mex. cum syn. *C. rufa* (sphalm. '*rusa*').

Cassia chamaecrista sensu Sessé & Mociño, Fl. Mex. ed. 2, 99. 1894.—"Habitat in praedio Tospae juxta Cordovam Veracruz."—Spm. authent., *Hb. S. & M. 1193,* MA! = F Neg. *44247.*

Commonly suffrutescent, rarely to 2 m, highly variable in length, density and orientation of vesture (as described for the whole species); gland trumpet- or tack-shaped, less often claviform, (0.4–)0.5–1(–1.2) × (0.35–)0.4–0.8(–1.1) mm; longer lfts to 6–13 × 1.2–2.2(–2.6) mm; otherwise as given in key.—Collections: 58.

Openings or clearings in pine-oak and more mesic or wet tropical woodlands, disturbed thickets, and becoming an abundant colonial weed of roadsides, pastures, and milpa, (300–)870–1800, perhaps locally to 2500 m, common on the Gulf slope of Sa. Madre Oriental in Mexico from s.-w. Tamaulipas to Puebla, thence s. interruptedly through highland Oaxaca and Chiapas to Guatemala (Huehuetenango) and n.-w. Honduras (Cortés); apparently disjunct in s. Costa Rica (Puntarenas), and collected once (*Bourgeau in 1865*) on the Mexican Plateau in Distrito Federal, perhaps introduced.—Fl. V–III.

The feature of var. *rufa* most directly and strikingly seen as variable is the vesture of stems and leaf-stalks, which may be simply incurved-puberulent (*C. rufa,* sens. str.), simply setose (*Ch. jalapensis,* sens. str.) or a variably proportioned mixture of the two (e.g. *G. L. Smith 1548,* p.p. NY). These variants occur within a few kilometers of each other on the Gulf slope of Mt. Orizaba, sometimes together, and are of random occurrence outside of Veracruz, both north and southward. Independently of pubescence the petiolar gland varies substantially in size and in elevation of its stipe, which varies from subfiliform to coarsely club-shaped. In the material examined we have detected no major separation of self-perpetuating races.

First collected in Mexico in late XVIII century, probably by Sessé and collaborators (Nueva España, hb. Pavon., BM, which probably = *Cassia chamaecrista* Sessé & Mociño, l.c. supra).

43b. Chamaecrista rufa (Martens & Galeotti) Britton & Rose var. **exsul** Irwin & Barneby, var. nov., a var. *rufa* imprimis florum parvulorum sepalis 5.5–7.5 (nec 8.5–13.5) mm, petalo abaxiali majori 8–9.5 (nec 10–19) mm longis, foliolis fere semper magis numerosis 32–45 (nec 18–32)-jugis usque, patriaque procul distanti diversa.—BRAZIL. Mato Grosso: Os Dardanelos, Rio Aripuanã, 15.VI.1974 (fl, fr), *Manoel dos R. Cordeiro 68.*—Holotypus, NY.

Suffruticose becoming shrubby, at anthesis 5–20 dm, the branchlets and lf-stalks incurved-puberulent and sometimes incipiently setulose with incurved-ascending hairs; gland 0.4–0.8 mm diam, either stipitate or subsessile, the short stipe slender or columnar, the whole in profile drum-shaped, obconical, or squatly tack-shaped, 0.4 mm shorter to 0.2 mm longer than the head; longer lfts of major lvs inserted 2–3 mm apart along rachis, up to 10–15 × 1.2–2.2(–2.4) mm; otherwise as given in key.—Collections: 4.

Scrub savanna and secondarily in capoeira replacing high Amazonian forest on terra firme, below 300 m, known only from tributaries of the upper Madeira (Jaciparaná; Rio Jamari) and Aripuanã rivers in Terr. Rondônia and n.-w. Mato Grosso, lat. ±9–10°S, Brazil.—Fl. IV–X.

The var. *exsul,* so named for its occurrence in places and environments unexpectedly remote from the main focus of *Ch. rufa* in Mexico and Central America, is the only Brazilian chamaecrista other than *Ch. flexuosa* with as many as 32–45 pairs of very narrow leaflets, and readily distinguished from that by the asymmetric venulation of the blades and by the small supra-axillary flowers. It simulates puberulent states of var. *rufa* closely, but differs nearly always in the more numerous leaflets of larger leaves and consistently, so far as our data go, in flower size. Moreover the midrib of the leaflets tends to be slightly less displaced, becoming intramarginal by 0.2–0.5 mm. But rare individual plants of var. *rufa* from Veracruz (e.g. *Purpus 1882,* GH) have up to 39 pairs of leaflets in the largest leaves and we find in the same region some overlap in the position of the midrib relative to the distal margin. The major secondary venules arising from the midrib on the blade's proximal side are generally 5–9 in var. *exsul,* 2–6 (but mostly 3–5) in var. *rufa.* It is especially interesting to find among the four known collections of var. *exsul,* certainly conspecific in a very narrow sense and sympatric with no kindred species from which alien characters could be derived by introgression, a substantial variation in the petiolar glands, the short stipe of which varies from slender to stout and columnar, yielding in profile a tack-shaped, drum-shaped or intermediately obconical figure. Here certainly, as probably elsewhere, variation in the gland is inherent in the constitution of the chamaecristae verae.

43c. Chamaecrista rufa (Martens & Galeotti) Britton & Rose var. **polyphlebia** Irwin & Barneby, var. nov., ab aliis speciei formis foliolis secus rachin confertissimis utrinque argute ac confertim multistriatis, ulterius a var. *rufa,* quoad flores majusculos simili, foliorum majorum foliolis magis numerosis 30–44 (nec 18–32)-jugis, a var. *exsuli,* quoad foliolorum numerum comparabili, costa exacte marginali floribusque magnis, ab ambabus patria procul aliena diversa. A *Ch. glandulosa* var. *tristicula* in vicinis Colombiae montibus obvia foliolis confertis angustissimis 0.8–1.5 mm tantum latis exacte dimidiatis multivenulosis facilius separatur.—COLOMBIA. Cundinamarca: grassy bank, 1400–1500 m, ''Monte Redondo'' to Quetame, 7.IX.1917 (fl), *F. W. Pennell 1869.*—Holotypus, NY; isotypus, US.

Diffuse herbaceous or erect frutescent, at anthesis 2.5–10(–?+) dm, subappressed-strigulose; petiolar gland 0.45–0.8 mm diam, elevated on a stout columnar stipe slightly constricted distally, the whole 0.6–1 mm tall, in profile 0–0.35 mm longer than diam of head; fls of var. *rufa*; otherwise as given in the key.—Collections: 7.

Open grassy or rocky hillsides, 800–1790 m, local, known only from the e. slope of Cordillera Oriental on the Upía and Guatiquía forks of río Meta (±lat. 4–5°N) in Cundinamarca and adjoining Boyacá, Colombia.—Fl. VIII–XII, III.

The type-collection of var. *polyphlebia* was at first mistaken by Britton & Killip (1936, p. 184) for *Ch. tristicula* (our *Ch. glandulosa* var. *tristicula*), but Britton later recognized the error and assigned to *Pennell 1869* (NY) a manuscript binomial, which however remained unpublished at his death in 1934, and was not taken up by his editor Killip. We agree with Britton that the large-flowered chamaecristas with strongly asymmetric leaflets found on the west and east slopes of Cordillera Oriental in Cundinamarca are sharply distinct entities. We differ only in our estimate of their affinities, referring *Ch. tristicula* to varietal status under *Ch. glandulosa,* which see for discussion of differential characters, while we describe the eastern one as a new variety of *Ch. rufa.*

As brought out in the varietal key, var. *polyphlebia* differs from other forms of *Ch. rufa,* all remotely allopatric either to the north of Panama or in Amazonian Brazil, in its very numerous, exactly dimidiate leaflets crowded along the rachis at intervals of no more than 0.8–1.5 mm, with blades coarsely and densely multivenulose from the marginal costa, the venules all of about equal length and strength. The var. *exsul* has about as many pairs of leaflets but these less crowded and different in the intramarginal costa giving rise to fainter and more distant secondary venulation; the var. *rufa* ordinarily at once fewer and less crowded leaflets consistently more simply and remotely venulose.

44. Chamaecrista chamaecristoides (Colladon) Greene, Pittonia **4**: 20. 1899. *Cassia chamaecristoides* Colladon, Hist. Cass. 134. 1816.—Typus sub. var. *chamaecristoide* indicatur.

Robust or slender, commonly small-lvd but large-fld, always psammophilous herbs, either monocarpíc or perennant and suffruticose, the freely, often divaricately branching main stems diffuse or decumbent, forming mats or low thickets (3–)5–18 dm diam, when engulfed by shifting sand sometimes adventitiously rooting, the annotinous branchlets and lf-stalks sparsely to densely and cinereously strigulose or pilosulous with all incumbent, partly incumbent and partly ascending, or all loosely ascending hairs up to 0.2–0.6(–0.9) mm, the foliage concolorous, the lfts as variably pubescent as the stems, often glabrous or glabrate above, rarely quite glabrous beneath, ciliolate.

Stipules erect submembranous, pale green turning brownish-stramineous, from oblique base lanceolate or ovate-acuminate (2–)3–7 × 0.7–1.7(–2.2) mm, rather weakly 5–11-nerved, glabrous to densely puberulent dorsally, persistent.

Lvs usually widely arcuate-spreading, in expanded outline broadly to narrowly ovate, the major cauline ones up to 1.5–5 cm, on some lateral branchlets often all smaller; petiole with wrinkled pulvinus (2–)2.5–5(–6.5) mm, narrowly thick-margined, at middle 0.35–0.6 mm diam, the shallow sulcus open at gland; gland 1 (if 2–3 then very small) situated near middle of petiole proper, urceolate or subdiscoid (0.2–)0.25–0.7(–0.9) mm diam, subsessile or shortly stipitate, in profile obconic to shortly trumpet-shaped, (0.2–)0.25–0.8 mm long, varying from 0.4(–0.5) mm longer to 0.3 mm shorter than diam of head; lfts of major lvs up to 8–17(–19) pairs inserted along rachis 0.6–0.7 mm apart, decrescent both up- and downward from below middle of rachis, in outline obliquely or dimidiately linear or linear-lanceolate or the distal ones, sometimes (var. *cruziana*) all, semi-oblong or -obovate, obliquely mucronulate, often subfalcately incurved, the larger up to 3–12 × 0.6–2.8 mm, at base cordate on proximal and cuneate or cuneately rounded on distal side, from base 3–5-nerved by a midrib intramarginal by 0.1–0.6(–0.7) mm and dividing the blade 1:2+–8, the midrib giving rise on proximal side to 2–7 secondaries, the first of these (or the innermost primary next to midrib) incurved-ascending beyond middle of blade, the venulation immersed or almost so on the (when dry sometimes striately wrinkled) upper face, bluntly, sometimes indistinctly prominulous beneath.

Peduncle and raceme-axis together 2–13(–20) mm, adnate through (0–)1–10(–15) mm, 1–3-fld; pedicels 5–15 mm; sepals brownish or yellowish, ovate- or lance-acuminate up to (6–)7–11; petals usually very unequal, 3 shorter oblong or oblong-obovate beyond the claw, 1 abaxial flabellate up to 9–17(–19) mm, the falcately oblong cucullus ± as long; androecium 10-merous, the anthers mostly red, the longer ones up to (5.5–)6.5–9(–11) mm; ovary white-pilosulous, pilosulous along the sutures only, or glabrous; style filiform 3.5–6 mm; ovules 10–14.

Pod (25–)30–65 × (3.5–)4.5–6 mm, the purplish-castaneous, ultimately nigrescent valves thinly pilosulous or glabrous; seeds ±3.4–4.3 mm, the dull fuscous testa sometimes paler at hilum or buff-ocher overall, finely shallow-pitted.

The close relationships of *Ch. chamaecristoides,* as we define it in these pages, lie in two directions: on the one hand toward *Ch. rufa,* which replaces it at elevations of 850 m upward on the Gulf slope of Cordillera Oriental, and *Ch. pedicellaris,* lowland but fully disjunct on Hispaniola and eastern Cuba; and on the other to *Ch. fasciculata,* into which it imperceptibly fades northward by way of the essentially indefinable entity here called var. *cruziana.* On grounds of overall morphological similarity a strong case could be made for combining *Ch. chamaecristoides* and *Ch. rufa* into one major species. Typical var. *chamaecristoides* and typical var. *rufa* are hardly different in shape and venation of the leaflets and differ principally in proportion of flower to leaf, var. *chamaecristoides* retaining the large longistylous flower of *Ch. rufa* alongside of xeromorphic reduction, at once in length and complexity, of the leaves. But in practice the var. *chamaecristoides* is so easily recognized and so sharply delimited ecologically that we feel it preferable to maintain the traditional specific separation. The differential characters of *Ch. pedicellaris* are emphasized under that heading.

More persuasive arguments might be developed for the subordination of *Ch. chamaecristoides* sens. lat. to *Ch. fasciculata,* between which we can find no discontinuities, either morphological or ecological. While var. *chamaecristoides,* sens. str., is separable easily enough by the syndrome of a perennial root and narrow, highly asymmetric or dimidiate leaflets, this is linked to the pubescent monocarpic phases of *Ch. fasciculata* with subsymmetric leaflets by a series of populations along the Gulf coast northward from Tampico in which the prostrate growth-form and dune habitat coincide with an annual or at best feebly perennant root and broader, moderately (variably) asymmetric leaflets. Texan botanists have long been aware that the distinctive plant of the Gulf dunes southward approximately from Aransas Bay, here called *Ch. chamaecristoides* var. *cruziana,* which Turner (1959, p. 79) recognized as *Cassia fasciculata* var. *ferrisiae* and which Isely (1975, p. 93) could accept at no higher level than that of unnamed variant of *C. fasciculata* var. *puberula,* intergrades freely and insensibly on the south Coastal Plain with erect, obligately annual *Ch. fasciculata.* Indeed it is difficult to find two populations in this area in which the vesture of the leaflets, the displacement of the midrib, the form and stipitation of the petiolar glands and the number of leaflets are strictly identical. The taxonomic problem posed by this mass of intergradient material is of the sort faced by a surgeon required to separate a pair of Siamese twins with one heart in common; irreparable damage to the integrity of one twin is inevitable. It appears to us equally rational to follow Isely and refer all the intergradient Texan forms, even those with highly asymmetric leaflets, or dune habitat, or both, to *Ch. fasciculata,* which is left with only the single differential character of an annual root; or to interpret the prostrate dune ecotype as a northern extension of *Ch. chamaecristoides* in which a shorter life span coincides with slightly broader and, in consequence, less asymmetric leaflets. The decision to adopt the latter course is one of several unpalatable but pragmatic ones made in the course of this revision. A third solution to the problem, a return to Otto Kuntze's concept of a multiracial megaspecies incorporating not only the taxa under immediate discussion but also *Ch. glandulosa* and *Ch. nictitans* sens. lat., may well be the soundest in a biological sense, but has been rejected as taxonomically unacceptable in practice. In shifting the emphasis from the *fasciculata*-like elements of the intergrades to those nearer *Ch. chamaecristoides* sens. str., we find it desirable to take up the epithet *cruziana* rather than the more familiar *ferrisiae,* herein referred to the synonymy of *Ch. fasciculata* sens. lat. The isotype of *Chamaecrista ferrisiae* cited above represents one of the erect pubescent forms of *Ch. fasciculata* found on coastal prairies inland from the Gulf itself, not the diffuse coastal form with short leaves which we interpret as *Chamaecrista cruziana.*

Our concept of *Ch. chamaecristoides* is further expanded to accommodate the controversial *Chamaecrista brandegeei,* which seems to represent the species very locally on the shore of the Gulf of California in Sinaloa. Known to us from only one collection, lacking mature fruits, obviously, therefore, inadequate to support definitive taxonomic conclusions, *Ch. brandegeei* was placed by Britton & Rose, because of its supposedly sessile but in reality short-stipitate petiolar glands, in ser. *Patellariae,* that is among monocarpic relatives of *Ch. nictitans* to none of which it appears closely akin. On the other hand it seems to differ from var. *chamaecristoides* sens. str., which can have the same gland, only in the somewhat smaller stipules, slightly less displaced midrib of the leaflets (in which it resembles var. *cruziana*), and shorter corolla, androecium and style. The habit of growth, so far as can be judged from the specimens available, seems to be that of var. *chamaecristoides,* and the habitat may well be so. A distributional disjunction between the east and west shores of the continent is not exceptional; compare, for example the range of *Dalea scandens* (Mill.) Clausen, like *Ch. chamaecristoides* based on a Houston collection from Veracruz and abundant around the coastal plain between Tamaulipas and Campeche, which is represented locally on the Pacific coast by an endemic var. *occidentalis* (Rydb.) Barneby.

Key to the Varieties of *Ch. chamaecristoides*

1. Perennial; entirely Mexican, on the Gulf coast s.-ward from inter-tropical Tamaulipas to Campeche, local on the Pacific coast from Sinaloa to Michoacán.
   2. Midrib of larger lfts intramarginal by 0.1–0.25(–0.4) mm, dividing the blade 1:(3.5–)4–8; secondary nerves from midrib 2–4; fl relatively large, the sepals 7–11, the longer petals 13–19, the long anthers up to 6.5–11, the style 4.5–6 mm; Gulf of Mexico; coastal Jalisco and Michoacán. Pod 5–6 mm wide. 44a. var. *chamaecristoides* (p. 772).
   2. Midrib of lfts intramarginal by 0.2–0.4 mm, dividing the blade 1:3.5–4; secondary nerves from midrib 4–7; fl smaller, the sepals 6–7, the longer petals 9–10, the long anthers ±5.5, the style 3.5 mm; Gulf of California (Sinaloa). Pod (little known) narrower, ±3–4 mm wide. 44b. var. *brandegeei* (p. 773).
1. Monocarpic (sometimes over-wintering); coastal s. Texas, Tamaulipas or extreme n. Veracruz between Aransas Bay and vicinity of Tampico. Midrib intramarginal by ±0.2–0.7 mm, dividing the blade 1:±2–4; fl of var. *chamaecristoides*; pod mostly 4–5 mm wide. 44c. var. *cruziana* (p. 773).

44a. Chamaecrista chamaecristoides (Colladon) Greene var. **chamaecristoides**. *Cassia chamaecristoides* Colladon, 1816, l.c., sens. strict.—"*Mill. Dict. n. 17*. Hab. in nova Hispania circa Vera-Crux."—Holotypus, the description of *17. Cassia chamaecrista* Miller, Gard. Dict. ed. 8. 1768 (non Linnaeus, 1753)! Typotypus, labelled '*Senna spuria mimosae foliis frutescens et procumbens, flore maximo, siliquis glabris*. Vera cruz, *Houston,* BM (hb. Mill.; cf. Britten & Baker, J. Bot. **35**: 232. 1897)!

Cassia cinerea Chamisso & Schlechtendal, Linnaea **5**: 599. 1830.—"In littore maris arenoso inter Tecolutla et Villa rica [Veracruz, Mexico]."—Holotypus, B; isotypi, *Schiede & Deppe s.n.*, BM = *Schiede 711*, K (hb. Hook.)! = NY Neg. *1423*. *Chamaecrista cinerea* (Chamisso & Schlechtendal) Pollard ex A. Heller, Cat. N. Amer. Pl. ed. 2, 5. 1900.

Chamaecrista chamaecristoides (Colladon) Rose, Contrib. U.S. Nat. Herb. **12**: 267. 1909, pro comb. nov.

Cassia cinerea sensu Bentham, 1871, p. 581.

Characters as given in key.—Collections: 36.

Coastal dunes, beaches and stabilized sand flats behind barrier dunes, below 5 m, locally abundant around the Gulf of Mexico between the mouth of río Pánuco in extreme s.-e. Tamaulipas and Isla del Carmen in Campeche; disjunct on coasts of s. Jalisco and w. Michoacán.—Fl. irregularly through the year.

Within var. *chamaecristoides* as delimited by facies and dune habitat we encounter the familiar pattern of variable pubescence, the foliage varying from bright green to gray-strigulose, the stems from incurved-puberulent to pilosulous, and the pod from glabrous to pilosulous. The typus of *Cassia cinerea* is simply a gray-pubescent extreme in a continuous series of variants. This was already well understood by Bentham, whose account is at fault only in the nomenclature, due to misinterpretation of the type of *C. chamaecristoides,* which he transferred to the synonymy of the endemic west Indian *C. pygmaea*. Houston's plant in the Miller herbarium (BM), which stands behind Miller's description of *C. chamaecrista* and thus in turn behind *C. chamaecristoides* Colladon, is certainly, as first pointed out by Rose (1909, p. 267) and confirmed by Britton & Rose (1930, p.

267), the plant still common on shores around and near the city of Veracruz. Greene's false equation *C. chamaecristoides* = *Ch. depressa* (Pittonia **4**: 29. 1899) introduced the epithet *chamaecristoides* temporarily into the literature of southern United States, but once rectified by Britton & Rose (l.c.) has left no further imprint.

44b. Chamaecrista chamaecristoides (Colladon) Greene var. **brandegeei** (Britton & Rose) Irwin & Barneby, stat. nov. *Chamaecrista brandegeei* Rose ex Britton & Rose, N. Amer. Fl. **23**(5): 295. 1930.—"Altata, Sinaloa, September 2, 1901, *T. S. Brandegee*.—Holotypus, UC! clastotypi (fragm) + phototypi, NY, US!

Characters as given in key to varieties and in brackets under the species.—Collection: 1.

Dunes and sandy flats behind barrier beaches, to be sought along the immediate shore line of Sinaloa near lat. 24–25°N, known only from the neighborhood of the type-locality at Altata.—Fl. IX–X(?).

44c. Chamaecrista chamaecristoides (Colladon) Greene var. **cruziana** (Britton & Rose) Irwin & Barneby, stat. nov. *Chamaecrista cruziana* Britton & Rose, N. Amer. Fl. **23**(5): 288. 1930.—"Vicinity of Pueblo Viejo, Veracruz, 2 km south of Tampico, June 1, 2, 1910, *E. Palmer 534*."—Holotypus, NY! isotypi, CAS, GH, K, MO, US!

Characters as given in key to varieties.—Collections: 18.

Dunes, beaches, and associated sand flats, below 5 m, locally abundant along the immediate coast of the Gulf of Mexico from Aransas Bay, Texas, s. to the mouth of Río Pánuco on the Veracruz-Tamaulipas border.—Fl. VI–X(–XII).

45. Chamaecrista pedicellaris (DeCandolle) Britton, Bull. Torr. Club **44**: 12. 1917. *Cassia pedicellaris* DeCandolle, Prod. Syst. Veg. **2**: 504. 1825.—Typus infra sub var. *pedicellari* indicatur.

Perennial from a blackish taproot or from the divisions of a shortly forking pluricipital caudex, the fine wiry, often atrocastaneous, simple or irregularly forking stems at anthesis mostly (0.5–)1–6(–7.5) dm, rarely frutescent and up to 1.5 m, commonly procumbent, when long flexuously trailing, when short tufted, when suffruticose sometimes assurgent or suberect, very variably strigulose or pilose with all short straight appressed, all short incumbent, partly incumbent and partly short but spreading, all short and spreading, or mixed incurved short and longer setiform hairs up to 0.2–1(–1.2) mm, the vesture usually gray, sometimes faintly lutescent distally, the subconcolorous foliage dull olivaceous drying brownish, the lfts varying from glabrous or glabrous but ciliolate to pilosulous beneath only or equally overall.

Stipules erect, narrowly triangular to lance-acuminate from an oblique or obliquely dilated, then 1–2 auriculate-cordate base, (2–)2.5–9 × 0.7–2.3 mm, from base (7–)8–15(–16) nerved, the green or anthocyanic dorsally glabrous or pubescent blades becoming papery brownish, persistent after fall of lf.

Major cauline lvs (deciduous late in season) 1.6–5(–6) cm, the expanded blade in outline ovate to broadly lanceolate; petiole with usually discolored but seldom much swollen pulvinus 1.5–4(–5) mm, at middle (0.3–)0.35–0.6(–0.7) mm diam, narrowly grooved and margined; gland 1(–2) near middle of petiole proper, urceolate, discoid or rarely low-convex (0.15–)0.2–0.7(–0.75) mm diam, stipitate but sometimes only shortly so, in profile varying from squatly or slenderly tack-shaped to trumpet- or pin-shaped, exceptionally obconic, (0.25–)0.3–1.2(–1.4) mm tall, usually as tall or up to 1 mm taller than diam of head, exceptionally a trifle shorter; lfts of major lvs (8–)10–22(–24) pairs, the more distant 0.8–2.5(–3) mm apart along rachis, decrescent both up- and downward from below middle of lf-stalk, in outline linear, linear-oblong, or dimidiately linear-lanceolate (the distal ones relatively broader), mostly obliquely acute or acuminulate, sometimes obtuse mucronulate, straight or falcately incurved, the larger up to (4.5–)5–12(–14) × (0.7–)0.8–2(–2.5) mm, at base cordate or cordately angulate on proximal and narrowly cuneate on distal side, the strongly displaced costa intramarginal by 0–0.3(–0.4) mm, (exceptionally by 0.4–0.6 mm), dividing the blade 1:(3–)3.5–10(+), the either narrowly or widely divergent venulation of ±4–18 major or major and minor secondary venules either parallel and camptodromous to margin of blade or brochidodromous intramarginally, the venation prominulous beneath, more faintly so or immersed above.

Peduncles solitary or rarely 2 serial from an axil, (1–)1.5–7(–11) mm, adnate through (0.5–)1–6(–8) mm, 1–2(–3)-fld; pedicels at anthesis subfiliform a little thicker in fruit, the longer ones (11–)13–32(–40) mm, some arising from subterminal axils or high on a raceme shorter, bracteolate (1–)2–8(–10.5) mm below calyx; buds ovoid-acuminate, thinly puberulent, rarely pilosulous or glabrate; sepals submembranous often reddish, ovate- or lance-acuminate up to (5.5–)6–9.5 mm; petals yellow, not very strongly dissimilar, 4 obovate- or flabellate-cuneate up to (8–)8.5–11.5(–12.5) mm, the cucullus about as long, falcately oblong; androecium 10-merous, the short inner cycle yellow and the outer red or red-tipped, sometimes all concolorous of either color, the longest (5–)5.5–7.2 mm; style linear commonly 2.6–4.6 mm, of some fls abortive and shorter; ovary strigulose; ovules (7–)8–12(–14).

Pod (17–)22–42(–46) × (3–)3.3–4.5(–4.8) mm, the purplish-castaneous, ultimately nigrescent valves thinly strigulose or incumbent-pilosulous; seeds 2.7–3.4 mm, the testa dull ochraceous with dark transverse line across distal end, faintly pitted.

This delicately ornamental chamaecrista is distinguished by a combination of perennial, commonly herbaceous or at best only weakly suffruticose life-form, nearly always very narrow, almost or actually dimidiate leaflets crowded on a short leaf-stalk, and relatively long-pedicelled flowers of moderate size and moderate asymmetry. It resembles in habit the Porto Rican *Ch. glandulosa* var. *mirabilis,* though quite different in the venulation and outline of the leaflets. It appears closely akin to the continental *Ch. rufa,* which is only precariously separable at the rank of species by normally coarser growth, large stipules, and at least on average longer primary leaves with more numerous pairs of leaflets. The purely Mexican, obligately maritime *Ch. chamaecristoides* has almost the foliage of *Ch. pedicellaris,* but a distinctively coarser habit of growth and larger, more asymmetric flowers borne on pedicels little or not longer than themselves. These three taxa can be maintained in their traditional taxonomic category only by emphasis on their perfect eco-geographical disjunction, not by any substantial morphological differences.

Our concept of *Ch. pedicellaris,* expanded to admit the expected variation in vesture and in height and diameter of the petiolar glands, readily accommodates the type-collections of *C. strigillosa, Ch. enneryana* and *Ch. haitiensis,* and further, with some strain, those of *C. buchii* and the almost exactly similar *Ch. leonardae,* different in their slightly wider leaflet-blades and consequently less strongly displaced midrib. When Bentham described *Cassia strigillosa* he had only three unrepresentatively depauperate samples from eastern Cuba for comparison with the fragmentary typus of *C. pedicellaris* from Hispaniola. These indeed differ in size and length of the glands, but this single difference has dwindled to insignificance in the light of subsequent collections from Haiti. The typus of *Ch. enneryana* is notable for relatively long, dense and spreading yellowish pubescence; we interpret it as a minor variant equivalent to the yellow-pilose variants of *Ch. fasciculata* in United States. During his herborizations around Ennery in January–February 1926, E. C. Leonard collected ten examples (all US) of what we here refer to *Ch. pedicellaris* var. *pedicellaris*; of these ten, two are pilosulous, the remainder puberulent and most of them different from *Ch. enneryana* only in this one attribute. Of the eight puberulent, six have narrow leaflets with submarginal midrib, two slightly wider blades with narrowly intramarginal midrib. These last correspond very closely with typical *C. buchii* and *Ch. leonardae,* described from single gatherings in the Moustique-Terreneuve Mountains at points 35 and 50 km respectively northwest of Ennery, in which the midrib of some or of all leaflets becomes intramarginal by as much as 0.4–0.6 mm, so as to divide the blade ±1:3–4. It is conceivable that this modification of the leaflet expressed genetic pollution of *Ch. pedicellaris* by the sympatric *Ch. nictitans* var. *glabrata,* ubiquitous in Hispaniola and encountered at Ennery by Leonard; and this might further account for the unusual combination in *Leonard 8802* (US) of the foliage of tyical *Ch. pedicellaris* with some (but not all) atypically abbreviated pedicels.

Within *Ch. pedicellaris* sens. lat. we have only one type of variation strongly and positively correlated with dispersal and this, because it involves a modification of the venation pattern in the leaf-blades, a relatively stable feature in the chamaecristas, is accepted as taxonomically significant.

Key to the Varieties of *Ch. pedicellaris*

1. Secondary venulation arising from proximal side of midrib consisting of 3–6 major with or without as many or twice as many minor intercalary venules ascending at narrow (less than 45°) angles and, at least the major, sometimes all, produced to margin of blade, when the major and intercalary venules all equally strong the lft appearing multistriate; Cuba, Haiti and extreme w. Dominican Republic. 45a. var. *pedicellaris* (p. 775).
1. Secondary venulation arising from midrib on proximal side consisting of ±4–6 major venules ascending at ±45–50° and brochidodromous within the margin of the blade, the loops so formed often enclosing minor intercalary venules, contributing to an irregular reticulum; Cordilleras Central and Septentrional and Yaque valley in centr. and n.-w. Dominican Republic. 45b. var. *adenosperma* (p. 776).

45a. Chamaecrista pedicellaris (DeCandolle) Britton var. **pedicellaris**. *Cassia pedicellaris* DeCandolle, 1825, l.c., sens. str.—". . . in Sancto-Domingo. Bertero . . . (v.s. in herb. Balb[is].)."—Holotypus, perhaps a fragment of a fuller sheet returned to TO, one stem + fl in packet, G-DC!

Cassia strigillosa Bentham, Trans. Linn. Soc. **27**: 581. 1871.—"Cuba, *Linden,* n. 1708, *Wright,* n. 147; St. Domingo, *herb. Mus. Petrop.*"—Lectoholotypus, *Wright 147,* "prope villam *Monte Verde* dictam, Cuba Orientali . . . 1859," K (hb. Hook.)! = NY Neg. *1430*; isotypi, BR, G, GH, LE! paratypi, *Linden 1708,* BR, FI (hb. Webb.), G, K, LE!—*Chamaecrista strigillosa* (Bentham) Britton, Bull. Torr. Club **44**: 12. 1917.

Cassia buchii Urban, Symb. Antill. **5**: 361. 1908.—"Hab. in Haiti prope Gonaives in pratis montanis siccis ad Terreneufe, 350 m. alt., m. Junio flor. et fruct.: Buch n. 91."—Holotypus, presumably †B; isotypus, GH!—*Chamaecrista buchii* (Urban) Britton, Bull. Torr. Club **44**: 12. 1917.

Chamaecrista enneryana Britton ex Britton & Rose, N. Amer. Fl. **23**(5): 275. 1930.—"Vicinity of Ennery, l'Artibonite, Haiti, February 23, 1926, *E. C. Leonard 10041.*"—Holotypus, NY! isotypus, US!—*Cassia enneryana* (Britton & Rose) Jiménez, Rhodora **62**: 236. 1960.
Chamaecrista haitiensis Britton ex Britton & Rose, N. Amer. Fl. **23**(5): 276. 1930.—"Hills above Riviere Froide, Haiti, March 5, 1925, *Gerrit S. Miller, Jr.*"—Holotypus, *Miller 227,* NY! isotypus, US!—*Cassia froidensis* Jiménez, Rhodora **62**: 236. 1960 (non *C. haitiensis* (Britton) Jiménez, 1960).
Chamaecrista leonardae Britton ex Britton & Rose, N. Amer. Fl. **23**(5): 281. 1930.—"Moustique Mountains, vicinity of Bassin Bleu, Haiti, April, 1929, *Emery C. Leonard & Genevieve M. Leonard 14982.*"—Holotypus, NY! isotypus, US!—*Cassia leonardae* (Britton) Jiménez, Rhodora **62**: 236. 1960.
Cassia holguinensis Borhidi, Acta Bot. Acad. Sci. Hung. **22**: 299. 1977.—"Cuba; prov. Oriente: Lomas de Holguin. Leg.: M. Lopez Figueiras 942! 3. apr. 753 . . ."—Holotypus, SV, not seen; isotypus, US! isoparatypus, *Léon 15690,* NY!
Chamaecrista pedicellaris sensu Britton & Rose, 1930, p. 276.
Chamaecrista strigillosa sensu Britton & Rose, 1930, p. 275.
Chamaecrista buchii sensu Britton & Rose, 1930, p. 276.

As described for the species, except for venulation of lfts as described in key to varieties.—Collections: 62.

Savannas, outcrops, open hillsides and pineforest glades, sometimes abundant along paths or in fallow farmland, ascending from sand or coral beach near sea level up to 800(–1200) m in the interior, there equally at home on limestone, laterite, or serpentine, common almost throughout Haiti (all departments) and extending w. into e. Cuba (Oriente, w. to Holguín) and e. into w. Dominican Republic (Dajabon, Santiago Rodríguez, San Juan, Independencia, Barahona, and Pedernales).—Fl. irregularly throughout the year.

45b. Chamaecrista pedicellaris (DeCandolle) Britton var. **adenosperma** (Urban) Irwin & Barneby, stat. nov. *Cassia adenosperma* Urban, Symb. Antill. **5**: 362. 1908.—"Hab. in Sto. Domingo in Sierra del Palo Quemado ad rupes calcareas, 440 m. alt., m. Majo flor. et fruct.: Eggers n. 1886."—Holotypus, presumably †B; no isotypus found.—*Chamaecrista adenosperma* (Urban) Britton, Bull. Torr. Club **44**: 12. 1917.

Chamaecrista adenosperma sensu Britton & Rose, 1930, p. 275.

As described for the species, but never (so far as known) yellow-pilosulous, the lfts venulose as described in key to varieties.—Collections: 29.

Sunny hillsides, fields, thickets, and pineforests, not known to reach the immediate seashore, 20–625 m, widespread in the basin of Río Yague del Norte, n.-w. Dominican Republic in Monte Cristi, Valverde, Santiago Rodríguez, Santiago and n.-w. La Vega, extending n. over the crest of Cordillera Septentrional to the Atlantic slope in Puerto Plata.—Fl. almost throughout the year.

46. Chamaecrista pygmaea (DeCandolle) Britton, Bull. Torr. Club **44**: 10. 1917. *Cassia pygmaea* DeCandolle, Mém. Soc. Phys. Genève **2**(2): 131. 1824.—Typus infra sub var. *pygmaea* indicatur.

Herbaceous or weakly suffruticulose perennial herbs with slender, pliantly humifuse or weakly assurgent, reportedly sometimes erect, simple or diffusely branched stems radiating from the crown of an at first slender, ultimately woody, sometimes enlarged napiform, blackish root, the young stems and lf-stalks minutely puberulent or pilosulous with incumbent, subappressed or incurved-ascending short hairs sometimes mixed with or replaced by longer fine spreading or spreading-incurved setuliform ones up to 0.6–1.2 mm, the small, often distant sensitive lvs olivaceous concolorous drying brownish, the lfts varying from glabrous minutely ciliolate to pilosulous only beneath or on both faces, when pilosulous sometimes in addition weakly setulose-ciliate.

Stipules erect appressed, ovate acute to lance-acuminate from a subsymmetrically rounded or strongly oblique and then semi-cordate-auriculate base, (1.5–)2–5(–5.5) × 0.7–1.4(–1.6) mm, from base 7–14-nerved, the submembranous green or purplish blades either glabrous or puberulent dorsally, ciliolate or weakly setulose-ciliate, becoming papery brown persistent.

Lvs either subisomorphic and all short, or heteromorphic, the distal ones then longer, more distant and more complex than the proximal ones, in expanded outline ovate to lance-ovate, the major adult ones up to 8–28(–31) mm; petiole with discolored but little swollen pulvinus (0.9–)1.2–2.5(–3) mm, at middle (0.2–)0.25–0.4 mm diam, very narrowly wing-margined; gland 1(–2) near middle of petiole proper, discoid or shallowly urceolate round 0.2–0.45(–0.5) mm diam, shortly stipitate or rarely sessile, in profile shortly trumpet- or tack-shaped, less commonly squatly obconic or depressed into the groove of petiole, 0.15–0.45(–0.5) mm tall, (0.2–)0.15 mm taller to 0.1(–0.2) mm shorter than diam of head; lfts of adult lvs up to (6–)8–21 pairs, crowded along the rachis at points up to 0.6–1.5 mm apart, in outline linear- or lance-oblong, narrowly oblong-elliptic or the smaller distal ones obliquely semiobovate, all obtuse mucronulate, up to (2–)2.5–5.5(–6) × 0.7–1.2(–1.4) mm, at base cordate on proximal and cuneate on distal side, from base 3–4(–5)-nerved by the strongly displaced or submarginal midrib with on its broad side 2–3(–4) short incurved-ascending primary venules, the slender midrib dividing the blade 1:2.5–5 and giving rise on its proximal side to 2–4 (exceptionally with 1–2 scarcely perceptible intercalary) secondaries, the venulation on upper face immersed or faintly raised, on the lower finely prominulous.

Peduncles 1–2(–3)-fld, 0.5–20 mm, adnate to stem through (0–)0.5–20 mm; pedicels 3–18(–20) mm, bracteolate 0.5–3.5 mm below calyx; buds ovoid short-acuminate, thinly pilosulous, puberulent or rarely glabrate; sepals lance- and ovate-acuminate, up to (2.4–)3–5.8 mm; petals yellow, 2 oblong-oblanceolate smaller, 2 obovate or obovate-flabellate longer, one of these distal and 4.7–8(–9.2) mm, the cucullus semiobovate crenulate shorter; androecium 9–10-merous, the filaments very short but filiform, the anthers either yellow or reddish, 2–3 longer up to 2.2–4 mm; ovary densely strigulose, exceptionally glabrous; style 1.1–2 mm; ovules (6–)7–16.

Pod linear (16–)18–40(–50) × 2.7–4 mm, the usually purplish-castaneous, ultimately nigrescent valves thinly pilosulous or strigulose, exceptionally glabrous; seeds (few seen) ±3 mm, the testa orange-ochraceous dull, shallow-pitted, with faint transverse dark line across distal end.

A delicately small-leaved, small-flowered chamaecrista, known certainly only from western Cuba, *Ch. pygmaea* resembles some forms of *Ch. nictitans,* especially its Caribbean var. *diffusa,* from which it differs in the strongly displaced or submarginal midrib of the leaflets and commonly also in the longer pedicels. The asymmetric leaflets are not essentially different from those of *Ch. nictitans* var. *ramosa,* but this is a more coarsely leafy, normally monocarpic plant with very shortly pedicellate flowers and distally dilated (not filiform) style. The nearest ally of *Ch. pygmaea* is probably *Ch. pedicellaris,* endemic to eastern Cuba and Hispaniola, sometimes very similar in growth-habit but different in the normally longer leaves and leaflets and larger flowers with distinctly longer androecia and styles. Except in the last two features there is some overlap in the recorded measurements and the biological reality might be better expressed if *Ch. pedi-*

cellaris and *Ch. pygmaea* were treated as vicariant aspects of one major specific type. See also discussion of the kindred, but Continental and substantially larger-flowered *Ch. chamaecristoides* and *Ch. rufa*.

Even within its relatively restricted range *Ch. pygmaea* is notably polymorphic in habit of growth, in vesture of stems and foliage, length and complexity of adult leaves, and flower-size. The pubescence, in the usual chamaecrista mode, may consist entirely of minute incurved hairs, entirely of longer spreading ones, or of the two sorts mixed in variable proportions; it is firmly correlated with no other variable quality or with dispersal, and is thus, as usual in the group, of no practical use whatever as a differential character. The ontogeny of the foliage exhibits two, usually quite distinct patterns. In all cases the early leaves (at anthesis sometimes already shed) are short (to 1–1.5 cm) and composed of few (±6–10) pairs of leaflets. In one developmental pattern leaves as short or almost as short are repeated indefinitely upward along the mature flowering and fruiting stems, and tend to be separated by internodes shorter than themselves. In the other pattern the upper leaves become rather abruptly longer and multifoliolate, at the same time separated by longer internodes. With short leaves there goes a tendency for diffuse branching, short peduncles only briefly adnate to shortened internodes, relatively small flowers, and short, few-ovulate pods. With longer adult leaves the stems tend to be few, simple or weakly branched, the peduncles become longer and further adnate to longer internodes, the flower a little larger and the pod on average longer. None of these characters, however, is firmly linked with the foliage patterns, and even in this we discern no really decisive discontinuity. We are therefore unable to accept *Ch. pygmaea* and *Ch. savannarum* as anything but diverse aspects of a single species.

Key to the Varieties of *Ch. pygmaea*

1. Peduncles 0.5–5(–6) mm long, adnate to the stem through (0–)0.2–0.5(–0.6) mm; lvs subhomomorphic, the lfts of longest adult ones 8–12(–15) pairs. 46a. var. *pygmaea* (p. 778).
1. Peduncles mostly 6–11(–20) mm long, adnate to stem through 5–15(–21) mm; lvs heteromorphic, the early ones (often shed by anthesis) as above, but the later ones, associated with fls, longer, with (13–)15–21 pairs of lfts. 46b. var. *savannarum* (p. 779).

46a. Chamaecrista pygmaea (DeCandolle) Britton var. **pygmaea**. *Cassia pygmaea* DeCandolle, 1824, l.c., sens. strict.—". . . trouvé à St.-Domingue par. M. Bertero."—Holotypus, G-DC!

Chamaecrista micrantha Britton, Bull. Torr. Club. **43**: 463. 1916.—". . . Cuba . . . near San Pedro, Isle of Pines (*Britton & Wilson 14294*)."—Holotypus, NY! isotypus, US!—*Cassia minutiflora* León, Contrib. Ocas. Mus. Hist. Nat. Colegio de la Salle **9**: 11. 1950 (non *C. micrantha* Guillemin, 1832).

Chamaecrista diffusissima Britton ex Britton & Rose, N. Amer. Fl. **23**(5): 288. 1930.—"Pinar, near Las Veguas, San Diego de los Baños, Pinar del Rio, Cuba, August 18, 1914, *Brother León 4394*.—Holotypus, NY!—*Cassia diffusissima* (Britton & Rose) León, Contrib. Ocas. Mus. Hist. Nat. Colegio de la Salle **9**: 11. 1950.

Chamaecrista roigii Britton ex Britton & Rose, N. Amer. Fl. **23**(5): 289. 1930.—"Cerro de Cabras, Pinar del Rio, Cuba, July 27, 1929, *J. T. Roig*."—Holotypus, NY!—*Cassia roigii* (Britton & Rose) León, Contrib. Ocas. Mus. Hist. Nat. Colegio de la Salle **9**: 12. 1950.

Chamaecrista pygmaea sensu Britton & Rose, 1930, p. 288.

Lvs as described in key; vesture pilosulous, puberulent, rarely almost 0; pedicels 5–18(–20) mm; fls commonly small, the sepals (2.4–)3–5(–5.6) mm, the longest petal 4.7–7 mm; ovules (6–)7–10(–14); pod (16–)18–30(–35) mm.—Collections: 21.—Fig. 49.

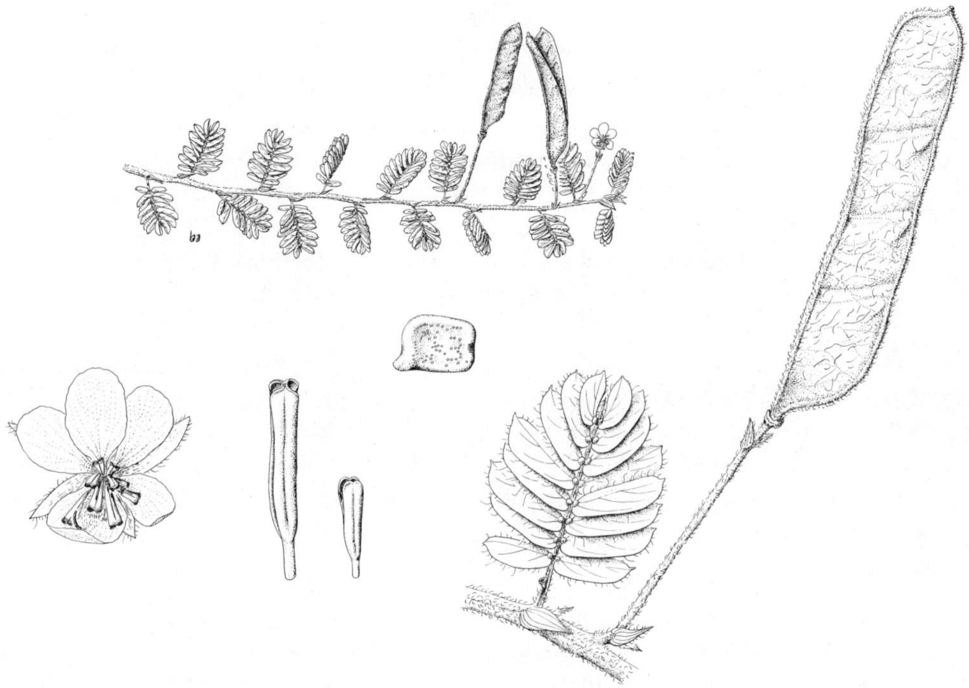

Fig. 49. *Chamaecrista pygmaea* (DC.) Britt. var. *pygmaea*. Flowering branchlet ×1; leaf + pod ×4; flower ×4; 2 stamens ×8; seed ×6 (*Britton et al. 6378*).

Savannas and open pinewoods, on sandy soils mostly below 100 m, w. Cuba: widespread and locally common in Pinar del Río; local on Isla de Pinos (Nueva Gerona, San Pedro); one record from w. Matanzas (Guanabana). Described from Santo Domingo, but not since collected on Hispaniola.—Fl. IV, VII–XII, perhaps intermittently through the year.

The three Brittonian segregates here reduced to synonymy are individually characterized by slightly different combinations of vesture, habit (due partly to maturity of the individual) and gland-size: *Ch. micrantha,* thinly pilose, supposedly with erect or ascending stems, and petiolar glands ±0.25–0.3 mm diam; *Ch. diffusissima,* glabrate except for remnants of pilose vesture on margins of stipules and leaflets, humifuse stems, and glands of the preceding; *Ch. roigii,* both pilose and incumbent-puberulent, habit undecipherable from the type-specimen, glands up to 0.5 mm diam. It may be worth noting that *Britton et al. 6760* from San Diego de los Baños, an approximate topotypus of *Ch. diffusissima* identified (NY) by Britton as *Ch. micrantha,* has the certainly diffuse stems of *Ch. diffusissima* combined with the vesture of *Ch. micrantha* and the gland of *Ch. roigii*. Britton's key to ser. *Pygmaea* (Britton & Rose, 1930, p. 272) bears little relation to reality even as expressed by material studied by Britton himself.

46b. Chamaecrista pygmaea (DeCandolle) Britton var. **savannarum** (Britton) Irwin & Barneby, stat. nov. *Ch. savannarum* Britton, Bull. Torr. Club **43**: 463. 1916.—". . . Cuba . . . near Siguanea, Isle of Pines (*Britton & Wilson* 14379)."—Holotypus, NY! isotypi, GH, US!—*Cassia sa-*

vannarum (Britton) Urban, Feddes Repert. **15**: 312. 1918, in obs., citatione manca; Macbride, Contrib. Gray Herb., n. ser. **59**: 26. 1919.

Chamaecrista savannarum sensu Britton & Rose, 1930, p. 288.

Lvs as decribed in key; vesture commonly incurved-puberulent, sometimes pilosulous; pedicels 3–10 mm; fls slightly larger than in var. *pygmaea,* the sepals 4–5.8 mm, the longest petal 6–7.5(–9.2) mm; ovules 9–16; pod 27–42(–50) mm.—Collections: 18.

Savannas, on white or brown sand, sometimes with thin pine forest or *Coccothrinax* palm, mostly below 150 m, w. Cuba: Isla de Pinos, widespread over the island; local in Pinar del Río (Cd. Pinar del Río, La Coloma).—Fl. almost throughout the year.

47. Chamaecrista glandulosa (Linnaeus) Greene, Pittonia **4**: 286. 1899, sens. magnopere ampliat. *Cassia glandulosa* Linnaeus, Syst. Nat. ed. 10, 1017. 1759.—Typus infra sub var. *glandulosa* indicatur.

Perennial, either potentially frutescent or arborescent (even though precociously flowering the first season as soft stemmed herbs) or truly herbaceous but the stems then arising from the crown of a pluricipital and vertical (never rhizome-like) lignescent root; lfts of larger lvs (8–)9–24 pairs (sometimes slightly fewer in rare Andean vars. *andreana, andicola* and *balsasana* and in Mexican var. *parralensis,* but these at once emphatically perennial or fruticose and large-fld), the midrib centric or moderately (in var. *tristicula* strongly) displaced; peduncles at least shortly adnate to stem; pedicels 7–31(–50) mm, the longer ones nearly always at least 10 mm, at least in fruit; fls mostly large or of moderate size, the long abaxial petal (8–)8.5–19(–23) mm; style linear (3–)3.4–8.7 mm (but sometimes abortive and shorter in some individual fls). Habit, pubescence, petiolar glands and venulation of lfts highly variable, described separately for each variety.—Centr. and s. Mexico and West Indies; Orinoco Valley in Venezuela; Andes from n.-w. Venezuela and n. Colombia to s. Peru and Bolivia; s.-e. Brazil; naturalized and cultivated elsewhere.—Fig. 2 (petiolar nectary).

The foregoing description of *Ch. glandulosa* is formulated in general terms such as to cover all but three of the perennial or fruticose Neotropical chamaecristas in which supra-axillary pedicels coincide with relatively large, long-pedicelled, dolichostylous flowers and scarcely or (except in var. *tristicula*) moderately asymmetric leaflets with midribs varying from centric to displaced by about one third of the blade's breadth. The exceptions noted are *Ch. deeringiana* of southeastern United States, distinguished by its unique horizontal root-system, the local Andean *Ch. pedemontana* and polymorphic West Indian *Ch. lineata,* each different from sympatric forms of *Ch. glandulosa* in the relatively few leaflets. The species as here interpreted is precariously distinguished from extratropical *Ch. fasciculata* by the perennial life-span, and the more precariously and artificially so because many of its forms flower in their first year from soft, essentially herbaceous stems, while races of *Ch. fasciculata* native to frostless peninsular Florida may survive the winter to become bushy-branched and basally lignescent. Duration of the root is again the only deciding difference between *Ch. glandulosa* sens. lat. and *Ch. nictitans* sens. lat., but all of the latter except its subsp. *disadena* are distinguished further by smaller or shorter-pedicellate flowers. Leaflets with strongly displaced or marginal midrib furnish the only ultimately diagnostic character of the group of *Ch. rufa, Ch. chamaecristoides* and *Ch.*

pedicellaris, and this difference is blurred by *Ch. glandulosa* var. *tristicula,* which we find inseparable from var. *flavicoma* except in this one feature. As here defined, *Ch. glandulosa* is coextensive, if taxa discovered after 1871 are disregarded, with Bentham's *Cc. flavicoma, glandulosa, virgata,* and *chamaecrista* var. *brasiliensis* combined. Bentham separated the first three primarily on characters of pubescence now found to have very little biological significance or taxonomic utility; he seems to have associated var. *brasiliensis* with North American *Ch. fasciculata* under the misapprehension that it was herbaceous. In the light of material gathered since Bentham's revision the discontinuities sundering *Ch. glandulosa* from the larger-flowered races of *Ch. nictitans* subsp. *disadena* on the one hand and from *Ch. fasciculata* on the other, have become so nebulous and moreover so unsatisfactory theoretically, that a return to Otto Kuntze's concept of an all-embracing *Cassia chamaecrista,* at least as this applies to the genuine *Chamaecristae* of the New World, must appear very attractive. The price, however, is an elaborately cumbrous taxonomic hierarchy even more forbidding and difficult to use than that proposed in this revision.

Due to the basic instability of those few phenetic characters which we have found helpful or diagnostic in defining geographically acceptable races within *Ch. glandulosa,* a key to its component varieties is extraordinarily difficult to construct, indeed is probably unattainable in the dependable form that one would wish. This reflects, of course, the contemporary stage of evolutionary activity and racial immaturity everywhere visible in the supra-axillary chamaecristas. While disapproving any key based primarily on dispersal patterns rather than morphology, we have been reduced to the expedient of two varietal keys, one for the endemic West Indian varieties of *Ch. glandulosa,* the other for its continental ones.

Key to the West Indian Varieties of *Ch. glandulosa*

1. Major lvs distinctly petioled, the petiole proper including pulvinus 2.5–6 mm; lft-blades plane; widespread, but not on Hispaniola.
    2. Lfts dorsally pilosulous in lower ⅓–⅔ at least on proximal side, often on both sides, of midrib; vesture of stems lutescent; Jamaica. 47a. var. *glandulosa* (p. 782).
    2. Lfts dorsally glabrous or minutely puberulent along midrib; vesture of stems colorless; widespread outside Jamaica.
        3. Larger lvs 3–12 cm, their larger lfts up to 7–25 mm; habit and habitats various, but seldom prostrate and if so allopatric; Puerto Rico, Virgin Is., and Antilles. 47c. var. *swartzii* (p. 784).
        3. Larger lvs 1.5–4(–4.5) cm, their larger lfts up to 3–7.5 mm; prostrate or assurgent, on white sands of the n. coastal plain of Puerto Rico. 47d. var. *mirabilis* (p. 787).
1. Major lvs sessile or almost so, the pulvinus with obsolete or very short and obscure petiole 1.5–3 mm; margin of lfts ± revolute, the blades hence at least shallowly concave dorsally; Hispaniola. 47b. var. *picardae* (p. 783).

Key to the Continental North and South American Varieties of *Ch. glandulosa*

1. Stems truly herbaceous, at anthesis 1–3.5 dm, arising annually from and dying back annually to the crown of the woody root or its cespitose pluricipital caudex; local either on the Mexican Plateau or in s.-centr. Colombia.
    2. Secondary venulation of lfts composed of 6–10 pairs of major with 1–2 intercalary venules; Colombia (Cundinamarca and Tolima to Nariño). 47k. var. *andreana* (p. 796).
    2. Secondary venulation of lfts composed of 3–5 pairs of major with or without faint intercalary venules; Mexico (Chihuahua to Guanajuato). 47l. var. *parralensis* (p. 797).
1. Stems herbaceous only in the first season of growth, early lignescent basally and at anthesis mostly 4.5–40 dm (sometimes dwarfed but still obviously frutescent in high Peruvian Andes);

widely dispersed from s. and e. Mexico (Guerrero and Yucatán) to Brazil, Bolivia and s. Peru, most highly developed in the Andes.

3. Lfts of major lvs 6–12 pairs; local varieties either of coastal s.-e. Brazil or (very local) n. Peru, in the latter case the venulation of upper face of lfts immersed.
   4. Petiolar gland sessile patelliform much wider than tall; lfts thin-textured, the venulation of both faces evident; s.-e. Brazil (Rio de Janeiro and Guanabara). 47m. var. *brasiliensis* (p. 798).
   4. Petiolar gland short-stipitate or subsessile, obconic or shortly trumpet-shaped, as tall or taller than diam of head; lfts thick-textured, the upper face smooth, apparently veinless; Marañón valley near 6°45′S in n. Peru. 47h. var. *balsasana* (p. 794).
3. Lfts of major lvs 12–24 pairs, (sometimes fewer in high-Andean var. *andicola*); widespread, but if Peruvian then lfts venulose on both faces.
   5. Pod slender, 3.5–4 mm wide; rare varieties of middle Orinoco valley and adjoining periphery of Guayana Highland in Venezuela and adjacent Colombia.
      6. Fl large, the sepals up to 12–16 mm, the long abaxial petal 15–23 mm; style 6.5–8.5 mm. 47i. var. *rapidarum* (p. 794).
      6. Fl small, the sepals up to 7–10 mm, the long abaxial petal 8–11 mm; style not over 6 mm. 47j. var. *crystallina* (p. 795).
   5. Pod relatively stout, 4.5–9 mm wide; allopatric.
      7. Midrib of lfts centric or moderately displaced, dividing the blade 1:1–2.5; widespread in the Andes; disjunct in Mexico.
         8. Relatively dwarf, diffuse or assurgent subshrubs of great elevations in the s. Ecuadorian, Peruvian and Bolivian Andes, at (2200–)2400–3300 m; major lvs up to 1.5–4 cm. 47f. var. *andicola* (p. 791).
         8. Taller assurgent or erect shrubs and treelets at anthesis (3–)5–40 dm; major lvs up to 3.5–12 cm; widespread from Mexico to Bolivia, but in the Andes from centr. Peru s.-ward mostly below 2000 m (higher in Colombia). 47e. var. *flavicoma* (p. 788).
      7. Midrib of lfts strongly displaced, dividing the blade 1:3–5; local in Cundinamarca, Colombia. 47g. var. *tristicula* (p. 793).

47a. Chamaecrista glandulosa (Linnaeus) Greene var. **glandulosa**. *Cassia glandulosa* Linnaeus, 1759, l.c. & Sp. Pl. ed. 2, 542. 1762, sens. str., exclus. syn. Breyn.—"Habitat in Jamaica."—Holotypus, acquired by Linnaeus from P. Browne, LINN *528/32*!—*Dialanthera glandulosa* (Linnaeus) Rafinesque, Sylva Tell. 127. 1838.

Cassia virgata Swartz, Prod. 66. 1788.—"Jamaica." & Fl. Ind. Occ. 2: 728. 1800.—"Habitat in *Jamaica, Hispaniola*."—Lectoholotypus, labelled "Jamaica, *Swartz* (hb. Swartz.)," S! = NY Neg. *9235*.—*Cassia glandulosa* var. (*β*) *ramosa* Grisebach, Fl. Brit. W. Ind. 211. 1864. *Chamaecrista virgata* (Swartz) Greene, Pittonia **4**: 31. 1899; Pollard, Field Mus., Bot. 2: 48. 1900, pro comb. nov. *Ch. glandulosa ramosa* Millspaugh, Field Mus., Bot. **1**: 493. 1902, technically nomen nudum, based on a non-existent "*C. ramosa* Linn.," doubtless belongs here.

Cassia stricta Schrank, Hort. Monac. **1**: pl. 34. 1819.—"Jamaica."—Described from a plant cultivated under glass at München, not known to survive. Lectoholotypus, the cited plate!—A plant grown at M under this name in 1836 appears to be *Ch. fasciculata*.—*C. chamaecrista* var. *stricta* (Schrank) DeCandolle, Prod. **2**: 503. 1825.—*C. glandulosa* var. *stricta* (Schrank) G. Don, Gen. Hist. Dichl. Pl. **2**: 447. 1832; Grisebach, Fl. Brit. W. Ind. 211. 1864, pro comb. nov.

Hepteireca glandulosa Rafinesque, Sylva Tell. 127. 1838, based on *Cassia glandulosa* sensu Hooker, Bot. Mag. t. 3435. 1835, non Linnaeus, 1753.—Holotypus, the cited plate!

Chamaecrista cowellii Britton ex Britton & Rose, N. Amer. Fl. **23**(5): 286. 1930.—"Between Corozal and Pedro Miguel, Panama [Canal Zone], March 14, 1905, *J. F. Cowell 406*."—Holotypus, NY!

Cassia glandulosa sensu Bentham, 1871, p. 577, minore ex parte. *Chamaecrista glandulosa* sensu Britton, 1917, p. 8; Britton & Rose, 1930, p. 286.

Suffruticose becoming shrubby, of rapid growth and early maturity, at anthesis (3–)5–16(–20) dm, puberulent with short incurved and commonly some or many longer spreading-ascending, yellowish hairs to 0.4–0.8 mm, the concolorous lfts pilosulous dorsally in lower ⅓–⅔ on both or on only the proximal side of midrib.

Stipules lance-acuminate 3.5–7(–8) × 0.65–1.4 mm.

Major cauline lvs to 4.5–9 cm, the petiole (2–)2.5–4.5 mm; petiolar glands 1(–3), the solitary or largest one urceolate 0.45–0.9 mm diam, coarsely long-stipitate, 0.9–2(–2.3) mm tall, in profile 0.3–0.9(–1.8) mm taller than diam of head; lfts of larger lvs up to 11–18(–19) pairs, oblong to lance-oblong-elliptic, the largest (9–)11–17 × (2.2–)2.8–4.4 mm, the pinnate venulation of 8–12 pairs of major secondary veins with as many intercalary ones.

Pedicels at anthesis mostly 9–14 mm, in fruit 10–16 mm (in some early and very late fls sometimes a trifle shorter); sepals up to (6–)7–10 mm; long abaxial petal 11.5–15 × 7–14.5 mm; style 3.6–4.4 mm; ovules 8–11.

Pod 32–43 × 4.5–5 mm, thinly puberulent.—Collections: 18.

Dry stony banks, thickets, and ledges along streams, sometimes on serpentine, 280–1300 m, locally abundant, coastal hills and interior mountains of Jamaica.—Fl. III–VI, and intermittently through the year.

We follow Britton & Rose (1930) and Adams (1972, p. 327) in the belief that typical *Ch. glandulosa,* described from specimens collected on Jamaica by Patrick Browne, is endemic to the island. It differs from the vicariant var. *picardae,* endemic to Hispaniola, in the petiolate leaves and plane leaflets, while resembling it in the golden pubescence of the young stems. The var. *swartzii,* of Puerto Rico and eastern Antilles, very closely simulates it, differing in the usually smaller glands, dorsally glabrous or subglabrous leaflets, and colorless vesture, or by some combination of these characters.

The typus of *Ch. cowellii,* from the lowlands of the Canal Zone a few miles inland from Panama City, matches Jamaican *Ch. glandulosa* at every point, the supposedly dense pubescence not really masking the venation of the leaflets, as claimed, except in very young leaves. If correctly labelled, *Ch. cowellii* must have been a cultivated plant. The species was overlooked by Schery in his treatment of *Cassia* for Flora of Panama (1951).

47b. Chamaecrista glandulosa (Linnaeus) Greene var. **picardae** (Urban) Irwin & Barneby, comb. nov. *Cassia pinetorum* var. *picardae* Urban, Feddes Repert. **15**: 311. 1918.—"Hab. in Haiti: Picarda no. 599."—Holotypus, presumably †B; no isotypus found.

Chamaecrista pinetorum Britton, Bull. Torr. Club **44**: 7. 1917.—"Pine woods, near Constanza, Santo Domingo, at 1,200 m. alt. (Tuerckheim 2887)."—Holotypus, NY! isotypi, A, G, GH, K, US!—*Cassia pinetorum* (Britton) Urban, Feddes Repert. **15**: 311. 1918.

Cassia selleana Urban, Feddes Repert. **15**: 311. 1918.—"Hab. in Haiti in Morne la Selle in sylvis apricis 1000 m, m. Jan. flor.: Buch no. 1148."—Holotypus, presumably †B; neoholotypus, former isotypus, GH!—*Chamaecrista selleana* (Urban) Britton ex Britton & Rose, N. Amer. Fl. **23**(5): 279. 1930.

Cassia cahosiana Urban & Ekman, Ark. Bot. **22A**: 33. 1928.—"HAITI. Massif des Cahos, Las Caobas prope Belladère in collibus cr[etaceis] 600 m alt., m. Febr. flor. et fruct.: [*E. L. Ekman*] n. H 5577."—Holotypus, S! = NY Neg. *9236*; paratypus, S!—*Chamaecrista cahosiana* (Urban & Ekman) Britton & Rose, N. Amer. Fl. **23**: 281. 1930.

Chamaecrista pinetorum sensu Britton & Rose, 1930, p. 280.

Early fruticose and sometimes ultimately subarborescent, at anthesis (3–)5–27(–35) dm, the young stems, lf-stalks and axes of inflorescence strigulose with straight appressed or less often incumbent whitish hairs up to 0.2–0.45 mm and often in addition thinly to densely pilosulous with widely spreading yellowish or whitish setules up to 0.4–1(–1.2) mm, with either type of vesture the lfts glabrous or glabrous and ciliolate, or with the shaggy yellowish vesture often pilosulous on the lower or on both faces, the hairs there commonly basally dilated and the lf-face consequently appearing papillate.

Stipules lance-acuminate from an obliquely dilated base, (3–)4–8.5 × (0.6–)1–2.1 mm, 8–15-nerved.

Major cauline lvs (deciduous during dry periods) up to 3–5.5(–7) cm, the pulvinus and obsolete or very short true petiole together 1.5–3 mm; petiolar gland solitary situated immediately adjacent to first pair of pulvinules, often immediately above the pulvinus, 0.7–1.6(–2) × 0.4–0.9 mm, in profile slenderly trumpet- or stoutly tack-shaped, (0–)0.1–1.2 mm taller than the discoid or urceolate head, seldom smaller glands on each segment of rachis; lfts of major lvs up to (8–)9–18(–19) pairs, up to (7–)8–15(–17) × (1.6–)2–4(–4.5) mm, the margins at least incipiently, often strongly revolute and the dorsal face therefore concave, the pinnate venulation of 5–9 major pairs, the intercalary ones either immersed or prominulous, especially so above.

Pedicels (7–)8–24 mm; sepals up to (7–)8–11(–12) mm; long petal up to 9–15 × (7–)8–12 mm (occasional abnormally small fls late in season); style (3–)3.5–5.5(–6) mm; ovary either strigulose or pilosulous; ovules 8–11.

Pod 30–50 × 5–7(–8) mm, the dark reddish-brown valves thinly pilosulous, incurved-puberulent, or glabrate.—Collections: 39.

Pinewoods and brush succeeding destroyed pine forest, on limestone, sandstone, laterite and bauxite, 600–1800(–2000) m, locally abundant in the mountains of s.-e. Haiti (Dépt. de l'Ouest) and of s.-w. and centr. Dominican Republic (e. to Cord. Central in prov. La Vega, s. to Sa. de Baoruco in prov. Barahona and Pedernales).—Fl. irregularly throughout the year.

We have assembled under the name var. *picardae* all representatives of the *C. glandulosa* complex found on Hispaniola, the endemic entity so constituted differing weakly from Jamaican var. *glandulosa* and its more eastern Antillean counterpart var. *swartzii* in the abbreviated or almost obsolete true petiole and in the marginally revolute, therefore dorsally concave leaflets. The variety consists of several minor races, corresponding with *C. selleana, Ch. pinetorum, C. pinetorum picardae, C. cahosiana,* and others intermediate between them. These are minutely distinguishable by combinations of leaflet-number and pubescence, and have some geographic significance. The plants of the Massif de La Selle in Haiti and of its eastward continuation, as Sa. de Baraoca, into southwest Dominican Republic, as also those of Sa. de Neiba immediately north of the last, have yellow-pilosulous branchlets and up to 12–18 pairs of leaflets in the primary leaves (*C. selleana,* sens. str.). In Cordillera Central of Prov. La Vega and Peravia the stems are appressed-puberulent with colorless hairs and the leaflets are on the average slightly larger and fewer, to 8–12 pairs in larger leaves (*C. pinetorum,* sens. str.). Near the Dominican frontier in northeastern Dépt. de l'Ouest, Haiti, the prevailing type combines the leaflets of *C. selleana* with vesture intermediate to that of *Ch. pinetorum* (var. *picardae,* sens. str.). The type of the inconsiderable *C. cahosiana* was segregated from *C. selleana* and *C. pinetorum* simply because of a relatively broad pod, 7–8 rather than commonly 5–7 mm wide. These incipient racial expressions of diversity seem in the context of sect. *Chamaecrista* insufficiently emphatic to deserve taxonomic rank.

Our interpretation of var. *picardae,* of which the holotypus was destroyed at Berlin, is based on the protologue and on *Ekman 1065* so identified by Urban himself.

47c. Chamaecrista glandulosa (Linnaeus) Greene var. **swartzii** (Wikstrom) Irwin & Barneby, comb. nov. *Cassia glandulosa* var. *swartzii* (Wikstrom) Macbride, Contrib. Gray Herb., n. ser. **59**: 26. 1919, based on *C.*

swartzii Wikstrom, Vetensk. Acad. Handl. **1825**: 430. 1826.—"Hab. in Isula S:ti Bartholomaei: Forsström."—Holotypus, labelled by Swartz "*Cassia* n. sp.," S! = NY Neg. *9234*; presumed isotypus, labelled "Ins. Sti. Bartholomaei: *Forsström*," S!—*Chamaecrista swartzii* (Wikstrom) Britton, Bull. Torr. Club **44**: 9. 1917.

Cassia otterbeinii G. F. W. Meyer, Prim. Fl. Essequib. 169. 1818.—"[Guyana:] in fruticetis, [collected by Heinrich Jakob Otterbein and sent to Meyer by Mertens]."—Holotypus, labelled "Essequebo, d. Otterbein, 1815," LE (hb. Mertens.)! = NY Neg. *9231*; caret in hb. GOETT (fide G. Wagenitz in litt., 4.IX.75).

Cassia polyadena DeCandolle, Mém. Soc. Phys. Genève **2**(2): 132. 1824.—". . . découverte à la Guadeloupe par . . . M. Bertero."—Holotypus, labelled "Guadel[oupe], Bertero, misit Balbis 1820," G-DC! = F Neg. *33460*.—*Chamaecrista polyadena* (DeCandolle) Britton, Bull. Torr. Club **44**: 10. 1917.

Cassia parkeriana DeCandolle, Prod. Syst. Veg. **2**: 504. 1825.—"circa Demerari."—Holotypus, labelled "Envoi de Demarari [Guyana], Mr. Parker. 1826," G-DC! = F Neg. *7001*; isotypus, labelled "Demarara, cultd.," K!

Cassia chamaecrista subsp. (*β*) *glandulosa* fma. *glabrescens* O. Kuntze, Rev. Gen. 169. 1891.—"St. Thomas, [West Indies]."—Holotypus, *O. Kuntze 45,* collected II.1874, US! NY caret!

Chamaecrista complexa Pollard ex Millspaugh, Field Mus., Bot. **2**: 47. 1900.—". . . on roadsides at Charlotte Amalia, St. Thomas, January 17–18, [1899], ([*Millspaugh*] *375*)."—Holotypus, F! clastotypus, (fragm), NY!—Reduced to *Ch. swartzii* by Britton, 1917, p. 9.

Chamaecrista swartzii sensu Britton, Addisonia **13**: 55, t. 444. 1928; Britton & Rose, 1930, p. 287.

Chamaecrista polyadena sensu Britton & Rose, 1930, p. 287; vix sensu Bentham, 1871, p. 575, nisi minima ex parte.

When adult suffruticose, fruticose, exceptionally subarborescent, at anthesis (1.5–)3–24(–40) dm, erect in sheltered sites, diffuse or prostrate in exposed ones, the stems and lf-stalks puberulent with incurved or some spreading whitish hairs to 0.2–0.8 mm, the lfts glabrous above, often so beneath, concolorous or paler beneath and then often red-margined when fresh.

Stipules triangular to lance-acuminate, 1.5–8(–11) × 0.4–1.7 mm.

Major cauline lvs 3–12 cm, in depauperate states only 2–3 cm, the petiole 2.5–6(–7) mm; petiolar gland 1(–3), urceolate or discoid, 0.15–0.9 mm diam, stipitate but sometimes shortly so, in profile 0.2–1.1(–1.5) mm tall, (0.3–)0.1 mm shorter to 0.8(–0.95) mm longer than diam of head; lfts of larger lvs (5–)8–23 pairs, linear-oblong to oblong-obovate, the largest (6–)7–25 × (1.2–)1.5–6.5(–9.5) mm, the secondary pinnate venulation of (5–)6–12 pairs of major secondaries with or without solitary or paired intercalary minor ones.

Pedicels at anthesis (6–)7–28 mm, in fruit (7–)8–28 mm; sepals up to (7–)8–13.5 mm; long abaxial petal (9–)10–18 × (6–)8–17.5(–21) mm; style (3–)3.4–4.2 mm; ovules (8–)9–11.

Pod (22–)27–52(–60) × (3–)3.5–5.5 mm, thinly puberulent.—Collections: 160.—Fig. 37 (androecium).

Open grassy banks, thickets, forest margins, dunes, sometimes on coral limestone and colonizing recent lava-flows, becoming dwarfed and prostrate on cliffs, shores, and windy summits, 1–600 m, widespread and locally abundant throughout the chain of islands forming the e. and n.-e. periphery of the Caribbean: Puerto Rico (with islas Vieques and Culebras); Virgin Is. (St. Thomas, St. Jan, Tortola, Virgin Gorda, Anegada, St. Croix); Leeward Is. (Anguilla, Saba, St. Eustatius, St. Kitts, Barbuda, Antigua, Monserrat, Guadeloupe, Marie Galante); Windward Is. (Dominica, Martinique, St. Lucia, St. Vincent, the Grenadines, Grenada); Barbados. Reported from Cuba, Hispaniola and Trinidad, but apparently by error.—Fl. most profuse X–V, but sporadically throughout the year.

The var. *swartzii,* as defined herein the most widespread Antillean form of *Ch. glandulosa,* is highly adaptable ecologically and phenologically plastic, subject to two, sometimes parallel modalities of variation, one adaptive, the other seasonal. A multitude of minor races, each embodying its own range of minor variations, and different one from the next in stature, in amplitude, coloration and venation of foliage, in size and number of petiolar glands, and in size of flower, occupy a great range of natural and disturbed habitats, from dunes and coral reefs within reach of salt spray up into the edges of mesic woodland or rain-forest, and thence out into hedges, pastures, and fallow ground. In protected sites the plants, even though apt to flower precociously as simple herbs, when adult become erect, branching shrubs up to two or rarely four meters tall, with relatively long primary leaves and many leaflets, superficially very different from the diffuse or prostrate, small-leaved and generally thrifty life-form imposed by exposure to wind or by barren soils. But where the seasons are well marked, the amply leafy plant, of whatever stature, discards its larger leaves during periods of water scarcity, nevertheless often continuing to flower from axillary brachyblasts clothed in the short, simple foliage that from the first and permanently characterizes the xeromorphic ecotypes.

While much of the polymorphism noted in var. *swartzii* can be accounted for in these terms, some must be attributed to the sort of genetic diversity that arises naturally under conditions of insular dispersal. The diameter and number of petiolar glands and the development of their stipes appear rather stable within a given population and sometimes within the confines of some smaller islands, as does the outline of the leaflets, regardless of their size. While the total dispersal of gland-types appears to be random and capricious, it is otherwise with the foliage. From Puerto Rico eastward through the Virgin Islands (except St. Croix) and smaller northern Leewards the average leaflet of a plant's most ample leaf is relatively narrow (±1.5–3 mm) and concolorous; on St. Croix and the length of the Windward chain to Grenada and Barbados the comparable leaflet is commonly broad (±3–6.5 mm) and paler beneath than above. These wider and often at the same time longer leaflets tend to have more elaborate and more sharply raised secondary (pinnate) venulation, with at least one and sometimes two distinct intercalary venules between each successive major one. Correlation between glands, leaflets, and venulation is not strong enough, however, to sustain any credible infraspecific taxa.

The typus of *Cassia polyadena,* a species maintained (under *Chamaecrista*) by Britton & Rose (1930) even though Britton had earlier (1917, p. 10) expressed lack of confidence in its distinctness from *Ch. swartzii,* represents a form with few (±14), broad, bicolored leaflets combined with several small, short-stalked rachis-glands, matched by collections from several of the Windward Islands and from Barbados. However, sympatric plants with the same leaf-type may have long-stipitate glands, either solitary on the petiole or also along the rachis, and the small glands of *C. polyadena* are encountered on Puerto Rico in conjunction with narrow leaflets. The typus of *Ch. complexa* was early recognized as being exactly Wikstrom's *C. swartzii.* Those of *Cassia parkeriana* and *C. otterbeinii,* referred by Bentham (1871) to *C. riparia* and *C. glandulosa* respectively, are typical narrow-leaflet var. *swartzii,* the first certainly and the second most probably only cultivated in Guyana, where Parker recorded the vernacular name Franchepan. The ten supposedly sterile but unequal and apically notched filaments described by Meyer in the protologue of *C. otterbeinii* are, of course, not filaments but the elongate anther-sacs characteristic of *Chamaecrista.*

The range of var. *swartzii* may possibly extend to Trinidad, but we have seen no complete specimen from that island. Fragments of *Hunt 1891* (=*Trinidad Herb. 5864*) at NY are labelled *Chamaecrista swartzii* in Britton's hand, but are too fragmentary for confirming identification.

47d. Chamaecrista glandulosa (Linnaeus) Greene var. **mirabilis** (Pollard) Irwin & Barneby, stat. nov. *Ch. mirabilis* Pollard, Proc. Wash. Acad. Sci. **15**: 19. 1902.—"Type . . . collected at Rio Piedras, Porto Rico, by Mr. & Mrs. A. A. Heller, February 27, 1899 (No. 642)."—Holotypus, NY!—A specimen numbered *Heller & Heller 642,* US, collected the following day at another place, is not isotypic.—*Cassia mirabilis* (Pollard) Urban, Symb. Antill. **4**(2): 276. 1905.

Chamaecrista mirabilis sensu Britton, 1917, p. 11; Addisonia **10**: 29, pl. 335. 1925; Britton & Rose, 1930, p. 286.

Prostrate or assurgent, the slender wiry, freely branched stems at anthesis (1–)2–8 dm, puberulent with fine incurved hairs to 0.2–0.5 mm, the stems sometimes subglabrous, the foliage appearing so but the lf-stalk always at least thinly hairy.

Stipules 1.5–5 × 0.4–1 mm, (4–)5–8-nerved.

Major lvs (1–)1.3–4.5 cm; petiole 1.5–5 × 0.2–0.3 mm; petiolar gland slenderly tack-shaped, 0.3–0.6 × 0.2–0.3 mm, in profile 0.05–0.3 mm taller than diam; often 1–2 smaller rachis glands present; lfts up to 11–18 pairs, linear or linear-elliptic, the largest of a plant (3–)3.5–7.5 × 0.7–1.4 mm, the midrib giving rise to (3–)4–7 pairs of secondary veins, intercalary ones not or scarcely perceptible; pedicels at anthesis (7–)9–15 mm, in fruit (7–)11–21 mm; sepals 6–9 mm; 3 short petals and cucullus 7–10.5 mm, the long abaxial petal amply flabellate obcordate 8.5–12.5 × 7–11 mm; 3 long red anthers up to 5.5–7 mm; style 3–3.6 × 0.2–0.25 mm; ovules 9–15.

Pod (18–)20–40 × 3–3.8 mm; seeds 2.2–2.9 mm.—Collections: 8.

White sand savannas below 150 m, local, known only from the north coastal plain of Puerto Rico between Río Manati and Río Pedras (66°–66°30′W).—Fl. XII–III.

This variety is very closely related to var. *swartzii,* the common suffruticose chamaecrista of Puerto Rico, differentiated only by its slender habit, on the average shorter leaves, and small leaflets. It may well consist merely of a set of ecotypes from sterile white sand habitats, similar to the dwarfed or depauperate states of var. *swartzii* found on exposed coral banks in the Virgin and Leeward Islands, but a trifle further reduced. It was originally compared by Pollard not with *C. glandulosa* var. *swartzii,* of which it has the pedicels, leaf-venulation, and long-styled asymmetric flower, but with the entirely different *C. aeschinomene,* and the wonderful feature to which the epithet alludes is merely the presence, but only in some leaves, of a few small supernumerary nectaries on the upper leaf-stalk. Such glands are, of course, frequent in other forms of *Ch. glandulosa,* and even if constant in *Ch. mirabilis,* which they are not, could provide no differential criterion. Britton & Rose (1930, p. 272) attributed to *Ch. mirabilis* a diagnostically narrow pod, 2–2.5 mm wide; but we have found not one single pod among the many seen by Britton which is actually less than 3 mm wide, all on the contrary falling well within the variational range of Puerto Rican var. *swartzii*. Comparison of Plates 335 (*Ch. mirabilis*) and 444 (*Ch. swartzii*) in *Addisonia,* painted by Mrs. F. W. Horne in the field, emphasizes the essential similarity in everything except size of leaflets. The miniature extreme of var.

mirabilis represented by *Britton 8696* (G, NY, US) is suggestive of some form of the Cuban *C. pygmaea,* but this differs decisively in its strongly asymmetric leaflets.

47e. Chamaecrista glandulosa (Linnaeus) Greene var. **flavicoma** (Humboldt, Bonpland & Kunth) Irwin & Barneby, comb. nov. *Cassia flavicoma* Humboldt, Bonpland & Kunth, Nov. Gen. & Sp. **6**(qu): 366. 1824.—"*Crescit prope fluvium Orinoci.*"—Holotypus, P (hb. H.B.K.)! isotypi, B (hb. Willd. *7986*)! P (hb. Bonpl.)!—*C. glandulosa* var. *flavicoma* (Humboldt, Bonpland & Kunth) Vogel, Syn. Gen. Cass. 63. 1837. *C. chamaecrista* fma. *flavicoma* (Humboldt, Bonpland & Kunth) O. Kuntze, Rev. Gen. 169. 1891. *Chamaecrista flavicoma* (Humboldt, Bonpland & Kunth) Greene, Pittonia **3**: 242. 1897.

Cassia pavoniana G. Don, Gen. Hist. Dichl. Pl. **2**: 447. 1832.—"Native of Peru. (v.s. in herb. Lamb[ert.])."—Holotypus, collected presumably by Dombey, Ruiz or Pavón and acquired by Lambert from the latter, not located; probable isotypi, labelled "*Cassia mimosoides.* Huánuco, 1787," MA (hb. Ruiz. Pav. *14/12*)! and distributed by Pavon as "*C. mimosoides,*" BM, FI, G!—Referred by Bentham, 1871, p. 557, to *C. glandulosa.*

Cassia stipulata G. Don, Gen. Hist. Dichl. Pl. **2**: 448. 1832.—"Native of Peru (v.s. in herb. Lamb[ert.])."—Holotypus not recognized as such, but perhaps "*Cassia stipulacea* sp. nov. del Peru" ex hb. Pavon., BM! isotypus, labelled "Peruvia. Herb. Pavon.," G!—Correctly equated with *C. flavicoma* by Bentham, 1871, p. 576.

Chamaecrista yucatana Britton & Rose, N. Amer. Fl. **23**(5): 287. 1930.—"Buena Vista Xbac, Yucatán, *G. F. Gaumer 1113.*"—Holotypus, NY! isotypi, GH, LE, US!—*Cassia yucatana* (Britton & Rose) Lundell, Phytologia **1**: 215. 1937.

Chamaecrista arborescens Britton & Killip, Ann. N.Y. Acad. Sci. **35**: 186. 1936.—". . . Culagá Valley, near Tapatá, north of Toledo, Norte de Santander, Colombia, 1800–2000 m. altitude, March 4, 1927, *Killip & Smith 20146* . . ."—Holotypus, NY! isotypi, A, US!

Chamaecrista dendrina Britton & Pittier, Bol. Soc. Venezol. Ci. Nat. **7**: 145. 1941.—"*Mérida*: Alrededores de Mérida . . . Febrero 5, 1928 (Pittier 12.864, tipo); Tabay . . . (Gehriger 348)."—Holotypus, *Pittier 12864,* NY! isotypi, A, K, MO, US, VEN! = NY Neg. *6333*; paratypi, *Gehriger 348,* A, G, NY, US!—*Cassia dendrina* (Britton & Pittier) Pittier, Cat. Fl. Venezol. 373. 1945, pro syn.

Chamaecrista multivenia Pittier, Cat. Fl. Venezol. 374. 1945, nom. nud.—Based on *Tamayo 2411* from Edo. Tachira, US, VEN!—*Cassia multivenia* Pittier, l.c., pro syn.

Highly variable in stature, flowering sometimes precociously as a single-stemmed suffruticulose herb but shrubby when adult and in favorable circumstances arborescent, at anthesis (5–)7–40 dm, the usually erect, rarely diffusely assurgent branches, lf-stalks, and inflorescence most commonly both puberulent with short incurved hairs and setulose-pilose with incurved-ascending, horizontally patent, or subretrorse, usually yellowish or golden hairs up to 0.7–1.5 mm, but either type of vesture predominant or either absent, both variable in density, the densely and usually amply leafy annotinous branchlets varying from densely shaggy-setose-pilose to thinly pilosulous or glabrate, the concolorous foliage likewise variably pubescent, the membranous lfts pilosulous on both faces, or only beneath, or glabrate on the faces and ciliolate, exceptionally glabrous.

Stipules ovate-acuminate to narrowly lance-attenuate or linear-acuminate, those subtending major cauline lvs up to (3–)4–15 × (0.6–)1–3 mm, the thin-textured blades green often becoming dry and brown, from the slightly asymmetric to shallowly semi-cordate (exceptionally obliquely cordate-amplexicaul) base prominently (on Sa. Nevada de Sta. Marta subobsoletely) (5–)9–19-nerved, persistent (in s. Mexico exceptionally deciduous).

Major cauline lvs up to (3.5–)6–12(–14.5) cm, the petiole (including often strongly dilated pulvinus) 2–7(–10) mm, at middle 0.5–1.3 mm diam; gland 1(–3) situated at, above or below middle of petiole proper, urceolate or shallowly saucer-shaped, round or elliptic (0.35–)0.4–1.1 mm diam, stipitate or subsessile, the glabrous or exceptionally puberulent stipe very variable in length and thickness, linear or dilated upward, the whole in profile usually club- or tack-shaped but when short-stipitate becoming squatly obconic and when very long-stipitate pin-shaped, (0.3–)0.4–2.8(–3.5) mm tall, 0.3 mm shorter to 2.3(–2.9) mm longer than diam of head, smaller glands often present on some distal segments of rachis; lfts of major cauline lvs up to (10–)12–22(–24) pairs, varying in outline from broadly oblong to oblong-elliptic, narrowly lance-oblong, or sublinear, mostly obtuse mucronate or aristulate, sometimes subacute, (7.5–)12–22(–25) × (1.6–)2–5.5(–6) mm, the blades plane or often subrevolute, from base 5–9-nerved by the slender centric or moderately displaced, straight or gently incurved midrib with on the narrow side 0–2, on the broad one 3–6 primary nerves, the secondary pinnate venulation of (4–)5–11 pairs of major secondaries with, in each interval, 0–2 minor ones, these sometimes, however, almost as strong as the major ones, the lfts then becoming multistriate, the venulation varying in prominence, sometimes sharper above than beneath, sometimes the reverse.

Peduncles usually solitary, sometimes 2 superposed from each axil, together with often well-developed raceme-axis (1.5–)3–20(–22) mm, adnate through 2–14(–18) mm, (1–)2–5(–7)-fld; pedicels mostly 12–27(–50, the longest commonly at least 15) mm, some late ones or those distal to a several-fld raceme shorter; sepals (7–)8–14(–16) mm; long abaxial petal (9–)11–17(–19) mm; ovary strigulose or loosely white-pilose; style filiform 3–5(–7) mm; ovules 9–16(–17).

Pod (30–)35–65(–70) × (4.5–)5–7.5 (in n. Colombia sometimes 8–9) mm, the valves either pilosulous or subappressed-puberulent; seeds (3.2–)3.4–4.7 (in n. Colombia sometimes up to 5–5.8) mm, the testa dull ocher-brown, the pits concolorous or often fuscous, or fuscous overall.—Collections: 117.

Open rocky or grassy hillsides, roadside hedges, thickets, fallow fields and rough mountain pastures, sometimes a weed of cafetal, in wooded country only in natural glades, cut-over forest, or on stream-banks or sand-bars, mostly between 900 and 2250 m but in n. Colombia up to 3100 m, interruptedly widespread through the Andes from Sa. Nevada de Sta. Marta, Sa. de Perijá and Cordillera de Mérida in n. Colombia and n.-w. Venezuela s. through the Cauca valley and Cordillera Occidental s.-ward from Antioquia, highland Ecuador, and inter-Andean valleys of the Amazon slope in Peru to the Upper Urubamba valley and, perhaps disjunctly, to the Yungas valleys in deptos. Larecaja and La Paz, Bolivia; reappearing in slightly differentiated form (=*C. pavoniana*) on the Pacific slope of centr. and n. Peru, on open hillsides and dry streambeds of the coastal lomas and Andean foothills mostly below 1000 m, in valleys of Ríos Chancay, Chillón, Rimac and Lurín in Dept. Lima (prov. Chancay, Lima, Huarochirí) and Dept. Cajamarca (prov. Hualgayoc), this ecotype apparently naturalized, at least formerly, in the Canary Islands (Palma); in Colombia absent from Magdalena Valley, Cordillera Central and e. slope of Cordillera Oriental, where replaced by subsp. *tristicula*; remotely disjunct in s. Mexico and adj. Guatemala; in pine-oak woodland near 1700 m in Sa. Madre del Sur, Guerrero; and on karst limestones and in scrub forest below 150 m in Yucatán, Quintana Roo, Campeche, and Petén.—Fl. in Venezuela and Colombia mostly (VII–)VIII–III(–IV), from Ecuador s.-ward X–VI, in Guerrero IX–II, in Yucatán IX, VI.

By expanding our concept of var. *flavicoma* to embrace all (except the far-southern var. *andicola*) large-flowered, long-pedicelled chamaecristas with sub-centrically costate leaflets known to occur in the Andes between Sa. Nevada de Santa Marta and Bolivia, we sweep into one net a polymorphic shoal of minor races which Bentham would have or (1871, and in annotation of specimens) actually did distribute between the ineffectively differentiated *C. flavicoma, C. glandulosa* and *C. virgata*. For Bentham, as later for Britton, *C. flavicoma* was characterized by a dense yellow pilosulous vesture, but an entity defined by this one feature would now be as heterogeneous in stature, stipules, glands, and width or venulation of the leaflets as the parallel, often sympatric analogues with gray or white vesture, whether sparse or dense. The large-flowered Andean chamaecristas are clearly in a stage of evolutionary ferment and present a taxonomically baffling series of minutely, or sometimes grossly, distinguishable forms, most of which can be precisely characterized only if we narrow them down to one or few populations, these not necessarily correlated with credible geographic patterns. The alternative to a comprehensive var. *flavicoma* is a return to the Brittonian method of applying names to individual populations.

Analysis of the petiolar glands of all available Andean var. *flavicoma* has brought to light a great range in size and form, this running sometimes in harmony, sometimes in discord with other modalitites of variation, but nevertheless showing some geographic sorting. In Venezuela, Colombia and northern Ecuador (Pichincha) the prevailing gland is stoutly stipitate, in profile claviform and mostly 0.8–1.5 mm tall. From southern Ecuador (Chimborazo) south to northern Cuzco (prov. Convención) in Peru, the average gland is shorter or, if no shorter, then slenderly stipitate, in profile tack-shaped or obconical; but the typus of *C. stipulata* and plants resembling it from the headwaters of Río Huallaga have the coarse glands of the northern region. Southward from the upper Urubamba valley in Cuzco to the Yungas in Bolivia the gland becomes rather abruptly longer and slender-stipitate, often pin-shaped in profile and 1–3.5 mm tall with head only 0.4–0.7 mm diameter. However in these three segments of the Andean range the other phenetic characters are extremely flexible in their expression, the mutual proportions of the duplex vesture, the color of hairs and of anthers (these red or yellow) the number of leaflets, the prominence of the secondary venulation, and the size of the pod each varying independently of the rest.

The most notable Andean variant known to us is that found on Sa. Nevada de Sta. Marta and Sa. de Perijá, in Magdalena and northern Norte de Santander in Colombia (to be expected in adjoining Venezuela), unlike all others in the massive pod 8–9 mm wide and correspondingly immense seeds about 5–6 mm long. But even this is not internally uniform, the plants of Sta. Marta having uniquely thin-textured stipules with scarcely visible venation, whereas those of Sa. de Perijá, which are *Chamaecrista arborescens* sens. str., have the ordinary striate stipules of var. *flavicoma*. The latter variant differs from the now well-known representative of var. *flavicoma* on Cordillera de Mérida immediately to the east, which has been segregated (if arborescent, but not when shrubby) as *Ch. dendrina*, in the fruit alone (pod valves 6–7.5, not 8–9 mm wide; seeds proportionately smaller) and could only be sustained as a distinct taxon by special license.

The cismontane Andean form of var. *flavicoma* segregated long ago as *C. pavoniana* G. Don is often recognizable, at least to the experienced eye, as distinct from most transmontane ones in the syndrome of thinly puberulent, distantly leafy stems, relatively small 6–9-nerved stipules mostly 4–9(–10) mm long, and narrow acutish, often falcately incurved leaflets up to 7.5–15 × 1.6–2.8(–3.2) mm. This form is taxonomically embarrassing, for it has no single gross morphological feature that cannot be matched, not only somewhere in var. *flavicoma*, but also somewhere within Antillean var. *swartzii*. In view of its relatively local range, mostly within a radius of 70 km of Callao, a port linked since European settlement of Peru with Spanish possessions in the Caribbean, it is possibly not truly indigenous, even though we have records going back to the late XVIII century, but actually is naturalized var. *swartzii*. The fact that what appears to be identically the same form of *C. glandulosa* was found naturalized in the Canary Islands a century ago (*Salzmann in 1869,* P) is evidence of its weedy potential. Nevertheless we have found no single individual plant of West Indian var. *swartzii* that matches simultaneously at every point this Peruvian expression of *C. glandulosa* and therefore, with hesitancy and lack of conviction, refer it provisionally to var. *flavicoma*.

This seems the appropriate place to mention three collections from Peru which we are at present unable to account for: without locality, *hb. Pavon*. (BM); without locality, *Mathews 918* (K); Huambos, prov. Chota, dept. Cajamarca, 2000 m, 11.IX.56 (fl, fr), *Soukup 4529* (US). Collectively these plants closely resemble some Caribbean var. *swartzii,* differing from all var. *flavicoma* in having short stipules, abbreviated leaves with no more than 9–12 small leaflets in the larger leaves. The material is insufficient to show the full stature, which appears to be, however, distinctly fruticose. Possibly these represent ecotypes of var. *flavicoma* from unfavorable microhabitats, but this is pure surmise. Bentham identified *Mathews 918* as *C. glandulosa* and we agree that it must represent a form of that species.

The disjunct Mexican populations here referred to var. *flavicoma* pose a real problem in classification, being not precisely matched by any particular Andean one, and being themselves internally heterogeneous. The plants of Yucatán and Campeche differ collectively from nearly all Andean var. *flavicoma* by the very narrow linear-attenuate stipules (6–11 × 0.6–1.1 mm), but have two types of foliage, one (*Ch. yucatana,* sens. str.) that compares favorably with some Andean plants with broad leaflets, the other (e.g. *Lundell 1427,* GH, NY, *7244,* US) with narrow and fewer leaflets, forcibly reminiscent of Caribbean var. *swartzii,* and possibly more nearly related in that direction than southward. One of the two known collections of var. *flavicoma* from Sa. Madre del Sur in Guerrero (Coalcomán distr., *Hinton 12708,* G, K, NY, US) is visually arresting for the syndrome of long narrow (×±2 mm) leaflets, extremely long (to 5 cm) pedicels, and extraordinary stipules, which are asymmetrically cordate-amplexicaul and early deciduous. We would certainly have described this if we had not seen *Hinton 10698* (G, K, NY, US) from district Minas a short distance to the east, in which the leaflets are the same but the persistent stipules and the pedicels comparable with Peruvian or Bolivian var. *flavicoma*. For the present we interpret *Hinton 12708* as an exceptional individual variant, one which, like the also profusely large-flowered variant from Santa Marta, mentioned above, would be a welcome addition to the list of chamaecristas in tropical cultivation.

Our statement of dispersal for var. *flavicoma* deliberately omits as implausible the supposed type-locality on the Orinoco. As Humboldt and Bonpland were generally specific and precise in their locality data, the vagueness of the phrase "near the Orinoco River" is itself negatively suggestive. The parts of the type collection that we have seen match most closely in general aspect, and in details of gland, vesture, leaflets, and venation, recent and contemporary collections of var. *flavicoma* from the middle and upper Cauca valley in Colombia, especially those from the environs of Popayán. We believe that *C. flavicoma* must have been collected first in the western cordillera of Colombia, not on the Orinoco.

47f. Chamaecrista glandulosa (Linnaeus) Greene var. **andicola** Irwin & Barneby, var. nov., ab arcte affini var. *flavicoma* statura tota deminuta, foliis majoribus (1–)1.5–4(–5, nec 5–12) cm usque longis, foliolisque minoribus saepiusque minus numerosis (6–)8–16(–18, nec 12–22)-jugis, 5–12 (nec 12–25) mm usque longis sed corolla ei var. *flavicomae* subaequimagna diversa.—PERU. Cuzco, prov. Anta: Mollepata, valle del Río Apurimac, 2800 m, I.1927 (fl, fr jun), *F. L. Herrera 1184*.—Holotypus, GH = NY Neg. *9197*.

(?)*Cassia calvens* Macbride, Field Mus., Bot. **13**, part 3(1): 159. 1943.—"Peru: Specimen in Field Museum without data, *Weberbauer 5496* . . ."—Holotypus, F! = NY Neg. *9182*.

Diffuse or assurgent, early suffruticose, at anthesis 1–5 dm, the slender, flexuous or straight young branchlets and lf-stalks puberulent with incurved whitish hairs up to 0.2–0.4 mm, sometimes canescently so, the lfts glabrous above, ciliolate, beneath glabrous, puberulent along midrib, or puberulent overall.

Stipules ovate-acuminate or lance-attenuate 2.5–8 × 0.7–1.7 mm, 7–12-nerved, either glabrous or puberulent dorsally.

Major lvs up to (1–)1.5–4(–5) cm, the petiole (1–)2–5 × 0.4–0.7(–0.8) mm; gland 1 at or well above middle of petiole proper, discoid or shallowly urceolate 0.2–0.5 mm diam, slenderly stipitate, in profile (0.3–)0.5–2 mm tall, commonly pin-shaped and much taller than diameter of head but sometimes tack-shaped and a trifle shorter; lfts of major lvs up to (6–)8–16(–18) pairs, linear-oblong to oblong-elliptic or -obovate obtuse mucronulate up to 5–12 × 1.5–2.8(–3.5) mm, the midrib centric or slightly displaced, the secondary venulation of 3–6 major pairs with between each 1(–2) minor intercalary ones, the venulation variably prominulous on either or both faces.

Pedicels 8–22 mm, some in fruit at least 10 mm; sepals 8–10 mm; long abaxial petal (usually broadly flabellate) 12–18(–20) mm; anthers red, the long ones up to 6.2–8.6 mm; ovary pilose or subappressed-puberulent; style linear 2.5–5.2 mm; ovules 8–15.

Pod (little known) 32–46 × 4.5–5.5, the valves commonly pilose, sometimes incurved-puberulent; seeds unknown.—Collections: 20.

Open rocky hillsides, low thickets, sometimes on road-cuts, 2400–3300 m, best known and typically developed on the Andean plateau and drier inter-Andean valleys drained by the Mantaro and Apurimac rivers in s.-e. Peru (Huancavelica, n. Ayacucho, Apurimac and centr. Cuzco); apparently disjunctly in the Marañón drainage in centr. Cajamarca and highland s. Ecuador (Loja); and atypically, transient to var. *glandulosa,* on the e. slope of the Andes in n.-w. Bolivia (La Paz).—Fl. IX, I–V.

A wiry, diffusely branching shrublet with the handsome, strongly asymmetric flower of *Ch. glandulosa* but foliage condensed and simplified, var. *andicola* is very likely a recent modification of the vicariant var. *flavicoma* adapted by selection to the severe climate of the Andean plateau. It is also extremely close to the form (*C. pavoniana*) of var. *flavicoma* found on the Pacific slope, in fact not separable by any one infallible differential character, but generally recognized at sight by the crowding of the obtuse leaflets along a shorter leaf-stalk and physically separated by a gap of some 1000 m elevation. The petiolar gland of var. *andicola* is highly variable, most often of the slenderly elongate pin-shaped type prevalent in var. *glandulosa* of lower elevations or moister sites in valleys within and near the same latitudes, but occasionally shortly tack-shaped. The Bolivian material referred to the variety has for the most part a less pronounced individual facies than the Peruvian, as it were fading into var. *flavicoma.* A fragmentary collection from an unidentified locality on the Bolivian plateau (La Granja, 2600 m, *Bro. Julio 119,* US), provisionally called var. *andicola,* is notable for a dense whitish puberulent vesture and minute petiolar glands. We have seen nothing quite like it.

We suspect, but cannot be absolutely sure, that the depauperate holotypus of *C. calvens,* not exactly matched by any other collection from Peru, represents starveling var. *andicola*. The condensed foliage, with up to six pairs of leaflets, closely resembles that of *Ochoa 707* (NY) from Ayacucho, but the flower is a little smaller (long petal ±11.5 mm). Until similar specimens can be associated with a known locality and environment the disposition of *C. calvens* must remain provisional.

47g. Chamaecrista glandulosa (Linnaeus) Greene var. **tristicula** (Humboldt, Bonpland & Kunth) Irwin & Barneby, stat. nov. *Cassia tristicula* Humboldt, Bonpland & Kunth, Nov. Gen. & Sp. **6**(qu): 367. 1824.—". . . in regno Novo-Granatensi, prope Fusagasugam, alt. 940 hex . . ."—Holotypus, P (*hb. H.B.K.*)! isotypi, B (hb. Willd. 7988)! = F Neg. *1760,* P (*hb. Bonpl.*)!—*Chamaecrista tristicula* (Humboldt, Bonpland & Kunth) Greene, Pittonia **3**: 242. 1897; Britton & Rose, Ann. N.Y. Acad. Sci. **35**: 184. 1936, pro comb. nov.

Cassia tristicula sensu Bentham, 1871, p. 581, p.p., exclus. syn.

Like vicariant forms of var. *flavicoma* except for the greater asymmetry of the lfts, herbaceous becoming shrubby, at anthesis 5–20 dm, the new branchlets, lf-stalks and pedicels pilosulous with stiff, widely ascending or horizontal, white or faintly lutescent hairs to 0.6–1.2 mm, sometimes in addition incurved-puberulent, the lfts glabrous ciliolate, or pilose (especially along midrib) beneath, or pilosulous on both faces.

Stipules lance-, rarely ovate-acuminate, (4–)6–14 × 1–2.3(–2.7) mm, coarsely 6–11-nerved, glabrous or pilosulous dorsally.

Primary cauline lvs (3.5–)4–9 cm, the petiole including swollen pulvinus 3–6 × 0.4–0.6(–0.75) mm; gland 1, at or above middle of petiole proper, urceolate 0.4–0.8 mm diam, stoutly stipitate, in profile claviform (0.5–)0.8–1.8 mm long, ±2–3 times as tall as diam of head; lfts up to 14–22 pairs, inserted along rachis at points 2–3.5 mm apart, obliquely or subdimidiately oblong or linear-oblong, obtuse aristulate, straight or subfalcate, up to 9–19 × 1.5–4.5 mm, the straight or slightly incurved costa distant 0.3–0.7(–1) mm from distal margin dividing the blade 1:3–5(–6), the secondary pinnate venulation subobsolete on narrow side, of 3–5(–6) major and often as many fainter intercalary ones on the broad side, the venulation finely prominulous on both faces.

Inflorescence of var. *flavicoma,* the pedicels (13–)15–30 mm; sepals to 9.5–12 mm; long abaxial petal to 11–13.5 mm; ovary pilose; style 3.5–4.5 mm, pilosulous below middle; ovules 10–15.

Pod (35–)40–67 × 5.5–6.5 mm, the valves spreading-pilosulous; seeds (few seen) ocher-brown, 3.7–4 mm.—Collections: 13.

Open or cleared woodland, rocky hillsides, becoming a weed of rough meadows, hedgerows and coffee orchards, 1300–2300 m, locally plentiful but of restricted range on the w. slope of Cordillera Oriental in Cundinamarca, Colombia, between Guaduas and Fusagasuga (lat. 5°5′–4°15′N).—Fl. X–V(–VII), perhaps irregularly throughout the year.

In contradiction to Bentham we interpret *Cassia tristicula* as a local, purely Colombian variety of *Ch. glandulosa* which replaces var. *flavicoma* on the west slope of the Eastern Cordillera in Cundinamarca; and we restore *C. rufa,* which formed the greater part of Bentham's concept of *C. tristicula,* to specific rank. The var. *tristicula* as reinterpreted is extremely close to those races of *Ch. glandulosa* var. *flavicoma* found in the Cauca valley and Western Cordillera in the same and more southern latitudes, differing solely in the greater displacement of the midrib of most leaflets, which divides the blade in proportions of 1:3–6, not 1:1–2.5. It appears to us much more distantly related to the only form of *C. rufa* currently known from Colombia, the var. *polyphlebia* locally endemic on the Meta slope of the same Cordillera in Boyacá and Cundinamarca, decisively separable as follows:

1. Lfts of major lvs up to 14–22 pairs, their blades 1.5–4.5 mm wide, the costa intramarginal by 0.3–1 mm, dividing the blade 1:3–5(–6); secondary venulation of broad side of blade consisting of 3–5(–6) major and often as many fainter intercalary nerves. *Ch. glandulosa* var. *tristicula*.
1. Lfts of major lvs up to 30–44 pairs, their blades 0.8–1.5 mm wide, the costa exactly marginal; secondary venulation from costa consisting of 10–20 subcontiguous, equally long and equally coarse nerves, the ventral surface of blade appearing multistriate. *Ch. rufa* var. *polyphlebia*.

47h. Chamaecrista glandulosa (Linnaeus) Greene var. **balsasana** Irwin & Barneby, var. nov., habitu toto var. *flavicomae* proxima, foliolis attamen crassis subcoriaceis venatione immersa facie utraque laevissimis parum sed satis diversa.—PERU. Cajamarca, prov. Celendín: along Río Marañón opposite Balsas, 800 m, 29.V.64 (fl, fr. mat.), *P. C. Hutchinson & J. K. Wright 5428*.—Holotypus, NY; isotypi, F, K, P, US.

Slender shrubs to 5 dm, with ± abruptly flexuous wiry branchlets, the castaneous bark of old stems splitting lengthwise, glabrous except for minutely puberulent annotinous branchlets, lf-stalks, and ovaries, the foliage bicolored, the thick-textured, apparently almost veinless lfts brownish-olivaceous above, paler beneath.

Stipules lanceolate or triangular-lanceolate 3–7 × 0.7–1.6 mm, from subsymmetric base (5–)6–11-nerved, persistent.

Major cauline lvs (3–)3.5–7 cm, the petiole 3.5–6 mm; petiolar gland 1, shortly stout-stipitate or subsessile, orbicular 0.4–0.8 mm diam, in profile trumpet-shaped or obconic, 0.5–1.2 mm tall; several smaller glands on distal segments of rachis; lfts (5–)6–12 pairs, lance- or linear-oblong, the larger up to 12–22 × 2.5–4 mm, the slender pallid, moderately excentric midrib with 3–5 pairs of pinnate secondary venules visible but scarcely prominulous beneath, fully immersed above.

Peduncles adnate through 2.5–12 mm, 1-3-fld; pedicels (10–)13–24 mm; sepals (glabrous) to 13–15 mm; long abaxial petal ±14–16 mm; style 6.5–8.5 mm, linear; ovules ±10–11.

Pod 30–45 × 4–5.3 mm, the castaneous sublustrous valves sparsely puberulent; seeds unknown.—Collections: 2.

Rock slides, 800 m, known only from the inter-Andean valley of Río Marañón, on the Cajamarca-Amazonas boundary near 6°45′S, in n. Peru.—Fl. ±IV–VI.

In almost all respects resembling the form of var. *flavicoma* found on the Pacific slope of the Andes in the latitudes of Chiclayo and Lima, but notable for the relatively few leaflets of such thick texture that the venulation of the upper face is fully immersed.

47i. Chamaecrista glandulosa (Linnaeus) Greene var. **rapidarum** Irwin & Barneby, var. nov., var. *flavicomae* praesimilis sed glandula petiolari crassa, florum maximorum sepala 12–16 (nec 8–12) mm usque, petalum abaxiale planum 15–23 (nec 11–15) mm longa, legumine angustissimo 3.5–4 (nec 4.7–6) mm lato, patriaque procul distanti diversa.—VENEZUELA. Bolívar: savanna edge at Laja Caruto, 80–90 m, e. bank of

Río Orinoco just below Raudal Marimare, 18.XII.1955 (fl, fr), *J. J. Wurdack & J. V. Monachino 39975.*—Holotypus, GH; isotypi, NY, P.

Shrubby or subshrubby to 2 m, thinly strigulose-pilosulous with appressed or incumbent hairs to 0.25–0.5 mm, the concolorous olivaceous lfts usually remotely strigulose beneath, ciliolate.

Stipules (4–)4.5–14 × 0.5–1 mm, 6–8-nerved.

Major lvs up to 5–11 cm, the petiole (4–)5–9 mm; glands 1–2(–3), broadly elliptic, convex or shallowly dimpled, 0.8–2 mm diam, subsessile, in profile squatly drum-shaped or obconic, 0.6–1 mm tall, (0.2–)0.4–1 mm shorter than diam of head; lfts of major lvs up to 16–24 pairs, oblong obtuse mucronulate to 12–20 × 2–4 mm, the midrib giving rise to 4–6(–7) pairs of major with or without faint intercalary secondary venules, these faintly prominulous on both faces.

Peduncles 6–17(–30) mm adnate through 2–15(–21) mm, 1–4(–5)-fld; pedicels up to 12–24 mm; sepals up to 12–16 mm; long abaxial petal 15–23 × 10–15; style 6.5–8.7 mm filiform; ovules 16–21.

Pod 40–60 × 3.5–4 mm, incumbent-pilosulous.—Collections: 4.

Habitat little known, sometimes at savanna margins, below 120 m, apparently very local along the banks of río Orinoco near and immediately below the mouth of río Meta, in n.-e. Vichada, Colombia, and adjoining Bolívar (presumably also Apure), Venezuela.—Fl. VIII–I.

Notable for the syndrome of extremely large flower, coarse subsessile petiolar gland, and narrow elongate pod, var. *rapidarum* cannot well be assimilated into any described form of *Ch. glandulosa.* In the diagnosis we have compared it to Peruvian var. *flavicoma,* similar in general appearance but different in the smaller flowers and broader pods. It is, however, very likely more closely related to its near neighbor var. *crystallina,* similar in the proportions of glands and pods but very much smaller-flowered, the sepals only 7–10 (not 12–16) and long abaxial petal only 8–11 (not 15–23) mm long.

The var. *rapidarum* was collected first somewhere on the banks of the Orinoco by the French naturalist-explorer Jean Chaffanon in 1886 (P).

47j. Chamaecrista glandulosa (Linnaeus) Greene var. **crystallina** Irwin & Barneby, var. nov., vix nisi foliolorum venatione simpliciori, venulis secundariis fortioribus (intercalariis, si adsunt, praetermissis) 2–5-jugis a var. *swartzii,* procul allopatrica in insulis Antillanis minoribus endemica, absimilis.—VENEZUELA. Terr. Amazonas: crystalline laja 100–120 m, 1–1.5 km e. of Hotel Amazonas, Puerto Ayacucho, 7.XI.1953, *B. Maguire, J. J. Wurdack & G. S. Bunting 36035.*—Holotypus, NY; isotypi, MA, P, US.

Shrubs and subshrubs at anthesis (2–)5–15 dm, freely branching distally, with brown woody trunks and atrocastaneous branchlets, finely or minutely pilosulous almost throughout with incumbent hairs to 0.2–0.5 mm, the lfts pubescent on both faces, beneath only, or subglabrous ciliolate.

Stipules 3–8 × 0.4–1.2 mm, (5–)6–10-nerved.

Major lvs (early shed) up to (3.5–)4–6 cm; petiole including pulvinus 3–5(–6) mm; glands 1(–2) round discoid or shallowly concave 0.6–1.2 mm diam, stoutly short-stipitate, in profile obconic, squatly drum- or tack-shaped, 0.7–1.0 mm tall, 0.2 mm longer to 0.2 mm shorter than diam of head; lfts of major lvs to 12–18 pairs (those of some branchlets fewer) narrowly oblong or oblong-elliptic to 9–13 × 2–3 mm, from base 4–6-nerved by midrib with on broad side 2–4, on narrower side 1 primary vein, the midrib giving rise to 2–5 major pairs with or without fainter intercalary secondaries, the venation faintly prominulous on both faces or only beneath.

Pedicels (7–)10–17 mm; sepals to 7–9.7 mm; short petals 3, 4–9.8 mm, the long plane abaxial one 8–11 × 5–7 mm, the cucullus a little longer, or shorter, 7.3–13 × 5–8 mm; long anthers 6.2–7.5 mm; style up to 6 mm, but often abortive in the material seen; ovules (9–)11–16.

Pod 28–47 × 3.5–4 mm, strigulose or pilosulous.—Collections: 5.

Crystalline outcrops, 100–550 m, local, around the n.-w. periphery of the Guayana Highland between Pto. Ayacucho and Cerro Bolívar in Terr. Amazonas and Edo. Bolívar, Venezuela, and n.-e. corner of Vaupés, Colombia.—Fl. IX–XII, III–IV.

This is a somewhat nebulous and precariously defined entity, having no one particular character that cannot be matched in Antillean *Ch. glandulosa* var. *swartzii,* although we can find nothing from the West Indies that is like it at all points simultaneously. Apparently confined to outcrops of the crystalline rock that forms the Guayana shield, var. *crystallina* may be visualized as existing on rock islands in a sea of forest, just as var. *swartzii* is broken up into a multitude of little races in conventional archipelagic fashion. The relatively simple venulation of the leaflets, although not unknown elsewhere in *Ch. glandulosa,* is immediately suggestive of *Ch. nictitans* var. *disadena*; but the present variety is reported by all collectors as a shrub, sometimes a genuinely woody plant as shown by *Maguire et al. 36008* (NY), not merely a basally lignescent herb of short duration. For this reason we refer var. *crystallina* to the *Ch. glandulosa* complex, although we are aware that in this as in var. *disadena* we have in hand the links that unite *Ch. nictitans* sens. lat. and *Ch. glandulosa* sens. lat. into a continuous fabric of variations.

A possibly distinctive feature of var. *crystallina* is the proportionately large cucullus that slightly exceeds in length the matching abaxial petal, in var. *swartzii* normally, perhaps always, the longest of the five. The perianth is, however, so variable in its irregularity, that we would place little reliance on a difference of this sort.

47k. Chamaecrista glandulosa (Linnaeus) Greene var. **andreana** (Britton & Killip) Irwin & Barneby, stat. nov. *Chamaecrista andreana* Britton & Killip, Ann. N.Y. Acad. Sci. **35**: 186. 1936.—". . . between Piedras and Ibagué, eastern base of Mt. Quindío, Tolima, Colombia, March 3, 1876, *André 1974* . . ."—Holotypus, NY! = NY Neg. *4968*; isotypus, K!—*Cassia andreana* (Britton & Killip) García-Barriga & Forero-González, Cat. Ill. Pl. Cundinamarca **3**: 52. 1969.

Chamaecrista pennellii Britton & Rose ex Britton & Killip, Ann. N.Y. Acad. Sci. **35**: 185. 1936.—". . . between Fusagasugá and Pandi, Cundinamarca, Colombia, November 30, 1917, *Pennell 2749.*"—Holotypus, US! isotypi, GH, NY! = NY Neg. *4950*.

Diffuse or prostrate perennial herbs from blackish woody roots, the weak simple or flabellately branching stems at anthesis 1–3 dm, the branchlets, lf-stalks and margins of stipules and lfts either simply puberulent with incumbent or curly hairs up to 0.2–0.4 mm or both puberulent and setulose with straight spreading ones up to 0.45–0.9 mm, the faces of the concolorous lfts glabrous except for random ciliation of midrib beneath.

Stipules lanceolate from subsymmetric base, 4.5–9 × 0.8–1.5 mm, 8–11-nerved from base.

Lvs 1.5–4 cm, ovate in expanded outline; petiole with pulvinus 1.5–3.5 mm, at middle 0.45–0.7 mm diam; gland 1, urceolate orbicular 0.3–0.7 mm diam, sessile or shortly stipitate, in profile scutellate, squatly turbinate or shortly tack-shaped, 0.2–0.6 mm tall, 0.5 mm wider to 0.2 mm narrower than high; lfts (3–)5–9(–11) pairs, oblong, oblong-elliptic or (distally) -obovate obtuse or emarginate, strongly mucronate, the larger of adult lvs up to (6–)8–14 × (2–)2.5–5 mm, from base 6–8-nerved by the excentric midrib with on the narrow side 2 very short and on broad side 3–5 primary veins, the midrib thence penninerved with (5–)6–10 pairs of major and between each of these 1–2 slightly or much fainter intercalary secondary nerves, the venulation prominulous on both faces.

Peduncles 1-2-fld; pedicels (10–)12–28 mm; buds thinly puberulent or setulose; sepals up to 7–10 mm; petals (little known) yellow, 4 obovate-cuneate up to 10–14(–17) mm, the dimidiate fifth about as long; long anthers 6–8 mm; ovary densely strigulose; style linear 3–6 mm; ovules 12–15.

Pod (little known) 30–40 × 3.5–4.5 mm, puberulent; seeds unknown.—Collections: 3.

Pastures and sandy grasslands, 1200–1680 m, apparently local, known only from centr. and s. Colombia, from middle Magdalena valley in Cundinamarca and Tolima s. to the upper Patia valley in Nariño.—Fl. XI–III (–VI).

A prostrate herbaceous chamaecrista with no strong differential character proper to itself, difficult to separate from juvenile forms of Antillean var. *swartzii*. It appears distinct from var. *flavicoma* only in lowly diffuse herbaceous growth-form, shorter stipules and fewer leaflets, the venulation of which is essentially identical. Sympatric var. *tristicula* is also normally taller and when adult suffruticose, and well defined by the strongly displaced midrib of the leaflets. Other Colombian chamaecristas have smaller flowers or, in the case of habitally similar *Ch. cordistipula,* basally cordate stipules combined with axillary peduncles.

The var. *andreana* occurs in two pubescence forms, one (typical *Ch. andreana*) with incurved-puberulent stems, leaf-stalks and flower-buds, the other (typical *Ch. pennellii*) with double vesture, in part puberulent, in part pilosulous-setulose. The type-localities of the two supposed species lie 100 km apart in the same latitude, respectively on the west and east sides of the Magdalena valley. Only the puberulent form is known from Nariño.

47l. Chamaecrista glandulosa (Linnaeus) Greene var. **parralensis** (Irwin & Barneby) Irwin & Barneby, Phytologia **44**(7): 500. 1979. *Cassia parralensis* Irwin & Barneby, Sida **6**(1): 18. 1975, based on *Chamaecrista goldmanii* Britton & Rose, N. Amer. Fl. **23**(5): 285. 1930 (non *Cassia goldmanii* Rose, 1906).—"Type from Parral, Chihuahua, September 19, 1898, *E. A. Goldman 98.*"—Holotypus, NY! isotypi, GH, US!

Herbaceous from a stout taproot and ultimately multicipital caudex, the simple or proximally several-branched stems decumbently radiating or weakly assurgent (1–)1.5–3.5 dm, like the lf-stalks and pedicels puberulent in variable density with short incumbent hairs and often, with margins of stipules and lfts, pilosulous or finely setose with widely spreading-ascending hairs up to (0.3–)0.5–1.4 mm, the concolorous, dull olivaceous, when dry brownish lfts glabrous on both faces or thinly pilose beneath.

Stipules obliquely ovate- or lance-acuminate, (2–)3–6 × 0.6–1.5 mm, (5–)6–10-nerved.

Major lvs up to (1.5–)2–4 cm, the petiole with pulvinus 1.5–4 × 0.3–0.65 mm; gland 1 at or above middle of petiole proper, shallowly urceolate 0.2–0.5 mm diam, commonly slender-stipitate and in profile tack- or slenderly trumpet-shaped, sometimes subsessile, 0.2–0.8(–1) mm tall, 0.1 mm shorter to 0.5(–0.6) mm longer than diam of head; lfts of major lvs up to 7–14(–16, of early lvs often 3–7) pairs, narrowly oblong to obliquely oblong-obovate obtuse mucronate up to 6–12 × 1.5–3(–3.5) mm, from base (3–)4–6-nerved by the slightly displaced midrib with on the broad side 2–4 and on the narrow side 0–1 primary nerves, the midrib thence penninerved with (2–)3–5 pairs of simple major and sometimes as many intercalary minor secondaries, the venulation pallid and prominulous on lower face, less so on the upper.

Pedicels (7–)9–18 mm, some fruiting ones at least 10 mm; sepals dark reddish, up to 7.5–11(–12) mm; long abaxial petal (broadly oblanceolate to flabellate) (10–)10.5–14(–15) mm; ovary gray-pilosulous or -strigulose; style linear 3.5–5 mm; ovules 12–19.

Pod linear 30–60 × 4 mm, the reddish-brown valves puberulent; seeds unknown.—Collections: 9.

Stony knolls, outcrops and gullied draws in arid grassland, ±1400–2400 m, apparently uncommon and scattered along the e. piedmont of Sa. Madre Occidental, both in parkland within the pine-belt and on the pure grasslands below it, on the sources of ríos Conchos and Nazas, from w.-centr. Chihuahua to s.-w. Durango (29°30′–24°S) thence e. to s.-centr. San Luis Potosí and, perhaps disjunctly, s. to the middle Lerma valley in Guanajuato (near 21°S), to be sought and expected in intervening Zacatecas and Aguas Calientes.—Fl. VIII–X.

Assessed in the narrow context of the Mexican Plateau flora, *Chamaecrista goldmanii* emerges as a well-defined entity, distinguished from the habitally similar *Ch. serpens* var. *wrightii,* which replaces it on the west slope of Sa. Madre, by the supra-axillary inflorescence, and from sympatric *Ch. nictitans* var. *mensalis* by the strong perennial rootstock and much larger, dolichostylous flower. It was in this provincial setting that we accepted it (l.c.) as an independent species, *Cassia parralensis*. Subsequently reevaluated in relation to the whole picture of multiracial North and South American *Ch. glandulosa,* this same taxon dwindles to one of a multitude of local expressions of a single specific type, modified scarcely or not at all in the organs of reproduction, very slightly so in details of foliage and its venation, but selectively adjusted to thrive in the pronounced seasonal regime of the Sierra Madrean grasslands. Climates leading to savanna climax vegetation, where a premium is set on survival during a prolonged dry season, favor the evolution of strong rhizomes, taproots or xylopodia, and it is reasonable to suppose that the phenetically plastic *Ch. glandulosa,* if it should find a foothold in such environment, would undergo rapid selection for this sort of life-form. Similar events must be responsible for the appearance in mutually remote localities of such closely similar but not identical taxa as *Ch. glandulosa* var. *parralensis* in Mexico, var. *andreana,* scarcely different except in the more elaborate venulation of the leaflets, in Colombia, and *Ch. deeringiana,* pseudo-rhizomatous rather than tap-rooted, on the porous coral limestones of peninsular Florida.

47m. Chamaecrista glandulosa (Linnaeus) Greene var. **brasiliensis** (Vogel) Irwin & Barneby, comb. nov. *Cassia chamaecrista* var. *brasiliensis* Vogel, Syn. Gen. Cass. 63. 1837.—"In Brasilia: Sellow leg. pr. Rio [de] Jan[eiro] in collibus siccis."—Holotypus, presumably †B; no isotypus found, but protologue and locality decisive.

Chamaecrista baileyorum Rose ex Bailey, Gentes Herb. **2**: 202, fig. 105. 1930.—"Collected by L. H. Bailey and Ethel Zoe Bailey on Morro da Viuva, in Rio de Janeiro, Brazil, March 8, 1924 (791)."—Holotypus, *L. H. & E. Z. Bailey 781* (nec *791*), US! clastotypus (fragm) + photogr., NY!

Cassia chamaecrista var. *brasiliensis* sensu Bentham, 1870, p. 172, pl. XLV, fig. 2, exclus. syn. velloz., specimin. Regnellianis, goiasensibus bolivianisque.

Slender, freely branching shrubs 8–15 dm, the young stems and lf-stalks puberulent with incumbent hairs to 0.2–0.5 mm, the stems sometimes cinereous, the pedicels in addition often weakly pilosulous with longer fine hairs up to 0.6–1 mm, the dull dark-olivaceous foliage appearing glabrous but the lfts ciliolate at least near base, often puberulent dorsally.

Stipules narrowly lance-attenuate (4.5–)6–12 × (0.5–)0.6–1.3 mm, from subsymmetric base 5–8-nerved, at first erect, becoming dry and sinuously recurving, persistent.

Major cauline lvs up to 4–8.5 cm, the petiole 3.5–7(–8) mm; petiolar gland 1 sessile patelliform, round or elliptic (0.5–)0.6–1.5 mm diam, in profile 0.3–0.5 mm tall; lfts of larger lvs 8–12(–14) pairs, linear-oblong to narrowly oblong-obovate, the largest of a plant 13–18(–21) × 3–4.4 mm, the pinnate venulation of 4–8(–9) pairs of major secondaries and between each of these 1–2 minor intercalary ones.

Pedicels at anthesis 13–27(–31) mm, in fruit (14–)18–35 mm; sepals up to (6.5–)7–12 mm; long abaxial petal 11–16 mm; style linear straight or almost so 4.5–6 mm, often puberulent to middle; ovules (8–)11–18.

Pod (40–)45–75 × 5–6.5 mm, the purplish-castaneous valves puberulent or minutely pilosulous.—Collections: 24.

Brushy declivities of morros facing the bay or ocean, and in thickets of restinga on the coastal sands, locally common along the coast of Rio de Janeiro and Guanabara, s.-e. Brazil, from Cabo Frio w. to Baia de Sepetiba (42–44°W long.).—Fl. almost throughout the year.

The var. *brasiliensis* is feebly characterized by the syndrome of frutescent stature, sessile patelliform petiolar glands and, as remarked by Vogel in the protologue, exceptionally narrow stipules. None of these features is unique to the variety, but no practical difficulty in recognising it is anticipated, for it is the only chamaecrista with long-pedicelled flowers known to occur naturally in eastern Brazil. An old specimen (*Isabelle in 1835*, G), purportedly collected on the frontier of Uruguay and Rio Grande do Sul, has been ignored in composing the statement of range. It was perhaps a cultivated plant brought to Missão São Borja from Rio de Janeiro. Branchlets of var. *brasiliensis* severed from their frutescent base are to an astonishing degree like whole plants of *Ch. fasciculata* severed at the root.

Bentham (1870, l.c.) confused genuine var. *brasiliensis* with *C. hypnotica* Vell. (=our *Ch. nictitans* var. *paraguariensis*) which, while similar in the form of the gland, is very easily separable by its monocarpic wandlike growth-habit and short-pedicelled flowers, characters explicit in Velloso's plate.

48. Chamaecrista pedemontana Irwin & Barneby, sp. nov., cum *Ch. nictitante* subsp. *brachypoda* hucusque confusa sed manifeste fruticosa nec monocarpica, stipulis parvis 2.5–7 (nec 7–17) mm longis, foliis adultis plerisque 2.5–4.5 (nec 4.5–9) cm longis, foliolis 5–7 (nec 6–12)-jugis patriaque subandina certe diversa, a *Ch. repenti* var. *repenti* quoad foliolorum numerum, formam necnon venationem arcte comparabili habitu glandulaeque petiolaris forma ut videtur distantior.—PERU.

San Martín: Salinas de Pilluana, I.1903 (fl), *E. Ule 6723.*—Holotypus, HAMB = NY Neg. *9195.*

Fruticose erect to 5(–?) dm with virgate leafy hornotinous branchlets, these and lf-stalks incurved-puberulent or densely pilosulous almost or quite throughout with fine spreading or spreading-incurved hairs up to 0.35–0.5 mm, the foliage subbicolored, the firmly chartaceous multistriate lfts either glabrous ciliolate or dull olivaceous and pubescent beneath and above either dull pubescent or glabrous sublustrous.

Stipules erect triangular- or lance-acuminate 2.5–7 × 0.7–1.2 mm, subsymmetrical at base, finely 5–9-nerved, persistent.

Major lvs widely ascending 2.5–4.3 cm, ovate in expanded outline; petiole with discolored but little swollen pulvinus 2.5–4.5 mm, at middle 0.5–0.7 mm diam, narrowly margined; gland 1, near middle of petiole proper, round cupuliform 0.45–0.8 mm diam, stoutly short-stipitate, in profile squatly tack- or funnel-shaped 0.5–0.8 mm tall, 0–0.2 mm longer than diam of head; similar glands on some distal segments of rachis; lfts of adult lvs 5–7 pairs, oblong or the shorter distal pair obliquely obovate, all obtuse or subobtuse mucronate, the larger up to 11–16 × 3–4 mm, at base angulate-cordate on proximal and rounded subcordate on distal side, from base 7–11-nerved by the slightly excentric midrib with proximally 4–7 and distally 2–3 short incurved-ascending primary veins, the midrib giving rise on each side to 5–9 major with, between each, 1–2 scarcely weaker (or sometimes faint) intercalary secondary venules, the venulation finely prominulous on both faces, but more sharply so above.

Peduncles 4–7 mm, adnate their whole length or almost so, 1–2-fld; pedicels (few seen) 9–12(–17) mm, bracteolate 2.5–4 mm below calyx; buds slenderly ovoid-acuminate, pilosulous; sepals broadly lance-acuminate up to 9 mm; petals yellow fading orange, 4 oblong-obovate-cuneate not longer than the sepals, the dimidiately oblong-obovate cucullus 13 mm; androecium 10-merous, the 2 long anthers ±8.5 mm; ovary pilosulous; style filiform ±3.5 mm; ovules 10–12.

Pod linear ±(28–)35–45 × 4–4.5 mm, the reddish-castaneous valves pilosulous; seeds ±3.5 mm, the testa dull ochraceous, lineolately brown-pitted.—Collections: 3.

Habitat not recorded, but to be expected in open rocky places, possibly on river-banks, 450–1700 m, apparently local on the e. slope of the Peruvian Andes in the valley of río Huallaga, deptos. San Martín (Lamas; Pilluana) and Huánuco (Huánuco).—Fl. VI–VII, XII–II(–?).

The earliest known collection (*Spruce s. n. in 1856,* K) of this certainly distinct but taxonomically ambiguous entity was referred by Bentham (1871, l.c.) to *Cassia brachypoda* but, as shown decisively by the type-collection, the plant is an erect shrub, not a monopodial and monocarpic herb, and differs further from genuine *C. brachypoda* in its short stipules and leaves and, as a consequence, in its overall facies. Its close affinity lies, we suspect, in the direction of *Ch. glandulosa* sens. lat., of which the only nearly sympatric element, var. *flavicoma,* is instantly separable by the longer multifoliolate leaves. In its relatively few (5–7) pairs of leaflets *Ch. pedemontana* technically resembles the wholly North American *Ch. lineata,* but we suspect that the similarity is coincidental. The one flower of *Ch. pedemontana* available for dissection showed a corolla, with four short petals not longer than the calyx and an exserted cucullus, unlike anything normally encountered in *Ch. lineata,* in which the petal paired with the cucullus is longest. But this single flower may well not be representative. In describing *Ch. pedemontana* we have been obliged to choose, arbitrarily, between: a) subordinating it to *Ch. lineata,* which would distort our concept of natural affinities; b) subordinating it to *Ch. glandulosa,* which would effectively delete the last tenuous distinction between that species and *Ch. lineata*; and c) presenting it as an autonomous species, although we are aware that it is too feebly characterized for that status. As elsewhere in sect. *Chamaecrista,* the alternatives all seem about equally unpalatable and the choice is ultimately governed by convenience.

49. Chamaecrista deeringiana Small & Pennell, Bull. Torr. Club **44**: 345. 1917.—". . . near Silver Palm, Dade County, Florida, collected in flower and fruit June 22, 1915, *J. K. Small, C. A. Mosier & G. K. Small 6454* . . ."—Holotypus, NY!—*Cassia deeringiana* (Small & Pennell) Macbride, Contrib. Gray Herb., n. ser. **59**: 24. 1919.

Chamaecrista deeringiana sensu Small, Addisonia **4**(1): pl. 121. 1919; Britton & Rose, 1930, p. 284. *Cassia deeringiana* sensu Pullen, 1963, p. 55, 79 (map); Isely, 1975, p. 78, map 26.

Perennial herbs with simple or sparingly branched stems erect or ascending, solitary or few together, from knots or short vertical caudices arising from shallowly buried, horizontally running woody rhizomelike roots, at anthesis 3–9 dm, except for the subterranean organs and perennial life-span closely resembling sympatric forms of *Ch. fasciculata,* appearing glabrous, but lf-stalks at least thinly and often also the stems distally, pedicels and ovary minutely appressed- or incurved-puberulent, the lfts glabrous, prominently venulose.

Stipules (ovate-) lance-acuminate (2.5–)4–10.5 × 0.8–1.9 mm, from oblique or shallowly subcordate base prominently 7–13-nerved, glabrous, persistent.

Adult lvs 3–8.5 cm; petiole with pulvinus 3–7(–9) mm; gland 1, near middle of petiole proper, round or elliptic, shallowly concave or discoid (0.5–)0.6–1.6(–2) mm diam, sessile or stoutly short-stipitate, in profile 0.25–1 mm tall, 0.25–1 mm shorter than diam of head; lfts of adult lvs up to 8–18(–20) pairs, narrowly lance- or linear-oblong, obtuse mucronulate or subacute, up to 9–17(–20) × 2–3.3(–"3.7") mm, from base 5–7-nerved by the centric or slightly displaced midrib with on its broad side 3–5, on its narrow one usually 1 shortly incurved-ascending primary veins, the midrib thence penninerved with (3–)4–6(–7) major and as many weaker or almost as strong intercalary secondaries, the venulation usually sharply prominulous on both faces.

Inflorescence and fl as in *Ch. fasciculata,* but the petals reportedly (Isely, 1975, p. 79) richer yellow, all or only 2–3 red-blotched at the claw, the long abaxial petal (12–)14–19 × 7–16 mm; androecium 10-merous, the longer 2–3 of the yellow, dull reddish or particolored anthers 8–10.5 mm; style linear 6–8 mm; ovules 12–19.

Pod 35–70(–85) × 4.5–6 ("–7.4" fide Pullen) mm; seeds little known, apparently not different from those of *Ch. fasciculata.*—Collections: 11.

Thin pine and pine-oak woodlands below 200 m, in deep sands or thin sandy humus overlying coral limestone, locally abundant but of discontinuous range in Florida and adjoining states: centr. and (more plentifully) s. peninsular Florida, from Hernando County s. to Key Largo; w. Florida, from the Apalachicola w. to the Escambia River, there extending n. just into Alabama; and apparently disjunct on the upper Flint River in w.-centr. Georgia (Taylor and Peach counties).—Fl. III–VII and in s. Florida sporadically through the fall and winter.

A species of enigmatic origin and oddly discontinuous dispersal, recognized by all modern students of the group as closely akin to *Ch. fasciculata,* from glabrescent sympatric forms of which it differs only in the cordlike horizontal perennial root, a structure duplicated nowhere in sect. *Chamaecrista.* Isely (1975, l.c.) finds unbranched stems and red anthers helpful in identifying specimens of *Cassia deeringiana* severed from their characteristic understructure, but these features are fallible, the more vigorous stems being by no means always simple, and the anthers, as noted by Small (1919, l.c.) varying from red to yellow. Pullen (1963, l.c.) raises the question as to whether *C. deeringiana,* in virtue of its perennial life-span and interrupted dispersal, may not represent a relict stock ancestral to *Ch. fasciculata* and perhaps other annual kindred. We have no empirical data with which to oppose this hypothesis, but think it a weak one if examined in the context of sect. *Chamaecrista* as a whole. The horizontal root-system, which appears to fulfill in the shallow soils of fire-prone pine savannas of Florida the function of a xylopodium in the campo cerrado of eastern Brazil, is (unlike a vertical taproot) a highly specialized one, and moreover coincides, in *Ch. deeringiana,* with the also specialized supra-axillary inflorescence of the "mimosoid" chamaecristas. We think it unlikely that a pseudo-rhizomatous ancestor of an enormous, highly diversified group of species dispersed throughout the tropics of the world should survive only in a small extratropical enclave in southeastern United States, the more so when it is known to resemble most closely the one member of that group which is itself entirely extratropical. We would suggest, to the contrary, that *Ch. deeringiana* arose in situ from sympatric *Ch. fasciculata,* of which it preserves intact every gross morphological feature above the root; and, further, that it is possibly biphyletic, a circumstance that would explain its bicentric range and curious absence from northern peninsular Florida (cf. Isely, 1975, map 26, where incidentally the Georgia stations are overlooked).

The species was named in honor of Charles Deering, landholder of Cutler, Dade County, Florida, who promoted J. K. Small's botanical exploration of the Everglades and southern peninsula.

50. Chamaecrista fasciculata (Michaux) Greene, Pittonia **3**: 242 (lapsu "*fascicularis*"). 1897. *Cassia fasciculata* Michaux, Fl. Bor. Amer. **1**: 262. 1803.—"Hab. in Pensylvania et Virginia."—Holotypus, P (hb. Michaux.)!

Cassia chamaecrista Linnaeus, Sp. Pl. 379. 1753, exclus. syn. Breyn. locisque caribaeis.—"Habitat in Jamaica, Barbados, Virginia."—Holotypus, ex hort. Upsal., LINN *528/30*!—*C. pulchella* Salisbury, Prod. stirp. hort. Chapel Allerton 326. 1796, nom. illegit. *Grimaldia chamaecrista* (Linnaeus) Schrank ex Link, Handbuch **2**: 141. 1831. *Xamacrista triflora* Rafinesque, Sylv. Tellur. 127. 1838, nom. illegit. *Chamaecrista chamaecrista* (Linnaeus) Britton, Bull. Torr. Club **44**: 12. 1917, tautonym. illegit.

Cassia triflora Jacquin, Pl. rar. hort. Schoenbrun. **4**: 40, t. 480. 1804.—"Patriam ignoro."—No typus found at W (1980); lectoholotypus, the cited plate!—Correctly referred by Bentham, 1871, p. 576, to *Cassia chamaecrista,* but the specimen seen by him in hb. Candoll. appears to be a form of *Ch. glandulosa.*

Cassia venosa Castiglione ex Zuccagni in Roemer, Collectanea 140. 1806.—Named from plants cultivated in Florence, presumably grown from seeds brought from United States by Castiglione. The description of pyxidate gland and red flare on upper petals confirms the identity. A possibly earlier description by Castiglione in "Auct. ad syn. hort. reg. Florent. 1795, no. 87," quoted by Zuccagni, has not been verified. If properly published, the epithet would have priority over *fasciculata.*

Cassia chamaecrista α *normalis* fma. *pubicaulis* O. Kuntze, Rev. Gen. 169. 1891.—"St. Thomas, [West Indies], U. St.: Charleston."—Lectoholotypus, *O. Kuntze 2862,* collected at Charleston, South Carolina, 9.IX.1874 (fl, fr), NY! The paratypus from St. Thomas, *O. Kuntze 62,* NY! = *C. glandulosa* var. *swartzii.*

Cassia chamaecrista α *normalis* fma. *hirsuta* O. Kuntze, Rev. Gen. 169. 1891.—"U. St.: St. Louis."—Holotypus, *O. Kuntze 2773,* collected 6.IX.1874 at St. Louis, Missouri, NY (2 sheets)!

Cassia chamaecrista var. *robusta* Pollard, Bull. Torr. Club **21**(5): 218. 1894.—". . . a single specimen in the Columbia College Herbarium collected [in 1840] by Dr. C. W. Short in the mountains of Kentucky . . ."—Holotypus, NY!—*C. robusta* (Pollard) Pollard, Bull. Torr. Club **24**: 150. 1897. *Chamaecrista robusta* (Pollard) Pollard ex A. Heller, Cat. N. Amer. Pl. ed. 2, 5. 1900. *Cassia fasciculata* var. *robusta* (Pollard) Macbride, Contrib. Gray Herb., n. ser. **59**: 24. 1919. *Chamaecrista fasciculata* β Pennell, Bull. Torr. Club **44**: 351. 1917. *Ch. fasciculata* var. *robusta* (Pollard) Moldenke, Boissiera **7**: 2. 1943. *Cassia fasciculata* var. *fasciculata* variant 3 Isely, Mem. N.Y. Bot. Gard. **25**(2): 92, map 30. 1975.—*Ch. robusta* sensu Britton & Rose, 1930, p. 296; *C. fasciculata* var. *robusta* sensu Pullen, 1963, p. 63.

Cassia mississippiensis Pollard, Bull. Torr. Club **21**(5): 219. 1894.—Type in herbarium of Columbia College, collected by Miss K. Skeehan, 1889, at Ocean Springs, Mississippi."—Holotypus, NY!—*Chamaecrista mississippiensis* (Pollard) Pollard ex A. Heller, Cat. N. Amer. Pl. ed. 2, 5. 1900.—*Ch. mississippiensis* sensu Britton & Rose, 1930, p. 296; equated with *C. fasciculata* var. *tracyi* by Pullen, 1963, p. 66, with var. *puberula* by Isely, 1975, p. 92, with var. *fasciculata* by Isely, 1975, p. 212.

Cassia depressa Pollard, Bull. Torr. Club **22**: 515, p. 251, 252 (fig. 6). 1895.—". . . River Junction, Gadsden Co., Florida. G. V. Nash, September 5, 1895 (no. 2571)."—Holotypus, US! isotypi, LE, NY!—*Chamaecrista depressa* (Pollard) Greene, Pittonia **3**: 242. 1897. *Ch. fasciculata* γ Pennell, Bull. Torr. Club **44**: 352. 1917. *Cassia fasciculata* var. *depressa* (Pollard) Macbride, Contrib. Gray Herb., n. ser. **59**: 25. 1919.—*Ch. depressa* sensu Britton & Rose, 1930, p. 296.—Referred to *C. fasciculata* var. *robusta* by Pullen, 1963, p. 63; to var. *fasciculata* by Isely, 1975, p. 89.

Chamaecrista bellula Pollard, Proc. Biol. Soc. Wash. **15**: 19. 1902.—"Type in [US], collected by Prof. S. M. Tracy at St. Vincent, Florida, September 9, 1899 (No. 6,326)."—Holotypus, US! isotypus, NY!—Referred to *Ch. fasciculata* by Pennell, 1917, p. 347; Britton & Rose, 1930, p. 296; to *C. fasciculata* var. *fasciculata* by Pullen, 1963, p. 58; overlooked by Isely, 1975.

Chamaecrista littoralis Pollard, Proc. Biol. Soc. Wash. **15**: 20. 1902.—"Type, No. 371,572 in [US], collected by Prof. S. M. Tracy and Prof. F. E. Lloyd on Breton Island, La., August 17, 1900 (No. 198)."—Holotypus, US! isotypus, NY!—*Chamaecrista mississippiensis* β Pennell, Bull. Torr. Club **44**: 354. 1917. *Cassia fasciculata* var. *littoralis* (Pollard) Macbride, Contrib. Gray Herb., n. ser. **59**: 25. 1919. *C. littoralis* (Pollard) Cory, Rhodora **38**: 406. 1936.—*Ch. littoralis* sensu Britton & Rose, 1930, p. 297. *C. fasciculata* var. *littoralis* sensu Pullen, 1963, p. 68. Referred by Isely (1975, p. 95) to *C. fasciculata* var. *puberula*.

Chamaecrista brachiata Pollard, Proc. Biol. Soc. Wash. **15**: 20. 1902.—"Type, No. 330,115, in [US], collected by Charles L. Pollard and G. N. Collins at Miami, Dade County, Florida, April 4–7, 1898 (No. 245)."—Holotypus, US! isotypus, NY!—*Cassia brachiata* (Pollard) Macbride, Contrib. Gray Herb., n. ser. **59**: 24. 1919. *C. fasciculata* var. *brachiata* (Pollard) Pullen [The *Cassia fasciculata* Complex 62. 1963 (Doctoral dissertation, unpublished)] ex Isely, Mem. N.Y. Bot. Gard. **25**: 87, 202, maps 29, 31, 32. 1975.—*Ch. brachiata* sensu Pennell, 1917, p. 345; Britton & Rose, 1930, p. 295.

Chamaecrista tracyi Pollard, Proc. Biol. Soc. Wash. **15**: 21. 1902.—"Type in [US], collected by Prof. S. M. Tracy at Koshtaw, Miss., September 15, 1898."—Holotypus, *Tracy 4914*, US!—*Cassia fasciculata* var. *tracyi* (Pollard) Macbride, Contrib. Gray Herb., n. ser. **59**: 25. 1919.—*Ch. fasciculata* var. *tracyi* sensu Pullen, 1963, p. 66.—Reduced to *Ch. mississippiensis* by Pennell, 1917, p. 353; Britton & Rose, 1930, p. 296; to *C. fasciculata* var. *puberula* variant 3 by Isely, 1975, p. 95.

Chamaecrista camporum Greene, Pittonia **5**: 108. 1903.—". . . collected by myself at Monticello, Illinois, 7 August, 1899."—Holotypus, ND (*herb. Greene 35144, 5, 6*)!—Referred to *Ch. fasciculata* by Pennell, 1917, p. 348; Britton & Rose, 1930, p. 296; to *C. fasciculata* var. *fasciculata* by Pullen, 1963, p. 58; overlooked by Isely, 1975.

Chamaecrista puberula Greene, Pittonia **5**: 134. 1903.—"On Galveston Island, [Texas,] 23 Sept. 1901."—Holotypus, *S. M. Tracy 7797,* ND! isotypi, NY, US (2 sheets)!—*Cassia fasciculata* var. *puberula* (Greene) Macbride, Contrib. Gray Herb., n. ser. **59**: 25. 1919. *C. greenei* Standley, Field Mus., Bot. **11**: 159. 1936.—*Ch. puberula* sensu Pennell, 1917, p. 355; Britton & Rose, 1930, p. 297; *C. fasciculata* var. *puberula* sensu Isely, 1930, p. 92, maps 33–36 (exclus. variant l). Reduced to *C. fasciculata* var. *tracyi* by Pullen, 1963, p. 66.

Chamaecrista rostrata Wooton & Standley, Contrib. U.S. Nat. Herb. **16**: 135. 1913.—"Type in [US] no. 660032 . . . at Logan, [Quay County, New Mexico], October 5, 1910, by Mr. Geo. L. Fisher."—Holotypus, US! isotypus, NY!—*Cassia fisheri* Cory, Rhodora **38**: 404. 1936. *C. rostrata* (Wooton & Standley) Tidestrom ex Tidestrom & Kittell, Fl. Ariz. New Mex. 157. 1941, homonym. poster. (non *C. rostrata* F. S. Voigt, 1828). *C. fasciculata* var. *rostrata* (Wooton & Standley) B. L. Turner, Field & Lab. **23**: 88. 1955.—*Ch. rostrata* sensu Britton

& Rose, 1930, p. 295. *C. fasciculata* var. *rostrata* sensu Pullen, 1963, p. 65; Isely, 1975, p. 96, maps 29, 38.

Chamaecrista ferrisiae Britton ex Britton & Rose, N. Amer. Fl. **23**(5): 285. 1930.—"Forty miles north of Edinburg on road to Falfur[r]ias, [in n. Hidalgo Co.], Texas, August 7, 1921, *Roxana S. Ferris & Carl D. Duncan 3202.*"—Holotypus, NY! isotypus, CAS!—*Cassia fasciculata* var. *ferrisiae* (Britton) B. L. Turner, Field & Lab. **23**: 90. 1955.—*C. fasciculata* var. *ferrisiae* sensu Turner, 1959, p. 79, map 40; reduced by Isely, 1975, p. 92, map 34, to *C. fasciculata* var. *puberula* variant 1.

Cassia fasciculata fma. *jenseni* Palmer & Steyermark, Ann. Missouri Bot. Gard **22**(3): 574. 1935.—"MISSOURI: Gray Summit, Franklin Co., Aug. 20, 1928, *L. P. Jensen* . . ."—Holotypus (not seen), MO.—*C. fasciculata* var. *fasciculata* fma. *jenseni* sensu Steyermark, Fl. Missouri 875. 1963.—Omitted by Pullen, 1963, and by Isely, 1975.

Cassia fasciculata var. *macrosperma* Fernald, Rhodora **42**: 455, pl. 635, fig. 1, 2. 1940.—". . . southeastern VIRGINIA: . . . along Kittewan Creek, Weyanoke, Charles City County, September 18, 1939, *Fernald & Long,* no. 11,350 . . ."—Holotypus, GH! isotypi, NY, US! paratypi, *Fernald & Long 11349, 11351, 11573, 11574, 11576,* all GH, some NY!—*C. fasciculata* var. *macrosperma* sensu Pullen, 1963, p. 62; referred to *C. fasciculata* var. *fasciculata* by Isely, 1975, p. 89.

Cassia fasciculata fma. *transmutata* Fernald, Rhodora **44**: 404, p. 724. 1942.—"Norfolk County, VIRGINIA: . . . north of Cornland, October 17, 1941, *Fernald & Long,* no. 14,955 . . ."—Holotypus, GH!—Omitted by Isely, 1975.

Cassia fasciculata fma. *mutata* Fernald, Rhodora **44**: 405, pl. 724, fig. 3. 1942.—"Nansemond County, VIRGINIA: . . . north of Whitemarsh School, July 19, 1939, *Fernald & Long,* no. 10685 . . ."—Holotypus, GH!—Omitted by Isely, 1975.

Cassia chamaecrista sensu Colladon, 1816, p. 129, max. ex parte; Vogel, 1837, p. 62, exclus. var. & ; Bentham, 1871, p. 576, exclus. var. *brasiliensi*; auct. bor.-amer. fere omnes ad annum 1917 usque.

Chamaecrista fasciculata sensu Pennell, 1917, pp. 347–352; Britton & Rose, 1930, p. 296. *Cassia fasciculata* sensu B. L. Turner, 1959, p. 78–79 (exclus. var. *ferrisiae*), maps 39–40; Pullen, 1963, pp. 56–69 (exclus. var. *ferrisiae*), figs. 7–8 (maps); Isely, 1975, pp. 84–96, 202 (exclus. var. *puberula* variant 1), maps 29–38.

Monocarpic but below or near the frost-line in s. Florida sometimes enduring into winter and weakly frutescent basally, either simple and virgately erect, or virgate and branched from near or above middle becoming bushy in age, or less often lacking a strong central axis and branched from near base, the stems then assurgent or subprostrate, at full anthesis (0.5–)1–11(–14) dm, the stems, lf-stalks, pedicels and buds puberulent with short incurved or subappressed hairs 0.2–0.8 mm and often in addition pilose with fine divaricate or widely incurved-ascending setae up to 0.8–1.8(–2.2) mm, but either type of vesture predominating or either, sometimes both, entirely lacking, the rich-green or olivaceous, sometimes glaucescent, dull or (living) sublustrous lfts either concolorous or often somewhat paler beneath, either glabrous, or glabrous ciliolate, or finely pilosulous above or on both faces, the vesture whitish or rarely lutescent.

Stipules erect lanceolate or lance-attenuate (2.5–)4–14(–15) × (0.7–)1–2.2(–2.6) mm, from oblique or (when broad) shallowly semicordate base (6–)7–13-nerved, the thinly herbaceous blades becoming papery persistent, either glabrous, glabrous ciliolate, or dorsally puberulent and remotely long-ciliate.

Lvs (disregarding smaller lower ones, commonly shed by anthesis) up to (2–)2.5–9(–11) cm, the blade in expanded outline ovate-oblong or -elliptic; petiole with pulvinus (2–)2.5–9 mm, at middle (0.25–)0.4–0.7(–0.8) mm diam; glands 1–2(–3, commonly only 1) near or slightly below middle of petiole proper, either discoid, or shallowly to deeply cupular, or low-convex, round or when relatively large perpendicularly dilated and elliptic in outline, (0.3–)0.4–2.3(–3.3, commonly 0.6–1.6) mm diam, either sessile or shortly stipitate, the stipe when present varying from much narrower than to as wide as the head, the whole in profile obconic to shallowly saucer-, tub-, squatly trumpet-, tack- or drum-shaped, 0.3–1.1 mm tall, almost always at least a trifle shorter, exceptionally up to 0.3 mm longer than diam of head; lfts of adult lvs up to (7–)8–21(–26) pairs (in fma. *mutata* lvs imparipinnate 5-foliolate), linear-oblong, oblong, oblong-oblanceolate or -elliptic, mostly obtuse mucronulate, sometimes acuminulate or (when narrow) subacute, the larger of largest lvs up to (5.5–)7–20(–23) × (1.3–)1.5–5(–6) mm, at base rounded or cordate on proximal side, cuneate to rounded on distal one, from base 4–6-nerved by the centric or moderately displaced midrib with on the broad side 2–4, on the narrow one (0–)1 incurved-ascending primary veins, the midrib dividing the blade 1:1–2.2 and giving rise to 2–5(–6) pairs of major with or without fainter or almost as strong intercalary secondary veins, these simple or almost so, the venulation prominulous on both faces, or only beneath, or faint and subimmersed.

Peduncles (1–)2–28 mm, adnate through 1–22(–26) mm, 1–4(–6)-fld; pedicels ascending and in fruit commonly spreading and distally incurved or geniculate under the pod, the longer of each raceme up to (6–)8–22(–26) mm, bracteolate above middle; buds ovoid-acuminate, puberulent, pilosulous, pilose, or glabrous; sepals submembranous, usually greenish, narrowly ovate- or lance-acuminate or -caudate up to 6–14 mm; petals yellow fading brown or orange, exceptionally (fma. *jenseni*) whitish, either concolorous or commonly 2–4 (but never the cucullus) maculate at the claw with an entire or 2-lobed red flare, the 3 adaxial homo- or erratically heteromorphic, symmetrically or obliquely oblanceolate, oblong-oblanceolate or obovate, up to (7–)7.5–16 × 4.5–13 mm, 1 abaxial similar but a little, often much, larger, mostly obovate-cuneate or flabellate, very obtuse or emarginate 10–23 × 7–18.5 mm, the cucullus as long or commonly shorter, obliquely obovate or subfalcately reniform, the margin often irregularly crenate; androecium 10-merous, the outer whorl of anthers longer, (2–)3 of these longest, up to (5.5–)6–10.5 mm, the sacs all yellow, or (some or all of them) yellow red-tipped, or (all or some) red, deep reddish-brown or red-violet; ovary strigulose, pilosulous, less often glabrous, sometimes shaggy-pilose; style filiform, gently incurved distally, (3.5–)4–7.2 mm; ovules (7–)9–21.

Pod linear-oblong, straight or curved outward, (25–)30–75(–85) × (3–)3.5–6.5 (in local "var. *macrosperma*" 6.5–10) mm, the valves usually reddish, castaneous or atropurpureous but paler where expressed over the seeds, sometimes green and red-margined, sometimes green throughout, when fully ripe nigrescent, either glabrous, or incurved-pilosulous or thinly pilose; seeds dark brown or black, dull or sublustrous, (2.8–)3.2–4.8 (in "var. *macrosperma*" 4.7–6.3) mm, finely lineolate-pitted; $n = 8$.—Collections: 332.—Fig. 50.

Open woodlands, fields, roadsides, coastal and inland dunes, quickly entering and sometimes dominating unstable microhabitats in a variety of climax communities, extending s.-ward into pine-savanna and coastal prairie, n.-e.-ward into disturbed hardwood forest, n.-w.-ward into tall-grass prairie, far w.-ward into sand-pockets and sandy gulches in short-grass prairie, usually in dry light soils or sands but occasional in waterlogged clays of ditches, stream-banks and tidally inundated estuarine flats, widespread and locally abundant below 1000 m over most of e., centr. and s.-e. United States, from s.-e. South Dakota and s. Minnesota to Pennsylvania and Massachusetts, s. to peninsular Florida and s. Texas, thence extending w. and up to 1350 m along the upper Cimarron, Canadian and Brazos valleys to s.-w. Kansas, the Oklahoma and Texas panhandles, and adjacent extreme e. New Mexico, on the coastal plain of s. Texas fading imperceptibly into *C. chamaecristoides*.—Fl. VI–X and around the Gulf coastal lowlands sometimes into I–III.

Our concept of *Ch. fasciculata,* as formulated in the foregoing description and

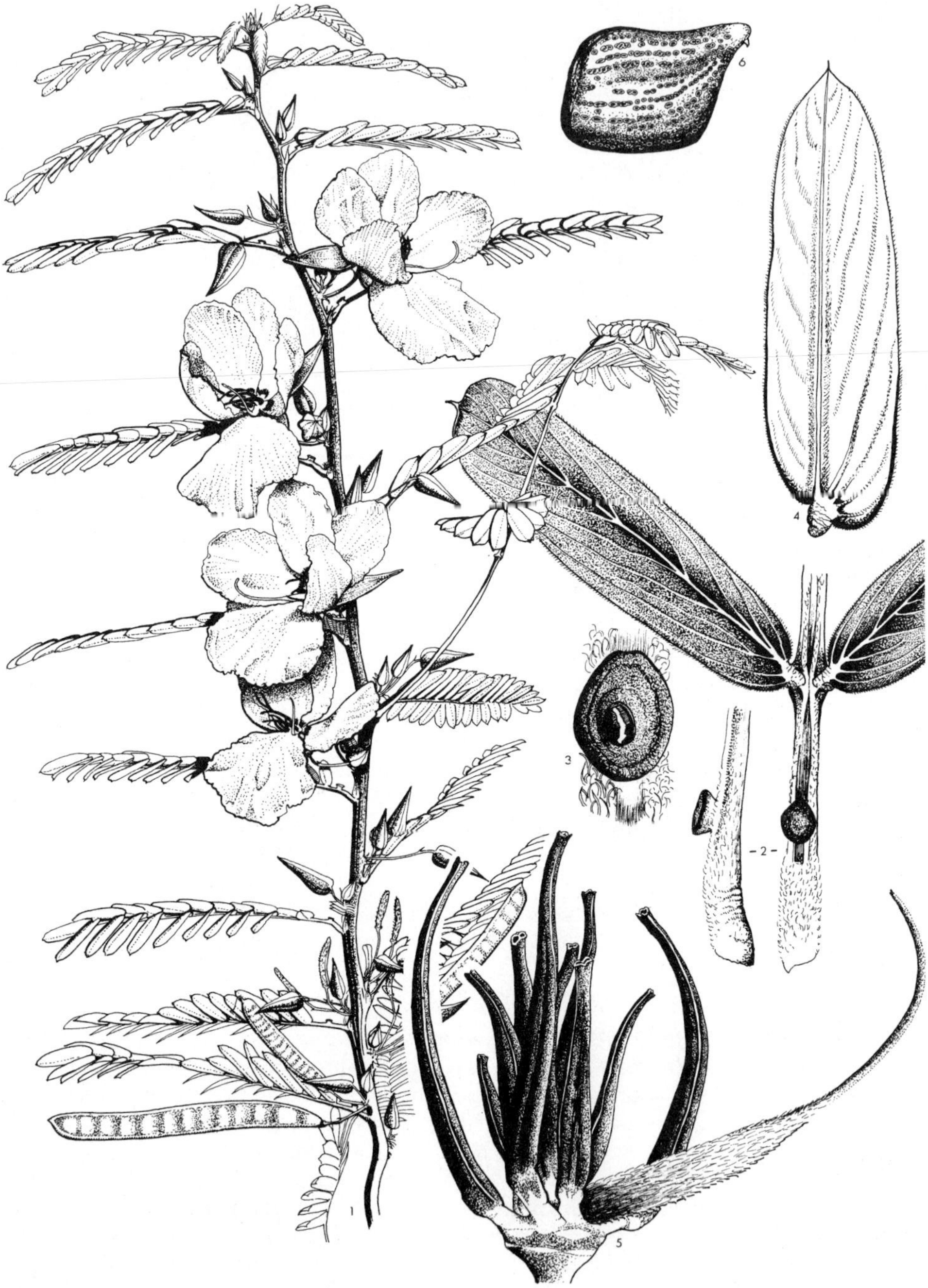

Fig. 50. *Chamaecrista fasciculata* (Michx.) Greene. 1) Habit ×1; 2) petiole + gland ×5; 3) gland ×20; 4) dorsal face of leaflet ×5; 5) androecium + pistil ×5; 6) seed ×5 (Drawn from life by W. S. Moye III).

synonymy, is coextensive with those of *Cassia fasciculata* presented by Turner (1959), Pullen (1963) and Isely (1975), but with these differences: 1) while recognizing the internal diversity of the species we find this of the same order, and no more extreme, than encountered in related "mimosoid" chamaecristas of the West Indies and Brazil, and in consequence we are unable to maintain formal subspecific categories; and 2) we transfer to *Ch. chamaecristoides* var. *cruziana* much of what has passed in Texas as *C. fasciculata* var. *ferrisiae*. As redefined, *Ch. fasciculata* embraces all those extratropical North American chamaecristas in which a monocarpic root coincides with subcentrically costate leaflets, adnate peduncles and relatively large, longistyled flowers. It is only feebly and artificially separated, on grounds of convenience, from 1) *Ch. glandulosa* by the annual life-span; from 2) large-flowered races of monocarpic *Ch. nictitans* by the more nearly sessile petiolar gland combined with temperate (and fully discrete) dispersal; and from 3) *Ch. chamaecristoides,* with which it freely intergrades in southern Texas, by the greater symmetry of the leaflets. Of three sympatric species, the widespread *Ch. nictitans* (including its var. *aspera*) is instantly separable by the small, short-pedicelled 5–8-androus flowers, and the local, Floridian *Ch. deeringinana* and *Ch. lineata* var. *keyensis* by perennial life-form, the former rhizomatous, the latter tap-rooted and differing further in relatively few (5–9) pairs of leaflets.

In the past 80 years *Ch. fasciculata* sens. lat. has been repeatedly analyzed, alternately from synthetic and analytic viewpoints, but all of the more thorough and detailed studies—we regretfully exclude from their number Britton & Rose's account in *North American Flora*—have treated the problems of internal diversity as though soluble without reference to the parental stock of tropical *Chamaecrista,* that hydra of which *Ch. fasciculata* itself is not more than an imperfectly severed head. For seen in relation to all of the kindred Neotropical chamaecristas the species emerges as merely a northern extension of a virtually continuous fabric of circum-Caribbean and Central American forms. The selective forces of a temperate climate have compressed the life-cycle of *Ch. fasciculata* into an obligately annual span, but have left the foliage, perianth and reproductive organs essentially unaltered, the flowers of *Ch. glandulosa* sens. lat., *Ch. fasciculata* sens. lat., *Ch. chamaecristoides, Ch. rufa,* and the largest-flowered races of *Ch. nictitans* var. *disadena* being structurally identical, as are the pods and seeds. We have much sympathy for the extreme synthetic position of Otto Kuntze (1894) who, following a voyage round the world and sundry encounters with supra-axillary chamaecristas in places as far apart as the Virgin Islands, New York, St. Louis, Trinidad, Bolivia, Annam and the Deccan, saw all of them as expressions of one multiracial species. The situation had been clearly perceived by Bentham (1871, p. 576) when he remarked that ". . . the whole group of the *Chamaecristae verae* seem to pass so gradually one into the other, that the delimitation of species here given must be admitted to be very unsatisfactory." But a logical application of Kuntze's principle to the taxonomy of the group leads to so cumbrous and elaborate an hierarchical system as to defeat the purposes of classification. The specific definition of *Ch. fasciculata* is as unsatisfactory to us as it was to Bentham but we think that in practice it is unavoidable.

Collation of the dense and extensive literature concerning *Ch. fasciculata* sens. lat. cited in our paragraphs of synonymy shows that really only a few variable qualities have ever been found taxonomically useful or significant, even though these qualities may occur in several character-states and thereby appear more numerous than they are. They fall into the categories of 1) growth-habit; 2) vesture; 3) petiolar glands; with, in particular cases, 4) leaflet-number; 5) color of anthers; and 6) width of pod. Each category deserves comment.

1) *Growth-habit:* The commonest form, prevalent throughout the range of the species, is virgate erect, simple when small or young, branched upward when vigorous or old, and extensively branched where mild climate permits a long season of growth. Less often the central axis is weak or inhibited and the growth-impulse is diverted into lateral branches, which may be either incurved-ascending or almost prostrate. Pullen (1963, p. 32) found that branching patterns are readily modified under cultivation; apparently they are not genetically controlled.

2) *Vesture:* A duplex vesture of shorter incurved or appressed and longer spreading hairs is characteristic not only of *Ch. fasciculata* but of all close Central American, South American and West Indian kindred, as also of the related Palaeotropical *Ch. mimosoides* (Brenan, 1967, p. 100–105). Throughout this group there is random emphasis now on one type of hair, now on the other, either preponderant or either (even both) lacking, the long hairs moreover sometimes lutescent but more often not. Long hairs (setae) may occur on all stem-axes and leaf-stalks as well as on the pedicels and sepals, but sometimes only on the latter. Short hairs may clothe only the stems (overall or in lines), leaf-stalks and pedicels, or may extend to one or both faces of the leaflets. The whole gamut of variation in length, orientation and dispersal of hairs is complex but essentially continuous and is very imperfectly correlated with other morphological characters or geography. Its visual impact is nevertheless often great, and has provoked description of several taxonomically inconsiderable forms. The full range of variation in pubescence is found mainly around the Gulf coastal plain between eastern Texas and western Florida. Northward and westward pilose stems and facially pubescent leaflets become independently rarer. Five of Pullen's six varieties of *C. fasciculata* were characterized principally in terms of pubescence, even though it was admitted that two of these may occur closely associated within a given population and that intermediate forms difficult to assign to any one of them are frequent. Except in the case of var. *puberula,* in which the leaflets are pubescent, Isely (1975) abandoned this position, substituting numbered *variants* for named varieties, although somewhat paradoxically insisting on sparse pubescence, provided that this coincides, in Florida (but not elsewhere), with large petiolar glands, as a differential character of var. *brachiata*. The cause of the inherent instability of pubescence in sect. *Chamaecrista* is unknown; but once recognized as a virtually ubiquitous phenomenon in the group it falls into perspective and may be dismissed as a useless or meaningless taxonomic character.

3) *Petiolar glands:* The parameters of observed variation in the petiolar gland of *Ch. fasciculata* will be found in our description of the species; we consider here only their possible bearing on the intraspecific taxonomy, their significance, that is, as markers of genuine genetic differentiation. The petiolar gland is a specialized outgrowth of the leaf-stalk and as such it is found in very large measure to vary in size with the vigor of its parent leaf. Glands on the first cauline leaves of a young plant, which are generally smaller and simpler than adult ones, are predictably smaller than glands of the mature foliage. The larger glands, 1 mm or more in diameter, are associated with relatively robust leaves and coarse leaf-stalks, and this irrespective of the development of the stipe, which varies independently of the head's diameter both in length and thickness. A very small gland, 0.3–0.6 mm diameter, associated as one would expect with slender-stalked leaves and relatively few and small (±7–10-jugate) leaflets, has been considered diagnostic of *Cassia fasciculata* var. *rostrata* of southern and Panhandle Texas, at least so long as combined with glabrous foliage and red anthers; but the same

gland in Massachusetts (*MacKeever 770,* NY), Florida (*Tracy 6536,* NY) or Mississippi (holotypus of *Ch. mississippiensis*) is passed over as insignificant, possibly because combined in the first case with partly yellow anthers, in the second with up to 12 pairs of leaflets, and in the last with puberulent foliage. Likewise a massive scutellate or trough-shaped gland, said to characterize, when linked in Florida with glabrescent foliage and stems, a robust endemic var. *brachiata,* can be matched readily, associated either with puberulent or with setose stems, in plants from as far north and west as Iowa and Missouri. If the petiolar gland is interpreted as a vegetative organ, subject to the sorts of emphasis and decline that are accepted as normal in the ontogeny of leaf-blades, its diagnostic value dwindles abruptly.

4) *Leaflet number:* As already mentioned in relation to the petiolar gland, the number of leaflets is, at least often, related to the leaf's position on the parent axis, whether this be the primary stem or a branch of the second or later orders, the earlier leaves constantly being shorter and simpler. In unfavorable or marginally tolerable habitats the ontogenetic sequence from simpler shorter to longer plurifoliolate leaves fails to find expression, a neotenic effect which Vassil'čenko (1965) has very convincingly attributed to environmental stress. This seems to be as true of typical *Ch. rostrata* on the high Plains of western Kansas and Texas as of phenotypically similar forms occurring as far east as New England (*Vail in 1888,* NY). The so-called *C. fasciculata* var. *rostrata* of the lower Rio Grande valley contradictorily combines in one plant the few leaflets and small gland of the elsewhere juvenile leaf with a vigorous, freely branching habit of growth. Isely (1975, p. 96, map 38), accepting var. *rostrata* as this had been defined by Turner (1959, map 39) and Pullen (1963, p. 65), suggests that it is a polyphyletic concept consisting of "several distinct to ambiguous [geographically] peripheral phases of [*C. fasciculata*]" in which small neat glabrous foliage and small glands happen to coincide. And indeed the average plant from the Staked Plains differs from that of the south Texas lowlands in its diminished stature and broader pod. Plants of the lowland race cultivated at Austin (*Irwin 1412,* NY) were found incompatible with other races of *C. fasciculata,* but since Pullen obtained only 20 fruits from 340 artificial intraspecific crosses in *C. fasciculata,* more evidence of genuine sexual barriers within the complex is desirable.

5) *Anther color:* Variation in anther color, noted first by Torrey & Gray (1840, p. 395–396), has been capably analyzed for *Ch. fasciculata* by Isely (1975, p. 90), who found a general drift toward yellow eastward from the Mississippi valley and toward red or red-violet westward, but so many populations with anthers of mixed color as to render the character useless in practice. There is no known link between anther-color and insect pollinators, but some may nevertheless exist.

6) *Pod:* While the pod of *Ch. fasciculata* sens. lat. varies greatly in length (with ovule number) and breadth (with seed size), there seems to be no discoverable discontinuity in the whole range of variation and hardly more than a hint of correlation with dispersal. The longest pods are found in peninsular Florida connected with so-called var. *brachiata,* but pods hardly shorter have been collected here and there throughout the Middle West. The narrowest pods, already mentioned as characterizing most of the south Texan forms that have been referred to so-called var. *rostrata,* are duplicated in parallel but not identical slender phenotypes in coastal Mississippi and elsewhere. The remarkable broad pod and massive seeds of so-called var. *macrosperma* are borne on a plant which at anthesis cannot by any detail be distinguished from the common robust glabrescent partridge pea of the Atlantic Coastal Plain. And among paratypes of var. *macrosperma* listed by Fernald in the protologue we encounter some fruits that

can be matched for size in populations from far inland. Nevertheless it would be interesting to see this *macrosperma* phase planted outside its estuarine habitat and studied cytogenetically.

In summary it may be said that 1) intraspecific diversity in *Ch. fasciculata* is greatest in the warm temperate and subtropical crescent suspended between the Everglades in Florida and the lower Rio Grande in Texas, and shows some incipient geographic segregation toward each end of this arc; and that 2) diversity decreases north and westward, as though there had occurred a progressive elimination of biotypes and simplification of the gene-pool as the species expanded into newly available habitats. While we agree with Isely (1975, passim) that potentially weedy *Ch. fasciculata* doubtless does exploit the ever-increasing disturbance of its environment and that the chance of reconstructing its pre-colonial dispersal is gone forever, we see no necessity to invoke introgressive hybridization between formerly better defined entities as prime or efficient causes of the random and reticulate patterns of phenotypic variation that meet the eye today. Similar erratic variation patterns are encountered, under quite different, insular circumstances in *Ch. lineata,* and were discovered in *Ch. flexuosa* and *Ch. serpens* by the earliest botanical travellers in Brazil, long before any significant modification of the primeval vegetation had taken place. A student of tropical *Chamaecrista* has the choice of accepting as a fact of life that sort of intraspecific variability that inspired the dissection of *Ch. fasciculata* or of retreating to the position of Britton & Urban who, driven by logic rather than sense, found a species in every particular combination of gland and hairiness that fell in their way. But in passing over the observed variation as taxonomically insignificant it is well to bear in mind that its cause and its biological significance remain a mystery.

Each of the named segregates from *Ch. fasciculata* listed in the synonymy was based on a particular plant or small populational sample the majority of which, like figures in an oriental carpet, had a particular individuality even while repeating a stated theme. This account would be incomplete without a record of their real or supposed differential characters:

i. *Cassia chamaecrista* of herb. Linn.—Virgate, incurved-puberulent; petiolar gland obconic ±0.8 mm diam; leaflets to 10 pairs; long petals to 14 mm.

ii. *C. triflora* Jacqu.—Apparently exactly as the preceding, except leaflets 6–8 pairs. A plant cultivated at Wien, not seen by us, the low leaflet number suggesting *C. lineata,* but said to be monocarpic.

iii. *C. fasciculata* Michx.—Exactly = 1. Michaux (or Richard) interpreted *C. chamaecrista* as a larger-flowered form, either puberulent or hirsute (hb. Michx., P!).

iv. fma. *pubicaulis* O. Kze.—Diffuse, the stems and lf-stalks densely pilosulous; petiolar gland tack-shaped, ±0.4 mm diam; lfts to 9 pairs, glabrous ciliolate.

v. fma. *hirsuta* O. Kze.—Probably virgate, the stems, lf-stalks, pedicels and ovary spreading-pilose; gland squatly drum-shaped ±0.6 mm diam; lfts 9–10 pairs, glabrous ciliolate.

vi. var. *robusta* Poll.—Exactly as v. except the glands substipitate saucer-shaped, 0.8–1.5 mm diam and lfts to 11 pairs.

vii. *C. mississippiensis* Poll.—Slender, diffusely branched from base, the stems, lf-stalks and dorsal face of lfts puberulent; gland either short-stipitate or subsessile, 0.3–0.5 mm diam; lfts 8–13 pairs, very small, to 4.5–5.5 mm, few-veined (secondaries 2–3 pairs); fls small, the long petal ±10 mm.

viii. *C. depressa* Poll.—Slender, diffusely branched from base, the stems thinly pilosulous and incurved-puberulent in lines, the foliage except for lf-stalk glabrous; gland subsessile, 0.45–0.65 mm diam; lfts up to 8–11 pairs; long petal ±14–16 mm.

ix. *Ch. bellula* Poll.—As i, except stems freely branching from below middle due to injury of primary axis.

x. *Ch. littoralis* Poll.—Diffusely branched from base, the stems puberulent but pedicels and ovary pilose, the lfts up to 13 pairs, puberulent on both faces; gland cupular-obconic 0.5–0.7 mm diam; long petal ±12.5 mm.

xi. *Ch. brachiata* Poll.—Tall erect branched distally, glabrous except for minutely puberulent ovary; gland sessile scutellate 0.9–1.1 mm diam; lfts to 13 pairs; long petal to (?)13.5 mm.—Described from an ageing plant, still flowering in April.

xii. *Ch. tracyi* Poll.—Puberulent throughout, the lf-stalk and ovary in addition subappressed-pilosulous; otherwise as ix.

xiii. *C. camporum* Greene—Virgate, subglabrous except for minutely ciliolate lfts and pilosulous sepals and pod; lfts ample 9–13 pairs.

xiv. *C. puberula* Greene—Puberulent throughout, freely branched upward, the attitude of stem not obvious from the specimens; gland shortly stipitate, 0.45–0.8 mm diam; lfts to ±11 pairs, the rather strongly displaced midrib together with pedicels only ±7–8 mm long in fruit suggesting passage into *Ch. chamaecristoides*.

xv. *Ch. rostrata* Woot. & Standl.—Erect, simple or few-branched, juvenile plants in fl and fr at 1–1.5 dm, except for glabrous lfts puberulent almost throughout; gland sessile or substipitate, 0.35–0.5 mm diam; lvs all short and of juvenile type, up to 2–3 cm and with only 6–8 pairs of lfts; long petal ±11–14 mm; style sometimes subpersistent, this suggesting the epithet.

xvi. fma. *jenseni* Palmer & Steyerm.—Not seen, but evidently a rare mutant with whitish petals, seen by the senior author in Texas.

xvii. var. *macrosperma* Fern.—Habit, vesture, gland, and lfts of vi, but pod ±7.5 mm wide.

xviii. fma. *transmutata* Fern.—A teratological form in which the raceme becomes densely branched like a miniature witch's-broom and the flower parts are transformed into bracteiform blades. Similar states, some less fully realized, are known in *Ch. nictitans* var. *pilosa* (*Irwin 2370*) and subsp. *brachypoda* (*Irwin et al. 16500*), both from Brazil.

xix. fma. *mutata* Fern.—A bizarre freak in which the axis of the leaf is contracted and the lfts reduced to 3 pairs, the distal pair completely fused into an entire blade, the leaf becoming in consequence imparipinnate 5-foliolate. An apparently analogous form from Hispaniola is described under *Ch. pedicellaris* var. *adenosperma*. Only a single individual plant of fma. *mutata* was found.

51. Chamaecrista nictitans (Linnaeus) Moench, Meth. pl. hort. bot. Marburg. 272. 1794, quoad descr., exclus. syn., hic magnopere ampliata.—*Cassia nictitans* Linnaeus, Sp. Pl. 380. 1753.—Typus infra sub var. *nictitanti* indicatur.

Monocarpic (under special circumstances over-wintering and weakly lignescent basally but never durably perennial or fruticose); peduncles adnate to stem, the pedicels appearing supra-axillary; all other phenetic characters labile and subject to wide ranges of variation.

We have brought together here into one multiracial *Ch. nictitans* all but one of the American monocarpic *Chamaecristae Mimosoideae* which, as inferentially predicted by Bentham a century ago, are found to form a continuous fabric of variation that yields only very reluctantly to taxonomic analysis. Allowance being made for conspicuous individual variation in habit, vesture, glands, and flower-size, the constituent races of *Ch. nictitans* share with *Ch. glandulosa* sens. lat. and *Ch. fasciculata* a monotonous sort of common facies which points, in our judgment, to a very close relationship among all and a presumably recent, doubtless continuing differentiation from common antecedents. The species is here arbitrarily separated off from *Ch. glandulosa* by monocarpic duration, which may well have arisen independently from several fruticose immediate ancestors, so that we cannot be sure that *Ch. nictitans* is a phyletically pure assembly. The one other American monocarpic ally already mentioned as excluded from our comprehensive concept of *Ch. nictitans* is, of course, *Ch. fasciculata* itself. While this can be distinguished by multiple characters from similar, distantly allopatric, large-flowered, longistylous *Ch. nictitans* subsp. *disadena* and subsp. *brachypoda,* it is no longer separable from the whole of *Ch. nictitans* by any single morphological difference. However, since *Ch. fasciculata* is endemic to the United States and sympatric only with small-flowered brachystylous elements of subsp. *nictitans,* the objections to its exclusion will be more theoretical than practical.

Key to the Subspecies of *Ch. nictitans*

1. Androecium 10-merous (if petiolar glands taller than diam of head here coincide with pedicels only 1–6 mm, see subsp. *nictitans* var. *diffusa*); West Indies and centr. Mexico to Argentina, not in U.S.
    2. Petiolar gland sessile (scutellate to obconic or squatly drum-shaped) or if stipitate only shortly so and in profile squatly tack-shaped, not taller than diam of head; pedicels 1.5–10 mm.
        3. Fls relatively ample and dolichostylous, the longer abaxial petal (8–)9–16 mm coinciding with style 4–7 mm; lfts of adult lvs 6–12(–14) pairs. Brazilian Planalto and adjac. Paraguay and Bolivia. 51/I. subsp. *brachypoda* (p. 812).
        3. Fls commonly both smaller and brachystylous, with longer abaxial petal less than 8 mm or style less than 2.5 mm or both; but if petal, style and dispersal in agreement with the preceding then the lfts 20–40 pairs. 51/II. subsp. *patellaria* (p. 814).
    2. Petiolar gland stipitate, in profile urceolate, tack- or pin-shaped, as tall or taller than diam of head; pedicels up to 6–14 mm. 51/III. subsp. *disadena* (p. 825).
1. Androecium 2–8(–9)-merous. Petiolar gland mostly as long or longer than diam of head, sometimes shorter only in var. *nictitans* of e. U.S.; fls always short-pedicelled, small, and brachystylous, the pedicels 1–6 mm, the long adaxial petal 3.5–8(–9) mm, the style 0.8–2 mm. 51/IV. subsp. *nictitans* (p. 832).

51/I. Chamaecrista nictitans (Linnaeus) Moench subsp. **brachypoda** (Bentham) Irwin & Barneby, stat. nov. *Cassia brachypoda* Bentham in Martius, Fl. Bras. **15**(2): 172. 1870.—"Habitat in prov. Minas Geraës campis ad S. João del Rey et versus Rio S. Francisco: Martius; ad Caxoeira do Campo: Claussen; in eadem provincia: Weddell; ad Montes Claros: Pohl; in prov. Goyaz collibus siccis ad Conceição et in virgultosis prope Arrayas: Gardner n. 3684, 3686.; ad Goyaz et Porto Imperial ad fl. Tocantins: Burchell. Crescit etiam in Boliviae prov. Chiquitos: D'Orbigny."—Lectoholotypus, *Gardner 3686,* collected at Arrayas, II.1840, fl, fr, K (hb. Benth.)! = NY Neg. *9190*; isotypus, K (hb. Hook.)! = IAP Neg. *1129*; paratypi, omnes manu Benth. determinati: *Weddell 1056,* G, P; *Claussen 14.155,* K; *Pohl 1116,* K! = NY Neg.

1540; *Burchell 6686, 8774,* K! *Martius s.n.* e S. João d'El Rey, M!—*D'Orbigny s.n.* e Bolivia, P, = *Ch. repens.*

Relatively coarse, amply leafy, monocarpic, with usually only 1 erect stem at anthesis (4–)5–16(–20) dm, simple or strictly few-branched near or beyond middle, except for often glabrous upper face of lfts puberulent throughout with forwardly or incumbently incurved hairs to 0.3–0.9 mm, the stems and lf-stalks sometimes in addition finely, sometimes densely pilose with widely ascending or horizontal setae to 1–2.4 mm, the foliage dull olivaceous concolorous, the lfts usually closely lineolate-venulose.

Stipules erect, narrowly lance-acuminate (6–)7–17 mm, at oblique base (1–)1.2–2.3(–2.5) mm wide, the blades becoming thinly papery but persistent, from base (7–)8–13(–15)-nerved, always pubescent or pilosulous dorsally.

Major cauline lvs to (4–)4.5–9(–10) cm, short-petiolate, the expanded blade oblong or ovate-oblong in outline; petiole including pulvinus (3–)4–9(–10) mm, above the gland open-sulcate, narrowly margined, 0.4–0.8 mm diam; petiolar glands 1(–2), situated near or above middle of petiole proper, usually sessile, scutellate or shallowly patelliform but highly variable in size and outline, (0.5–)0.7–3(–3.6) mm diam, when less than 1 mm diam round, when longer elliptic, locally (especially in Mato Grosso) smaller, tack-shaped or turbinate and only 0.3–0.5 mm diam, 0.3–0.6 mm tall; often additional 1–several smaller tack-shaped or urceolate glands on distal one or all segments of rachis; lfts of major lvs 6–12(–14) pairs, well-spaced along rachis, in outline oblong, oblong-elliptic, rarely ovate-oblong, or the distal pair sometimes obliquely obovate, commonly obtuse mucronulate or aristulate, rarely deltately subacute, the longer of each plant to 12–26 × 3.5–9 mm, at base broadly cordate on proximal and cuneate to rounded on distal side, from base (4–)5–7(–9)-nerved by straight midrib with on its wide side 3–5, on its narrow one 1(–2) short incurved-ascending primaries, the subcentric or moderately excentric costa dividing the blade 1:(1–)1.1–1.75, penniveined with 5–11(–12) pairs of major secondary venules and often as many prominulous intercalary ones, when all venulation prominulous the blades becoming closely lineolate, the venation commonly prominulous both above and below but often more sharply so beneath.

Peduncles 6–30 mm, adnate to stem for 5.5–28 mm, (1–)2–4(–5)-fld; pedicels at anthesis 3–9(–11) mm, in fruit incurved-erect 4–12(–15) mm, bracteolate 1.5–3 mm below calyx; bracteoles lance-acuminate 2.5–6 mm; buds ovoid-acuminate pilosulous; sepals often fuscous or reddish externally, lance-acuminate to ovate-acute, the longer ones 8.5–14.5 mm; petals yellow, often not strongly differentiated, the 3 adaxial (or all but the cucullus) obovate-cuneate to elliptic-oblanceolate, (7–)8–15 × 5–8 mm, the longer abaxial petal up to 8–14(–16) mm, the cucullus obliquely semi-obovate or broadly falcate-oblanceolate 7.5–14 × 7–13 mm; stamens 10, the long anthers to 6–9.5 mm; ovary densely appressed-pilosulous; style linear or almost so, (3–)4–7 × 0.3–0.5 mm; ovules (9–)11–21.

Pod erect linear, usually slightly curved outward, (30–)35–65(–78) × 4–5(–5.5) mm, the green, reddish or atrocastaneous, finally nigrescent valves finely pilosulous or (especially along sutures) setose with spreading-ascending, rarely incurved hairs up to 0.4–1.4(–2.1) mm; seeds 2.6–3.7 mm, the testa fuscous or black, dull to moderately lustrous, lineolate-pitted.—Collections: 67.

Thickets in cerrado, sandy, sometimes seasonally flooded campo, and becoming weedy in disturbed cerradão, on river-banks, in rough pastures and along highways, (200–)400–1100 m, widespread over the Brazilian Planalto from the middle Tocantins valley near 8°S in s.-e. Pará s. and s.-e. to the headwaters of Rio São Francisco in s.-centr. Minas Gerais and the valley of Rio Tieté in centr. São Paulo, s.-w. and w. through Sa. do Roncador and Sa. de Maracajú to extreme s.-e. Mato Grosso and adjacent Paraguay; and interruptedly to s.-e. Bolivia (near 62°W in prov. Santa Cruz).—Fl. X–VI(–VII).

Of all forms of *Ch. nictitans* found to the south of the Amazonian Hylaea, subsp. *brachypoda* is perhaps the most easily recognized by a combination of one simple or weakly branched stem rising wandlike from a monocarpic root, dorsally pubescent stipules, in the largest leaves only few (to 12, rarely 14) pairs of ample, subsymmetrically veiny leaflets, and longistyled, in context of the species as a whole, large flowers. In habit and flower-size it has everything in common with sympatric large-flowered races of subsp. *patellaria* var. *paraguariensis* which, however, differs in its externally glabrous stipules and in comparable adult leaves much more numerous (at least 20, and up to 40) pairs of leaflets. In foliage it resembles *Ch. repens* var. *repens,* found immediately to the south and west and perhaps marginally sympatric around this periphery of its range; but var. *repens* has several, shorter stems arising from a perennial root, again dorsally glabrous stipules, and a petiolar gland raised on a pediment wider than itself. *Chamaecrista venturiana,* hardly different from *Ch. repens* in life-form and stipules, has the gland almost of subsp. *brachypoda* but once again fewer and in addition more oblique and more simply venulose leaflets.

The petiolar gland of subsp. *brachypoda* is extraordinarily polymorphic. Most commonly it takes the same form as that of var. *paraguariensis,* that is a shield or shallow dish, round when small but often greatly dilated lengthwise, and either strictly sessile within the expanded sulcus of the petiole or resting as it were directly on the petiolar wings, no stipe being visible in profile. In south-central Minas Gerais and again far distant in east-central and southern Mato Grosso, the gland abruptly becomes much smaller, turbinate or tack-shaped and about as broad as tall, sometimes distinctly even though shortly stipitate, and especially in Sa. do Roncador in Mato Grosso is accompanied on one or all segments of the leaf-rachis by further glands of similar form. Size and number of glands are not correlated with any other morphological feature so that here, as in the case of *Ch. glandulosa,* we can derive from them no taxonomic benefits or insights. Most of the populations of subsp. *brachypoda* known from Sa. do Roncador and vicinity are further peculiar for a relatively dense and long setose pubescence not equalled elsewhere in the range of the subspecies. One of these (*Irwin et al. 16500*) has bizarre teratological flowers with many perianth and some androecial members transformed into leaflike and stemlike structures.

51/II. Chamaecrista nictitans Linnaeus subsp. **patellaria** (Colladon) Irwin & Barneby, comb. nov. *Cassia patellaria* DeCandolle ex Colladon, Hist. Casses 125, t. 16. 1816, sens. ampliat.—Typus infra sub var. *ramosa* citatur.—*C. glandulosa* subsp. ( ) *patellaria* (Colladon [''DeCandolle'']) O. Kuntze, Rev. Gen. 169. 1891.

Fl short-pedicellate, small and brachystylous except in some races of var. *paraguariensis,* the pedicels 1.5–10(–11) mm, the longest petal mostly 3.5–10 mm and style 0.8–2.2 mm (in var. *paraguariensis* up to 14 mm and 4 mm respectively); androecium (9–)10-merous; petiolar gland sessile or almost so, scutellate, shallowly patelliform, squatly tub- or drum-shaped, or turbinate-obconic, if incipiently stipitate the stipe stout, the whole in profile shorter than diam of head.—Ubiquitous through the Neotropics.

We have assembled under the title of subsp. *patellaria* a series of short-pedicelled, small-flowered, brachystylous and decandrous forms which run parallel, as it were, to subsp. *nictitans,* but differ in the sessile or subsessile glands. The gland-type is not, unfortunately, an absolute criterion, for a gland in no respect different from that of subsp. *patellaria* var. *glabrata* is known in subsp. *nictitans* var. *nictitans*; but in the latter case associated with a degraded androecium of less than ten members. Moreover where the ranges of the two subspecies interfinger or coincide the difference in the glands is pronounced. While small-flowered forms of var. *paraguariensis* cannot well be divorced from subsp. *patellaria,* its larger-flowered ones are inconveniently similar to sympatric Brazilian subsp. *brachypoda,* differing principally in high leaflet-number. The var. *ramosa* is unusual in the species as a whole for the almost always emphatically displaced midrib of the leaflets, but in reality we find a complete series of intermediate states between the narrow, truly dimidiate leaflet of this variety in its most pronounced form and a wider, merely asymmetric leaflet usual in var. *glabrata*.

Key to the Varieties of subsp. *patellaria*

1. Midrib of lfts centric or moderately displaced, dividing the blade 1:1–2.2; style linear or almost so.
    2. Adult lvs up to 7–15(–21) cm, with up to (19–)20–38(–44) pairs of subsymmetric lfts, their midrib centric or almost so; secondary venulation from midrib of 4–9 major with commonly well defined intercalary venules. Vesture never setose or pilosulous, of short appressed or incurved hairs only. 51/IIa. var. *paraguariensis* (p. 815).
    2. Adult lvs commonly up to 3–9 cm with up to 10–26 pairs of at least moderately asymmetric lfts, their midrib displaced to divide the blade ±1:2; secondary venulation from midrib of 2–6 pairs of major with faint or 0 intercalary venules. Vesture highly variable, pilose, setose, puberulent, or 0.
        3. Sutures of pod wingless or essentially so, rarely produced along each side into an obscure flange up to 0.4 mm wide appressed to the face of the valve; widespread through West Indies, local in s. Mexico and n.-e. Brazil (Pará), a scattered weed in S. America, Hawaii, and Old World Tropics. 51/IIb. var. *glabrata* (p. 822).
        3. Sutures of pod winged lengthwise on each side, the herbaceous, finally papery wings at maturity 0.75–1.6 mm wide; local in s. Mexico, Cuba, Trinidad, and lower Orinoco and Amazon valleys in Venezuela and Brazil. 51/IIc. var. *praetexta* (p. 821).
1. Midrib of lfts very strongly displaced or submarginal, dividing the blade 1:2.5+; style dilated distally, obliquely truncate at the stigma. 51/IId. var. *ramosa* (p. 818).

51/IIa. Chamaecrista nictitans (Linnaeus) Moench subsp. **patellaria** (Colladon) var. **paraguariensis** (Chodat & Hassler) Irwin & Barneby, comb. nov. *Cassia flavicoma* var. *paraguariensis* Chodat & Hassler in Bull. Herb. Boiss. II, **4**: 827. 1904.—"[Paraguay:] . . . in campo pr. Igatimí, Nov., [*Hassler*] n. 5420."—Holotypus, G! = NY Neg. *9196*; isotypi, K, P (2 sheets).

(?)*Cassia gemina* Vellozo, Fl. Flum. Ic. **4**: t. 74. 1827 & p. 161. 1881.—"Habitat . . . campis, seu fruticetis excultis maritimis ad Sacellum C. Antonii Joary dictum."—Holotypus, the cited plate!

Cassia hypnotica Vellozo, Fl. Flum. Ic. **4**: t. 75. 1827 & p. 161. 1881.—"Habitat eodem loco." i.e. as the preceding.—Holotypus, the cited plate!—*C. chamaecrista* var. *hypnotica* (Vellozo) N. Mattos, Loefgrenia **70**: 2. 1976.

Chamaecrista simplex Standley, Contrib. U.S. Nat. Herb. **18**: 103. 1916.—"Type in the [US] no. 679815, collected in the Sabana de Dormisolo, near Chepo. Province of Panama, Panama . . . October, 1911, by H. Pittier (no. 4655)."—Holotypus, US! isotypi, NY (2 sheets)!—*Cassia simplex* (Standley) Standley ex Pittier, Cat. Fl. Venezol. 374. 1945, pro syn.

Chamaecrista browniana Britton & Rose, N. Amer. Fl. **23**(5): 293. 1930.—"Type from between Quetame and Susumuco, Cundinamarca, Colombia, August 24, 1917, *Pennell 1351* . . ."—Holotypus, NY! = NY Neg. *4960*; clastotypus, US!—*Cassia pennelliana* Amshoff, Meded.

Bot. Mus. & Herb. Rijks. Univ. Utrecht **52**: 28. 1939. *C. browniana* (Britton & Rose) Philipson, Mutisia **24**: 2. 1956, homon. poster. (non *C. browniana* Kunth, 1822–1824).

Erect, obligately monocarpic herbs at anthesis (2.5–)3.5–15("–25") dm, commonly simple wandlike, sometimes (when robust) strictly few-branched distally, branched at base only when cut or injured, the stems and lf-stalks strigulose with subappressed, incumbent, or straighter, narrowly ascending hairs up to 0.25–0.7(–0.8) mm, nowhere setose, the concolorous foliage except for minutely scabrid-ciliolate lfts glabrous, only the pod loosely pilosulous, exceptionally glabrous throughout.

Stipules erect, lance-acuminate, those associated with larger adult lvs and flowers (6.5–)7.5–21 mm, at subsymmetric base (1.2–)1.5–3 mm wide, from base strongly 10–17-nerved, the blades glabrous dorsally, minutely ciliolate.

Lvs ascending and often incurved, heteromorphic, the early ones relatively short and simple, those (alone described hereafter) at middle and fertile nodes up to (5.5–)7–15(–21) cm, short-petioled, the expanded blade in outline narrowly lance-elliptic to linear-oblong; petiole including pulvinus (4–)4.5–9.5(–14) mm, above the gland narrowly coarsely margined, openly sulcate and 0.5–1.2 mm diam; glands 1–2, situated near middle of petiole proper, scutellate or patelliform sessile, in outline nearly always elliptic longer than wide, but when small subcircular, the one or the larger of a pair up to (0.7–)0.9–3.2(–4) mm diam, in profile always much wider than tall; lfts up to (19–)20–38(–44) pairs well spaced along rachis, in outline narrowly oblong, narrowly oblong-elliptic, or ovate-elliptic, obtuse mucronate or subacuminulate, the largest of larger lvs of a given plant up to (5.5–)6.5–20(–22) × 1.3–4.6 mm, at base angulately cordate on the broad and cuneate to rounded on the narrow side, from base 6–9-nerved by the costa with on the broad side 4–6, on the narrow one 1–2 primary nerves, the costa thence penniveined, with 4–9 major pairs of secondaries and commonly but not always as many intercalary ones, the latter varying from faint or subobsolete (externally) to almost as strong as the major set, the venulation therefore appearing either laxly or closely parallel-lineolate, usually prominulous on both faces but often more sharply so and pallid beneath.

Peduncles at most nodes paired, the inner (4–)6–24 mm, adnate its whole length, mostly 2–4, rarely 5(–6)-fld, the exterior one very short or obsolete, becoming axillary, in some slender individuals represented by mere rudiments of bracts; pedicels at anthesis 2.5–6(–9) mm, in fruit ascending, thickened 3.5–8(–11) mm, bracteolate 1–2.5 mm below calyx; bracteoles ovate- or lance-acuminate (1.5–)2–5 mm; buds slenderly ovoid-acuminate, especially along midrib of outer sepals thinly strigulose-pilosulous, rarely glabrous; sepals ovate- and lance-acuminate up to (4.5–)6–14.5 mm; petals yellow, seldom widely expanding, all very variable in length and amplitude, the 3 adaxial oblanceolate to broadly obovate-cuneate 5.5–15 × 4.5–8 mm, the long abaxial one either similar to the last or longer, broadly oblanceolate to flabellate (5.5–)6.5–14(–16) mm, the cucullus obliquely obovate or transversely dilated-reniform, usually a little shorter than the last, 5.5–13(–14) × 5–10 mm; long anthers (4.4–)5.5–9.5 mm; ovary densely strigulose-pilosulous, exceptionally glabrous; style linear or when short slightly dilated distally, never very strongly incurved, 0.8–4(–5.5) × 0.3–0.55 mm; ovules (9–)11–22.

Pod erect, straight or slightly curved outward, (27–)30–60(–66) × 3.5–5.2 mm, the castaneous or purplish-red, finally nigrescent valves densely pilosulous with fine ascending hairs; seeds 2.7–3.3 mm, the fully ripe testa sublustrous, fuscous black-pitted or soot-black overall, the pits shallow, sometimes faint.—Collections: 95.—Fig. 36 (petiolar nectary).

Seasonally wet grasslands (savanna, campo, llano), thickets in scrubland, disturbed woods, cerrado, sometimes becoming weedy in hedges, roadsides, fallow fields and orchards, in Central America ascending into the pine-oak belt, mostly between 450 and 1000 m, but in Panama and on the llanos of e. Colombia at 20–450 m, in the inter-Andine valleys of Colombia to 1500–1700 m, and in s.-e. Brazil to 1200 m, discontinuously widespread both n. and s. of the Amazonian Hylaea: s. Mexico (Tabasco, Guerrero, Oaxaca, Chiapas) to Colombia (s. in Cauca valley to Cauca, in Magdalena valley to Tolima, e. to Int. Meta) and rarely to Venezuela (Cord. Costanera in Aragua and D.F.; Gran Sabana in Bolívar); disjunctly on the Brazilian Planalto from s. Ceará s., between Rios São Francisco and Xingú, to w. Minas Gerais, centr. and w. São Paulo, and w. Paraná, thence e. rarely into Rio de Janeiro and commonly w. through s. Mato Grosso to e.-centr. Paraguay and the Yungas in trans-Andine Bolivia.—Fl. n. of the Equator mostly VI–II(–IV), to the s. mostly XI–V.

We assemble here under the name of *Ch. nictitans* var. *paraguariensis* all those obligately monocarpic *Chamaecristae Mimosoideae* characterized by simple wandlike stems, sessile patelliform petiolar glands, narrowly elongate multifoliolate adult leaves and short-pedicelled flowers, which have passed in Central America and Colombia as either *C. simplex* or *Ch. browniana* (=*C. pennelliana*) and in Brazil either as *C. subtriflora* (which really = *Ch. repens multijuga*) or some variety of *C. chamaecrista* or *C. flavicoma*. There is little new in our consolidation of names, for Woodson & Schery (1951, p. 62) have already recognized the taxonomic synonymy of *C. simplex,* described from lowland Panama, and *C. pennelliana,* described from middle elevations in Colombia and El Salvador; while Britton long ago (in annotation, NY) admitted the latter to the flora of Bolivia in the shape of *Bang 549* and *Rusby 2854,* plants certainly conspecific with the typus of Paraguayan *C. flavicoma* var. *paraguariensis*. Thus the chain of synonymies, established in advance, only requires positive assertion. For want of type specimens some residual doubt clings to the identity of *C. gemina* and *C. hypnotica* Vell., one of which we believe to be the oldest of all applicable names at the specific level, the more so because the plant has been found only once (*W. Hoehne 6007,* NY) in modern times on the Atlantic slope of southern Brazil and was probably therefore not native "*in excultis maritimis*" within the range of *Flora Fluminensis*. But since, like so many other chamaecristas, var. *paraguariensis* is often at least potentially weedy, considerations of range do not weigh heavily here. Vellozo's plate of *C. hypnotica* agrees in all details with large-flowered Brazilian forms of var. *paraguariensis,* showing the annual root, the simple stem, the sessile glands, the many (to 22 pairs) small ovate-oblong leaflets with subcentric, penniveined costa, and the short pedicels that together characterize no other chamaecrista known to us, certainly none of the three taxa (*Ch. nictitans* subsp. *patellaria* var. *ramosa,* subsp. *disadena* var. *pilosa* and *Ch. glandulosa* var. *brasiliensis*) encountered today in coastal Rio de Janeiro or Guanabara. Bentham's reduction of *C. hypnotica* to his own *C. chamaecrista* var. *brasiliensis* (=*Ch. glandulosa brasiliensis*), a shrub with no more than 12 pairs of leaflets in the largest leaves, is difficult to comprehend.

Multiple similarities in habit and morphological detail point to a very close relationship between var. *paraguariensis,* var. *glabrata* and subsp. *brachypoda*. The latter, identical in habit, differs essentially only in the dorsally pubescent stipules and the absolutely fewer (at most 14, not 20–40) pairs of leaflets, which tend to be larger though identical in venulation. On the other hand it is the more excentric midrib and simpler secondary venulation of the leaflets that alone ultimately distinguishes var. *glabrata*. The petiolar glands of each of these three related cassias are quite variable in size and offer no firm differential characters. The coarse multifoliolate (sometimes polyploid) race of var. *glabrata* found in the Amazon delta region and described by Bentham as *C. patellaria* var. *longifolia* resembles some individual small-flowered var. *paraguariensis* in everything but the leaflet venation. The middle Amazonian glabrate form of *Ch. nictitans disadena* is another habital doppelganger for var. *paraguariensis* but nevertheless recognized at once by the venulation and the coarsely turbinate, not shield- or dish-shaped petiolar glands.

The perianth of var. *paraguariensis* is extremely unstable in size, the petals together with long anthers and styles varying together in what appears to be a continuous series only imperfectly correlated with geographical dispersal and, furthermore, subject to seasonal modification on a given plant. The largest and smallest flowers are not sympatric, however. The former, with styles 2–4(–5.5) mm long, occur only on the Brazilian Planalto, within the range of the related subsp. *brachypoda,* from which it seems possible that genes for large, outcrossing flowers may have been derived. The smallest, probably autogamous flowers, with styles only 0.8–1.7 mm long, often associated with the slender habit of growth and small leaflets which are the sole peculiarities of the ecotype known as *Cassia simplex,* are found only northward from Guayana and Colombia. But flowers of moderate size have been encountered at both extremities of the variety's range, in Mexico and in southern Brazil and Paraguay. A noteworthy character of var. *paraguariensis,* very unusual in *Ch. nictitans* sens. lat., is the relative uniformity in vesture, the setose type of trichome being quite suppressed. A minor variant (*Anderson 11330,* CURI, NY), completely hairless even to the ovary, superficially resembles the glabrate Cuban form of var. *glabrata* segregated by Britton as *Ch. glaberrima*.

51/IIb. Chamaecrista nictitans (Linnaeus) Moench subsp. **patellaria** (Colladon) var. **ramosa** (Vogel) Irwin & Barneby, comb. nov. *Cassia patellaria β ramosa* Vogel, Syn. Gen. Cass. 66. 1837.—"In Brasilia: Sellow leg. pr. Rio [de] Jan[eiro]."—Holotypus, presumably †B; neoholotypus, *Sellow s.n.,* K (ex hb. berol.)! isotypi, P, W!

Cassia patellaria DeCandolle ex Colladon, Hist. Casses 125, t. 16. 1816.—"Hab. in Cayenna et ad Isthmum Panama."—Lectoholotypus (Britton, 1917, p. 10, by reference to type-locality), from Cayenne, Thibaud, 1815, G-DC! = F Neg. *7002*; paratypus, mounted with the preceding, from Panama, comm. Lagasca, 1807, no. *39,* G-DC!—*Chamaecrista patellaria* (DeCandolle) Greene, Pittonia **4**: 32. 1899.

Cassia biglandulosa Bertoloni, Opuscoli Sci. Bologna **3**: 409, t. XIX, fig. 1. 1819.—"Habitat in provincia *di Rio Janeiro* Brasiliae. Floruit in horto botanico Bononiensi . . ."—No typus known to survive, but the elaborate protologue decisive.

Cassia patellaria ? var. *longifolia* Bentham in Martius, Fl. Bras. **15**(2): 175. 1870.—". . . in prov. Paraensi ad Serra de Majuré prope Santarem: Spruce."—Holotypus, *Spruce 771**, K (hb. Benth., adnotavit Sandwith.)! = NY Neg. *9189*; isotypus, P!

Chamaecrista villosissima Britton & Rose, N. Amer. Fl. **23**(5): 290. 1930.—"Table land about Ocuilapa, Chiapas, August 21, 1895, *E. W. Nelson 3014*."—Holotypus, NY! isotypi, GH, US!—*Cassia villosissima* (Britton & Rose) Lundell, Phytologia **1**: 215. 1937.

Cassia patellaria sensu Vogel, 1837, p. 66, exclus. var. *glabrata*; Bentham, 1870, p. 174 & 1871, p. 578, exclus. var. *Chamaecrista patellaria* sensu Britton & Rose, 1930, p. 294.

Monocarpic herbs of rapid maturation but sometimes of more than one season's duration, usually coarsely and abundantly leafy, highly variable in habit and adult stature, with either one stem simple erect and wandlike, one erect and branched distally, or branched from the earliest lf-axils and the lateral stems then incurved-ascending, at early anthesis (1–)1.5 dm upward, in fruit to 1.1(–1.6) m, pubescent usually almost throughout with short incurved and (especially lf-stalks and stems distally) longer widely spreading or spreading-incurved setae up to 0.6–1.3(–1.6) mm, the lfts commonly pilosulous on both faces and the hairs dorsally arising from minutely pustular base (but these sometimes filiform), the pubescence often lacking from 1 or both faces and persistent only as a minute ciliolation of the margin, the venation prominulous only beneath, the foliage dull olivaceous concolorous, the very small, fleeting fls supra-axillary, short-pedicellate.

Stipules erect persistent lanceolate or lance-acuminate or -caudate, except for some small lowest ones (4–)5–14(–16) × (0.6–)1.2–2 mm, at base rounded and slightly contracted on contrapetiolar side, the blades herbaceous, prominently (5–)7–11-nerved.

Lvs, disregarding some lowest short ones, (2–)2.5–8(–8.5) cm, shortly petioled, the stout lf-stalk usually incurved-ascending, the expanded blade ovate or broadly lanceolate in outline; petiole including the scarcely dilated pulvinus 3–6(–7) mm, openly sulcate-marginate; petiolar glands 1–2, if 2 then the distal commonly larger, this situated immediately below first pair of lfts, the second near or below middle of petiole proper, in outline orbicular or elliptic, plane and scutellate or by involution of the thick, pallid or rubescent margin becoming urceolate, either quite sessile or raised on a pediment or stout stipe to 0.4 mm, becoming then either shallowly tub-shaped or shortly trumpet-shaped or pileiform; rachis ±2–7 cm, the interfoliolar segments narrowly sulcate-marginate, the margins contracted at insertion of lfts but not upwardly dilated; lfts of adult lvs 10–23, of some larger lvs at least 12 pairs, in outline linear-oblong or obscurely linear-oblanceolate, the small subterminal ones often dimidiately narrow-obovate, straight or subfalcate, at apex obtuse or acute but aristate by excurrent costa, at base on proximal side obtusely angulate-auriculate, the midrib displaced by ±⅓–¼ width of blade, slender but sharply prominulous beneath, stouter than the 2–3 weak basal nerves on its broad side, on narrow side faintly or obsoletely penninerved, on the broad side often more strongly so.

Peduncles mostly 4–22(–27) mm, adnate almost their whole length to stem, solitary or commonly paired, the outer then much shorter, sometimes subaxillary, (1–)2–3(–4)-fld; pedicels at anthesis 0.5–3.5 mm, in fruit erect thickened 2–5.5(–6) mm, bracteolate near or above middle; bracteoles lance-subulate, 1.5–3 mm, persistent; buds slenderly ovoid-acuminate, at least thinly pilosulous; sepals submembranous greenish ovate- or lance-acuminate to 4–6.2 mm; petals yellow, fading orange by noon of first day, unequal, the 3 adaxial smaller obovate to spatulate 2.6–4.5 mm, the 2 abaxial larger, 1 subsymmetrically obovate, 3.6–5.5 mm, incurved over the gynoecium and its associated long anther, the cucullus of ± same length but oblique; stamens 9–10, the inner whorl always small, with short or abortive anthers, at least 2 of the outer whorl larger, with anthers (2.4–)2.5–3.8 mm; ovary densely white-pilose with appressed or ascending hairs; style 0.8–1.4(–1.6) mm, strongly incurved, ± enlarged upward and obliquely truncate at the stigma; ovules 8–13.

Pod erect, linear-oblong, straight or slightly incurved, 2–4.5 × 0.3–0.42 cm, the valves brownish or deep reddish-purplish-tinged, in age nigrescent, thinly or densely pilosulous; seeds subquadrate 2.1–3.2(–3.4) mm, the testa fuscous overall, fuscous beyond the pallid base, or ochraceous and fuscous-lineolate, shallow-pitted, dull, mucilaginous when wetted.—Collections: 301.—Fig. 2 (petiolar nectary).

Native probably of both upland and seasonally flooded lowland savannas, llanos, campos, and of exposed unsettled habitats along rivers and shores, now colonial in disturbed or cut-over woodland, along highways, in suburban and pastoral waste places, and a weed of gardens and plantations (cafetal, cocal, bananal), mostly between sea level and 600 m but ascending in Mexico to 1100, in Venezuela to 1300, in s.-e. Brazil to 1550, in Colombia to 1600, and in Bolivia to 1700 m, interruptedly dispersed over much of the Neotropical zone e. of the Andes and outside the Amazonian Hylaea: Cuba (all provinces) and Jamaica; Gulf slope and coastal plain of Mexico from extreme s. San Luis Potosí to Tabasco and n. Chiapas; Belize; Costa Rica and Panama; very abundant in the Cauca and Magdalena valleys in Colombia, from Magdalena s. to Cauca and Huila, in Valle de Cauca extending to the Pacific coast (Buenaventura) and in Boyacá and Meta to the trans-Andine llanos; local in Venezuela (Maracaibo Basin, Cord. Costanera, one record each from Guárico and Bolívar) and the Guianas; local in Atlantic Amapá and lower Amazon valley and delta in Pará, Brazil; again locally plentiful over e. Brazil from s. Maranhão and Pernambuco to São Paulo and interior Rio Grande do Sul, s.-w. to Paraguay, n. Argentina (Formosa, Chaco, Corrientes, Misiones), and along the e. foothills of the Andes from San Martín, Peru, to the Yungas in Bolivia.—Fl. and fr throughout the year except where drought-inhibited.—Introduced (De Wit, 1955, p. 287) in Malaysia.

The syndrome of sessile glands, subdimidiate leaflets and very small asymmetric flowers which traditionally characterized var. *ramosa* (=*Cassia patellaria* sensu Bentham and all subsequent authors) is a strong one and it is with some reluctance that we here reduce the species to varietal status. In the context of the whole complex of Neotropical annual small-flowered chamaecristas the single unique character of *C. patellaria* is, however, the emphatic displacement of the midrib, and even this is not a fixed quality, but varies with the breadth of the leaflet, the midrib becoming in relatively broad ones hardly more excentric than in narrower leaflets of var. *glabrata*. Furthermore some ambiguous Puerto Rican plants (e.g. *Sintenis 3899,* NY, US; *Underwood & Griggs 241,* NY), with the venation of the latter on the same plant with the flower of *C. patellaria,* fully bridge any gap between the two. In the narrower context of our subsp. *patellaria* the var. *ramosa* is distinguished by the dilated style-tip which in the bud is incurved toward the orifice of one long anther and generally receives pollen from it before expansion of the corolla. But this device, which promotes a probably obligate autogamy, is repeated in subsp. *nictitans* var. *jaliscensis* and subsp. *disadena* var. *pilosa,* thereby losing some of its diagnostic weight. The closely related var. *paraguariensis* which, under different names, has often been confused with var. *ramosa,* largely no doubt because of its similar glands, is more easily distinguished by the truly subcentric midrib and consequently subsymmetric leaflets. These in sleep tend to fold face-to-face in a plane parallel to the leaf-stalk, completely concealing this with a linear series of vertically overlapping blades, rather than ascending, as in var. *ramosa,* at an acute angle from the rachis, in a closed rank of comblike tines.

Four sheets in the *Flora Brasiliensis* herbarium (M) which represent a coarsely leafy form of var. *ramosa* were cited by Bentham in the protologue of *C. subtriflora* Mart. ex Benth. Although one of these (=IPA Neg. *1390,* F Neg. *6259*) is labelled 'São Gonçalo d'Amarante, Piauí' and the other three 'Barra do Rio Negro' (i.e. Manaus), they seem to be duplicates of one collection, precisely similar in details of ontogeny and superficial discoloration of tissues. We believe them to be the plants intended by Martius to bear the name *C. subtriflora* and the cited photographs were taken under the misapprehension that they are holotypic for that species. Bentham's concept of *C. subtriflora,* however, even though he cited these small-flowered plants of var. *ramosa,* was substantially and explicitly based on material of the large-flowered, longistyled *Ch. repens* var. *multijuga* (q.v.).

The earliest collection of var. *ramosa* known to us is a plant sent from Veracruz by Dr. Houston in 1730 (BM, hb. Mill.!), accompanied by polynomial diagnoses by Houston himself and by Philip Miller. These phrase-names do not correspond, however, with any published in the eighth edition of *Gardeners' Dictionary* (1768).

51/IIc. Chamaecrista nictitans (Linnaeus) Moench subsp. **patellaria** (Colladon) var. **praetexta** (Vogel) Irwin & Barneby, stat. nov. *Cassia praetexta* Vogel, Syn. Gen. Cass. 67 & Linnaea **11**: 712, descr. ampliat. 1837.—"Hbt. In Guyana! (v.s. in Hrb. Kth.)."—Holotypus, labelled "Guiane française, Poiteau ded. 1824," †B = F Neg. *1735*! neoholotypus, former isotypus, collected at Cayenne by Poiteau, 1819–1821, G! isotypi, K, LE, P!

Chamaecrista lucesiae Pittier in Bol. Soc. Venezol. Ci. Nat. **8**: 135. 1943.—"*Anzoátegui*: Sabanas anegadas del río Caní, entre Guanipa y Cantaura . . . Agosto 24, 1942 (Pittier 15112, tipo)."—Holotypus, VEN! = NY Neg. *6328*; isotypus, US!

Chamaecrista nelsonii Britton & Rose, N. Amer. Fl. **23**(5): 291. 1930.—"Between Niletpec [=Niltepec] and Zanatepec, Oaxaca, July 15, 1895, *E. W. Nelson 2805*."—Holotypus, collected at 400 ft, in early anthesis, US! clastotypus, NY!

Cassia praetexta sensu Bentham, 1870, p. 175; 1871, p. 579.

Monocarpic herbs (1.5–)4–10 dm, with usually 1 erect stem strictly few-branched from near or beyond middle, or less often branched from near root-crown and the lateral stems incurved-ascending, appearing glabrous below the coarsely ciliolate pod but the young internodes and pedicels minutely or thinly puberulent with incurved hairs, the concolorous foliage brown when dry, the lft-margins either glabrous or minutely ciliolate.

Stipules erect, submembranous, lance-acuminate 4–10 × 1.1–2 mm, from base striately 7–13-nerved, persistent.

Adult lvs 3–9 cm, shortly petiolate, the expanded blades lance-elliptic in outline; petiole including pulvinus 3–8 mm, at middle 0.25–0.8 mm diam, very narrowly margined; gland 1, situated near or above middle of petiole proper, sessile, in profile tub- or drum-shaped, rarely turbinate, slightly wider than high, 0.25–0.9 mm tall, the round or elliptic head 0.35–1.2 mm diam; lfts 10–21 pairs, in outline narrowly oblong to broadly linear, obtuse and mucronate or acuminulate, straight or slightly falcate, 5–16 × 1–2.7 mm, at base cordate on proximal and cuneate on distal side, the moderately excentric costa dividing the blade ±1:2, from base (3–)4–5-nerved, the lateral nerves all short, 3 on broad and 0–1 on narrow side, the costa remotely 3–6-penninerved, the venulation above immersed, beneath finely prominulous.

Peduncles adnate to stem through 4–15 mm; racemes 1–4(–5)-fld, the axis 0–4 mm; pedicels at anthesis 1.5–4.5 mm, in fruit 3.5–6.5 mm, bracteolate 0.5–2 mm below calyx; buds ovoid-acuminate, glabrous or almost so; sepals lance-acuminate, 5–6.5 mm; petals yellow, the 3 adaxial broadly oblanceolate to obovate-cuneate 5–6.5 × 2.5–3.5 mm, the 2 abaxial longer or broader, one broadly obovate 6.5–8 × ±6 mm, the cucullus obliquely reniform ±6 mm long, 7 mm wide; stamens 10, the long anthers 3.6–4.4 mm; ovary glabrous on sides, densely pilosulous along sutures; style straight linear ±1.5 mm; ovules 8–13.

Pod erect, linear slightly curved outward, 2.5–4.5(–5.1) cm long, the thin-textured glabrous valves 3–4 mm wide, the thick sutures winged lengthwise on either side, the thinly herbaceous, commonly reddish wings coarsely ciliolate, becoming 0.75–1.6 mm wide; seeds 2.6–3.2 mm, the testa atrofuscous above the fulvous base, dull, sharply lineolate-pitted.—Collections: 28.

Savannas, mostly below 500 m, apparently rare and nowhere abundant, scattered around the n.-e. shoulder of S. America, from the lower Amazon valley (downstream from near mouth of Rio Trombetas) in Pará, through coastal French Guiana and Guyana (not yet recorded from Surinam) to Trinidad (Piarco Savanna) and the lower Orinoco valley in Venezuela (Guárico and Anzoátegui); one record, perhaps adventive, from suburban Manaus in Amazonas; disjunctly in Cuba (Oriente and Camaguey) and the Pacific lowlands of Isthmus of Tehuantepec in s.-e. Oaxaca, Mexico.—Fl. mostly X–IV, but along the Amazon River perhaps irregularly through the year.

Until the characteristic sutural wings develop with the maturing pod, the plant body of var. *praetexta* is virtually identical with that of those glabrescent or glabrous Cuban forms of var. *glabrata* which Britton & Rose (1930) described as *Ch. glaberrima*; and except for smaller, shorter-styled flowers and laterally glabrous ovary mimics in exact detail the glabrescent forms of subsp. *disadena* var. *disadena* encountered on savannas of the central Amazon valley and interior Guyana. In both of these forms, moreover, the sutures of the ripe pod sometimes carry obscure but definite herbaceous flanges up to 0.4 mm wide, narrower, it is true, than those of var. *praetexta,* which become up to 0.8–1.6 mm wide, but of precisely the same essential structure. We interpret var. *praetexta* as a derivative of the *Ch. nictitans* stock inhabiting northeastern South America and the southern Caribbean region in which one single vegetative feature, already present in facultative form, has become exaggerated. In the absence of supporting differential characters we confidently reduce *C. praetexta* to varietal status. Exaggerated sutural wings appear to have developed in parallel style in the genus *Senna,* the local and rare *S. pentagonia* (P. Miller) hardly differing otherwise from widespread and variable *S. obtusifolia* (Linnaeus) Irwin & Barneby.

51/IId. Chamaecrista nictitans (Linnaeus) Moench subsp. **patellaria** (Colladon) var. **glabrata** (Vogel) Irwin & Barneby, comb. nov. *Cassia patellaria* var. *glabrata* Vogel, Syn. Gen. Cass. 66. 1837.—"(*C. aristata* Hb. W[illdenov.] 8000). Hbt. In Para Brasiliae."—Holotypus, *Hb. Willd. 8000/1,* collected by Sieber and sent to Willdenow by Hoffmannsegg, B!—*Cassia aristata* Willdenow ex Walpers, Repert. **1**: 836. 1842, pro syn.

Cassia aeschinomene DeCandolle ex Colladon, Hist. Casses 127, t. 17. 1816.—"Hab. in Sto. Domingo . . ."—Holotypus, labelled 'Brotero, misit Balbis,' G-DC!—*C. mimosoides* var. *aeschinomene* (Colladon) Bentham, Trans. Linn. Soc. London **27**: 579 ("*Aeschynomene*"). 1871. *Chamaechrista aeschinomene* (Colladon) Greene, Pittonia **4**: 32. 1899.

Cassia lechenaultiana DeCandolle, Mém. Soc. Phys. Genève **2**(2): 132. 1824.—". . . communiquée par . . . M. Lechenault, qui l'a découverte au Bengale."—Holotypus, labelled 'Bengal, M. Leschenault, 1821,' G-DC! cf. De Wit, 1955, p. 280–282.—*C. mimosoides* var. *lechenaultiana* (DeCandolle) Ohashi, J. Jap. Bot. **50**: 308. 1975.

Cassia smaragdina Macfadyen, Fl. Jam. **1**: 347. 1837.—". . . common in the mountain districts [of Jamaica]."—Holotypus not found at BM, but description decisive in context of origin.—*Chamaecrista smaragdina* (Macfadyen) Britton ex Britton & Rose, N. Amer. Fl. **23**(5): 293. 1930.

Cassia chamaecrista Linnaeus subsp. (α) *normalis* var. *diffusa* O. Kuntze, Rev. Gen. 169. 1891.—"Portorico: Caguas."—Holotypus, *Kuntze 292,* collected 10.III.1891 (fl, fr), NY!—*C. chamaecrista* var. *diffusa* fma. *incana* O. Kuntze, l.c., based on the same typus.

Chamaecrista millspaughii Pollard, Field Mus., Bot. **2**: 47. 1900.—". . . near Bayamon, Porto

Rico, January 11 [1899] ([*Millspaugh*] *312*) . . ."—Holotypus, F! clastotypus (fragm), NY!—Reduced to *Ch. aeschinomene* by Britton & Rose, 1930, p. 294.

Chamaecrista fasciata Britton, Bull. Torr. Club **37**: 352. 1910.—". . . between Bath and Cuna-Cuna Gap [Jamaica], at 300 m. altitude (*Britton 3500,* September, 1908)."—Holotypus, NY!—*Cassia fasciata* (Britton) Fawcett & Rendle, Fl. Jam. **4**(2): 115. 1920.

Chamaecrista dussii Britton, Bull. Torr. Club 44: 9. 1917.—". . . Trou-Vaillant, Parnasse, Martinique (*Père Duss 1121*)."—Holotypus, NY!

(?)*Cassia martinicensis* Urban, Feddes Repert. **15**: 315. 1918.—"Hab. in Martinique: Duss no. 1121b."—Holotypus, presumably †B; no isotypus found.—*Chamaecrista martinicensis* (Urban) Britton & Rose, N. Amer. Fl. **23**(5): 294. 1930.

Chamaecrista glaberrima Britton ex Britton & Rose, N. Amer. Fl. **23**(5): 291. 1930.—"Type from Triscornia, Havana, [Cuba,] April 5, 1905, *H. A. Van Hermann 805*."—Holotypus, NY!—*Cassia glaberrima* (Britton) León, Contrib. Ocas. Mus. Hist. Nat. Collegio de la Salle (Habana) **10**(=Fl. Cuba **2**): 265. 1951.

Chamaecrista rovirosana Britton ex Britton & Rose, N. Amer. Fl. **23**(5): 292 ("*Rovirosana*"). 1930.—". . . Ribera de Yxtacomitan, Tabasco, [Mexico], August 27, 1888, *J. N. Rovirosa 253*."—Holotypus, NY!—*Cassia rovirosana* (Britton) Lundell, Carnegie Inst. Wash. Publ. **478**: 212. 1937.

Chamaecrista tabascoensis Britton & Rose, N. Amer. Fl. **23**(5): 299. 1930.—"Santa Anita, Tabasco, [Mexico,] February 5, 1890, *J. N. Rovirosa 709*."—Holotypus, US! clastotypus (fragm + photo), NY! isotypus, K!—*Cassia tabascoënsis* (Britton & Rose) Lundell, Carnegie Inst. Wash. Publ. 478: 212. 1937.

Chamaecrista aeschinomene sensu Britton & Rose, 1930, p. 294.

Cassia lechenaultiana sensu Degener, New Ill. Fl. Hawaii, anno 1934, t. sine num. ("*leschenaultiana*"); Fl. Taiwan **3**: 201, pl. 535. 1977.

Cassia smaragdina sensu Adams, 1972, p. 328.

Chamaecrista fasciata sensu Britton & Rose, 1930, p. 292. *Cassia fasciata* sensu Adams, 1972, p. 328.

Closely resembling var. *ramosa,* but on the average less densely pubescent, the lfts always glabrous above, sometimes on both faces, commonly ciliolate, but the whole plant occasionally glabrous to the ovary or even throughout, longer setose hairs sometimes present on lf-stalks and young stems.

Lvs (2–)2.5–9(–14) cm, the lf-stalk either straight, slightly incurved, or slightly decurved; petiolar glands 1–2, commonly 1, varying from almost plane and sessile to saucer- or cup-shaped and obconic in profile, (0.35–)0.4–1 (rarely 1.2) mm diam, 0.2–1 mm tall, either much wider or slightly narrower than high; lfts 10–26(–31) pairs, linear or linear-oblanceolate, 4–21(–24) × 1–3.5(–4) mm, the longest of a plant mostly 9–19 mm.

Pedicels at anthesis subobsolete to 8 mm, in fruit erect thickened, 2–10(–12) mm; sepals 5–7.5 mm; petals 2–3 narrowly obovate or oblanceolate 5–7 mm, either shorter than or about as long as sepals, 2 anterior larger, well-exserted, one (6–)6.3–10 mm, obovate-cuneate, the cucullus of nearly same length but oblique; stamens 10, 2–3 longer, their anthers (3.5–)3.8–5.4(–6) mm; ovary densely white-pilose or white-strigose; ovules (7–)8–14; style 1.2–2.1 mm, not or scarcely thickened distally.

Pod (2.2–)2.5–5 × (0.35–)0.4–0.5 cm, the wide sutures often reddish, sometimes bordered by a narrow wing up to 0.5 mm, the valves usually pilose with spreading hairs, sometimes puberulent with incurved hairs, exceptionally glabrous.—Collections: 264.

Lowland habitats of var. *ramosa,* mostly below 400 but in Haiti and Jamaica up to ±800 m, perhaps originally an element of the natural savanna flora but now largely weedy, abundant throughout Cuba, Hispaniola, Puerto Rico and Jamaica, but on the Lesser Antilles known only from Guadeloupe and (one old record) St. Thomas, w. sporadically into s. Mexico (Tabasco; Veracruz; n. and w. Oaxaca; Guerrero), Honduras, and Nicaragua, s. to lower Orinoco valley in Venezuela

(Guárico; Monagas), coastal Guyana (one old record), and the Amazon delta region in Pará, Brazil; disjunct (probably adventive) in n. Peru (Madre de Dios; Amazonas) and upper Amazonian Brazil (Rondônia). Long naturalized in s. India, Ceylon and Java; Fiji, Samoa, and Hawaiian Is. (Oahu; Kauai; Maui); Mauritius.—Fl. almost throughout the year unless drought-inhibited.

The var. *glabrata* is the commonest weedy monocarpic chamaecrista of the larger West Indian islands, ubiquitous in Cuba, Hispaniola and Puerto Rico, common on Jamaica, but curiously lacking from the outer Antilles other than Guadeloupe. Its continental range is discontinuous and probably largely adventitious, except possibly in Veracruz, Mexico, and in the lower Orinoco and Amazon valleys, where it is represented by seemingly endemic races notable for large glands and ample foliage, one known to be polyploid. Outside the West Indies, where to the highest degree variable in habit and vesture, var. *glabrata* becomes relatively uniform. We suspect that the Pacific and Indomalayan populations are descended from one or few ancient introductions from the Caribbean region and are in consequence biotypically impoverished.

The variety as a whole seems almost equally related to subsp. *nictitans* var. *diffusa* as it is to subsp. *aeschinomene* var. *ramosa*. From the former it is distinguished ultimately only by the sessile or subsessile, not slenderly stipitate and tack- or pin-shaped petiolar gland; from the latter ultimately by the strongly asymmetric leaflets and less dilated style. As a general rule the disciform or shallowly dish-shaped petiolar glands of var. *ramosa* can be contrasted with the turbinate or cup-shaped glands of var. *glabrata,* but the exceptions are too numerous to make the criterion effective. The flower of var. *ramosa* is usually smaller and its corolla more strongly asymmetric, but here again we find no practical division. None of these characters were mentioned by Britton & Rose who (1930, p. 273, in clave) separated *Ch. patellaria* (=var. *ramosa*) from *Ch. aeschinomene* by relative density of pubescence. Annotations by Britton (NY) show that this was indeed the practical differential character employed. As a consequence *Ch. patellaria* was thought to extend north into Puerto Rico, but all such material that we have examined proves to be relatively more pubescent var. *glabrata*. In our definition of the latter we permit enough latitude in vesture and petiolar glands to admit the types of six Brittonian segregates, the particular characters of which are mentioned directly. Pollard's *Ch. millspaughii,* already reduced by Britton himself, requires no further comment.

i. *C. smaragdina:* at least as interpreted by Adams (l.c.), a form with procumbent or assurgent stems and puberulent, not pilose, ovary and pod. Said to occur in Jamaica higher than *Ch. fasciata,* an erect form pilose with at least some patent setae.
ii. *Ch. fasciata:* ovary shaggy-pilose giving rise to pod pilose with hairs up to 2.5 mm. Similar plants occur in Cuba; and even in Jamaica do not replace typical var. *glabrata* with subappressed-pilosulous ovary.
iii. *Ch. dussii:* supposed to differ by slightly larger petals, but not significantly larger-flowered than several collections annotated by Britton as *Ch. aeschinomene*.
iv. *Ch. glaberrima:* based on a plant that really is hairless in all parts, even ovary and pod. Other collections from La Habana province referred here by Britton have glabrous foliage but puberulent stems, or leaf-stalks, or pubescent ovary. We regard *Ch. glaberrima* as a minor variant in which pubescence is greatly reduced but seldom fully suppressed. Until the pod matures, this phase of var. *aeschinomene* is virtually indistinguishable from var. *praetexta*.

v. *Ch. rovirosana:* another negligible variant with subglabrous foliage, but with ovary pilose and pod persistently so along the sutures.

vi. *Ch. tabascoënsis:* based on a sheet of fragmentary specimens, four stems severed within 2 dm of apex and described, for no obvious reason, as "shrubby." Collected at end of the growing season (February 5), these stems bear proximally a few malformed pods at nodes from which the leaves have disjointed and distally a few newly expanded leaves and flowers. They have the characteristic aspect of plants that have survived a period of drought and put forth a little fresh growth following a winter rain. Emphasizing "subsessile" flowers and 1–2-seeded pods, Britton & Rose referred *Ch. tabascoënsis* to a monotypic series. However, the short flowering pedicel is not unusual in either var. *glabrata* or var. *ramosa,* and a dissection showed eight ovules in one of the new flowers, showing that the pod is only few-seeded by abortion. In fact no fully formed seeds are present in the semi-abortive fruits. A vigorous leafy plant (*Rovirosa 253,* NY) collected near the type-locality of *Ch. tabascoënsis* but not acquired until after Britton's death, manifests the normal vernal aspect.

51/III. Chamaecrista nictitans (Linnaeus) Moench subsp. **disadena** (Steudel) Irwin & Barneby, stat. nov. *Cassia disadena* Steudel, Flora (Regensb.) **26**: 760. 1843.—Typus infra sub. var. *disadena* citatur.

Cassia stenocarpa Vogel, Syn. Gen. Cass. 68 & Linnaea **11**: 713 (descr. ampliat.). 1837.—"Hbt. in Brasilia: Sellow leg.; Luschnath leg. pr. Bahia."—The syntypi, lost at B, surely represent one or both of the varieties of subsp. *disadena* recognized below; but which one is doubtful (cf. discussion infra).—*Chamaecrista stenocarpa* (Vogel) Standley, Contrib. U.S. Nat. Herb. **18**: 104. 1916.

Pedicels in fruit up to 6–14(–21) mm, potentially but not always actually longer than in other subspp. of *Ch. nictitans,* the corolla either subrotate or strongly asymmetric, the long abaxial petal up to (5.5–)6–15 mm; style either linear and 3–5.5 mm or dilated upward and only 1.4–3.3 mm; androecium 10-merous; petiolar gland stipitate, in profile tack-, trumpet-, urn-shaped or broadly claviform, taller than diam of the head.—S. Mexico to s.-e. Brazil and Paraguay.

Of all our sometimes arbitrarily delimited infraspecific taxa in *Ch. nictitans,* the present subspecies is the most difficult to define or to defend from criticism. It accommodates those monocarpic chamaecristas with emphatically stipitate petiolar glands which differ from subsp. *nictitans* in the longer pedicels and larger, or if scarcely larger then more regular, subrotate corolla. It is approximately equivalent to *C. stenocarpa* Vog. as understood by Bentham and many subsequent writers, although when large-flowered and represented by severed stems unaccompanied by data on duration, partly equated by Bentham with *C. glandulosa.* The subspecies stands in a critical relation to the whole complex of Neotropical *Chamaecristae Mimosoideae* in the sense that the monocarpic duration of *Ch. nictitans* sens. lat. is here combined with the flower of *Ch. glandulosa,* confirming Otto Kuntze's doubtless biologically real while taxonomically monstrous concept of a huge multiracial species within which no substantial discontinuities have yet developed. We divide subsp. *disadena* into two intergradient and largely sympatric varieties, a longistyled presumably outcrossing var. *disadena* which is sometimes almost impossible to disentangle from shrubby or at least strongly perennial *Ch. glandulosa,* and a brachystyled, presumably autogamous var. *pilosa* which in Central America fades imperceptibly into subsp. *nictitans* var. *jaliscensis.*

Key to the Varieties of subsp. *disadena*

1. Style linear or almost so, 3–5.5 × 0.2–0.4 mm. 51/IIIa. var. *disadena* (p. 826).
1. Style dilated upward and shorter, 0.8–3(–3.3) × 0.4–0.7 mm.
   2. Petals heteromorphic but not strongly graduated in length, the 2 abaxial subequal, mostly 6–12 mm, the inner perianth subrotately campanulate; androecium always 10-merous, the longer anthers commonly 3.7–7 mm; style 1.4–3.3 mm. 51/IIIb. var. *pilosa* (p. 829).
   2. Petals strongly graduated, 2 abaxial much longer than the rest, up to 3.5–7(–8) mm; androecium 4–10-merous, the longer anthers 1.7–4.5 mm; style 0.8–1.6(–2) mm. Cf. subsp. *nictitans* var. *jaliscensis*.

51/IIIa. Chamaecrista nictitans (Linnaeus) Moench subsp. **disadena** (Steudel) var. **disadena** (Steudel) Irwin & Barneby, stat. nov. *C. disadena* Steudel, 1843, l.c., sensu str.—"[Surinam, *Hostmann*] nr. 1179."—Holotypus not found at P; isotypi, G (3 sheets), GH (2 sheets), FI (hb. Webb.), LE, NY!

Chamaecrista seleri Rose ex Britton & Rose, N. Amer. Fl. **23**(5): 287. 1930.—"Guatemala . . . Barranca del Rio de las Vacas, Salida de Isabel [properly Izabal], October 15, 1896, *C. & E. Seler 2286*."—Holotypus, NY! isotypi, GH, US (2 sheets + photo)!—*Cassia seleri* (Britton & Rose) Lundell, Phytologia **1**: 215. 1937.

Chamaecrista chiriquensis Britton & Rose, N. Amer. Fl. **23**(5): 287. 1930.—"Vicinity of San Felix, eastern Chiriqui, Panama, 1912, *Pittier 5216*."—Holotypus, NY! isotypus, US!

Erect or rarely diffuse, essentially herbaceous and commonly monocarpic but in age basally suffruticose herbs at anthesis (3–)5–16(–20) dm, commonly both puberulent with short incurved hairs and the stems distally with leaf-stalks and sometimes the pedicels hirsute with spreading fine setae up to 1–2.4 mm, but either type of vesture sometimes sparse or 0, the subconcolorous lfts often dorsally, exceptionally also ventrally puberulent, rarely pilosulous, the margins either minutely ciliolate or randomly setulose, exceptionally glabrate.

Stipules lance-acuminate or attenuate, those at flowering nodes (7–)8–14(–17) mm, at base (1.3–)1.5–2.6(–2.8) mm wide, from base (7–)9–15-nerved.

Adult lvs (3–)3.5–10.5(–13.5) cm, shortly petioled, the expanded blades lanceolate or narrowly ovate in outline; petiole (including pulvinus) 3–5.5(–8) mm, at middle 0.4–0.75(–0.8) mm diam, narrowly margined; petiolar glands 1–2(–3), the first (or only) one situated at, below, or above middle of petiole proper, in profile trumpet-, urn-, or broadly club-shaped, sometimes coarsely urceolate-turbinate (0.5–)0.6–1.6 × (0.25–)0.35–0.8(–1.2) mm, at least a trifle (often much) taller than the diam of the circular head; sometimes smaller glands on rachis below 1–5 distal pairs of lfts; lfts of adult lvs (10–)12–24(–26) pairs, linear to narrowly oblong or linear-elliptic, mostly obtuse and abruptly apiculate or aristate, the larger up to (7–)9–22(–25) × 1.4–3.4(–3.7) mm, at base on broad side cordate or obtusely angulate, on narrow side broadly cuneate to subcordate, the costa varying from centric to moderately displaced, dividing the blade 1:1–2, from base on broad side shortly 3–4- and on narrow side 1–2-nerved, the costa penninerved with (3–)4–6(–7) major secondary venules with sometimes additional intercalated ones, then appearing 8–12-nerved on either side, the venulation commonly impressed on upper and finely prominulous on lower face, exceptionally subimmersed on both.

Peduncles adnate through (1–)2–14(–26) mm, (1–)2–5-fld; pedicels at anthesis weak subfiliform 4–11 mm, in fruit spreading-ascending thickened straight or incurved distally, (4–)6–14(–21) mm; buds ovoid-acuminate, thinly pilosulous or incurved-puberulent, sometimes glabrate distally; sepals lance-acuminate (6–)6.5–12 mm; petals yellow, 3 adaxial oblanceolate to narrowly obovate-cuneate, 6–10.5(–11) mm, 1 long petal obovate-flabellate (7–)8–14(–15) × 5.5–10.5 (–12) mm, the cucullus obliquely reniform 5–10(–12) mm; stamens 10, the anthers greenish-yellow, yellow, or red (drying blackish), the 2–3 long ones (5–)5.5–10.5(–12.5) mm; ovary either strigulose or pilosulous; style linear, gently incurved or straight, 3–5.5 × 0.2–0.4 mm, not or only very obscurely dilated distally; ovules (11–)13–26.

Pod linear, slightly arcuate, (30–)35–63(–70) × (2.5–)3–4(–4.4) mm, the green, brown or deep reddish- or purplish-castaneous valves shortly incurved-pilosulous or -puberulent, less often thinly setose; seeds fulvous or fuscous, dull, 2.6–3.7 mm, the testa lineolate-pitted.—Collections: 122.

Savannas at low and middle elevations, in Central America ascending into the pine-oak belt up to 2100 m, in Guayana and within the Amazonian Hylaea occupying scattered sunny habitats along rocky stream banks, islands in river-rapids, rock-outcrops, beaches, and becoming weedy in capoeira or cultivated ground, in coastal Bahia a common and locally abundant weed of roadsides and cocoa plantations but ascending inland to cerrado habitats up to 1200 m, of discontinuous range: Central America from Oaxaca (Cerro Zempoaltepetl, near 17°N, 86°W) s.-e. to w. Panama (Chiriquí); llanos e. of Cordillera Oriental in n.-e. Colombia (Norte de Santander; Int. Meta); Cordillera Costanera (Distrito Federal, perhaps a waif) and middle Orinoco valley (Cerro San Borja, Bolívar) in Venezuela; interior Guyana, Surinam, and adjacent Amapá and Terr. do Roraima in Brazil, from the upper Mazaruni, Rupununi and Courantyne valleys to the Wilhelmina and Tumucumaque ranges on the Jarí-Oiapoque watershed; somewhat disjunct along middle Amazon valley in Amazonas between Manacapurú and Purequequare (59°40′–60°40′W) with one record from upper Rio Negro (Vaupés) near 67°W; again disjunct along the Atlantic lowlands of n.-e. Brazil in Ceará and Bahia, extending occasionally inland to Chapada Diamantina (Sa. do Sincorá) and thence s. into n. Sa. do Espinhaço near 17°S in n.-centr. Minas Gerais; one record from n.-e. Paraguay (Río Apa, near 22°S).—Fl. in Central America mostly VIII–I, in the Guianas, Amazonia and e. Brazil almost throughout the year.

Like other discontinuously dispersed varieties of *Ch. nictitans,* the var. *disadena* consists of numerous races minutely distinguishable by proportions and coarseness of the petiolar glands, relative proportions and position of the duplex vesture, and length of the pedicels and flower parts. The nomenclaturally typical form, from Surinam, closely matched by modern collections from the Wilhelmina Mts. and Sa. Tumucumaque and from the headwaters of Rio Branco in Terr. do Roraima, has usually two slenderly stalked petiolar glands, setose stems, and long narrow puberulent pod containing some 14–26 seeds; but the flower varies in size from one collection to the next, the feature common to all being the linear style, on which we ultimately depend for separating var. *disadena* from the very closely related var. *pilosa*. The populations found in coastal Bahia, largely derived from roadsides and cocais, are essentially typical, but have usually only one petiolar gland. In riparian and disturbed habitats along the middle Amazon valley in Amazonas var. *disadena* is represented by a robust form with thinly puberulent or glabrate, when dry brownish foliage and stems, coarsely urceolate glands, and a thinly setose, not puberulent pod of which the sutures are sometimes dilated into incipient wings. This form, collected first by Spruce near Manaus and, in slightly more pubescent form by the Schomburgks in Guyana, was misinterpreted by Bentham (1871, p. 577) as *C. riparia* H.B.K. (of which the typus represents the short-styled var. *jaliscensis*) and treated by him as specifically distinct from "*C. stenocarpa.*" At anthesis it closely mimics subsp. *patellaria* var. *praetexta,* which replaces it on savannas of the lower Amazon valley; this differs in the smaller flower with short-styled ovary setose on the sutural angles but not on the faces, and when adult, of course, in the remarkable wing-margined pod. While it is readily recognized in Amazonas, on the savannas of interior Guyana it fades imperceptibly into typical var. *disadena* found in similar habitats immediately to the east and west. The one collection of this glabrescent form from the upper Rio Negro is peculiar for the presence of glands on several distal segments of the leaf-rachis, a phenomenon now known to occur in many distantly related forms of supra-axillary chamaecrista.

The var. *disadena* of southern Mexico and Central America, found largely (but not exclusively, e.g. typus of *Ch. chiriquensis*) in relatively undisturbed upland habitats associated with pine- or oak-forest, has the vesture and (1 or 2) glands of typical *disadena,* but a fewer (mostly 11–15- not 14–26)-ovulate and in consequence shorter pod, suggesting that of the entirely West Indian *Ch. glandulosa* var. *swartzii,* different in its fruticose adult stature and short stipules. The also closely similar *Ch. glandulosa* var. *flavicoma,* allopatric in the Andes (and in disjunct stations in Yucatan and southern Mexico) is only arbitrarily separable by the once again potentially shrubby or arborescent life-form and the usually very long, at anthesis flexuous pedicels.

One collection of var. *disadena* from Ceará (20 mi inland from Fortaleza, *G. Bolland s.n.,* K) deserves special mention. With the habit of the coarse glabrate Amazonian race mentioned above but setose stems and leaf-stalks it has exceptionally large flowers (calyx and long abaxial petal to 15 mm and long anthers 12.5 mm) and the calyx lobes are prominently several-striate, not indeed so closely parallel-venulose as those of *Ch. calycioides,* but more prominently than those of any other New World chamaecrista known to us. Since *Ch. calycioides* was collected with it (*Bolland s.n.,* K) it may not be fanciful to suppose an exchange of genes between the two, however unlikely that may appear from the assumed difference in chromosome number.

In adopting for this varietal taxon the epithet *disadena,* we deliberately pass over the earlier and probably relevant *stenocarpa* for two reasons: we have failed to typify *C. stenocarpa* Vogel with assurance; while its shifting application by Bentham, Britton and others to related members of the *Ch. nictitans* complex, particularly to what we here call var. *pilosa,* var. *jaliscensis,* and var. *disadena* in part only, have eroded its utility in practical taxonomy. The protologue of *Cassia stenocarpa* is based on two plants lost in the Berlin disaster, one collected by Sellow, of which we have not knowingly seen any duplicate, and one by Luschnath from Bahia, unfortunately not further identified as to number, date or place. Bentham equated *C. stenocarpa* with the Luschnath collection, supposedly from Ilhéus, that was distributed as Martius Herb. *No. 720.* A duplicate of *Martius 720* at Berlin survives as Field Negative *1749,* but if this is the Luschnath specimen seen by Vogel there is nothing on the label to prove it. Bentham (1870, p. 174) cited as *C. stenocarpa* a specimen collected at Cruz de Cosme by Martius; the duplicate in Bentham's herbarium has short but linear styles. So far these all appear to represent our var. *disadena.* We have seen also, however, in the Vienna herbarium a *Luschnath 193* and at Kew *Luschnath 834* from the same locality, the former dated August 1835, both of which represent the short-styled relative var. *pilosa.* There is nothing decisive in the protologue pointing to one or other of the two entities known to occur in the Luschnath collection. By contrast the type-set of *C. disadena* is widely dispersed, unambiguous, and of excellent quality.

51/IIIb. Chamaecrista nictitans (Linnaeus) Moench subsp. **disadena** (Steudel) var. **pilosa** (Bentham) Irwin & Barneby, comb. nov. *C. riparia* var. *pilosa* Bentham in Martius, Fl. Bras. **15**(2): 174. 1870.—". . . ad Lagoa Santa prov. Minas Geraës: Warming."—No holotypus found at K; presumed isotypus, *Warming 207,* collected 26.II.1866, NY!

Cassia propinqua Humboldt, Bonpland & Kunth, Nov. Gen. & Sp. **6**(folio): 369. 1824.—"Crescit prope Caripe [edo. Monagas, Venezuela] et in monte Cocollar, alt. 400 hex."—Holotypus, labelled "n. *288,* Cocollar, Caripe," P-HBK! isotypi, dated Vindem., an 8, P (hb. Bonpland.)! = NY Neg. *9185,* and labelled "Neu Anadalusien, *Humboldt.,*" B (*hb. Willd. 8009*)!—*C. erecta* Willdenow ex Steudel, Nom. Bot. ed. 2, 305. 1840, pro syn., taken from *hb. Willdenov. 8009.*

Chamaecrista rekoi Britton & Rose, N. Amer. Fl. **23**(5): 289. 1930.—"Cerro Espino, Oaxaca, 600 m. altitude, September, 1917, *B. P. Reko 3608.*"—Holotypus, US! clastotypus (fragm), NY!

Chamaecrista tonduzii Britton & Rose, N. Amer. Fl. **23**(5): 290. 1930.—"Guatemala. Type collected July, 1921, *A. Tonduz 649.*"—Holotypus, NY! isotypus, US!—*Cassia mayana* Lundell, Phytologia **1**: 215. 1937 (non *C. tonduzii* Standl., 1919).

Chamaecrista fenixensis Britton & Rose, N. Amer. Fl. **23**(5): 291. 1930.—". . . Fenix, [Sa. Madre del Sur], Chiapas, April, 1925, *Purpus 10553.*"—Holotypus, NY! isotypus, misnumbered *Purpus 10008,* US!—*Cassia fenixensis* (Britton & Rose) Lundell, Phytologia **1**: 214. 1937.

Chamaecrista comayaguana Britton & Rose, N. Amer. Fl. **23**(5): 291. 1930.—". . . vicinity of Siguatepeque, Comayagua, Honduras, February 14–27, 1928, *Standley 56253.*"—Holotypus, NY! isotypus, US!—*Cassia comayaguana* (Britton & Rose) Lundell, Phytologia **1**: 214. 1937.

Chamaecrista conzattii Britton & Rose, N. Amer. Fl. **23**(5): 293. 1930.—"Cerro San Antonio, 1600 m. altitude, Oaxaca, September 8, 1921, *C. Conzatti 4228.*"—Holotypus, NY! isotypus, US!

Chamaecrista salvadorensis Britton ex Britton & Rose, N. Amer. Fl. **23**(5): 293. 1930.—"Vicinity of Ahuachapán, El Salvador, January 9–27, 1922, *Standley 19819.*"—Holotypus, NY! isotypi, GH, US!

Chamaecrista stenocarpoides Britton ex Britton & Rose, N. Amer. Fl. **23**(5): 293. 1930.—". . . vicinity of San Sebastián, San José, Costa Rica, February 1, 1924, *Standley 32701.*"—Holotypus, NY! isotypus, US!—*Cassia stenocarpoides* (Britton) Lundell, Phytologia **1**: 215. 1937. *C. stenocarpa* var. *stenocarpoides* (Britton) Schery, Ann. Mo. Bot. Gard. **38**[=Fl. Panama **5**(3)]: 60, fig. 123. 1951.

Monocarpic but often long-enduring, suffrutescent in age and perhaps sometimes perennant, plastic in stature and deportment, the stems commonly erect and incurved-ascending, usually branched but sometimes simple, less often, both naturally and after browsing or cutting, diffuse or humifuse, at anthesis (1–)2–10(–13, reportedly –25) dm, commonly at once incurved-puberulent and pilose with fine spreading or spreading-incurved setae up to 0.7–2.6 mm, the lfts usually glabrous on both faces, sometimes finely pilosulous beneath and exceptionally above also, their margins minutely scabrid-ciliolate or charged with random setae up to 0.7–1 mm.

Adult lvs (sometimes produced only subsequent to the early fls) up to (2–)2.5–8.5(–11) cm; petiole 2–6 mm; petiolar gland 1, greatly variable in form, either stipitate or subsessile, the shallowly concave head 0.2–0.8 mm diam, the whole in profile varying from broadly to narrowly turbinate or urn-, trumpet- or tack-shaped, 0.2–1 mm tall and from 0.3 mm taller to 0.3 mm shorter than head's diam; lfts of adult lvs (12–)13–26(–29) pairs, the longer ones linear to oblanceolate, straight or subsigmoid, up to (5–)7–20(–23) × (1–)1.2–3.5(–4) mm.

Pedicels 1–3, at anthesis slenderly flexible, the solitary one or the first of a fascicle up to 5–9(–11) mm, in fruit thickened, sigmoidally spreading-incurved and up to 7–12(–14, exceptionally –18) mm; sepals up to (5–)5.3–8.3(–9.3) mm; petals usually little graduated in length, forming when fully expanded a campanulate perianth, the cucullus and long petal only slightly longer though both or at least the former much wider than the 3 adaxial ones, up to (5.5–)6–12 mm; sta-

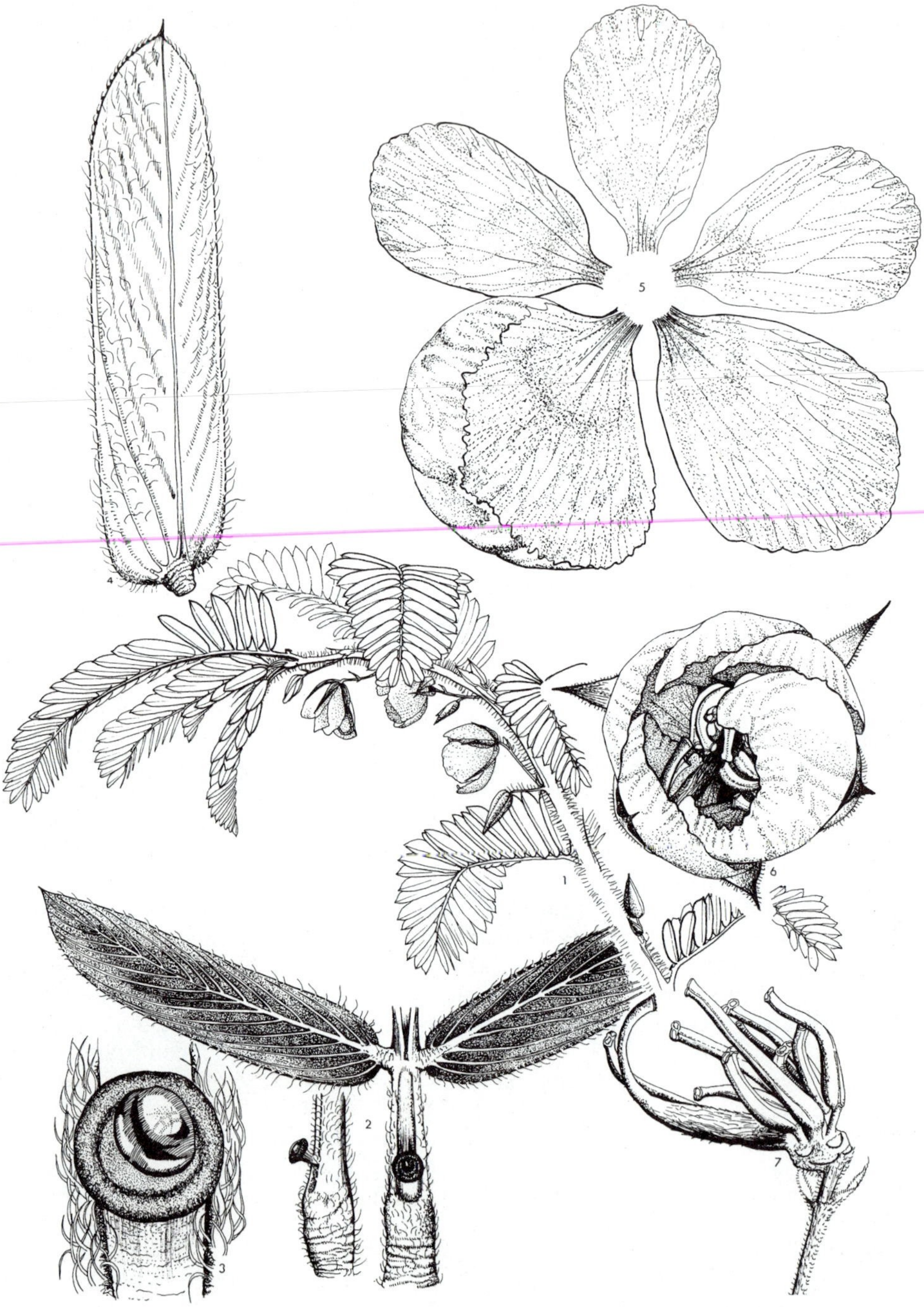

Fig. 51. *Chamaecrista nictitans* (Linn.) var. *pilosa* (Benth.) I. & B. 1) Flowering branchlet ×1; 2) petiole and gland ×5; 3) gland ×20; 4) dorsal face of leaflet ×5; 5) flower ×5; 6) petals ×5; 7) androecium + pistil ×5 (Drawn from life by W. S. Moye III).

mens 10, the 3 longest anthers to (3.3–)3.7–6.8(–8) mm; ovary strigulose or pilosulous; style strongly incurved, (1.4–)1.6–3(–3.3) mm, dilated at apex to 0.4–0.7 mm diam; ovules (7–)11–24.

Pod linear in outline, straight or slightly sigmoid-curved, (20–)27–67 × 2.7–4.5 mm, the castaneous or purplish, finally nigrescent valves either incurved-puberulent or less often thinly weakly pilose; seeds pale brown or pale chestnut, often fuscous-suffused or -lineolate, dull or sublustrous, 2.4–3.2 mm.—Collections: 133.—Fig. 51.

Thickets, shores, savanna-campos, disturbed woodlands, ascending from sea level up into the pine-oak- and rainforest, now commonest as a colonial weed along highways, in waste places, plantations and gardens, ascending in Mexico and Central America to 1200–2000, in Venezuela to 2300, and on the Brazilian Planalto to 1500 m, discontinuously widespread through much of tropical America outside the West Indies, the Amazonian Hylaea, the Guayana Highland, and most of the Andes: frequent in Central America from w. Panama (Chiriquí) to e. and s. Mexico (Tamaulipas; s. Veracruz; Nueva Galicia, Sa. Madre del Sur in Oaxaca and Guerrero); absent from the rest of Panama and Colombia, but again rather frequent in the uplands of n. Venezuela from Cord. de Mérida e. to Monagas; local in centr. Ecuador (Cotopaxi) and on the Pacific slope in n. Peru (Lambayeque); reappearing in e. Brazil, where highly local in n. Maranhão and (weedy) e. Pernambuco and coastal Bahia (Ilhéus), thence s. on the Atlantic slope to Rio Grande do Sul and inland to the Paraguay drainage in s.-w. Mato Grosso and Paraguay, extreme s.-e. Bolivia, and common in centr. and s. Sa. do Espinhaço s.-ward from ±16°30′.—Fl. almost throughout the year, except where inhibited or killed by drought.—*Ta hata ha guazu* (Guarani)

The essential features of our newly defined var. *pilosa,* originally separated from *Cassia riparia* and *C. stenocarpa* by a particular type of vesture now seen to have no real diagnostic significance, are the relatively long weak flowering pedicels, the perianth of moderate size and asymmetry, and the dilated style-tip. It is very closely related to var. *jaliscensis,* which has the same style, but smaller, usually more asymmetric corolla and substantially shorter pedicels; and to var. *disadena,* which has essentially the same flower and pedicels, but a linear and on the average longer style.

Variation within var. *pilosa* follows the pattern set by other members of the *Ch. nictitans* complex, involving great plasticity of life-form and stature, emphasis and loss of either element of the duplex vesture, and modification of the petiolar gland both in absolute size and in proportion of stalk to head. The Brittonian segregates listed in the synonymy are based on individual combinations of these variable factors:

- i. *Ch. rekoi:* Densely pilosulous throughout, the lfts subappressed-pilose on both faces; gland trumpet-shaped, long-stalked.
- ii. *Ch. tonduzii:* Copiously setose and puberulent, of average stature; gland tack-shaped, short-stalked.
- iii. *Ch. fenixensis:* Specimens mutilated by browsing or cutting and consisting of old stems laterally regenerating, the leafy shoots bearing only immature leaves; pubescence of the last but gland smaller.
- iv. *Ch. comayaguana:* Gland of *Ch. tonduzii,* but growth-habit slender, diffuse, leaves relatively small, setose pubescence much reduced.
- v. *Ch. conzattii:* Habit nearly of the last, but setose pubescence copious and petiolar gland small, tack-shaped, as *Ch. fenixensis*; lfts relatively broad, narrowly oblong.

vi. *Ch. salvadorensis:* Erect, copiously setose; gland almost of *Ch. tonduzii*; leaflets linear-oblong.
vii. *Ch. stenocarpoides:* Essentially as the last, but the leaflets of *Ch. conzattii*.

The only one of the proposed segregates which may have some genetic substance is *Ch. comayaguana,* a slender diffuse small-leaved form from forest habitats above 1000 m. However the diffuse habit of growth is encountered in coarser plants at lower elevations, and typical erect var. *pilosa* is found also at montane elevations. The opposite extreme, a giant phase with long leaves and large flowers, reportedly reaching (in Mérida, Venezuela) a stature of 2.5 m but usually not more than half as tall, occurs at random points within the Central American range of var. *pilosa* (cf. *Breteler 3161* from Venezuela; *Skutch 2880* from Costa Rica). Its ploidy level should be investigated.

51/IV. Chamaecrista nictitans (Linnaeus) Moench subsp. **nictitans**. *Cassia nictitans* Linnaeus, 1753, l.c., sens. ampliat.—Typus infra sub var. *nictitanti* citatur.

Fl short-pedicellate, small, brachystylous (probably autogamous), the pedicels 1–6(–8) mm, the longest petal 3.5–8 mm, the style 0.8–1.6(–2) mm; androecium 4–10-merous; petiolar gland commonly stipitate and in profile taller than diam of head, but sometimes (in var. *nictitans* of e. United States only) shorter. Mostly North and Central American and West Indian, 1 extending to n.-w. Peru and (disjunctly, perhaps adventive) to w. Argentina.

Disregarding exceptions noted in the description, we define subsp. *nictitans* as a closeknit group of monocarpic forms characterized by a small autogamous flower and stipitate petiolar glands prevailingly taller than the diameter of the head. It differs ideally but, in the nature of the whole species, indecisively from subsp. *patellaria* in the petiolar gland and from subsp. *disadena* in the prevailingly shorter pedicels combined with a shorter (or if not shorter more asymmetric), shorter-styled flower. A depleted androecium of less than 9 members is common in subspecies *nictitans,* but while found in this subspecies alone is not characteristic of it. By way of var. *diffusa* the subspecies passes directly, in the West Indies, into subsp. *patellaria* var. *glabrata* and by way of var. *jaliscensis* in s.-centr. Mexico into subsp. *disadena* var. *pilosa*. Its morphological boundaries are therefore quite arbitrary, even though we believe that it corresponds to a biological reality.

Key to the Varieties of subsp. *nictitans*

1. Style linear or very slightly dilated distally, at apex 0.15–0.3 mm diam, rarely to 0.4 mm diam in var. *nictitans* of e. United States.
    2. Corolla subrotate, the abaxial plane petal and cucullus commonly broader but scarcely longer than the rest; androecium (9–)10-merous; circum-Caribbean n. (feebly) into the Bahamas. 51/IVa. var. *diffusa* (p. 833).
    2. Corolla usually highly asymmetric, the plane abaxial petal opposed to the cucullus usually much the largest, but if only slightly longer than the adaxial petals then the androecium 4–8-merous, or the dispersal different, or both.
        3. Lfts ciliolate with at least a few relatively long hairs 0.4 mm or longer; ovules mostly 8–19; w. United States and Mexico, from trans-Pecos Texas to centr. Arizona, s. to Querétaro, Jalisco, and s. Baja California. 51/IVc. var. *mensalis* (p. 837).
        3. Lfts marginally glabrous or ciliolate-scaberulous distally with hairs not over 0.35 mm; ovules mostly 5–10; widespread in e. United States, from Illinois to New England, s. to e. Texas and peninsular Florida, thence s. to w. Cuba, Jamaica, and Honduras.
            4. Stems distally with leaf-stalks and pods pilose with yellowish setae commonly up

to 1–2.5 mm; Atlantic coastal plain from s.-e. South Carolina to peninsular Florida and s.-ward. 51/IVd. var. *aspera* (p. 838).

4. Stems, lf-stalks and pods puberulent or pilosulous with hairs only up to 0.2–0.7 mm, or sometimes fully glabrous; endemic to United States, n.-ward from s. Florida. 51/IVe. var. *nictitans* (p. 840).

1. Style dilated distally, at apex (0.3, but only when very short) 0.35–0.7 mm diam; Mexico, Cuba, Jamaica, s. to n.-w. Peru; w. Argentina. Androecium 4–10-merous. 51/IVb. var. *jaliscensis* (p. 834).

51/IVa. Chamaecrista nicititans (Linnaeus) Moench subsp. **nictitans** var. **diffusa** (DeCandolle) Irwin & Barneby, stat. nov. *Cassia diffusa* DeCandolle, Mém. Soc. Phys. Genève **2**(2): 130. 1824.—". . . née dans le jardin de Genève de graines recueillies dans l'île de Porto-Ricco, par . . . Bertero, et . . . communiquées par . . . Balbis."—Holotypus, dated 1822, G-DC!—*Chamaecrista diffusa* (DeCandolle) Britton, Ann. Mo. Bot. Gard. **2**: 41. 1915.—Non *C. glandulosa* var. *diffusa* O. Kuntze, 1891.

Cassia brachycarpa Urban, Feddes Repert. **15**: 312. 1918.—". . . in Sto. Domingo prope Constanza in pinetis 1190 m alt., fl. June.: v. Türckheim no. 3336."—Holotypus, presumably †B; isotypi, M, NY!—*Chaemaecrista brachycarpa* (Urban) Britton & Rose, N. Amer. Fl. **23**(5): 291. 1930.

(?)*Cassia exunguis* Urban, Feddes Repert. **15**: 312. 1918.—"Habitat in Portorico inter Aibonito et Coamo . . . m. Nov. flor. et fruct.: Sintenis no. 2831."—Holotypus, †B; no isotypus located.—*Chamaecrista exunguis* (Urban) Britton ex Britton & Rose, N. Amer. Fl. **23**(5): 289. 1930.

Chamaecrista chamaecrista sensu Pennell, 1917, p. 348; sensu Britton & Rose, 1930, p. 290, exclus. syn. Linn., Moench., Rafin., necnon var. *jaliscensi*.

Chamaecrista pavonia (sub tabula "*pavonis*") *Americana, siliqua multiplici* Breynius, Exot. Pl. Cent. Prima 66, t. 24. 1678.—"In Americana Insula *Curacao* dicta, oritur."

Heteromorphic in growth-habit, either erect and simple from a shallow annual root in maturity becoming branched distally, or diffusely to humifusely branched from a usually monocarpic but sometimes short-lived perennial root, the stems at anthesis (0.3–)0.5–6(–7) dm, puberulent with fine incurved hairs up to 0.2–0.7 mm and the stems distally with 1f-stalks in addition often pilose with longer more widely spreading-ascending ones up to 0.7–1.7 mm, the lifts commonly puberulent beneath and glabrous above, sometimes glabrous on both faces, always at least minutely ciliolate, sometimes pilose-ciliate.

Adult lvs either associated with fls or produced only beyond the early fls (in starveling or very young individuals sometimes absent) 1.5–6.5(–7) cm, early (rarely all) lvs shorter and simpler; petiole 2–5 mm; petiolar gland 1, circular, shallowly saucer-shaped or almost discoid, (0.15–)0.2–0.5(–0.6) mm diam, stipitate, the stipe either slenderly linear or stouter and gradually dilated upward, (0.2–)0.3–1 mm tall, the whole longer than or as long as diam; lfts of early lvs and those of many lateral branchlets 5–8, of most adult lvs (7–)8–20(–22) pairs, linear or narrowly oblong, up to (3–)5–15(–18) × 1–2.6 mm.

Pedicels 1–3, at anthesis 0.5–4(–4.5), in fruit (1.5–)2–6 mm; sepals up to (3.5–) 3.8–6(–6.2) mm; long abaxial petal and cucullus of about equal length, 3.7–6 (–6.2) mm, 0.8 mm shorter to 0.2(–0.5) mm longer than the longer sepals; stamens usually 10, 5 much smaller, 1 rarely staminodal or wanting, 3 larger with anthers up to (1.8–)2.1–4.2(–4.5) mm; ovary either strigulose or pilosulous; style 0.9–1.6 mm, not or scarcely dilated, at apex 0.2–0.3 mm diam.

Pod (14–)17–37 × 2.4–3.7(–4) mm, the fuscous, reddish-castaneous, or nigrescent valves either incurved-puberulent or finely spreading-pilosulous; seeds 1.9–2.5 mm, the testa overall light brown, or this fuscous-pitted, or fuscous overall, dull.—Collections: 95.

Rocky and sandy seashores, savannas, streamside sands and gravels, sometimes on dunes or in pinewoods, becoming weedy in waste places, fallow fields, gardens, and in packed soil of paths and highway margins, mostly below 300 but in Hispaniola up to 1200 and in n. Venezuela to 1600 m, locally common but discontinuously dispersed around the e. and centr. Caribbean basin, from Jamaica through e. Cuba (Oriente), Hispaniola, Puerto Rico, Virgin Is. (St. Thomas, St. John, St. Croix), Leeward Is. (St. Barthélemy, St. Kitts), Windward Is. (Guadeloupe, St. Vincent, Grenada) to the Lesser Antilles (Aruba, Curaçao, Bonaire) and n. Venezuela (Isla Margarita, seaward slopes of Miranda, Lara and Distrito Federal, and Penísola de Paraguaná in Falcón); extending feebly n. from Haiti to the Bahamas (South Caicos and New Providence), there sympatric with var. *aspera*; apparently disjunct near 250 m in the middle Orinoco (Atabapo) valley in n.-e. Vaupés, Colombia.—Fl. primarily X–III, but in weedy circumstances or granted moist subsoil intermittently through the year.

Our concept of var. *diffusa* is nearly equivalent to Britton's *Chamaecrista chamaecrista,* and includes a diversity of growth- and pubescence forms collectively characterized by small stipitate petiolar glands, ciliolate leaflets, a tiny decandrous flower with two larger abaxial petals which are subequal in length and shorter to barely longer than the sepals, and a linear style. It is of interest to note in passing that Britton in this case permitted within one species both erect annual and prostrate more enduring life-forms, doubtless because he had himself found them growing together on the islands of St. Kitts and Curaçao, as they have been encountered elsewhere by others (see the suite of specimens, NY, US, collected by Leonard on Ile de la Tortue, Haiti, nos. *11605, 11708* erect, nos. *13339, 14066* prostrate).

As characterized herein, var. *diffusa* differs from the sympatric var. *glabrata,* the dwarfer forms of which are virtually identical in habit, in the smaller flower, the tack-shaped rather than sessile or obconic petiolar gland, and the generally narrower pod; from mostly allopatric vars. *aspera* and *nictitans* in the ciliolate leaflets and usually perfect decandry, from the first further in the more dilated gland and from the latter, but less consistently than from marginally sympatric var. *aspera,* in having the two carinal petals of subequal length. The sometimes habitally inseparable var. *jaliscensis* differs principally only in the dilated style, while distantly allopatric var. *mensalis,* which has the same style, differs in the more reduced androecium and more unequal carinal petals.

The var. *diffusa* was described first by Breynius from Curaçao in 1678, and his figure, excellent except for the fancifully misrepresented androecium, was cited by Linnaeus in the protologue of *Cassia chamaecrista.* Our reasons for rejecting this element of the mixed *C. chamaecrista* as alone typical, the choice of Pennell (1917) and all subsequent students of the group, were set out under *C. chamaecrista* Linn. emend. Benth. in an earlier paper (Irwin & Barneby, 1976, p. 437–439).

51/IVb. Chamaecrista nicititans (Linnaeus) Moench subsp. **nictitans** var. **jaliscensis** (Greenman) Irwin & Barneby, comb. nov. *C. leptadenia* var. *jaliscensis* Greenman, Proc. Amer. Acad. Sci. **41**: 239 ("*jaliscense*") 1905.—"MEXICO. State of Jalisco: Guadalajara, September, 1886, *Dr. Ed. Palmer,* no. 460 . . . State of Guerrero: vicinity of Acapulco, October, 1894, to March, 1895, *Dr. Ed. Palmer,* no. 23 . . ."—Lectoholotypus, *Palmer 23*, GH! isotypi, MO, NY, US! paratypi, intergradient to var. *mensalis, Palmer 460,* GH, NY, US!

Cassia riparia Humboldt, Bonpland & Kunth, Nov. Gen. & Sp. **6**(qu): 369. 1824.—". . . in ripa fluminis Magdalenae, prope Mompox [=Mompós, dept. Bolívar, Colombia]."—Holotypus, P-HBK!—*Chamaecrista riparia* (Humboldt, Bonpland & Kunth) Britton, Bull. Torr. Club **44**: 11. 1917.
Cassia broughtonii Fawcett & Rendle, J. Bot. (London) **55**: 37. 1917.—"*Hab.* Jamaica, . . . *Broughton*! *Macfadyen*! King's House Grounds . . . *Harris* 6949."—Lectoholotypus, *Broughton Herb. III/228,* BRISTM (not seen) = photo mounted with *Harris 6949,* BM! paratypi, *Harris 6949,* BM, NY!
Chamaecrista wilsonii Britton & Rose, N. Amer. Fl. **23**(5): 290. 1930.—"Honduras . . . shore of Laguna Quemada, March 3, 1903, *Percy Wilson 634.*"—Holotypus, NY! isotypus, US!—*Cassia wilsonii* (Britton & Rose) Standley, Field Mus., Bot. **10**: 219. 1931, technically illegitimate, wanting citation.
Chamaecrista longicoma Britton & Rose, N. Amer. Fl. **23**(5): 292. 1930.—"Rio Florido, Sinaloa, *M. P. Dehesa 1505.*"—Holotypus, US!
Chamaecrista nicoyana Britton ex Britton & Rose, N. Amer. Fl. **23**(5): 292. 1930.—"Near Nicoya, Costa Rica, December, 1899, *Tonduz 13556.*" —Holotypus, NY! isotypi, K, US!—*Cassia nicoyana* (Britton) Standley, Field Mus., Bot. 18: 516. 1937.
Chamaecrista stenocarpa sensu Britton & Rose, N. Amer. Fl. **23**(5): 289. 1930; non *C. stenocarpa* Vog.
Chamaecrista chamaecrista sensu Britton & Rose, op. cit. 290, p.p.; non *C. chamaecrista* Linnaeus.
Cassia stenocarpa sensu Schery, Ann. Mo. Bot. Gard. **38**[=Fl. Panama **5**(3)]: 59. 1951.

Monocarpic but highly variable in stature and vigor, at anthesis (0.4–)1–9(–12) dm, the stem either simple and erect, or branched from near or only well above base (the secondary axes then incurved-ascending), or less often diffuse or humifuse, the most robust forms basally lignescent in age, the whole highly variable in pubescence, the stems and lf-stalks usually both puberulent with short incurved-subappressed hairs and pilose with spreading, pallid or yellowish setae up to 1–2(–2.3) mm, but either type of hair sometimes lacking, the lfts commonly glabrous on both faces, sometimes puberulent beneath, their margins usually either minutely scabrous-ciliolate to ciliate with random longer setae, or rarely smooth, the pod nearly always loosely pilose, exceptionally incurved-puberulent.
Adult lvs (1.5–)2.5–9.5(–11) cm; petiole 2–5(–6) mm; petiolar glands 1–2, orbicular, cupulate or almost discoid, (0.3–)0.4–1(–1.1) mm diam, stipitate, the stipe either slenderly linear or dilated upward, the whole 0.5–1.7 mm tall, varying in profile from slenderly trumpet-shaped or pileiform to tack-shaped, rarely broadly subuliform, and at least as tall, usually taller than diameter of head; lfts of larger lvs up to (9–)12–25(–28) pairs, linear or narrowly oblong, straight or slightly falcate, up to 6–21 × (1–)1.3–3 mm.
Pedicels at anthesis (0.5–)1–3.5, in fruit 2–6 mm; longer sepals 3.5–6.8(–8) mm; petals strongly asymmetric, 2(–3) adaxial much shorter than the rest, the longer (1–)2 and the cucullus (2.8–)3.5–6.7(–7, exceptionally –8) mm; stamens (4–)5–10, the longer anthers 1.7–4.5(–4.8) mm; ovary densely subappressed-pilose or barbate; style 0.8–1.6(–2) mm, dilated upward and hooked, at apex (0.3–)0.35–0.7 mm diam; ovules (7–)8–15(–16).
Pod (20–)24–52 × 2.5–4.6(–5) mm, the valves usually reddish or castaneous, sometimes green with red sutures, finally nigrescent, loosely pilose or less often merely puberulent; seeds 2.5–3.3(–3.5) mm, the testa fuscous or brownish-nigrescent, not or scarcely lustrous, lineolate, the pits often darker-colored than the ground.—Collections: 174.—Fig. 36 (petiolar nectary).

Open sandy and gravelly hills, savannas, glades in light or disturbed woodland, shores and beaches, becoming a weed along highways, paths and in gardens or orchards, mostly below 700 but in Guatemala up to 1200, in s. Mexico up to 1740, in Ecuador up to 2200, in n.-w. Argentina reportedly up to 3000 m, widespread and common in Central America from the Isthmus of Tehuantepec in Mexico to the lower Magdalena valley in Colombia, extending n. in Mexico into the Balsas Depression and (gradually passing into var. *mensalis*) to the Lerma-Santiago valley in Jalisco and (somewhat disjunct) to the Pacific coast and foothills of s. Sinaloa, also n.-e. around the Caribbean slope to s. Tamaulipas, the Yucatan peninsula, and feebly represented on w. Cuba (Pinar del Río; La Habana) and Jamaica; s. highlands (Loja) and Pacific lowlands (Guayas) of Ecuador; lowland n.-w. Peru (Piura and Tumbes); highly disjunct in n.-w. Argentina (Salta and Jujuy), where perhaps introduced.—Fl. in Mexico and Centr. America VIII–II, in Argentina XII–III, but intermittently through the year if moisture permits, and capable of abrupt regeneration following unseasonable storms.

Our concept of var. *jaliscensis* encompasses a horde of growth- and pubescence-forms in which a small, highly asymmetric flower with apically thickened style coincides with evidently but variably stipitate petiolar glands as tall or taller than the diameter of the head. Its diverse elements differ collectively from partly sympatric varieties of subsp. *patellaria* in gland-type, from the latter's var. *ramosa* further in the less asymmetric midrib of the leaflets but not in the flower, and from its var. *glabrata* further in the asymmetric flower but not in the leaflets. From its closer kin within ssp. *nictitans* it is finally distinguished only by the dilated style, and from subsp. *disadena* var. *pilosa,* which has the same style, by the strongly asymmetric corolla and (often) reduced androecium.

The more immediately notable minor variants of var. *jaliscensis* are those in which the ordinarily abundant long setae of stem, leaf-stalk and stipule-margin are few, small, or wanting, and those which, due probably to edaphic accident, fail to develop a strong erect primary axis and in consequence assume a diffuse or trailing habit. The flower, always very small relative to the foliage, varies considerably in length and amplitude, but this is imperfectly correlated with vigor of growth. Number of stamens is clearly linked to amplitude of the perianth, a 9–10-merous androecium being found commonly in the larger flower that predominates on the Caribbean slope in Mexico and southward, dwindling in Mexico northwestward from Chiapas (except locally in Sinaloa) into a 5–7-merous one. At the same time the style-tip, emphatically dilated southward from Guerrero, fades by insensible steps northwestward into the linear style of var. *mensalis,* ultimately distinguished by this feature alone. The girth of the pod varies from quite as slender as that of var. *mensalis,* which is commonly 2.5–3.5 mm, up to 4.5 or even 5 mm in the robust forms which correspond to Britton's misconception of *Cassia stenocarpa* Vog. The named segregates which we refer to var. *jaliscensis* are characterized by particular syndromes of the variable features:

i. *Cassia riparia:* Described from an ageing plant in fruit only, but apparently representing the form of var. *jaliscensis* common in lower Magdalena valley; pod hirsutulous.
ii. *Cassia broughtonii:* Robust and erect, puberulent but scarcely setose below the pod. This was equated even by Britton with the widespread plant of the Central American states which he called *Ch. stenocarpa*.
iii. *Chamaecrista wilsonii:* As the last, but the pod incurved-puberulent, not pilose.
iv. *Chamaecrista longicoma:* Pubescent as *C. broughtonii* but peculiar in var. *jaliscensis* for the relatively large flower, the long petals up to ±8 mm. A closely similar plant with slightly taller glands (*Montes & Salazar 507,* US), also from Sinaloa, was referred by Britton to *Ch. stenocarpa*. Both have the long-barbate ovary common in Panamanian var. *jaliscensis*, 9 or 10 stamens,

and the dilated style characteristic of their kind, in these characters differing widely from sympatric var. *mensalis* of the Sinaloan foothills. Fruiting specimens (*Mexia 380,* CAS) from the same region have pilose pods ±5 mm wide and probably represent the same localized large-flowered race.

v. *Chamaecrista nicoyana:* Slender, procumbent, short-leaved, either densely or sparsely setose below the pilose pod. The type, from "sommets gazonnés" suggests an ecotype from extremely poor stony soil, comparable to plants flowering at a stature of 5 cm which were collected in river sand in El Salvador (*Standley 21192,* US). Elsewhere the prostrate or diffuse-incurved growth-habit is associated either with small leaves and flowers (*Breedlove & Raven 13244* from Chiapas; *Purpus 7850* from Veracruz), or with relatively large leaves and flowers, this in Honduras either inland (*Molina 24654,* NY) or on the ocean beach (*Schipp 720,* GH, NY).

51/IVc. Chamaecrista nictitans (Linnaeus) Moench subsp. **nictitans** var. **mensalis** (Greenman) Irwin & Barneby, Phytologia **44**(7): 501. 1979. *Cassia leptadenia* var. *mensalis* Greenman, Proc. Amer. Acad. **41**: 238. 1905.—"MEXICO. State of Chihuahua: Hacienda San Miguel, near Batopilas, September, 1885, *Dr. Ed. Palmer,* no. 175 . . . State of Sonora: vicinity of Guaymas, on mesas among grass, October, 1887, *Dr. Ed. Palmer,* no. 242 . . ."—Lectoholotypus, *Palmer 242,* GH! isotypi, NY, US! paratypi, *Palmer 175,* GH, NY, US!

Cassia leptadenia Greenman, Proc. Amer. Acad. **41**: 238. 1905.—"UNITED STATES. Texas: . . . 48 km east of El Paso, May to October, 1849, *Chas. Wright,* no. 154 . . ."—Lectoholotypus (Pennell, 1917, p. 355), *Wright 154,* GH! isotypus, US!—Greenman cited 11 collections; all conspecific; the later lectotypification by Pullen (1963, p. 69), *Bigelow s.n.* from Comanche Spring, is not sustained.—*Chamaecrista leptadenia* (Greenman) Cockerell, Muhlenbergia **4**: 68. 1908.

Chamaecrista leptadenia sensu Britton & Rose, 1930, p. 297. *Cassia leptadenia* sensu Isely, 1975, p. 105, map 44 (U.S.A. only).

Decisively monocarpic but variable in vigor and stature, at anthesis (0.5–)1–6(–8) dm, erect or exceptionally diffuse, when small simple or few-branched from base or middle upward, when stouter bushily branched from near or well above base, the stems distally and lf-stalks simply puberulent with short incurved hairs or both puberulent and thinly setulose with spreading-incurved ones up to 0.4–1.1(–1.3) mm, the lfts usually glabrous on both faces, ciliolate with at least a few, but sometimes random setules 0.4+ mm.

Adult lvs (2–)2.5–8(–9) cm; petiole (1.5–)2–5(–6) mm; petiolar gland 1(–2), always stipitate, 0.25–0.9 mm diam, in profile tack- or slenderly trumpet-shaped and 0.3–1.3 mm long, almost always taller than diam; lfts of adult lvs (7–)10–22(–25) pairs, linear or narrowly oblong, up to (3–)4–14 × ±1–3 mm.

Pedicels at anthesis 0.5–4, in fruit 2–8 mm; sepals 3–5.6(–6) mm; petals often fading pinkish, 4 equalling or shorter than the longer sepals, the fifth, opposed to the short cucullus, commonly but not always exserted, 3.5–7(–7.5) mm; stamens (4–)5–8, the long anthers ±2–3 mm; ovary strigulose or rarely pilosulous; style 0.8–1.5 mm, linear or at tip very slightly dilated, at the stigma 0.2–0.3 mm diam; ovules (6–)8–19.

Pod (17–)20–48(–56) × 2.5–3.8(–4) mm, the reddish- or purplish-castaneous, rarely green valves commonly puberulent, sometimes loosely pilose; seeds 2.5–3.2 mm, fuscous or blackish, dull, the lineoles often appearing as lines of fuscous-margined craters.—Collections: 68.

Arid grasslands, thorn-forest, sandy and gravelly desert slopes and washes, roadsides, and openings in pine-oak- or oak-forest, near sea level on Gulf of California up to 2000 m in n. Sa. Madre Occidental, widespread and following summer rains locally abundant, s.-w. United States from trans-Pecos Texas w. through s.-w. New Mexico to the middle Gila valley and Baboquivari Mts. in s.-centr. and s. Arizona, s. into Mexico along the w. slope of Sa. Madre and around margins of Sonoran Desert in Sonora, through Sinaloa to Nayarit and n.-e. Jalisco; s. Baja California (Sa. de la Giganta and Cape region); and from Texas s. more sparingly along the e. slope of Sa. Madre and margins of the Chihuahuan Desert to Zacatecas and Querétaro, rarely s. to n. Michoacán where passing insensibly into var. *jaliscensis*.—Fl. VIII–XI, into II s.-ward.

Observers in the United States are accustomed to think of var. *mensalis* as the western species *Cassia leptadenia,* admittedly related to eastern *C. nictitans* but conveniently separated from it in Texas by a gap of some 350 miles which approximately corresponds with the limestone caprock of Edwards Plateau (cf. Turner, 1959, maps 33, 35). However, in the larger view these two entities, which are, or can be, essentially identical in habit, foliage, flower-structure and pod, and differ only slightly in features of pubescence and average pod size, are seen to be northerly outliers of a tropical American complex to which each is inseparably connected by intermediate forms, *C. nictitans* southward through the West Indies and *C. leptadenia* southward through western Mexico. The present geographical discontinuity suggests that they arose independently from different immediate southern antecedents and have followed separate paths northward on either side of the Continental Divide; if so, they exhibit a remarkable morphological parallelism, which extends to the often much degraded androecium.

The two differential characters traditionally ascribed to *C. leptadenia* are a relatively narrow, many-seeded pod, and long-ciliolate leaflets. To these Pennell (1917, p. 341) added a supposed difference in relative length of the odd, long petal, but this has proved a delusion. Turner found (1959, p. 73) that in Texas the stipules of *C. nictitans* are longer than those of *C. leptadenia,* but this difference evaporates south of the border. While the pod of typical var. *nictitans* is indeed, on the average, shorter, wider, and fewer-seeded than that of its western analogue, all recent authors allow some overlap in measurements, and Pullen (1963, p. 55) was reduced to adopting marginal pubescence of the leaflets as the sole reliable criterion. Theoretically feeble though the difference is, especially in the context of sect. *Chamaecrista,* it is the only absolute one known. Isely (1975, l.c., maps 44, 49) separated the two species by geography alone and so evaded the need for morphological characters.

Internally var. *mensalis* varies greatly in stature, the stressful climates of the Sonoran Desert and Sierra Madrean arid grasslands having apparently selected thrifty strains which mature very rapidly, like associated desert annuals, and may bear fruits at axils beginning only three or four leaves removed from the cotyledons. But such plants can continue to vegetate and flower so long as moisture and temperature permit. Southward in Mexico coarser and more leafy strains predominate, some of which may perhaps winter-over, although essentially monocarpic. Southward from the valley of Río Grande de Santiago in Jalisco var. *mensalis* is replaced by var. *jaliscensis,* different in the dilated style-tip, and in the area of overlap and intergradation commonly more densely setose.

51/IVd. Chamaecrista nictitans (Linnaeus) Moench subsp. **nictitans** var. **aspera** (Elliott) Torrey & Gray. *Cassia aspera* Muhlenberg ex Elliott, Sketch Bot. S. Car. & Ga. **1**: 474. 1821.—"On Eding's island near Beaufort

[Beaufort Co., South Carolina], common . . ."—Holotypus, the protologue! Not represented (Weatherby, Rhodora **44**: 249, sequ. 1942) in hb. Elliott., CHARL.—*C. nictitans β aspera* (Elliott) Torrey & Gray, Fl. N. Amer. **1**: 396. 1838. *Nictitella aspera* (Elliott) Rafinesque, Sylva Tell. 128. 1838. *Chamaecrista aspera* (Elliott) Greene, Pittonia **3**: 243. 1897.

Cassia simpsoni Pollard, Bull. Torr. Club **21**: 221. 1894.—"Florida—Big Pine Key, [J. H.] Simpson (May, 1891), no. 174."—Holotypus, US!—*Chamaecrista simpsoni* Pollard ex A. Heller, Cat. N. Amer. Pl., ed. 2, 5. 1900. *Cassia aspera* var. *simpsoni* (Pollard) Macbride, Contrib. Gray Herb., n. ser. **59**: 25. 1919.

Chamaecrista confusa Britton ex Britton & Rose, N. Amer. Fl. **23**(5): 292. 1930.—". . . pinelands, Harold Road, New Providence, Bahamas, September 1, 1904, *Britton & [L. J. K.] Brace 427.*"—Holotypus, NY!—*Cassia caymanensis* C. D. Adams, Phytologia **20**(5): 310. 1970 (non *C. confusa* Philippi, 1894).

Chamaecrista aspera cum var. *β* sensu Pennell, 1917, p. 361; Britton & Rose, 1930, p. 298. *Cassia aspera* sensu Pullen, 1963, p. 70; Isely, 1975, p. 67, map 19. *C. nictitans* var. *aspera* sensu Gillis, Phytologia **35**(2): 89. 1976.

Chamaecrista simpsoni sensu Britton & Rose, 1930, p. 298.

Chamaecrista riparia sensu Britton, Bull. Torr. Club **44**: 11. 1917 (basionym. exclus.). *Cassia caymanensis* sensu Adams, 1972, p. 328.

Annual or sometimes over-wintering monocarpic, perhaps randomly perennial, at anthesis 1.5–7.5(–10) dm, the stem simple erect, simple and bushy-branching from either below or near middle, or procumbent and humifuse, either throughout or distally minutely puberulent and also, with lf-stalks, pilose with divaricate yellowish setae up to 1.1–2.5 mm, but the puberulent or pilose vestures, exceptionally both, sometimes greatly reduced or lacking, the lfts glabrous on both faces, at least at tip minutely scabrous-ciliolate with forwardly directed hairs less than 0.35 mm.

Adult lvs (1.5–)2–7.5(–9) cm; petiolar glands 1(–2), discoid or shallowly concave 0.2–0.7 mm diam, stipitate, the stipe commonly dilated upward, in profile either narrowly claviform or sometimes narrowly trumpet- or tack-shaped, 0.3–0.9(–1.1) mm tall, as long or longer than diam of gland; lfts of adult lvs up to (12–) 16–28(–32) pairs, linear or linear-oblong, the largest 4–14 × 1–2.3 mm, some often (especially of later lvs, and outside Florida) becoming falcately twisted and appearing obliquely lanceolate.

Pedicels at anthesis 1–2.5 mm, in fruit 2–4.5 mm; sepals 3.3–5.4 mm; long abaxial petal 5–8(–9) mm, ±1.3–2.1 times as long as the oblique cucullus; stamens 5–8(–"9"), the long anthers 1.6–2.8 mm; ovary loosely pilose (exceptionally glabrous); style linear 0.9–1.3 mm, at tip 0.15–0.25 mm diam; ovules 5–9(–13).

Pod (15–)18–32(–36) × (3.5–)4–5.5 mm, the valves purplish-castaneous, paler over the seeds, finally nigrescent, almost always pilose with fine spreading setae 1.2–2.4 mm; seeds fuscous-brown or blackish, dull or sublustrous, finely lineolate, (2.2–)2.4–3.3 mm.—Collections: 52.

Low seasonally moist sandy pinewoods, beaches, dunes, sometimes on limestone pavement or coral detritus, becoming weedy and locally abundant in old fields, along paths, ditches and railways, mostly below 50 m, widespread over inland and coastal peninsular Florida n.-ward from the Keys and Everglades, extending n.-w. to Gadsden County and along the Atlantic coast into the s.-e. corner of South Carolina; n. Bahama Is. (Grand Bahama, Abaco, Andros, New Providence and Eleuthera); w. Cuba (La Habana); Grand Cayman I; Jamaica (mostly western); Honduras (Corozal distr., a weed, and intermediate to var. *jaliscensis*).—Fl. in U.S. mostly VIII–I, in W.I. interruptedly through the year.

Our concept of var. *aspera,* only slightly, insofar as it includes West Indian

Ch. confusa, more comprehensive than that of *C. aspera* current in United States (Pullen, 1963, p. 70; Isely, 1975, p. 67), stands midway between vars. *nictitans* and *jaliscensis,* differing from the first principally in the long setose pubescence of the upper stems, leaf-stalks and pods, and from the latter, often similar in pubescence, principally in the linear style. To distinguish it from distantly allopatric var. *mensalis* we are obliged to fall back once again on the inconsiderable morphological difference of shorter cilia on the leaflets, but the pod is generally broader, shorter, and fewer-seeded. The prostrate, short-leaved form of var. *aspera* (=*C. simpsoni*), found in southern peninsular Florida and the Keys, closely resembles in habit and foliage the West Indian var. *diffusa,* but differs in the strongly asymmetric corolla, nearly always reduced androecium, and again a broader pod.

The differences supposed by Britton to exist between Bahamian *Ch. confusa* and *Ch. aspera* of Florida and western Cuba scarcely emerge from his account in *North American Flora,* the first, often annual in fact, being keyed out in the supposedly perennial group *Patellariae* whereas the latter, along with the admittedly sometimes perennant *Ch. simpsoni,* appears among the annual group *Fasciculatae*. We find them identical except for one feature, the gland being in continental *aspera* generally claviform as seen in profile and not more than 0.35 mm diameter at tip, in the plants of Bahamas and Jamaica more definitely tack-shaped and up to 0.4–0.6 mm diameter. Both types of gland occur, however, in Cuba, and the whole variation recorded is less than that unanimously accepted within continental var. *nictitans*.

The plants called *Ch. simpsoni,* treated by Pennell (1917, p. 361) as an unnamed minor variant of *Ch. aspera* and by Pullen (1963, p. 72, in *Cassia*) as "nothing more than an ecotype" of the same, stand in relation to var. *aspera* exactly as the prostrate, short-leaved *Ch. nicoyana* stands to erect or ascending var. *jaliscensis,* repeating a pattern of habital variation found throughout the whole specific complex of *Ch. nictitans*. While the contrast between the most robust, erect, bushy-branching and the most slender, diffuse or humifuse extremes of each variety is visually arresting, these are connected in each case by a chain of intermediate types. It may be noted in this particular case that Adams (1972, p. 328, as *C. caymanensis*) recognizes both growth forms in Jamaican var. *aspera*. Isely (l.c.) draws attention to individual variants of the *simpsoni* type with merely puberulent or even fully glabrous (cf. *Small 3787,* p.p., NY) stems, foliage, and pod; these are particularly suggestive of West Indian var. *diffusa* although retaining the asymmetric flower and few stamens of var. *aspera*.

51/IVe. Chamaecrista nictitans (Linnaeus) Moench subsp. **nictitans** var. **nictitans**. *Ch. nictitans* Moench, 1794, l.c., exclus. syn., emend. Greene, Pittonia **3**: 243. 1897, as originally published a technical (by retroactive ruling) synonym of *Cassia chamaecrista* L., but the unique epithet obviously derived from *C. nictitans* L. and the accompanying description of a pentandrous androecium decisive.—*Cassia nictitans* Linnaeus, 1753, l.c., sens. str.—"*Habitat in* Virginia."—Lectotypus (Pennell, 1917, p. 356): Hortus Cliffortianus t. 36. 1737! a sterile typotypus survives in Hb. Cliffort., *Cassia No. 1* (exclus. flor.), BM!—*Nictitella amena* Rafinesque, Sylva Tell. 128. 1838, nom. illegit. *Cassia chamaecrista* ζ *nictitans* (Linnaeus) O. Kuntze, Rev. Gen. 169. 1891.

Cassia procumbens Linnaeus, Sp. Pl. 380. 1753.—"Habitat in Indiis," but based on *Cassia americana procumbens herbacea, mimosae foliis* . . . J. Ammann, Acad. Petrop., Commentarii **12**: 240. 1740.—". . . in horto Academico . . . ex seminibus . . . circa Philadelphiam urbem in Pennsylvania, Americae septentrionalis . . . collectis."—Lectoholotypus, the quoted description of Ammann.—The specimen LINN *528/34*! labelled *C. procumbens*, identified by Bentham, 1871, p. 578, with *C. pygmaea* DC., is not known to have been in herb. Linn. in 1753. Cf. Pennell, 1917, p. 357.—*Chamaecrista procumbens* (Linnaeus) Greene, Pittonia **4**: 28, quoad nom. 1899.

Cassia chamaecrista ζ *nictitans* fma. *pubicaulis* O. Kuntze, Rev. Gen. 169. 1891.—"U[nited] St[ates]: Garrison am Hudson-fluss und Sickestown." Lectotypus, *Kuntze 2475*, collected at Garrison, N.Y., 12 August 1874, NY!

Cassia multipinnata Pollard, Bull. Torr. Club **22**: 515, pl. 250, pl. 252, fig. 3. 1895.—"Near Jacksonville [Florida]. A. H. Curtiss, North American Plants. No. 512 [properly 712], distributed as *C. nictitans* and *C. nictitans* var. *aspera*; second distribution, No. 5157, September 15, and October 27, 1894. Tallahassee. G. V. Nash, No. 2403, August 12, 1895."—Lectoholotypus (Pennell, 1917, p. 359), *Curtiss 5157*, US! paratypi, *Curtiss 712*, NY (3 sheets), US!—*Chamaecrista nictitans* β Pennell, l.c. *Cassia nictitans* var. *multipinnata* (Pollard) Macbride, Contrib. Gray Herb., n. ser. **59**: 25. 1919.—Reduced to *C. nictitans* by Pullen, 1963, p. 75 and by Isely, 1975, p. 113.

Cassia multipinnata var. *nashii* Pollard, Bull. Torr. Club **22**: 515. 1895.—". . . River Junction, Gadsden Co., Florida, by Mr. [G. V.] Nash, September 5, 1895 (n. 2577)."—Holotypus, NY (ex hb. Columbia Coll.)! isotypi, LE, US!—Reduced to *Ch. multipinnata* by Pennell, 1917, l.c.

Cassia aspera var. *mohrii* Pollard, Bull. Torr. Club **24**: 150. 1897.—"Type in the herbarium of the U.S. Geol. Surv. of Alabama, collected in Mobile in 1878 by Dr. Mohr . . ."—No typus seen.—*Chamaecrista aspera* var. *mohrii* (Pollard) Pollard ex A. Heller, Cat. N. Amer. Pl. ed. 2, 5. 1900. *Ch. nictitans* γ Pennell, Bull. Torr. Club **44**: 360. 1917. *Cassia nictitans* var. *mohrii* (Pollard) Macbride, Contrib. Gray Herb., New Ser. **59**: 25. 1919. *Chamaecrista mohrii* (Pollard) Small ex Britton & Rose, N. Amer. Fl. **23**(5): 298. 1930.—Maintained by Pullen, 1963, p. 77, as a var. of *C. nictitans* but reduced to it by Isely, 1975, p. 113.

Chamaecrista nictitans var. *commixta* Pollard & Maxon, Proc. Biol. Soc. Wash. **14**: 163. 1901.—"Type . . . collected by Charles L. Pollard and William R. Maxon . . . along the New River at Quinnimont, W. Va., August 21, 1899. (No. 31)."—Holotypus, US!—*Cassia nictitans* var. *commixta* (Pollard & Maxon) Millspaugh, W. Va. Geol. Surv. **5**: 283. 1913.—Reduced by Pennell, 1917, p. 357, to *Ch. nictitans*.

Cassia nictitans var. *hebecarpa* Fernald, Rhodora **38**: 423. 1936.—"VIRGINIA . . . bluff along Chesapeake Bay, Old Town Neck, Northampton Co., October 13, 1935, *Fernald, Long & Fogg*, no. 5316 . . ."—Holotypus, GH!—Maintained by Gleason, 1952, p. 386 and by Pullen, 1963, p. 76, but reduced to *C. nictitans* by Isely, 1975, p. 113.

Cassia nictitans var. *leiocarpa* Fernald, l.c. 1936.—"KENTUCKY. Pine Mountain, Bell Co., September, 1893, *T. H. Kearney*, no. 496."—Holotypus, GH!—*Chamaecrista nictitans* var. *leiocarpa* (Fernald) Moldenke, Boissiera **7**: 2. 1943.—Maintained by Gleason, 1952, l.c. but reduced to *C. nictitans* by Pullen, 1963, p. 75, and by Isely, 1975, p. 113.

Cassia chamaecrista sensu Schrank, Pl. rar. hort. Monac. **1**(4): t. 33. 1819.

Annual but sometimes (especially s.-ward) basally lignescent in age, at anthesis 1–7 dm, the stem either erect, when varying from simple to distally branched, or diffusely branched from base, commonly puberulent or distally, with the leafstalks, pilosulous with incumbent or spreading-incurved hairs up to 0.2–0.7 mm, the lfts glabrous or less often puberulent on both faces, their margins smooth or minutely scabrous-ciliolate with hairs not over 0.15 mm, occasionally glabrous throughout.

Lvs (disregarding short simpler early ones) 2–7 cm; petiole 2–7 mm; gland usually solitary, rarely 2, discoid or shallowly saucer-shaped, red, mostly orbicular in outline, 0.3–1 mm diam, usually stipitate but the stipe varying from slender to stout and distally dilated, the whole 0.3–1 mm tall and either much longer or much shorter than diam of the head; lfts (6–)8–23(–26) pairs, linear-oblong or narrowly oblong-elliptic (4–)6–26 mm, the midrib slightly excentric, the secondary venation either faint or prominulous.

Pedicels at anthesis up to 2 mm, in fruit 1–3.5 mm; sepals 3.2–5.3 mm; long petal obovate-cuneate or spatulate, 3.5–6.5(–7) mm, either much or only a trifle longer than the oblique cucullus; stamens 4–5, ±0–2(–3) interior staminodia or

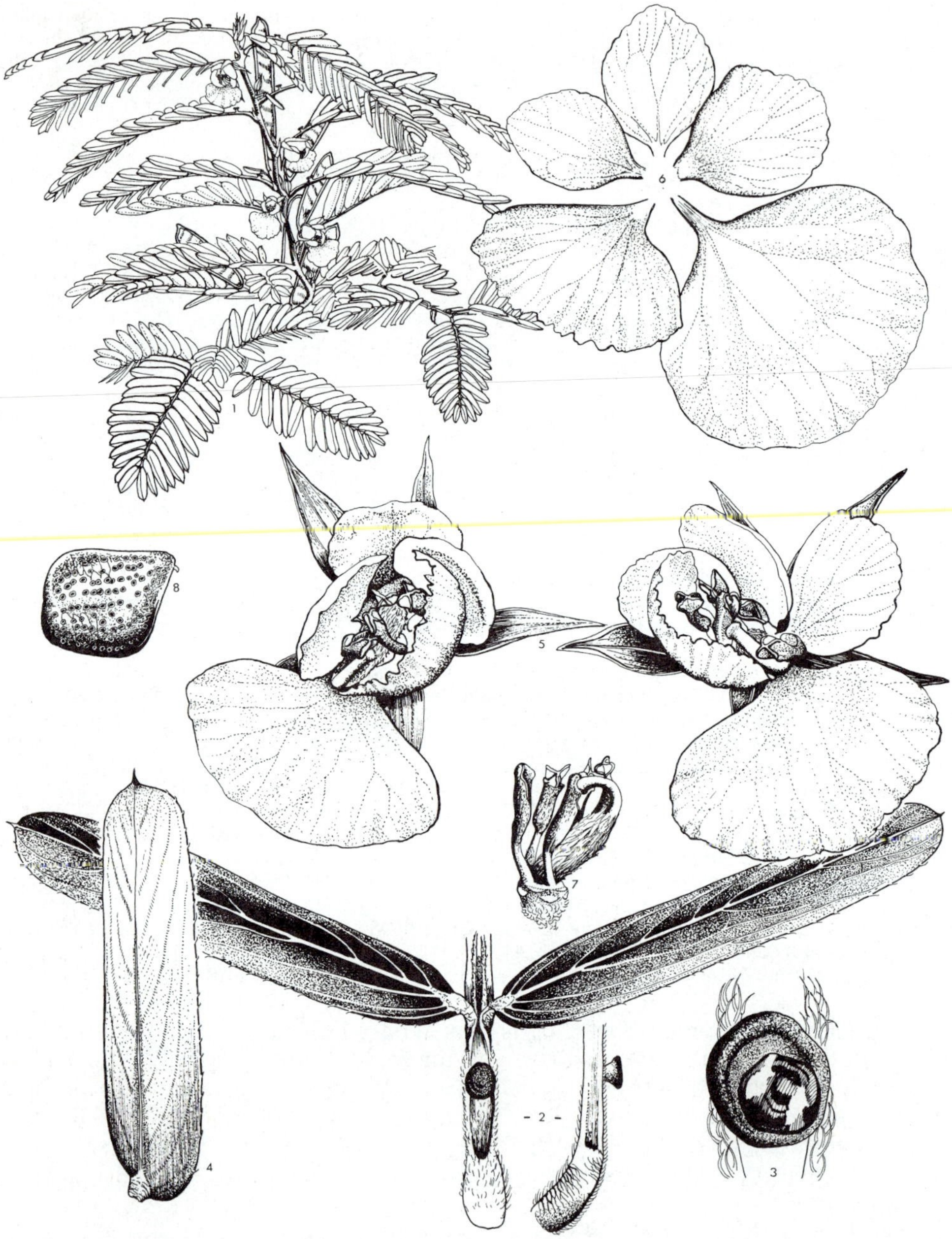

Fig. 52. *Chamaecrista nictitans* (Linn.) Moench var. *nictitans*. 1) Flowering branchlet ×1; 2) petiole and gland ×5; 3) gland ×20; 4) dorsal face of leaflet ×5; 5) a pair of flowers, showing enlarged abaxial petal alternately left and right ×5; 6) petals ×5; 7) androecium + pistil ×5; 8) seed ×5 (Drawn from life by W. S. Moye III).

degraded stamens, the long anthers (1.4–)1.7–2.8 mm; style 1–1.4 mm, slightly dilated upward and 0.25–0.4 mm diam at apex; ovary usually strigulose, sometimes minutely pilosulous or quite glabrous; ovules (4–)5–10.

Pod (unless deformed) 18–43 × (2.8–)3.2–5.5(–5.8) mm, the valves commonly puberulent, less often shortly pilosulous or fully glabrous; seeds dark brown, sometimes paler toward the hilum, sublustrous, faintly pitted, 2.7–3.4 mm.—Collections: 249.—Fig. 52.

Open woods, thickets, sandy fields and pastures, shores, hedgerows, and waste places, locally abundant and an early invader of land cleared for agriculture or by logging, widespread over e. and s.-e. United States but n. of the Tennessee V. and s. New York discontinuously scattered and perhaps only immigrant since Colonial times, Massachusetts and s.-w. Vermont to centr. Illinois, s. to centr. Florida, the Gulf Coast, s.-centr. Oklahoma, and Texas e. of Edwards Plateau; cf. Pullen, 1963, fig. 9; Turner, 1959, map 35; Isely, 1975, map 49.—Fl. (VII–) VIIII–frost, or s.-ward well into winter.

The var. *nictitans,* endemic to United States, has been subject to intense analysis in recent decades (Pullen, 1963; Isely, 1975). Internal variation, as we perceive it, is documented by the foregoing description, while differential characters distinguishing it from allied forms have been noted under vars. *aspera* and *mensalis* immediately preceding. The var. *nictitans* stands in relation to the whole neotropical *Ch. nictitans* complex as does *Cassia nomame* (Sieb.) Kitagawa to the paleotropical complex of *Ch. mimosoides* (L.) Greene. Each has migrated northward into a cold-winter climate and acquired the ability to survive severe frost in the form of seeds.

Numbered Exsiccatae of sect. *Chamaecrista*

(Figures in parentheses refer to taxa as numbered in the text)

Abbott, R. Q. 389 (51/IVb), 1679 (45a), 1699 (45a)
Abbott, W. L. 495 (51/IId)
Abraham, A. A. 69 (15/IIIa)
Acuña, J. 4405 (51/IId)
Acuña, J. & Br. Alain 2274 (32a)
Adams, P. 18 (21a)
Agostini, G. 658 (28a), 663 (51/IIb)
Alain, Bro. (Liogier) 1036 (40a), 1124 (28a), 1548 (32a), 1849 (51/IId), 2137 (46b), 4069 (51/IId), 4075 (28a), 6398 (40c), 9136 (40c), 9908 (40c), 11073 (45b), 12424 (45b), 12448 (45b), 12465 (45a), 13262 (45b), 14645 (45a), 15137 (45b), 15166 (45b), 15620 (45b), 15865 (45b), 15898 (45b), 16261 (45b), 16364 (45b), 16610 (45b), 18833 (45b), 20800 (45b).
Alain, Bro. & E. P. Killip 2143 (31)
Alain, Bro. & G. R. Proctor 6173 (31)
Albuquerque, B. W. de 1336 (51/IIIa)
Albuquerque, B. W. de & L. Coelho 385 (51/IIIa)
Alexander, E. 183 (31)
Allard, H. A. 13116 (51/IVa), 14907 (45b), 14923 (45b), 16064 (51/IId), 16481 (47b), 16489a (47b)
Allem, A. 69 (26a), 139 (24), 870 (33)
Allen, C. 26 (26a), 264 (29), 653 (21a)
Allen, G. M. 56 (10b)
Allen, P. H. 659 (51/IIIb), 1049 (31), 2772 (51/IIb), 5044 (32a)
Alston, A. H. G. 5190 (28a), 5494 (51/IIIa), 5597 (51/IIb)
Alvarado, A. 77 (51/IIIb)
Alvaranga, M. 18 (21a)
Anderson, W. R. 6419 (33), 6498 (32a), 6783 (51/I), 6826 (31), 6840 (21a), 6913 (27), 6964 (32a), 6977 (31), 6988 (21a), 7113 (31), 7391 (51/I), 7703 (33), 7823 (32a), 7888 (21a), 7964 (24), 8376 (7c), 8377 (7c), 8669 (3), 8716 (7c), 8745 (33), 8807 (51/IIb), 9000 (3), 9127 (30), 9283 (26a), 9285 (21a), 9297 (32a), 9361 (24), 9544 (31), 9781 (50/IIa), 9858 (33), 11182 (21a), 11186 (47m), 11223 (26a), 11230 (51/IIb), 11232 (32a), 11272 (27), 11280 (33), 11282 (21a), 11296 (51/I), 11325 (24), 11330 (51/IIa), 11340 (51/IIb), 11345 (26a), 11351 (27), 11378 (33), 11385 (24), 11393 (51/IIb), 11433 (24), 11435 (27), 11456 (33), 11458 (51/I), 11461 (31), 11526 (50/IIb), 11542 (15), 11563 (7a), 11567 (2), 11587 (3), 11590 (7a), 11602 (5a)
Anderson, W. R. et al. 26060 (3), 34688 (51/I), 35125 (32a), 35126 (51/IIb), 35431 (7a), 35811 (18c), 36061 (15), 36123 (4), 36263 (15), 36406 (32a), 36407 (35b), 36628 (35b), 36631 (35b), 36955 (37), 37022 (32a), 37027 (28b)
Andrade, A. G. & M. Emmerich 529/521 (26b)
André, E. F. 1398 (47g), 1919 (26a), 1974 (47k), 2514 (47e), 2814 (47e)
Apolinar Angel, Bro. 674 (51/IIb), 719 (27)
Appun, C. F. 1231 (1), 1535 (51/IVa)
Araujo, D. 497 (21a)
Arbo, M. M. 1730 (29), 1770 (29)
Archer, W. A. 358 (51/IIb), 790 (26a), 1031 (47e), 3303 (47g), 3303a (47g), 3609 (32a), 4742 (21a), 4803 (51/IIIb), 4870 (21a), 5027 (15), 7660 (50/IId)
Archer, W. A. & Melho Barreto 4907 (4), 4942 (4)
Arguelles, J. 4 (26e), 137 (51/IVc)

Aristeguieta, L. 2221 (47j), 2382 (32a), 2412 (47e), 3374 (47e), 4433 (51/IIc), 5859 (28a), 6324 (29), 7595 (51a), 7602 (32a), 7730 (51/IIIb)
Arnoldo-Broeders, M. 1878 (51/IVa), 3854 (51/IVa)
Arsène, Bro. 5647 (51/IVc), 6293 (34a)
Asplund, E. 1684 (47e), 16677 (32b)

Backer, C. A. 34534 (51/IId)
Badillo, V. 1981 (51/IIIb), 2112 (51/IIIb), 2611 (51/IIIb), 3656 (32a), 3705 (21a), 4608 (28a)
Bailey, L. H. & E. Z. 159 (51/IIIb), 171 (47c), 254 (47c), 781 (47m), 1214 (32a)
Baker, C. F. 466 (21a), 590 (26a)
Baker, J. G. 1560 (51/IIb), 3536 (51/IIb)
Balansa, B. 1414 (29), 1415 (21a), 1416 (51/IIb), 1416a (39), 1417, 1417a, 1417b (26b), 1418 (32a), 3105 (32a)
Baldwin, J. T., Jr. 4529 (51/IId)
Balls, E. K. 5478 (43a)
Bang, M. 549 (51/IIa), 613 (47e), 2081 (51/IIb)
Barbosa, H. 1161 (12)
Barclay, A. S. 1711 (51/IVc), 3535 (51/IIb)
Barclay, G. W. 283 (47e), 2794 (51/IId)
Barkley, F. A. 13833 (21b), 14498 (34a)
Barlow, F. D. 15/8 (44a)
Barneby, R. C. 17783 (47c)
Barnes, C. R. & W. J. G. Land 304 (26e), 482 (51/IVb)
Bartlett, H. H. 10139 (51/IIIb), 10245 (34a), 10554 (34a), 10932 (51/IIIb) 11600 (31), 11620 (51/IIIb), 17508 (51/IId), 17369 (47b)
Basilio Augusto, Bro. 1131 (51/IId)
Beard, J. S. 264 (47c), 361 (47c), 1004 (47c), 1257 (47c)
Beck, S. G. 6548 (51/IIa)
Bélanger, C. P. 89 (47c), 306 (47c)
Belém, R. P. 468 (33), 470 (24), 476 (33), 489 (33), 1315 (50/IIIa), 1245 (51/IIIa), 1464 (32a), 1535 (32b), 1666 (51/IIIa), 1777 (51/IIIa), 1780 (50/IIIa).
Belém, R. P. & Mendes Magalhães, G. 584 (51/IIIa), 701 (51/IIIa), 704 (51/IIb).
Belém, R. P. & J. M. Mendes 105 (35b), 151 (35b), 154 (26b), 158 (26b), 353 (35b), 370 (35b), 377 (1), 378 (1), 380 (1), 381 (1), 385 (37), 425 (35b), 445 (51/IIIb)
Belém, R. P. & R. S. Pinheiro 2079 (51/IIIa), 2232 (51/IIIa), 3049 (21a)
Bello, W. 126 (7b)
Benítez de Rojas, C. E. 482 (51/IIb), 490 (29)
Benoist, R. 1344 (51/IIc), 3975 (47e)
Berlandier, J. L. 173 (44a), 230 (44c), 850 (34a), 997 (21b), 1069 (51/IVb), 2427 (21b), 2479 (21b)
Bermúdez, L. A. 18 (51/IIb)
Bernoulli, G. 61 (51/IIIb), 1192 (51/IVb)
Berro, M. P. 2753 (35a)
Berry, P. E. 82 (51/IVa)
Bertoni, M. S. 1511 (35a), 2047 (32a), 2271 (35a)
Bhorai, M. 863 (51/IIc)
Biolley, P. 8997 (51/IIIb)
Black, G. A. 8733 (21a), 9926 (51/IIb), 12508 (26a), 14023 (26a), 16920 (21a)
Blake, S. F. 7602 (51/IVb), 7655 (26a), 7751 (31)
Blanchet, J. S. 252 (51/IIIa), 375 (32a), 842 (51/IIIb), 1046 (51/IIIb), 2747 (30), 3485 (26a), 3711 (35b), 3844 (32b)
Boldingh, I. 2324 (47c), 3288B (41), 7162 (51/IVa)
Bonpland, A. 475 (35a)
Bossio, H. 21 (26b)
Botteri, M. 445 (43a), 668 (51/IIb)
Bourgeau, E. 87 (43a), 2289 (31), 2673 (51/IIb), 2866 (43a), 3352 (31)
Bovell, J. R. 48 (47c)
Box, H. E. 615 (47c), 783 (47c)
Boyan, R. 110 (1), 206 (51/IIIa)
Brace, L. J. K. 1843 (40a), 1971 (40a), 3463 (40a), 3498 (40a), 3985 (10b), 4007 (40a), 4290 (40a), 4565 (40a), 4944 (40a), 4972 (10a), 5044 (40a), 5098 (40a), 5231 (10a), 6753 (40a), 6922 (40a), 6962 (40a), 7080 (40a)
Brade, A. C. 17717 (21a), 28048 (22)
Brandegee, T. S. 175 (51/IVc)
Braun, O. 2 (26b)
Breckon, G. J. & D. A. Kolterman 2342 (34c)
Breedlove, D. 6125 (43a), 7456 (43a), 7754 (43a), 9162 (51/IIa), 11818 (51/IIIa), 11864 (26e), 12862 (43a), 13805 (51/IIIa), 14842 (43a) 20310 (51/IVb), 21950 (21a), 26629 (51/IVb), 27632 (51/IIIa), 28250 (51/IVb), 28548 (28a), 37763 (31)
Breedlove, D. & P. H. Raven 13244 (51/IVb)
Breedlove, D. & R. F. Thorne 18418 (26a)
Brenan, J. P. M. 14320 (51/IIIb)
Brenes, A. M. 5(51/IIIb), 1910 (21a), 3676 (31), 4645 (51/IIIb), 4654 (32a), 4830 (31), 5195 (31), 5990 (51/IIIb), 6354 (51/IIIb), 13237 (51/IIIb), 14321 (32a), 14322 (51/IIb), 18590 (51/IIb), 17292 (32a), 18618 (31), 18811 (32a), 23162 (51/IIIb), 23189 (31)
Breteler, F. J. 3082 (51/IIIb), 3161 (51/IIIb), 3302 (47e), 4140 (47e), 4743 (32a), 4791 (26a)
Bridges, T. 125 (47e)
Bristol, M. L. 199 (40a)
Britton, N. L. 20 (47a), 24 (40a), 42 (47c), 670 (51/IId), 811 (40e), 1112 (40a), 1136 (51/IVa), 1520 (51/IVa), 1676 (51/IVa), 1787 (51/IId), 2395(40a), 2513 (40a), 2965 (51/IIb), 2969 (28a), 3159 (51/IVa), 3186 (51/IVa), 3291 (47a), 3294 (51/IIb), 3423 (51/IId), 3448 (10a), 3482 (51/IId), 3500 (51/IId), 3659 (51/IId), 4016 (51/IId), 4341 (42a), 5721 (51/IId), 5758 (51/IId), 5771 (51/IVa), 5772 (51/IVa), 5831 (51/IId), 5867 (51/IId), 5883 (51/IId), 5893 (47c), 5919 (51/IVa), 6188 (40a), 6229 (51/IId), 6376 (32a), 6378 (46a), 6384 (51/IId), 6406 (51/IId), 6420 (32a), 6437 (51/IId), 6441 (51/IId), 6521 (51/IId), 6663 (51/IIb), 6760 (46a), 6843 (47d), 6916 (51/IVa), 6965 (47c), 7050 (46a), 7085 (51/IId), 7315 (47c), 7470 (51/IId), 7506 (51/IId), 7526 (47c), 7812 (51/IId), 7954 (47d), 8006 (47d), 8329 (40c), 8672 (51/IId), 8696 (47d), 8741 (51/IId), 8825 (47c), 8835 (51/IVa), 8990 (47d), 9074 (47c), 9110 (42a), 9122 (51/IVa), 9210 (47c), 9211 (51/IId), 9212 (51/IVa), 9218 (42a), 9226 (51/IId), 9394 (42a), 9813 (51/IId), 9837 (42a), 10047

(51/IVa), 10095 (28a), 12780 (51/IId), 12865 (51/IId), 13224 (51/IId), 14005 (40a), 14937 (40a), 15096 (32a), 15100 (28a), 15791 (46b)
Britton, N. L. & K. R. Boynton 288 (42a), 300 (47c)
Britton, N. L. & L. J. K Brace 397 (40a), 678 (10a)
Britton, N. L. & J. F. Cowell 19 (47c), 20 (47c), 127 (47c), 273 (51/IVa), 280 (51/IVa), 314 (47c), 408 (51/IId), 1243 (51/IId), 1490 (27c), 1542 (51/IVa), 3837 (47c), 4032 (42a), 4705 (40c), 4997 (40c)
Britton, N. L., J. F. Cowell & W. E. Hess 1740 (40c)
Britton, N. L. & F. S. Earle 6542 (28a), 6581 (46a)
Britton, N. L. & T. E. Hazen 1581 (51/IIb)
Britton, N. L. & W. E. Hess 2712 (42a), 2770 (47c)
Britton, N. L. & A. Hollick 2055 (40a), 2125 (40a)
Britton, N. L. & D. W. Marble 1222 (47c)
Britton, N. L. & C. F. Millspaugh 2163 (40a), 2505 (40a), 2615 (40a), 2670 (10b), 2774 (10b), 2841 (10b), 2916 (40a), 3041 (10b), 5416 (40a), 5895 (10a), 5910 (10b), 6009 (10b)
Britton, N. L. & J. N. Rose 1406 (42a)
Britton, N. L. & J. A. Shafer 250 (47c), 283 (51/IVa), 285 (47c), 671 (47c), 802 (47c), 1500 (51/IId), 1714 (51/IVa), 1915 (40c), 2053 (51/IId), 2919 (51/IVa), 2988 (51/IVa)
Britton, N. L. & W. M. Wheeler 49 (47c)
Britton, N. L. & P. Wilson 6 (51/IId), 185 (51/IIb), 278 (51/IId), 315 (51/IId), 374 (51/IId, 5648 (40a), 14294 (46a), 14313 (32a), 14379 (46b)
Broadway, W. E. 1844 (51/IVa), 3062 (51/IIb), 5045 (28a), 7398 (28a)
Brown, S. 85 (51/IVa)
Bryant, E. M. 117 (47c)
Buch, W. 91 (45a), 1148 (47b), 1358 (47b), 1421 (45a), 1862 (47b)
Buchtien, O. 213 (47e), 776 (51/IIb), 778 (47e), 1786 (51/IIb), 2424 (51/IIb), 4109 (51/IIa), 4334 (51/IIb), 5421 (47e), 6166 (51/IIb), 7319 (50/IIb)
Bunting, G. S 468 (31), 2518 (47e), 3546 (21a), 3594 (51/IIIa), 3636 (32a)
Burch, D. 6667 (40c)
Burchell, W. J. 1610 (32a), 1205 (32a), 4276 (32a), 4288 (21a), 4487 (51/IIa), 5707 (33), 5756 (8), 6618 (27), 6774 (27), 6840 (32a), 6867 (51/IIa), 7582 (17), 8055 (33), 8235 (24), 8303 (31), 8414 (32a), 8483 (30), 8548 (30), 8551 (21a), 8552 (27), 8554 (30 & 32a), 8626 (27), 8660 (27), 8664 (28b), 8679 (27, 30 & 31), 8747 (28b), 8824 (51/I), 8925 (51/IIb), 9000 (30), 9605 (51/IId), 10011 (51/IId), 10022 (21a)
Burger, W. C. & M. Burger 8454 (51/IIb)
Burkart, A. 18717 (35a)
Burt, R. L. & H. W. Koster 104 (31), 131 (32a)
Byrne, R. 288 (40a)

Cabot, T. D. 19 (47e)
Cabrera, E. 77 (51/IVb), 149 (47e), 199 (47e), 261 (51/IVb), 536 (51/IVb), 578 (47e), 721 (51/IVb), 749 (47e), 1337 (47e)
Calderón, S. 1014 (28a), 1018 (31), 1036 (21a), 1066 (51/IIa), 1202 (51/IIIb)
Callaway, R. 388 (51/IVb), 396 (32)
Calzada, J. I. 374 (44a)
Camp, W. H. 3004 (47e), 3129 (47e), 3541 (51/IVb)
Campbell, E. 6020 (26a)
Campos Novães, J. de 227 (21a)
Campos Porto, P. 1233 (15), 2564 (21a)
Carauta, J. P. P. 1242 (51/IIb), 3346 (47m)
Carcerelli, C. 83 (51/IIb)
Cárdenas, M. 3559 (47e), 4524 (21a), 5376A (32b)
Cárdenas de Guevara, L. 804 (21a), 1001 (32a), 1003 (21a)
Carriker, M. A. 31 (47e)
Carrillo, E. (47e)
Carter, A. 4827 (51/IVc), 5100 (51/IVc), 5297 (51/IVc)
Carvalho da Silva, L. M. 16 (51/IIIb)
Casaretto, G. 1591 (32a), 2622 (7a), 2940 (4)
Cassen, P. M. 591 (28a), 611 (28a)
Castellanos, A. 24505 (35a), 26279 (21a)
Castillón, L. 9011 (36)
Cavalcante, P. 2092 (21a)
Cedillo T., R. 496 (51/IIIa)
Chaffanon, J. 128 (47i)
Chagas, J. 6979 (51/IIIa)
Chaves, D. 407 (51/IVb)
Chevalier, N. 153 (40a)
Choussy, F. 86 (51/IIa)
Chiang, F. 834 (26f)
Clarke, O. F. 219 (51/IIb), 308 (51/IIb)
Claussen, P. 86 (15), 148 (12), 208 (18c), 295 (7a), 694 (51/I), 733 (51/I), 854 (12), 861 (51/I), 1717 (7a)
Clemente, Bro. 550 (51/IId), 5777 (51/IId), 7135 (51/IId)
Clos, E. C. 1914 (51/IIb), 3229 (32a)
Clover, E. U. 930 (21b), 1102 (21b)
Coelho do Moraes, J. 2119 (25), 2119 (26b & 28a), 2161 (21a)
Coker, W. C. 10 (40a)
Collins, G. N. 3 (47a)
Combs, R. 167 (32a), 368 (51/IIb), 505 (40a)
Conrad, J. 2294 (51/IIIb)
Constantino, D. 137 (51/IIb)
Conzatti, C. 858 (51/IIIb), 1701 (32a), 1870 (51/IIIb), 3754 (51/IVb), 4228 (51/IIIb)
Cook, O. F. & G. N. Collins 370 (51/IId), 372 (51/IId), 598 (51/IId), 624 (51/IId)
Cook, O. F., C. S. Schofield & C. B. Doyle 189 (51/IId), 190 (51/IId)
Cooley, G. R. 8263 (47c), 8535 (47c), 8793 (47c)
Cooper, J. J. 170 (51/IIIb), 5766 (51/IIIb), 5766 (51/IVb)
Coradin, L. 107 (51/IIb), 136 (21a), 762 (26a)
Cordeiro, M. dos R. 68 (42b), 727 (51/IId), 1335 (51/IIIa)
Correll, D. S. 12308 (51/IId), 40463 (40a), 40725 (40a), 14424 (40a), 41430 (40a), 41657 (40a), 42088 (40a), 43007 (40a), 43165 (40a), 43167 (10c), 43188 (10c), 43386 (40a), 43460 (40a),

43705 (40a), 44054 (40a), 44467 (40a), 44763 (40a), 46112 (10a), 46321 (40a), 45263 (40a), 46563 (10c), 46616 (40a), 46676 (40a), 48365A (40a), 48314 (40a), 49292 (40a), 49942 (40a), 50253 (10a), 51211 (40a)
Cory, V. L. 55285 (21b)
Coulter, J. M. 3337 (43a)
Cowan, R. S. 1310 (51/IIb), 1661 (47c)
Croat, T. 11 (47c), 12827A (51/IVb), 12910 (51/IVb), 24010 (21a), 24128 (51/IVb), 38364 (51/IIIb), 38448 (51/IIIb), 39165 (34a), 43941 (43a), 45239 (32a)
Crosby, M. R., H. A. Hespenheide & W. R. Anderson 35 (40a), 246 (47a), 338 (40e)
Cuatrecasas, J. 1869 (47e), 3755 (51/IIa), 3867 (32a), 4028 (47i), 4056a (31), 4086 (47i), 13181 (51/IIIa), 18605 (51/IIb), 19759 (51/IIb), 23007 (26a), 24699 (47e), 24872 (26a), 26220 (47e), 27475 (51/IIb)
Cuming, H. 989 (47e)
Curran, H. M. 563 (32b)
Curran, H. M. & Haman 179 (51/IVa)
Curtiss, A. H. 7 (51/IVa), 180 (40a), 259 (32a), 616 (26a)
Cutler, H. 7014 (32a)

Dahlgren, B. E. & E. Sella 498 (51/IId)
Daniel, Bro. 898 (51/IIb), 1462 (47e), 1482 (47e), 3818 (47e)
D'Arcy, W. G. 262 (47c), 2613 (47b), 6301 (31), 6308 (51/IIa), 9771 (31)
Dash, J. S. 48 (47c)
Davidse, G. 1371 (51/IIIb), 3926 (51/IVb), 4237 (26a), 4551 (32a), 5136 (24), 5639 (51/IIb), 10341 (51/IIIb), 10895A (21a), 11439 (51/IIIb), 11629 (36b), 15492 (26a), 15497 (21a), 15588 (21a)
Davidson, M. E. 700 (32a)
Davis, D. H. 42 (21a), 759 (51/IIIa), 805 (26a), 806 (21a)
Davis, P. H. 59747 (32a)
Dawe, M. T. 831 (47e)
Dawson, E. Y. 14535 (31), 14912 (27), 26505 (40a), 26559 (40a), 26564 (40a), 26681 (40a)
Degener, O. 2488 (51/IId), 7284 (51/IId), 10875 (51/IId), 26651 (51/IIIa)
Dehesa, M. P. 1536 (32a)
Delascio, F. & R. Liesner 6865 (28a), 6872 (21a)
Delgado, A. A. de 22 (51/IVb)
Delgado S., A. 53 (51/IIIb), 688 (47e), 759 (26a), 951 (51/IId)
Dias da Rocha, – 3951 (32a)
Dodge, C. W. & A. A. Hunter 8606 (31)
Dodge, C. W. et al. 16895 (51/IIa), 16911 (31)
Dombey, J. 838 (47e)
Dombrowski, L. Th. 7006 (51/IIb), 7007 (32a), 7055 (32a), 7068 (39), 7219 (32a)
Dorrien-Smith, – 134 (32a), 193 (33), 194 (23)
Dressler, R. L. & Q. Jones 192 (51/IId)
Drouet, F. 2041 (51/IId)
Drushel, J. A. 9329 (34a)
Dryander, E. 316 (47e)
Duarte, A. P. 1294 (21a), 2146 (2), 2676 (4), 2802 (8), 2833 (21a), 4211 (51/IIIb), 5060 (32a), 6501 (18c), 7471 (21a), 7795 (30), 7861 (32a), 8646 (21a), 8728 (7b), 8983 (32b), 9065 (2), 10301 (35b), 10430 (7c), 13997 (32a), 14011 (18a), 14140 (35b)
Duarte, A. P. & G. Barroso 7478 (35b)
Duarte, A. P. & A. Castellanos 365 (51/IIIb), 382 (35b)
Duarte, A. P. & E. Pereira 8946/9960 (7a)
Ducke, A. 160 (51/IIIa), 1600 (51/IIIa), 1892 (51/IIIa)
Dugand, A. 4418 (28a), 5206 (28a), 5451 (51/IVb), 6168 (28a), 7113 (28a), 7140 (28a)
Dugand, A. & R. Jaramillo 3832 (51/IIb)
Dugès, A. 257bis (47l)
Duke, J. A. 5924 (51/IIa), 6028 (51/IIa), 6204 (51/IIb), 9177 (51/IIIb), 9213 (51/IIIb), 13652 (31)
Duque-Jaramillo, J. M. 2631 (47e)
Dusén, P. 165 (21a), 7590 (39), 9220 (32a), 9454 (39), 16357 (39), 16410 (32a)
Duss, A. 397 (47c), 1117 (47c), 1121 (51/IId), 1892 (51/IVa), 2627 (47c), 2625 (51/IId)
Dutra, J. 569 (35a)
Dwyer, J. D. 79 (21a), 329 (21a), 7617 (51/IIb), 9355 (51/IVb), 11523 (31), 11603 (21a), 12146 (21a), 12152 (31)
Dziekanowski, C. 1693 (34a), 1737 (34a), 3159 (51/IIa)

Eames, A. J. 151 (51/IId)
Earle, F. S. & G. F. Baker 2440 (51/IId)
Ebinger, J. E. 768 (51/IIIb), 1018 (21a)
Eggers, Baron H. 33 (47c), 119 (47c), 257 (47c), 906 (47c), 1085 (51/IIb), 1419 (51/IIb), 3880 (40a), 3914 (40a), 3985 (40a), 5507 (51/IIb), 6247 (47c), 7232 (47c)
Egler, W. A. 1485 (51/IId), 1485 (51/IIb), 23892 (21a)
Egler, F. E. 39/242 (47c)
Eiten G. & L. 1691 (32a), 1712 (21a), 1798 (32a), 1885 (32a), 1888 (32a), 1889 (51/IIb), 2270 (33), 2530 (21a), 2604 (51/IIb), 2621 (21a), 2644 (21a), 2660 (51/IIb), 2700 (21a), 2901 (32a), 2910 (21a), 3220 (21a), 3504 (21a), 3572 (31), 3592 (32a), 3761 (27), 3762 (32a), 3765 (30), 3805 (27), 3999 (51/IIb), 4003 (30), 4020 (37), 4174 (37), 4236 (21a), 4348 (25), 4546 (25), 4716 (32a), 5840 (21a), 5844 (51/IIb), 9548 (33), 9744 (33), 9823 (33), 9994 (27), 10005 (51/I), 10647 (25), 10815 (21a), 10871 (26a)
Ekman, E. L. 830 (51/IId), 1065 (47b), 1752 (51/IIb), 1753 (50/IIb), 1805 (45a), 1900 (47b), 2430 (35a), 2558 (40a), 4009 (45a), 4162 (10c), 4748 (51/IId), 5577 (47b), 5961 (45a), 6603 (40a), 7243 (45a), 8042 (51/IId), 9087 (51/IVa), 9486 (40a), 9847 (51/IId), 10223 (40d), 10430 (32a), 10453 (51/IVa), 10711 (28a), 10958 (32a), 11670 (42a), 12685 (45b), 14301 (51/IId), 14498 (45b), 14928 (51/IIb), 14929 (51/IId), 14956 (51/IIc), 15554 (32a), 15772 (40d), 15901 (51/IVa), 16158 (45b), 16173 (40a), 16220 (45b), 16697 (40a), 16960 (51/IId), 17020 (46a), 17268 (46a), 17900 (46b), 18101 (26a)

Elias, Bro. 275 (51/IIa), 328 (51/IIa), 398 (28a), 787 (28a)
Elias, T. S. 751 (31)
Enríquez, A. 55 (51/IVc)
Enríquez, O. G. 789 (51/IVb)
Ervandberg, L. C. 14 (51/IIb)
Eupunino, A. 385 (51/IIIa)

Falcão, J. A. et al. 801 (21a)
Faris, J. A. 220 (51/IVa)
Fawcett, W. 8051 (28a)
Fay, J. J. & J. I. Calzada 913 (44a)
Fendler, A. 319 (51/IVa), 321 (51/IIIb & 51/IVa), 321b (51/IIa), 336 (51/IIb), 1855 (32a), 1853 (51/IIb)
Fereira, A. in Richards 6816 (33)
Fernández, A. 59 (51/IIIb), 467 (51/IIb), 481 (26a), 527 (51/IIIb), 630 (28a)
Fernández Yépez, F. 866 (28a)
Ferrari, G. 483 (47e)
Ferreira, V. F. 798 (32a), 842 (7c), 1503 (14a)
Ferreyra, R. 328 (47e), 2759 (47f), 3094 (51/IIIb), 5903 (51/IVb), 6028 (51/IVb), 7786 (51/IIb), 11023 (47e), 17841 (51/IIa)
Ferris, R. 5904 (51/IVc), 6198 (21b)
Fevereiro, P. C. 106 (32a)
Fiebrig, K. 39 (21a), 50 (51/IIb), 171 (51/IIb), 292 (29), 306 (32a), 626 (21a), 1290 (51/IIb), 1450 (26a), 4045 (29), 4455 (21a), 5033 (32a), 5058 (35a), 5164 (29), 5365 (51/IIb)
Finch, R. G. 27 (47c)
Fisher, G. L. 144 (44a), 158 (51/IIb), 35434 (40f), 35487 (21a)
Fishlock, W. C. 8 (47c), 13 (47c), 268 (47c), 283 (47c)
Flores, J. S. 484 (51/IVb)
Florschütz, J. & P. A. 1663 (21a)
Foldats, E. 3577 (32a)
Fonseca, M. 108 (37)
Fosberg, F. R. 22125 (26a), 38807 (21a)
Fournet 296 (51/IVa), 1760 (41)
Fredholm, A. 3182 (51/IIb)
Freire Alemão, F. & M. de Cysneiros 471 (35b), 472 (30)
Freitas Campos, E. & J. M. de 3410 (32a)
Freitas Leitão Filho, H. de 25 (51/IIb)
Fries, R. E. 1540 (51/IIIb)
Fróes, R. L. 23668 (21a), 34041 (51/IIIb)
Fromm, E. & E. Santos 1262/1284 (32a)
Fromm, E. et al. 1343 (51/IIIb)
Fuertes, M. 496 (45a), 684 (47b), 1707 (51/IId), 1713 (51/IId)
Fujimoto, Y. 10 (51//IIb)
Funck, N. 279 (47e), 491 (47e)
Funck, N. & J. Schlim 80 (47c)

Galeotti, H. 3183 (26a), 3291 (31), 3311 (43a), 3408 (44a), 3462 (21a), 3470 (43a)
Garcés O., C. 15 (26a)
García, H. 5185 (51/IIa)
García Barriga, H. 7517 (51/IIb), 8199 (32b), 17211 (43c), 18525 (51/IIb)
Gardner, G. 176 (51/IIIa), 749 (47m), 787 (32b), 967 (32a), 989 (21a), 1573 (35b), 1569 (30), 2127 (35b), 2128 (26a), 2129 (30), 2131 (21a), 2145 (25), 3683 (33), 3684 (51/I), 3685 (21a), 3686 (51/I), 3689 (24), 4528 (7a), 4529 (5a), 6005 (32a)
Garnier, Bro. A. 535 (51/IIIb), 3072 (51/IVb)
Gärtner, A. 20 (47e)
Gaudichaud-Beaupré, C. 127 (47m), 219 (51/IIIb), 887 (21a), 888bis (32a)
Gaumer, G. F. 122 (32b), 885 (51/IVb), 1113 (47e), 1410 (47e), 1531 (47e), 2182 (47e), 2383 (47e), 23458 (51/IVb), 23474 (47e)
Gay, C. 2072 (47e)
Geay, F. 1892 (28a)
Gehriger, W. 53 (26a), 348 (47e), 615 (51/IIIb)
Gehrt, G. 8291 (51/I), 8293 (21a)
Gentle, P. 833 (51/IVb), 974 (51/IIIb), 976 (31), 977 (26a), 3080 (51/IVb), 3674 (31), 3904 (51/IIIb), 4176 (31), 4177 (21a), 8237 (51/IIb)
Gentry, A. 1632 (31), 5786 (51/IIa), 6714 (51/IVb), 7654 (21a), 19150 (47e), 20130 (51/IVb), 21437 (21a)
Gentry, H. S. 1710 (51/IVc), 1750 (26e), 2816 (26e), 5104 (32a), 5284 (26d), 6366 (26e), 6535 (51/IVc), 6714 (51/IVb), 6720 (34a), 17813 (51/IVc), 17838 (26e), 20340 (34c), 20411 (43a)
Geyger, E. 1 (47e)
Ghiesbreght, A. B. 87 (43a), 170 (43a), 252 (43a)
Gibbs, P. E., H. F. Leitão Filho et al. 3458 (21a)
Gillis, W. T. 7759 (40a), 7829 (40a), 8178 (47c), 10321 (21b), 11678 (40a), 11802 (10c), 11803 (10c), 11908 (40a), 11936 (40a), 12279 (40a)
Gilly, C. L. 272 (51/IIb)
Ginés, Hno. 3054 (51/IIIb), 3872 (28a), 4417 (51/IVa)
Glaziou, A. 4992 (32a), 4993 (47m), 6502 (51/IIIb), 10662 (30), 12618 (30), 12620 (47m), 14637 (32a), 19052 (12), 19082 (2), 19083, in part (5a), 19084 (6), 19086 (12), 19087 (7a), 19088 (27), 19088bis (27), 19089 (7a), 19094 (14a), 19103 (18c), 20257 (12), 20268 (7a), 20269 (7b), 20980 (24), 20987 (33), 20988 (31 & 32a), 20989 (32a), 21006bis (8)
Glocker, E. F. von 156 (26a), 468 (32a), 599 (32a)
Goldman, E. A. 98 (47l)
Goll, G. W. 470 (51/IId), 478 (51/IId), 801 (51/IId), 838 (51/IId), 952 (51/IId)
Gomes, I. 838 (37)
González, E. 7 (51/IVb)
González, L. C. 18C064 (51/IVb)
Goodland, R. 211 (21a), 219 (21a), 512 (51/IIb), 671 (51/IIc), 764 (26a)
Goodspeed, T. H. 35397 (47e)
Goodwin, G. G. & B. W. 17 (47c)
Gottsberger, I. 6 (51/IIb), 48 (21a), 18271 (51/I)
Grant, M. L. 10955 (47e)
Grant, V. 956 (51/IIIb)
Gregg, J. 1078 (26e)
Greenman, J. M. 13 (44a), 5081 (51/IVb)
Grether, R. 509 (44a), 920 (47e), 1227 (26g)
Grisol, M. 11 (28a), 30 (26a), 31 (28a)
Grosourdy, – 13 (26a)
Guerrero C., B. 684 (44a)
Guillemin, J. B. A. 745 (51/IIb)

Gutiérrez, G. 1470 (26a)
Guyon, H. 168 (51/IId)

Hackett, H. E. 190 (40a)
Hahn, L. 197 (47c)
Hallberg, B. 1035 (51/IIIa)
Hammen, Th. van der 1129 (47e)
Hanbury-Tracy, J. 34 (47e)
Harley, R. 15038 (21a), 15078 (35b), 15204 (37), 15214 (1), 15215 (1), 15787 (51/IIIa), 15875 (35b), 15886 (21a), 16860 (37), 17481 (51/IIIa), 18742 (12), 18977 (37), 18988 (21a), 19021 (37), 19103 (30), 19123 (38), 19264 (37), 19301 (21a), 19415 (21a), 19843 (32b), 20141 (32b)
Harmon, W. E. 3434 (51/IIIa), 4435 (51/IIIa)
Harris, W. 6911 (26a), 6949 (51/IVb), 8906 (28a), 9615 (40e), 9652 (40a), 10641 (51/IId), 11725 (40a), 12126 (40e), 12268 (28a), 12838 (51/IIb)
Harrison, S. G. 1025 (21a), 1038 (51/IIIa), 1093 (26a), 1174 (51/IIIa), 1288 (21a), 1340 (51/IIb), 1408 (21a), 1445 (1)
Hart, J. 664 (51/IId), 1043 (51/IId)
Hartweg, C. T. 961 (47g)
Hassler, E. 153 (51/IIIb), 1024 (51/IIIb), 1487 (26b), 1561 (21a), 1578 (21a), 1589 (51/IIb), 1785 (21a), 2437a (51/IIb), 2759 (51/IIb), 3776 (51/IIb), 3813 (26b), 4033 (21a), 4478 (21a), 4626 (32a), 5003 (21a & 23), 5258 (51/IIb), 5420 (51/IIa), 6113 (21a), 6608 (26b), 6609 (29), 6015 (51/I), 6239 (32a), 6386 (26b), 6387 (29), 6643 (51/IIIb), 7037 (32b), 7713 (35a), 7994 (29), 8060 (51/IIIb), 8462 (32b), 9073 (51/IIb), 9116 (51/IIa), 9933 (51/IIa), 10088 (23), 10778 (32b), 10999 (51/IIIa), 11004 (35a), 11011 (39), 12372 (26b)
Hatschbach, G. 3594 (39), 6345 (32a), 6859 (21a), 7836 (39), 7900 (51/IIb), 13212 (32a), 15202 (21a), 17927 (21a), 20855 (21a), 21444 (51/IIIb), 21820 (51/IIb), 21824 (26a), 23579 (51/IIb), 23597 (51/I), 23665 (26a), 23701 (32b), 23749 (26b), 23828 (32a), 23926 (39), 26027 (23), 26168 (26a), 26994 (18a), 27025 (18b), 27271 (2), 27350 (15), 27393 (7a), 27443 (7c), 27498 (5a), 27919 (18b), 29518 (29), 29841 (15), 30058 (4), 30077 (18c), 30215 (19), 31274 (32a), 31498 (33), 32010 (51/I), 32134 (51/IIIb), 33701 (21a), 33806 (26a), 33809 (51/IIb), 33817 (51/I), 33953 (21a), 34033 (24), 34057 (51/IIb), 34067 (51/IIb), 34092 (31), 34230 (33), 34277 (51/IIa), 35266 (15), 35678 (39), 36800 (31), 36875 (21a), 37654 (21a), 37692 (32a), 37721 (21a), 38018A (32a), 38018B (51/IIb), 38022 (51/I), 38034 (24), 38178 (27), 38228 (51/I), 38230 (51/I), 38248 (27), 38254 (27), 38258 (51/IIa), 38274 (31), 38138 (21a), 38323 (27), 38347 (21a), 38386 (21a), 38496 (30), 38600 (26a), 38664 (51/IIb), 39114 (26b), 39117 (32b), 39382 (26b), 39407 (35b), 39438 (35b), 39495 (1), 39512 (26a), 39520 (30), 39545 (1), 39559 (22), 39572 (37), 39659 (32b), 40901 (14a), 41209 (51/I), 41237 (35b), 41340 (32a), 41627 (16), 42022 (21a), 42063 (35b), 42082 (37), 42151 (27), 42450 (26a), 42475 (35b) 42791 (27), 43050 (8), 48404 (21a)
Haught, O. 2147 (51/IIb), 2292 (29), 3575 (21a)
Hayes, S. 108 (51/IVb)
Heller, A. A. & E. G. 57 (47c), 242 (51/IId), 259 (51/IVa), 642 (47d), 1175 (47c), 1178 (47c), 1376 (51/IId), 1969 (51/IId), 2615 (42a), 6183 (51/IId), 6362 (47c)
Herbst, D. 2305 (51/IId)
Heriberto, Bro. 231 (31), 260 (51/IIb), 417 (51/IVb)
Heringer, E. P. 919 (26a), 3374 (21a), 3813 (27), 6425 (33), 9064 (33), 10221 (35b), 10440 (51/IIIa), 11042 (27), 13053 (51/I), 14908 (32b), 15782 (30), 15869 (21a), 15955 (24)
Heringer, E. P. et al. 79 (26a), 98 (35b), 380 (25), 408 (26a), 811 (21a), 954 (51/IIb), 1002 (32a)
Heringer, E. P. & A. Castellanos 6061 (32a)
Hermann, F. J. 10974 (32b)
Hermann, H. A. van 101 (51/IIb), 184 (51/IId), 805 (51/IId)
Hernández M., R. 587 (42a)
Herrera, F. L. 1184 (47f)
Herrera, G. S. 938 (51/IIb)
Herter, W. G. 1328 (32a)
Herzog, Th. 1172 (36), 1304 (32a)
Hess, L. 1238 (51/IIIb)
Hess, W. E. 242 (51/IId), 1371 (51/IId), 1599 (51/IId), 3390 (42a), 3580 (51/IId)
Heyde, A. W. & E. Lux 3312 (28a), 3728 (31), 4169 (51/IVb), 5154 (51/IIa), 6133 (28a)
Hill, S. R. 1773 (51/IIb), 2270 (40a), 2372 (40a), 2408 (40a)
Hinton, G. B. 1593 (32a), 1656 (26a), 1766 (51/IVb), 1864 (32a), 4575 (21a), 4663 (51/IVb), 5972 (51/IVc), 6122 (21b), 6754 (32a), 6862 (26e), 9070 (21a), 9542 (26e), 6614 (51/IVb), 9660 (51/IVb), 10000 (21a), 10536 (26e), 10688 (51/IIIb), 10698 (47c), 10841 (51/IId), 11385 (51/IIIb), 11388 (31), 12708 (47e), 14660 (51/IIa), 15239 (32a)
Hioram, Bro. 88 (47c), 4868 (40d)
Hitchcock, A. S. 20218 (51/IVb)
Hitchcock, C. L. & L. Stanford 6845 (34a)
Hodge, W. H. 593 (47c), 594 (47c), 1257 (47c), 2176 (47c), 2544 (47c), 2633 (47c), 3277 (47c)
Hoehne, F. C. 160 (32a), 2659 (24), 2666 (29), 4729 (29), 5058 (12)
Hoehne, W. 5537 (32b), 5547 (32a), 6007 (51/IIa)
Holdridge, L. R. 879 (47b), 1126 (45a)
Holm, Th. 119 (51/IId), 139 (42a)
Holm-Nielsen, L. et al. 2056 (51/IVb), 2162 (51/IVb), 3163 (51/IIIb)
Holmquist, J. de D. 13 (1)
Holt, E. G. & W. Gehriger 54 (32a), 211 (26b)
Holton, I. F. 987 (47g)
Hostmann, F. W. 374a (21a), 451 (51/IIb), 950 (51/IIb), 959 (51/IIb), 1179 (51/IIIa)
Howard, R. A. 3052 (47c), 5218 (32a), 5711 (51/IId), 6008 (40a), 8119 (47b), 8200 (47b), 8780 (47b), 8961 (45a), 9262 (47b), 9535 (51/IVa), 9798 (51/IId), 10015 (40a), 10659 (47c), 11033 (47c), 11034 (47c), 11084 (47c), 11345 (47c), 12399 (45a), 12448 (45a), 17005 (32a)

Huber, O. 548 (21a), 683 (21a), 685 (26b), 729 (26b), 881 (26b), 889 (21a), 1004 (21a), 1088 (21a), 1145 (51/IIc), 1266 (26b), 1487 (26b), 2186 (21a)
Humbert, H. 27138 (43c), 27176 (27)
Humboldt, F. W. von & A. Bonpland 288 (51/IIIb), 1182 (26b)
Hunnewell, F. W. 15279 (51/IVa), 16025 (47e), 18854 (40a), 20130 (47c)
Hunt, D. R. 3 (21a), 32 (31), 239 (51/IIIb), 5785 (27)
Hunt, D. R. & Ferreira 4139 (51/I), 5590 (51/I)
Hunt, D. R. & J. F. Ramos 6271 (32a)
Hunter, A. A. & P. H. Allen 18 (31), 76 (51/IVb), 754 (31)
Hutchinson, P. C. & J. K. Wright 3402 (51/IIIb), 5428 (47h), 6833 (47e)

Ibarrola, C. T. 166 (26a), 256 (26a), 356 (32a), 848 (26a), 1126 (26b) 1295 (26b), 1387 (35a), 1399 (26b), 1582 (32a), 1663 (35a), 1769 (35a), 1925 (35a), 2056 (21a), 2146 (21a), 2350 (35a), 2354 (32a), 2715 (35a), 2832 (32a), 3065 (32a), 3671 (26a), 3974 (26a)
Imray, J. 311 (47c)
Irwin, H. S. 303 (51/IIIa), 475 (1), 592 (21a), 707 (26a), 1208 (21b), 1209 (21b), 1250 (51/IVc), 1257 (26e), 1270 (50/IVb), 1255 (32a), 1274 (26e), 1356 (51/IVb), 1392 (51/IVb), 1396 (34a), 2012 (51/IIb), 2013 (32a), 2014 (51/IIIb), 2108 (51/IIIb), 2140 (51/IIIb), 2356 (21a), 2358 (32a), 2369 (51/IIIb), 2370 (51/IIIb), 2378 (21a), 2380 (51/IIIb), 2386 (27), 2392 (12), 2405 (12), 2411 (27), 2428 (20), 2446 (2), 2447 (18c), 2448 (3), 2464 (18b), 2467 (2), 2468 (14a), 2470 (7a), 2472 (3), 2475 (7c), 2483 (7a), 2494 (2), 2499 (51/IIb), 2502 (12), 2515 (27), 2519 (32a), 2520 (8), 2532 (51/I), 2534 (31), 2548 (27), 2549 (29), 2551 (51/I), 2552 (27), 2554 (33), 2558 (51/I), 2559 (27), 2569 (24), 2571 (31), 2582 (33), 2587 (51/IIb), 2588 (27), 2608 (21a), 2613 (35b), 2614 (26a), 2819 (51/IIb)
Irwin, H. S. et al. 7851 (33), 8394 (32a), 8812 (32a), 8925 (21a), 8941 (33), 8966 (51/I), 9182 (24), 9185 (33), 9468 (33), 9540 (32a), 9566 (32a), 9620 (33), 9709 (24), 9801 (8), 10000 (8), 10073 (32a), 10074 (21a), 10115 (33), 10287 (35b), 10575 (32a), 10576 (21a), 10778 (33), 10860 (32a), 11021 (27), 11068 (33), 11080 (21a), 11103 (24), 11111 (24), 11365 (33), 11508 (51/IIa), 11610 (24), 11955 (51/I), 12107 (32a), 12146 (51/I), 12195 (24), 12223 (33), 12873 (33), 13007 (33), 13025 (33), 13076 (33), 13133 (51/IIb), 13578 (21a), 13619 (33), 13691 (33), 13757 (32a), 13819 (27), 13887 (33), 13955 (24), 14186 (33), 14287 (51/I), 14382 (31), 14406 (31), 14644 (32a), 14648 (35b), 14693 (35b), 14898 (35b), 14920 (35b), 14937 (21a), 14953 (30), 15062 (27), 15078 (27), 15218 (32a), 15301 (27), 15303 (51/IIa), 15353 (32a), 15492 (33), 15617 (32b), 15996 (51/IIa), 16084 (33), 16157 (21a), 16222 (51/IIa), 16335 (51/I), 16409 (33), 16410 (32b), 16500 (51/I), 16534 (21a), 16632 (27), 16938 (21a), 16945 (32b), 16964 (51/I), 16989 (51/IIa), 17408 (51/I), 18501 (27), 18508 (21a), 18771 (24), 18798 (32a), 18800 (31), 18870 (51/I), 18978 (33), 19053 (51/I), 19081 (31), 19421 (51/IIa), 19460 (33), 19650 (32a), 19781 (51/IIIb), 20022 (15), 20197 (3), 20231 (51/IIIb), 20356 (20), 20426 (32a), 20500 (18c), 20624 (4), 20635 (2), 20653 (5a), 20885 (37), 20921 (51/IIIb), 21086 (31), 21147 (51/IIb), 21154 (27), 21155 (32a), 21246 (21a), 21310 (27), 21336 (51/I), 21344 (21a), 21409 (51/IIb), 21447 (51/I), 21491 (31), 21626 (27), 21627 (31), 21881 (18c), 21881a (18b), 21956 (32a), 22154 (2), 22216 (7a), 22313 (3), 22462 (21a), 22530 (51/IIIb), 22619 (21a), 22733 (6), 22821 (32a), 22881 (51/IIIb), 22901 (7c), 22927 (14a), 22965 (35b), 23007 (51/IIIa), 23107 (14b), 23330 (21a), 23409 (11), 23535 (16), 23599 (32a), 23792 (51/IIb), 23897 (8), 23912 (33), 24253 (33), 24339 (31), 24405 (33), 24450 (51/I), 24479 (27), 24502 (32a), 24650 (31), 24687 (33), 24767 (32a), 24786 (32a), 24874 (31), 25075 (21a), 25076 (33), 25103 (27), 25178 (51/IIa), 25291 (32a), 25344 (24), 25467 (32a), 25482 (8), 25721 (32a), 25725 (21a), 25774 (27), 25857 (51/I), 25865 (33), 25872 (27), 25956 (27), 26050 (51/IIa), 26124 (26a), 26189 (51/IIb), 26247 (51/IIb), 26324 (51/I), 26415 (33), 26428 (51/I), 26457 (31), 26604 (32a), 26642 (33), 26778 (51/IIb), 26816 (27), 26895 (21a), 26970 (51/I), 27157 (51/I), 27253 (33), 27413 (5a), 27432 (21a), 27586 (32a), 27635 (5a), 27682 (3), 27721 (5a), 27848 (7a), 27927 (21a), 27960 (5a), 28101 (7a), 28107 (7a), 28119 (21a), 28124 (51/IIIb), 28165 (14a), 28262 (7a), 28269 (6), 28589 (7a), 29012 (4), 29343 (12), 29584 (21a), 29594 (51/IIb), 29680 (12), 30378 (12), 30559 (51/IIb), 30560 (32a), 30666 (37), 30871 (35b), 30887 (21a), 31193 (1), 31262 (21a), 31279 (31a), 31280 (26a), 31281 (30, 31293 (31), 31306 (35b), 31659 (37), 31802 (51/I), 31902 (33), 32051 (27), 32329 (35b), 32442 (32b), 32446 (21a), 32585 (37), 32631 (22), 32641 (35b), 33107 (21a), 34612 (21a), 34613 (17), 34698 (27), 34703 (17), 34780 (32a), 35026 (31), 36566 (21a), 36787 (21a), 37199 (40a), 37200 (40a), 37284 (50/IId), 37286 (51/IId), 37287 (51/IId), 37288 (51/IId), 37289 (51/IId), 37290 (47b), 37291 (47b), 37292 (51/IVa), 37293 (51/IId), 37295 (51/IId), 37297 (45b), 37298 (51/IId), 37304 (51/IId), 37305 (51/IVa), 37307 (51/IVa), 37308 (51/IId), 37310 (51/IId), 37311 (51/IVa), 37312 (51/IId), 37313 (51/IVa), 37316 (42b), 37317 (51/IId), 54495 (51/IIIa), 55330 (51/IIIa)
Irwin, H. S., J. M. Pires & L. Y. Th. Westra 47939 (51/IIb)
Irwin, H. S. & T. R. Soderstrom 5080 (24), 5081 (33), 5340 (21a), 5367a (24), 5457 (21a), 5503 (51/IIIb), 5599 (24), 5632 (32a), 5697 (24), 6213 (21a), 6623 (33), 6624 (51/I), 7021 (33), 7113 (31), 7116 (32a), 7229 (24), 7433 (33)

Jack, J. G. 6182 (51/IId), 7508 (40a), 7779 (32a), 7889 (32a), 8098 (32a)
Jäger, B. 294 (45a)
Jahn, A. 64 (47e), 504 (47e), 1268 (51/IVa)
Janzen, D. H. 1100 (51/IVb), 1107 (51/IVb)
Jenman, G. S. 5513 (51/IIc), 6209 (26a)
Jennings, O. E. 299 (21a), 521 (40a)
Jesús, J. O. de 95 (51/IIIa)
Jiménez M., A. 1386 (32a), 2391 (51/IIb), 827 (51/IId)
Jiménez, J. de J. 868 (45b), 1488 (51/IId), 1506 (51/IVa), 1552 (47b), 1567 (47b), 1666 (45b), 1704 (51/IId), 1711 (40c), 2049 (45b), 2148 47b), 2387 (32a), 2511 (51/IId), 2875 (45b), 3218 (40c), 3263 (47b), 3285 (47b), 3874 (51/IId), 4411 (51/IId), 6147 (45b), 8502 (45b), 8531 (51/IVa), 8532 (51/IId), 8645 (45b), 27247 (40c)
Jobert, C. 178 (51/IIb), 1067 (25)
Johnson, C. D. 1178 (32a), 1186 (51/IVb), 1287 (32a), 1398 (32a), 1525 (51/IVb), 1556 (51/IIIb), 1807 (47e), 1827 (47e), 2035 (51/IVb), 2045 (51/IVb), 2070 (51/IVb), 1377 (51/IVa)
Johnston, I. M. 928 (51/IVb), 7199 (34a), 7865 (51/IVc), 8135 (51/IVc)
Johnston, J. R. 1057 (28a)
Johnston, M. C. 2435 (21b), 5073 (21b), 5353 (21b), 5383B (34b), 5445 (34a), 5459 (34a), 10317 (34a)
Joly, A. B. et al. 255 (2), 260 (18c), 1864 (18c), 2198 (4), 2901 (15)
Jones, M. E. 27192 (32a), 27327 (51/IVb)
Jönsson, G. 195a (39)
Jörgensen, P. 3594 (32a), 4616 (35a)
Julio, Bro. 119 (47f)
Junk, W. 279 (51/IIIa)
Jurgensen, C. 505 (51/IVb), 759 (43a)
Juzepczuk, S. 1909 (31)

Kamb, P. 2037 (51/IIa)
Kappler, A. 1578 (51/IIb)
Karwinski von Karwin, W. F. von 164 (44a), 1176 (31)
Kearney, T. H. & R. H. Peebles 14470 (26e), 14865 (26e)
Kegel, A. H. H. 1103 (21a)
Kellerman, W. A. 7619 (31), 7699 (51/IIIa)
Kenoyer, L. A. 1625 (51/IIIb), 2422 (47l)
Kerber, E. 63 (43a), 339 (31)
Killip, E. P. 3023 (51/IIa), 3243 (31), 3324 (51/IIa), 4154 (31), 11535 (51/IIb), 32278 (32a), 32311 (21a), 32317 (32a), 32354 (46a), 33230 (51/IIb), 35620 (51/IIb), 37154 (51/IIb), 37645 (21a), 42536 (32a), 42828 (46b), 42943 (46b), 43016 (40a), 43662 (46b), 43864 (46b), 43975 (46b), 44032 (46b), 44141 (46b), 44697 (32a), 45020 (28a), 45121 (31), 45142 (46b), 45196 (51/IId), 45483 (31), 45589 (40a), 45627 (28a), 45636 (46b), 45728 (51/IId), 45741 (51/IIb), 45817 (31), 45887 (46b)
Killip, E. P. & J. Cuatrecasas 38425 (26a), 38426 (51/IIb), 39017 (51/IIb)
Killip, E. P. & A. C. Smith 14462 (51/IVb), 20146 (47e), 20777 (51/IIa), 20876 (51/IIb), 22315 (47e), 23108 (47e), 23181 (47e), 23223 (47e), 24108 (47e), 24239 (47e), 30242 (51/IId), 30418 (21a)
Kimber, C. 904 (47c)
King, R. M. 1828 (26a), 1888 (21a), 1913 (31), 1915 (26a)
Kirkbride, J. H., Jr. 2483 (26a)
Kishler, L. 296 (47l)
Klevens, M. J. & F. A. Barkley 170319 (51/IIb)
Koczicki, C. 286 (12)
Krapovickas, A. 11762 (32a), 11798 (51/IIb), 12095 (32a), 12127 (26b), 12512 (51/IIb), 16529 (36a), 16686 (26b), 29056 (35a), 29142 (26b), 29154 (21a), 30379 (51/IVb), 30869 (51/IVb), 31574 (32a), 31800 (32a), 32224 (51/I), 32284 (26a), 32869 (21a), 33347 (32a), 34778 (32a), 34876 (51/IIb), 35083 (21a), 36164 (51/I), 36177 (31), 36311 (32a)
Kruckeberg, A. A. 4813 (34a)
Kuhlmann, J. G. 2201 (32a), 2336 (47m), 3187 (21a), 3235 (51/IIIa)
Kummrow, R. 497a (51/IIa), 518 (51/IIb), 1093 (32a)
Kuntze, O. 45 (47c), 62 (47c), 292 (51/IId), 1078 (51/IIb)
Kupper, W. 123 (51/IIIb)

Lakela, O. 29039 (26a), 32172 (32a)
Langlassé, E. 544 (26d), 990 (21a),
Langman, I. 3929 (44a)
Lasseigne, A. 4303 (51/IIb), 4312 (51/IIIb), 4896 (44), 4937 (51/IIIb), 4944 (51/IVb)
Lasser, T. 828 (28a), 1542 (51/IIa)
Laughlin, R. M. 1120 (26e), 1540 (51/IIIa), 1822 (51/IIIa), 2570 (51/IIIa)
Lavastre, Bro. Basilio Augusto 459 (40c), 539 (45b), 672 (47b), 962 (45b), 1711 (40c), 2273 (45b)
Leavenworth, W. C. 235 (34a)
Leblond, J. B. 611 (51/IIc)
Lechler, W. 2014 (47e)
LeDoux 2171 (26e)
Le Gallo, C. 693 (41), 2523 (51/IVa)
Lehmann, F. C. 96 (47e), 2373 (32b), 2505 (47g), 5365 (51/IIa), 8479 (32b)
Lehto, E. 16346 (51/IId)
Leite, J. E. 1983 (35a), 2253 (51/IIb)
León, Hno. 149 (26a), 1393 (51/IId), 1707 (51/IVb), 3801 (51/IId), 4393 (28a), 4394 (46a), 4639 (40a), 6659 (40a), 6961 (21a), 6967 (46a), 8050 (40a), 8549 (32a), 8607 (51/IId), 8665 (40a), 8612 (51/IIb), 8878 (51/IIb), 8881 (40a), 8884 (51/IId), 9371 (51/IId), 9543 (40a), 9591 (40a), 9676 (51/IId), 9935 (51/IId), 10564 (51/IId), 11778 (40a), 12782 (28a), 13131 (40a), 13134 (40a), 13686 (51/IIb), 13702 (51/IId), 13974 (51/IId), 14809 (46b), 15077 (51/IId), 15078 (28a), 15364 (46a), 15690 (45a), 15756 (40a), 16335 (40a), 16697 (40a), 17525 (40a), 19590 (21a), 20661 (40a), 22397 (51/IId), 22400 (32a).
León, Hno. & Hno. Hioram 4361 (32a), 4473 (46a), 4594 (32a)
León, Hno. & E. P. Killip 17022 (46a)

Leonard, E. C. 3460 (45a), 3822 (47b), 4073 (45a), 7032 (51/IId), 7153 (45a), 7341 (51/IId), 7349 (51/IVa), 7559 (45a), 7728 (51/IId), 7857 (51/IId), 7977 (51/IId), 8044 (45a), 8049 (45a), 8136 (51/IId), 8193 (51/IId), 8420 (45a), 8600 (51/IId), 8631 (51/IId), 8750 (51/IId), 8802 (45a), 8873 (45a), 8939 (45a), 8983 (51/IId), 9013 (45a), 9059 (45a), 9073 (45a), 9180 (51/IId), 9387 (51/IId), 9626 (51/IId), 9720 (45a), 9953 (45a), 10041 (45a), 10042 (45a), 10048 (45a), 11278 (51/IVa), 11605 (51/IVa), 11699 (40a), 11708 (51/IVa), 11925 (45a), 11937 (45a), 12129 (51/IId), 12181 (51/IId), 12187 (51/IId), 12370 (40a), 12649 (51/IVa), 12838 (51/IId), 12857 (45a), 13156 (45a), 13360 (45a), 13396 (45a), 13412 (45a), 13422 (45a), 13479 (51/IId), 13714 (51/IId), 13770 (45a), 14059 (40a), 14065 (51/IVa), 14066 (51/IVa), 14081 (40a), 14180 (51/IVa), 14403 (51/IId), 14423 (40a), 14849 (45a), 14982 (45a), 15231 (45a), 15740 (45a).

LeSueur, H. 691 (51/IVc), 693 (26e)

Lévy, P. 181 (51/IVb)

Lewis, W. H. 3151 (51/IIa), 3339 (51/IIb)

Lewton, F. L. 178 (21b)

Liebmann, F. M. 27 (43a), 28 (44a), 31 (51/IVb), 5036 (31)

Liesner, R. L. 152 (51/IVb), 5376 (51/IIIb), 5616 (32a), 6180 (31), 7108 (31)

Liesner, R. L. & J. D. Dwyer 1413 (31), 1526 (51/IVb)

Lillo, M. 7290 (36)

Lima, A. 53/1287 (51/IIc), 53/1787 (51/IIIa), 58/2907 (22), 58/2923 (35b), 58/3019 (21a), 60/3534 (35b), 61/3716 (21a), 61/3821 (12), 61/3894 (18b), 61/3875 (14a)

Lima, N. 154 (32a)

Lindeman, J. C. 4297 (51/IIIa)

Lindeman, J. C. & J. H. de Haas 106 (21a), 3295 (51/IIIb), 3650 (35a), 4579 (35b), 5003 (39), 13330 (21a)

Linden, J. 80 (47c), 346 (47e), 686 (43a) 1312 (31), 1708 (45a)

Liogier, A. 1740 (51/IId), 9247 (42a), 9893 (42a), 9955 (51/IId), 10238 (32a), 10294 (51/IId), 10514 (42a), 10650 (51/IId), 11026 (51/IId), 11435 (51/IId), 11540 (47b), 12052 (51/IId), 12491 (47b), 13176 (51/IId), 13684 (47b), 13729 (47b), 14002 (47b), 14134 (51/IVa), 15192 (51/IId), 15359 (51/IVa), 15460 (51/IId), 15469 (47b), 15743 (51/IId), 15820 (51/IId), 16093 (47b), 16535 (42b), 17579 (40c), 18309 (51/IId), 18399 (47b), 18673 (40c), 18678 (40c), 18756 (51/IVa), 19396 (47b), 19543 (47b), 19686 (47b), 19898 (51/IVa), 19964 (51/IId), 20631 (42a), 20962 (47b), 23580 (51/IId), 25198 (40c), 25988 (51/IId), 26079 (40c), 27383 (40c), 27755 (28a), 29574 (42)

Lisbôa, P. 156 (51/IIIa)

Lloyd, F. E. 567 (47c), 571 (47c), 116 (47c), 691 (47c)

Lobb, W. 243 (47e)

Lobo de Anhanguera, U. 22 (33)

Löfgren, A. 170 (21a), 376 (30), 446 (27), 752 (35b), 829 (25)

López, A. & A. Sagástegui 4257 (47e), 4299 (47e), 5298 (47e)

López Figueiras, M. 431 (51/IId), 660 (51/IId), 942 (45a), 1368 (40a), 2892 (40a)

López-Guillén, – 202 (47e)

López Miranda, A. 260 (47e)

Lossen, W. 558 (32a)

Lourteig, A. 1926 (32a), 2032 (35a)

Luederwaldt, H. 999 (51/IIa)

Luetzelburg, P. von 166 (3), 257 (3), 305 (21a), 12284 (30), 12377 (37), 12387 (32a), 20507 (26a), 20844 (26a), 20971 (26a), 21476 (1), 23683 (27), 26976 (51/IIIa), 27068 (30)

Luna, A. 181 (40a)

Luna Peixoto, A. 756 (51/I)

Lund, P. W. 191 (33)

Lundell, C. L. 682 (51/IIIb), 1427 (47e), 6695 (21a), 6915 (31), 6983 (31), 6990 (21a), 7285 (34a), 7444 (47e), 8055 (21b), 14943 (21b)

Lyonnet, E. 487 (21a), 2820 (51/IVb)

Luschnath, B. 832 (21a)

Maas, P. J. M. & L. Y. Th. Westra 3582 (21a), 3680 (26a), 3629 (21a), 4042 (32b)

Macbride, J. F. 2853 (47e), 3541 (47e), 3813 (47e)

MacBryde, B. 423 (51/IVb)

McDaniel, S. 14753 (51/IVb)

MacDougall, W. B. H517 (51/IVb)

Macedo, A. 187 (33), 1606 (33), 1704 (32a), 1749 (51/IIa), 1766 (51/IIb), 2101 (51/IIb), 3787 (12)

McGill, L. & D. Keil 7876 (34a)

Machado de Campos, S. 67 (21a)

McKee, H. S. 11187 (51/IVb), 10590 (51/IId), 10794 (26a), 10897 (31), 10906 (44a)

McVaugh, R. 20378 (32a)

Magallanes, A. S. 54 (26f), 1169 (44a), 1263 (51/IVb)

Maguire, B. 22755 (51/IIb), 32006 (21a), 35888 (47j), 36008 (47j), 36011 (21a), 36035 (47j), 36056 (31), 36057 (32a), 36058 (21a), 36059 (26b), 40094 (1), 46257 (21a), 49002 (15), 49019 (15), 49056 (18c), 49123 (5a), 49142 (18b), 49210 (11), 49230 (14b), 54038 (51/IIIa), 54192 (51/IIIa)

Makrinius, E. 485 (21a)

Mahler, W. 5880 (32a)

Malme, G. O. A. 2244 (24), 3245 (29), 3379 (51/IIIb)

Manning, W. E. & M. S. 53163 (34a)

March, W. T. 657 (47a)

Marie-Victorin, Bro. 22070 (40a)

Markgraf, F. 3537 (12)

Marsh, E. G. 52 (34a), 506 (34a), 1754 (34a)

Marshall, S. A. et al. 6355 (51/IVb), 6480 (51/IVb)

Martinelli, G. 1348 (21a), 2584 (15), 2585 (15), 2859 (51/IIIb), 5186 (51/IIIb)

Martinet, M. 896 (47e)

Martínez-Calderón, G. 349 (31), 1005 (51/IId), 1061 (51/IId), 1512 (31), 1722 (51/IId), 2021 (51/IId), 2022 (51/IIb)

Martius, K. F. P. von 223 (29), 586 (21a), 720 (51/

IIIb), 806 (51/IIIa), 1077 (51/IIb), 1078 (51/IIb), 1268 (35b), 1271 (32a), 3272 (37)
Marulanda, L. 59 (32b)
Mathews, A. 917 (47e), 918 (47f), 1576 (51/IIb), 1925 (47e), 3271 (51/IVb)
Mattos, J. R. & M. F. 8295 (21a), 8346 (21a), 8438 (21a), 8501 (32a), 13551 (21a)
Mattos Silva, L. A. 217 (35b), 252 (32b)
Matute, G. 280 (31)
Matuda, E. 23 (43a), 45 (28a), 1978 (43a), 2122 (51/IVb), 31231 (21a), 32266 (26e)
Maxon, W. R. 1623 (47a), 2805 (47a), 5131 (32a), 8586 (51/IIb), 8637 (47a), 8771 (47a), 8784 (47a).
Maxon, W. R. & E. P. Killip 306 (28a), 1403 (47a)
Mélinon, – 26 (51/IIb)
Mello Barreto, M. 1433 (51/IIIb), 3211 (27), 4776 (7b), 5971 (12), 5975 (15), 7700 (12), 10063 (7c)
Mello Filho, 975 (47m), 1690 (26a), 3368 (15), 3523 (4), 3605 (5b)
Mendes Magalhâes, G. 92 (15), 4718 (21a), 5432 (21a), 17667 (3), 17679 (26a), 19933 (8), 25736 (2)
Mexia, Y. 380 (51/IVb), 790 (26e), 5182 (51/IVc), 648 (51/IVc), 745 (32a), 825 (51/IVc), 952 (51/IVc), 4382 (32a), 5740 (5a)
Meyer, T. 3557 (51/IIb), 10838 (35a)
Miers, J. 3204 (47m), 4662 (32a)
Mikan, J. C. 5659 (47m)
Miller, G. S. 37 (47c), 172 (45a), 227 (45a), 303 (45a)
Miller, O. O. & J. R. Johnston 210 (51/IVa)
Miller, W. de W. & L. Griscom 55 (51/IIIb), 99 (51/IIIb)
Millspaugh, C. F. 314 (51/IId), 375 (47c), 601 (47c), 2281 (40a), 2354 (40a), 9064 (40a), 9174 (40a), 9232 (51/IVa), 9297 (40a), 9303 (10c)
Mimura, I. 289 (32a), 299 (32a)
Miranda, F. 2827 (26f)
Mitchell, E. R. 29 (40a)
Mizoguchi, K. 377 (21a)
Mocquerys, – 1094 (47e), 1254 (51/IIIb)
Mohr, C. 94 (43a), 667 (43a)
Molina, J. A. & F. A. Barkley 18Va241 (51/IVa)
Molina R., A. 1614 (51/IIIb), 2619 (26c), 8741 (51/IIIa), 1167 (26c), 3623 (51/IVb), 11536 (43a), 13065 (51/IIIa), 13440 (31), 13528 (51/IIIb), 14610 (26c), 14631 (26c), 14726 (51/IIIa), 14769 (26a), 15264 (51/IIIa), 18548 (51/IIIa), 20638 (32b), 20723 (51/IVb), 22239 (51/IIIb), 22254 (51/IIIa), 22411 (31), 22799 (51/IVb), 24000 (31), 24654 (51/IVb), 24902 (51/IIIa), 24919 (31), 25193 (31), 25198 (51/IVb)
Montaldo, P. 3752 (51/IIb)
Monteiro da Costa, R. C. 286 (21a)
Monteiro Neto, H. 169 (25)
Montes, J. E. 462 (35a), 1599 (51/IIb), 2049 (35a), 4736 (35a), 9650 (51/IIb), 14924 (51/IIb), 10324 (35a)
Moore, H. E. 2997 (43a)
Moreira, H. & M. 489 (51/IIIb)
Morel, T. 944 (51/IIb), 4573 (51/IIb)
Moreno, P. 305 (47e), 368 (51/IVb)
Moreno, Z. 260 (47e)
Mori, S. A. 9271 (51/IIIa), 9371 (51/IIIa), 9514 (35b), 9513 (37), 9921 (32b), 9926 (32b), 11081 (32b), 11604 (30), 12456 (3), 12606 (12), 12965 (12)
Morillo, G. & M. Ishikawa 3609 (32b), 3652 (31)
Moritz, J. W. K. 1771 (51/IIb)
Morong, T. 171 (32a), 938 (51/IIb)
Morton, C. V. 4073 (51/IIb), 4379 (51/IId), 4791 (47c), 4870 (47c), 5173 (47c), 5612 (47c), 5664 (47c)
Morton, C. V. & J. Acuña 2930 (51/IId), 3079 (51/IId), 3242 (51/IIb), 3890 (51/IId)
Müller, F. 81 (44a), 104 (43a), 586 (31), 1014 (44a), 1114 (43a), 1628 (43a)
Mueller, C. H. & M. T. 517 (34a), 3422 (51/IVc)
Mutis, C. 2388 (43c), 4282 (32b)
Myers, J. G. 5509 (26a)

Nagelkerken, W. P. 148 (51/IVa)
Narváez Montes, M. & A. E. Salazar 507 (51/IVb), 622 (32a)
Nash, G. V. 122 (40a), 629 (51/IId), 780 (45a), 787 (45a), 1433 (45a)
Nash, G. V. & N. Taylor 910 (10c), 1261 (10c), 1301 (40a), 3829 (10c)
Nealley, G. C. 288 (21b)
Nee, M. 6868 (51/IIa)
Neill, D. A. 1132 (51/IVb), 2239 (51/IVb), 2761 (51/IVb), 2869 (51/IVb), 2923 (51/IIb), 6355 (51/IVb)
Nelson, Cirilo 209 (51/IVb)
Nelson, E. W. 790 (43a), 853 (51/IIa), 2310 (32a), 2805 (51/IIc), 2850 (21a), 3014 (51/IIb), 3608 (43a)
Nienstedt, E. F. 408 (31)
Norris, D. O. 57 (51/IIa), 116 (51/IIb), 159 (32a), 184 (26a), 197 (32a), 292 (51/IIb), 300 (51/IIa), 333 (21a), 383 (32a), 411 (21a)
Northrop, J. I. 393 (40a), 638 (10a)
Novais, C. 229 (32a)
Novelo, A. 447 (44a)

Occhioni, P. 1163 (21a), 5647 (5a)
Ochoa, C. 286 (47e), 434 (47e), 707 (47f)
Oliveira, E. 1474 (27), 3287 (21a), 6088 (51/IIb), 6438 (51/IIb)
Oliveira Torres, J. de 5 (12)
Onishi, E. & S. Fonseca 34 (32a)
Orbigny, A. d' 966 (35a)
Orcutt, C. R. 1218 (34a), 3098 (44a), 3295 (31), 5538 (44a)
Ortega, J. G. 5999 (32a), 6088 (26e), 7035 (51/IVc), 7288 (26e)
Ortíz, R. T. 650 (31)

Pabst, G. F. J. 8736 (21a), 9575 (51/IIb)
Padilla, S. A. 136 (51/IIIb), 441 (31)
Palmer, E. 29 (26e), 31 (26e), 125 (32a), 156 (34a), 175 (51/IVc), 231 (26a), 242 (51/IVc), 254 (44a), 267 (21a), 283 (34a), 395 (32a), 405 (27l), 460 (51/IVb), 529 (47l), 534 (44c), 610 (26a), 654 (51/IVc), 1053 (51/IIb), 1107 (32a), 1606 (51/IVc)

Palmer, W. & J. H. Riley 396 (51/IIb), 1078 (46b)
Parra, M. 439 (34c)
Partch, M. 69/55 (31)
Passarelli, A. 81 (51/IIIb)
Peck, M. E. 289 (51/IVb), 603 (51/IVb), 691 (21a)
Pedersen, T. M. 66 (32b), 205 (51/IIb), 3003 (35a), 3073 (26a), 3283 (35a), 3726 (21a), 4492 (21a), 5244 (51/IIIb), 5302 (35a), 5832 (51/IIb), 6476 (26a), 8082 (35a), 9441 (21a), 9494 (26b), 9619 (32a), 9454 (51/IIb)
Peebles, R. H., Harrison & Kearney 4624 (26e), 4707 (26e)
Pennell, F. W. 1351 (51/IIa), 1869 (42c), 2702 (47g), 2749 (47k), 2886 (26a), 2934 (51/IIa), 3465 (51/IIa), 3674 (51/IIa), 4009 (29), 4029 (51/IVb), 5477 (47e), 5614 (51/IIb), 5642 (51/IIb), 5661 (26a), 9553 (40b), 10690 (47e), 10698 (51/IIb), 10839 (51/IIb), 12071 (28a) 12072 (26a), 1454 (51/IIa), 1499 (51/IIb), 1509 (51/IIb), 17926 (42a), 18148 (47l), 19535 (26e), 20252 (51/IVc), 2773 (51/IIb)
Pennell, F. W. & E. P. Killip 5961 (51/IIa), 6040 (51/IIb)
Perazzo Barbosa, V. 34 (51/IIIb), 171 (51/IIIb), 261 (51/IIIb)
Peredo, I. 305 (32b)
Pereira, E. 283 (21a), 346 (21a), 1487 (7c), 1598 (5a), 1649 (21a), 2035 (35b), 2037 (12), 2046 (21a), 2133 (37), 5016 (21a), 5841 (14a), 5754 (47m), 7329 (21a), 8843 (32a), 8925 (15), 9827 (32b)
Pereira, E. & G. F. W. Pabst 2893/3729 (2), 2895/3731 (4), 8333/9444 (21a), 8395/9506 (35b), 8578/7853 (35a), 9218/8219 (4), 9524/8413 (1), 9188/8189 (15)
Pérez Arbaláez, E. 344 (51/IIb), 2016 (47g), 6427 (51/IIb)
Perkins, J. R. 1029 (47a)
Pfeifer, H. W. 1340 (51/IIIb), 1495 (31), 1778 (51/IIa)
Philcox, D. 3074 (32a), 3176 (32a), 8209 (51/IIc)
Philcox, D. & A. Ferreira 3371 (27), 3662 (33), 3748 (27), 3959 (33), 3968 (21a), 4139 (51/I)
Philipson, W. R. 755 (51/IId)
Philipson, W. R. & J. M. Idrobo 1341 (51/IIa)
Picarda, L. 599 (47b)
Pickel, D. B. 2470 (51/IIb)
Pilger, R. K. F. 285 (21a)
Pilz, G. E. 1319 (26c)
Pinheiro, R. S. & T. S. Santos 2290 (32b)
Pinkus, A. S. 19 (51/IIIa)
Pires, J. M. 75/19 (51/IId), 1403 (21a), 9433 (21a), 9907 (32a), 9988 (43b), 14656 (26a), 51072 (21a)
Pires, J. M. & G. A. Black 1905 (31), 1974 (32a), 2018 (31), 2033 (26a), 2056 (32a), 2285 (21a)
Pires, J. M. & P. Cavalcante 52113 (51/IIc), 52113A (51/IIb)
Pires, J. M. & N. T. Silva 10516 (51/IId)
Pires, J. M. & L. Y. Th. Westra 47938 (51/IIIa)
Pittier, H. 2104 (51/IVb), 3311 (32a), 4633 (31), 4655 (51/IIa), 4749 (51/IVb), 4828 (51/IIb), 4854 (31), 5216 (51/IIIa), 6874 (51/IVb), 7170 (51/IIIb), 7238 (51/IIIb), 7243 (26a), 7416 (26a), 7542 (51/IIIb), 7982 (26a), 7981 (28a), 8699 (26a), 8702 (28a), 8704 (32a), 9444 (51/IIIb), 9549 (51/IIIb), 9681 (26a), 9811 (26a), 9844 (51/IIa), 10592 (32b), 10606 (21a), 10807 (26a), 11613 (51/IIIb), 12864 (47e), 12868 (51/IIIb), 14283 (21a), 14751 (21a), 15078 (32a), 16338 (32a)
Plee, A. 107 (29)
Plowman, T. et al. 8383 (27), 8976 (51/I), 9019 (27), 9165 (21a), 9218 (27), 9287 (30), 9288 (31), 9289 (32a), 9295 (51/IIb), 9324 (30), 9376 (31), 9609 (21a), 9669 (21a)
Poeppig, E. F. 1629 (47e), 1630 (47e)
Poggie, J. J. 39 (44a)
Pohl, J. G. 449 (51/I), 546 (7b), 787 (33), 907 (32a), 949 (7b), 997 (21a), 1116 (51/I), 1395 (27), 1442 (51/IIa), 1521 (51/I), 2126 (17), 3135 (14b), 3152 (14b), 5296 (12)
Pollard, C. L. 80 (51/IVa)
Pollard, C. L. & W. Palmer 332 (51/IIb)
Ponce, M. & G. D. Rodríguez 13 (21a)
Porter, D. M. 1367 (21b)
Potter, D. 5501 (47c)
Powell, A. M. & B. L. Turner 2719 (34a)
Prance, G. T. et al. 2094 (30), 4248 (21a), 4496 (1), 5193 (43b), 5414 (21a), 5737 (21a), 6027 (21a), 7205 (51/IIIa), 11647 (51/IIIa), 18501 (43c), 20195 (51/IIIa), 24394 (51/IIc)
Pringle, C. G. 1234 (47l), 1596 (47l), 1820 (32a), 2079 (34a), 5418 (26e), 6805 (44a), 7839 (43a), 8337 (43a), 9731 (26e), 10053 (51/IVb), 11402 (34a), 11405 (26e), 11406 (32a)
Prior, A. 1050 (51/IId)
Proctor, G. R. 3370 (40e), 10858 (47b), 16066 (51/IId), 16845 (47c), 17502 (47c), 18178 (47c), 18413 (28a), 18441 (26a), 18449 (28a), 18563 (47c), 1888 (47c), 19563 (51/IVa), 19803 (47a), 19816 (28a), 19821 (51/IId), 19908 (47c), 20283 (47c), 20469 (47c), 20602 (51/IIb), 20729 (26a), 20808 (47a), 21526 (28a), 21798 (47c), 22872 (28a), 22941 (28a), 23242 (40a), 24833 (47a), 25882 (47c), 25971 (47c), 32729 (28a), 33179 (40a)
Proctor, G. R. & W. Gillis 33283 (10c), 34033 (10c), 34078 (10c)
Pulle, A. 512 (51/IIIa)
Purpus, C. A. 1194 (34c), 1880 (31), 1882 (43a), 5884 (51/IIb), 6086 (44a), 7201 (21a), 7850 (51/IVb), 7850 (21a), 8432 (34b), 9135 (21a), 9269 (51/IVb), 9384 (40f), 10003 (21a), 10004 (51/IIIb), 10010 (21a), 10472 (51/IVb), 10553 (51/IIIb), 10581 (51/IVb), 10592 (26e)
Pursell, R. A. 8068 (32a), 8868 (51/IIIb), 9297 (32a), 9302 (21a), 9474 (51/IId)

Quelch, J. J. & F. McConnell 140 (1), 719 (1)
Questel, A. 533 (51/IVa), 534 (51/IVa), 541 (51/IVa), 584 (47c), 594 (47c), 2320 (51/IId), 2321 (47c), 2322 (47c), 4676 (51/IId), 5169 (41), 5182 (41)

Rambo, B. 39081 (21a), 40587 (51/IIb), 40587 (51/IIb), 51945 (35a)

Ramia, M. 2576 (21a), 2599 (32a), 2638 (21a), 3104 (21a)
Ramón de la Sagra—see Sagra
Ramos, J. 309 (51/IIa)
Ramsammy, J. R. 99 (47c)
Ratter, J. A. et al. 1010 (33), 1169 (51/IIa)
Reed, E. 398 (51/IIIb), 432 (51/IIIb)
Regnell, A. F. 49 (32a), I/59 (51/IIa), III/484 (32a), III/485 (33), III/487 (27), III/488 (21a)
Reineck, E. M. 134 (35a)
Reiss, R. 144 (32a)
Reko, B. P. 3608 (51/IIIb), 4553 (26e), 5014 (32a)
Ricksecker, A. E. 112 (51/IVa), 166 (47c), 194 (47c)
Riedel, L. 20 (47m), 30 (32a), 34 (51/IIb), 36 (21a), 113 (47m), 60 (21a), 114 (51/IIb), 119 (51/IIb), 164 (21a), 177 (23), 178 (51/IIa), 179 (21a), 198 (4), 216 (26a), 218 (39), 248 (7b), 411 (39), 468 (23), 477 (21a), 481 (24), 483 (9), 484 (29), 558 (4), 559 (3), 562 (18b), 571 (33), 622 (31), 674 (32a), 678 (32a), 765 (35b), 802 (31), 808 (29), 992 (24), 1074 (2), 1162 (32a), 1207 (18b), 1239 (9), 1379 (33), 1582 (51/IIa), 1771 (32a), 2031 (33), 2032 (51/I), 2383 (26a), 2449 (8), 2901 (12)
Ripley, H. D. & R. C. Barneby 14830 (26a)
Robert, A. 399 (31)
Robinson, J. W. L. 33 (51/IIb)
Rodrigues, N. 2279 (51/IIIa)
Rodrigues, W. 397 (51/IIIa)
Rodrigues, W. & A. Aubréville 666 (21a), 707 (32a)
Rodríguez, F. M. 281 (51/IIb)
Rodríguez, H. 261 (28a)
Rodríguez, J. V. 3525 (51/IVb)
Rodríguez, L. 2537 (51/IId), 2983 (51/IId), 4682 (51/IId)
Rodríguez, S. 21 (47g)
Roe, K. 1296 (44a)
Roig, J. T. 1797 (40a), 3840 (28a), 3859 (21a), 5590 (39a)
Rojas, T. in Hassler 2590 (36), 2590a (36), 10038 (33), 12672 (32a)
Romero C., R. 1193 (51/IIIa), 1254 (51/IVa)
Romero Castañeda, R. 4311 (51/IVb)
Rosa, N. A. & M. R. Santos 2003 (21a)
Rosas, M. 10 (43a), 511 (43a)
Rose, J. N. 2253 (26e), 2469 (26e), 2471 (51/IVc), 3169 (47c), 3482 (26e), 4252 (40c), 11253 (34c), 21699 (51/IIIb), 22205 (47e), 23851 (47e)
Rose, J. N., W. R. Fitch & P. G. Russell 3249 (47c), 3295 (47c), 3804 (51/IId), 4316 (51/IId)
Rose, J. N. & R. Hay 5888 (34c), 6093 (43a)
Rose, J. N. & J. H. Painter 6971 (51/IVb), 7335 (32a), 7363 (51/IVc), 7363 (51/IVb), 7481 (26e), 9500 (51/IVc), 10241 (51/IVb)
Rose, J. N., P. C. Standley & P. G. Russell 13046 (26e), 13796 (32a), 13824 (26e)
Rose, J. N. & P. G. Russell 19927 (21a), 20038 (37), 20276 (51/IIIb), 20689 (21a), 20718 (21a)
Roth, L. 1833 (21a), 1834 (51/IIb), 1840 (3), 1845 (15)
Rothrock, J. T. 683 (26e)
Rovirosa, J. N. 45 (31), 253 (51/IId), 709 (51/IId), 725 (44a)
Rozynski, H. W. von 599 (43a)
Rudd, V. 499 (21a), 2061 (51/IId)
Ruiz, Thirza 3574 (51/IVb)
Ruiz Teran, L. 1580 (47e)
Runyon, R. 1481 (21b)
Rusby, H. H. 1022 (47e), 2854 (51/IIa)
Rusby, H. H. & F. W. Pennell 1077 (26a)
Rzedowski, J. 5410 (34b), 5561 (34b), 6668 (47l), 11375 (47l), 20007 (44a), 21196 (51/IVb)

Sachet, M.-H. 370 (51/IVd)
Saer d'Héguert, J. 205 (51/IVa), 905 (26a)
Sagástegui, A. et al. 7144 (47e)
Sagra, R. de la 619/1 (40a)
St. Hilaire, A. de 471 (5a), 822 (4), 2032ter (3), 2346 (32b), A1/27 (32a), B1/347 (12), B1/612 (27), B/700 (51/IIIb), B1/882 (4), B1/905 (2), B1/1742 (14b), B1/1997 (14a), B1/1998 (14a), B2/2153 (12), Ba/2265ter (12), B2/2399bis (7b), B2/2415bis (27), B2(2)/252 (21a), B2(2)/258 (35b), B2(2)/296 (32b), C1/312 (4), C1/655 (33), C1/933 (8), C1/1171 (39), C2/1297 (26b), C2/1383 (39), C2/1383 (51/IIb), C2/1463bis (51/IIIb), C2/2490 (35a), C2/2672 (51/IIb), D/659 (33)
Salvosa, F. M. 762 (40a)
Salzmann, P. 2046 (21a)
Sampaio, A. J. de 3 (3), 508 (51/IIa), 2998 (51/IIb), 3007 (51/IIIb), 3311 (51/IIIb), 3371 (51/IIIb), 3908 (51/IIIb), 6084 (51/IIb), 6349 (51/IIIb), 6439 (51/IIIb), 8053 (21a)
Sandeman, C. 3574 (47e), 5526 (47e), 4464 (47e)
Santoro, J. 358 (51/IIb), 423 (21a), 434 (21a)
Santos, E. 1177 (32a), 1598 (32b)
Santos, E. & J. C. Sacco 1919/2180 (26a), 1926/2187 (51/IIIb), 1947/2208 (51/IIIa), 1948/2209 (37), 1985/2247 (26a)
Santos, R. S. & A. Castellanos 24153 (21a), 24298 (14b)
Santos, T. S. dos 717 (21a), 3046 (51/IIIa)
Saravia, C. 2222 (26a), 2537 (28a), 2881 (28a), 2907 (29)
Saravia, C. & D. Johnson 494 (26a), 564 (26a)
Sargent, F. H. 37 (42a), 696 (47c)
Sartorius, C. 35 (43a)
Sastre, C. 1509 (51/IIIa)
Sastre, C. & J. Jérémie 1969 (41)
Saunders, S. G. E. 105 (51/IIb), 740 (47f)
Saunders, G. B. 555 (44a)
Schaller, G. 91 (32a), 105 (51/I)
Schiede, C. W. J. 219 (31), 709 (43a), 711 (44a)
Schiede, C. W. J. & Deppe 708 (31)
Schiller, I. 185 (partim) (21b), 955 (21b)
Schimpff, H. J. F. 454 (47e), 1025 (51/IVb)
Schinini, A. 12575 (21a), 12556 (32a), 12742 (51/IIb), 13312 (32a), 15097 (32a)
Schipp, W. A. 490 (51/IVb), 637 (21a), 638 (31), 720 (51/IVb)
Schlim, J. 106 (51/IVb), 106 (51/IIa), 174 (51/IIb), 281 (29)
Schnee, L. 495 (51/IIIb)
Schomburgk, Richard 840 (1)

Schomburgk, Robert 57 (21a), 59 (21a), 176 (51/IIIa), 387 (32b), 398 (51/IIIa), 441 (51/IIb), 498 = 771 (32b), 531 (26a), II/582 (1), 683 (51/IIc), 720 (51/IIIa), 722 (21a), 840 (26a)
Schott, A. 105 (21b), 290 (21b), 976 (47e)
Schreiter, R. 3974 (36), 9233 (36), 11095 (51/IVb)
Schroeder, A. H. 147 (34a)
Schubert, B. G. 1712 (47e), 1842 (44a)
Schubert, B. G. & M. Sousa 1946 (32a), 1978 (26e)
Schumann, W. 193 (471)
Schultes, R. E. & López 9264 (31)
Schulz, C. L. 399 (35a), 664 (32a)
Schulz, E. D. 429 (21b)
Schulz, J. P. 600 (51/IIb), 707 (31)
Schunke, C. A105 (51/IIb)
Schunke V., J. 3021 (51/IVb)
Schwacke, A. W. 1023 (30), 6317 (47m)
Schwarz, G. J. 313 (32a), 1810 (35a), 449 (35a), 482 (35a), 8992 (35a)
Schwindt, E. 461 (35a), 1210 (35a), 2430 (35a), 2763 (51/IIb)
Seaton, H. E. 140 (43a), 436 (51/IIb)
Seemann, B. 217 (31)
Segadas Vianna, F. & J. Loredo 1105 (2), 3604 (21a), 4006 (47m)
Seibert, R. J. 1772 (51/IVa)
Seler, E. 1977 (21a), 2286 (51/IIIa), 2411 (31), 3721 (44a), 4238 (51/IVb), 4476 (44a)
Semir, J. 2028 (2), 2356 (2), 2374 (18c), 2731 (18c), 4099 (2), 5032 (4)
Senn, H. A. 113 (51/IId)
Sessé & Mociño Herb. 1005 (31), 1146 (51/IId), 1146bis (51/IVb), 1147 (21a), 1193 (43a), 1199 (26d)
Seymour, F. C. 2357 (51/IVb), 3683 (31), 3990 (51/IVb), 5999 (21a), 6443 (51/IVb)
Shafer, J. A. 58 (47c), 125 (51/IId), 143 (32a), 151 (47c), 330 (51/IId), 375 (51/IIb), 466 (51/IId), 498 (40a), 653 (40a), 720 (40a), 1184 (45a), 1269 (45a), 1719 (40a), 2362 (51/IId), 2471 (51/IId), 2582 (40a), 2778 (40a), 2793 (42a), 2780 (47c), 2874 (47c), 2895 (51/IVa), 2899 (51/IId), 3029 (47c), 3785 (40a), 7733 (51/IIb), 7806 (40a), 8414 (40a), 9067 (51/IId), 10404 (28a), 10434 (28a), 10583 (21a), 10709 (46a), 10881 (21a), 10971 (26a), 11062 (28a), 11649 (28a)
Shannon, W. C. 546 (31)
Sharp, A. J. 4629 (31), 441192 (43a), 50329 (34a)
Shepherd, G. J. et al. 3950 (19)
Shimoya, C. 3578 (51/I)
Shreve, F. 6663 (51/IVc)
Shreve, F. & E. R. Tinkham 9818 (34a), 9566 (34a)
Sieber, F. W. 110 (47c), 251 (51/IId)
Silva, I. H. 10 (51/IIb)
Silva, M. 704 (21a), 1287 (21a), 1134 (51/IIc), 1628 (21a), 1850 (51/IIIa)
Silva, M. G. 2924 (21a), 4506 (24)
Silva, M. G. & R. Bahia 3027 (27)
Silva, M. G. & A. Pinheiro 4135 (32a), 4156 (24)
Silva, J. M. de 202 (51/IId), 263 (51/IId)
Silva, N. T. 4779 (26b), 4873 (26b), 58288 (51/IIIa), 59732 (21a)
Silveira, A. & H. de Magalhaes 42 (7b), 402 (51/IIb), 527 (51/I), 556 (27)
Silva Manso, A. L. P. da 219 (29)
Sintenis, P. 367 (51/IId), 631 (40c), 1165 (47c), 1967 (47c), 3312 (51/IVa), 3450 (40c), 3746 (51/IVa), 4472 (47c), 4775 (51/IId), 4926 (42a), 5079 (51/IId), 5402 (47c), 6716 (47c)
Skutch, A. F. 1009 (43a), 1349 (51/IVb), 5278 (51/IVb)
Small, J. K. 8914 (40a)
Small, J. K. & J. J. Carter 8514 (10a), 8761 (40a), 8853 (10a)
Smith, A. C. 2389 (51/IIIa), 3256 (21a), 10188 (47c), 10578 (47c)
Smith, Austin A/588 (51/IIIb)
Smith, C. E. et al. 3765 (34c)
Smith, C. L. 1905 (31)
Smith, H. H. 66 (29), 67 (26a), 274 (51/IIb), 275 (51/IVb), 275b (51/IVb), 276 (28a), 707 (51/IIb), 709 (27), 782 (51/IVa), 1187 (47c), 2537 (51/IVb)
Smith, J. Donnell 1028 (51/IIIa), 2293 (51/IVb), 3711 (51/IIIa), 5212 (51/IIa)
Smith, L. B. 2069 (21a), 3139 (40a), 6422 (21a), 6614 (21a), 6933 (32a)
Smith, L. B. & A. C. Brade 2326 (51/IIIb)
Smith, L. B. & Klein 10643 (39), 12113 (39), 14758 (32a)
Smith, L. B. & P. R. Reitz 6062 (21a)
Smith, L. C. 754 (51/IVc), 1548 (43a)
Smith, S. G. 1122 (32a)
Snethlage, E. H. 47 (51/IId)
Sneidern, K. von 204 (51/IIa), 602 (51/IIb), 1363 (51/IIa), 1158 (51/IIb), 4607 (51/IIb)
Soukup, J. 4030 (47f), 4529 (47f), 5436 (47f), 5468 (47f)
Sousa, M. 2202 (50/IId), 2848 (51/IId), 3300 (44a), 5390 (34c), 5787 (26f), 5843 (21a), 5844 (51/IIa), 5855 (21a), 5872 (21b), 5875 (31), 6314 (51/IIIb), 6471 (32a), 7109 (26g), 7330 (21a), 7597 (51/IId), 7723 (26f), 7735 (34c), 7951 (51/IIa), 8244 (51/IIIb), 8360 (51/IVb), 8382 (51/IIIb), 8434 (32a), 8439 (51/IVb), 8485 (51/IVb), 8497 (21a), 8634 (26g), 8795 (31), 8811 (51/IId), 8986 (32a), 9110 (26g), 9112 (21b), 9206 (31), 9233 (51/IId), 9443 (51/IIIb), 9499 (26g), 9554 (51/IVb), 9579 (31), 9580 (26a), 9581 (31), 9592 (21a), 9743 (51/IIIb), 9890 (21a), 9948 (51/IIIb), 10124 (26g), 10125 (21b), 10270 (51/IId), 10286 (51/IId), 10314 (34c), 10442 (34c), 10589 (21a), 11025 (47e), 11681 (51/IVb)
Spellenberg, R. 3850 (51/IVc)
Spellman, D. L. 1388 (51/IIIa)
Sprague, T. A. 257 (51/IIa)
Spruce, R. 551* (51/IIc), 771 (51/IIb), 771* (51/IIb), 1292 (51/IIIa), 3003 (31), 3589 (26a)
Stafford, D. 1220 (47e)
Stahel, G. & Gonggrijp 852 (51/IIb)
Standley, P. C. 2880 (51/IIIb), 10702 (51/IIIb), 14433 (51/IVb), 14645 (26c), 19597 (51/IVb), 19819 (51/IIIIb), 19859 (51/IIIb), 20419 (21a),

21192 (51/IVb), 23246 (26c), 24176 (51/IVb), 24285 (51/IVb), 24433 (26a), 24660 (51/IVb), 25194 (51/IIa), 25231 (51/IVb), 25838 (31), 26031 (51/IVb), 26289 (31), 26342 (51/IIa), 26600 (51/IIa), 26636 (31), 26663 (26c), 26798 (51/IIb), 26826 (26c), 26827 (51/IIIb), 27142 (51/IVb), 27240 (51/IVb), 27266 (51/IVb), 27337 (26c), 27390 (51/IIb), 27452 (51/IIIb), 27657 (51/IVb), 27753 (31), 28014 (51/IIa), 28337 (51/IVb), 28730 (51/IIIb), 29015 (51/IVb), 29095 (31), 29166 (51/IIb), 29239 (51/IVb), 29277 (26c), 29416 (51/IVb), 30053 (51/IIb), 31811 (31), 32154 (51/IIb), 32580 (51/IIIb), 32701 (51/IIIb), 32868 (51/IIIb), 33291 (51/IIIb), 35825 (51/IIIb), 40019 (31), 41325 (51/IIIb), 53807 (51/IVb), 54084 (51/IId), 54218 (51/IVb), 54775 (51/IVb), 54983 (51/IVb), 56253 (51/IIIb), 56279 (31), 57941 (51/IIIa), 58047 (51/IIIa), 60301 (51/IIIa), 72705 (51/IVb), 73593 (51/IVb), 73909 (26a), 74295 (51/IVb), 77386 (51/IIIa), 77839 (51/IVb)

Stanford, L. R. et al. 871 (34b), 2038 (34b)

Steere, W. C. 3100 (21b)

Stehlé, H. 446 (41), 2818 (51/IId), 3503 (47c), 4314 (47c), 4322 (47c)

Stein, S. 4 (51/IIb)

Steinbach, J. 3787 (31), 5388 (32b), 5332 (51/IIb), 5389 (21a), 5628 (21a), 6165 (21a), 6336 (21a), 6354 (32b), 7366 (51/IIb), 7466 (51/I)

Steinbach, R. F. 396 (51/IIb)

Stevens, F. L. 1750 (51/IId), 6192 (40c), 6233 (40c), 6284 (40c)

Stevens, W. D. 448 (26e), 3975 (51/IIIb), 4420 (51/IVb), 4512 (51/IIIb), 4930 (51/IVb), 5189 (51/IVb), 5238 (51/IVb), 5385 (51/IVb), 6139 (51/IIIb), 7654 (31), 7687 (31), 7827 (21a), 7836 (31), 8127 (51/IIa), 8623 (51/IIa), 9300 (51/IIa)

Stevenson, J. A. 85 (51/IId), 1114 (51/IId), 1693 (47c), 2050 (47c)

Steward, W. C. 247 (21a)

Stewart, R. M. & I. M. Johnston 3023 (51/IVc)

Steyermark, J. A. 38433 (31), 42428 (51/IIIa), 49583 (51/IIa), 50463 (43a), 51295 (43a), 52191 (28a), 52198 (51/IVb), 86227 (21a), 88318 (21a), 88319 (51/IIb), 88901 (32a), 97215 (47e), 99393 (51/IIIb), 100059 (51/IIb), 100107 (51/IIb), 102746 (31), 107669 (51/IIIb), 112149 (51/IIIb), 115238 (28a)

Stimson, W. R. 408 (51/IId), 778 (40a), 1068 (40a), 3144 (51/IId)

Stoffers, A. L. 308 (51/IVa), 2135 (51/IVa), 4715 (47c)

Stork, H. E. & O. B. Horton 10103 (47e), 10444 (47f)

Strang, H. & A. Castellanos 1095/26678 (15), 1096/26679 (2)

Stuckert, T. 2284 (36), 6011 (36), 6099 (36)

Stuessy, T. 1028 (47b)

Sucre, D. 97 (51/IIa), 526 (32a), 1081 (47m), 1279 (47m), 2092 (51/IIIb), 3857 (21a), 4092 (51/IIIb), 9572 (47m), 9851 (51/IIIb)

Sucre, D. & P. I. S. Braga 4623/1509 (32b)

Sumichrast 667 (43a), 1726 (43a)

Swallen, J. 3685 (35b), 9464 (32a), 10730 (21a), 10949 (31), 10950 (26c)

Tamayo, F. 804 (51/IVa), 1066 (51/IVa), 1167 (51/IVa), 1474 (32a), 1725 (26a), 2411 (47e), 2567a (51/IIIb), 2623 (27a), 3537 (21a), 4015 (31), 4095 (31), 4139 (51/IIb), 4356 (47e), 4601 (51/IIc)

Tate, G. H. H. 28 (21a), 86 (21a), 121 (32b), 176 (1), 1018 (51/IIb)

Tavares, S. 536 (37), 637 (51/IIIb), 657 (51/IIIb)

Taylor, A. A. 124 (46b)

Taylor, J. & C. 7363 (51/IVb)

Taylor, N. 87 (51/IVa), 330 (51/IIb)

Tejera, E. 238 (32b)

Télez, O. 279 (34c), 1145 (47e), 1187 (51/IVb), 1295 (47e), 1337 (47e), 1376 (47e), 1813 (47e), 2217 (47e), 2486 (47e), 3503 (47e), 3747 (47e), 3783 (51/IVb)

Terorgh, J. 100 (47b)

Thompson, J. B. 175 (47c), 541 (47c), 723 (47c), 882 (47c)

Thurber, G. 1060 (51/IVc)

Thurm, D. 199 (51/IId)

Thurn, E. F. im A/71 (1)

Tomlin, S. 8 (51/IIIb)

Ton, A. S. 1145 (43a), 1388 (43a)

Tonduz, A. 649 (51/IIIb), 1566 (51/IIIb), 4718 (43a), 4990 (31), 7198 (51/IIIb), 7274 (51/IIIb), 10920 (32a), 13556 (51/IVb)

Toro, R. A. 777 (51/IIb), 783 (47e)

Topping, D. LeR. 3036 (51/IId), 3069 (51/IId)

Tovar, A. 4100 (51/IIb)

Tovar, O. 287 (47f), 1915 (47f), 2113 (47f)

Traill, J. W. H. 177 (51/IIb), 178 (51/IIc)

Trejos, A. V27 (41/IIb)

Triana, J. 4389 (47e), 4392 (47g)

Trinta, Z. A. & E. Fromm 567/1643 (51/IIIb), 794/1870 (21a), 1432/2435 (51/IIIb)

Trott, S. 247 (51/IIa)

Trujillo, B. 742 (51/IIa), 1030 (51/IIIb), 1755 (51/IVb), 1817 (51/IIa), 2126 (32a), 2356 (51/IIb), 2369 (28a), 2371 (32a), 2492 (28a), 2494 (32a), 2611 (51/IIIb), 2925 (51/IVa), 3979 (26a), 4947 (51/IVa), 5994 (21a), 7278 (28a), 10851 (51/IIb), 11627 (21a), 13098 (32a), 13101 (32a), 13106 (51/IIb)

Tucker, J. M. 656 (51/IIIa), 737 (51/IIIb), 2554 (51/IVc)

Tuerckheim, H. von 61 (51/IIIb & 51/IVb), 2887 (47b), 3336 (51/IVa), 3571 (31), 3680 (42a), 8666 (51/IVb)

Tutin, T. G. 1591 (32a)

Tweedie, J. 1269 (47m), 11068 (32a)

Ugent, D. 5373 (47e)

Ule, E. 2BA (37), 389 (51/IIIa), 411 (29), 2868 (29), 4820 (47m), 6952 (32b), 7435 (25), 7527 (21a), 7526 (37), 7771 (26a), 8137 (32b), 8140 (21a), 8141 (51/IIIa)

Underwood, L. M. & F. S. Earle 1618 (51/IId), 1679 (51/IVa)

Underwood, L. M. & R. F. Griggs 7 (51/IId), 43 (47c), 241 (51/IId), 661 (42a)

Uribe Uribe, L. 698 (47e), 798 (26a), 1279 (51/IIa)
Usteri, A. 57 (51/IIa), 61 (32a), 77b (32a), 113b (39)

Valeur, E. J. 30 (51/IId), 824 (51/IId)
Vargas, C. 211 (47f), 1320 (51/IIb), 2511 (47e), 2586 (51/IIb), 2720 (47e), 12265 (51/IIb)
Vauthier, – 72 (51/IIb), 152 (51/IIb), 153 (7a), 154 (5a), 155 (51/I), 157 (32a), 161 (18b)
Velasco, R. 109 (51/IIIb), 117 (51/IIb), 162 (51/IIa), 170 (51/IIa), 184 (51/IIa), 316 (51/IIIb)
Vivas, J. M. 127 (51/IIb)
Vélez, I. 1846 (40c), 2291 (26b), 2662 (32b), 2954 (40c), 3029 (47c), 3273 (47c), 3412 (47c), 3541 (40c)
Velloso, H. P. 7 (51/IIIb)
Ventura A., F. 9084 (32a)
Venturi, S. 1471 (36), 4898 (36), 7625 (51/IVb), 8031 (36), 8217 (36), 8326 (51/IVb), 8772 (36), 9106 (26b), 9143 (32a)
Vianna Freire, C. 243 (51/IIIb)
Vidal, J. 1476 (27), 5157 (51/IIIb), I-307 (27), I-313 (27), I-412 (51/IIa), I-476 (27), II-5366 (51/IIIb), II-5816 (51/IIIb)
Viereck, H. W. 360 (34b), 484 (34b), 1053 (34a)
Villareal de Puga, M. 1213 (44a)
Vogel, C. 504 (51/IIIb)

Waby, J. F. 64 (47c)
Wagner, R. J. 7 (47c), 1080 (47c), 1603 (40c)
Walker, R. 1224 (31)
Wallace, Dunn & LeDoux 514 (51/IVb)
Warming, E. 11 (51/IIc), 193 (33), 195 (21a), 197 (51/I), 201/2 (27), 205 (51/IIb), 207 (51/IIIb), 263 (51/IIIb)
Watson, S. 413a (31)
Wawra von Fernsee, R. 330 (21a), 1097 (43a)
Weaver, J. N. 644 (34a)
Weberbauer, A. 1445 (47e), 5496 (47f), 5633 (47e)
Webster, G. L. 10353 (40a)
Weddell, H. A. 491 (32a), 584 (51/IIb), 1056 (51/I), 1860 (27), 1868 (27), 1915bis (24), 2286 (26a), 2398 (21a), 2600 (51/I), 3159 (51/IIb & 51/IId), 3311 (32a), 4150 (47f), 4540 (47e), 5111 (32a)
Weston, A. 5012 (21a)
White, O. E. 1548 (51/IIa)
White, S. S. 2462 (51/IVc), 2540 (51/IVc), 2620 (51/IVc), 2738 (26e), 3012 (26e), 3272 (26e), 3503 (51/IVc), 3623 (26e), 3927 (26e), 3973 (51/IVc)
Wieder, – 5 (34a)
Wiggins, I. L. 7201 (51/IVc), 7209 (51/IVc), 7357a (26a)
Wiggins, I. L. & R. C. Rollins 375 (51/IVc), 410 (26e)
Wight, A. E. 53 (40a), 165 (32a)
Wilbur, R. L. 7512 (47c), 7633 (47c), 7748 (47c), 8045 (47c), 8140 (47c), 10821 (51/IVb), 7297 (47c), 20534 (51/IIb), 22935 (21a)
Williams, Ll. 11930 (21a), 13433 (26b), 13462 (26b), 13812 (26b), 15925 (21a)
Williams, L. O. 5306 (51/IIIb), 5498 (32a), 16900 (26c), 17295 (26c), 22154 (43a), 26348 (15/IIb), 27271 (51/IIIa), 42802 (26c)
Williams, L. O. & V. Assis 5235 (27), 5523 (32a), 5839 (51/IIb), 7506 (15), 7534 (21a)
Williams, L. O. & A. Molina R. 10579 (26c), 12405 (51/IVb), 16949 (51/IIIb)
Williams, R. S. 105 (31), 114 (21a), 115 (51/IIb), 308 (51/IIb), 506 (27), 518 (51/IIb), 742 (51/IIb)
Wilson, P. 634 (51/IVb), 1288 (51/IIb), 7384 (40a), 7656 (10c), 7687 (10c), 7800 (40a), 7849 (40a), 7895 (10b), 9200 (32a), 9214 (51/IIb)
Windler, D. R. 2880 (32a), 2912 (32a), 3004 (44a), 3005 (31)
Woodson, R. E. 1659 (51/IIa)
Woolston, A. L. 381 (51/IIb), 493 (32a)
Woytkowski, F. 474 (47e), 5589 (47e), 7227 (51/IIb), 34010 (47e), 35194 (51/IIb)
Wright, A. E. 226 (10a)
Wright, C. 146 (40c), 147 (46b), 154 (51/IVc), 297 (26c), 1034 (26e), 1599 (51/IId), 1600 (51/IIb), 2374 (32a), 2375 (28a), 2377 (40d), 2378 (21a), 2379 (46a), 2380 (46a)
Wurdack, J. J. & J. V. Monachino 39817 (51/IIIa), 39829 (32a), 39928 (26b), 39975 (47i), 39981 (26b), 40907 (31), 40948 (21a)
Wydler, H. 33 (47c), 132 (51/IId)
Wynd, F. L. & C. H. Muller 332 (34a)

Yuncker, T. G. 5702 (26a), 8368 (51/IVb)

Zehntner, – 765 (37), 766 (37)
Zizumbo, D. & P. Colunga 322 (51/IVb)

E. Sectio CALICIOPSIS Irwin & Barneby

Chamaecrista Moench sect. **Caliciopsis** Irwin & Barneby, sect. nov., inter sect. *Chamaecristam* et *Xerocalycem* quasi intermedia, ab illa sepalis more *Xerocalycis* striatim parallele multinervosis, ab hac sepalis inter se subaequilongis (nec insigniter gradatim imbricatis) necnon chromasomatum numero (x = 8, nec 7) diversa.—Sp. typica: **Ch. calycioides** (Colladon) Greene.

Cassia sect. *Caliciopsis* Irwin & Barneby, Mem. N.Y. Bot. Gard. **30**: 3 (in clave), 1977, nom. nud.
Cassia sect. *Chamaecrista* subsect. *Xerocalyx* sensu Bentham, 1871, p. 569, min. ex parte.
Chamaecrista ser. *Deeringianae* sensu Britton & Rose, 1930, p. 272 (in clave), min. ex parte.

Resembling in morphology and chromosome number herbaceous members of sect. *Chamaecrista* and in the close parallel striate venation of the sepals the sect. *Xerocalyx,* but differing from the latter in the subequal sepals, the pluri (not exactly 1–2)-jugate, few-nerved lfts and chromosome number ($x = 8$, not 7).—Spp. 2, of tropical and warm-temperate North and South America.

The ambiguous character of *Ch. calycioides* relative to sect. *Chamaecrista* and sect. *Xerocalyx* was first pointed out by Irwin (1964, passim) and sectional status in *Cassia* was informally proposed for it by Irwin & Barneby in 1977. As brought out in the foregoing description, it combines characters of both sections and, if included in either, would fatally weaken the discontinuities between them. Whether sect. *Caliciopsis* stems independently from the matrix that gave rise to both *Chamaecrista* and *Xerocalyx* or represents a recombination of genetic material from both of them is unknown, but for practical reasons the group is now best treated as an autonomous unit of the same rank. A second species, known to us only from the protologue, appears to have the same essential characters as *Ch. calycioides*.

Key to the Species of sect. *Caliciopsis*

1. Stems diffuse or prostrate 1–9 dm; stipules 2–7.5 mm; lfts 5–12(–3) pairs, 0.7–2.3 mm wide; petals 5–14 mm; pod (0.7–)1–4.5 cm long; range of the section. 1. *Ch. calycioides* (p. 858).
1. Stems stiffly erect-ascending 10–15 dm; stipules 8–17 mm; lfts 12–24 pairs, 3–4 mm wide; peduncles briefly supra-axillary; petals 25+ mm; pod 4–8 cm long; known only from Ceará, Brazil. 2. *Ch. duckeana* (p. 861).

1. Chamaecrista calycioides (Colladon) Greene, Pittonia **4**: 32. 1899. *Cassia calycioides* DeCandolle ex Colladon, Hist. Casses 125, t. 20, fig. B. 1816.—Typus infra sub var. *calycioide* indicatur.

Precociously flowering perennial herbs from dark brown or blackish, at length fusiformly thickened roots, the simple or few-branched stems diffusely or humifusely radiating from the root-crown or the divisions of a short suffrutescent caudex, at anthesis 1–9 dm, commonly puberulent with short incurved and pilose with fine, widely divergent or incurved-ascending setae up to 1–3 mm, but the setae sometimes and the shorter vesture seldom, both rarely, lacking, the dull olivaceous concolorous lfts glabrous above or on both faces or less often puberulent throughout, the margins ciliolate, the stems, blades of stipules and sepals (between the close-set venation), petals (especially toward the claw) and valves of pod speckled with minute vertical violet lineoles.

Stipules erect, broadly lance- to ovate-acuminate, (1.5–)2–7.5 × 0.7–2 mm, at subsymmetric base rounded or when broader incipiently cordate, prominently 7–11-nerved, persistent.

Adult lvs mostly 1–4, rarely 3.5–6 cm, the expanded blade ovate to lanceolate, short-petiolate; petiole 2–4.5 mm, at middle 0.25–0.7 mm diam, coarsely wing-margined; petiolar glands 1(–3) stipitate, urn- or almost disc-shaped 0.15–0.9 mm diam, the stipe varying from filiform to stout, the whole in profile pin- or tack-shaped, 0.25–1.3 mm tall and varying from a little shorter to much taller than diameter of head; lfts of adult lvs (4–)5–12 pairs, the lowest and uppermost smaller, in outline linear to narrowly oblong-elliptic, sharply obliquely aristulate, at base cordate on proximal side, the larger up to 5–13 × 0.7–2.3 mm, straight or subfalcate, the firmly membranous blades beneath prominently, above faintly venose, from base (4–)5–6-nerved, the midrib displaced to divide the blade 1:3–6 accompanied on narrow (×0.1–0.4 mm) side by 1 parallel submarginal nerve and on the broad side by 1 adjacent produced almost or quite to apex and commonly anastomosing into the mucro, a second nearly as strong but not reaching apex, with 1–2 exterior and shorter, the stronger nerves weakly forking distally.

Peduncles exactly axillary, 0–1(–3.5) mm, commonly 10-, rarely 1–2(–3)-fld; pedicels at anthesis obsolete up to 1(–2.5) cm, in fruit thickened, 0.5–12(–28) mm, bracteolate 1–3(–12) mm below calyx; bracteoles resembling smaller stipules, lanceolate or ovate, 1–3 mm; buds lance-acuminate in outline; sepals narrowly lance-acuminate, mostly 5–11(–12.5), in one rare var. only 2.6–3.1 mm, externally puberulent or rarely glabrate; petals yellow fading orange, 4 subhomomorphic varying from spatulate and narrowed to a slender claw to obovate-cuneate, mostly 5–14, rarely 2.4–3 mm, the cucullus as long or a trifle shorter, usually obliquely reniform-obovate; stamens 3–10; ovary usually densely, sometimes thinly strigose-pilose; style (0.5–)1–5 mm, linear or when short somewhat dilated at apex; ovules 2–14.

Pod commonly linear, rarely shortened and oblong (cf. key to vars.), the pale-green or sometimes purple-tinged, violet-dotted valves usually weakly long-pilose, rarely only puberulent or glabrous; seeds (poorly known) ±2.4–3.1 mm, the testa dull ochraceous marked along distal keel by a fuscous transverse line, mucilaginous when wetted.

Chamaecrista calycioides consists of several geographically disjunct races, but only one of these is thought to deserve taxonomic status. Discussion of minor internal variation follows description of var. *calycioides*.

Key to the Varieties of *Ch. calycioides*

1. Fl relatively large, the sepals 5–12.5, the petals 5–14 mm; stamens 8–10; ovules 7–16; pod linear in profile, 2.5–4.5 cm, 7–11 times longer than wide; range of the species (Texas to Argentina) 1a. var. *calycioides* (p. 859).
1. Fl minute, the sepals and petals ±2.5–3 mm; stamens 3–5; ovules 2–3; pod broadly oblong in outline, 7–10 ± 4 mm, ± twice longer than wide; local in Oaxaca, Mexico. 1b. var. *lentiformis* (p. 861).

1a. Chamaecrista calycioides (Colladon) Greene var. **calycioides.** *Cassia calycioides* DeCandolle ex Colladon, 1816, l.c., sens. str.—"Hab. in Cayenna."—Holotypus, without further data, collector not recorded (but perhaps J. B. Patris), G-DC! = F Neg. *7003*.

Chamaecrista aristellata Pennell, Bull. Torr. Club **44**: 342. 1917.—"Berlandier 2036 . . . the precise locality . . . unrecorded" but according to isotypic label at GH "Texas in sylvis arenosis prope el rio Medina [probably in the present Bexar or Medina county]. Julio 1829."—Holotypus, NY! isotypi, GH, K, MO, P, US (2 sheets)!—*Cassia aristellata* (Pennell) Cory ex Cory & Parks, Bull. Tex. Agric. Exper. Sta **550**: 56. 1938 (sphalm. "*aristatella*").
Cassia pedersenii Burkart, Darwiniana **15**: 519, fig. 5, 6. 1969.—"ARGENTINA, prov. Corrientes: depto. Lavalle, km 753 de la ruta 12, leg. T. M. Pedersen, no. 5824 . . . el 27.II. 1961 . . ."—Holotypus, SI (not seen); isotypi, K, NY, US!
Cassia calycioides sensu DeCandolle, Prod. **2**: 503. 1825; Vogel, 1837, p. 64; Bentham, 1870, p. 160, t. XLIII, fig. 2; 1871, p. 569; Pollard, 1894, p. 217; Isely, 1975, p. 72, map 21.
Chamaecrista aristellata sensu Britton & Rose, 1930, p. 285. *Cassia aristellata* sensu Turner, 1959, p. 78, map 35.

Habit and vesture of the species; lvs 1.5–6 cm; gland(s) 0.2–0.9 mm diam, 0.3–1.3 mm tall, the stipe usually filiform, sometimes stout; lfts of adult lvs 5–12 pairs; peduncles 0–3.5 mm, 1(–3)-fld; fruiting pedicels 2–12(–25) mm; sepals 5–12.5 mm; petals 5–14 mm, varying from ±1 mm shorter to 3 mm longer than sepals, the cucullus strongly differentiated, commonly ± twice as wide as the rest; style 1–5 mm, either filiform or (especially when short) a little dilated and at stigma up to 0.3–0.4 mm diam; stamens 8–10, the long anthers 3.3–6(–7.2) mm; pod as described in key, the valves usually pilose, sometimes merely puberulent, exceptionally glabrate.—Collections: 57.

Savanna, campo, pampa, usually in sandy soils, below 600 m, widely scattered through the tropics and subtropics of the Americas, apparently nowhere common; Coastal Plain of s. Texas (Turner, l.c. sub *C. aristellata,* map 35; Isely, l.c., map 21) and adjacent Tamaulipas; Isthmus of Tehuantepec, Oaxaca, Mexico; n. Venezuela (Carabobo and Cojedes to Monagas, mostly on the Orinoco slope of the coastal cordillera); foothills of Sa. Pacaraima in Terr. Roraima (Sas. do Mel and de Xiriri), Brazil and probably adjacent Guyana; French Guiana; scattered over n.-e. and e. Brazil from the lower Amazon valley in Pará to Rio Grande do Norte and e.-centr. Goiás; and again occasional in widely scattered stations within the lower Paraná-Paraguay basin from n.-centr. Paraguay (Río Apa) to s. Brazil (n.-centr. Sâo Paulo) and n. Argentina (Santiago del Estero and Corrientes).—Fl. in Texas and Mexico V–X, in Venezuela III–VIII, in Brazil nearly throughout the year, in far south X–III.

The immense distances intervening between major populational centers and remotely scattered smaller ones of var. *calycioides,* such as savanna-enclaves on the lower Amazon, are such as must divide the whole gene pool into decisively isolated segments and therefore favor or even promote morphological differentiation. In the circumstances, however, it is the essential homogeneity, not the internal diversity of the species that is more remarkable. Apart from composition and density of the duplex vesture, unstable everywhere in *Chamaecrista,* the variable phenetic characters are: number of leaflets; diameter and height of petiolar glands; length of perianth; and length of style. In Texas and adjoining Mexico the leaflets of adult leaves are mostly 7–9 pairs, the gland very small (0.2–0.4 mm diam) and elevated on a filiform stalk, while sepals and petals are relatively short (±5.5–7 mm) and the stout style only 1.1–2 mm long. The typus of *C. calycioides* from Cayenne and all the samples known from Brazil between Pará and Rio Grande do Norte are essentially like the Texan plant except that the leaflets are slightly more numerous, up to 9–12(–13) pairs in adult leaves. In Venezuela and Oaxaca we find the same gland and leaflets as the last combined with a larger flower, the sepals up to 8–11 mm, the petals to 8.5–14 mm, and a linear style 2.5–4.5 mm long. The southern foothills of Pacaraima harbor a form of var. *calycioides,* perhaps taller and more erect than others, in which the flower and leaflets of the last are combined with a coarse gland 0.5–0.9 mm diameter elevated on a stout stipe no longer than the diameter of the head. In the Paraná-Paraguay basin the leaflets are once again relatively few as in Texas, mostly 5–9 pairs, but with this type of leaf the flower varies greatly in amplitude and the style from 1 to 5 mm in length. Here the gland, while small-headed, tends to be shorter-stalked than northward. In the context of compatriot forms such as represented by *T. Meyer 6074* (A) from Ituzaingó, Corrientes, the typus of *C. pedersenii* is unique in one feature only, the pedicels, elsewhere never above 12 mm long, being greatly elongated, up to 2.5 cm, carrying the flower well beyond the tip of the short subtending leaf. Despite its visually striking nature we are obliged to reduce *C. pedersenii* to the status of a taxonomically negligible minor variant. Our one collection of var. *calycioides* from Santiago del Estero, with the foliage and gland of *C. pedersenii* but the small, short-pedicelled flower prevalent in eastern Brazil or Texas, is scarcely less well-marked.

The name *Cassia aristellata* was used first in manuscript by Asa Gray, but never published by him, the plant in question, *Berlandier 2036,* having in the meantime been recognized by Bentham (1871, l.c.) as conspecific with *C. calycioides*. As defined by Pennell, *Chamaecrista aristellata* was thought to differ from genuine *C. calycioides* in fewer leaflets and 1- rather than 2–3-flowered peduncles. But these distinctions, as noted by Isely (1975, p. 72), have lost importance as more material has accumulated.

The epithet *calycioides,* which might have been better spelled "*calicioides,*" refers to the petiolar gland, likened to the fruiting body of the lichen *Calicium*.

1b. Chamaecrista calycioides (Colladon) Greene var. **lentiformis** Irwin & Barneby, var. nov., a var. *calycioide* floribus minutis sessilibus, androecio reducto 3–5-mero, et legumine abbreviato late oblongo 2–3 (nec lineari 7–16)-spermo primo intuitu diversa.—MEXICO. Oaxaca: San Geronimo, 5.X.1933 (fl, fr), *C. D. Mell 2109*.—Holotypus, US; isotypi, NY, US.

Humifuse, 1.5–4 dm, puberulent and sparsely setulose; adult lvs 1–2 cm; gland 0.15–0.2 mm diam, 0.3–0.45 mm tall, the stipe filiform; lfts 5–8 pairs; peduncles obsolete, 2-fld; pedicels at anthesis scarcely evident, becoming 0.5–1 mm long in fruit; sepals 2.6–3.1 mm; petals 2.4–3 mm, 4 alike and ±1.4 mm wide, the cucullus scarcely differentiated; style 0.5 mm; stamens 3–5, the long anthers up to 1.1–1.3 mm; pod as described in key, the valves thinly pilose.—Collection: 1 (typus).

Habitat not recorded, but to be sought in savanna on sandy ground, probably near 100 m, known only from the type-locality, surmised to be San Jerónimo Ixtepec on the Isthmus of Tehuantepec railroad n. of Juchitán.

A remarkable little plant, with all essential characters of *C. calycioides* but too far modified in its diminished flower, depauperate androecium and abbreviated pod to be classed among the minor variants of var. *calycioides* described above. The small pod might suggest a history similar to that of *Chamaecrista tabascoënsis,* which we have interpreted as a drought-modified *C. nictitans* var. *diffusa,* the short pod of which becomes few-seeded only by abortion of several ovules. The three complete plants of var. *lentiformis* which form the type-series are in full vigor of anthesis and early fruit, and two dissected ovaries contained 2 and 3 ovules respectively, no more numerous than the maturing seeds. We assume that var. *lentiformis* arose as a mutation from var. *calycioides* which also occurs, but rarely, on the Isthmus lowlands.

2. Chamaecrista duckeana (Bezerra & Fernandes) Irwin & Barneby, comb. nov.—*Cassia duckeana* Bezerra & Fernandes, Bradea **11**(50): 337, fig. 1, 2. 1979.—"Specimina ab A. Fernandes lecta in Serra Meruoca, Ceará, loco Lages, 08.07.1957—HOLOTYPUS . . . HPB n. 1714 . . . et loco Santo Antonio, 15.07.57, HPB n. 1728."—Photo in protologue!

Stiffly erect herb or subshrub 1–1.5 m, spreading pilosulous.

Stipules narrowly lance-subulate 8–17 mm striate, spreading-pilosulous especially toward margins.

Lf-stalks 3–9 cm; gland below proximal pair of lfts ± stipitate; lfts 14–24 pairs, dimidiately linear-oblong acute or subobtuse mucronulate 8–20 × 3–4 mm, either pilosulous or glabrous, ± ciliolate, the venation apparently as in *Ch. calycioides*.

Peduncles ± connate to stem, more than 1-fld, the pedicels 5–15 mm; sepals slightly unequal, the outer 12–14 mm; petals ±25 mm (±1 inch).

Pod 4–8 × 0.3–0.5 cm, pubescent or glabrate.—No collections seen, the description compiled from protologue.

Habitat not recorded, known from 2 collections from Serra Meruoca, n.-e. of Sobral in n.-e. Ceará, Brazil.—Fl. VI–VII(–?).

This seems clearly and closely related to *Ch. calycioides,* as assumed by the describers, but falls beyond the limits of variation observed elsewhere in that species, the greater stature, number of leaflets, supra-axillary fascicles of flowers and ample petals contributing to a strong differential character. Nevertheless, since typical *Ch. calycioides* has been long known from Ceará, the possibility that *Ch. duckeana* represents a marginally extreme form of it must be borne in mind. Reassessment of its status will be required as material for comparison accumulates.

Numbered Exsiccatae of *Ch. calycioides* var. *calycioides*

Aristeguieta, L. 4271
Belém, R. P. & J. M. Mendes 155
Benoist, R. 1330, 1348
Berlandier, J. L. 2036
Black, G. A. 20104
Burchell, W. J. 9001
Cory, V. L. 11807, 20505
Ducke, A. 16036
Eiten, G. & L. 3646
Fiebrig, K. 4217
Freire Allemâo, L. & M. de Cysneiros 472
Gardner, G. 2130
Gentry, A. 11154
Glaziou, A. 10662
Goodland, R. 536
Hatschbach, G. 39418
Heringer, E. P. et al. 27
Irwin, H. S. 1210
Jobert, C. 1023
Johnston, M. C. 4548
Jones, F. B. 493
King, R. M. 1816
Labouriau, M. 169
Luetzelburg, P. von 20842, 21496, 26471
Metz, Sr. Mary 304
Meyer, T. 6074
Monteiro da Costa, R. C. 4
Murça Pires, J. 5067
Parks, H. B. 5067
Pedersen, T. M. 5824
Pittier, H. 7923, 9008, 11698
Plowman, T. et al. 9321
Pursell, R. A. 8559
Ramia, M. 2581, 2598, 2604
Rodríguez, S. 19, 20
Schomburgk, Robert 473
Silva, M. 1529
Spruce, R. 851
Tavares, S. 419, 1007
Trujillo, B. 2501
Ule, E. 8139
Venturi, S. 7613
Vidal, J. 956

F. Sectio XEROCALYX (Bentham) Irwin & Barneby

Chamaecrista Moench sect. **Xerocalyx** (Bentham) Irwin & Barneby, comb. nov. *Cassia* sect. *Chamaecrista* subsect. *Xerocalyx* Bentham in Martius, Fl. Bras. **15**(2): 155. 1870 & Trans. Linn. Soc. London **27**: 568. 1871.—Sp. lectotypica: *Cassia diphylla* Linnaeus ≡ **Chamaecrista diphylla** (Linnaeus) Greene.—*Cassia* sect. *Lasiorhegma Xerocalyx* Vogel, Syn. Gen. Cass. 53. 1837, statu haud indicato. *C.* sect. *Xerocalyx* (Bentham) Irwin, Mem. N.Y. Bot. Gard. **12**(1): 44. 1964.

Lfts 1–2 pairs sessile, the pulvinule reduced to a transverse line or crescent around base of blade; inflorescence axillary (or very shortly supra-axillary), 1 (rarely 2–3)-fld; sepals strongly graduated, firm-textured, closely parallel-nerved, appearing multistriate; fl resupinate, 1 structurally abaxial (alternately right and left) petal becoming functionally adaxial and incurved over the 10-merous androecium; anthers very short-stalked, in outline narrowly lanceolate or linear, unequal, 2–3 structurally abaxial of the outer cycle largest; seeds narrowly pyri- or claviform, finely lineolate-pitted; $n = 7$.—Herbs, subshrubs and shrubs, the herbs sometimes monocarpic and the shrubs sometimes precociously flowering as herbs.

Circumscription of sect. *Xerocalyx* is adopted without modification from the last revision (Irwin, 1964), in which the morphological, chromosomal and chemical characters of the group were defined. The substantial differences between the taxonomy and nomenclature of the last and the present revisions are due to four circumstances:

a) Ten years of botanical exploration in Brazil, with particular emphasis on *Cassia,* have provided a vastly greater data-base in the shape of specimens, ecological information, and experience of the living plants.

b) New perspectives gained through comparative studies of related groups of *Chamaecrista,* in particular of sect. *Chamaecrista,* have modified our views of the optimal specific unit taxonomically useful in a group of organisms not separated by geographical or precise morphological discontinuities.

c) Types not accessible to the senior author while working at Austin, Texas, have since been located and reinterpreted.

d) Transfer of sect. *Xerocalyx* from *Cassia* to *Chamaecrista* imposes new names for most taxa.

In its entirety sect. *Xerocalyx* forms an extremely distinct, instantly recognizable type, superficially resembling the paucifoliolate members of sect. *Chamaecrista* but different in the resupinate flower (not easily observed in herbarium specimens), strongly graduated, multistriate sepals, claviform seeds, and chromosome number. Internally it is as rich in minor phenetic variation as it is uniform in basic features. The flower and pod, granted some latitude in size, present no tangible differential characters. As a result the taxonomist is now, as in the past, obliged to fall back, for descrimination of taxa, on characters of growth-habit, foliage (including petiolar glands), pubescence, and length of pedicel, all highly labile and hence unsatisfactory for his purpose. In the biological sense the section is probably best viewed as one enormous macrospecies in which complex evolutionary processes are currently active but not yet so advanced as to give rise to truly discrete units. Even the loss of one pair of leaflets, which in practice satisfactorily sets off *Ch. diphylla* sens. lat. from the remainder, cannot be regarded, in context of the genus and in particular of subgen. *Lasiorhegma,* as any great evolutionary step; it was unaccompanied by any major floral or ecological modification. The tetraphyllous members of the group, in which Bentham (1871, with repeated reservations) recognized nine species (with three varieties) and Irwin (1964) fourteen (with fifteen varieties), form a series, or more accurately a network, of intergradient types between which we, following Amshoff (1939), have sought vainly for pronounced morphological discontinuities. Nevertheless we recognize that a specific concept so wide as to include such extremes as amply leafy, large-flowered *Cassia latistipula* alongside of minute-leaved *C. curvifolia* and small-flowered forms of *C. desvauxii* becomes a taxonomic monstrosity. Compromising between biological naturalness and taxonomic convenience we propose to treat tetraphyllous *Xerocalyx* as composed of two species, or two aggregates of forms conventionally called species, which we separate primarily on amplitude of foliage and length of petiole. Arbitrary as it is in fact, this division separates at the specific level not only the extreme foliage types already mentioned, but also several narrowly sympatric pairs of microphyll (*Ch. ramosa*) and macrophyll (*Ch. desvauxii*) populations which either in a given locality or over a more extensive common range maintain sharply distinct individual facies without signs of intergradation.

Phenetic variation in tetraphyllous *Xerocalyx* involves relatively few factors, which may be analyzed as follows:

1. *Stature:* The adult plant may be a) an erect wand, simple or with few branch-

es from the middle upward, these narrowly ascending, or plagiotropic, or even pendulous; b) a bushy shrub or shrublet with several primary trunks from the rootstock, branching distally into a rounded head; a diffuse but wiry shrublet, sometimes with adventitiously rooting stems; or a prostrate, weakly suffrutescent herb with either stiff or pliantly humifuse stems. Some wand-forms seem to be monocarpic; but others are all perennial and potentially fruticose.

2. *Vesture:* Soft pubescence of accumbently incurved or of weakly erect hairs may clothe the whole plant or be completely lacking, but these extremes are rare. Pubescence appears first, or persists longest, as a line of hairs on stem and pedicel and on sutures of the pod; it extends thence to dorsal keel of petiole, margin of leaflets and stipules, the whole circumference of stem and pedicel, the sepals, and the dorsal or finally both faces of leaflets. As in sect. *Chamaecrista,* vesture is almost never firmly correlated with any other morphological feature.

3. *Stipules:* The stipule blades, always cordate at base and when broad amplexicaul, when long concealing the internodes, vary greatly in size, outline and duration. Usually they are ovate or ovate-oblong and acute, sometimes narrowly oblong or narrowly triangular and acuminate, sometimes broadly ovate and obtuse, mostly persistent after fall of the associated leaf but in one variety of each species caducous. Amplitude of stipules varies either independently of leaf-blades or synchronously with them, their relative proportions contributing materially to the adult plant's facies.

4. *Petiolar gland:* The gland, usually saucer- or cup-shaped, exceptionally scutellate, varies from 0.2 to 2 mm diameter and may be either sessile and, if small enough, concealed in the petiole's ventral groove, or raised on a stout pediment shorter than the diameter of the head, or distinctly stipitate, when becoming stud- or finally slenderly pin-shaped as viewed in profile. While the overall type of gland is more or less characteristic of a given variety, it is highly unstable in fine detail, almost always so between plants of a population, often between leaves of a given stem.

5. *Leaf:* The leaf-stalk varies from about 1 to 15 mm in length, this connected generally but inexactly with the lengths of the whole leaf and of individual leaflets. The distal pair of leaflets varies from the same size as the proximal one up to twice as long, rarely a trifle shorter. Individual leaflets vary from 1 up to 45 mm in length and in outline from broadly obovate to linear, passing insensibly through ever more narrowly obovate or obovate-elliptic forms into lance- or oblance-elliptic ones, the broader types varying from obtuse to obtuse-mucronate or subacute, the narrower from obtuse to sharply acute or acuminate and sometimes falcate. Texture of the blades varies from stiffly to thinly chartaceous, independently of amplitude; and independently of either size or texture from olivaceous to blue-green, sporadically glaucous.

6. *Pedicels:* The pedicels vary from much shorter up to 2–3 times the length of the sepals, and independently from shorter up to about twice as long as the subtending leaf. The relative lengths of leaf and pedicel greatly affect the gross aspect of the plant. Some plants which bear only long-pedicelled flowers at early and middle anthesis may produce late in the growing season some short-pedicelled ones, these often smaller and sometimes sterile.

7. *Perianth:* The longest of the five sepals varies from 6 to 20 mm in length, but falls mostly in the range between 8 and 16 mm. The longer petals vary from about one third shorter up to one third longer than the long sepal. The cucullus, which by resupination of the flower forms a shed roof over the androecium, is often red-streaked or -dotted towards the claw, the size and density of the dots,

which occasionally run together into a scarlet flare, being highly variable even within small populations of otherwise similar plants.

By ringing changes on the foregoing set of variable factors it is possible to assemble all known phenotypes of the tetraphyllous group. The number theoretically possible is, of course, legion, and of those known to exist many occur, so far as present evidence goes, only as individual plants or as single populations. Relatively few have achieved a precarious stability over extensive ranges and probably these alone deserve taxonomic status. On the Brazilian Planalto, where sect. *Xerocalyx* forms a virtually ubiquitous element of the cerrado and high campo floras, two, three, or even four clearly different sorts may be encountered within an area of a few hectares, each maintaining locally a pronounced individual character, but occasionally, as on Sa. do Cipó in Minas Gerais, melding into a complex swarm of hybrids. In the recent past it was conjectured that the taxonomic difficulties in the group might be satisfactorily explained by free and widespread gene-flow between initially discrete species. It is now clear, however, that geographically isolated populations can exhibit a degree of internal variation of the same sort and degree as those conveniently situated to receive alien pollen. We now feel that much (but not all) of the phenetic polymorphism is due to an inherently unstable genome. Some of the strange disjunctions in range (cf. *Ch. desvauxii* var. *langsdorfii*) could in these conditions be explained as coincidental effects of random sorting: identical phenotypes (doppelgänger) making independent appearances in widely separated places. Controlled breeding experiments are required to settle these questions, experiments best pursued in planaltine Brazil, whose curious soils support a flora not readily cultivated elsewhere. We have no choice at present but to adopt for *Xerocalyx* a pragmatic classification involving many arbitrary judgments. The definitions and keys in the following pages should be read in this light.

Key to the Species of sect. *Xerocalyx*

1. Lfts 2 pairs.
   2. Relatively macrophyllous, the petiole together with pulvinus of most larger lvs (3–)3.5–8.5 mm, if less than 3.5 mm the distal pair of lfts in larger lvs >15 mm long. 1. *Ch. desvauxii* (p. 865).
   2. Relatively microphyllous, the petiole together with pulvinus of most larger lvs 0.6–3.5(–4) mm, if over 3.3 mm the distal pair of lfts of larger lvs <15 mm long. 2. *Ch. ramosa* (p. 884).
1. Lfts 1 pair. 3. *Ch. diphylla* (p. 891).

1. Chamaecrista desvauxii (Colladon) Killip, Brittonia **3**(2): 165. 1939, sensu hic magnopere ampliato. *Cassia desvauxii* Colladon, Hist. Casses 131. 1816.—Typus infra sub var. *desvauxii* indicatur.

Habit and fls of sect. *Xerocalyx,* but relatively large-lvd, either the petiole (including pulvinus) 3.5+ mm or the distal lfts of larger lvs 15+ mm, or both; North and South American (Mexico to Argentina, lacking in West Indies.

Key to the Varieties of *Ch. desvauxii*

1. Petiolar gland sessile or only shortly stipitate, the whole in profile shorter (or no longer) than diam of head.
   2. Stipules deciduous with or even before the associated lf, the annotinous branchlets naked; scattered around n., w., and s. edges of Amazon basin. 1g. var. *piptostegia* (p. 874).
   2. Stipules persistent, either entire or as pallid marcescent shreds, along the leafless annotinous branchlets.

3. Stipules 2–5 × 1–1.5 mm; local on rocks along middle Orinoco river in Venezuela and Colombia. 1j. var. *triumviralis* (p. 876).
3. Stipules (5–)6–26 mm long; allopatric.
  4. Pedicels 14–55 mm, shorter only in some late-season fls.
    5. Lfts narrowly linear on linear-acuminate grasslike, up to 18–25 × 1–1.7 mm, 8–16 times longer than wide; local in Chapada Diamantina and Sa. do Espinhaço, Bahia and Minas Gerais. 1m. var. *graminea* (p. 878).
    5. Lfts proportionately broader, much so if sympatric.
      6. Proximal pair of lfts slightly larger than the distal pair, concealing them when closed together in sleep; Surinam, French Guiana, and Amapá. 1i. var. *saxatilis* (p. 875).
      6. Proximal pair of lfts shorter than the distal pair, these exserted from between the proximal when closed in sleep.
        7. Fl very small, the longer petals 6–8.5 mm; cf. long-pedicellate forms of 1p. var. *modesta* (p. 882).
        7. Fl of moderate or large size, the longer petals 10–27 mm.
          8. Stipules at once very large and rounded at apex, 15–26 × 10–20; Brazilian Planalto to centr. Paraguay and extreme n. Argentina. 1a. var. *glauca* (p. 867).
          8. Stipules smaller, 5–16 × 2.5–10 mm, commonly acute or abruptly deltate at apex; widespread and partly sympatric with the last.
            9. Stems pliantly humifuse herbaceous; foliage usually pallid-glaucescent and gray-pilose; local on crest of Sa. do Espinhaço in centr. Minas Gerais. 1h. var. *malacophylla* (p. 874).
            9. Stems either erect or diffuse, but if the latter suffruticose and stiff; foliage glabrous or pilosulous, but if the latter not pallid-glaucescent; where sympatric with the last distinctly shrubby.
              10. Larger lfts narrowly elliptic, oblong-elliptic, linear-oblanceolate or sublinear, (12–)15–26 × 2–6(–6.5) mm, mostly 3.5–8 (–10) times longer than wide; mostly erect wandlike plants with single main stem branched distally; Brazilian Planalto s. to Paraná and Paraguay; remotely disjunct in Colombia (Huila). 1n. var. *langsdorfii* (p. 879).
              10. Larger lfts oblanceolate to broadly obovate, obtuse, mostly 11–34 × 4.5–16 mm, ±2–4 times longer than wide; plants variable in habit and stature but mostly several-stemmed from base and frutescent when adult; widespread within and beyond range of last.
                11. Plant prostrate or diffuse, the stems often adventitiously radicant; pedicels mostly 25–55 mm; Brazil only, along the Atlantic coast from n. Bahia to Paraná, inland over the Planalto to s.-centr. Mato Grosso and s. Goiás. 1e. var. *desvauxii* (p. 872).
                11. Plant erect or ascending, bushy-branched when adult; pedicels mostly 15–35(–42) mm; widely and interruptedly dispersed from w. and s. Mexico to Belize, Colombia, the Guayana Highland, lower Amazon valley, and over the Brazilian Planalto to Paraguay and n. Argentina, but not coastal s. of Amazon delta.
                  12. Lfts membrano-chartaceous, not cartilaginous-margined; widespread over the range just given. 1c. var. *mollissima* (p. 868).
                  12. Lfts subcoriaceous, margined with a pallid cordlike nerve 0.25–0.4 mm diam; local in n.-w. Bahia. 1d. var. *circumdata* (p. 871).
  4. Pedicels 1–9(–12) mm.
    13. Midrib of lfts straight or almost so, the blade (beyond the oblique attachment) symmetrically obovate, oblanceolate, linear-elliptic, or linear.
      14. Lfts narrowly linear-acuminate or linear 15–27 × 1–2 mm, 10–16 times as long as wide; local in Pernambuco and w. Bahia. 1l. var. *linearis* (p. 878).
      14. Lfts absolutely or at least proportionately broader, much so in range of preceding.

15. Fls very small, the longer petals 6–8.5 mm; lfts linear to linear-oblanceolate 2–4 mm wide, 4.5–10 times as long as wide; slender erect, mostly monocarpic and wandlike; local on Brazilian Planalto from e.-centr. Mato Grosso and s. Goiás to São Paulo. 1p. var. *modesta* (p. 882).

15. Fls larger, the longer petals 11–25 mm; lfts elliptic-oblanceolate to obovate, mostly (3.5–)4–15 mm wide and 2–5 times as long as wide; variable in habit, either bushy-branched or wandlike; sympatric with the last but also much more widely dispersed, from Honduras s. interruptedly to e. Colombia and n.-e. Peru, e. to the Guayana Highland, the coastal plain of Surinam and adjoining Brazil, thence s. to interior Bolivia, n. Paraguay, and Goiás. 1k. var. *brevipes* (p. 876).

13. Midrib of lfts incurved, the blades falcately oblanceolate to obliquely obovate, the margin bordered by a raised pallid cartilaginous nerve; local in centr. and s.-w. Mato Grosso. 1q. var. *chapadicola* (p. 883).

1. Petiolar gland slenderly stipitate, in profile pin-shaped, taller than diam of head. Local in e. and s.-e. Brazil and in Paraguay.

16. Stipules obovate obtuse 6–14 mm wide; blade of larger lfts (7–)9–16 mm wide; gland in profile 1.5–2.4 mm tall; s.-centr. Goiás (Sa. Dourada). 1b. var. *peronadenia* (p. 868).

16. Stipules lance-ovate subacute 3.5–7 mm wide; blade of larger lfts 4–12 mm wide; gland in profile 0.6–1.3 mm tall; allopatric.

17. Plant diffuse; larger lfts obovate 8–12 mm wide; local in Piauí and Bahia. 1f. var. *latifolia* (p. 873).

17. Plant erect; larger lfts oblanceolate 4–8 mm wide; w.-centr. Paraguay. 1o. var. *pirebebuiensis* (p. 881).

1a. Chamaecrista desvauxii (Colladon) Killip var. **glauca** (Hassler) Irwin & Barneby, comb. nov. *Cassia latistipula* var. *glauca* Hassler, Feddes Repert. Sp. Nov. **8**: 129. 1910.—"Paraguay: Sierra de Amambay, in campis pr. Estrella, flor. et fruct. mens. Jan.; Hassler no. 10080, leg. Rojas."—Holotypus, collected 27.VII.1910, G (4 sheets)! isotypus, K!

Cassia latistipula Bentham in Martius, Fl. Bras. **15**(2): 156. 1870.—". . . prope Riacho Fundo prov. Ceara [in reality the Riacho Fundo in foothills of Sa. do Cipó, s.-centr. Minas Gerais]: Riedel; ad Caxoeira do Campo prov. Minas Geraës: Claussen; ad Uberava prov. S. Paulo [now w. Minas Gerais]: Regnell coll. III. n. 468; in . . . Sertao de Amaro-Leite prov. Goyaz: Weddell."—Lectoholotypus (Irwin, 1964, p. 56): *Riedel 570*.—The holotypus could not be found at LE in 1976; isotypus, K = NY Neg. *1514*!

Cassia arlindo-andradei Hoehne, Arq. Bot. Estado S. Paulo, n. ser. form. major. **2**(1): 16, 17 (photo), t. 4 (optima!). 1944.—"Campo Grande, Mato Grosso. No. 39.070 Oswaldo Handro . . . cultivada de sementes trazidas de Mato Grosso, em 1936, por F. C. Hoehne e A. Gehrt . . ."—Isotypus, SP!

Foliage (so far as known) glabrous, bright green, blue-green, or pallidly glaucescent; larger stipules contiguous or imbricated, concealing the stem, 15–26 × 10–20 mm, at apex broadly rounded to broadly deltate-subacute, persistent; lf-stalk 7–20(–24) mm; gland 0.4–1.7 mm diam, sessile or almost so; larger lfts up to (20–)25–38(–46) × (10–)12–21(–23) mm; pedicels 22–42(–50) mm; long petals (14–)16–21(–25) mm; pod 40–65(–70) × (6–)7–9.5(–10.5) mm; ovules (15–)16–21(–23).

Virgin and disturbed cerrado, sometimes weedy along roads and in rough pasture or coppiced woodland, mostly 300–1200 m, ascending on Chapada dos Veadeiros in e.-centr. Goiás to 1600 m and descending along the lower Paraguay and Paraná rivers to ±100 m, widespread over the Brazilian Planalto from Chapada Diamantina (Sa. do Sincorá) in Bahia s. along Sa. do Espinhaço to s. Minas Gerais (Caxambú) and e. São Paulo (Guararema) and w. and s. over s. Goiás and s. Mato Grosso to extreme s.-e. Rondônia, centr. Paraguay, n.-e. Paraná, and extreme n. Argentina (Ituzaingó, prov. Corrientes).

Collections of var. *glauca,* equivalent of *C. latistipula* sensu Irwin (1964, p. 56, fig. 338, q.v. for discussion of variability), have doubled since 1964, filling out the already known range in planaltine Brazil and extending it north into Bahia and south into Paraná. Vouchers (NY) for the new limital records are: BAHIA: Sa. do Sincorá, 19 km n.-e. of Ibicoara, 1.II.74 (fl, fr), *Harley et al. 15767.* RONDÔNIA: Vilhena, 3.I.79 (fl, fr), *M. G. Silva & A. Pinheiro 4120*. SÃO PAULO: Guararema, on roadside, 24.IX.67 (fl, fr), *Moreira de Souza* in herb. I. A. C. *19562*. PARANÁ: Nova Esperança, 3.IX.69 (fl, fr), *Hatschbach 22163*.

The variety may be visualized ideally as a coarsely leafy, always glabrous version of var. *mollissima* with very large obtuse stipules upwardly imbricated to conceal the stem. Alike in Goiás and Minas Gerais var. *glauca* fades by way of populations with stipules either broadly acute at apex or shorter than the internodes, or both, into amply leafy var. *mollissima*.

1b. Chamaecrista desvauxii (Colladon) Killip var. **peronadenia** Irwin & Barneby, var. nov., adspectu toto et praesertim stipulis magnis obtusis var. *glaucae* (Hassler) Irwin & Barneby praesimilis sed glandula petiolari parvula eximie stipitata 1.5–2.4 mm usque longa foliolisque rachi valde abbreviato subpalmatim dispositis diversa.—BRAZIL. Goiás: frequent among sandstone rocks about summit of Sa. Dourada, 800 m, 20 km s.-e. of Goiás Velho, 18.I.1966 (fl, fr), *Irwin, Souza & Reis dos Santos 11696*.—Holotypus, UB; isotypi, F, GH, K, LE, NY, RB, S, US.

Foliage glabrous; larger stipules forwardly imbricated or nearly as long as internodes, obovate obtuse up to 10–21 × 6–14 mm; lf-stalk of larger lvs 5–8 mm, the rachis only 1–1.5 mm; gland urceolate 0.4–0.7 mm diam, slenderly stipitate, in profile pin- or slenderly trumpet-shaped 1.5–2.4 mm tall, at least 1 mm longer than diam of head; larger lfts cuneate-obovate obtuse up to 15–27 × (7–)9–16 mm; pedicels 17–26 mm, ± as long as lf; longest sepal 14–18 mm; longer petals 16–20 mm; pod (little known) ±40 × 7 mm, glabrous or sparsely ciliolate along sutures; ovules 14–18.—Fig. 2 (petiolar nectary).

Stony slopes in cerrado and about sandstone outcrops, locally abundant on caprock of Sa. Dourada near 800 m, known only from the type-locality s.-e. of Goiás Velho near 50°W, 16°15′S in s.-centr. Goiás.—Fl. XII–II.

A striking variation on the theme of var. *glauca,* notable for the pin-shaped gland whose stipe is two to three times as tall as its head (whence the epithet) and for the subpalmately disposed leaflets separated by a rachis only 1–1.5 mm long. The var. *latifolia,* similar in the gland, has the habit, stipules and foliage of var. *desvauxii* and, like var. *pirebebuiensis,* which has the foliage of var. *langsdorfii,* is interpreted as a parallel but independent mutant. Additional material of var. *peronadenia,* all from the summit of Sa. Dourada (NY): 10.I.65 (fl), *Heringer 10857*; 18.XII.68 (fl), *Harley & Barroso 11320 = Barroso et al. 790*.

1c. Chamaecrista desvauxii (Colladon) Killip var. **mollissima** (Bentham) Irwin & Barneby, comb. nov. *Cassia desvauxii* var. (B) *mollissima* Bentham in Martius, Fl. Bras. **15**(2): 157. 1870.—"in provincia S. Paulo: Weir, Burchell, Silva Manso."—Lectoholotypus (Irwin, 1964, p. 101): *Weir 205,* K! = NY Neg. *1515*.—*C. tetraphylla* var. *mollissima* (Bentham) Irwin, Mem. N.Y. Bot. Gard. **12**(1): 101, fig. 350 (map). 1964.

Cassia lanceolata Persoon, Syn. Pl. **1**: 456. 1805.—"Hab. in Cajenne. (Herb. Lamarck.)."—Holotypus not seen, not found in P-LAMK by Amshoff (1939, p. 25) who, however, describes a possible isotypus in hb. Persoon. (L) and refers it to *C. tetraphylla* sensu Amshoff.—*C.*

persooni Colladon, Hist. Casses 119. 1816 (non *C. lanceolata* Forskål, 1775).—The only other tetraphyllous *Xerocalyx* with lfts as long known from Cayenne is var. *saxatilis* with obovate lfts.—*Chamaecrista persoonii* (Colladon) Greene, Pittonia **4**: 31. 1899.

Cassia pulchra Humboldt, Bonpland & Kunth, Nov. Gen. & Sp. **6**(folio): 362. 1823.—"Crescit prope San Carlos del Rio Negro."—Holotypus *Humboldt & Bonpland 1181,* P-HBK! isotypus, B (*hb. Willd. 7922*)!—*Chamaecrista pulchra* (Humboldt, Bonpland & Kunth) Britton & Rose ex Britton & Killip, Ann. N.Y. Acad. Sci. **35**: 184. 1936.

Cassia tecta Vogel, Syn. Gen. Cass. 56 & Linnaea **11**: 705, descr. ampliat. 1837.—"Hbt. In Brasilia: Sellow leg."—Holotypus, *Sello 5435,* †B = F Neg. *1754*!—*C. tecta* sensu Irwin, 1964, p. 84, fig. 347 (map).

Chamaecrista madrensis Rose ex Britton & Rose, N. Amer. Fl. **23**(5): 274. 1930.—"Between Aguacate and Dolores, Tepic [=Nayarit, Mexico], August 6, 1897, *J. N. Rose 2023.*"—Holotypus, US; clastotypus, NY!—*Cassia madrensis* (Britton & Rose) Irwin, Mem. N.Y. Bot. Gard. **12**(1): 88, fig. 348 (map). 1964.

Cassia bartlettii Standley, Field Mus., Bot. **11**(4): 132. 1932.—"British Honduras: Mountain Pine Ridge, El Cayo District, February 21, 1931, *H. H. Bartlett 11649* . . ."—Holotypus, F (no. *658353*); isotypi, K, MICH!—*C. bartlettii* sensu Irwin, 1964, p. 87, fig. 348 (map).

Cassia tetraphylla var. *longifolia* Amshoff, On S. Amer. Papilionaceae 26. 1939.—"Pará: H[erbario do] J[ardim] B[otânico do] R[io de Janeiro] 1780 . . ."—Holotyus, U.

Cassia tetraphylla var. *colombiana* Irwin, Mem. N.Y. Bot. Gard. **12**(1): 104. 1964.—". . . Colombia, Santander, '18 km s of Socorro . . . alt. 1200 m.' *H. St. John 20538.* 28 June 1944."—Holotypus, NY! isotypi, MICH, UC, US!

Cassia desvauxii (typica) sensu Bentham, 1870, p. 157, 1871, p. 568; *C. tetraphylla* var. *tetraphylla* sensu Irwin, 1964, p. 89, fig. 349 (map); non *C. desvauxii* Colladon, nec *C. tetraphylla* Desv. sens. str.

When adult frutescent or suffruticose, branched from base or distally, the stems erect or incurved-ascending (2.5–)4–14(–25) dm, sometimes precociously flowering from a single stem branched distally; foliage usually olivaceous, rarely glaucescent, either glabrous, or glabrous and pilosulous-ciliolate, or sparsely to densely, even velutinous-pilosulous on both faces, sometimes the proximal lvs glabrous but those of distal branchlets pilosulous, the vesture pallid or rarely fulvescent; larger stipules (3–)5–15(–18) × 2.5–7(–8) mm, when relatively small ovate-acute or -acuminate, half or less than half as long as internodes, when larger ovate-oblong or broadly lanceolate acute or deltately subacute, subequally to slightly longer than internodes, persistent; lf-stalk of larger lvs (4–)5–11(–12) mm, those of some late-season branchlets sometimes shorter, the petiole (3–)4–8(–11), the rachis 1–3 mm; gland scutellate or saucer-shaped (0.3–)0.5–2 mm diam, mostly sessile but (especially in s.-centr. Goiás) sometimes shortly stout-stipitate, then in profile shortly peg-, tack- or trumpet-shaped to 0.7(–1) mm tall but not or scarcely longer than diam of head; larger lfts obovate to oblanceolate obtuse or obtuse mucronulate 13–34 × (3.5–)4.5–14(–16) mm, ±(1.5–)2–4 times longer than wide, the distal pair ±$^1/_5$–½ longer than the proximal; pedicels (12–)15–35(–42) mm, commonly as long or longer, rarely shorter than leaf, mostly as long or longer, rarely shorter than longer sepals; long sepal (12–)13–21(–24) mm; long petal (12–)14–24(–27) mm; pod 30–50 × 6–8(–10) mm, the sutures always at least puberulent, the valves varying from glabrous to strigulose or pilosulous, sparsely or densely so, the vesture pallid or yellowish; ovules (10)12–20(–24).

Moist or seasonally dry campo, savanna, and brushy grassland transitional to woodland, in Mexico and Central America entering open pine-forest, ascending from near sea level on the Atlantic Coastal Plain of the Guianas where found on sandy low-level terraces from just above clay alluvium, and the lower Amazon valley up to ±1000 m in Mexico, to 1500 m on the Eastern Cordillera in Colombia and on the Guayana Highland, and up to 1300 m on the Brazilian Planalto (Chapada dos Veadeiros and Sa. do Espinhaço), interruptedly widespread over much of tropical South America e. of the Andes and disjunct n.-ward in Belize and Mexico: very local in Mexico on Sas. Madre Occidental (Nayarit) and Sur (Guerrero) and on the coastal plain of Tabasco (near Huimanguillo); on the Maya Mts., Belize and probably adjoining Guatemala; Cordilleras Oriental and Central in Colombia (Norte de Santander, Santander, Caldas); Guayana Highland in Venezuela (Amazonas, Bolívar) and adjacent Guyana and Terr. do Roraima; coastal and interior (campina) sands in Surinam, French Guiana and lower Amazonian Brazil (Pará); and s.-ward over the Brazilian shield from the upper Guaporé in Rondônia, Sa. do Roncador in Mato Grosso, Sa. do Cachimbo in s.-w. Pará, and s. Ceará to Paraguay, n. Argentina (Misiones) and n.-e. Paraná.—Fl. in equatorial latitudes almost throughout the year, s.-ward mostly XII–IV, n.-ward mostly X–III, especially during and after rains.

Our present comprehensive concept of the often misleadingly named var. *mollissima* accommodates not only the original pubescent type described from São Paulo but also the otherwise identical glabrous or subglabrous one which passed until recently as *Cassia tetraphylla* var. *tetraphylla* (or var. *longifolia* Amsh.). It is equivalent to the broad-leaved, large-flowered *C. desvauxii* sensu Bentham deprived of its var. *brevipes* which, differing essentially only in the abbreviated pedicels, has from the very beginning included both glabrous and pubescent phases. In the last revision (Irwin, 1964, p. 101) var. *mollissima* was so construed as to appear endemic to southeastern Brazil and adjoining Paraguay and Argentina, the similar plants from the Guayana Highland being interpreted as minor variants of sympatric glabrous var. *tetraphylla*. It remains true in the light of intensive new exploration that whole populations of pilosulous plants are much more frequent from central Goiás southward than elsewhere in the range of var. *mollissima* sens. lat.; but it now seems theoretically illogical and is in practice impossible to distinguish these from the parallel form of Guayana mentioned by Irwin (1964, p. 101, variant 1), unless the source of the plant is known. Individual plants in which glabrous young foliage gives way in lateral mature branchlets to densely pilosulous foliage (as in the holotypus of *C. piauhiensis*, which = *Ch. desvauxii* var. *latifolia*) demonstrate the plastic nature of pubescence in this, as in related species. A hypothesis of introgression between purely glabrous and purely pubescent species is in context of sect. *Xerocalyx* or the whole genus *Chamaecrista* no longer required to explain sudden loss or gain of trichomes.

Much of the phenetic variation encountered in var. *mollissima* sens. lat. was analyzed in detail by Irwin (1964, sub var. *tetraphylla*) and needs no repetition. Ample new collections from the Brazilian Planalto confirm the passage, imperceptibly gradual, from broad-leaved, large-flowered var. *mollissima* to narrow-leaved, slightly smaller-flowered *C. langsdorfii*, herein reduced to coordinate varietal rank within *Ch. desvauxii*. An exceptionally large-flowered and ample-leaved glabrous var. *mollissima* is now known from several collections from Sa. do Roncador (e.g. *Ratter et al. 1396*; *Irwin et al. 16215*); this simulates var. *glauca*, but has stipules of var. *mollissima* or var. *langsdorfii*. Serra dos Pireneus and vicinity has yielded a glabrous var. *mollissima* with distinctly though stoutly stipitate petiolar glands (cf. *Irwin et al. 18537, 18649, 18706*) suggestive of a trend that culminated in var. *peronadenia* on Sa. Dourada, but the gland, shorter and coarser, and here associated with stipules and leaves of var. *mollissima*, not of var. *glauca*. However, the variety of such minor modifications present on the Planalto seems virtually inexhaustible.

Our reduction to synonymy of taxa recognized in 1964 requires commentary: a) *Cassia tecta*, already known to Vogel in gray-pubescent (typical) and brownish-pubescent (β) phases, cannot be maintained on basis of hair-color. It may be visualized as a form intermediate between ideal var. *mollissima* and var. *langsdorfii*, resembling the latter in the thatch of relatively long, often but not invariably overlapping stipules that covers the stems (whence the epithet). On Sa. do Cipó typical *C. tecta* is associated with a habitally similar and obviously closely related glabrous plant with the same stipules but larger flowers, this indistinguishable technically from glabrous var. *mollissima* sens. lat. These two forms as observed in the field seemed to contribute, as also sympatric var. *malacophylla,* pubescent like *C. tecta* but prostrate, to a complex introgressive community within which no taxonomic lines could be drawn. The situation on Sa. do Cipó will not be understood without detailed biosystematic studies. b) *Cassia madrensis* was described from an almost flowerless plant collected the first week of August, that is at the beginning of the rainy season in Sa. Madre Occidental, and we think it probable that its microphyllous aspect, which can be matched on late season branchlets of var. *mollissima* from Venezuela and Brazil, is due to this circumstance. No tetraphyllous xerocalyx has been collected subsequently in Nayarit and the only other gathering from Mexico (Guerrero, *Hinton 11192,* collected in flower in January) is exactly duplicated by material of var. *mollissima* from the Guayana Highland (e.g. *Altson 538,* NY). Ovate as opposed to narrowly ovate to lanceolate bracteoles, used to separate *C. madrensis* and *C. bartlettii* collectively from *C. tetraphylla* sens. lat., no longer provide an effective differential character. For note on disjunct dispersal of *C. madrensis* see under var. 1e. c) *C. bartlettii* seemed at the time of description safely distinct from anything described in the then new revision of *Cassia* in *North American Flora* and was not compared with South American *C. desvauxii,* of which it seems to us certainly only a series of displaced populations. Similar disjunctions of range between Mexico, Belize and northern South America are encountered in *Ch.* (*Absus*) *zygophylloides* (Guayana Highland, Belize, Oaxaca) and *Ch. viscosa* (Guerrero, Colombia) and must reflect past migrations of relatively xerophytic floras at a period when tropical rain-forest was in retreat.

1d. Chamaecrista desvauxii (Colladon) Killip var. **circumdata** Irwin & Barneby, var. nov., a var. *mollissima,* cum qua foliolorum ambitu adspectuque toto bene convenit, foliolis subcoriaceis nervo marginali crasso cartilagineo albicanti circumdatis, a var. *langsdorfii* imprimis foliolis late ellipticis 6–8 (nec 2–6) mm latis multo crassius marginatis diversa.—BRAZIL. Bahia: ±45 km w. of Vanderlei along BR-242 (near 12°S, 44°30′W, 7.IV.1976 (fl, fr), *G. Davidse, T. P. Ramamoorthy & D. M. Vital 12040.*—Holotypus, US = NY Neg. *10495.*

Glabrous throughout, the foliage dull olivaceous, the stiffly chartaceous lfts paler beneath, all margined with a pallid cordlike nerve 0.25–0.4 mm diam; stipules 9–16 × 6–9 mm, a little shorter than the internodes, at apex obtusely deltate, persistent; lf-stalk stout 4.5–5.5 mm; gland 0.7–0.9 mm diam, obscurely stipitate, broader than tall; larger lfts ±20–24 × 6–8 mm; pedicels 25–30 mm; longer petals ±16 mm; ovules 12.—Collections: 1. Cerrado, in sandy soils near 800 m, known only from the type-locality in the middle Rio São Francisco valley in n.-w. Bahia.—Fl. II–IV.

This striking variant on the theme of *Ch. desvauxii* resembles the glabrous state of var. *mollissima* in general aspect and in the broadly elliptic outline of the leaflets, but these differ in their stiff texture and in the pallid nerve that forms a blunt raised rim around the blade. Occasional leaflets of var. *langsdorfii* show a similar but less pronounced marginal thickening, but are then of substantially narrower outline.

1e. **Chamaecrista desvauxii** (Colladon) Killip var. **desvauxii**. *Cassia desvauxii* Colladon, Hist. Casses 131. 1816, sens. str., based on *C. tetraphylla* Desvaux, Jour. Bot. **3**: 72. 1814 & Ann. Sci. Nat. I, **9**: 428. 1826 (sub *C. desvauxii*), descr. ampliat.—"Habitat *in* America calidiore."—Holotypus, annotated by Amshoff in 1938 and described by her, 1939, p. 25, P (hb. Desvaux.)! = F Neg. *39819*.—Non *C. tetraphylla* Martyn, 1797; nec *C. tetraphylla* var. *tetraphylla* sensu Irwin, 1964, p. 89, quae = *Chamaecrista desvauxii* var. *mollissima* sens. lat.—*Chamaecrista tetraphylla* (Desvaux) Britton & Rose ex Britton & Killip, Brittonia **3**(2): 165. 1939.

Cassia adscendens Martius ex Colla, Herb. Pedem. **2**: 281. 1834.—"Brasilia, cap: Frio [Rio de Janeiro]."—No typus known to survive. Overlooked by Bentham (1871), but the description and locality decisive.

Cassia uniflora γ (?) *glaucescens* Vogel, Syn. Gen. Cass. 54. 1837.—Collected by Sellow, but no locality recorded; presumed isotypus, *Sellow s.n.*, P!

Cassia langsdorfii var. *longipedicellata* Irwin, Mem. N.Y. Bot. Gard. **12**(1): 65. 1964.—"Brazil, Goiás, 'Porto Real.' *Burchell 8490* . . ."—Holotypus, K! isotypus, GH!

Cassia tetraphylla var. *littoralis* Irwin, Mem. N.Y. Bot. Gard. **12**(1): 102. 1964.—"Brazil, Guanabara, 'vicinity of Cabo Frio.' *J. N. Rose & P. G. Russell* 20695. 8 Aug. 1915."—Holotypus, NY! isotypus, US!

Prostrate or weakly ascending, the branches often adventitiously rooting where in contact with ground, the pedicels ascending toward the meridian; foliage (so far as known) glabrous, the stems, lf-stalks and pedicels varying from glabrous to pilosulous, the lf-stalks sometimes setulose, the lfts bright green or blue-green, not known to be glaucous; larger stipules (near middle of stems) ovate to broadly lance- or oblong-ovate acute up to (6–)7–16(–18) × (3–)4–6(–10) mm, half as long to a little longer than internodes, those of terminal branchlets late in season smaller; lf-stalk of larger lvs 4–12 mm, the rachis 2–4 mm; gland 0.3–1.6 mm diam, either sessile and scutellate to urceolate or stoutly short-stipitate, then in profile peg- or funnel-shaped and up to 0.7(–1) mm tall; larger lfts broadly oblanceolate to broadly obovate obtuse up to 11–25 × 5–10 mm; pedicels (2–)2.5–5.5 cm; longest sepal 9–18 mm; longest petal of early fls 11–15(–20) mm, of some late fls only 7–10 mm; pod 30–55 × (5–)5.5–7.5(–8) mm, glabrous or pilosulous; ovules 12–22.

Interruptedly widespread over s.-e. Brazil but of bicentric range: on dunes and in coastal restingas along the Atlantic coast from n. Bahia to Paraná; and on sandy or gravelly soils of campo and campo cerrado at 300–1100 (in centr. Minas Gerais to 1300) m, scattered over the Planalto from centr. Bahia (Sa. do Tombador) to centr. Minas Gerais, s.-w. across centr. and s. Goiás to n.-e. and s.-centr. Mato Grosso (Sas. do Roncador and São Gerónimo).—Fl. year round along the coast, mostly X–III inland.

Our concept of var. *desvauxii* is coextensive with the combination *Cassia tetraphylla* var. *littoralis* + *C. langsdorfii* var. *longipedicellata* of the last revision (Irwin, 1964) and consists of those races of *Ch. desvauxii* in which a prostrate, often adventitiously rooting habit coincides with broad leaflets, sessile or only short-stipitate glands, and flowers of moderate size. It may very likely embrace at least two, possibly more numerous, phyletically independent mutations, but the coastal and interior phases are certainly no longer morphologically separable, the distinction between puberulent pedicels (*C. tetraphylla*) and glabrous ones (*C. langsdorfii*) being no longer tenable. On the Atlantic coast and in interior Bahia var. *desvauxii* is sympatric with *Ch. ramosa*, different in the small leaflets and emphatically stipitate glands, and on the Planalto in Goiás and Mato Grosso with vars. *mollissima, langsdorfii,* and *glauca,* collectively different in their more or less upright, bushy or wandlike habit of growth. Plants of the form prevalent in the interior are notable in the field for the blue-green tone of the foliage and the attitude of the pedicels, ascending toward the meridian from the prostrate stems.

Reinterpretation of original *Cassia desvauxii* as equivalent of the maritime southern plant latterly known as *C. tetraphylla* var. *littoralis* and not the Amazonian one to which the name has been applied is based on comparison of the holotypus, not available to Irwin in 1964, with our copious material from restingas of coastal Rio de Janeiro and Guanabara. These maritime populations are subject to seasonal dimorphism, an amply leafy aspect prevalent in wet fall months giving way to a more or less microphyllous one in the dry season. Desvaux's plant corresponds in minute detail with the latter phase, and is matched especially closely with the distal branchlets of *Fuad Atala 285* (NY) which conveniently retains toward the base of the stems some of the ample early foliage. There is no exact record of origin for Desvaux's specimen, but it could well have been among the plants collected at Rio de Janeiro in 1767 by Commerson on Bougainville's expedition around the world.

Vouchers for limital records of var. *desvauxii* are: Bahia: Sa. do Tombador, *Irwin et al. 30522, 30602* (NY). Mato Grosso: Sa. do Roncador, *Irwin et al. 16403, 16625* (NY); w. of Coxim, *Nienstedt 48* (NY). Paraná: Caioba to Matinhos, *Lindeman & de Haas 2622, 5065* (NY).

1f. Chamaecrista desvauxii (Colladon) Killip var. **latifolia** (Bentham) Irwin & Barneby, comb. nov. *Cassia uniflora* var. (γ) *latifolia* Bentham in Martius, Fl. Bras. **15**(2): 158. 1870.—". . . in campis mediterraneis prov. Bahiensis: Martius."—Holotypus, *Martius 2210,* ipso manu Bentham. determinatus, "in campis et ad sylvas Catingas inter V. Caxoeira et Feira da Conceição . . . Febr. [1819]," M! = NY Neg. *9307*; isotypi, M (3 sheets)!

Cassia piauhiensis Irwin, Mem. N.Y. Bot. Gard. **12**(1): 61, fig. 341 (map). 1964.—"Brazil, Piauhí, 'Sud-Piauhy, Campo Grande. Caatinga,' *Dr. v. Luetzelburg 366.* May 1914."—Holotypus, NY!

Exactly as var. *desvauxii* except for the stipitate petiolar gland, this 0.25–0.5 mm diam, in profile pin-shaped 0.7–1.3 mm tall; foliage either glabrous or the distal lvs sometimes densely short-pilosulous on both faces.

In caatinga 40–500 m apparently rare, known only from centr. and and w.-centr. Bahia and from one station each in n. Minas Gerais (Sa. do Calixto) and s. Piauí (Campo Grande)—Fl. I–II, V.

The var. *latifolia* was originally described as intermediate between *C. uniflora* sens. Bentham (largely our *Ch. ramosa*) and *C. desvauxii* (our var. *mollissima*), approaching the latter in the relatively large flower and, implicitly, in the ampler foliage. The stipitate gland, which is the one, obviously precarious character separating var. *latifolia* from our var. *desvauxii,* was not described by Bentham and was overlooked when the essentially identical *C. piauhiensis* was segregated.

A manuscript description, *Martius Observatio 2210, Cassia glabrata* Mart. (nom. nud.), is so far at variance with the associated specimen in the characters of glandular pubescence and subequal sepals that it must have been misplaced in mounting and could only refer to some member of sect. *Absus.*

1g. Chamaecrista desvauxii (Colladon) Killip var. **piptostegia** Irwin & Barneby, var. nov., var. *mollissimae* (Bentham) Irwin & Barneby quam maxime affinis sed stipulis cito (diu ante folium lapsum) deciduis, caulibus annotinis ideo nudis abstans.—COLOMBIA. Vaupés: sabanas sobre piedras areniscas, ±450 m, Cerro Yapoboda, río Kuduyari, 5–6.X.1951 (fl), *Schultes & Cabrera 14362*.—Holotypus, NY; isotypus, US.

When adult erect, bushy-branched distally, up to 7–12 dm, the purplish-castaneous stems pilosulous when young, the olivaceous lfts either glabrous or glabrous ciliolate; stipules submembranous, lance-oblong, obtuse or abruptly acute, up to 6.5–14 × 2.5–6.5 mm, all deciduous before the associated lf; lf-stalk 5–11 (–12) mm, the petiole 3–7 mm, the rachis 1–2 mm; petiolar gland 0.5–1.1 mm diam, sessile or elevated on a low mounded stipe as wide or slightly wider than head; lfts oblanceolate or narrowly obovate-elliptic obtuse or subacute, the larger of larger lvs 20–42 × 5–12(–14) mm, the distal pair often much longer than the proximal one and often a trifle arched backward; pedicel, fl and pod of var. *mollissima*.

Campo or savanna, on white sand, quartzite, or sandstone, ±250–500 m, interruptedly dispersed around the n., w. and s. periphery of the Amazon Basin: sources of Río Vaupés (ríos Kubiyu and Kuduyari) in Vaupés, Colombia; middle Orinoco valley in n.-w. Amazonas, Venezuela; Guayana Highland in Bolívar, Venezuela (Gran Sabana near Uriman); Sa. Tumucumaque and vicinity, along the frontier in Surinam, Brazil (Pará, Amapá) and probably French Guiana; and on middle and upper Rios Xingú and Araguaia in s.-e. Pará and n.-e. Mato Grosso (Parque Nacional do Xingú; Sa. do Roncador).—Fl. apparently the year round.

Readily distinguished from glabrous forms of var. *mollissima,* which it otherwise closely resembles, by early loss of the membranous stipules, the year-old stems in consequence naked and not clothed in gray marcescent tatters. Like other discontinuously dispersed forms of *Ch. desvauxii,* the var. *piptostegia* consists of major geographic populations that are minutely different and technically separable, possibly indeed independent but parallel modifications of the protean ancestral stock of tetraphyllous xerocalyx. The Colombian plants (holotypus; *García Barriga et al. 16043*; *Schultes & Cabrera 18321*) have glabrous leaflets, sessile glands, and relatively long petioles; those from northeastern Brazil and adjoining countries (*Steyermark 75288*; *Oldenburger ON-166*; *Fittkau & Coêlho* in INPA *12835*; *Cavalcante 2512*; *Pires & Cavalcante 52023*) have ciliolate leaflets, glands raised on a mounded pediment, and proportionately very long distal leaflets bent backward in the style of the narrow-leaved form of *Ch. diphylla* that has been known as *C. cultrifolia*; while the Xingú-Araguaia populations (*D. Coêlho* in INPA *15877*; *Fróes 29821*; *Argent et al. 6640*; *Santos et al. in Ratter 1496*) combine ciliolate but ampler leaflets with sessile glands.

1h. Chamaecrista desvauxii (Colladon) Killip var. **malacophylla** (Vogel) Irwin & Barneby, stat. nov. *Cassia malacophylla* Vogel, Syn. Gen. Cass. 55. 1837.—"In Brasilia: Sellow leg."—Holotypus, †B (hb. Kunth. e "Serra do S. Antonio [=Sa do Cipó; oct.] 1818" = F Neg. *1707*; isotypus, K (hb. Hook.)! = NY Neg. *1519*.—*C. malacophylla* sensu Bentham, 1870, p. 159, 1871, p. 569; sensu Irwin, 1964, p. 84, fig. 347 (map).

Essentially as var. *desvauxii,* but the stems weakly pliantly humifuse and like the pallidly glaucescent stipules and foliage softly pilose with vertical or subretrorse hairs to 0.7–1.8 mm; stipules ovate acute, the larger 5–14 × 3.5–7 mm, as long to half as long as internodes; petiole 3–7 mm, rachis 1.5–4 mm; gland subsessile or slightly elevated, urceolate 0.3–0.5 mm diam, in profile shorter than diam of head; larger lfts 11–20 × 5–10 mm; pedicels ascending toward the meridian 24–50 mm; long sepal 9.5–11 mm; long petals 10–16 mm; pod 25–31 × 5–7 mm; ovules 9–13.

High campo, on quartz sands and about outcrops, 1000–1300 m, local on and near the crest of Sa. do Espinhaço between Diamantina and Sa. do Cipó in lat. ±18°–19°30′S, centr. Minas Gerais.—Fl. VIII–III, perhaps intermittently through the year.

A visually striking form, notable for foliage at once pallidly glaucescent and gray-pilose, but otherwise different from typical var. *desvauxii* only in the weaker, pliantly humifuse stems that run together to form wide patches in the thin campos of the serra crest. A collection from Datas at 1300 m (*Irwin et al. 22539*), within the range of var. *malacophylla,* has almost its habit but the glabrous foliage of var. *desvauxii*. On Serra do Cipó var. *malacophylla* is associated with both glabrous and pilose phases of var. *mollissima* and may be involved there in complex hybrid swarms.

1i. Chamaecrista desvauxii (Colladon) Killip var. **saxatilis** (Amshoff) Irwin & Barneby, comb. nov. *Cassia tetraphylla* Desvaux var. *saxatilis* Amshoff, On S. Amer. Papilionaceae 27. 1939.—"Suriname, Upper Litanie Riv. (mount Knopaiamoi, Rombouts 809 . . .)."—Holotypus, U!—*C. saxatilis* (Amshoff) Irwin, Mem. N.Y. Bot. Gard. **12**:(1): 86, fig. 348 (map). 1964.

Slender erect, usually single-stemmed from shallow roots, paniculately branched distally 3–10 dm, the branches either strictly ascending or plagiotropic, the foliage olivaceous glabrous or glabrous ciliolate; stipules submembranous broadly lance-ovate acute or acuminate, persistent, the larger ones 6–14 × 2–4 mm, as long or only ⅓ as long as internodes; longer lf-stalks 4–8 mm, the petiole (3–)3.5–6 mm, the rachis 1–2.5; petiolar gland scutellate or shallowly concave 0.5–1.4 mm diam subsessile or shortly stipitate, in profile obconic or squatly stud-shaped, 0.2–0.8 mm tall; lfts subdimidiately obovate or obliquely broadly oblanceolate, the proximal pair usually slightly larger (in sleep covering the distal pair), the larger ones 10–19(–23) × 4–8(–9) mm; pedicels very slender wiry 2–3.3(–4) cm, surpassing lf; longer sepals (7–)9–11 mm; long petal (5–)7–11(–14) mm; pod 17–27(–38) × 4.5–5.5(–7) mm; ovules (6–)7–9(–11).

Sand savanna and outcrops in forest or along rivers, a pioneer colonist of temporary river islands, soil-filled crevices and depressions in granite rock faces, 10–550 m, interior Surinam, French Guiana, and Amapá, Brazil, from the middle Coppename River (Voltz Berg) e. to Rio Araguarí, s. to s. slope of Sa. Tumucumaque on the headwaters of Rio Jarí, in French Guiana and Amapá coming out to the coast (at Cayenne) or the coastal plain (downstream from Pôrto Platon).—Fl. year around.

Readily distinguished from sympatric var. *mollissima* by the slender habit, small flower, and short few-ovulate pod. The variety is more closely related to the distantly allopatric var. *triumviralis* which occupies similar microhabitats on granitic knobs and associated savannas of the middle Orinoco valley, but differs in diminished leaf- and stipule-blades, smaller sessile glands, and shorter pedicels.

1j. Chamaecrista desvauxii (Colladon) Killip var. **triumviralis** Irwin & Barneby, nom. et stat. nov. *Cassia gracilis* Kunth, Mimoses 120, t. XXXVI. 1819.—"Crescit in ripa fluminis Orinoci, prope San Borja . . . près de la Mission de San Borja, au sud de l'embouchure du Meta."—Holotypus, ticketed "n. 1176, San Borja Orinocensium," P-HBK! isotypus, mislabelled "Rio Negro," B (*hb. Willd. 7923*)!—*Chamaecrista gracilis* (Kunth) Pittier, 3rd Conf. Interamer. Agric. Caracas 373. 1945.

Cassia gracilis sensu Humboldt, Bonpland & Kunth, Nov. Gen. & Sp. 6(qu): 362, 1823; Bentham, 1870, p. 159 and 1871, p. 569, amb. p.p., exclus. pl. austro-brasil.; Irwin, 1964, p. 58, fig. 339 (map).

Habit of var. *saxatilis,* the foliage sometimes glaucescent, always glabrous; larger stipules 2–5 × 1–1.5 mm, persistent; lf-stalks 3–6 mm, the petiole 2.5–4.5 mm, the rachis 0.7–1.4 mm; gland sessile 0.4–0.7 mm diam; larger lfts obliquely oblanceolate 5–10 × 2–5 mm; pedicels 8–24 mm; long sepals 6–7.5 mm; long petal 6–11 mm; pod 16–27 × 4.5–6.5 mm, strigulose; ovules (4–)5–7.

Granitic (laja) outcrops and associated quartz sand savannas at ±80–300 m, known only from the immediate banks of Río Orinoco from the mouth of Río Atabapo downstream to the mouth of Río Meta, in Amazonas, Venezuela and Vichada, Colombia (lat. 4°–6°20′N).—Fl. IV—XII, but probably irregularly throughout the year.

While instantly distinguished from sympatric vars. *mollissima* and *piptostegia* by its delicately diminished foliage and small flowers, var. *triumviralis* is only precariously separable, unless by remotely allopatric dispersal, from the habitally similar var. *saxatilis* by generally (not absolutely) smaller stipules, leaves and flowers. There is considerable variation in leaf-size, the smallest neatest foliage (e.g. *Maguire et al. 36044*) strongly suggesting that of some south Brazilian races of *Ch. ramosa* var. *erythrocalyx* (=*C. langsdorfii tenuis*), with which Bentham indeed (as discussed by Irwin, 1964, p. 59) confused it. The var. *triumviralis* spans the artificially conceived discontinuity between macrophyllous *Ch. desvauxii* and microphyllous *Ch. ramosa,* and is here subordinated to the former chiefly because of its strong morphological and ecological similarity to var. *saxatilis*. It is interesting that the granite outcrops which furnish the insular primary habitats of var. *saxatilis* in the Guianas and var. *triumviralis* on the Orinoco also harbor related members of sect. *Chamaecrista,* in the first case *Ch. nictitans* var. *disadena,* in the second *Ch. glandulosa* vars. *crystallina* and *rapidarum*. All these obligately heliophilous cassias which can survive in the setting of tropical high forest only on rocky river banks and barren monadnocks, must have reached their present circumscribed niches at a period when broad paths of open savanna opened up across the Amazon Basin in dry eras.

The change of epithet from *gracilis* to *triumviralis* (for H.B.K.), permissible with change of taxonomic rank, is aimed at eliminating confusion among the all-too-similar varietal epithets *tenuis, gracilis,* and *tenuicaulis* current in the tetraphyllous complex of sect. *Xerocalyx*.

1k. Chamaecrista desvauxii (Colladon) Killip var. **brevipes** (Bentham) Irwin & Barneby, comb. nov. *Cassia desvauxii* var. (β) *brevipes* Bentham in Martius, Fl. Bras. **15**(2): 157. 1870.—". . . in provincia Goyaz, e. gr. prope Curralinho et in Serra dos Christaës: Pohl; ad Natividade: Gardner n. 2125.; inter Funil et S. João: Burchell. In Guiana quoque plantam legit Schomburgk."—Lectotypus (Irwin, 1964, p. 99): *Pohl 843* e "Rio Lage in Serra dos Chrystaës"; isotypus, K! paratypi, *Gardner*

2125, BM! *Burchell 8943, 9038,* K! *Robt. Schomburgk 190,* K!—*Cassia tetraphylla* var. *brevipes* (Bentham) Irwin, Mem. N.Y. Bot. Gard. **12**(1): 99 (exclus. syn. *Chamaecrista lehmannii*), fig. 350 (map). 1964.

Cassia brevipes DeCandolle ex Colladon, Hist. Casses 119, t. 9, fig. A. 1816.—"Hab. in Panamaide (Lagasca)."—Holotypus, ticketed "no. 65, Lagasca, 1807," G-DC! = F Neg. *33459.*—*Chamaecrista brevipes* (DeCandolle ex Colladon) Greene, Pittonia **4**: 31. 1899. *Cassia tetraphylla* var. *aurivilla* Irwin, Mem. N.Y. Bot. Gard. **12**(1): 96, fig. 350 (map). 1964 (non *C. tetraphylla* var. *brevipes* (Bentham) Irwin, 1964).

Cassia desvauxii var. *stipulacea* Pilger, Engler, Bot. Jahrb. **30**: 158. 1901.—"Mattogrosso: zerstreut am Rande des Uferwaldes am oberen Ronuro ([*Hermann Meyer*] n. 621.—Fruchtend im Mai 1899)."—Holotypus, †B; isotypus, US!

Cassia brevipes sensu Bentham, 1871, p. 568; Schery, 1951, p. 58.

Habit, foliage, vesture and fl of var. *mollissima*; stipules shorter than internodes; petiolar gland always sessile, 0.6–2 mm diam; pedicels (1–)2–8(–12) mm, at once shorter than subtending lf and the longer sepals; pod (17–)20–35 (–40) × (5–)6–12.5(–14) mm, the valves strigose or pilosulous, the vesture gray, sordid, or yellow; ovules (8–)9–15.

Savanna habitats of var. *mollissima,* scattered within the latter's range but less widespread and vicariantly allo- or locally sympatric: Gulf coastal plain of Honduras (Cortés); Mosquito Coast of Nicaragua (Zalaya); upland w. and Pacific Costa Rica (San José and Puntarenas); Panama (Chiriquí, up to 1550 m, Coclé and Panamá); n.-w. Venezuela (Zulia, Lara); llanos of e. Colombia s. to Cordillera de Macarena; n.-e. Peru (Loreto); Guayana Highland in Venezuela (Amazonas, Bolívar) and adjoining Rupununi District of Guyana and Terr. do Roraima in Brazil; coastal plain of Surinam and Amapá, Brazil; and reappearing s. of the Amazonian Hylaea from s. Ceará, n. Goiás, n.-e. Mato Grosso and the upper forks of ríos Beni and Mamoré in Bolivia s. to s.-w. Goiás, s. Mato Grosso and immediately adjacent Paraguay (Sa. de Amambay).—Fl. in Central America X–III, in equatorial latitudes irregularly through the year, on the Brazilian Planalto mostly (X–)I–VI.

The implausibly interrupted dispersal of var. *brevipes* coincident with an internal range of variability that often, but not always, runs parallel to that of immediately vicariant or even sympatric var. *mollissima* suggests that the entity as defined herein is not phyletically natural but rather an assembly of minor variants that independently have acquired abbreviated pedicels. This however is a striking feature and maintenance of a traditional var. *brevipes,* if modified to admit a pod so variable in length, outline and vesture as to include var. *aurivilla* of the last revision (=*C. brevipes* Bentham), serves a useful taxonomic purpose. Correlation between a short, proportionately broad pod and yellow pilose vesture of the valves, well sustained in Costa Rica and Panama, fails to define two short-pedicelled forms in South America. Yellow pilose vesture occurs on short and narrow, long and narrow, and broad short pods, and these cannot be sorted in meaningful geographic patterns. In fact most of the South American material that has been referred to var. *aurivilla* because of the loose lutescent vesture of the pod or ovary has mature pods 3–4 times longer than wide, except for vesture exactly as in var. *mollissima.* The wide-valved pod, whether long or short, that could accommodate relatively long seeds does not in fact do so, and length of pod is only weakly correlated with ovule-number.

Over most of its range var. *brevipes* has the adult fruticose or suffruticose habit of var. *mollissima.* On the Orinoco-Amazon divide in Venezuela and Colombia and again in northern Goiás a possibly distinct entity may be characterized by a simple or distally few-branched wandlike habit that coincides with relatively narrow leaflets suggestive of var. *langsdorfii*. On the Esmeralda savannas in Amazonas this is sympatric with frutescent var. *ventuarensis,* and in Vaupés with var. *piptostegia*.

A Peruvian locality for var. *brevipes* mapped by Irwin (1964, l.c.), based on a Pearce collection from Pata, needs confirmation; the plant may have originated at Pata in Bolivia. The variety is known in Peru definitely only from Loreto (*W. Fox 54,* K).

1l. Chamaecrista desvauxii (Colladon) Killip var. **linearis** (Irwin) Irwin & Barneby, comb. nov. *Cassia tetraphylla* var. *linearis* Irwin, Mem. N.Y. Bot. Gard. **12**(1): 94 (exclus. pl. Blanchet.), fig. 350 (map). 1964.—"Brazil, Pernambuco, 'banks of the Rio Preto.' *Gardner 2828*. Sept. 1939."—Holotypus, K!

Either single-stemmed erect, then simple or few-branched distally and wandlike, or commonly when adult diffusely branched from crown of slender root, at anthesis (1–)1.5–4(–"10") dm, puberulent throughout or the lfts glabrous facially and ciliolate; stipules triangular-lanceolate (4–)5–7.5 × 1–2 mm; lf-stalk 3–4.5(–5) mm, the petiole 2–3.4 mm, the rachis 1–1.6 mm; gland sessile, 0.3–0.6 mm diam; lfts linear or linear-acuminate, the proximal pair only a trifle shorter than the distal, these of larger lvs 15–27 × 1–2 mm, 10–16 times longer than wide; pedicels 2–5 mm, much shorter than the sepals; long sepal 8–10 mm; longer petals 9–14 mm; pod 18–24 × 4–6.5 mm; ovules 6–10.—Collections: 2.

Campo or brejo, in Bahia near 750 m, apparently rare and local, known only from the type-locality in interior Pernambuco and from Espigão Mestre in w. Bahia ±100 km w.-s.-w. of Barreiras.—Fl. VIII–IX, I–III, perhaps throughout the rainy season.

Originally based on two collections, the typus from Pernambuco with pedicels shorter than the sepals, and *Blanchet 2536* from Jacobina, in Chapada Diamantina, here transferred, because of the long pedicels, to a newly described var. *graminea*. As here circumscribed, var. *linearis* very closely simulates the more slender races of var. *brevipes,* differing essentially only in the shorter petiole and narrower, grasslike leaflets. An ample collection from western Bahia (*Anderson et al. 36723*) includes with diffusely branched older plants that match the holotypus in all details others that are erect and wandlike, either simple or few-branched distally. These last are so compellingly suggestive of var. *brevipes* that we suspect that var. *linearis* is little more than an extreme narrow-leaved form of it, perhaps not taxonomically separable. For this reason we have referred var. *linearis* along with the seemingly closely allied var. *graminea* to *Ch. desvauxii* sens. lat. rather than to *Ch. ramosa,* in spite of the often abbreviated petiole.

1m. Chamaecrista desvauxii (Colladon) Killip var. **graminea** Irwin & Barneby, var. nov., a var. *lineari* (Irwin) Irwin & Barneby imprimis pedicellis elongatis 14–17 (nec 2–5) mm longis abstans.—BRAZIL. Minas Gerais: 30 km by road s.-w. of Gouveia, at km 60 on road to Curvelo, 1150 m, 11.IV.1973 (fl, fr jun), *W. R. Anderson 8639*.—Holotypus, R; isotypi, F, GH, K, LE, MBM, NY (2 sheets), S, UB (not seen), UC, US.

Very slender, flexuously procumbent herbs 1–4 dm, appearing glabrous below the strigulose ovary but the stems and lf-stalks sometimes puberulent or pilosulous, the very narrow grasslike lfts glabrous or ciliolate; stipules narrowly lance-attenuate 6–11 × 1.5–2.5 mm, persistent; lf-stalks 3–4.5 mm, the petiole 1.5–3.5 mm, the rachis 1–1.4 mm; gland sessile or shortly stipitate, 0.25–0.9 mm diam, in profile (0.1 mm longer) 0–0.4 mm shorter than diam of head; lfts linear-lanceolate acute or acuminate, the proximal pair only a trifle shorter than the distal, these up to 15–25 × 1–1.7 mm, 8–16 times longer than wide; pedicels subfiliform 14–27 mm; long sepal 12–17 mm; longer petals 13–16 mm; ripe pod unknown; ovules 8–10.—Collections: 3.

Sandy campo near 1000–1200 m, apparently rare and local, known from only three stations, one in centr. Sa. do Espinhaço near 18°40′S in centr. Minas Gerais and two in Chapada Diamantina near 11°20′–13°S (Jacobina; Mucujé) in centr. Bahia.—Fl. in Minas Gerais IV–V, in Bahia VI–VIII, the full range unknown.

The foregoing description is based primarily on the ample type-series which, while resembling var. *linearis* in the extremely narrow grasslike leaflets, differs in the elongate pedicels which elevate the proportionately large fls to or beyond the level of the fulcrant leaf. With this we associate *Blanchet 2536* from Jacobina, mentioned by Bentham (1870, p. 158) as one of the ambiguous forms connecting *Cassia langsdorfii* with *C. gracilis* and referred by Irwin (1964) to var. *linearis,* from which it differs in the developed pedicels. While it might appear a priori preferable to stretch definition of var. *linearis* to encompass both long- and short-pedicelled flowers, this will remain logically indefensible unless we abandon the same distinction between the otherwise inseparable vars. *mollissima* and *brevipes*.

A unicate specimen (*Irwin et al. 32602a,* NY) from Sa. do Tombador, in Chapada Diamantina about 100 km southwest of Jacobina, is perhaps referable to var. *graminea* but differs from the collections cited above in the rather stiffer (but still procumbent, sometimes adventitiously rooting) stems and slightly wider (to 2.5 mm), oblanceolate subobtuse rather than lance-attenuate leaflets. More ample collections from the highlands of Bahia are needed before the full range of variation in var. *graminea* can be known.

1n. Chamaecrista desvauxii (Colladon) Killip var. **langsdorfii** (Kunth ex Vogel) Irwin & Barneby, stat. nov. *Cassia langsdorfii* Kunth ex Vogel, Syn. Gen. Cass. 55 & Linnaea **11**: 703, descr. ampliat. 1837.—"In Brasilia: Sellow leg."—Holotypus, B (hb. Kunth.) = F Neg. *1697.—Chamaecrista langsdorfii* (Kunth ex Vogel) Britton ex Pittier, 3rd Conf. Interamer. Agric. Caracas 373. 1945.

Cassia borbonioides Vogel, Syn. Gen. Cass. 55 & Linnaea **11**: 703, descr. ampliat. 1837.—"In Brasilia: Sellow leg."—Holotypus, †B; no isotypus seen by Bentham or subsequently but the protologue unambiguous (cf. comment below).

Chamaecrista lehmannii Britton & Rose ex Britton & Killip, Ann. N.Y. Acad. Sci. **35**: 184. 1936.—"Guagua, Huila, Colombia, *Lehmann B. T. 1095* . . ."—Holotypus, NY!

Cassia chartacea Irwin var. *chartacea* Irwin, Mem. N.Y. Bot. Gard. **12**(1): 79 (exclus. syn.), fig. 346 (map). 1964.—"'Brasil,' *Martius 1076* . . ."—Holotypus, M! isotypi, BM, K, NY (2 sheets), S!

Cassia langsdorfii sensu Bentham, 1870, p. 158 ("*Langsdorffii*"), max. ex parte. *C. langsdorfii* var. *langsdorfii* sensu Irwin, 1964, p. 62 ("*langsdorffii*"), fig. 342 (map).

Erect, at first simple and few-branched distally or (rarely) few-stemmed from base, the lateral stems stiffly ascending, when adult frutescent and stiffly branched upward, the branches either ascending or plagiotropic, the whole at anthesis (2–)3–16(–20) dm, the lvs mostly abruptly divaricate, the olivaceous or pallidly subglaucescent foliage commonly glabrous, the stipules and lfts less often ciliolate, the stems varying from glabrous to strigulose or pilosulous, the petioles often dorsally setulose; stipules ovate- or lance-acuminate or abruptly acute, almost always as long or longer than internodes, persistent, the larger up to 12–24 × 2.5–8(–9) mm; lf-stalks of longer lvs 4–10(–11) mm, the petiole 3–8.5, the rachis ±1–2 mm; petiolar gland scutellate or shallowly concave (0.3–)0.4–1.6 mm diam, usually sessile and when both small and sessile sometimes depressed into the sulcus and inconspicuous, less often raised on a short stout stipe and becoming in profile drum- or squatly tack-shaped, but shorter (exceptionally a trifle longer) than diam of head; lfts narrowly elliptic, oblong-elliptic, linear-oblanceolate or sublinear, acute or obtuse mucronulate, the distal pair usually a trifle longer than the proximal, the larger lfts up to (12–)15–26 × 2–6(–6.5) mm, (3–)3.5–8(–10) times longer than wide, the blades varying from submembranous to stiffly chartaceous, the stiffer ones bordered by a raised pallid marginal nerve; pedicels 8–30(–38) mm, mostly as long or shorter, rarely a little longer than lf, either glabrous, or puberulent on one or both faces; long sepal (11–)12–19(–20) mm; long petal (10–)11–18 mm; pod (26–)30–52 × (5.5–)6–7.5(–9) mm, thinly strigulose or the faces sometimes glabrous; ovules (9–)11–18(–22).

Cerrado and campo, sometimes about outcrops and often locally abundant, along with microphyllous Melastomataceae, on seeps or moist rocky stream banks, mostly 500–1100, on Chapada dos Veadeiros up to 1400 m, widespread and locally common over the centr. highlands of the Brazilian Planalto from e.-centr. to s.-w. Goiás, w. across Rio S. Francisco to the w. slope of n. Sa. do Espinhaço (Montes Claros) in Minas Gerais, thence s. along the Serra to Sa. do Itabirito and along Sa. Geral into the Triângulo, with outlying stations on the headwaters of Rios Paraiba and Paranapanema respectively in far e. São Paulo and n.-e. Paraná, on Sa. de Amambay in e.-centr. Paraguay, in centr. Mato Grosso (near Cuiabá), and distantly disjunct at ±1200 m on upper Río Magdalena in Huila, Colombia (near 3°N).—Fl. mostly X–IV, in moist places nearly throughout the year.

While the range of dispersal (except for the Colombian extension, due to taxonomic decision) of the taxon *langsdorfii* has not been extended over that mapped by Irwin (1964, l.c.), the amount of material available for comparative study has tripled, compelling us to modify quite radically the former concept. Before going further, we have to correct the misinterpretation of *Cassia langsdorfii* sens. str. of the last revision. Vogel described *C. langsdorfii* (the spelling was altered by Bentham to "*langsdorffii*") and *C. borbonioides* as close kindred different principally in pubescence, the former having ciliolate leaflets with variably pubescent pedicels whereas the latter was glabrous below the ovary. Obviously a circumscription and key that attributed to *C. langsdorfii* a prime differential character of glabrous pedicels was an unhappy error. We are now satisfied, however, that pubescence of the pedicels has neither taxonomic nor biological significance in sect. *Xerocalyx,* so that in effect we are able, following modification of the definition, to maintain the epithet *langsdorfii* for the same material, with which we now associate the pubescent analogue segregated as *C. chartacea*.

The central and prevailing expression of the newly defined, more heterogeneous var. *langsdorfii* described above is an erect plant, at first single-stemmed with either ascending or plagiotropic branches, lance-oblong stipules normally concealing the internodes, and petioles, of the length found in var. *mollissima,* abruptly divaricate from the stem; but different from var. *mollissima* in the narrower, mostly linear-elliptic or narrowly oblong-elliptic, acute or less often obtuse leaflets consistently glabrous facially but either ciliolate or not so, coinciding wth flowers of moderate size. While most mature plants of this nature attain a stature of less than one meter, some old ones, especially where growing in thickets at the edge of gallery woodland or along streams, become shrubby and up to two meters tall; but whether these are genetically distinct or merely aged individuals of the common wandlike type we do not know. The gland of var. *langsdorfii* is now seen to vary enormously in size, but is only rarely elevated on a stout stipe and even then not taller than the diameter of the head. When very small it may be concealed between the ventral wings of the petiole but usually projects beyond them. The leaflets vary in texture from firm, with pallid thickened (always glabrous) marginal nerve, to chartaceous with sharp (either glabrous or ciliolate) margins, but we can no longer expect correlation between texture of leaf-blades and size of glands. As a consequence *C. chartacea* var. *chartacea,* which seemed in the light of a small sample specifically distinct in its syndrome of thin-textured leaflets, small glands, and pubescent pedicels, has faded to the status of a poorly differentiated minor variant. The type-collection of *C. chartacea,* of which the precise origin is unknown, is somewhat unusual in having, along with the supposed differential characters already mentioned, obtuse leaflets and relatively small stipules that do not overlap along the stems as an imbricated thatch. The few precisely similar modern collections come from the northwest corner of Distrito Federal near Sobradinho, Planaltina and Formosa, and we suspect that Martius may have collected the typus somewhere on the headwaters of Rio Paranã close beyond the present confines of the District.

The var. *langsdorfii* occupies a position crucially central at once to the whole curve of phenetic variation in tetraphyllous xerocalyx and to its focus of maximum diversity on the Brazilian Planalto. In southern Goiás it is in places narrowly sympatric with *Ch. desvauxii* var. *glauca,* var. *mollissima,* var. *brevipes* and var. *modesta* and with *Ch. ramosa* var. *parvifoliola,* var. *lucida* and var. *erythrocalyx*; and we have specimens ambiguously intermediate to all of them except var. *glauca,* with which it is connected however by way of var. *mollissima.* Our division of the tetraphyllous complex into two specific series primarily on the basis of petiole length casts var. *langsdorfii* (with *C. chartacea* var. *chartacea*) on the side of *Ch. desvauxii* and the very closely related and intergradient var. *parvifoliola* and var. *erythrocalyx* (with *C. chartacea* var. *tenuicaulis*) on that of *Ch. ramosa.* The artificiality of this surgical division is obvious, but we can find no more natural line of separation.

The typus of *Ch. lehmanii,* interpreted by Irwin (1964) as *Ch. brevipes* with paradoxically long pedicels, cannot be distinguished from some Brazilian var. *langsdorfii* and would be noteworthy in southern Goiás only for the relatively short stipules, a feature already encountered, as noted above, in the type-collection of *C. chartacea.* A second collection (*Fosberg 19812*) from Huila shows the characteristic wand habit of var. *langsdorfii,* not manifest in the typus.

1o. Chamaecrista desvauxii (Colladon) Killip var. **pirebebuiensis** (Chodat & Hassler) Irwin & Barneby, stat. nov. *Cassia pirebebuiensis* Chodat & Hassler, Bull. Herb. Boiss. II, **4**: 825. 1904.—"Paraguay: . . . in colle Tobaty, Sept., Hassler n. 6118 . . . eodem loco, Mart., n. 3983."—Lectoholotypus (Irwin, 1964, l.c. infra, by implication): *Hassler 6118,* G! isotypi, NY, S! paratypi, *Hassler 3983,* G, NY!

Cassia pirebebuiensis sensu Irwin, 1964, p. 59, fig. 340 (map).

Erect, 1–few-stemmed, at anthesis 2.5–6 dm, except for strigulose ovary glabrous; stipules broadly lanceolate to oblong-ovate 8–16 × 3.5–7 mm, as long or little longer than internode; lf-stalk 4.5–7 mm, the petiole (3.5–)4–5.5 mm, the rachis 0.7–1.4 mm; gland 0.3–0.5 mm diam, slenderly stipitate, in profile pin-shaped 0.6–1.2 mm tall, 0.3–0.8 mm taller than diam of head; lfts oblanceolate, the proximal pair ±1/5 shorter than the distal, these up to 16–25 × 4–8 mm, 3–4 times longer than wide; pedicels 24–37 mm; longest sepal 12–14 mm; longer petals 10.5–15 mm; pod 42–53 × 7–9 mm; ovules 15–25.

Open stony places, ±250–400 m, apparently known only from prov. Caraguatay in w.-centr. Paraguay (Tobatí; Caacupe; Pirebebuy).—Fl. VIII–III.

The var. *pirebebuiensis* differs essentially from var. *langsdorfii* only in the pin-shaped petiolar gland. This is the gland of distantly allopatric var. *latifolia* which, apart from its range, is only precariously distinguished by diffuse rather than erect stems and proportionately broader stipules and leaflets. We maintain them as distinct in the belief that they must be parallel independent modifications derived from var. *langsdorfii* and var. *desvauxii* respectively, whereas var. *peronadenia,* with the gland yet further etiolated, appears to have developed from *C. desvauxii* var. *glauca*.

1p. Chamaecrista desvauxii (Colladon) Killip var. **modesta** Irwin & Barneby, var. nov., habitu adspectuque var. *langsdorfii* simulans, imprimis floribus minutis, sepalo longiori 8–11 (nec 12–19) mm, petalo longiori 6–8.5 (nec 11–18) mm longis abstans.—BRAZIL. Goiás: wet ground of gallery margin, 900 m, ±58 km n.-e. of Catalão in Contraforte Geral de Goiás, 26.I.1970 (fl, fr jun), *Irwin et al. 25425*.—Holotypus, UB; isotypi, F, GH, K, LE, NY, RB, S, US.

Slender erect from an apparently monocarpic root, simple or few-branched distally, 2–8(–"15") dm, except for strigulose ovary and dorsally barbate petiole glabrous, or the stem and pedicels sometimes unilaterally puberulent, the lfts always glabrous; stipules lanceolate 7–15 × (2–)2.5–4.5 mm, mostly ± as long as internode, persistent; lf-stalks divaricate or widely ascending (4–)4.5–7 mm, the petiole (3–)3.5–5.5 mm, the rachis 1–1.5 mm; gland scutelliform sessile or almost so 0.6–1.4 mm diam; lfts subsymmetrically linear, linear-elliptic or narrowly oblanceolate, the proximal pair ±¼ shorter than the distal, these of longer lvs 14–20 × 2–4 mm, 4.5–10 times longer than wide; pedicels very slender (5–)7–17(–20) mm; longest sepal 8–11 mm; long petals 6–8.5 mm, shorter than the longer sepals; pod 25–35 × 5–8 mm; ovules 10–14.—Collections: 12.

Moist campo and brejo within the cerrado climax woodland, 500–1200 m, locally plentiful in the valleys of rios Corumbá, São Marcos and upper Paranaíba in s.-w. Goiás and adjacent Distrito Federal and Minas Gerais; apparently disjunct on headwaters of Rio Araguaia in e.-centr. Mato Grosso and in s.-centr. São Paulo.—Fl. X–IV.

The essentially uniform series of twelve collections, described here as *Ch. desvauxii* var. *modesta,* differs from other races of *Ch. desvauxii* in the very small flower, comparable in size only with that of *Ch. diphylla*. In wandlike habit of growth, sparse vesture, firm-textured narrow leaflets and sessile, usually large petiolar glands, var. *modesta* closely resembles var. *langsdorfii* and occupies an enclave within the latter's more ample range. It might be thought to consist merely of a selection of small-flowered minor variants or seasonal states of var. *langsdorfii,* but as most of our material is in full vigor of growth and at young anthesis, the flower's small size cannot be attributed to either drought or senescence. In addition to the type-series we have seen the following collections (all NY):

Goiás: Anápolis, 22.III.76 (fl,fr), *Hatschbach 38251*; 5–6 km s. of Cristalina, 2.XI.65 (fl,fr), *Irwin et al. 9755* and 4.III.66 (fl,fr), *13431*. Distrito Federal: Parque do Gama, 20 km s. of Brasília, 28.III.73 (fl,fr), *W. R. Anderson 7983*; 25 km s.-w. of Brasília, *Hatschbach, Anderson et al.* 11422; 15 km e. of Brasília, 30.I.66 (fl), *Irwin et al.12155*. Minas Gerais: 29 km w. of Uberlândia, 1.II.59 (fl), *Irwin 2533*; 41 km s. of Carmo do Paranaíba, 10.II.59 (fl,fr), *Irwin 2617*; 35 km n.-w. Paracatú, 8.II.70 (fl), *Irwin et al. 26318*; Morro das Pedras n.-e. of Patrocinio, 29.I.70 (fl), *Irwin et al. 2566*. Mato Grosso: Xavantina, 30.XII.67 (fl), *Philcox & Fereira 3780*. São Paulo: n. of Botucatu, 8.II.72 (fl,fr), *I. Gottsberger 97R-8272*.

1q. Chamaecrista desvauxii (Colladon) Killip var. **chapadicola** Irwin & Barneby, var. nov., a var. *langsdorfii* habitu foliolisque interdum coriaceis craspedo-neuris simili forsan propiuscule affinis sed foliolorum ambitu valde obliquorum nervo mediano falcatim incurvo pedicellisque abbreviatis 3–9 (nec 8–30) mm longis distincta; a *Ch. ramosa* var. *lucida* foliolis obliquis incurvis pedicellisque brevibus comparabili imprimis petiolis plerisque elongatis 4–6 (nec 1–2.5) mm longis abstans.—BRAZIL. Mato Grosso: Salto das Sete Quedas on Rio Verde, ±15 km s.-w. of Rio Verde, 8.II.1975 (fl, fr), *Anderson, Hatschbach, Barneby & Gates 11268*.—Holotypus, MBM; isotypi, K, NY, US.

Erect, sometimes wandlike but at least distally branched subshrubs at anthesis 5–9 dm, with densely leafy, either ascending or plagiotropic branchlets, either glabrous below the always strigulose ovary or the stem sometimes unilaterally puberulent and the petioles ventrally ciliolate, the foliage always glabrous, olivaceous or glaucescent, the firmly chartaceous leaflets prominently cartilaginous-margined; stipules broadly ovate-cordate acute 7–11 × 4–6.5 mm, a little shorter or longer than internode, persistent; lf-stalks (3.5–)4.5–7 mm, the petiole (3–)4–6 mm, the rachis 0.8–1.4 mm; gland 0.7–1.3 mm diam, sessile or almost so, in profile depressed-drum-shaped or broadly obconic, 0.2–0.7 mm shorter than diam of head; lfts falcately oblanceolate to obliquely obovate, at apex obtuse mucronulate or broadly deltate, the proximal pair ±¼ shorter than the distal, these of larger lvs (9–)11–18 × 4–6 mm, 2.5–3.5 times longer than wide; pedicels 3–9 mm, shorter than stipules; long sepal 9–12 mm; longer petals 9–14 mm; pod 26–38 × 6–8 mm; ovules 10–15.—Collections: 5.

Cerrado on sandstone, 450–700 m, locally plentiful on both slopes of the Amazon-Paraguai divide in centr. and s.-w. Mato Grosso.—Fl. IX–II.

The earliest known collection of var. *chapadicola*, made about a hundred years ago in "western Brazil" by the Austrian naturalist Tamberlik (*s.n.*, NY), was referred by Irwin (1964, p. 73) to *Ch. curvifolia* var. *lucida* but particularly noted as atypical, simulating *C. langsdorfii* in the coriaceous texture and strong primary and marginal nervation of the leaflets. Now better known, this striking plant emerges as a definite entity morphologically intermediate between *Ch. ramosa* var. *lucida* and *Ch. desvauxii* var. *langsdorfii*, having the incurved leaflets and short pedicels of the former combined with stiffly margined leaflets and longer petioles of the latter. It is one of several tetraphyllous *Xerocalyx* cassias linking the microphyllous and macrophyllous series which we arbitrarily distinguish by petiole length, the critical differential character throwing var. *chapadicola* on the side of *Ch. desvauxii*.

Additional material, all from Mato Grosso (all at NY): Xavantina-Cachimbo road, km 223, 21.I.68 (fl), *Philcox & Fereira 4138*; Chapada dos Guimarães, 13.II.75 (fl, fr), *Anderson et al. 11355*; ibid., 14.XI.75 (fl), *Hatschbach 37552*.

2. Chamaecrista ramosa (Vogel) Irwin & Barneby, comb. nov. *Cassia ramosa* Vogel, Syn. Gen. Cass. 55. 1837.—Typus infra sub var. *ramosa* indicatur.

Habit and fls of sect. *Xerocalyx* but relatively small-lvd, the petiole (including pulvinus) 0.6–4 mm, the lfts scarcely ever more than 15 mm; South American only, from Venezuela to s.-e. Brazil.

Key to the Varictics of *Ch. ramosa*

1. Midrib of lfts straight or almost so, the blades (beyond the oblique attachment) symmetrically obovate, oblanceolate, elliptic, or linear; if proximal pair of lfts incipiently incurved, then pedicels much over 10 mm.
  2. Stipules persistent, either entire or as pallid marcescent tatters, on leafless annotinous branchlets.
    3. Petiolar gland distinctly stipitate, in profile as tall or taller than diam of head; lfts obovate or oblanceolate; lower Orinoco valley s. to Pará, thence along the Atlantic coast to Rio de Janeiro, inland only to centr. Bahia. 2a. var. *ramosa* (p. 884).
    3. Gland sessile or short-stipitate, in profile usually shorter than diam of head, if as long then lfts linear-oblong-elliptic and range different.
      4. Lfts elongately linear up to 15–27 × 1–2 mm, 8–16 times longer than wide; cf. 1b. *Ch. desvauxii* var. *graminea* (if pedicels 14–27 mm) or 1k. *Ch. desvauxii* var. *linearis* (if pedicels 2–5 mm).
      4. Lfts shorter or, if (rarely) as long as last, then proportionately wider, 2.5–7 times longer than wide.
        5. Plants of middle Orinoco and Orinoco-Amazon divide in Amazonas, Venezuela, adjoining Colombia and Terr. do Roraima, Brazil; pedicels 3–10 mm and stipules 6–12 mm, concealing the internodes (if pedicels over 10 mm or stipules 2–5 mm much shorter than internodes, cf. 1j. *Ch. desvauxii* var. *triumviralis*). 2b. var. *ventuarensis* (p. 885).
        5. Plants of Brazilian Planalto s. from s.-w. corner of Pará; pedicels at least 11, mostly over 14 mm long. 2c. var. *parvifoliola* (p. 886).
  2. Stipules deciduous with or before associated lf, the annotinous branchlets naked; highlands of Brazilian Planalto. 2d. var. *erythrocalyx* (p. 887).
1. Midrib of lfts incurved, the blades falcately lanceolate, elliptic, or obliquely to dimidiately obovate; pedicels 2–10 mm.
  6. Distal pair of lfts 9–17 mm; Goiás, adjoining Mato Grosso, and s.-w. Pará. 2e. var. *lucida* (p. 888).
  6. Distal pair of lfts 2.5–6.5(–7.5) mm; e.-centr. Goiás to Bahia, Ceará and lower Amazonia. 2f. var. *mollissima* (p. 890).

2a. Chamaecrista ramosa (Vogel) Irwin & Barneby var. **ramosa**. *C. ramosa* Vogel, 1837, l.c., sens. str.—"In Brasilia: Sellow legit inter Victoria [Espirito Santo] et Bahia [=Salvador]; in itinere a Minas ad S. Paulo; Luschnath pr. Bahia in campis pr. mare."—Lecto-isotypus (Irwin, 1964, p. 73): *Sello s.n.*, K! = NY Neg. *1518*.—*Cassia uniflora* Sprengel var. (β) *ramosa* (Vogel) Bentham in Martius, Fl. Bras. **15**(2): 158. 1870. *C. uniflora* var. *parvifolia* Bentham, Trans. Linn. Soc. **27**: 568. 1871, nom. illegit. *C. tetraphylla* var. *ramosa* (Vogel) Amshoff, On S. Amer. Papilionaceae 27. 1939.

Cassia savannensis Miquel, Ann. & Mag. Nat. Hist. **11**: 15. 1843.—"*Hab.* Surinami, locis arenosis, in Tooden-Savanna, prope fl. Suriname Sup . . ."—Holotypus, *Focke 454*, U!—*C. savannensis* sensu Miquel, Stirp. Surinam, Select. 16 ("*savanensis*"). 1850.—Equated by Bentham, 1871, p. 568, with *C. uniflora* var. *parvifolia* = *C. ramosa* Vog.

Chamaecrista ramulosa Killip & Pittier ex Pittier, Bol. Soc. Venezol. Ci. Nat. **7**(48): 146. 1941.—". . . *Anzoátegui*: Lomas de los Caños, valle del Río Carí, al sur de El Tigre . . . Marzo 23 y Deciembre 15, 1940 (Pittier 14.318 y 14.476, tipo) . . . el No. 14.476 se ha escogido como tipo."—Holotypus not seen; paratypus (*Pittier 14318*), VEN! = NY Neg. *6336*.—*Chamaecrista rusbyi* Britton & Rose ex Pittier, l.c., pro syn.—Equated with *C. ramosa* by Irwin, 1964, p. 74.

Cassia ramosa var. *maritima* Irwin, Mem. N.Y. Bot. Gard. **12**(1): 77, 1964.—"Brazil, Guanabara, 'Vicinity of Cabo Frio.' *J. N. Rose & P. G. Russell 20688,* 8 Aug. 1915."—Holotypus, NY! isotypus, US!

Cassia ramosa sensu Irwin, 1964, p. 73, max. ex parte, exclus. syn. *C. uniflora utiarityi* locisque nonnullis austrobrasiliensibus, fig. 345 (map, inclus. var. *maritima*).

Erect, assurgent or diffusely trailing, stiffly much-branched shrublets, at anthesis (1–)2–11(–15) dm, varying (both within and between populations) from glabrous except for unilaterally puberulent stems and pedicels or even fully glabrous to densely finely velutinous-puberulent or -pilose throughout, the foliage olivaceous or less often glaucescent, the stems or trunklets often adventitiously rooting when in contact with the ground or when engulfed in sand; stipules ovate-cordate (2–)2.5–8(–9) × (1.5–)2–4.5 mm, varying from much shorter up to as long as or, on some terminal congested branchlets, longer than the internode, all becoming papery, persistent; lf-stalk (1.6–)2–4.5(–5) mm, the petiole 1.2–3(–3.5) mm, the rachis (0.4–)0.6–1.5 mm; petiolar gland ureceolate (0.1–)0.15–0.55(–0.8) mm diam, stipitate but the stipe variable in length and coarseness, the whole in profile pin-, stud- or rarely coarsely trumpet-shaped, 0.4–2 mm tall, 1.6–0.1 mm longer, exceptionally to 0.1 mm shorter than diam of head; lfts obliquely obovate to oblanceolate obtuse or mucronulate-subacute, the larger (distal) pair ¼–½ longer than the proximal, the longest lfts up to 5–13(–15) × 1.5–5.5(–7) mm, 1.8–3.5(–4.5) times longer than wide; pedicels slender wiry, of most fls 12–32(–36) mm, of some late ones shorter; longest sepal 9–14(–15) mm; longest petal (10–) 11–18 mm, nearly always a little longer than longest sepal; pod 24–44 × 4.5–6(–7) mm, strigulose, ciliolate along sutures, or glabrous; ovules 11–23.

Savannas, coastal dunes and restingas, ascending s.-ward and inland into cerrado-caatinga, mostly from near sea level up to 450 m, on Guayana Highland up to 1500 and in interior Bahia to 1000 m, interruptedly widespread over e. S. America: middle and lower Orinoco valley in n.-w. and n. Bolívar, Guárico and Anzoátegui, Venezuela; Guayana Highland and lowland periphery in Bolívar, Venezuela, Guyana, and Terr. do Roraima, Brazil; white or brown sands of the coastal plain of the Guianas and Pará; along immediate coast of n.-e. Brazil from Rio Grande do Norte to Rio de Janeiro; disjunctly inland on highlands of interior Bahia (Sas. do Tombador and Agua de Rega).—Fl. interruptedly throughout the year, especially during and just after rains.

2b. Chamaecrista ramosa (Vogel) var. **ventuarensis** (Irwin) Irwin & Barneby, comb. nov. *Cassia tetraphylla* var. *ventuarensis* Irwin, Mem. N.Y. Bot. Gard. **12**(1): 95, fig. 349 (map). 1964.

Erect shrublets, at first wandlike, when adult paniculately branched distally and up to 1–2 m, the annotinous branches densely leafy, the stems, lf-stalks and pods pilosulous, the lifts glabrous or glabrous ciliolate; stipules broadly lanceolate up to 6–12 mm, ± as long as internodes, papery-marcescent; lf-stalks 2–4(–5) mm, the petiole 1–3 mm, the rachis ±0.8–1.4 mm; gland elliptic 0.3–0.5 mm diam, sessile and concealed in petiolar groove; lfts obliquely oblanceolate to narrowly obovate obtuse, the distal pair slightly larger, these of larger lvs 10–19 × 3.5–5 mm, 2.5–4 times longer than wide; pedicels 3–10(–18) mm; fls of var. *ramosa*; pod 2.5–3.2 × 6–8 mm; ovules 7–9.

Seasonally wet savannas, river-banks and shrubby banks, ±250–1600 m, scattered around the s. and s.-w. margins of the Guayana Highland: forks of Río Ventuari (4–6°N, 65–67°W) in n. Amazonas, Venezuela; savannas of Rio Xeriuini (near lat. 0°, 62°W) in s. Território do Roraima, Brazil.—Fl. XII–IV, VII, perhaps almost throughout the year.

The var. *ventuarensis* has the leaflets and sessile glands of sympatric *Ch. desvauxii* var. *mollissima* with the short pedicels of sympatric *Ch. desvauxii* var. *brevipes*, but the short leaf-stalks of *Ch. ramosa*. It is thus morphologically intermediate between the macrophyllous and microphyllous series of sect. *Xerocalyx*. Following our arbitrary separation of the two series by length of leaf-stalk it falls with *Ch. ramosa*, differing from the vicariant var. *ramosa* in the relatively large leaflets, sessile glands, and shorter pedicels. Known in 1964 only from the Ventuari basin in Venezuela, but now disjunctly from a west affluent of Rio Branco in Território do Roraima, Brazil (vouchers: *Pires et al. 13936, 13990*, NY).

2c. Chamaecrista ramosa (Vogel) var. **parvifoliola** (Irwin) Irwin & Barneby, comb. nov. *Cassia langsdorfii* var. *parvifoliola* Irwin, Mem. N.Y. Bot. Gard. **12**(1): 68, fig. 342 (map, the dot on Ceará requiring confirmation). 1964.—"Brazil, Minas Gerais, 'Serra do Cipó, 15 km E. of S. José do Almeido, on road to Morro [do] Pilar.' *H. S. Irwin 2429*. 12 Jan. 1959."—Holotypus, TEX! isotypi, MICH, NY (2 sheets), R, UC, US, VIC!

Cassia chartacea var. *tenuicaulis* Irwin, Mem. N.Y. Bot. Gard. **12**: 82, fig. 346 (map). 1964.—"Brazil, São Paulo, 'in campis arenosis, Araracoara.' *Riedel 2198*."—Holotypus, collected V.1834, US!

Slender erect, diffuse or procumbent shrublets freely branching from base or from well below middle, at anthesis (2–)3–8(–10) dm, commonly except for unilaterally puberulent stems and pedicels and for strigulose ovary glabrous, sometimes glabrous throughout or the dull olivaceous to glaucescent foliage ciliolate, less commonly (in either mixed or pure populations) pilosulous on dorsal or both faces; stipules ovate- or triangular-lanceolate to lance-oblong-acuminate 3–10 (–13) × 1.2–2.7(–3.5) mm, varying from a little longer to less than ½ as long as internode, becoming pallid-papery, persistent; lf-stalks commonly widely ascending or horizontal, of larger lvs 2–4.4(–6) mm, the petiole 1.6–3.4(–4) mm, the rachis 0.5–1.3(–2) mm; gland shallowly concave 0.2–1(–1.5) mm diam, commonly sessile, if shortly stipitate becoming in profile obconic or peg-, tack-, or drum-shaped, 0.4–0.8(–1) mm tall, mostly much shorter, rarely up to 0.4(–0.5) mm longer than diam of head; lfts linear-oblong or -elliptic to narrowly oblanceolate, acute or obtuse mucronulate, the distal pair ±½–1 times longer than the proximal, these sometimes incipiently incurved, the distal lfts of larger lvs up to (5–)6–15 (–18) × (1–)1.5–3.5(–4) mm, 3.5–7(–8) times longer than wide; pedicels very slender wiry, mostly (12–)13–30(–33) mm and longer than lf, of some late-season fls shorter, either glabrous or puberulent along one or both sides; long sepal (8–)9–15(–16) mm; longer petals (8–)9–15 mm, a little longer or shorter than long sepal; pod 20–40(–44) × 4.5–7 mm; ovules 8–17(–21).

Stony campo and outcrops in cerrado, s.-ward mostly 250–800 m, in centr. Minas Gerais and highland Goiás mostly 550–1370 m, locally abundant along the crest and w. slope of Chapada Diamantina and Sa. do Espinhaço from Sa. do Sincorá in s.-centr. Bahia s. to Belo Horizonte (13°–20°S) in centr. Minas Gerais, s. interruptedly along the n. slope of Sa. da Mantiqueira and through e. São Paulo (sources of the Rio Tieté and formerly around the city of São Paulo) to upper Iguaçu valley in e. Paraná (26° S), w. in Minas through the Triângulo to s. Mato Grosso, thence n. in scattered stations through e. Goiás (Sa. dos Cristais; Distrito Federal) to Chapada dos Veadeiros; apparently disjunct in upper Araguaia valley in n.-e. Mato Grosso (Barra do Garças n. through Sa. do Roncador) and on Sa. do Cachimbo in s.-w. Pará.—Fl. X–VI.

The var. *parvifoliola* might be visualized and interpreted from either of two angles: as a diminished, microphyllous and bushy-branched derivative of *Ch. desvauxii* var. *langsdorfii* (Irwin, 1964, l.c.); or as one of two Planaltine representatives of *Ch. ramosa* in which the petiolar gland is sessile or only stoutly and shortly stipitate. Over much of its range it is sympatric with *Ch. ramosa* var. *erythrocalyx,* similar in ramulose habit of growth but different in the small, early caducous stipules, and standing in relation to this as does *Ch. desvauxii* var. *mollissima* to var. *piptostegia*. In some populations of var. *parvifoliola* the smaller lower pair of leaflets are incipiently falcate and these plants scarcely differ, except in relatively long pedicels, from *Ch. ramosa* var. *lucida*. Two collections from Alto Garças on the upper Araguaia in east-central Mato Grosso (*Hatschbach 34164*; *Anderson 11394*) and one from central Minas Gerais (Biribiry, *Hatschbach 27971*), in which the foliage of var. *parvifoliola* is combined with the short pedicels of var. *lucida,* present a further ambiguously intermediate set of characters in which we cannot yet recognize a stabilized taxonomic entity.

In the light of ample contemporary collections the distinction between a relatively southern *Cassia chartacea* var. *tenuicaulis* (of São Paulo and Paraná) with puberulent pedicels and a more widely dispersed *C. langsdorfii* var. *parvifoliola* with glabrous pedicels has broken down completely. The plant of the southern campos does tend to have weaker stems and smaller stipules than much of var. *parvifoliola* from middle and high elevations in Sa. do Espinhaço, but it is impossible to define two entities in these terms. Bentham's treatment (1870, p. 159) of the early collections from São Paulo (*Burchell 4280*; *Lund* from Araraquará = *Riedel 2198,* later to become typus of var. *tenuicaulis*) as unnamable forms intermediate between *C. langsdorfii* and *C. gracilis* is evidence of the essential similarity of all tetraphyllous *Xerocalyx* and the difficulty of attaining a realistic and viable taxonomy.

An enigmatic plant from coastal Bahia near 17°S (Pôrto Seguro, 11.VI.62, *A. P. Duarte 6665,* NY), from a locality where var. *ramosa* alone could be expected, has all technical characters of var. *parvifoliola* except for genuinely prostrate stems radiating from the root-crown. It is tentatively referred here but because of uncertainty as to its true status the locality is omitted from the statement of range.

2d. Chamaecrista ramosa (Vogel) var. **erythrocalyx** (Martius ex Bentham) Irwin & Barneby, comb. nov. *Cassia gracilis* var. *erythrocalyx* Martius ex Bentham in Martius, Fl. Bras. **15**(2): 161. 1870.—". . . in montibus districtus Adamantium provinciae Minas Geraës: Martius."—Holotypus, *Martius 1268,* M! isotypus, M!

Cassia langsdorfii var. *tenuis* Irwin, Mem. N.Y. Bot. Gard. **12**(1): 67, fig. 342 (map). 1964.—"Brazil, Goiás, 'Serra de Chrystaes' *Pohl 90.* 1818."—Holotypus, NY! isotypi F, GH, M, MO!

Erect or diffuse, freely branching subshrubs, at anthesis (1.5–)2.5–10 dm, the olivaceous or exceptionally glaucescent foliage either glabrous or finely pilosulous, the purplish-castaneous annotinous branchlets naked; stipules ovate-cordate to broadly lanceolate (1.5–)1.8–8 × (1–)1.3–3 mm, usually shorter than the internode, deciduous with or before the lf; lf-stalks of larger lvs (1.5–)2–4.5(–5) mm, the petiole 1.5–3.5(–4), the rachis 0.2–1 mm; gland urceolate (0.2–)0.3–0.8(–0.9) mm diam, sessile or shortly stipitate, if the latter then in profile obconic to tack-, peg- or shortly trumpet-shaped, 0.3–0.8 mm long, from 0.7 mm shorter to 0.25 mm longer than diam of head; lfts linear-oblanceolate to narrowly obovate obtuse, obtuse-mucronulate, rarely acute, the larger (distal) pair ±¼–½ longer than the proximal, up to (6–)7–15(–19) × 1.5–4.5(–6) mm, (3–)3.5–6(–9) times longer than wide; pedicels (13–)15–30(–35) mm, glabrous, unilaterally puberulent, or puberulent all around; long sepal (8–)10–16 mm; longest petal (10–)11–16(–17) mm, either a little longer or shorter than the longest sepal; pod 22–48 × 5.5–7.5(–8.5) mm, finely strigulose or glabrous; ovules 10–19.—Fig. 53.

High campo sujo, stony or sandy meadows, rocky stream-banks and about outcrops, 950–1400, on Chapada dos Veadeiros up to 1750 m, locally abundant along the crest of Sa. do Espinhaço from Grão Mogol s. to Sa. do Cipó (±16°30′–19°30′S) in centr. Minas Gerais, and on the caprock of Chapada dos Veadeiros and Sa. dos Cristais in e.-centr. and s.-e. Goiás.—Fl. (X–)XI–III(–V).

The well known phase of var. *erythrocalyx* corresponding to *Cassia langsdorfii* var. *tenuis,* which occurs on Serra dos Cristais both in common pilosulous and rarer glabrate forms, is a slender erect shrublet some 5–15 dm tall with wiry naked trunks and gracefully drooping branchlets clothed in neat glaucescent foliage. The stipules are ovate-cordate and much shorter than the internodes, and the petiolar glands are shortly stipitate, trumpet-shaped in profile. An almost exactly similar except slightly larger-flowered race (*Irwin et al. 23326*) occurs on Sa. Grão Mogol, far distant in northern Minas Gerais, and this in turn very closely resembles the type-collection of *C. gracilis* var. *erythrocalyx,* obtained by Martius in the same general region. The latter, mentioned inconclusively in the last revision (Irwin, 1964, p. 68) as a probable synonym of var. *tenuis,* is now known to us through the holotypus and imposes an earlier varietal epithet.

The populations of var. *erythrocalyx* on the crest of Chapada dos Veadeiros like most of those known from Sa. do Espinhaço southward from Diamantina differ in longer or at least proportionately longer, lanceolate stipules often coinciding with a more bushy growth-form and sometimes (but not always) with sessile glands. The immediate environs of Diamantina have yielded in addition two visually arresting forms, one (*Mexia 5828*; *Irwin 2474*; *Irwin et al. 22629*) notable for relatively ample foliage, lanceolate stipules and stud-shaped glands, closely simulating in general aspect small-leaved forms (or late-season specimens) of *Ch. desvauxii*; the other (*Glaziou 19091*; *Irwin et al. 21964, 27677*) remarkable for tiny ovate stipules, minute foliage and relatively short and broad sepals. The one character common to all these is the early loss of the stipules and in consequence a framework of naked wiry atrocastaneous old stems supporting the leafy annotinous branchlets. This, too, is the one feature useful in practice to distinguish var. *erythrocalyx* from sympatric var. *parvifoliola* and its otherwise virtually identical but more southern form described as *C. chartacea* var. *tenuicaulis*.

2e. Chamaecrista ramosa (Vogel) var. **lucida** (Bentham) Irwin & Barneby, comb. nov. *C. curvifolia* var. (β) *lucida* Bentham in Martius, Fl. Brasil. **15**(2): 160. 1870.—". . . in campis pascuis montanis provinciarum Minas Geraës: Martius; Goyaz: Gardner no. 4120., Burchell; Piauhy: Gardner no. 2550 . . ."—Lectoholotypus, *Gardner 4120,* collected V.1840 (fl, fr) between Santos Domingos and Possé in e.-centr. Goiás, K! isotypi, BM, BR, NY (hb. Meisner.), S! paratypi, *Martius s.n.,* M (cited by Irwin, 1964, p. 72, as holotypus, but seen only in photograph); *Burchell*

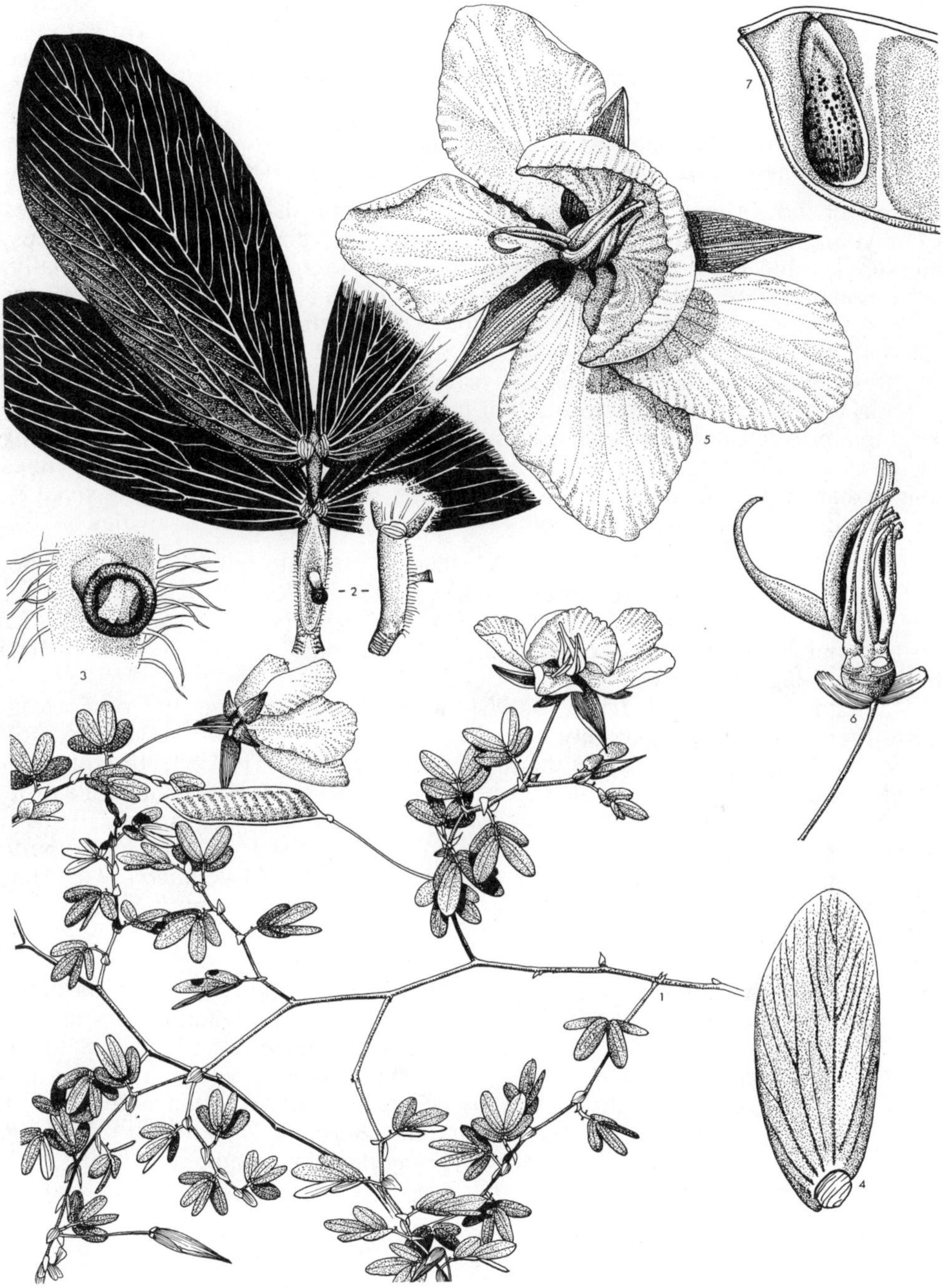

Fig. 53. *Chamaecrista* (*Xerocalyx*) *ramosa* (Vog.) var. *erythrocalyx* (Benth.) I. & B. 1) Flowering branchlet ×1; 2) leaf + petiolar gland ×5; 3) gland ×20; 4) dorsal face of leaflet ×5; 5) flower ×5; 6) androecium + pistil ×5; 7) seed ×5 (Drawn from life by W. S. Moye III).

8256, from between Natividade and Chapada, K!—*Gardner 2550* = var. *mollissima*.

Cassia uniflora var. *utiarityi* Hoehne, Comm. Linh. Telegr. Estrat. de Matto Grosso ao Amazonas **45**(5) Bot. **8**: 47, estampa 144 (lapsu sub fma. *utiarityi*). 1919.—"[Hoehne] 2075 . . . Colhida [VI.1909, fl] no Utiarity, margens do Rio Papagayo [Mato Grosso, near 58°20′W, 13°S]."—Holotypus not seen, but the figure and protologue decisive.
Cassia curvifolia var. *lucida* sensu Irwin, 1964, p. 72, exclus. pl. Tamberlik., fig. 344 (map).

Erect or diffuse, stiffly branched subshrubs at anthesis 3.5–10 dm, sometimes when young wandlike and few-branched distally, the stems commonly pilosulous, the stipules and lfts commonly glabrous ciliolate, less often pilosulous facially, exceptionally quite glabrous; stipules ovate to lance-acuminate, the larger ones 5–13 × 2.5–6 mm, mostly as long or slightly longer than the internode; lf-stalks 1.6–3.4(–4) mm, the petiole 1–2.4(–3), the rachis 0.6–1.1 mm; gland subsessile or shortly stipitate, 0.3–0.7 mm diam, in profile shortly peg- or drum-shaped, 0.3–0.8 mm tall, 0.3 mm shorter to 0.5 mm longer than diam of head; lfts obliquely semi-ovate to falcately lance-elliptic, all acute, the proximal pair ± half as long as the distal, these of larger lvs up to 8–17 × 2–4 mm, the blades ± 3–5 times longer than wide, the midrib of all ± incurved, pedicels 2–10 mm; long sepal 8–12 mm; longer petals 9–13 mm; pod 19–37 × 5–7 mm, pilosulous; ovules 8–16.

Cerrado, commonly in sandy places, 300–950 m, scattered around the n. periphery of the Brazilian Planalto on the headwaters of Rios Tapajós and Tocantins, from n.-w. Mato Grosso (Utiariti, in Chapada dos Parecis) to s.-w. Pará (Sa. do Cachimbo) and n. and e.-centr. Goiás (Araguaina s. to Possé).—Fl. I–VIII, perhaps throughout the year.

Maintained here, as by Irwin in 1964, with some misgiving, and perhaps no more than a luxuriant, relatively large-leaved state of var. *mollissima*, next following, although apparently for the most part allopatric. The collection of Tamberlik, mentioned as perhaps distinct by Irwin (1964), is described above as *Ch. desvauxii* var. *chapadicola*. The var. *utiarityi*, referred by Irwin (1964) to var. *ramosa*, otherwise unknown from Mato Grosso, appears from the phototype to have the obliquely lanceolate leaflets and short pedicels of var. *lucida*. The Martius collection from Serro Frio cited in the protologue represents a form of *Ch. ramosa* var. *parvifoliola*.

2f. Chamaecrista ramosa (Vogel) var. **mollissima** (Bentham) Irwin & Barneby, comb. nov. *C. curvifolia* var. (γ) *mollissima* Bentham in Martius, Fl. Brasil. **15**(2): 160. 1870.—". . . in campis alpestribus prope Diamantina, in Serro Frio aliisque locis prov. Minas Geraës: Martius; in via a Rio Bagagem ad Cachoeiras, prov. Goyaz: Pohl; ad Cavalcante ejusdem prov.: Burchell."—Lectoholotypus, *Burchell 7690*, K! = NY Neg. *1521*; paratypi, *Martius s.n.* ("*C. vetula* Mart. ined."), M! *Pohl 1962*, M!

Cassia curvifolia Vogel, Syn, Gen. Cass. 55. & Linnaea **11**: 704, descr. ampliat. 1837.—"In Brasilia: Sellow leg."—Holotypus, a single branchlet ("ramulum tantum vidi"), †B; no isotypus known to survive.
Cassia curvifolia sensu Bentham, 1871, p. 569, max. ex parte. *C. curvifolia* var. *curvifolia* sensu Irwin, 1964, p. 70 (exclus. typ. indicat.), fig. 344 (map).
Cassia curvifolia var. *mollissima* sensu Irwin, 1964, p. 73, fig. 344 (map).

Either erect or diffuse, repeatedly, often stiffly ramified subshrubs at anthesis 3–12(–15) dm, the annotinous branchlets and pedicels puberulent or pilosulous, the stipules, lfts, sepals and ovary varying (independently) from glabrous or glabrous ciliolate to densely minutely pilosulous, exceptionally glabrous throughout; stipules ovate-cordate-acuminate, the larger up to 2.5–5 mm, ± as long, rarely shorter than internode; lf-stalks abruptly divaricate or widely ascending 1–2 mm, the petiole 0.6–1.3 mm, the rachis 0.2–0.8 mm; gland 0.1–0.8 mm diam, sometimes highly variable from lf to lf on a branchlet, either subsessile or shortly stipitate, the stipe coarse or slender, in profile shortly pin-, peg-, drum- or inversely cone-shaped, (0.1–)0.2–0.5 mm tall, 0.3 mm longer to 0.3(–0.5) mm shorter than diam of head; lfts obliquely, dimidiately or falcately obovate or oblanceolate, the distal pair ±½–⅓ longer than the proximal, these of larger lvs 2.5–6.5(–7.5) × 1.2–2.7 mm, the midrib of all, but especially of the shorter proximal pair, incurved, the apex strongly oblique, mucronulate; pedicels 2–10(–13.5) mm; sepals 8–11.5 mm; longer petals up to 9–11.5(–13) mm; pod 26–36 × 4.5–7.5 mm; ovules 8–14.

Campo and cerrado of interior uplands and sand-flats, restinga and stony or sandy savannas of the coastal plain and lowlands, 10–925 m, interruptedly widespread over e. Brazil between 2°N and 14°S, from the lower Amazon valley downstream from Rio Trombetas to the delta region in Amapá and Pará, e. to interior Ceará, s. to coastal n. and interior Bahia, there s. along Espigão Mestre to both slopes of the Rio Corrente-Paraná divide in s.-w. Bahia and e.-centr. Goiás; reports from Minas Gerais based wholly on collections of Martius supposedly from near Diamantina and Sêrro have not been confirmed by the many modern collectors visiting this region.

The most highly reduced and xeromorphic of the microphyllous series in sect. *Xerocalyx,* notable for the minute, sometimes almost ericoid foliage which is, however, only a trifle smaller than that of some genuine var. *ramosa* but then different in the more oblique, dimidiate or falcate, obliquely mucronulate leaflets combined with shorter pedicels. The more and less densely pubescent states, corresponding with var. *curvifolia* and var. *mollissima* (sens. str.) of Bentham (1870) and Irwin (1964), which are sympatric in Ceará, Maranhão and Bahia, can no longer be recognized taxonomically. The variety passes by insensible steps eastward across the Planalto into the coarser and slightly larger- (but similarly oblique-) leaved var. *lucida,* and through this is connected by intergrades with var. *parvifoliola* and with *Ch. desvauxii* var. *langsdorfii.*

The collections of Martius, supposedly from Diamantina (Tejuco) and Sêrro Frio, cited by Irwin (1964) as typi of var. *curvifolia* and var. *mollissima,* are here rejected as such. *Cassia curvifolia* was based wholly on a unicate collection of Sellow, now lost. The var. *mollissima* was based on several well-documented collections in Bentham's own herbarium in addition to that of Martius (M). Since the locality-data of the latter are in all probability inexact, we feel it appropriate to select a new lectotypus (q.v. supra).

3. **Chamaecrista diphylla** (Linnaeus) Greene, Pittonia **4**: 28. 1899. *Cassia diphylla* Linnaeus, Sp. Pl. 376. 1753.—''*Habitat in* India.''—Holotypus, LINN *528/1*!

Cassia cultrifolia Humboldt, Bonpland & Kunth, Nov. Gen. & Sp. **6**(qu): 363. 1823.—''Crescit ad ripam fluminis Orinoci.''—Holotypus, P-HBK!—The Humboldt specimen in hb. Willd. (no. *7918*), B! is labelled Cumaná and is therefore not certainly isotypic.—*Chamaecrista cultrifolia* (Humboldt, Bonpland & Kunth) Britton & Rose ex Britton & Killip, Ann. N.Y. Acad. Sci. **35**: 183. 1936.

Cassia diphylla sensu Bentham, 1870, p. 155; 1871, p. 568; Schery, 1951, fig. 122; Irwin, 1964, p. 52 (exclus. syn. ult.), fig. 337 (map). *Chamaerista diphylla* sensu Britton & Rose, 1930, p. 273.

Cassia cultrifolia sensu Bentham, 1870, p. 156; 1871, p. 568; Irwin, 1964, p. 50, fig. 336 (map).

Monocarpic or short-lived perennial, the stem either 1 simple erect, or commonly several assurgent or radiating from root-crown, simple or few-branched, at anthesis 1.5–11 dm, the foliage glabrous, glabrous ciliolate, or puberulent, olivaceous or glaucescent; stipules of sect. *Xerocalyx* 9–25 mm, mostly as long or little shorter than internodes; petiole 3–8 mm; gland 0.4–1 mm diam, sessile or shortly stipitate; lfts 1 pair, in outline varying from obliquely obovate or suborbicular to narrowly oblanceolate, linear-elliptic, or linear, 10–42 × (3–)4–20 mm, when narrow commonly falcately arcuate or dimidiate; pedicels (12–)15–37 mm; longest sepal ±8–15 mm, the longer petals equalling it or a little shorter; pod 25–52 × 4–7.5 mm; ovules (9–)12–20.—Fig. 14 (pod, seed).

A survey of the material of *Cassia diphylla* and *C. cultrifolia* accumulated at NY before and, particularly, since the last revision (1964) brings home the truth of Bentham's remark (1870, p. 156) that these species differ only in shape and (weakly correlated) texture of the leaflets. Criteria of pubescence and thickness of leaflet margins (Irwin, 1964, p. 47) are no longer diagnostic of two entities delimited by leaf-shape. Much of broad-leaved *C. diphylla* from Cuba has ciliolate leaflets, and leaflets glabrous and ciliolate occur in otherwise identical plants in single collections (e.g. *Matuda 3793,* NY). Thickness of the marginal nerve is an inverse function of the blade's amplitude, the wider, thinner, more mesophytic type thought proper to *C. diphylla* having, as expected, a sharper, less corneous edge than that of the nrrow, xeromorphic blade of *C. cultrifolia*. A winnowing by leaf-shape, putting in one pile those with semi-obovate or obliquely obovate blades either straight or convex along the abaxial margin, and in another those with oblanceolate or lunately narrow-elliptic abaxially straight to shallowly concave blades, yields an embarrassingly large residue of intermediates, difficult to assign either way, these in the past having been referred to *C. diphylla* if prostrate or from Central America and to *C. cultrifolia* if erect or from Panama or southward. The whole range of variation in leaf-shape seems to follow a unimodal curve, on which the typus of *C. cultrifolia* is situated close to one extreme, but thc typus of *C. diphylla* only a trifle beyond the mean on the opposing side. The dispersal of the two extreme types is not identical, the broad one alone being found in the West Indies, but extending s. into lower Amazonia and sporadically south in Brazil to Minas Gerais and Mato Grosso, whereas the *cultriformis* type, dominant on the Brazilian Planalto, extends interruptedly north to Nicaragua (*Molina 14882* etc., NY) and southern Mexico (*Matuda 3793,* already cited).

The collective range of *Ch. diphylla,* including *C. cultrifolia,* may be summarized: n. periphery of the Caribbean from Cuba through Puerto Rico and Hispaniola to St. Kitts; Mexico, mostly on the Gulf slope (Veracruz to Tabasco and Chiapas) but reaching the Pacific slope in Oaxaca; Central America from Guatemala and Belize to Panama; n.-e. Colombia to Trinidad and the Guianas; middle and lower Amazon valley in Brazil and thence s. over the Planalto to s. Minas Gerais, s.-w. Goiás, and centr. Mato Grosso; Atlantic coastal plain in Rio de Janeiro; formerly introduced in Phillippine Is. (Luzon).

The synonym *C. pentandria* Larranaga, listed in the revision (Irwin, 1964, p. 52) under *C. diphylla,* is removed with some confidence to *Ch.* (sect. *Chamaecrista*) *rotundifolia,* which better fits the epithet, by being pentandrous, and also the type-locality, which is in or near the city of Rio de Janeiro and not, as surmised, in Uruguay.

BIBLIOGRAPHY

Adams, C. D. 1972. *Cassia*. *In:* Flowering plants of Jamaica 321–329.

Ali, S. I. 1971. *Cassia*. *In:* Flora of West Pakistan **3**: 53–82.

Allen, O. N. & E. K. Allen. *Cassia* sens. lat. *In:* The Leguminosae. A source book of characteristics, uses, and nodulation 140–148.

Amshoff, G. J. H. 1939. On South American Papilionaceae.

Aubréville, A. 1968. *Cassia*. *In:* Flore du Caméroun **9**: 50–70, pl. 8–12.

Batka, J. B. 1849. Ueber Senna. Bot. Zeit. **7**(11): 185–194, t. II.

———. 1866. Monographie des Cassien gruppe *Senna*.

Bentham, G. 1870. *Cassia*. *In:* Martius, Flora Brasiliensis **15**(2): 82–176.

———. 1871. Revision of the genus *Cassia*. Trans. Linn. Soc. London **27**: 503–591, t. 60–63.

Bravo, L. D. 1978a. El género *Cassia* en la Argentina, I. Serie *Aphyllae*. Darwiniana **21**(2–4): 343–391, fig. 1–15.

———. 1978b. Estudio comparativo de las plántulas de las subespecies de *Cassia aphylla* Cav. Darwiniana **21**(2–4): 393–399.

———. 1981. El genero *Cassia* en la Argentina. III. Serie *Pachycarpae*. Darwiniana **23**: 257–277, fig. 1–5.

Brenan, J. P. M. 1958. New and noteworthy *Cassias* from tropical Africa. Kew Bull. **1958**: 231–252.

———. 1967. *Cassia*. *In:* Flora of tropical East Africa, Leguminosae, pt. **2**: 47–103, fig. 10–16.

Britton, N. L. & J. N. Rose. 1930. Caesalpiniaceae Cassieae. *In:* North American Flora **23**(4–5): 227–301.

——— **& E. P. Killip.** 1936. Mimosaceae and Caesalpiniaceae of Colombia. Ann. New York Acad. Sci. **35**(3): 101–208.

Buchmann, S. 1974. Buzz pollination of *Cassia quiedondilla* (Leguminosae) by bees of the genera *Centris* and *Melipona*. Bull. S. California Acad. Sci. **73**(3): 171–173.

Burkart, A. 1952. *Cassia*. *In:* Las Leguminosas Argentinas silvestres y cultivadas, 2da ed., 162–169, fig. 29.

Capitaine, L. 1912. *Cassiées*. *In:* Etudes des Légumineuses 354–361.

Capuron, G. 1968. Notes sur quelques Cassiées Malgaches. Adansonia II, **8**(1): 17–37.

Clos, D. apud C. Gay. 1954. *Cassia*. In Historia física y política de Chile. Botánica **2**: 234–245.

Colladon, L. Th. F. 1816. Histoire naturelle et médicale des Casses.

Corby, H. D. L. 1981. The systematic value of Leguminous root nodules. *In:* Polhill & Raven, Advances in Legume Systematics **2**: 657–676.

Cronquist, A. 1981. Caesalpiniaceae. *In:* An integrated system of classification of Flowering Plants 595–598, fig. 5.16.

Datta, R. M. 1933. Aneuploidy in the genus *Cassia*. Current Sci. **1**: 364.

DeCandolle, A. P. *Cassia*. *In:* Prodromus systematis naturalis regni vegetabilis **2**: 489–507.

Delgado Salinas, A. & M. Sousa S. Biologia floral del género *Cassia* en la región de Los Tuxtlas, Veracruz. Bol. Soc. Bot. Mex. **37**: 5–45.

DeWit, H. C. D. 1955. A revision of the genus *Cassia* as occurring in Malaysia. Webbia **11**: 197–292, fig. 1–3.

Dimitri y Rial Alberti. 1954. Las especies del género *Cassia* cultivadas en la Argentina. Revista de Invest. Agric. **8**(1): 5–34.

Don, G. 1832. *Cassia, Chamaefistula, Cathartocarpus*. *In:* A general history of Dichlamydeous Plants **2**: 438–454.

Dormer, K. J. 1953. A study of phyllotaxy in the genus *Cassia*. New Phytol. **52**: 313–315.

Ducke, A. 1939. As Leguminosas da Amazonia Brasileira.

Dulberger, Rivka. 1981. The floral biology of *Cassia didymobotrya* and *C. auriculata*. Amer. J. Bot. **68**(10): 1350–1360.

Fawcett, W. & A. B. Rendle. 1920. *Cassia*. *In:* Flora of Jamaica **4**: 99–116.

Fontenelle, G. B. 1979. Contribuição ao estudo de biologia floral de *Cassia silvestris* Vell.—Considerações Anatomicas. Leandra Ano VIII–IX: 49–84.

García-Barriga, H. & E. Forero González. 1968. *Cassia*. *In:* Catálogo Ilustrado de las plantas de Cundinamarca **3**: 52–64, fig. 15–20.

Ghesquière, J. 1934. Matériaux pour l'étude des *Cassia* de la province malgache. Rev. Zool. Bot. Africaines **26**: 125–154.

Gleason, H. A. *Cassia*. *In:* The new illustrated flora of NE. United States and adjacent Canada **2**: 384–386.

Gordon-Gray, K. D. & D. P. K. Schorn. 1975. Studies in the genus *Cassia* in South Africa: I. Taxonomic notes on species of the subgenus *Lasiorhegma*, section *Chamaecrista*. J. S. Afr. Bot. **41**(3): 133–162, fig. 1–9.

Greene, E. L. 1897. The genus *Chamaecrista*. Pittonia **3**: 238–243.

Humboldt, A. von, A. Bonpland & S. Kunth. *Cassia*. *In:* Nova genera et species plantarum **6**(qu): 336–371.

Hutchinson, J. 1964. *Cassia*. *In:* The genera of Flowering Plants **1**: 246–247.

Irwin, H. S. 1964. Monographic studies in *Cassia* (Leguminosae-Caesalpinioideae). I. Section *Xerocalyx*. Mem. New York Bot. Gard. **12**(1): 1–114, fig. 1–350.

———. 1970. *Cassia*. *In:* Corell, D. S. & M. C. Johnston, Manual of the vascular plants of Texas 785–792.

——— **& R. C. Barneby.** 1975. Notes preliminary to an account of *Cassia* in the Chihuahuan Desert. Sida **6**(1): 7–18.

——— **&** ———. 1976a. Notes on the generic status of *Chamaecrista* Moench (Leguminosae Caesalpinioideae). Brittonia **28**(1): 28–36.

——— **&** ———. 1976b. Nomenclatural notes on *Cassia* Linnaeus (Leguminosae Caesalpinioideae). Brittonia **28**(4): 435–442.

——— **&** ———. 1977. Monographic studies in *Cassia* (Leguminosae Caesalpinioideae). IV. Supplementary notes on section *Apoucouita* Bentham. Brittonia **29**(3): 277–290.

——— **&** ———. 1978. Monographic studies in *Cassia* (Leguminosae Caesalpinioideae). III. Sections *Absus* and *Grimaldia*. Mem. New York Bot. Gard. **30**: 1–300, pl. 1–26.

——— **&** ———. 1981. Tribe Cassieae Bronn. *In:* Polhill & Raven, Advances in Legume Systematics, pt. **1**: 97–106.

——— **& D. J. Rogers.** 1967. Monographic studies in *Cassia* (Leguminosae Caesalpinioideae). II. A taximetric study of section *Apoucouita*. Mem. New York Bot. Gard. **16**: 71–120, fig. 1–55.

——— **& B. L. Turner.** 1960. Chromosomal relationships and taxonomic considerations in the genus *Cassia*. Amer. J. Bot. **47**(4): 309–310.

Isely, D. 1974. Leguminosae of the United States: II. Subfamily Caesalpinioideae. Mem. New York Bot. Gard. **25**(2): i–viii, 1–168.

Kearney, T. H. & R. H. Peebles. 1951. *Cassia*. *In:* Arizona Flora 405–406.

Keay, R. W. J. 1958. *Cassia*. *In:* Flora of tropical W. Africa, ed. 2, **1**(2): 450–455, fig. 147–148.

Langman, I. K. 1964. A selected guide to the literature on the flowering plants of Mexico.

Larsen, K., S. S. Larsen & J. E. Vidal. 1980. Flore du Cambodge, du Laos, et Viêt-Nam **18** (Césalpinioidiées): 1–227.

Lasseigne, A. 1979a. Studies in *Cassia* (Leguminosae Caesalpinioideae) I. New taxa and combinations in section *Chamaefistula* series *Corymbosae*. Iselya **1**(1): 7–18.

———. 1979b. Studies in *Cassia* (Leguminosae Caesalpinioideae) II. External seed morphology. Iselya **1**(2): 63–80, fig. 1–110.

———. 1979c. Studies in *Cassia* (Leguminosae Caesalpinioideae) III. Anther morphology. Iselya **1**(3): 141–160, fig. 1–85.

———. 1979d. Studies in *Cassia* (Leguminosae Caesalpinioideae) IV. The seedling, with special emphasis on sleep movement. Iselya **1**(4): 187–199.

Macbride, J. F. 1943. *Cassia*. *In:* Flora of Peru. Field Mus., Bot. **13**, pt. 3(1): 143–184.

McMinn, H. E. 1939. *Cassia*. *In:* An illustrated manual of California shrubs 228–229.

Malick, K. C. & B. Krishna. 1978. Identification of seeds of Indian Senna of commerce. Bull. Bot. Surv. India **20**(1–4): 113–116, fig. 1–8.

Martius, C. 1957. Versuch einer Mongraphie der Sennesblätter.

Muñoz Pizarro, C. 1960. Las especies de plantas descritas por R. A. Philippi en el siglo XIX.

Okpon, E. N. U. 1969a. Morphological notes on the genus *Cassia*: I. Notes Roy. Bot. Gard. Edinb. **29**(2): 185–195, fig. 1–4.

———. 1969b. Morphological notes on the genus *Cassia*: II, III. Notes Roy. Bot. Gard. Edinb. **29**(3): 331–342, fig. 1–4.

Pantulu, J. 1940. A note on the chromosome numbers of *Cassia*. Current Sci. **9**: 416–417.

———. 1946. Chromosome number in *Cassia sophera* L. Current Sci. **15**: 77.

Pardha Saradhi, P. & H. Y. Mohan Ram. 1981. Some aspects of floral biology of *Cassia fistula* Linn. (The Indian Laburnum). Current Sci. **50**(18): 804–805.

Peixoto, A. L., B. Correa de Souza & M. P. Morim. 1978. Contribução ao estudo do género Cassia L. Observações biológicas. Rodriguesia **44**: 245–252.

Pennell, F. W. 1917. The genus *Chamaecrista* in the United States. Bull. Torrey Bot. Club **44**: 339–362.

Pittier, H. 1945. *Cassia* [and segregates]. *In:* Pittier et al. Cat. Fl. Venezol. **1**: 370–376.

Polhill, R. M. & P. H. Raven. 1981. Advances in Legume Systematics.

Pullen, T. M. 1963. The *Cassia fasciculata* complex. Doctoral Thesis, University of Georgia, unpublished.

Robbins, W. J., F. Kavanagh & J. D. Thayer. 1947. Antibiotic activity of *Cassia reticulata* Willd. Bull. Torrey Bot. Club **74**(4): 287–292.

Robertson, K. R. & Yin-Tse Lee. 1976. The genera of Caesalpiniaceae (Leguminosae) in the Southeastern United States. J. Arn. Arb. **57**(1): 1–53.
Schery, R. W. 1951. *Cassia. In:* Woodson, Fl. Panama **5**(3): 39–83, fig. 117–128.
Seaforth, C. E. 1962. *Cassia.* Tropical Science **4**(3): 159–162.
Serrato Valenti, G. 1971. Adumbratio fl. aethiopicae 22. Caesalpiniaceae gen. *Cassia.* Webbia **26**(1): 1–99.
Seymour, F. C. 1973. *Cassia* in Nicaragua. Phytologia **27**(5): 330–348.
Shreve, F. & I. L. Wiggins. 1964. *Cassia* and *Chamaecrista. In:* Vegetation and flora of the Sonoran Desert 614–621.
Standley, P. C. 1922. *Cassia. In:* Trees and Shrubs of Mexico. Contrib. U.S. Nat. Herb. **23**(2): 400–411.
———. 1937. *Cassia. In:* Flora of Costa Rica. Field Mus., Bot. **18**(1): 513–519.
——— **& J. A. Steyermark.** 1946. *Cassia. In:* Flora of Guatemala. Fieldiana, Bot. **24**(5): 105–132.
Steyaert, R. L. 1950. Notes sur des *Cassia* africains et asiatiques de la section *Chaemaecrista.* Bull. Jard. Bot. de l'Etat, Bruxelles **20**(2): 233–268, pl. VIII–X, fig. 23–26.
———. 1952. *Cassia. In:* Flore du Congo Belge et du Ruanda-Urundi **3**: 496–531, fig. 44, 45, pl. XXXV–XXXVIII.
Symon, D. E. 1966. A revision of the genus *Cassia* in Australia. Trans. Roy. Soc. S. Australia **90**: 73–146.
Takhtajan, A. L. 1980. Outline of the classification of Flowering Plants. Bot. Rev. (Lancaster) **46**(3): 1–225.
Teixeira Ormond, W., M. C. Bezerra Pinheiro [Rodrigues Correia] & R. Fernandes de Oliveira. 1976. Contribução a citologia de *Cassia ramosa* Vogel. Rev. Brasil. Biol. **36**(2): 503–506.
———, ——— & ———. 1977. Contribução a citologia de *Cassia tetraphylla* Desvaux. Rev. Brasil. biol. **36**(2): 503–506.
Thorp, R. W. & J. R. Estes. 1975. Intrafloral behavior of bees on flowers of *Cassia fasciculata.* J. Kans. Entom. Soc. **48**(2): 175–184.
Turner, B. L. 1955. The *Cassia fasciculata* complex (Leguminosae) in Texas. Field & Lab. **23**: 87–91.
———. 1959. *Cassia. In:* The Legumes of Texas 71–80.
Vassil'čenko, I. T. 1965. Neoteničeskie Ismeneniya i Rastenii. Leningrad, Akad. Nauk SSSR.
Venkatesh, C. S. 1956a. The form, structure and special ways of dehiscence of anthers of *Cassia*—I. Subgenus *Fistula.* Phytomorphology **6**: 168–176.
———. 1956b. The form, structure and special ways of dehiscence of anthers of *Cassia*—II. Subgenus *Lasiorhegma.* Phytomorphology **6**: 272–277.
———. 1957. The form, structure and special ways of dehiscence of anthers of *Cassia*—III. Subgenus *Senna.* Phytomorphology **7**: 253–273.
Vogel, Th. 1837. Synopsis Generis Cassiae. Berlin: Typis Nietackianis.
Wiggins, I. L. 1980. *Cassia* and *Chamaecrista. In:* Flora of Baja California 648–650, 652.

INDEX

Genera in LARGE CAPITALS; categories between genus and species in LARGE and SMALL CAPITALS; names of all ranks in **boldface** if accepted, *italicized* if synonymous, in plain roman if of doubtful status; pairs of synonyms are linked by ≡ if homotypic, by = if heterotypic. Note abbreviations: **C.** = **Cassia**; **Ch.** = **Chamaecrista**; **S.** = **Senna**.

ADIPERA = **S.** ser. **COLUTEOIDEAE**
arborea = **S. pendula advena**
arsenei = **S. × floribunda**
bicapsularis ≡ **S. bicapsularis bicapsularis**
bicolor = **S. × floribunda**
corymbosa ≡ **S. corymbosa**
indecora = **S. pendula advena**
jahnii = **S. santanderensis**
laevigata = **S. septemtrionalis**
ovalifolia ≡ **S. pendula ovalifolia**
purpusii ≡ **S. purpusii**
santanderensis ≡ **S. santanderensis**
spiciflora = **S. bicapsularis bicapsularis**
stahlii ≡ **S. pendula stahlii**
submontana = **S. pendula ovalifolia**

BACTYRILOBIUM = **CASSIA**
bacillare = **S. bacillaris bacillaris**
ferrugineum ≡ **C. ferruginea ferruginea**
fistula ≡ **C. fistula**
grande ≡ **C. grandis**
molle = **C. grandis**

CASSIA — p. 4
subgen. FISTULA ≡ **CASSIA**
subgen. LASIORHEGMA ≡ **CHAMAECRISTA**
subgen. SENNA ≡ **SENNA**
sect. ABSUS ≡ **CH.** sect. **ABSUS**
sect. APOUCOUITA ≡ **CH.** sect. **APOUCOUITA**
sect. CALICIOPSIS ≡ **CH.** sect. **CALICIOPSIS**
sect. CHAMAECRISTA ≡ **CH.** sect. **CHAMAECRISTA**
sect. CHAMAEFISTULA ≡ **S.** ser. **CORYMBOSAE**
sect. CHAMAESENNA ≡ **S.** ser. **PICTAE**
sect. EARLEOCASSIA ≡ **S.** ser. **BRACHYCARPAE**
sect. FISTULA ≡ **CASSIA** ser. **CASSIA**
sect. GRIMALDIA = **CH.** sect. **GRIMALDIA**
sect. HERPETICA = **S.** ser. **PICTAE**
sect. LEIOCALYX ≡ **CH.** sect. **CHAMAECRISTA**
sect. ONCOLOBIUM = **S.** ser. **BASIGLANDULOSAE**
sect. PROSOSPERMA = **S.** ser. **TRIGONELLOIDEAE**
sect. PSILORHEGMA ≡ **S.** sect. **PSILORHEGMA**
sect. XEROCALYX ≡ **CH.** sect. **XEROCALYX**
subsect. ADENOPHYLLUM ≡ **CH.** subsect. **ADENOPHYLLUM**

CASSIA (*Cont.*)
subsect. BASEOPHYLLUM ≡ **CH.** subsect. **BASEOPHYLLUM**
subsect. CHAMAEFISTULA = **S.** ser. **COLUTEOIDEAE**
subsect. CHAMAESENNA = **S.** ser. **PICTAE**
subsect. OTOPHYLLUM = **CH.** subsect. **OTOPHYLLUM**
subsect. XEROCALYX ≡ **CH.** sect. **XEROCALYX**
ser. ABSOIDEAE ≡ **CH.** ser. **ABSOIDEAE**
ser. ADENOPHYLLAE ≡ **CH.** ser. **ADENOPHYLLAE**
ser. **AMAZONICAE** — p. 16
ser. ANDROMEDEAE ≡ **CH.** ser. **ANDROMEDEAE**
ser. APHYLLAE ≡ **S.** ser. **APHYLLAE**
ser. ATROGLANDULOSAE ≡ **CH.** ser. **ATROGLANDULOSAE**
ser. BACILLARES ≡ **S.** ser. **BACILLARES**
ser. BASIGLANDULOSAE ≡ **S.** ser. **BASIGLANDULOSAE**
ser. BAUHINIANAE ≡ **CH.** ser. **BAUHINIANAE**
ser. **BIFARIAE** — p. 10
ser. BRACHYCARPAE ≡ **S.** ser. **BRACHYCARPAE**
ser. BRACHYCARPAE = **S.** ser. **SENNA**
ser. BRACTEOLATAE ≡ **CH.** ser. **BRACTEOLATAE**
ser. **CASSIA** — p. 13
ser. CATHARTICAE ≡ **CH.** ser. **CATHARTICAE**
ser. CHAMAECRISTAE VERAE ≡ **CH.** ser. **CHAMAECRISTA**
ser. CONFERTAE ≡ **S.** ser. **CONFERTAE**
ser. CONFERTAE ≡ **CH.** ser. **CONFERTAE**
ser. CORIACEAE ≡ **S.** ser. **CORIACEAE**
ser. CORIACEAE ≡ **CH.** ser. **CORIACEAE**
ser. CORYMBOSAE = **S.** ser. **COLUTEOIDEAE**
ser. ERICIFOLIAE ≡ **CH.** ser. **ERICIFOLIAE**
ser. EXCELSAE ≡ **S.** ser. **EXCELSAE**
ser. FLORIDAE ≡ **S.** ser. **FLORIDAE**

CASSIA (*Cont.*)

ser. GEMINATAE ≡ **CH.** ser. **GEMINATAE**

ser. GLUTINOSAE ≡ **CH.** ser. **GLUTINOSAE**

ser. GRACILLIMAE ≡ **CH.** ser. **GRACILLIMAE**

ser. **GRANDES** — p. 28

ser. HASSLERANAE ≡ **CH.** ser. **HASSLERANAE**

ser. HEDYSAROIDES ≡ **CH.** ser. **HEDYSAROIDES**

ser. **HETEROSPERMAE** — p. 35

ser. INCANAE ≡ **CH.** ser. **INCANAE**

ser. INCURVATAE ≡ **CH.** ser. **INCURVATAE**

ser. INTERGLANDULOSAE ≡ **S.** ser. **INTERGLANDULOSAE**

ser. JUGIGLANDULOSAE = **S.** ser. **BACILLARES**

ser. LAXIFLORAE ≡ **S.** ser. **LAXIFLORAE**

ser. LOMATOPODAE ≡ **CH.** ser. **LOMATOPODAE**

ser. LUCIDAE ≡ **CH.** ser. **LUCIDAE**

ser. MICROPHYLLAE ≡ **CH.** ser. **MICROPHYLLAE**

ser. MIMOSOIDEAE = **CH.** ser. **CHAMAECRISTA**

ser. **MOSCHATAE** — p. 32

ser. NIGRICANTES ≡ **CH.** ser. **NIGRICANTES**

ser. **OBOLOSPERMAE** — p. 46

ser. OCHNACEAE ≡ **CH.** ser. **OCHNACEAE**

ser. OLIGOSPERMAE ≡ **CH.** ser. **OLIGOSPERMAE**

ser. PACHYCARPAE ≡ **S.** ser. **PACHYCARPAE**

ser. PANICULATAE ≡ **CH.** ser. **PANICULATAE**

ser. PAUCIJUGAE ≡ **CH.** ser. **PROSTRATAE**

ser. PICTAE ≡ **S.** ser. **PICTAE**

ser. PINIFOLIAE ≡ **CH.** ser. **PINIFOLIAE**

ser. PROSTRATAE ≡ **CH.** ser. **PROSTRATAE**

ser. PTERIDOPHYLLAE ≡ **CH.** ser. **PTERIDOPHYLLAE**

ser. RIGIDULAE ≡ **CH.** ser. **RIGIDULAE**

ser. ROSTRATAE = **S.** ser. **INTERGLANDULOSAE**

ser. SECUNDAE ≡ **CH.** ser. **SECUNDAE**

ser. SESBANIANAE = **CH.** ser. **CHAMAECRISTA**

ser. SETOSAE ≡ **CH.** ser. **SETOSAE**

ser. SPINULOSAE ≡ **CH.** ser. **SPINULOSAE**

ser. STRICTIFOLIAE ≡ **CH.** ser. **STRICTIFOLIAE**

ser. SUBAPHYLLAE = **CH.** ser. **BAUHINIANAE**

CASSIA (*Cont.*)

ser. SUBCORIACEAE = **CH.** ser. **CORIACEAE**

ser. **TETRAPLEURAE** — p. 36

ser. THARPIA ≡ **S.** ser. **THARPIA**

ser. TORAE = **S.** ser. **TRIGONELLOIDEAE**

ser. TRACHYCARPAE ≡ **CH.** ser. **TRACHYCARPAE**

ser. TRIGONELLOIDEAE ≡ **S.** ser. **TRIGONELLOIDEAE**

ser. UNIJUGAE ≡ **CH.** ser. **UNIJUGAE**

ser. URSINAE ≡ **CH.** ser. **URSINAE**

absus (Linn.) ≡ **Ch. absus**

meonandra = **Ch. absus meonandra**

absus (S. & M.) = **S. insularis**

acanthoclada ≡ **S. acanthoclada**

acapulcensis = **S. pallida pallida**

acinacicarpa = **S. morongii**

acisperma — p. 54

acosmifolia ≡ **Ch. acosmifolia**

euryloba ≡ **Ch. acosmifolia euryloba**

oropedii = **Ch. acosmifolia oropedii**

aculeata ≡ **S. aculeata**

acuminata (Moench) ≡ **S. marilandica**

acuminata (Willd.) ≡ **Ch. apoucouita**

acunae — p. 54

acuruensis ≡ **S. acuruensis acuruensis**

acuta = **S. cumingii cumingii**

acutifolia = **S. alexandrina**

acutisepala ≡ **S. acutisepala**

adamantina ≡ **Ch. adamantina**

adenophora ≡ **Ch. adenophora**

adenophylla ≡ **Ch. adenophylla**

adenopoda = **Ch. ensiformis ensiformis**

adenosperma ≡ **Ch. pedicellaris adenosperma**

adiantifolia ≡ **Ch. adiantifolia**

adscendens = **Ch. desvauxii desvauxii**

advena ≡ **S. pendula advena**

aeschinomene = **Ch. nictitans glabrata**

affinis ≡ **S. affinis**

aiarana ≡ **Ch. aiarana**

alata (Linn.) ≡ **S. alata**

perennis = **S. alata**

rumphiana = **S. alata**

alata (S. & M.) = **S. polyantha**

albida = **S. multiglandulosa**

alcaparillo = **S. bicapsularis augusti**

alcaparra ≡ **S. cumingii alcaparra**

alexandrina = **S. alexandrina**

altoana ≡ **Ch. altoana**

amambaya ≡ **Ch. amambaya**

amazonica = **S. spectabilis spectabilis**

amiciella ≡ **Ch. amiciella**

amphibola ≡ **Ch. amphibola**

ampliflora = **S. multijuga multijuga**

anceps ≡ **Ch. anceps**

andersonii ≡ **Ch. andersonii**

andina (Phil.) = **S. arnottiana**

andina (Rose) ≡ **S. cushina**

andreana ≡ **Ch. glandulosa andreana**

andrieuxii ≡ **S. andrieuxii**

andromedea ≡ **Ch. andromedea**

angulata ≡ **S. angulata**

CASSIA (*Cont.*)
angustifolia = **S. angulata angulata**
miscadena ≡ **S. angulata miscadena**
angustifolia = **S. alexandrina**
angustisiliqua ≡ **S. angustisiliqua angustisiliqua**
anisopetala = **S. peralteana**
annunciata = **S. reticulata**
antillana = **S. nitida**
apaënsis = **Ch. repens repens**
aphylla = **S. aphylla aphylla**
rigida ≡ **S. rigida**
robusta ≡ **S. chacoënsis**
trichosepala ≡ **S. trichosepala**
apiculata ≡ **S. apiculata**
apoucouita ≡ **Ch. apoucouita**
appendiculata = **S. australis**
applanata = **S. pistaciifolia picta**
aprica = **Ch. cathartica paucijuga**
apsidoneura ≡ **S. apsidoneura**
arborea = **S. siamea**
arborescens (P. Mill.) = **S. atomaria**
arborescens (S. & M.) — p. 54
arborescens (Vahl) = **S. sulfurea**
arduinervis = **Ch. lineata brevipila**
brevipila ≡ **Ch. lineata brevipila**
arenaria = **Ch. flexuosa**
arequipensis ≡ **S. birostris arequipensis**
argentea ≡ **S. argentea**
arida ≡ **S. arida**
aristata (Benth.) ≡ **Ch. aristata**
aristata (Walp.) = **Ch. nictitans glabrata**
aristellata = **Ch. calycioides calycioides**
aristulifera = **Ch. pascuorum**
arlindo-andradei = **Ch. desvauxii glauca**
armata ≡ **S. armata**
arnottiana ≡ **S. arnottiana**
andina = **S. arnottiana**
arrojadoana = **Ch. mucronata**
articulata = **S. villosa**
aspera ≡ **Ch. nictitans aspera**
mohrii = **Ch. nictitans nictitans**
simpsonii = **Ch. nictitans aspera**
aspleniifolia ≡ **Ch. aspleniifolia**
astrochiton ≡ **Ch. astrochiton**
astroites = **S. villosa**
atomaria ≡ **S. atomaria**
glabrata ≡ **S. mollissima glabrata**
atroglandulosa ≡ **Ch. atroglandulosa**
augusti ≡ **S. bicapsularis augusti**
aurantia ≡ **S. aurantia**
auris-zerdae ≡ **Ch. auris-zerdae**
aurita — p. 54
aurivilla ≡ **Ch. aurivilla**
australis ≡ **S. australis**
aversiflora ≡ **S. aversiflora**
axillaris (Pitt.) = **S. oxyphylla oxyphylla**
axillaris (S. & M.) = **S. fruticosa**
azulana ≡ **Ch. azulana**
babylonica = **Ch. absus absus**
bacillaris ≡ **S. bacillaris**
bacillus = **C. javanica**
bahamensis = **S. ligustrina**
latifolia ≡ **S. mexicana latifolia**
bahiae ≡ **Ch. bahiae**

CASSIA (*Cont.*)
barbata ≡ **Ch. barbata**
bartlettii = **Ch. desvauxii mollissima**
basifolia ≡ **Ch. basifolia**
aristulata = **Ch. basifolia**
bauhiniifolia ≡ **Ch. rotundifolia grandiflora**
bauhinioides ≡ **S. bauhinioides**
arizonica = **S. bauhinioides**
pilosior ≡ **S. pilosior**
belemii ≡ **Ch. belemii**
paludicola ≡ **Ch. belemii paludicola**
benitoënsis ≡ **S. benitoënsis**
benthami ≡ **Ch. benthami**
benthamiana ≡ **Ch. benthamiana**
berlandieri = **S. fruticosa**
berteri = **S. bicapsularis bicapsularis**
berteriana ≡ **S. mexicana berteriana**
bicapsularis ≡ **S. bicapsularis bicapsularis**
aristata = **S. bicapsularis bicapsularis**
augusti ≡ **S. bicapsularis augusti**
chilensis ≡ **S. candolleana**
eriocarpa ≡ **S. pendula eriocarpa**
indecora = **S. pendula advena**
pilosa = **S. pendula paludicola**
pubescens = **S. pendula advena**
quadrijuga = **S. bicapsularis bicapsularis**
rostrata ≡ **S. rostrata**
tenuifolia ≡ **S. pendula tenuifolia**
biflora (Linn.) — vide **S. pallida** p. 531
rostrata ≡ **S. rostrata**
biflora (P. Mill.) = Aeschynomene sp.
bifoliola ≡ **Ch. bifoliola**
bifoliolata = **Ch. rotundifolia rotundifolia**
pentandra = **Ch. rotundifolia rotundifolia**
rotundifolia ≡ **Ch. rotundifolia rotundifolia**
biglandulosa = **Ch. nictitans ramosa**
bijuga = **S. macranthera nervosa**
bimarginata = **S. undulata**
birostris ≡ **S. birostris birostris**
blancheti ≡ **Ch. cytisoides blancheti**
bonariensis = **S. corymbosa**
bonplandia — p. 54
bonplandiana = **C. fistula**
borbonioides = **Ch. desvauxii langsdorfii**
botteriana = **S. pendula ovalifolia**
boyanii ≡ **Ch. boyanii**
brachiata = **Ch. fasciculata**
brachyblepharis ≡ **Ch. brachyblepharis**
brachycarpa = **Ch. nictitans diffusa**
brachypoda ≡ **Ch. nictitans brachypoda**
multijuga ≡ **Ch. repens multijuga**
brachyrachis ≡ **Ch. brachyrachis**
brachystachya ≡ **Ch. cytisoides brachystachya**
bracteata = **S. alata**
bracteolata ≡ **Ch. bracteolata**
bracteosa — p. 405
bradei ≡ **Ch. linearifolia bradei**
brasiliana = **C. grandis**
ferruginea ≡ **C. ferruginea**
tomentosa = **C. grandis**
brasiliensis = **C. grandis**
brevicalyx ≡ **Ch. brevicalyx brevicalyx**
elliptica ≡ **Ch. brevicalyx elliptica**
brevipes = **Ch. desvauxii brevipes**

CASSIA (*Cont.*)
brevipila ≡ **Ch. lineata brevipila**
brittoniana = **S. hayesiana**
brongniartii ≡ **S. brongniartii**
broughtonii = **Ch. nictitans jaliscensis**
browniana (Kunth) = **S. polyantha**
browniana (Philipson) = **Ch. nictitans paraguariensis**
bucherae ≡ **Ch. bucherae**
buchii = **Ch. pedicellaris pedicellaris**
bulbotricha = **Ch. cathartica paucijuga**
burchellii ≡ **Ch. burchellii**
burkartiana ≡ **S. burkartiana**
cachimboënsis = **S. kuhlmannii**
caeciliae = **S. guatemalensis guatemalensis**
caesia ≡ **S. corifolia caesia**
caespitosa ≡ **Ch. caespitosa**
cafesiana — p. 54
cahosiana = **Ch. glandulosa picardae**
caiapo ≡ **Ch. caiapo**
calixtana ≡ **Ch. calixtana**
calliantha = **S. multijuga multijuga**
calvens = **Ch. glandulosa andicola**
calycioides ≡ **Ch. calycioides calycioides**
campanae = **S. cruckshanksii**
campicola ≡ **Ch. campicola**
camporum ≡ **Ch. campestris**
cana (Nees & Mart.) ≡ **S. cana cana**
cana (Steud.) = **S. multiglandulosa**
canca = **S. sophera**
candolleana ≡ **S. candolleana**
canescens = **S. mollissima mollissima**
caracasana = **S. hirsuta hirsuta**
caracensis ≡ **Ch. caracensis**
caribaea ≡ **Ch. caribaea caribaea**
carnaval = **S. spectabilis spectabilis**
carobinha ≡ **Ch. carobinha**
caroliniana = **S. occidentalis**
carthaginensis = **S. bacillaris bacillaris**
catapodia ≡ **Ch. catapodia**
cathartica ≡ **Ch. cathartica cathartica**
floribunda ≡ **Ch. amambaya**
paucijuga ≡ **Ch. cathartica paucijuga**
catharticoides ≡ **Ch. catharticoides**
catiarae ≡ **Ch. catiarae**
catingae ≡ **S. acuruensis catingae**
caudata ≡ **S. caudata caudata**
cavalcantina ≡ **Ch. cavalcantina**
caymanensis = **Ch. nictitans aspera**
celiae ≡ **Ch. celiae**
centiflora ≡ **Ch. centiflora**
centijuga = **S. multijuga verrucosa**
cernua ≡ **S. cernua**
chacoënsis ≡ **S. chacoënsis**
chaetoblepharis = **Ch. obtecta**
chaetostegia ≡ **Ch. chaetostegia chaetostegia**
obolaria ≡ **Ch. chaetostegia obolaria**
chamaecrista (Linn.) = **Ch. fasciculata**
brasiliensis ≡ **Ch. glandulosa brasiliensis**
flavicoma ≡ **Ch. glandulosa flavicoma**
glandulosa = **Ch. glandulosa glandulosa**
glandulosa glabrescens = **Ch. glandulosa swartzii**
normalis = **Ch. glandulosa glandulosa**
normalis diffusa = **Ch. nictitans glabrata**

CASSIA (*Cont.*)
normalis hirsuta = **Ch. fasciculata**
normalis pubicaulis = **Ch. fasciculata**
pubicaulis = **Ch. nictitans nictitans**
robusta = **Ch. fasciculata**
stricta = **Ch. glandulosa glandulosa**
chamaecrista (S. & M.) = **Ch. rufa rufa**
chamaecristoides ≡ **Ch. chamaecristoides chamaecristoides**
chapadae ≡ **Ch. chapadae**
chapmanii ≡ **S. mexicana chapmanii**
chartacea = **Ch. desvauxii langsdorfii**
tenuicaulis = **Ch. ramosa parvifoliola**
chiapensis ≡ **S. guatemalensis chiapensis**
chloroclada ≡ **S. chloroclada**
choriophylla ≡ **Ch. choriophylla choriophylla**
latifolia ≡ **Ch. choriophylla latifolia**
chrysocarpa ≡ **S. chrysocarpa**
macrocarpa = **S. tapajozensis**
psilocarpa = **S. rizzinii**
chrysoclada = **S. multijuga lindleyana**
chrysogyne = **S. chrysocarpa**
chrysoloma — p. 54
chrysophylla = **S. atomaria**
chrysosepala ≡ **Ch. chrysosepala**
chrysotingens = **Ch. setosa detonsa**
ciliata (Hoffg.) = **S. uniflora**
ciliata (Raf.) = **S. occidentalis**
ciliolata ≡ **Ch. ciliolata ciliolata**
caprina ≡ **Ch. ciliolata caprina**
pulchella ≡ **Ch. ciliolata pulchella**
cinerascens ≡ **Ch. cinerascens**
cinerea (C. & S.) = **Ch. chamaecristoides chamaecristoides**
cinerea (Colla) — p. 54
cipoana ≡ **Ch. cipoana**
clarendonensis = **S. mexicana berteriana**
clarensis = **Ch. lineata lineata**
clausseni ≡ **Ch. clausseni clausseni**
cyclophylla ≡ **Ch. clausseni cyclophylla**
megacycla ≡ **Ch. clausseni megacycla**
closiana = **S. candolleana**
cobanensis ≡ **S. cobanensis**
cochabambae = **S. morongii**
collae = **S. bicapsularis bicapsularis**
coluteoides = **S. pendula glabrata**
comayaguana = **Ch. nictitans pilosa**
compitalis ≡ **Ch. compitalis**
conferta ≡ **Ch. conferta conferta**
machrisiana ≡ **Ch. conferta machrisiana**
simulans ≡ **Ch. conferta simulans**
virgata ≡ **Ch. conferta virgata**
confinis ≡ **S. confinis**
confusa — p. 55
conjugata = **S. brongniartii**
cookii = **S. mandoni**
coquimbensis ≡ **S. cumingii coquimbensis**
cordistipula ≡ **Ch. cordistipula**
coriacea ≡ **Ch. coriacea**
corifolia = **S. corifolia corifolia**
corymbosa (Gom.-Ort.) ≡ **S. × floribunda**
corymbosa (Lamk.) ≡ **S. corymbosa**
plurijuga ≡ **S. × floribunda**
cotinifolia ≡ **Ch. cotinifolia cotinifolia**
glaberrima ≡ **Ch. cotinifolia glaberrima**

CASSIA (*Cont.*)
leptodictya ≡ **Ch. cotinifolia leptodictya**
percoriacea ≡ **Ch. cotinifolia percoriacea**
covesii ≡ **S. covesii**
cowanii — p. 24
cowanii — p. 26
guianensis — p. 25
peruviana — p. 26
craspedoneura = **Ch. ochnacea purpurascens**
crassifolia = **S. corymbosa**
crassiramea ≡ **S. crassiramea**
crassisepala = **S. pendula glabrata**
crenulata ≡ **Ch. crenulata**
crista = **S. angustisiliqua angustisiliqua**
oligophylla = **S. angustisiliqua angustisiliqua**
cristalinae ≡ **Ch. cristalinae**
crommyotricha ≡ **Ch. crommyotricha**
crotalarioides ≡ **S. crotalarioides crotalarioides**
vogeliana = **S. crotalarioides crotalarioides**
cruckshanksii ≡ **S. cruckshanksii**
cubensis — p. 55
cultrifolia = **Ch. diphylla**
cumingii ≡ **S. cumingii cumingii**
cuneata = **Ch. lineata lineata**
cuneifolia = **Ch. viscosa major**
curvifolia = **Ch. ramosa mollissima**
lucida ≡ **Ch. ramosa lucida**
mollissima ≡ **Ch. ramosa mollissima**
cushina ≡ **S. cushina**
cuspidata = (?) **S. pistaciifolia glabra**
cytisoides ≡ **Ch. cytisoides cytisoides**
blancheti ≡ **Ch. cytisoides blancheti**
brachystegia ≡ **Ch. cytisoides brachystegia**
confertiformis ≡ **Ch. cytisoides confertiformis**
cytisoides ≡ **Ch. cytisoides cytisoides**
decora ≡ **Ch. cytisoides decora**
micrantha ≡ **Ch. cytisoides micrantha**
unijuga ≡ **Ch. cytisoides unijuga**
dalbergiifolia ≡ **Ch. dalbergiifolia**
dawsonii ≡ **Ch. dawsonii**
deamii ≡ **Ch. zygophylloides deamii**
debilis ≡ **Ch. debilis**
decipiens (Desv.) — p. 55
decipiens (Grsb.) = **S. stenophylla**
decrescens ≡ **Ch. decrescens**
decumbens ≡ **Ch. decumbens**
deeringiana ≡ **Ch. deeringiana**
demissa ≡ **S. demissa demissa**
dendrina = **Ch. glandulosa flavicoma**
densiflora = **S. fruticosa**
densifolia ≡ **Ch. densifolia**
felipensis ≡ **Ch. feliciana**
dentata ≡ **Ch. dentata**
depressa = **Ch. fasciculata**
desertorum ≡ **Ch. desertorum**
rugosula ≡ **Ch. rugosula**
desvauxii ≡ **Ch. desvauxii desvauxii**
brevipes ≡ **Ch. desvauxii brevipes**
mollissima ≡ **Ch. desvauxii mollissima**
stipulacea = **Ch. desvauxii brevipes**

CASSIA (*Cont.*)
didymobotrya ≡ **S. didymobotrya**
diffusa ≡ **Ch. nictitans diffusa**
diffusissima = **Ch. pygmaea pygmaea**
diphylla ≡ **Ch. diphylla**
disadena ≡ **Ch. nictitans disadena**
discolor = **S. pallida pallida**
dispar = **S. pallida** (sens. lat.)
disperma = Peltophorum sp.
distichoclada ≡ **Ch. distichoclada**
dombeyana = **S. aurantia**
domingensis ≡ **S. domingensis**
dormiens = **S. pendula glabrata**
doylei ≡ **S. multijuga doylei**
drepanophylla = **Ch. repens multijuga**
duartei ≡ **Ch. duartei**
duckeana ≡ **Ch. duckeana**
dumalis ≡ **Ch. dumalis**
dumetorum = **S. reticulata**
durangensis ≡ **S. durangensis durangensis**
iselyi = **S. durangensis iselyi**
dysophylla = **S. velutina**
echinocarpa ≡ **Ch. echinocarpa**
edulis (Posada) = **S. spectabilis spectabilis**
edulis (S. & M.) = **S. pendula advena**
egleri ≡ **Ch. egleri**
egregia = **S. kurtzii**
eitenorum ≡ **Ch. eitenorum eitenorum**
regana ≡ **Ch. eitenorum regana**
ekmaniana = **S. racemosa racemosa**
elachistophylla ≡ **Ch. elachistophylla**
elegans (H.B.K.) = **S. septemtrionalis**
elegans (Voigt) — p. 55
elliptica (H.B.K.) = **S. atomaria**
elliptica (S. & M.) = **S. atomaria**
emarginata (Clos) = **S. candolleana**
emarginata (Linn.) = **S. bicapsularis**
unijuga = **S. atomaria**
emarginata (P. Mill.) = **Ch. pilosa pilosa**
enneandra = **Ch. zygophylloides deamii**
enneaphylla = **S. sulfurea**
enneryana = **Ch. pedicellaris pedicellaris**
ensiformis ≡ **Ch. ensiformis ensiformis**
maranonica ≡ **Ch. ensiformis maranonica**
plurifoliolata ≡ **Ch. ensiformis plurifoliolata**
erecta = **Ch. nictitans pilosa**
eremobia ≡ **S. cumingii eremobia**
ericifolia ≡ **Ch. ericifolia**
esmeraldensis ≡ **S. gundlachii**
excelsa (H.B.K.) = **C. fistula**
excelsa (Schrad.) ≡ **S. spectabilis excelsa**
acutifolia = **S. spectabilis spectabilis**
exsudans ≡ **Ch. exsudans**
xanthadena ≡ **Ch. xanthadena**
exunguis = **Ch. nictitans diffusa**
fabaginifolia = **Ch. rotundifolia rotundifolia**
fabrisii ≡ **S. fabrisii**
fagifolia — p. 55
fagonioides ≡ **Ch. fagonioides**
falcata (Dum.-Cours.) = **S. corymbosa**
falcata (Linn.) = **S. occidentalis**
falcinella = **S. quinquangulata**
fasciata = **Ch. nictitans glabrata**

CASSIA (*Cont.*)
fasciculata ≡ **Ch. fasciculata**
brachiata = **Ch. fasciculata**
depressa = **Ch. fasciculata**
ferrisiae = **Ch. fasciculata**
jenseni = **Ch. fasciculata**
macrosperma = **Ch. fasciculata**
mutata = **Ch. fasciculata**
puberula = **Ch. fasciculata**
robusta = **Ch. fasciculata**
tracyi = **Ch. fasciculata**
transmutata = **Ch. fasciculata**
fastigiata (Nees) = **S. spectabilis excelsa**
fastigiata (Vahl) = **S. surattensis**
fastuosa — p. 40
calva — p. 43
fastuosa — p. 42
feliciana ≡ **Ch. feliciana**
felipensis = **Ch. nictitans disadena**
fenixensis = **Ch. nictitans pilosa**
ferruginea — p. 38
ferruginea — p. 40
velloziana — p. 40
filicifolia ≡ **Ch. filicifolia**
filipes = **Ch. rotundifolia grandiflora**
fisheri = **Ch. fasciculata**
fistula — p. 14
fistuloides = **C. fistula**
fitchiana = **S. angustisiliqua angustisiliqua**
flaccida = **S. cumingii coquimbensis**
flavicoma ≡ **Ch. glandulosa flavicoma**
paraguariensis ≡ **Ch. nictitans paraguariensis**
flexuosa (Linn.) ≡ **Ch. flexuosa flexuosa**
cuyabensis = **Ch. flexuosa flexuosa**
pubescens = **Ch. flexuosa flexuosa**
flexuosa (P. Mill.) = **Ch. serpens serpens**
flexuosa (DC.) = **S. candolleana**
floribunda ≡ **S. × floribunda**
elegans = **S. septemtrionalis**
herbertiana = **S. × floribunda**
scandens ≡ **S. pendula scandens**
florida = **S. siamea**
fockeana = **S. bacillaris bacillaris**
foederalis = **Ch. foederalis**
foetida (G. Don) = **S. stipulacea**
foetida (Pers.) = **S. occidentalis**
foetida (S. & M.) = **S. foetidissima foetidissima**
foetidissima ≡ **S. foetidissima**
foliolosa ≡ **S. pallida foliolosa**
foliosa (G. Don) — p. 55
foliosa (Phil.) — p. 55
foliosa (Pitt.) = **S. oxyphylla oxyphylla**
fragilis ≡ **Ch. fragilis**
fraxinifolia = **S. pistaciifolia glabra**
froidensis = **Ch. pedicellaris pedicellaris**
frondosa (Ait.) = **S. angustisiliqua angustisiliqua**
crista = **S. angustisiliqua angustisiliqua**
pubescens = **S. angustisiliqua angustisiliqua**
frondosa (H. & A.) = **S. candolleana**
fructicosa — p. 55

CASSIA (*Cont.*)
frutescens = **S. sophera**
fruticescens = **S. viminea**
fruticosa ≡ **S. fruticosa**
benthamiana ≡ **S. bacillaris benthamiana**
gatunensis ≡ **S. dariensis gatunensis**
fulgens (Macfad.) ≡ **S. angustisiliqua fulgens**
fulgens (Wall.) = **S. multijuga multijuga**
fulva = **S. andrieuxii**
fuscescens ≡ **Ch. fuscescens**
galegifolia — p. 56
galeottiana ≡ **S. galeottiana**
gardneri ≡ **S. gardneri**
garrettiana ≡ **S. garrettiana**
gaudichaudii ≡ **S. gaudichaudii**
gaumeri ≡ **S. pallida gaumeri**
gemina = **Ch. nictitans paraguariensis**
geminata ≡ **Ch. geminata**
geminiflora (Collad.) = **S. pallida geminiflora**
geminiflora (Schrank) = **S. sophera**
geniculata (G. Don) = **S. villosa**
geniculata (S. & M.) = **S. villosa**
geraldi ≡ **Ch. geraldi**
gigantea = **S. siamea**
gigantifolia ≡ **S. macrophylla gigantifolia**
gilliesii ≡ **Ch. gilliesii**
glaberrima = **Ch. nictitans glabrata**
glabra = **Ch. cytisoides cytisoides**
glandulosa = **Ch. glandulosa glandulosa**
flavicoma ≡ **Ch. glandulosa flavicoma**
ramosa = **Ch. glandulosa glandulosa**
stricta = **Ch. glandulosa glandulosa**
swartzii ≡ **Ch. glandulosa swartzii**
glauca = **S. sulfurea**
koenigii = **S. surattensis**
suffruticosa = **S. surattensis**
glaucescens (Benth.) = **S. urmenetae**
glaucescens (Hoffg.) — p. 56
glaucofilix ≡ **Ch. glaucofilix**
glaziovii ≡ **Ch. glaziovii**
goldmanii = **S. polyantha**
gonoclada ≡ **Ch. gonoclada**
gooddingii = **S. hirsuta glaberrima**
goyazensis ≡ **Ch. neesiana goyazensis**
gracilis ≡ **Ch. desvauxii triumviralis**
erythrocalyx ≡ **Ch. ramosa erythrocalyx**
gracillima ≡ **Ch. benthami**
grammica = **Ch. lineata brachyloba**
grandiflora = **S. × floribunda**
grandis — p. 30
grandistipula ≡ **Ch. rotundata grandistipula**
granulata (Macbr.) = **Ch. lineata brachyloba**
granulata (Rizz.) ≡ **Ch. rizzinii**
graveolens — p. 56
greenei = **Ch. fasciculata**
greggii ≡ **Ch. greggii greggii**
grisea = **S. atomaria**
gualanensis = **S. pallida pallida**
guaranitica ≡ **S. silvestris guaranitica**
latifolia = **S. silvestris guaranitica**
guatemalensis = **S. guatemalensis guatemalensis**
gundlachii ≡ **S. gundlachii**
gymnothyrsa ≡ **Ch. gymnothyrsa**

CASSIA (*Cont.*)
haitiensis = **S. angustisiliqua angustisiliqua**
hartwegii ≡ **S. oxyphylla hartwegii**
hassleri ≡ **Ch. hassleri**
haughtii ≡ **S. haughtii**
hayesiana ≡ **S. hayesiana**
hebecarpa ≡ **S. hebecarpa**
  longipila = **S. hebecarpa**
hedysaroides ≡ **Ch. hedysaroides**
helveola = **S. birostris huancavensis**
  huancavensis ≡ **S. birostris huancavensis**
hemicardia = **S. oxyphylla oxyphylla**
herbertiana = **S. × floribunda**
herpetica = **S. alata**
herzogii ≡ **S. herzogii**
hilariana ≡ **S. hilariana**
hintoni — p. 35
hioramii = **Ch. lineata brevipila**
hirsuta ≡ **S. hirsuta hirsuta**
hirsutissima = **Ch. aurivilla**
hirta — p. 56
hispida = **Ch. hispidula**
hispidula ≡ **Ch. hispidula**
hoffmanseggii = **S. latifolia**
  gardneriana ≡ **S. georgica georgica**
holguinensis = **Ch. pedicellaris pedicellaris**
holwayana = **S. holwayana holwayana**
homophylla — p. 56
hookeriana ≡ **S. birostris hookeriana**
hornemannii — p. 56
houstoniana = **Aeschynomene** sp.
huancabambae ≡ **S. huancabambae**
huidobriana — p. 57
humboldtiana = **S. spectabilis spectabilis**
humilis (M. & G.) = **S. cobanensis**
humilis (Pers.) = **S. obtusifolia**
humilis (Steud.) ≡ **Ch. kunthiana**
huntii ≡ **Ch. huntii huntii**
  correntina ≡ **Ch. huntii correntina**
hybrida = (?) **S. × floribunda**
hymenaeifolia ≡ **Ch. hymenaeifolia**
hypnotica = **Ch. nictitans paraguariensis**
hypoleuca ≡ **S. cana hypoleuca**
ignorata = **S. pilifera pilifera**
imbricans ≡ **Ch. imbricans**
inaequilatera (Balb.) = **S. bacillaris bacillaris**
inaequilatera (Ram. Goy.) = **S. hayesiana**
inaguensis ≡ **Ch. caribaea inaguensis**
incana ≡ **Ch. incana**
incarnata ≡ **S. incarnata**
incurvata ≡ **Ch. incurvata incurvata**
  zanclodes ≡ **Ch. incurvata zanclodes**
indecora = **S. pendula advena**
  advena ≡ **S. pendula advena**
  glabrata ≡ **S. pendula glabrata**
inflata = **S. bicapsularis bicapsularis**
insignis = **S. bacillaris bacillaris**
insularis ≡ **S. insularis**
isidorea ≡ **Ch. isidorea**
itabiritoana ≡ **Ch. itabiritoana**
itaculumiensis = **Ch. andromedea**
italica ≡ **S. italica**
itambana ≡ **Ch. itambana**
itzana = **Ch. flexuosa texana**

CASSIA (*Cont.*)
ixodes ≡ **Ch. ixodes**
jacobinea ≡ **Ch. jacobinea**
jahnii = **S. santanderensis**
jalapensis = **Ch. rufa rufa**
jamaicensis ≡ **Ch. lineata jamaicensis**
javanica — p. 46
  agnes = **C. javanica indochinensis**
  indochinensis — p. 50
  javanica — p. 50
  microcalyx — p. 51
  nodosa = **C. javanica indochinensis**
  pubifolia — p. 51
juruenensis ≡ **Ch. juruenensis**
keyensis ≡ **Ch. lineata keyensis**
killippii = **Ch. fagonioides fagonioides**
kirkbridei ≡ **Ch. kirkbridei**
kuhlmannii ≡ **S. kuhlmannii**
kunthiana ≡ **Ch. kunthiana**
kurtzii ≡ **S. kurtzii**
labouriaeae ≡ **Ch. labouriaeae**
laeta = **S. mollissima mollissima**
laevigata = **S. septemtrionalis**
  scandens ≡ **S. pendula scandens**
lamprosperma ≡ **Ch. lamprosperma**
lanceolata = **Ch. desvauxii mollissima**
langsdorfii ≡ **Ch. desvauxii langsdorfii**
  longipedicellata = **Ch. desvauxii desvauxii**
  parvifoliola ≡ **Ch. ramosa parvifoliola**
  tenuis = **Ch. ramosa erythrocalyx**
larensis = **S. macranthera quadrifoliolata**
latifolia (Desf.) — p. 56
latifolia (G. F. W. Mey.) ≡ **S. latifolia**
  falcistipula = **S. latifolia**
latistipula = **Ch. desvauxii glauca**
  glauca ≡ **Ch. desvauxii glauca**
latopetiolata = **S. versicolor versicolor**
lavradiiflora ≡ **Ch. lavradiiflora**
lavradioides ≡ **Ch. lavradioides**
laxiracemosa ≡ **Ch. neesiana laxiracemosa**
lechenaultiana = **Ch. nictitans glabrata**
leiandra — p. 17
  guianensis ≡ **C. cowanii guianensis**
  peruviana ≡ **C. cowanii peruviana**
leiantha = **Ch. hispidula**
leiophylla ≡ **S. leiophylla**
  pubescens = **S. cobanensis**
lentiscifolia ≡ **Ch. lentiscifolia**
leonardii = **Ch. pedicellaris pedicellaris**
leptadenia = **Ch. nictitans mensalis**
  jaliscensis ≡ **Ch. nictitans jaliscensis**
  mensalis ≡ **Ch. nictitans mensalis**
leptocarpa ≡ **S. hirsuta leptocarpa**
  glaberrima ≡ **S. hirsuta glaberrima**
  hirsuta ≡ **S. hirsuta hirta**
leptophylla — p. 43
leucopilis ≡ **Ch. leucopilis**
liebmannii ≡ **S. racemosa liebmannii**
ligustrina (Linn.) ≡ **S. ligustrina**
  eggersiana = **S. ligustrina**
  jaegeriana = **S. ligustrina**
ligustrina (P. Mill.) = **S. italica**
limensis = **S. bicapsularis bicapsularis**
lindheimeriana ≡ **S. lindheimeriana**

CASSIA (*Cont.*)
lindleyana ≡ **S. multijuga lindleyana**
linearifolia ≡ **Ch. linearifolia linearifolia**
bradei ≡ **Ch. linearifolia bradei**
linearis = **S. sophera**
lineata ≡ **Ch. lineata lineata**
brachyloba ≡ **Ch. lineata brachyloba**
littoralis = **Ch. fasciculata**
loefgreniana = **S. tropica**
lomatopoda ≡ **Ch. lomatopoda**
longicoma = **S. argentea**
longicuspis ≡ **Ch. longicuspis**
longiglandulosa ≡ **S. longiglandulosa**
longirostrata = **S. pallida longirostrata**
longisiliqua (Blanco) = **S. hirsuta puberula**
longisiliqua (Linn. f.) = **S. atomaria**
lorentzii = **S. arnottiana**
loretensis ≡ **S. loretensis**
lotoides = **Ch. hispidula**
lucayana ≡ **Ch. caribaea lucayana**
lucens = **S. silvestris silvestris**
luetzelburgii = **Ch. venulosa**
lundii ≡ **Ch. lundii**
lutescens = **S. multiglandulosa**
macdougaliana ≡ **Ch. greggii macdougaliana**
macedoi ≡ **Ch. macedoi**
machaeriifolia ≡ **Ch. machaeriifolia**
machrisiana ≡ **Ch. conferta machrisiana**
macradena = **S. occidentalis**
macranthera ≡ **S. macranthera macranthera**
macrocarpa = **S. cernua**
macrophylla ≡ **S. macrophylla macrophylla**
macropoda = **Ch. pilosa pilosa**
madrensis = **Ch. desvauxii mollissima**
magnifica = **S. multijuga lindleyana**
malacophylla ≡ **Ch. desvauxii malacophylla**
malacotricha = **Ch. machaeriifolia**
mandoni ≡ **S. mandoni**
manzanilloana = **S. pendula advena**
maranonica ≡ **Ch. ensiformis maranonica**
marginata (Roxb.) ≡ **C. roxburghii**
marginata (S. & M.) = **S. polyantha**
marginata (Willd.) = **S. polyphylla polyphylla**
marilandica ≡ **S. marilandica**
mariquitensis — p. 206
maritima = **S. pilifera subglabra**
martiana ≡ **S. martiana**
martinicensis = **Ch. nictitans glabrata**
marylandica ≡ **S. marilandica**
mattogrossensis = **S. pilifera subglabra**
mayana = **Ch. nictitans pilosa**
maxonii = **S. hayesiana**
medellinensis = **S. pendula advena**
medica = **S. quinquangulata**
medsgeri = **S. marilandica**
megalantha = **C. javanica**
melanocarpa = **S. viminea**
mensarum ≡ **Ch. serpens mensarum**
mensicola ≡ **S. mensicola**
mexicana = **S. mexicana mexicana**
grandiflora ≡ **S. foetidissima grandiflora**
moustiquensis = **S. mexicana mexicana**
micans ≡ **S. macranthera micans**
michoacanensis = **S. atomaria**

CASSIA (*Cont.*)
microphylla (Glaz.) ≡ **Ch. harmsiana**
microphylla (Willd.) = **Ch. mimosoides**
microsenna ≡ **Ch. microsenna**
midas — p. 60
milleri = **Ch. pilosa pilosa**
mimosoides ≡ **Ch. mimosoides**
aeschinomene = **Ch. nictitans glabrata**
lechenaultiana = **Ch. nictitans glabrata**
minutiflora = **Ch. pygmaea pygmaea**
mirabilis ≡ **Ch. glandulosa mirabilis**
misera = **S. brongniartii**
mississippiensis = **Ch. fasciculata**
mohrii = **Ch. nictitans nictitans**
mollicaulis ≡ **Ch. mollicaulis**
mollifolia ≡ **Ch. mollifolia**
mollis = **C. grandis**
mollissima ≡ **S. mollissima mollissima**
monaden = **S. macranthera macranthera**
monantha ≡ **S. uniflora**
monophylla = **Ch. rotundifolia rotundifolia**
monozyx ≡ **S. monozyx**
monserratensis = **Ch. serpens wrightii**
monticola ≡ **Ch. monticola**
morelensis = **S. wislizeni pringlei**
mornicola — p. 57
morongii ≡ **S. morongii**
moschata — p. 33
mucronata ≡ **Ch. mucronata**
mucronifera ≡ **S. mucronifera**
mucronulosa = **S. uniflora**
multiflora (M. & G.) ≡ **S. holwayana holwayana**
multiflora (Scheele) — p. 57
multiflora (Vog.) = **S. macranthera macranthera**
multifoliolata ≡ **S. multifoliolata multifoliolata**
multiglandulosa ≡ **S. multiglandulosa**
multijuga ≡ **S. multijuga multijuga**
multinervia ≡ **Ch. multinervia**
multipennis ≡ **Ch. multipennis**
multipinnata = **Ch. nictitans nictitans**
nashii = **Ch. nictitans nictitans**
multiseta ≡ **Ch. multiseta**
multisiliqua — p. 57
multivenia = **Ch. glandulosa flavicoma**
mutisiana ≡ **S. mutisiana**
myrophenges ≡ **Ch. myrophenges**
myrtifolia = **S. stipulacea anglorum**
nairobensis = **S. didymobotrya**
nana ≡ **S. nana**
nanodes ≡ **Ch. nanodes**
× **nealiae** — p. 16
neesiana ≡ **Ch. neesiana neesiana**
goyazensis ≡ **Ch. neesiana goyazensis**
laxiracemosa ≡ **Ch. neesiana laxiracemosa**
subnitida ≡ **Ch. neesiana subnitida**
neglecta ≡ **S. neglecta neglecta**
acuminata ≡ **S. hirsuta acuminata**
entreriana = **S. scabriuscula**
negrensis ≡ **Ch. negrensis negrensis**
albuquerquei ≡ **Ch. negrensis albuquerquei**
nelsonii = **S. skinneri**

CASSIA (*Cont.*)

nemorosa ≡ **S. pallida nemorosa**
nervosa ≡ **S. macranthera nervosa**
nicaraguensis ≡ **S. nicaraguensis**
nicoyana = **Ch. nictitans jaliscensis**
nictitans ≡ **Ch. nictitans nictitans**
commixta = **Ch. nictitans nictitans**
hebecarpa = **Ch. nictitans nictitans**
leiocarpa = **Ch. nictitans nictitans**
multipinnata = **Ch. nictitans nictitans**
mohrii = **Ch. nictitans nictitans**
niqueroënsis = **Ch. lineata lineata**
nitida ≡ **S. nitida**
nodosa = **C. javanica indochinensis**
nuda ≡ **Ch. nuda**
nudicaulis ≡ **S. nudicaulis**
nummulariifolia ≡ **Ch. nummulariifolia**
nutans = **S. mollissima glabrata**
obcordata ≡ **Ch. obcordata**
obliqua ≡ **S. obliqua**
obliquifolia = **S. occidentalis**
oblongifolia ≡ **S. oblongifolia**
oligophylla ≡ **S. neglecta oligophylla**
obovata = **S. italica**
obtecta ≡ **Ch. obtecta**
obtusa (Clos) = **S. candolleana**
obtusa (Roxb.) = **S. italica**
obtusifolia ≡ **S. obtusifolia**
occidentalis ≡ **S. occidentalis**
aristata = **S. occidentalis**
entreriana = **S. scabriuscula**
glabra = **S. sophera**
ovato-lanceolata ≡ **S. hirsuta puberula**
scabriuscula ≡ **S. scabriuscula**
sophera ≡ **S. sophera**
ochnacea ≡ **Ch. ochnacea ochnacea**
latifolia ≡ **Ch. ochnacea latifolia**
mollis ≡ **Ch. ochnacea mollis**
purpurascens ≡ **Ch. ochnacea purpurascens**
speluncae ≡ **Ch. ochnacea speluncae**
ochrosperma ≡ **Ch. ochrosperma**
olesiphylla ≡ **Ch. olesiphylla**
oligosperma ≡ **Ch. oligosperma**
opaca = **S. stipulacea stipulacea**
oppositifolia = **Ch. lavradioides**
orbiculata ≡ **Ch. orbiculata orbiculata**
cercidifolia ≡ **Ch. orbiculata cercidifolia**
trichothyrsus ≡ **Ch. orbiculata trichothyrsus**
ustulata ≡ **Ch. orbiculata ustulata**
orcuttii ≡ **S. orcuttii**
oreades = **S. urmenetae**
orenocensis ≡ **Ch. orenocensis**
organensis ≡ **S. organensis organensis**
ornata = **S. latifolia**
ornithopoides = **S. uniflora**
otterbeinii = **Ch. glandulosa swartzii**
ovalifolia (Colla) — p. 57
ovalifolia (M. & G.) ≡ **S. pendula ovalifolia**
oxyadena — p. 57
oxyphylla ≡ **S. oxyphylla oxyphylla**
pachycalyx = **Ch. cathartica cathartica**
pachycarpa = **C. grandis**

CASSIA (*Cont.*)

pachyclada ≡ **Ch. pachyclada**
pachyphylla ≡ **Ch. tragacanthoides rasa**
pachypoda ≡ **Ch. pachyclada**
pachyrrhiza ≡ **S. pachyrrhiza**
pallens = **S. mexicana latifolia**
pallida ≡ **S. pallida pallida**
pallidifolia = **S. ruiziana ruiziana**
pallidior = **S. pallida pallida**
palmeri = **Ch. serpens wrightii**
paniculata ≡ **Ch. paniculata**
papillosa ≡ **S. papillosa papillosa**
paposana — p. 57
papulosa — p. 58
paradictyon ≡ **S. paradictyon**
paraënsis ≡ **S. paraënsis**
parahyba Vell. = **Schizolobium** sp.
paraguensis = **Ch. viscosa paraguayensis**
paramariboënsis Miqu. = **Aeschynomene** sp.
paraunana ≡ **Ch. paraunana**
parkeriana = **Ch. glandulosa swartzii**
parralensis ≡ **Ch. glandulosa parralensis**
parviflora — p. 58
parvifolia — p. 58
parvistipula ≡ **Ch. parvistipula**
pascuorum ≡ **Ch. pascuorum**
patellaria ≡ **Ch. nictitans patellaria**
glabrata ≡ **Ch. nictitans glabrata**
longifolia = **Ch. nictitans ramosa**
patellaria = **Ch. nictitans ramosa**
ramosa ≡ **Ch. nictitans ramosa**
patula = **S. sophera**
pauciflora = **Ch. hispidula**
pavoniana = **Ch. glandulosa flavicoma**
pazensis = **S. versicolor versicolor**
pearcei = **S. pistaciifolia glabra**
pedersenii = **Ch. calycioides calycioides**
pedicellaris ≡ **Ch. pedicellaris pedicellaris**
pencana — p. 58
pendula ≡ **S. pendula pendula**
bracteata = **S. cajamarcae**
eriocarpa ≡ **S. pendula eriocarpa**
hemirostrata ≡ **S. pendula hemirostrata**
pennelliana = **Ch. nictitans paraguariensis**
pentagonia ≡ **S. pentagonia pentagonia**
pentandra = **Ch. rotundifolia rotundifolia**
pentandria = **Ch. rotundifolia rotundifolia**
peralteana ≡ **S. peralteana**
persoonii = **Ch. desvauxii mollissima**
peruviana ≡ **Chamaefistula elegans**
petensis = **S. peralteana**
petropolitana = **S. sulfurea**
philippi ≡ **Ch. philippi**
phyllostachya ≡ **Ch. phyllostachya**
piauhiensis = **Ch. desvauxii latifolia**
picachensis = **Ch. flexuosa flexuosa**
picta ≡ **S. pistaciifolia picta**
pilicarpa ≡ **Ch. pilicarpa**
pilifera ≡ **S. pilifera pilifera**
pilosa = **S. pilifera pilifera**
sericea = **S. pilifera pilifera**
subglabra ≡ **S. pilifera subglabra**
pilosa ≡ **Ch. pilosa pilosa**
pilosior ≡ **S. pilosior**

CASSIA (*Cont.*)
pinetorum = **Ch. glandulosa picardae**
picardae ≡ **Ch. glandulosa picardae**
pinoi ≡ **Ch. lineata pinoi**
pirebebuiensis ≡ **Ch. desvauxii pirebebuiensis**
pistaciifolia ≡ **S. pistaciifolia pistaciifolia**
glabra ≡ **S. pistaciifolia glabra**
pittieri = **S. quinquangulata**
planaltoana ≡ **Ch. planaltoana**
planifolia ≡ **Ch. planifolia**
planisiliqua (Lamk.) — p. 58
planisiliqua (Linn.) = **S. occidentalis**
plumieri = **S. occidentalis**
pohliana ≡ **Ch. pohliana**
poiretioides = **Ch. rigidifolia rigidifolia**
polita ≡ **Ch. polita**
polyadena = **Ch. glandulosa swartzii**
polyantha ≡ **S. polyantha**
polymorpha ≡ **Ch. polymorpha**
polyphylla ≡ **S. polyphylla polyphylla**
polystachya ≡ **Ch. polystachya**
portoricensis ≡ **Ch. portoricensis portoricensis**
callosa = **Ch. portoricensis portoricensis**
granulata = **Ch. lineata brachyloba**
porturegalensis = **S. italica**
potentilla ≡ **Ch. potentilla potentilla**
potosina ≡ **Ch. greggii potosina**
praetexta ≡ **Ch. nictitans praetexta**
pringlei ≡ **S. wislizeni pringlei**
proboscidea = **S. sophera**
procumbens (Linn.) = **Ch. nictitans nictitans**
procumbens (P. Mill.) = **Ch. hispidula**
prominens = **S. macranthera macranthera**
propinqua = **Ch. nictitans pilosa**
prostrata = **Ch. serpens serpens**
psammophila = **Ch. cordistipula**
pseudociliolata ≡ **Ch. dawsonii**
psoraleopsis ≡ **Ch. psoraleopsis**
pteridophylla ≡ **Ch. adiantifolia pteridophylla**
puberula = **S. bacillaris bacillaris**
pubescens = **S. hirsuta puberula**
pubigera = **S. hirsuta hirta**
pudibunda ≡ **S. macranthera pudibunda**
pulchella = **Ch. fasciculata**
pulcherrima — p. 58
pulchra = **Ch. desvauxii mollissima**
pumila = **S. cobanensis**
pumilio ≡ **S. pumilio**
punctata (Scheele) — p. 58
punctata (Vog.) ≡ **Ch. punctata**
punctulata ≡ **Ch. punctulata**
punctulifera ≡ **Ch. punctulifera**
purpusii ≡ **S. purpusii**
pycnophylla = **Ch. filicifolia**
pygmaea ≡ **Ch. pygmaea pygmaea**
quadrifoliolata ≡ **S. macranthera quadrifoliolata**
quebracho = **S. candolleana**
quiedondilla ≡ **S. pallida quiedondilla**
quinquangulata ≡ **S. quinquangulata quinquangulata**
racemosa (P. Mill.) ≡ **S. racemosa racemosa**
tenuifolia = **S. silvestris silvestris**
racemosa (Mutis) = **S. hayesiana**

CASSIA (*Cont.*)
ramiflora = **Ch. ensiformis ensiformis**
ramosa ≡ **Ch. ramosa ramosa**
maritima = **Ch. ramosa ramosa**
ramosissima — p. 58
reflexa ≡ **S. marilandica**
× **regia** — p. 32
reniformis ≡ **S. reniformis**
repens ≡ **Ch. repens repens**
reticulata ≡ **S. reticulata**
rhonhofiana — cf. **C. swartzioides**
richardiana = **S. multijuga multijuga**
riedelii — p. 59
rigida ≡ **S. rigida**
inermis = **S. rigida**
rigidicaulis ≡ **S. rigidicaulis**
rigidifolia ≡ **Ch. rigidifolia rigidifolia**
veadeirana ≡ **Ch. rigidifolia veadeirana**
riparia = **Ch. nictitans jaliscensis**
pilosa ≡ **Ch. nictitans pilosa**
ripleyana ≡ **S. ripleyana**
robiniifolia ≡ **S. robiniifolia**
robusta = **Ch. fasciculata**
roemeriana ≡ **S. roemeriana**
roigii = **Ch. pygmaea pygmaea**
rojasiana = **S. morongii**
roncadorensis ≡ **Ch. roncadorensis**
roraimae ≡ **Ch. roraimae**
rostrata ≡ **S. rostrata**
rotundata ≡ **Ch. rotundata rotundata**
angustifolia ≡ **Ch. cinerascens**
grandistipula ≡ **Ch. rotundata grandistipula**
rotundifolia ≡ **Ch. rotundifolia rotundifolia**
bauhiniifolia = **Ch. rotundifolia grandiflora**
grandiflora (Benth.) ≡ **Ch. rotundifolia grandiflora**
grandiflora (Micheli) = **Ch. rotundifolia grandiflora**
longipedicellata = **Ch. rotundifolia grandiflora**
rovirosana = **Ch. nictitans glabrata**
rowleeana = **S. papillosa papillosa**
roxburghii — p. 51
rubriflora — p. 26
rufa ≡ **Ch. rufa rufa**
rugosa ≡ **S. rugosa**
rugosula ≡ **Ch. rugosula**
ruiziana ≡ **S. ruiziana ruiziana**
sagotiana = **C. spruceana**
salvatoris ≡ **Ch. salvatoris**
santanderensis ≡ **S. santanderensis**
canarensis ≡ **S. lasseigniana canarensis**
parvianthera ≡ **S. lasseigniana lasseigniana**
sapindifolia ≡ **S. silvestris sapindifolia**
savannarum = **Ch. pygmaea savannarum**
savannensis = **Ch. ramosa ramosa**
saxatilis ≡ **Ch. desvauxii saxatilis**
scabra ≡ **Ch. scabra**
scabrella = **S. multijuga lindleyana**
scandens ≡ **S. scandens**
scarlatina ≡ **C. swartzioides scarlatina**
schlimii = **Ch. cordistipula**

CASSIA (*Cont.*)

sclerocarpa = **S. latifolia**
scleroxyla = **S. domingensis**
scleroxylon ≡ **Ch. scleroxylon**
secedens = **S. spinescens spinescens**
secunda ≡ **Ch. secunda**
seleri = **Ch. nictitans disadena**
seleriana = **S. nicaraguensis**
selleana = **Ch. glandulosa picardae**
selloi = **S. multijuga lindleyana**
semaphora ≡ **Ch. semaphora**
semifalcata = **S. multijuga lindleyana**
semperflorens = **S. angustisiliqua angustisiliqua**
senna = **S. alexandrina**
sennoides = **S. bicapsularis bicapsularis**
sensitiva = **S. uniflora**
sepium = **S. affinis**
septemtrionalis ≡ **S. septemptrionalis**
septentrionalis (S. & M.) = **S. septemtrionalis**
septentrionalis (Zuccagni) = **S. septemtrionalis**
sericea = **S. uniflora**
serpens ≡ **Ch. serpens serpens**
  grandiflora ≡ **Ch. serpens grandiflora**
sessilis = **Ch. andromedea**
seticrenata ≡ **Ch. seticrenata**
setosa ≡ **Ch. setosa setosa**
  detonsa ≡ **Ch. setosa detonsa**
  paucivenia ≡ **Ch. setosa paucivenia**
  subsetosa ≡ **Ch. setosa subsetosa**
shaferi ≡ **S. mexicana shaferi**
siamea ≡ **S. siamea**
  puberula = **S. siamea**
siliquosa = **S. mollissima mollissima**
silvestris ≡ **S. silvestris silvestris**
  ferruginea = **S. silvestris unifaria**
  olivacea = **S. silvestris unifaria**
simplex = **Ch. nictitans paraguariensis**
simplicifolia — p. 59
simpsonii = **Ch. nictitans aspera**
simulans = **S. foetidissima grandiflora**
sincorana ≡ **Ch. sincorana**
sipoana ≡ **Ch. cipoana**
skinneri ≡ **Ch. skinneri**
smaragdina = **Ch. nictitans glabrata**
solimoesensis = **Ch. apoucouita**
sophera ≡ **S. sophera**
  ligustrinoides = **S. sophera**
sophoroides ≡ **Ch. sophoroides**
souzana ≡ **Ch. souzana**
speciosa (H.B.K.) = **S. spectabilis spectabilis**
speciosa (Schrad.) = **S. macranthera micans**
  nervosa ≡ **S. macranthera nervosa**
spectabilis ≡ **S. spectabilis spectabilis**
spiciflora = **S. bicapsularis bicapsularis**
spinescens ≡ **S. spinescens spinescens**
spiniflora ≡ **S. spiniflora**
spinigera ≡ **S. spinigera**
spinulosa ≡ **Ch. spinulosa**
splendida ≡ **S. splendida splendida**
  angustifolia = **S. splendida splendida**
  striata ≡ **S. macranthera striata**
spruceana — p. 19

CASSIA (*Cont.*)

stahlii ≡ **S. pendula stahlii**
staminea = **C. ferruginea ferruginea**
standleyi = **S. hayesiana**
stellata = **S. villosa**
stenocarpa = **Ch. nictitans disadena**
  stenocarpoides = **Ch. nictitans pilosa**
stenocarpoides = **Ch. nictitans pilosa**
stenophylla ≡ **S. stenophylla**
stenosepala = **S. foetidissima foetidissima**
stillifera ≡ **Ch. stillifera**
stipulacea (Ait.) ≡ **S. stipulacea stipulacea**
  tenuistipula = **S. cruckshanksii**
stipulacea (Willd.) = **S. stipulacea**
stipulata = **Ch. glandulosa flavicoma**
striata ≡ **S. macranthera striata**
stricta = **Ch. glandulosa glandulosa**
strictifolia ≡ **Ch. strictifolia**
strictula ≡ **Ch. strictula**
strigillosa = **Ch. pedicellaris pedicellaris**
strobilacea = **S. reticulata**
subdecrescens ≡ **Ch. subdecrescens**
subelliptica = **S. mandoni**
submontana = **S. pendula ovalifolia**
subnitida ≡ **Ch. neesiana subnitida**
subpeltata ≡ **Ch. subpeltata**
subsetosa ≡ **Ch. setosa subsetosa**
subtriflora = **Ch. repens multijuga**
subulata ≡ **S. subulata**
succedanea = **S. marilandica**
suffruticosa = **S. surattensis**
sulcata (Blanco) = **S. sulfurea**
sulcata (DC.) = **S. cernua**
sulfurea ≡ **S. sulfurea**
supplex ≡ **Ch. supplex**
surattensis ≡ **S. surattensis**
  suffruticosa = **S. surattensis**
swainsoni ≡ **Ch. swainsoni**
swartzii ≡ **Ch. glandulosa swartzii**
swartzioides — p. 20
  scarlatina — p. 22
  swartzioides — p. 22
sylvestris ≡ **S. silvestris silvestris**
tabascoënsis = **Ch. nictitans glabrata**
tagera auct. amer. ≡ **Ch. kunthiana**
tapahuascana = **S. pallida quiedondilla**
tapajozensis ≡ **S. tapajozensis**
tarantan = **S. reticulata**
tarapacana = **S. birostris arequipensis**
taubertiana = **Ch. cathartica cathartica**
tecta = **Ch. desvauxii mollissima**
tenella = **Ch. serpens grandiflora**
tenuicaulis Jones = **Aeschynomene** sp.
tenuifolia ≡ **S. tenuifolia**
tenuisepala ≡ **Ch. tenuisepala**
tenuissima (Linn.) — vide **S. pallida** p. 531
tenuissima (S. & M.) = **S. polyphylla polyphylla**
tenuivenosa = **Ch. rotundifolia rotundifolia**
tephrosiifolia ≡ **Ch. tephrosiifolia**
tetrafoliata — p. 59
tetraphylla (Desv.) ≡ **Ch. desvauxii desvauxii**
  aurivilla ≡ **Ch. desvauxii brevipes**
  brevipes ≡ **Ch. desvauxii brevipes**

CASSIA (*Cont.*)
colombiana = **Ch. desvauxii mollissima**
linearis ≡ **Ch. desvauxii linearis**
littoralis = **Ch. desvauxii desvauxii**
longifolia = **Ch. desvauxii mollissima**
mollissima ≡ **Ch. desvauxii mollissima**
ramosa ≡ **Ch. ramosa ramosa**
saxatilis ≡ **Ch. desvauxii saxatilis**
ventuarensis ≡ **Ch. ramosa ventuarensis**
texana ≡ **Ch. flexuosa texana**
thonningii = **Ch. absus absus**
timoriensis ≡ **S. timoriensis**
tomentosa (Linn. f.) = **S. multiglandulosa**
albida = **S. multiglandulosa**
tomentosa (Wall.) = **S. hirsuta hirsuta**
tonduzii ≡ **S. tonduzii**
tora auct. amer. = **S. obtusifolia**
hirsuta = **S. pilifera pilifera**
humilis = **S. obtusifolia**
obtusifolia ≡ **S. obtusifolia**
toroides = **S. obtusifolia**
tortuënsis = **Ch. caribaea inaguensis**
tortuosa = **S. pallida trichocraspedon**
totonaca = **S. spectabilis spectabilis**
trachycarpa ≡ **Ch. trachycarpa trachycarpa**
acutifolia ≡ **Ch. trachycarpa acutifolia**
trachyclada = **Ch. viscosa major**
trachypus ≡ **S. trachypus**
tragacanthoides ≡ **Ch. tragacanthoides tragacanthoides**
trichocraspedon ≡ **S. pallida trichocraspedon**
trichoneura = **S. skinneri**
trichosepala ≡ **S. trichosepala**
trichopoda ≡ **Ch. trichopoda**
luxurians ≡ **Ch. pilosa luxurians**
trichothyrsus ≡ **Ch. orbiculata trichothyrsus**
triflora (Jacqu.) = **Ch. fasciculata**
triflora (Vahl) = **S. atomaria**
trinitatis = **S. spectabilis spectabilis**
tristicula ≡ **Ch. glandulosa tristicula**
tropica ≡ **S. tropica**
tuberculata — p. 59
tuerckheimii = **Ch. portoricensis portoricensis**
turquinae = **S. ligustrina**
unica = **S. quinquangulata**
uniflora (P. Mill.) ≡ **S. uniflora**
uniflora (Spreng.) — p. 59
glaucescens = **Ch. desvauxii desvauxii**
latifolia ≡ **Ch. desvauxii latifolia**
unijuga ≡ **S. unijuga**
urmenetae ≡ **S. urmenetae**
urophyllidia ≡ **Ch. urophyllidia**
ursina ≡ **Ch. ursina**
vargasii ≡ **S. vargasii**
vauthieri ≡ **Ch. vauthieri**
velutina (Gar.-Bar. & For.) ≡ **S. viarum**
velutina (Vog.) ≡ **S. velutina**
venatoria ≡ **Ch. venatoria**
venenifera = **S. hirsuta hirsuta**
venosa (Desv.) = **Ch. cytisoides cytisoides**
venosa (Hornem.) — p. 59
venosa (Zuccagni) = **Ch. fasciculata**
venulosa ≡ **Ch. venulosa**
venustula = **S. pallida pallida**

CASSIA (*Cont.*)
verbenensis = **S. tonduzii**
verdickii = **S. didymobotrya**
vernicifolia = **Ch. hedysaroides**
vernicosa = **S. septemtrionalis**
verrucosa ≡ **S. multijuga verrucosa**
versicolor ≡ **S. versicolor versicolor**
vestita ≡ **Ch. vestita**
viarum ≡ **S. viarum**
viciifolia ≡ **S. viciifolia**
villosa ≡ **S. villosa**
villosissima = **Ch. nictitans ramosa**
viminea ≡ **S. viminea**
virgata (A. Rich.) = **S. chrysocarpa**
virgata (Sw.) = **Ch. glandulosa glandulosa**
virginis ≡ **Ch. virginis**
viscosa ≡ **Ch. viscosa viscosa**
major ≡ **Ch. viscosa major**
paraguayensis ≡ **Ch. viscosa paraguayensis**
viscoso-pilosa = **Ch. viscosa major**
vogeliana = **S. crotalarioides**
weberbaueri = **S. mollissima mollissima**
wightiana = **S. multiglandulosa**
williamsii ≡ **S. williamsii**
wilsonii = **Ch. nictitans jaliscensis**
wislizeni ≡ **S. wislizeni wislizeni**
painteri ≡ **S. wislizeni painteri**
villosa ≡ **S. wislizeni villosa**
wrightii ≡ **Ch. serpens wrightii**
xanthadena ≡ **Ch. xanthadena**
xanthophylla = **S. mollissima glabrata**
xinguënsis ≡ **Ch. xinguënsis**
xiphoidea = **Ch. pallida pallida**
yucatana = **Ch. glandulosa flavicoma**
yucatensis = **S. pallida gaumeri**
zygophylloides ≡ **Ch. zygophylloides zygophylloides**
caribaea ≡ **Ch. zygophylloides caribaea**
colligans ≡ **Ch. zygophylloides colligans**
deamii ≡ **Ch. zygophylloides deamii**
CASSIEAE (tribus) — p. 1
CASSIINAE (subtribus) — p. 2
CATHARTOCARPUS = **CASSIA**
bacillaris ≡ **S. bacillaris bacillaris**
bacillus = **S. bacillaris bacillaris**
bicapsularis ≡ **S. bicapsularis bicapsularis**
brasilianus = **C. grandis**
erubescens = **C. grandis**
excelsus = **C. fistula**
ferrugineus ≡ **C. ferruginea**
fistula ≡ **C. fistula**
fistuloides = **C. fistula**
grandis ≡ **C. grandis**
humboldtianus = **S. spectabilis spectabilis**
javanicus ≡ **C. javanica javanica**
marginatus ≡ **C. roxburghii**
moschatus ≡ **C. moschata**
nodosus = **C. javanica indochinensis**
roxburghii ≡ **C. roxburghii**
speciosus = **S. spectabilis spectabilis**
trinitatis = **S. spectabilis spectabilis**
CHAMAECASSIA = S. ser. COLUTEOIDEAE
laevigata = **S. septemtrionalis**

CHAMAECRISTA — p. 636
sect. **Absus** — p. 644
sect. **Apoucouita** — p. 638
sect. **Caliciopsis** — p. 857
sect. **Chamaecrista** — p. 664
sect. **Grimaldia** — p. 664
sect. **Xerocalyx** — p. 862
subsect. **Absus** — p. 647
subsect. **Adenophyllum** — p. 647
subsect. **Baseophyllum** — p. 646
subsect. **Otophyllum** — p. 647
ser. **Absoideae** — p. 660
ser. **Adenophyllae** — p. 655
ser. **Andromedeae** — p. 651
ser. **Apoucouita** — p. 641
ser. **Atroglandulosae** — p. 648
ser. **Bauhinianae** — p. 727
ser. **Bracteolatae** — p. 650
ser. **Catharticae** — p. 647
ser. **Chamaecrista** — p. 738
ser. Chiapenses = **Ch.** ser. **Chamaecrista**
ser. **Confertae** — p. 658
ser. **Coriaceae** — p. 667
ser. Dimidiatae = **Ch.** ser. **Chamaecrista**
ser. **Ericifoliae** — p. 660
ser. Fasciculatae = **Ch.** ser. **Chamaecrista**
ser. **Flexuosae** — p. 695
ser. **Geminatae** — p. 651
ser. Glandulosae = **Ch.** ser. **Chamaecrista**
ser. **Glutinosae** — p. 658
ser. **Gracillimae** — p. 653
ser. **Greggianae** — p. 735
ser. **Hassleranae** — p. 656
ser. **Hedysaroides** — p. 656
ser. **Incanae** — p. 648
ser. **Incurvatae** — p. 659
ser. Lineatae ≡ **Ch.** ser. **Chamaecrista**
ser. **Lomatopodae** — p. 651
ser. **Lucidae** — p. 656
ser. **Microphyllae** — p. 651
ser. **Nigricantes** — p. 648
ser. **Ochnaceae** — p. 656
ser. **Oligospermae** — p. 648
ser. **Paniculatae** — p. 655
ser. Pilosae = **Ch.** ser. **Prostratae**
ser. **Pinifoliae** — p. 650
ser. **Prostratae** — p. 706
ser. **Pteridophyllae** — p. 641
ser. Pygmaeae = **Ch.** ser. **Chamaecrista**
ser. **Rigidulae** — p. 654
ser. Rotundifoliae = **Ch.** ser. **Bauhinianae**
ser. **Secundae** — p. 651
ser. **Setosae** — p. 650
ser. **Spinulosae** — p. 659
ser. **Strictifoliae** — p. 659

CHAMAECRISTA (*Cont.*)
ser. Tabascoënses = **Ch.** ser. **Chamaecrista**
ser. Tagera = **Ch.** ser. **Prostratae**
ser. **Trachycarpae** — p. 648
ser. Tristiculae = **Ch.** ser. **Chamaecrista**
ser. **Unijugae** — p. 659
ser. **Ursinae** — p. 650
ser. Yucatanae = **Ch.** ser. **Chamaecrista**
absus — p. 664
absus — p. 664
meonandra — p. 664
acosmifolia — p. 660
acosmifolia — p. 660
euryloba — p. 660
oropedii — p. 660
adamantina — p. 650
adenophora — p. 653
adenophylla — p. 656
adenosperma ≡ **Ch. pedicellaris adenosperma**
adiantifolia — p. 641
adiantifolia — p. 641
pteridophylla — 641
aeschinomene = **Ch. nictitans glabrata**
aiarana — p. 642
alamosensis = **Ch. serpens wrightii**
altoana — p. 654
amambaya — p. 656
amiciella — p. 661
amphibola — p. 651
amplistipulata = Ch. flexuosa flexuosa
anceps — p. 679
andersonii — p. 660
andreana ≡ **Ch. glandulosa andreana**
andromedea — p. 651
apoucouita — p. 642
arborescens = **Ch. glandulosa flavicoma**
arduinervis = **Ch. lineata brevipila**
aristata — p. 683
aristellata = **Ch. calycioides calycioides**
aspera ≡ **Ch. nictitans aspera**
commixta = **Ch. nictitans nictitans**
mohrii = **Ch. nictitans nictitans**
aspidiifolia — p. 642
aspleniifolia — p. 641
astrochiton — p. 650
atroglandulosa — p. 648
auris-zerdae — p. 651
aurivilla — p. 650
azulana — p. 655
bahiae — p. 642
baileyorum = **Ch. glandulosa brasiliensis**
barbata — p. 660
basifolia — p. 733
bauhiniifolia = **Ch. rotundifolia grandiflora**
belemii — p. 660
bellula = **Ch. fasciculata**
benthami — p. 654
benthamiana — p. 654
bifoliola — p. 658
bifoliolata = **Ch. rotundifolia rotundifolia**
boyanii — p. 642

CHAMAECRISTA (*Cont.*)
brachiata = **Ch. fasciculata**
brachyblepharis — p. 655
brachycarpa = **Ch. nictitans diffusa**
brachyrachis — p. 654
bracteolata — p. 650
brandegeei ≡ **Ch. chamaecristoides brandegeei**
brevicalyx — p. 660
  elliptica — p. 660
brevipes ≡ **Ch. desvauxii brevipes**
browniana = **Ch. nictitans paraguariensis**
bucherae — p. 647
buchii = **Ch. pedicellaris pedicellaris**
burchellii — p. 690
caespitosa — p. 658
cahosiana = **Ch. glandulosa picardae**
caiapo — p. 656
calixtana — p. 648
calycioides — p. 858
  calycioides — p. 859
  lentiformis — p. 861
campestris — p. 661
campicola — p. 651
camporum = Ch. fasciculata
caracensis — p. 659
cardiostegia — p. 678
caribaea — p. 680
  caribaea — p. 682
  inaguensis — p. 682
  lucayana — p. 682
carobinha — p. 661
catapodia — p. 659
cathartica — p. 647
  cathartica — p. 647
  paucijuga — p. 647
catharticoides — p. 647
catiarae — p. 658
cavalcantina — p. 648
celiae — p. 655
centiflora — p. 650
chaetostegia — p. 654
  chaetostegia — p. 654
  obolaria — p. 654
chamaecrista = Ch. fasciculata
chamaecristoides — p. 769
  brandegeei — p. 773
  chamaecristoides — p. 772
  cruziana — p. 773
chapadae — p. 661
chiapensis = **Ch. serpens wrightii**
chiriquensis = **Ch. nictitans disadena**
choriophylla — p. 691
  choriophylla — p. 693
  latifolia — p. 693
  rossicorum — p. 693
chrysosepala — p. 656
ciliolata — p. 654
  caprina — p. 654
  ciliolata — p. 654
  pulchella — p. 654
cinerascens — p. 672
cinerea = **Ch. chamaecristoides chamaecristoides**

CHAMAECRISTA (*Cont.*)
cipoana — p. 654
clarensis = **Ch. lineata lineata**
clausseni — p. 655
  clausseni — p. 655
  cyclophylla — p. 655
  megacycla — p. 655
comayaguana = **Ch. nictitans pilosa**
compitalis — p. 644
complexa = **Ch. glandulosa swartzii**
conferta — p. 658
  conferta — p. 658
  machrisiana — p. 658
  simulans — p. 658
  virgata — p. 658
confusa = **Ch. nictitans aspera**
conzattii = **Ch. nictitans pilosa**
cordistipula — p. 721
coriacea — p. 647
cotinifolia — p. 658
  coriacea — p. 658
  cotinifolia — p. 658
  glaberrima — p. 658
  leptodictya — p. 658
cowellii = **Ch. glandulosa glandulosa**
crenulata — p. 657
cristalinae — p. 653
crommyotricha — p. 658
cruziana ≡ **Ch. chamaecristoides cruziana**
cultrifolia = **Ch. diphylla**
cuneata = **Ch. lineata lineata**
cuprea — p. 749
cytisoides — p. 647
  blanchetii — p. 647
  brachystegia — p. 647
  confertiformis — p. 647
  cytisoides — p. 647
  decora — p. 647
  micrantha — p. 647
  unijuga — p. 647
dalbergiifolia — p. 653
dawsonii — p. 655
debilis — p. 647
decrescens — p. 653
decumbens — p. 654
deeringiana — p. 801
delicata ≡ **Ch. serpens delicata**
dendrina = **Ch. glandulosa flavicoma**
densifolia — p. 654
dentata — p. 658
depressa = **Ch. fasciculata**
desertorum — p. 657
desvauxii — p. 865
  brevipes — p. 876
  chapadicola — p. 883
  circumdata — p. 871
  desvauxii — p. 872
  glauca — p. 867
  graminea — p. 878
  langsdorfii — p. 879
  latifolia — p. 873
  linearis — p. 878
  malacophylla — p. 874
  modesta — p. 882

CHAMAECRISTA (*Cont.*)
mollissima — p. 868
peronadenia — p. 868
piptostegia — p. 874
pirebebuiensis — p. 881
saxatilis — p. 875
triumviralis — p. 876
didyma — p. 651
diffusa ≡ **Ch. nictitans diffusa**
diffusissima = **Ch. pygmaea pygmaea**
diphylla — p. 891
distichoclada — p. 674
duartei — p. 642
duckeana — p. 861
dumalis — p. 653
dussii = **Ch. nictitans glabrata**
echinocarpa — p. 659
egleri — p. 660
eitenorum — p. 642
eitenorum — p. 642
regana — p. 642
elachistophylla — p. 653
enneryana = **Ch. pedicellaris pedicellaris**
ensiformis — p. 642
ensiformis — p. 642
maranonica — p. 642
plurifoliolata — p. 642
ericifolia — p. 660
exsudans — p. 650
exunguis = **Ch. nictitans diffusa**
fagonioides — p. 661
fagonioides — p. 661
macrocalyx — p. 661
fasciata = **Ch. nictitans glabrata**
fasciculata — p. 802
robusta = **Ch. fasciculata**
feliciana — p. 654
fenixensis = **Ch. nictitans pilosa**
ferrisiae = **Ch. fasciculata**
filicifolia — p. 655
flavicoma ≡ **Ch. glandulosa flavicoma**
flexuosa — p. 696
flexuosa — p. 698
texana — p. 700
fodinarum — p. 661
foederalis — p. 653
fragilis — p. 658
fuscescens — p. 650
geminata — p. 651
geraldi — p. 657
gilliesii — p. 648
glaberrima = **Ch. nictitans glabrata**
glandulosa — p. 780
andicola — p. 791
andreana — p. 796
balsasana — p. 794
brasiliensis — p. 798
crystallina — p. 795
flavicoma — p. 788
glandulosa — p. 782
mirabilis — p. 787
parralensis — p. 797
picardae — p. 783
ramosa = **Ch. glandulosa glandulosa**

CHAMAECRISTA (*Cont.*)
rapidarum — p. 794
swartzii — p. 784
tristicula — p. 793
glaucofilix — p. 655
glaziovii — p. 650
glischrodes — p. 648
goldmanii ≡ **Ch. glandulosa parralensis**
gonoclada — p. 702
gracilis ≡ **Ch. desvauxii triumviralis**
grammica = **Ch. lineata brachyloba**
granulata = **Ch. lineata brachyloba**
greggii — p. 735
greggii — p. 737
macdougaliana — p. 737
potosina — p. 737
gymnothyrsa — p. 655
haitiensis = **Ch. pedicellaris pedicellaris**
harmsiana — p. 653
hassleri — p. 656
hatschbachii — p. 656
hedysaroides — p. 656
hioramii = **Ch. lineata brachyloba**
hispidula — p. 661
huntii — p. 653
correntina — p. 653
huntii — p. 653
hymenaeifolia — p. 641
imbricans — p. 653
inaguensis ≡ **Ch. caribaea inaguensis**
incana — p. 648
incurvata — p. 659
incurvata — p. 659
zanclodes — p. 659
isidorea — p. 653
itabiritoana — p. 656
itambana — p. 649
ixodes — p. 650
jacobinea — p. 661
jalapensis = **Ch. rufa rufa**
jamaicensis ≡ **Ch. lineata jamaicensis**
juruenensis — p. 660
keyensis ≡ **Ch. lineata keyensis**
kirkbridei — p. 659
kunthiana — p. 724
labouriaeae — p. 656
lagotois — p. 694
lamprosperma — p. 656
langsdorfii ≡ **Ch. desvauxii langsdorfii**
lavradiiflora — p. 658
lavradioides — p. 659
lehmannii = **Ch. desvauxii langsdorfii**
lentiscifolia — p. 658
leonardae = **Ch. pedicellaris pedicellaris**
leptadenia = **Ch. nictitans mensalis**
leucopilis — p. 650
linearifolia — p. 659
bradei — p. 659
linearifolia — p. 659
lineata — p. 752
brachyloba — p. 757
brevipila — p. 758
jamaicensis — p. 757
keyensis — p. 756

CHAMAECRISTA (*Cont.*)
lineata — p. 754
pinoi — p. 761
littoralis = **Ch. fasciculata**
lomatopoda — p. 651
longicoma = **Ch. nictitans jaliscensis**
longicuspis — p. 660
lucayana ≡ **Ch. caribaea lucayana**
lucesiae = **Ch. nictitans praetexta**
lundii — p. 655
macdougaliana ≡ **Ch. greggii macdougaliana**
macedoi — p. 654
machaeriifolia — p. 649
macropoda = **Ch. pilosa pilosa**
madrensis = **Ch. desvauxii mollissima**
martinicensis = **Ch. nictitans glabrata**
mazatlensis = **Ch. serpens wrightii**
mexiae = **Ch. serpens wrightii**
micrantha = **Ch. pygmaea pygmaea**
microsenna — p. 647
millspaughii = **Ch. nictitans glabrata**
mirabilis ≡ **Ch. glandulosa mirabilis**
mississippiensis = **Ch. fasciculata**
mohrii = **Ch. nictitans nictitans**
mollicaulis — p. 654
monticola — p. 659
mucronata — p. 684
multinervia — p. 685
multipennis — p. 654
multipinnata = **Ch. nictitans nictitans**
multiseta — p. 660
multivenia = **Ch. glandulosa flavicoma**
myrophenges — p. 659
nanodes — p. 654
neesiana — p. 653
goyazensis — p. 653
laxiracemosa — p. 653
neesiana — p. 653
subnitida — p. 653
negrensis — p. 642
albuquerquei — p. 642
negrensis — p. 642
nelsonii = **Ch. nictitans praetexta**
nicoyana = **Ch. nictitans jaliscensis**
nictitans — p. 811
aspera — p. 838
brachypoda — p. 812
diffusa — p. 833
disadena — p. 825, 826
glabrata — p. 822
jaliscensis — p. 834
mensalis — p. 837
nictitans — p. 832, 840
paraguariensis — p. 815
patellaria — p. 814
pilosa — p. 829
praetexta — p. 821
ramosa — p. 818
niqueroënsis = **Ch. lineata lineata**
nuda — p. 653
nummulariifolia — p. 654
obcordata — p. 762
obtecta — p. 651
ochnacea — p. 656

CHAMAECRISTA (*Cont.*)
latifolia — p. 656
mollis — p. 656
ochnacea — p. 656
purpurascens — p. 656
speluncae — p. 656
ochrosperma — p. 651
olesiphylla — p. 670
oligosperma — p. 648
onusta — p. 642
orbiculata — p. 655
cercidifolia — p. 655
orbiculata — p. 655
trichothyrsus — p. 655
ustulata — p. 655
orenocensis — p. 651
ortegae = **Ch. serpens wrightii**
pachyclada — p. 655
palmeri = **Ch. serpens wrightii**
paniculata — p. 650
papillata — p. 688
paraunana — p. 660
parvistipula — p. 704
pascuorum — p. 747
patellaria = **Ch. nictitans ramosa**
pedemontana — p. 799
pedicellaris — p. 773
adenosperma — p. 776
pedicellaris — p. 775
pennellii = **Ch. glandulosa andreana**
persoonii = **Ch. desvauxii mollissima**
philippi — p. 648
phyllostachya — p. 650
pilicarpa — p. 648
pilosa — p. 718
luxurians — p. 720
pilosa — p. 719
pinetorum = **Ch. glandulosa picardae**
pinoi ≡ **Ch. lineata pinoi**
planaltoana — p. 654
planifolia — p. 659
pohliana — p. 653
polita — p. 654
polyadena = **Ch. glandulosa swartzii**
polymorpha — p. 653
polystachya — p. 641
portoricensis — p. 763
atlantica — p. 765
granulata = **Ch. lineata brachyloba**
portoricensis — p. 765
potentilla — p. 673
potentilla — p. 674
specuum — p. 674
potosina ≡ **Ch. greggii potosina**
procumbens = **Ch. nictitans nictitans**
psoraleopsis — p. 653
puberula = **Ch. fasciculata**
pulchra = **Ch. desvauxii mollissima**
punctata — p. 657
punctulata — p. 661
punctulifera — p. 658
pygmaea — p. 776
pygmaea — p. 778
savannarum — p. 779

CHAMAECRISTA (*Cont.*)
ramosa — p. 884
erythrocalyx — p. 887
lucida — p. 888
mollissima — p. 890
parvifoliola — p. 886
ramosa — p. 884
ventuarensis — p. 885
ramulosa = **Ch. ramosa ramosa**
rekoi = **Ch. nictitans pilosa**
repens — p. 742
multijuga — p. 745
repens — p. 744
rigidifolia — p. 655
rigidifolia — p. 655
veadeirana — p. 655
riparia = **Ch. nictitans jaliscensis**
robusta = **Ch. fasciculata**
roigii = **Ch. pygmaea pygmaea**
roncadorensis — p. 661
roraimae — p. 669
rostrata = **Ch. fasciculata**
rotundata — p. 675
grandistipula — p. 676
interstes — p. 677
rotundata — p. 677
rotundifolia — p. 727
grandiflora — p. 732
rotundifolia — p. 731
rovirosana = **Ch. nictitans glabrata**
rufa — p. 765
exsul — p. 768
polyphlebia — p. 769
rufa — p. 767
rugosula — p. 660
rusbyi = **Ch. ramosa ramosa**
salvadorensis = **Ch. nictitans pilosa**
salvatoris — p. 660
savannarum ≡ **Ch. pygmaea savannarum**
scabra — p. 651
schlimii = **Ch. cordistipula**
scleroxylon — p. 642
secunda — p. 651
seleri = **Ch. nictitans disadena**
selleana = **Ch. glandulosa picardae**
semaphora — p. 659
serpens — p. 708
delicata — p. 713
grandiflora — p. 712
isthmogenes — p. 715
mensarum — p. 713
oaxacana — p. 715
serpens — p. 711
wrightii — p. 714
seticrenata — p. 656
setosa — p. 650
detonsa — p. 651
paucivenia — p. 651
setosa — p. 650
subsetosa — p. 651
simplex = **Ch. nictitans paraguariensis**
simplifacta — p. 694
simpsonii = **Ch. nictitans aspera**
sincorana — p. 653

CHAMAECRISTA (*Cont.*)
smaragdina = **Ch. nictitans glabrata**
sophoroides — p. 648
souzana — p. 661
spinulosa — p. 659
stenocarpa = **Ch. nictitans disadena**
stenocarpoides = **Ch. nictitans pilosa**
stillifera — p. 658
strictifolia — p. 659
strictula — p. 654
strigillosa = **Ch. pedicellaris pedicellaris**
subdecrescens — p. 653
subpeltata — p. 642
supplex — p. 722
swainsoni — p. 701
swartzii ≡ **Ch. glandulosa swartzii**
tabascoënsis = **Ch. nictitans glabrata**
tagera (auct.) = **Ch. kunthiana**
tenella = **Ch. serpens grandiflora**
tenuisepala — p. 707
tephrosiifolia — p. 648
tetraphylla ≡ **Ch. desvauxii**
texana ≡ **Ch. flexuosa texana**
tonduzii = **Ch. nictitans pilosa**
tortuënsis = **Ch. caribaea inaguensis**
trachycarpa — p. 648
acutifolia — p. 648
trachycarpa — p. 648
tracyi = **Ch. fasciculata**
tragacanthoides — p. 686
rasa — p. 688
tragacanthoides — p. 687
trichopoda — p. 716
tristicula ≡ **Ch. glandulosa tristicula**
tuerckheimii = **Ch. portoricensis portoricensis**
ulmea — p. 689
uniflora = **Ch. desvauxii**
urophyllidia — p. 648
ursina — p. 650
vauthieri — p. 657
venatoria — p. 648
venturiana — p. 746
venulosa — p. 671
vestita — p. 750
villosissima = **Ch. nictitans ramosa**
virgata = **Ch. glandulosa glandulosa**
virginis — p. 650
viscosa — p. 661
major — p. 661
paraguayensis — p. 661
viscosa — p. 661
wilsonii = **Ch. nictitans jaliscensis**
wrightii ≡ **Ch. serpens wrightii**
xanthadena — p. 650
xinguënsis — p. 642
yucatana = **Ch. glandulosa flavicoma**
zygophylloides — p. 660
caribaea — p. 661
colligans — p. 661
deamii — p. 661
zygophylloides — p. 661
CHAMAEFISTULA = **S.** ser. COLUTEOIDEAE
acuminata = **S. hayesiana**
anconis = **S. hayesiana**

CHAMAEFISTULA (*Cont.*)
angusta ≡ **S. papillosa angusta**
antillana = **S. nitida**
apiculata = **S. hayesiana**
astroites = **S. villosa**
axillaris = **S. oxyphylla oxyphylla**
bacillaris ≡ **S. bacillaris bacillaris**
barbinervis = **S. macrophylla gigantifolia**
berlandieri = **S. fruticosa**
birostris = **S. birostris birostris**
bolivarensis = **S. dariensis dariensis**
caudata ≡ **S. caudata caudata**
chiapensis = **S. hayesiana**
collinsii = **S. quinquangulata**
corymbosa ≡ **S. corymbosa**
crotalarioides ≡ **S. crotalarioides**
curranii = **S. papillosa papillosa**
dariensis ≡ **S. dariensis dariensis**
deficiens = **S. hayesiana**
densiflora = **S. fruticosa**
elegans — p. 205
fagifolia ≡ **C. fagifolia**
falcinella = **S. quinquangulata**
floribunda ≡ **S. × floribunda**
fluviatilis = **S. papillosa papillosa**
foliosa = **S. oxyphylla oxyphylla**
fruticosa ≡ **S. fruticosa**
gatunensis ≡ **S. dariensis gatunensis**
gigantifolia ≡ **S. macrophylla gigantifolia**
hartwegii ≡ **S. oxyphylla hartwegii**
hayesiana ≡ **S. hayesiana**
hemicardia = **S. oxyphylla oxyphylla**
herbertiana = **S. × floribunda**
huilana ≡ **S. huilana**
ignota = **S. fruticosa**
inaequilatera = **S. bacillaris**
indecora = **S. pendula advena**
inflata = **S. bicapsularis bicapsularis**
insularis ≡ **S. insularis**
klugii = **S. quinquangulata**
laevigata = **S. septemtrionalis**
lanata = **S. hayesiana**
larensis = **S. macranthera quadrifoliolata**
latifolia ≡ **S. latifolia**
macrophylla ≡ **S. macrophylla macrophylla**
maxonii = **S. hayesiana**
melanocarpa = **S. viminea**
membranacea = **S. hayesiana**
micradenia = **S. huilana**
nicaraguensis = **S. papillosa papillosa**
obliqua ≡ **S. obliqua**
oxyphylla ≡ **S. oxyphylla oxyphylla**
papillosa ≡ **S. papillosa papillosa**
pendula ≡ **S. pendula pendula**
puberula = **S. bacillaris bacillaris**
pumila = **S. oxyphylla oxyphylla**
quadrifoliolata ≡ **S. macranthera quadrifoliolata**
quinquangulata ≡ **S. quinquangulata quinquangulata**
rekoi = **S. quinquangulata quinquangulata**
rowleeana = **S. papillosa papillosa**
ruiziana ≡ **S. ruiziana ruiziana**
sinaloana = **S. fruticosa**

CHAMAEFISTULA (*Cont.*)
smithiana ≡ **S. smithiana**
sophera ≡ **S. sophera**
speciosa = **S. macranthera micans**
standleyi = **S. hayesiana**
subcoriacea = **S. hayesiana**
subpilosa = **S. hayesiana**
tocotana = **S. viminea**
tolimensis — p. 207
toroana — p. 207
tortiramis = **S. oxyphylla oxyphylla**
undulata ≡ **S. undulata**
valerioi = **S. hayesiana**
villosula = **S. quinquangulata quinquangulata**
viminea ≡ **S. viminea**
williamsii = **S. hayesiana**
CHAMAESENNA = S. ser. PICTAE
aculeata ≡ **S. aculeata**
colombiana = **S. pistaciifolia pistaciifolia**
nicaraguensis ≡ **S. nicaraguensis**
pistaciifolia ≡ **S. pistaciifolia pistaciifolia**
reticulata ≡ **S. reticulata**
velutina ≡ **S. viarum**
COWELLOCASSIA = S. ser. SAPINDIFOLIAE
domingensis ≡ **S. domingensis**
scleroxylon = **S. domingensis**

DESMODIOCASSIA ≡ S. ser. ASTROITES
villosa ≡ **S. villosa**
DIALANTHERA = CH. ser. CHAMAECRISTA
glandulosa ≡ **Ch. glandulosa**
DIALLOBUS = S. ser. TRIGONELLOIDEAE
uniflorus = **S. obtusifolia**
DISTEREPTA = CH. ser. PROSTRATAE
pilosa ≡ **Ch. pilosa pilosa**
DITREMEXA = S. ser. BASIGLANDULOSAE
confusa = **S. ligustrina**
hirsuta ≡ **S. hirsuta hirsuta**
marilandica ≡ **S. marilandica**
medsgeri = **S. marilandica**
nashii = **S. marilandica**
occidentalis ≡ **S. occidentalis**
sophera ≡ **S. sophera**

EARLEOCASSIA = S. ser. BRACHYCARPAE
argentea ≡ **S. argentea**
arida ≡ **S. arida**
bauhinioides ≡ **S. bauhinioides**
confinis ≡ **S. confinis**
covesii ≡ **S. covesii**
crotalarioides ≡ **S. crotalarioides**
demissa ≡ **S. demissa demissa**
durangensis ≡ **S. durangensis durangensis**
lindheimeriana ≡ **S. lindheimeriana**
roemeriana ≡ **S. roemeriana**
vogeliana = **S. crotalarioides**
ECHINOCASSIA ≡ S. ser. ACULEATAE
aculeata ≡ **S. aculeata**
EMELISTA = S. ser. TRIGONELLOIDEAE
mucronifera ≡ **S. mucronifera**
mucronulosa = **S. uniflora**
obtusifolia ≡ **S. obtusifolia**
pilifera ≡ **S. pilifera pilifera**

GAUMEROCASSIA = S. ser. ISANDRAE
andrieuxii ≡ **S. andrieuxii**
ekmaniana = **S. racemosa racemosa**
liebmannii ≡ **S. racemosa liebmannii**
peralteana ≡ **S. peralteana**
GRIMALDIA ≡ CH. sect. GRIMALDIA
absus ≡ **Ch. absus absus**
baileyorum = **Ch. fagonioides fagonioides**
chamaecrista = **Ch. fasciculata**
confusa = **Ch. desvauxii**
deamii ≡ **Ch. zygophylloides deamii**
decora = **Ch. fagonioides fagonioides**
hispidula ≡ **Ch. hispidula**
killipii = **Ch. fagonioides fagonioides**
opifera = **Ch. absus absus**
orenocensis ≡ **Ch. orenocensis**
punctulata = **Ch. punctulata**
standleyi = **Ch. viscosa viscosa**
viscosa ≡ **Ch. viscosa viscosa**

HEPTEIRECA = CH. ser. CHAMAECRISTA
glandulosa ≡ **Ch. glandulosa glandulosa**
HERPETICA = S. ser. PICTAE
alata ≡ **S. alata**

ISANDRINA ≡ S. ser. ISANDRAE
acapulcana = **S. mollissima glabrata**
arborescens = **S. bicapsularis bicapsularis**
emarginata = **S. bicapsularis bicapsularis**
maxonii = **S. atomaria**
xanthophylla = **S. mollissima glabrata**

LEONOCASSIA = S. ser. BASIGLANDULOSAE
stenophylla ≡ **S. stenophylla**

MIMOSA
nodosa = **S. bacillaris**

NICTITELLA = CH. ser. CHAMAECRISTA
aspera ≡ **Ch. nictitans aspera**
nictitans ≡ **Ch. nictitans nictitans**

OPHIOCAULON = CH. ser. PROSTRATAE
serpens ≡ **Ch. serpens serpens**

PALMEROCASSIA = S. ser. DESERTICOLAE
painteri ≡ **S. wislizeni painteri**
pringlei ≡ **S. wislizeni pringlei**
unijuga ≡ **S. unijuga**
villosa ≡ **S. wislizeni villosa**
wislizeni ≡ **S. wislizeni wislizeni**
PANISIA = S. ser. INTERGLANDULOSAE
biflora = nom dub.
PEIRANISIA = S. ser. INTERGLANDULOSAE
aristulata = **S. multijuga multijuga**
benitoënsis ≡ **S. benitoënsis**
botteriana = **S. pendula ovalifolia**
chiapensis ≡ **S. guatemalensis chiapensis**
clarendonensis = **S. mexicana berteriana**
crista = **S. angustisiliqua angustisiliqua**
deamii = **S. pallida pallida**
doylei ≡ **S. multijuga doylei**
falconiensis = **S. skinneri**

PEIRANISIA (*Cont.*)
fitchiana = **S. angustisiliqua angustisiliqua**
foliolosa ≡ **S. pallida foliolosa**
gaumeri ≡ **S. pallida gaumeri**
guatemalensis ≡ **S. guatemalensis guatemalensis**
haitiensis = **S. angustisiliqua angustisiliqua**
holwayana ≡ **S. holwayana holwayana**
inaguensis ≡ **S. angustisiliqua inaguensis**
jamaicensis = **S. angustisiliqua fulgens**
longirostrata ≡ **S. pallida longirostrata**
macrochlamys = **S. robiniifolia robiniifolia**
magniflora = **S. viciifolia**
multijuga ≡ **S. multijuga multijuga**
mutisiana ≡ **S. mutisiana**
orcuttii ≡ **S. orcuttii**
palmeri ≡ **S. pallida palmeri**
polyphylla ≡ **S. polyphylla polyphylla**
quiedondilla ≡ **S. pallida quiedondilla**
robiniifolia ≡ **S. robiniifolia robiniifolia**
saeri ≡ **S. saeri**
salvadorensis = **S. holwayana holwayana**
santanderensis = **S. robiniifolia robiniifolia**
scopulorum ≡ **S. guatemalensis scopulorum**
shaferi ≡ **S. mexicana shaferi**
simulans ≡ **S. foetidissima grandiflora**
tonduzii ≡ **S. tonduzii**
turquinae = **S. ligustrina**
velutina = **S. pallida pallida**
verbenensis = **S. tonduzii**
viciifolia ≡ **S. viciifolia**
williamsii ≡ **S. williamsii**
yucatensis = **S. pallida gaumeri**
PHRAGMOCASSIA = S. ser. SKINNERANAE
skinneri ≡ **S. skinneri**
PSEUDOCASSIA = S. ser. EXCELSAE
anisopetala = **S. peralteana**
petensis = **S. peralteana**
spectabilis ≡ **S. spectabilis spectabilis**
PSILORHEGMA ≡ S. sect. PSILORHEGMA
suffruticosa = **S. sulfurea**
PTEROCASSIA = S. ser. GALEOTTIANAE
galeottiana ≡ **S. galeottiana**
goldmanii = **S. polyantha**
polyantha ≡ **S. polyantha**
rubricaulis = **S. polyantha**

SCIACASSIA ≡ S. ser. FLORIDAE
siamea ≡ **S. siamea**
SCOLODIA = S. ser. BACILLARES
viminea ≡ **S. viminea**
SENNA — p. 64
subgen. TAGERA = CH. ser. PROSTRATAE
sect. ASTROITES — p. 452
sect. CHAMAEFISTULA — p. 82
sect. PARADICTYON — p. 484
sect. PEIRANISIA — p. 486
sect. PSILORHEGMA — p. 77
sect. SENNA — p. 455
ser. ACULEATAE — p. 478
ser. APHYLLAE — p. 567
ser. ARMATAE — p. 291
ser. BACILLARES — p. 102
ser. BASIGLANDULOSAE — p. 405

SENNA (*Cont.*)
ser. **Brachycarpae** — p. 260
ser. **Chlorocladae** — p. 563
ser. **Coluteoideae** — p. 345
ser. **Confertae** — p. 257
ser. **Coriaceae** — p. 220
ser. **Deserticolae** — p. 570
ser. **Egregiae** — p. 566
ser. **Excelsae** — p. 599
ser. **Floridae** — p. 97
ser. **Galeottianae** — p. 448
ser. **Harleyanae** — p. 209
ser. **Interglandulosae** — p. 488
ser. **Isandrae** — p. 578
ser. **Laxiflorae** — p. 225
ser. **Nanae** — p. 290
ser. **Pachycarpae** — p. 293
ser. **Pictae** — p. 456
ser. **Sapindifoliae** — p. 85
ser. **Senna** — p. 479
ser. **Skinneranae** — p. 217
ser. **Spinescentes** — p. 210
ser. **Spinigerae** — p. 476
ser. **Stipulaceae** — p. 324
ser. **Temperatae** — p. 443
ser. **Tharpia** — p. 288
ser. **Trigonelloideae** — p. 237
ser. **Trolliiflorae** — p. 100
acanthoclada — p. 570
aculeata — p. 478
acuparata — p. 202
acuruensis — p. 506
acuruensis — p. 508
catingae — p. 509
interjecta — p. 508
acutifolia = **S. alexandrina**
acutisepala — p. 140
affinis — p. 123
alata — p. 460
alexandrina — p. 481
alexandrina (Garsault) = **S. alexandrina**
andrieuxii — p. 585
angulata — p. 177
angulata — p. 178
miscadena — p. 178
angustisiliqua — p. 514
angustisiliqua — p. 515
fulgens — p. 516
inaguensis — p. 516
aphylla — p. 570
apiculata — p. 273
apiculata — p. 275
longipes — p. 275
apsidoneura — p. 246
araucarietorum — p. 370
arborescens (P. Mill.) = **S. atomaria**
arborescens (Vahl) = **S. sulfurea**
argentea — p. 272
arida — p. 275
aristeguietae — p. 502
armata — p. 292
arnottiana — p. 316
atomaria — p. 588
aurantia — p. 354

SENNA (*Cont.*)
aurata = **S. septemtrionalis**
australis — p. 235
aversiflora — p. 529
aymara — p. 310
bacillaris — p. 111
bacillaris — p. 113
benthamiana — p. 114
barnebyana = **S. aurantia**
bauhinioides — p. 286
benitoënsis — p. 418
bicapsularis — p. 399
augusti — p. 403
bicapsularis — p. 401
birostris — p. 337
arequipensis — p. 342
birostris — p. 344
campiana — p. 345
helveola — p. 343
hookeriana — p. 339
huancavensis — p. 341
totorae — p. 340
brongniartii — p. 322
burkartiana — p. 372
cajamarcae — p. 350
cana — p. 226
calva — p. 232
cana — p. 229
hypoleuca — p. 230
phyllostegia — p. 231
pilosa — p. 230
candolleana — p. 376
caudata — p. 197
caudata — p. 198
diadena — p. 198
centranthera — p. 468
cernua — p. 419
chacoënsis — p. 569
chloroclada — p. 565
chrysocarpa — p. 171
cobanensis — p. 250
collicola — p. 412
confinis — p. 278
corifolia — p. 221
caesia — p. 222
corifolia — p. 222
cornigera — p. 199
corymbosa — p. 397
covesii — p. 276
crassiramea — p. 569
crotalarioides — p. 268
cruckshanksii — p. 330
cuatrecasasii — p. 164
cumingii — p. 312
alcaparra — p. 315
coquimbensis — p. 315
cumingii — p. 314
eremobia — p. 314
cushina — p. 475
dariensis — p. 128
dariensis — p. 130
gatunensis — p. 131
hypoglauca — p. 131
smaragdina — p. 132

SENNA (*Cont.*)
demissa — p. 270
  demissa — p. 271
  radicans — p. 272
didymobotrya — p. 467
domingensis — p. 96
durangensis — p. 281
  durangensis — p. 282
  iselyi — p. 282
exigua = **Ch. absus**
fabrisii — p. 570
× **floribunda** — p. 360
foetidissima — p. 262
  foetidissima — p. 264
  grandiflora — p. 264
formosa — p. 501
fruticosa — p. 121
galeottiana — p. 450
gardneri — p. 192
garrettiana — p. 98
gaudichaudii — p. 80
georgica — p. 193
  bangii — p. 195
  georgica — p. 195
guatemalensis — p. 302
  calcarea — p. 306
  chiapensis — p. 305
  guatemalensis — p. 304
  hidalgensis — p. 306
  oligantha — p. 303
  scopulorum — p. 305
gundlachii — p. 95
harleyi — p. 209
haughtii — p. 587
hayesiana — p. 133
hebecarpa — p. 446
herzogii — p. 145
hilariana — p. 396
hirsuta — p. 425
  acuminata — p. 431
  glaberrima — p. 432
  hirsuta — p. 434
  hirta — p. 433
  leptocarpa — p. 431
  puberula — p. 429
  streptocarpa — p. 429
holwayana — p. 559
  absoidea — p. 561
  holwayana — p. 560
huancabambae — p. 561
huilana — p. 196
incarnata — p. 526
insularis — p. 162
italica — p. 482
itatiaiae — p. 349
koelziana — p. 298
kuhlmannii — p. 224
kurtzii — p. 566
lasseigniana — p. 353
  canarensis — p. 354
  lasseigniana — p. 354
latifolia — p. 143
lechriosperma — p. 236
leiophylla — p. 248

SENNA (*Cont.*)
ligustrina — p. 409
lindheimeriana — p. 266
longiglandulosa — p. 215
loretensis — p. 146
lourteigiana — p. 170
macranthera — p. 181
  andina — p. 188
  lindeni — p. 187
  macranthera — p. 183
  micans — p. 185
  nervosa — p. 184
  pudibunda — p. 186
  quadrifoliolata — p. 187
  striata — p. 185
macrophylla — p. 137
  gigantifolia — p. 139
  macrophylla — p. 140
malaspinae — p. 311
mandoni — p. 331
marilandica — p. 444
martiana — p. 465
mensicola — p. 284
mexicana — p. 414
  berteriana — p. 416
  chapmanii — p. 417
  latifolia — p. 417
  mexicana — p. 416
  shaferi — p. 417
mollissima — p. 592
  glabrata — p. 594
  mollissima — p. 595
monozyx — p. 575
morongii — p. 364
mucronifera — p. 247
multifoliolata — p. 295
  eurystegia — p. 298
  metaxi — p. 297
  mimetes — p. 297
  multifoliolata — p. 297
multiglandulosa — p. 357
multijuga — p. 492
  doylei — p. 500
  lindleyana — p. 497, 498
  multijuga — p. 495
  peregrinatrix — p. 499
  verrucosa — p. 496
mutisiana — p. 512
nana — p. 290
neglecta — p. 421
  furnicola — p. 425
  grandiflora — p. 423
  neglecta — p. 423
  oligophylla — p. 424
nicaraguensis — p. 463
nitida — p. 159
nudicaulis — p. 570
obliqua — p. 148
oblongifolia — p. 374
obovata = **S. italica**
obtusa = **S. italica**
obtusifolia — p. 252
occidentalis — p. 436
orcuttii — p. 268

SENNA (*Cont.*)
organensis — p. 326
  extratropica — p. 328
  friburgensis — p. 329
  heterandra — p. 328
  organensis — p. 328
oxyphylla — p. 117
  hartwegii — p. 121
  oxyphylla — p. 120
pachyrrhiza — p. 570
pallida — p. 531
  bahamensis — p. 548
  brachyrrhachis — p. 545
  cordillerae — p. 554
  delgadoana — p. 558
  foliolosa — p. 553
  gaumeri — p. 546
  geminiflora — p. 540
  goldmaniana — p. 547
  isthmica — p. 539
  lemniscata — p. 555
  longirostrata — p. 544
  macdougalliana — p. 556
  nemorosa — p. 543
  pallida — p. 537
  palmeri — p. 543
  quiedondilla — p. 549
  shreveana — p. 542
  trichocraspedon — p. 557
  triquetripes — p. 552
  tuerckheimiana — p. 551
papillosa — p. 125
  angusta — p. 128
  papillosa — p. 128
paradictyon — p. 485
paraënsis — p. 245
pendula — p. 378
  advena — p. 389
  ambigua — p. 385
  dolichandra — p. 387
  eriocarpa — p. 386
  glabrata — p. 382
  hemirostrata — p. 389
  indistincta — p. 387
  meticola — p. 391
  missionum — p. 384
  ovalifolia — p. 391
  paludicola — p. 393
  pendula — p. 388
  praeandina — p. 385
  recondita — p. 384
  scandens — p. 395
  stahlii — p. 392
  tenuifolia — p. 394
pentagonia — p. 255
  pentagonia — p. 256
  valens — p. 257
peralteana — p. 597
pilifera — p. 241
  pilifera — p. 243
  subglabra — p. 243
  tubata — p. 245
pilosior — p. 279
pinheiroi — p. 180

SENNA (*Cont.*)
pistaciifolia — p. 471
  glabra — p. 473
  picta — p. 473
  pistaciifolia — p. 472
pneumatica — p. 329
polyantha — p. 448
polyphylla — p. 517
  montis-christi — p. 519
  neglecta — p. 520
  polyphylla — p. 519
praeterita — p. 333
pumilio — p. 288
punoënsis — p. 362
purpusii — p. 319
quinquangulata — p. 153
  meizonoloba — p. 156
  quinquangulata — p. 155
racemosa — p. 580
  coalcomanica — p. 584
  liebmannii — p. 585
  moctezumae — p. 583
  racemosa — p. 582
  sororia — p. 584
reniformis — p. 223
reticulata — p. 458
rigida — p. 570
rigidicaulis — p. 569
riparia = **S. marilandica**
ripleyana — p. 285
rizzinii — p. 174
robiniifolia — p. 520
roemeriana — p. 282
rostrata — p. 528
rugosa — p. 188
ruiziana — p. 150
  micrandra — p. 152
  ruiziana — p. 152
rupununiensis — p. 166
saeri — p. 524
sandwithiana — p. 116
santanderensis — p. 355
scabriuscula — p. 435
scandens — p. 149
septemtrionalis — p. 365
siamea — p. 98
silvestris — p. 89
  bifaria — p. 92, 94
  guaranitica — p. 91
  sapindifolia — p. 91
  silvestris — p. 89
  unifaria — p. 93
  velutina — p. 93
skinneri — p. 218
smithiana — p. 132
sophera — p. 440
sousana — p. 596
speciosa = **S. surattensis**
spectabilis — p. 600
  excelsa — p. 604
  spectabilis — p. 603
spinescens — p. 213
  schultesiana — p. 215
  spinescens — p. 214

SENNA (*Cont.*)
spiniflora — p. 569
spinigera — p. 476
splendida — p. 190
gloriosa — p. 192
splendida — p. 191
stenophylla — p. 413
stipulacea — p. 334
anglorum — p. 336
stipulacea — p. 336
subtrijuga — p. 141
subulata — p. 403
sulfurea — p. 78
surattensis — p. 81
talpana — p. 301
tapajozensis — p. 175
tenuifolia — p. 179
timoriensis — p. 98
tonduzii — p. 525
toroides = **S. obtusifolia**
trachypus — p. 509
trianae — p. 163
trichosepala — p. 570
trolliiflora — p. 101
tropica — p. 368
uncata — p. 200
undulata — p. 157
uniflora — p. 258
unijuga — p. 576
urmenetae — p. 318
vargasii — p. 373

SENNA (*Cont.*)
velutina — p. 232
versicolor — p. 307
heterosperma — p. 309
versicolor — p. 309
viarum — p. 474
viciifolia — p. 522
villosa — p. 452
viminea — p. 160
weddelliana — p. 323
williamsii — p. 513
wislizeni — p. 571
painteri — p. 574
pringlei — p. 573
villosa — p. 575
wislizeni — p. 574
wurdackii — p. 167
SERICEOCASSIA ≡ **S.** ser. **CONFERTAE**
uniflora ≡ **S. uniflora**

THARPIA ≡ **S.** ser. **THARPIA**
pumilio ≡ **S. pumilio**

VOGELOCASSIA = **S.** ser. **TRIGONELLOIDEAE**
cobanensis ≡ **S. cobanensis**
leiophylla ≡ **S. leiophylla**

XAMACHRISTA = **CH.** ser. **CHAMAECRISTA**
triflora = **Ch. fasciculata**
XEROCASSIA ≡ **S.** ser. **ARMATAE**
armata ≡ **S. armata**